Harper & Row's

Complete Field Guide to North American Wildlife

Consultants

Birds
Lester L. Short
Curator of Ornithology
American Museum of Natural History

Mammals
Sydney Anderson
Curator of Mammalogy
American Museum of Natural History

Reptiles and Amphibians
Patricia Gardner Haneline
Andrew Mellon Fellow
Department of Biological Sciences, University of Pittsburgh

Fishes
Franklin C. Daiber
Professor
College of Marine Studies, University of Delaware

Mollusks
Harald A. Rehder
Zoologist Emeritus
National Museum of Natural History, Smithsonian Institution

Other Marine Invertebrates
Robert D. Barnes
Professor of Biology
Department of Invertebrate Zoology, Gettysburg College

Harper & Row's

Complete Field Guide to North American Wildlife

Eastern Edition

Covering 1500 species of birds, mammals, reptiles, amphibians, food and game fishes of both fresh and salt waters, mollusks, and the principal marine invertebrates occurring in North America east of the 100th meridian from the 55th parallel to Florida north of the Keys

Assembled by Henry Hill Collins, Jr.

Illustrations by Paul Donahue, Jennifer Emry-Perrott, Nancy Lou Gahan, John Hamberger, Michel Kleinbaum, Klarie Phipps, Stephen Quinn, Guy Tudor, Nina L. Williams, John Cameron Yrizarry

Harper & Row, Publishers

New York, Hagerstown, Philadelphia, San Francisc
Cambridge, London, Mexico City, São Paulo, Sydr

1817

Portions of this work originally appeared in somewhat different form in *Complete Field Guide to American Wildlife: East, Central and North.*

FIRST EDITION

Library of Congress Cataloging in Publication Data

Collins, Henry Hill, 1905–1961.
Harper & Row's complete field guide to North American wildlife, Eastern edition.
Includes index.
1. Zoology—North America. 2. Zoology—Atlantic States. 3. Animals—Identification. I. Title.
QL151.C62 591.974 80-8198
ISBN 0-06-181163-7

81 82 83 84 85 10 9 8 7 6 5 4 3 2 1

Contents

Harper & Row's
Complete Field Guide to North American Wildlife

Sponsoring Editor
Nahum Waxman

Project Editor
Marta Hallett

Production Editor
Bernard Skydell

Project Design Director
Abigail Moseley

Manufacturing Direction
Thomas Malloy

Manufacturing Coordination
Pat Slesarchik

Design Assistants
Sarah Hartman
Barbara Knight

Indexer
Sydney Wolfe Cohen

Copy Chief
Dolores Simon

Copy Editing
Carole Berglie
Duane Berry
Bernie Borok
Virginia Ehrlich
Janet Field
Rebecca Finnell

Composition: TriStar Graphics
Color Separation: Sterling Regal Incorporated
Printing and Binding: R. R. Donnelley & Sons Co.
Jacket Printing: Longacre Press, Inc.

Introduction

Harper & Row's Complete Field Guide to North American Wildlife takes its inspiration from *Complete Field Guide to American Wildlife,* compiled by the late Henry Hill Collins, Jr., and published by Harper & Row in 1959. This important book, covering the Eastern United States and Canada only, was the first single-volume field guide covering all of the major families of larger wildlife—birds, mammals, reptiles, amphibians, fishes, mollusks, and other marine invertebrates.

In the mid-1960s a decision was made at Harper & Row to carry out important revisions in Mr. Collins's field guide and to add a second volume covering the Western United States and Canada. To this end, Jay Ellis Ransom was engaged to prepare a text for the Western Edition, and, in due course, boards of consultants were assembled to review and edit in detail the texts of both volumes.

Working with Mr. Collins's and Mr. Ransom's compilations, the two boards, one for the eastern volume and one for the western, reviewed exhaustively each and every entry, revising, rewriting, and bringing the content into conformity with scientifically attested descriptive information and the latest taxonomic thinking.

In addition, the eastern volume, which did not include species of the southeastern United States, was expanded to cover that region.

Fifteen of the country's outstanding wildlife artists were brought in to work on the projects. All but a dozen of the plates from the original eastern volume were scrapped and replaced with totally new art. The western volume contains only new art, done especially for this work.

Now, after more than fifteen years in the making, this remarkable collaboration is complete and we are pleased to present it for the use of both professional and amateur field natural scientists.

The purpose of the Field Guides is field identification. Included are more than 3300 species, described by appearance, habitat, life zones, behavior, reproduction, and other important aspects. Covered are all species as follows:

In the East
East of the 100th meridian, north through temperate Canada (approximately the 55th parallel) and south to southern Florida (the 26th parallel, taking in the peninsula but not the Keys). Included are food and game fish species to a depth of 25 fathoms and mollusks and other marine invertebrates to a depth of 10 fathoms.

In the West
West of the 100th meridian, north through temperate Canada (approximately the 55th parallel) and south to the Texas-Mexico border, including all major Texas species but not necessarily specialized coverage of all desert species). Included are all food and game fish species to a depth of 25 fathoms and other marine invertebrates to 10 fathoms.

Names and Classification
The common and scientific names used in these books represent the viewpoint of the consultants concerned with each section. In general the consultants follow the prevailing scientific checklists in their field, and in the instructions to their sections cite the

authority they have used. In many individual cases these scientists have developed their own conclusions on certain points, and in these cases the terminology or the order of the materials may not conform to the checklists.

How to Use This Book

In trying to identify any species, note (and jot down, if possible) all the visible characteristics before the animal disappears from sight. If it is a bird, in addition to general color and size, try to determine the color of its beak, legs, feet, and undertail feathers. Notice whether it has wing bars, an eye-ring, white on its tail; stripes on its body; a patch on its cheek or a line over its eyes. Check the shape of its bill. It is important to remember that quick observation is an essential part of field identification, for frequently a species darts in and out of sight very rapdily. Thus, it is necessary for the observer to note visual characteristics quickly, and this can be achieved only if the observer knows what to look for in a particular group of animals.

For mammals, particularly important to note are color patterns—whether they be on the sides, back, or underparts. Color patterns and scales are the key to identification of snakes. So too the number and character of stripes for a lizard. In a fish, the color markings are of great import, but observe them, to the extent possible, while the fish is still alive, for, as with the marine invertebrates, once the animal dies, its colors will change and fade. Naturally, size, habitats, habits, and voice, if any, can be important clues to the identity of any species.

When you turn to the appropriate section of the book, check the ranges and eliminate all species not within your range and season. Check also habitat data, and eliminate those species which appear in a habitat not normally observed. Then turn to the illustrations, which combined with the text, will provide all necessary information for making an identification.

Be sure to note the following points in using these guides:

1. Entries are by common name, with the scientific (Latin) name below. Many marine invertebrate species, however, have no common names and are listed by scientific names only.

2. To the right, opposite each entry, is a set of boldface numbers providing illustration references. Numbers preceded by the designation *Fig.* refer to figures directly in the text. Such figures are generally no more than two or three pages away from the text entry they illustrate. Numbers not preceded by the designation *Fig.* and taking the form of two numerals separated by a colon (e.g., **32:5**) refer to an illustration on one of the plates, with the first half of the number indicating the plate and the second half indicating a specific figure on the plate (thus, **32:5** refers to figure 5 on Plate 32).

3. Plate captions include page references back to the text entries for the species illustrated, so that throughout this book cross-referencing in both directions is complete.

Some species may show color variations because of age, season, sex, environment, or molt—and, as a result, an animal seen in the field may not be colored exactly as the one illustrated. To the extent possible, this factor has been considered, so that in many cases the most prevalent color variations are illustrated. However, if this is not the case with the species sighted, read once again the text description to learn if the species seems to conform in every other respect. It is essential to use both the illustrations and the text for accurate identification. If there is doubt about the identity of a species, don't mark it as having been observed. Scrupulous accuracy in field identification is necessary for any field person.

Life List

This book can be used to keep a life list of species seen and identified. First, there is included in a section, just before the index, an alphabetical list of all the species in the book by category. Second, the reader may make notes in the index or in the margins of the book, right beside the text descriptions of each species identified. Some people may find it useful to specify the place and date of observation. In this way the guide can become a personal lifetime natural history diary of each new sighting.

Acknowledgments

For their aid in providing reference materials for these volumes, thanks go to Mildred Bobrovich, Assistant Librarian for Reference Services at The American Museum of Natural History; Jill Fairchild at the Sea Library in Santa Monica, California; and Mary Ann Nelson at the Duke University Marine Laboratory in Beaufort, North Carolina.

To the consulting editors particular thanks are due, for their exemplary efforts in guiding the plan of each chapter, nurturing and editing the text descriptions, and examining each and every piece of artwork.

And, of course, grateful acknowledgment to the artists, whose fine work and whose enthusiasm for the project have done so much to help make these volumes the accomplishment we believe they are. Special thanks to Guy Tudor, who contributed so much more than his paintings.

The Editors

Birds

Consulting Editor
Lester L. Short
Curator of Ornithology
American Museum of Natural History

Illustrations
Water Birds and Game Birds, Plates 1–7, 10–11, 17, 24–28
Guy Tudor

Flying Ducks, Plates 8–9 Michel Kleinbaum
Birds of Prey, Plates 12–16 Paul Donahue
Owls and Nightjars, Plates 29–30 Stephen Quinn

Shorebirds, Woodpeckers, Perching Birds, Plates 18–23, 31–48
John Cameron Yrizarry

Text Illustrations, Michel Kleinbaum, Guy Tudor,
John Cameron Yrizarry

Birds

Class Aves

A bird is an animal with feathers; most, but not all, birds fly. It is generally presumed that those that do not fly are descended from ancestors that did. For this ability, the bodies of birds are specially adapted. The bones are light but strong, and some contain air sacs connected with the lungs. Wings—adaptations of the vertebrate forelimb—are designed to propel the bird through the air. In flightless forms, they balance the bird as it runs, as in the ostrich, or propel it through the water, as in the penguin.

The tail is made entirely of feathers, with a bony base, in contrast to the mammals in which the bones run down the middle. The tail may be either long or short, and is sometimes forked in strong flying species. What looks like, and is generally called, the leg is more or less the equivalent of the human foot (tarsometatarsus), with toes at the end. What looks like a reversed knee is really a joint somewhat the equivalent of the human ankle, the true knee usually being hidden by feathers next to the bird's body. Most birds have four toes, three forward and one, the hallux, behind. Some birds, like the woodpeckers, parrots, and cuckoos, have two toes forward and two behind. In many birds the toes lock automatically around the perch when a resting bird lowers its body, allowing it to sleep without falling.

Birds have no teeth, so a bill adapted for the bird's preferred diet is used to tear, crush, or seize the flesh, seeds, or insects upon which a particular species subsists. Food is sometimes temporarily stored in a crop, an enlargement of the esophagus. The gizzard, or main stomach, is a tough organ, effectively replacing the teeth and heavy jaw muscles of other animals, and the process of digestion is often aided by stones or gravel, which the bird swallows for that purpose. The products of the digestive, excretory, and reproductive systems are all discharged through a common opening, the cloaca. Birds are warm-blooded, with a body temperature varying between 98°F and 112°F (36.7°–44.4°C).

Birds show a complex mixture of unmodifiable ("instinctive"), somewhat modifiable, and highly modifiable (or "learned") behavior. Among the most remarkable of the essentially unlearned patterns is the migration flight of some young shorebirds that leave their summer homes independently of their parents and follow, for thousands of miles to their winter homes, a route they have never before traveled.

Evolution

Birds arose from the reptiles; the feather is, in essence, a modified scale such as still occurs on their legs and feet. Birds are seldom preserved as fossils, and their record in the rocks is still scant compared with other groups, such as the mammals. It is known, however, that the gull-like *Ichthyornis* and the flightless, somewhat loonlike *Hesperornis* had already appeared by the Cretaceous period, and that the giant flightless land predator, the seven-foot (2.1-m) tall *Diatryma*, inhabited what is now Wyoming in the Eocene Epoch.

In the Pleistocene, in what is now Los Angeles, the Giant Condor, *Teratornis,* was occasionally engulfed in the La Brea tar pits. Almost into historic times the flightless Elephant Bird, *Aepyornis*, persisted on the island of Madagascar. One of the flightless moas of New Zealand, which reached a height of over ten feet (3.1 m), was contemporaneous with early humans, who exterminated the last of them.

Adaptation

Within the stringent limitations imposed by flight, birds have adapted in diverse ways to varied environments. Birds in their thousandfold varieties are the unchallenged champions of the air. Arctic Terns may fly 22,000 miles (35,405 km) a year in their extended migrations. Some swifts reach speeds of up to 200 miles per hour (322 km/hr). Hawks, vultures, and albatrosses can soar for long periods, hardly flapping a wing. And, because they are warm-blooded, birds can survive in climates from the Poles to the Equator.

Conservation

Birds have long served humans for game, food, and feathers, as well as in their predatory capacity as destroyers of insects and rodents. Because of the prodigious abundance of wildlife in the early days, the possibility of the permanent disappearance of any species seemed inconceivable. As a result, conservation came too late to save the Great Auk, Labrador Duck, Passenger Pigeon, and Carolina Parakeet, although the Passenger Pigeon was once probably the single most numerous bird species in the entire world.

In the 1890s the public was finally shocked into action at the disappearance of so great a number of birds. The National Audubon Society was formed and set out to save the egrets, which the hunters of plumes for women's hats had almost totally destroyed. Long, hard campaigns resulted in laws that protect virtually all birds. A recent step forward is the Federal Rare and Endangered Species Program.

Urbanization, industrialization, the draining of wetlands, and unwise agricultural practices are continually eliminating the habitats and food supplies of many species of birds, thus causing the greatest modern threat to birds. Also, the widespread use of pesticides has been implicated in the reduction in numbers of certain birds.

Habitat

Birds can be found almost anywhere—even in city parks and backyards. The best places, however, are where different types of habitat meet, such as the edges of woods, shorelines, and marshes. National, state, and county parks, wildlife refuges, and nature sanctuaries are good places to find birds. Gardens, parks, and open suburbs also may have a variety of species, particularly where sufficient undergrowth and cover have been left, and where dead trees and dead limbs (for nesting holes and exposed perches) have not all been cleared away. On the other hand, although deep woods, desert, mountains above timberline, and certain other places of uniform habitat are not rich in birds, the species found there include some not likely to be seen elsewhere.

The best time of day for bird-watching is the early morning, when birds start singing and actively feeding. During the heat of the day most land birds become inactive and silent, but some will start singing and feeding again toward evening. There may be considerable bird activity throughout the day when it is cloudy or raining lightly. Spring, when migrating birds in breeding plumage are hurrying north, is the best time of the year to see birds; however, the drawn-out fall migration, with the birds in nonbreeding or immature plumage, offers a greater challenge.

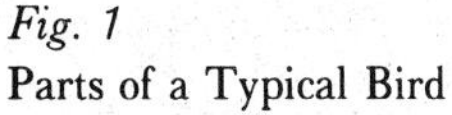
Fig. 1
Parts of a Typical Bird

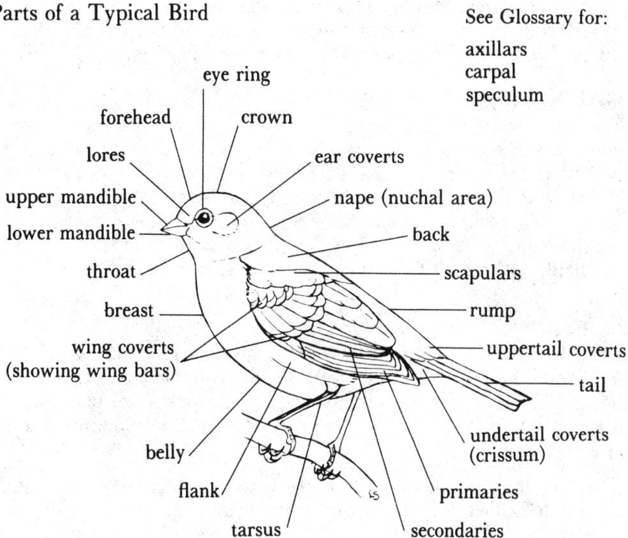

Voice
It is important to be able to identify birds by sound, especially if the bird in question is hidden by foliage or silhouetted against the sky. In some cases, it is the best means of distinguishing between species that visually may appear almost identical.

For each species discussed in the text, such data on voice are given as may be useful for identification. Whenever possible, the song or call has been rendered into English words, or into syllables that can be pronounced. Inadequate as any graphic attempt to portray sound must be, the use of English phrases is the most useful memory aid.

Food
The food a bird eats is correlated closely with the kind of bill, or beak, it has. A sparrow with a stout, conical beak is adapted to a diet of hard-coated seeds; a hawk with a sharp, hooked beak, to the tearing and shearing of flesh. Knowledge of a bird's diet is always of interest and may be of help in identification. Hence, brief descriptions of the principal food of each species are given.

Reproduction
All birds are hatched from eggs laid by the female, usually in a nest, sometimes on the bare ground. The eggs are incubated by the female or male, or by both alternately. In North America, the cowbirds are a notable exception to this practice; they lay their eggs in the nests of other birds.

Incubation takes from eleven to sixty-three days. The young of some species, such as ducks, grouse, and shorebirds, are able to run about almost as soon as they are hatched; these are termed *precocial*. But the *altricial* young of the majority of species must be fed in the nest for several days or weeks until they are fledged.

Mating is customarily preceded by a more or less elaborate courtship performance. When their plumages differ, males of the species, except in a few cases, are generally more brightly colored than the females. However, in many groups, such as herons, shorebirds, and sparrows, the sexes are alike in plumage. In the birds of prey the female is often larger than her mate.

Nest and Eggs

With a little practice, the nests and eggs of many species can be identified in the field. The concise descriptions given in the text will be of assistance, but the advanced student will want to use one of the several available guides to birds' nests and eggs. There is often much variation in the color and markings of birds' eggs, so the descriptions given are of typical examples only. When found in the field, a bird's nest should not be disturbed, nor robbed of its eggs.

Plumage

Most birds have more than one plumage; that is, covering of feathers. The variations that occur in the coloration of certain birds usually correspond to specific periods of the year (seasonal), sexual differentiation (male or female), or to stages of development (adult and immature).

There is almost always a difference between the immature, those incapable of breeding, and adult plumages. This is especially prolonged in the water birds and birds of prey. In some groups, gulls for example, there are distinctions between the juvenals and immatures as well. During the juvenal period, usually when the bird is first out of its nest in summer, a bird may exhibit one plumage; this may distinctly change to another plumage when the bird subsequently grows into its immature stage.

Changes that relate to a period of time in the life of the bird may also appear in adults, birds which are capable of reproduction. Generally, the terms *spring, fall, winter,* and *summer,* in relation to plumage, apply to the differences in adult breeding and nonbreeding plumages, rather than to the season of the year. *Spring* and *summer* usually apply to breeding plumages; *fall* and *winter* to nonbreeding. A spring bird, for example, is generally considered to be an adult in its bright breeding plumage. In this context, the term does not apply strictly to the spring season as we think of it (March to June), although it is loosely related to this time of year. The bird may be in this plumage somewhat before, and for a time after, the season as we know it. By the same definition, the "fall" plumage, as used herein, may be acquired by midsummer in a group such as the shorebirds.

The most common plumages of the individual species are discussed within the text when they are a distinct factor in identifying a species.

Range

The area of coverage for this chapter includes species found in eastern North America north of the 25th parallel, and west to the 100th meridian. The central prairie states along the axis of the 100th meridian constitute the "twilight zone," where the ranges of eastern and western birds converge and overlap, to some extent. Ranges apply to the species as a whole only within the geographic coverage of the chapter.

Most birds within the given range that are not permanent residents breed in the north in summer and migrate south for the winter.

Exceptions are a few petrels and shearwaters, which breed in the Southern Hemisphere and then cross the Equator to "winter" in North American waters during our summer. Both the normal breeding and wintering ranges of each species are given in the text.

Range boundaries are seldom clearly defined in the wild and tend to change slightly as time passes; hence the ranges given should be understood to be approximations only. Species may occasionally be found somewhat outside, or be absent from some space inside, the indicated range. Within a range area, of course, a species will normally occur only in its usual habitat.

Nomenclature
Except in a few cases in which recent studies indicate a change, the English and scientific names and the taxonomy of birds used in this chapter are those set forth in the 1957 *Check-list of North American Birds* by the American Ornithologists' Union, and supplements since published in *The Auk*. The families, genera, and species are arranged more or less in phylogenetic order; that is, the sequence employed by taxonomists to reflect their understandings of avian evolution, from more primitive to more advanced types. Because of gaps in knowledge and difficulties posed by any linear sequence, the taxonomic arrangement is subject to constant reinterpretation and change. To aid in identifying, the most specific characteristic appears in italics, preceded by a bullet (•), under the description heading for most species.

Illustrations
Virtually every bird that is discussed in this chapter is illustrated and, where appropriate, varying plumages are shown. All the birds are adults unless otherwise specified in the caption. If there is no gender symbol shown, it is because the sexes are identical, or nearly so.

A page of illustrations divided by a rule indicates that there is a change in scale. Otherwise, all the species shown are proportionately sized on the page. The sizes given are the average known sizes.

USEFUL REFERENCES

Books

American Ornithologists' Union. 1957. *Check-list of North American Birds.* 5th ed. Baltimore: American Ornithologists' Union.

Austin, O. L., Jr., and Singer, A. 1961. *Birds of the World.* New York: Golden Press.

Baker, J. H. 1941. *The Audubon Guide to Attracting Birds.* New York: Doubleday & Co.

Barton, R. 1955. *How to Watch Birds.* New York: McGraw-Hill.

Bent, A. C. 1919–68. *Life Histories of North American Birds.* 23 vols. Washington, D.C.: U.S. National Museum.

Berger, A. J. 1971. *Bird Study.* Reprint. New York: Dover Publications.

Bull, J. 1964. *Birds of the New York Area.* New York: Harper & Row.

Bull, J. *Birds of New York State.* New York: Harper & Row.

Bull, J., and Farrand, J. 1977. *The Audubon Society Field Guide to North American Birds, Eastern Region.* New York: Knopf.

Darling, L., and Darling, L. 1962. *Bird.* Boston: Houghton Mifflin.

Gilliard, E. T. 1958. *Living Birds of the World.* New York: Doubleday & Co.

Godfrey, W. E. 1966. *The Birds of Canada.* Bulletin 203. Ottawa: National Museum of Canada.

Harrison, H. H. 1975. *A Field Guide to Birds' Nests.* Boston: Houghton Mifflin.

Hoffmann, Ralph. 1927. *Birds of the Pacific States.* Boston: Houghton Mifflin.

Lanyon, W. E. 1963. *Biology of Birds.* Garden City: Natural History Press.

Leck, C. 1975. *Birds of New Jersey: Their Habits and Habitats.* New Brunswick: Rutgers University Press.

Ovington, R. 1976. *Birds of Prey in Florida.* St. Petersburg, Fla.: Great Outdoors.

Palmer, R., ed. 1962– . *Handbook of North American Birds.* Vols. 1, 2, 3. New Haven: Yale University Press.

Peterson, R. T. 1947. *A Field Guide to the Birds.* 2nd ed. Boston: Houghton Mifflin.

———. 1961. *A Field Guide to Western Birds.* Boston: Houghton Mifflin. 60 pl.

———. 1963. *The Birds.* New York: Time-Life Books.

Pettingill, O. S., Jr. 1953. *A Guide to Bird Finding West of the Mississippi.* New York: Oxford University Press.

Pough, R. H. 1946. *Audubon Land Bird Guide.* New York: Doubleday & Co. 48 col. pl.

———. 1951. *Audubon Water Bird Guide.* New York: Doubleday & Co.

Reed, C. A. 1965. *North American Birds' Eggs.* New York: Dover Publications. Illus.

Reilly, E. M., Jr. 1968. *The Audubon Illustrated Handbook of American Birds.* New York: McGraw-Hill.

Robbins, C. S.; Brunn, B.; Zim, H. S.; and Singer, A. 1966. *A Guide to Field Identification: Birds of North America.* New York: Golden Press.

Snyder, D. E., and Griscom, L. 1955. *Birds of Massachusetts.* Salem: Peabody Museum.

Stefferud, A., ed. 1966. *Birds in Our Lives.* Washington, D.C.: U.S. Department of Interior.

Van Tyne, J., and Berger, A. J. 1976. *Fundamentals of Ornithology.* New York: John Wiley.

Welty, J. C. 1975. *The Life of Birds.* New York: W. B. Saunders.

Wing, L. W. 1956. *Natural History of Birds.* New York: Ronald. Lists many state and local books and periodicals.

Periodicals

American Birds. Bimonthly. National Audubon Society, 950 Third Avenue, New York, New York.

Audubon Magazine. Bimonthly. National Audubon Society, 950 Third Avenue, New York, New York.

The Auk. Quarterly. American Ornithologists' Union, Allen Press, Inc., Lawrence, Kansas.

Birding. Bimonthly. American Birding Association, Inc., Box 4335, Austin, Texas.

The Condor. Quarterly. Cooper Ornithological Society, Allen Press, Inc., Lawrence, Kansas.

The Wilson Bulletin. Quarterly. Wilson Ornithological Society, Allen Press, Inc., Lawrence, Kansas.

Records and Tape Cassettes
Kellogg, P. P., and Allen, A. A. *A Field Guide to Bird Songs.* Boston: Houghton Mifflin. More than 300 eastern species, to accompany Peterson's *A Field Guide to the Birds.*

GLOSSARY

♂—Male
♀—Female

Axillar (*pl.* axillars) Feathers of the underwing, between the wing lining and the body ("armpits").

Breeding An adult bird in the plumage in which it reproduces.

Carpal Relating to the wrist (carpus).

Displaying A male bird in breeding display pose.

Form A distinct plumage type, such as of a subspecies, or color phase.

Immature A bird in its first year, usually the first fall and winter following juvenal plumage; occasionally in large birds may extend to second or third year.

Juvenal A bird in its first plumage after natal down during summer; followed by immature plumage.

Molt The loss of one feather plumage and acquiring of another.

Nonbreeding An adult bird that is in a plumage other than that in which it reproduces.

Phase A coloration other than the normal plumage, irrespective of sex, age, or season.

Race A subgroup (or subspecies) within a species.

Resident A species that breeds and winters in the same general latitude.

Sibling Species Closely related, look-alike species that coexist.

Speculum A patch of color, usually iridescent, on the secondaries; mainly waterfowl.

Sub-Adult A plumage between immature and adult; sometimes or regularly breeds in this stage.

Water and Game Birds and Birds of Prey

Loons

Order Gaviiformes

LOONS

Family Gaviidae

Loons are large water birds with strong, pointed bills and thick necks, midway in size between ducks and geese. They are dark above and white below, with white spots on the back in summer. Their wings are slender and pointed. They fly with rapid, fairly deep wingbeats (but slower than any duck), with necks extended and drooping, and with feet trailing. They dive rapidly, and can swim long distances under water. Loons are northern birds that feed on fish and are seen in most of the east only on migration and in winter.

COMMON LOON

Gavia immer **1:1**

Description
Size, 28–36 in. (71.1–91.4 cm). Heavy bill held straight. Summer: head dark, • *back evenly spotted with white squares*, but spots not visible at great distance. Winter: hindneck and back unspotted gray; throat, cheek, and underparts white.
Similarities
Arctic and Red-throated have thinner necks; Common's back is darker than smaller Red-throated.
Habitat
Prefers large, deep freshwater lakes for breeding; at other seasons more often found in fresh water than other loons; commonest loon in most of the east.
Habits
Swims low in water, sometimes with only neck and head visible; patters on water into wind to take flight; ungainly on land.
Voice
Wild and quavering, often compared to insane laughter.
Eggs
2; olive-brown with spots; 3.5 x 2.2 in. (8.9 x 5.6 cm).
Range
Breeds throughout N., from n. Canada, s. to N.H. and e. from Mass. to Minn.; winters s. to Gulf Coast.

ARCTIC LOON

Gavia arctica **1:2**

Description
Size, 23–29 in. (58.4–73.7 cm). • *Bill straight* or downcurved. Summer: • *head gray*; white spots above only on scapulars, • *2 patches* on each side. Winter (difficult identification): • *head and neck grayish*, back black (often with scaly appearance), body small and thin, bill thin. Breeding: throat sometimes turns black first

along lower edge, showing black line across lower throat (helpful but not diagnostic).

Similarities
In general, Arctic is similar to, but smaller than, Common; bill of Arctic is appreciably smaller; in winter, Arctic's body is smaller, thinner, and with thinner bill. Red-throated has upturned bill and a speckled brownish-gray back.

Habitat
Large lakes, ponds, tundra waters; in winter, most maritime of eastern loons.

Habits
Similar to Common.

Voice
Loud squalls and screams; seldom heard in winter.

Eggs
2; olive-buff with dark spots; 3.0 x 1.8 in. (7.6 x 4.6 cm).

Range
Breeds in Arctic, s. to Hudson Bay; winters mainly in W., but e. to Man., Ont., Que.; rarely farther s. and e.

RED-THROATED LOON
Gavia stellata **1:3**

Description
Size, 24–27 in. (61.0–68.6 cm). Head held at upward tilt when at rest. In eastern range, only loon with • *upturned bill*. Summer: head gray; nape striped; throat rusty-red; back brownish-black,• *speckled with white*. Winter: head and neck gray above, white below; back brownish-gray, more conspicuously speckled with white.

Similarities
Other loons are darker-colored and heavier with straighter bill.

Habitat
In summer, small ponds and other bodies of water, fresh or salt; other seasons, usually salt water.

Habits
Springs into the air without pattering on water; will often fly rather than dive; walks, but not with ease.

Voice
Not as loud or as wild as other loons; more complaining.

Eggs
2; brownish-olive, sparingly spotted; 2.8 x 1.8 in. (7.1 x 4.6 cm). Nest is muddy platform at edge of tundra lake or grassy pool, or on islet.

Range
Breeds on tundra along n. Atlantic Coast, s. to Newfoundland and w. to n. Man.; winters s. to Fla., w. to Gulf Coast.

Grebes
Order Podicipediformes

GREBES
Family Podicipedidae

These ducklike birds sit low in the water and have thin necks, sharp bills, and no apparent tails; their feet are very much to the rear. They are almost helpless on land, but are excellent swimmers and divers. They hold their heads erect when swimming and run along on the surface of the water before taking flight. They fly with neck inclined downward. Most eat aquatic invertebrates and fish.

RED-NECKED GREBE

Podiceps grisegena **1:4**

Description
Size, 18–22½ in. (45.7–57.2 cm). Short body; long neck; straight, • *yellow bill.* Summer: • *cheeks white,* neck red-brown. Winter: cap dark, neck red, body gray; • *2 white patches* on each wing and (except in some first-year birds) white crescent on cheek; dark upperparts and light underparts blend without sharp line of contrast. Neck thicker, head and bill heavier than in other grebes. Most loonlike of all grebes (but no loon has white wing patches).
Similarities
Common Loon has larger, longer body, shorter neck. Horned Grebe is smaller, neck white, bill dark. Red-throated (winter) Loon has whitish throat and upraised pale bill. Red-breasted Merganser in flight shows only 1 white wing patch, has faster wingbeat, and holds neck horizontally.
Habitat
Freshwater ponds in summer; fresh and salt water in winter.
Voice
Various loonlike wails; a single bray.
Eggs
4–5; bluish-white, usually heavily nest-stained; 2.3 x 1.4 in. (5.8 x 3.6 cm).
Range
Breeds in n. Canada, s. to N.H., w. to Minn.; winters s. to N.Y., coastally to Fla.

HORNED GREBE

Podiceps auritus **1:6**

Description
Size, 12–15¼ in. (30.5–38.7 cm). Summer: crown and cheeks dark green, ear tufts orange-buff, neck and flanks dark chestnut. Winter: gray above; cheeks, foreneck, and underparts clear white. In flight, shows white wing patch.
Similarities
See Eared Grebe.
Habitat
In summer, freshwater ponds and lakes; in winter, fresh and salt water.
Eggs
4–5; bluish- or olive-white, usually heavily nest-stained; 1.7 x 1.2 in. (4.3 x 3.0 cm).
Range
Breeds in n. Canada, s. to Wis.; winters along Atlantic Coast, s. to Gulf Coast.

EARED GREBE

Podiceps nigricollis **1:7**

Description
Size, 12–14 in. (30.5–35.6 cm). Summer: • *crest,* neck, and back • *black*; ear tufts golden; cheeks orange; flanks dark chestnut. Winter: gray above, light below; lower cheek and spot behind ear are whitish; • *sides of neck grayish. Bill appears upturned.*
Similarities
Horned Grebe shows entire cheek bright white and contrasting gray and white neck in winter; has wing patches.

Habitat
Fresh water; breeds in colonies in shallow marshy parts of lakes; many winter on salt water.
Voice
"Poo-eep, poo-eep; hick-rick-up, hick-rick-up" (Dawson).
Eggs
4–5; indistinguishable from those of Horned; 1.7 x 1.2 in. (4.3 x 3.0 cm).
Range
Breeds in W.; winters casually to Atlantic Coast.

WESTERN GREBE

Aechmophorus occidentalis **1:5**

Description
Size, 22–29 in. (55.9–73.7 cm). Large, long neck; head held high above water; dark gray above, white below, giving a two-toned, • *black-and-white appearance.* Crown and hindneck black; back gray; cheek, foreneck, wing patch, and underparts white; bill yellow, slightly upturned; eyes red; neck long, slender.
Similarities
Red-necked Grebe in winter is smaller, bill duskier yellow, plumage dingy gray, neck especially gray. Loons have shorter neck, no wing patches.
Habitat
Lakes and ponds.
Habits
Neck droops in middle in flight.
Voice
"A loud, double-toned, whistled *c-r-r-ee-ee-r-r-ee*" (Chapman).
Eggs
3–6; dull, bluish-white to olive-buff, usually nest-stained; 2.3 x 1.5 in. (5.8 x 3.8 cm).
Range
Breeds in W.; winters casually to Great Lakes and se. coastline.

PIED-BILLED GREBE

Podilymbus podiceps **1:8**

Description
Size, 12–15 in. (30.5–38.1 cm). Brown above, light below; bill short, heavy, whitish; no white wing patches; undertail white. Summer only: • *black throat and ring around bill.*
Habitat
In spring and summer on ponds and marshes; in winter on fresh and salt water.
Habits
On water, carries tail cocked, showing white below; can dive or submerge by sinking body until only head shows.
Voice
Series of harsh *cow-cow-cow* notes on 1 pitch, somewhat cuckoolike.
Eggs
5–7; indistinguishable from those of Horned Grebe; 1.7 x 1.2 in. (4.3 x 3.0 cm).
Range
Breeds throughout Canada and U.S.; winters s. from New England.

Fulmars, Shearwaters, and Petrels

Order Procellariiformes

These web-footed, powerfully winged water birds have nostrils in a tube on top of the bill. Their home is the open sea, but they breed on islands.

FULMARS AND SHEARWATERS

Family Procellariidae

For long periods, these birds glide on stiff wings close to the surface of the sea. They swim well. Frequently they collect in numbers around fishing boats for scraps and waste. Shearwaters are open-ocean birds. They are dark above and, except for the Sooty, white below. They alternate deep flaps with wheeling glides just over the waves. In gliding, the long slender wings are held straight out and stiff. In the East, they are summer or fall visitors, generally well offshore. Their food consists of small fish, crustaceans, and plankton.

NORTHERN FULMAR

Fulmarus glacialis **2:7**

Description
Size, 17–20 in. (43.2–50.8 cm). Light phase: back and mantle gray; head, tail, and underparts white; bill stubby, yellow (sometimes darker). Dark phase: dark gray all over, bill dusky-brown. Intermediates occur frequently. Tube-nosed bill and manner of flight distinguish it from gulls. Habitat is open ocean.
Similarities
Shearwaters do not flap as much; dark phases of Sooty similar to Fulmar, but darker.
Habits
Ship follower and scavenger; flaps more than most shearwaters.
Voice
Grunts, chuckles, cackles.
Eggs
1; white; 2.9 x 2.0 in. (7.4 x 5.1 cm); little or no nest on rock ledge.
Range
Breeds along coastline, s. to Newfoundland; winters at sea, s. to N.J.

CORY'S SHEARWATER

Puffinus diomedea **2:8**

Description
Size, 21 in. (53.3 cm). Largest eastern shearwater and the only one with • *yellow bill.* Brownish-gray above, white below; bill long, thick, yellow; • *cap same color as back* and blends into white of throat, sometimes shows indistinct white V at base of tail.
Similarities
Greater is darker, has conspicuous white V at base of tail, and black cap makes sharp line of contrast with white cheek.
Eggs
1; 3.0 x 2.0 in (7.6 x 5.1 cm).
Range
Breeds in N. Hemisphere, e. of N. America; appears during our summer off Atlantic Coast, s. to S.C.

GREATER SHEARWATER
Puffinus gravis **2:5**

Description
Size, 18 in. (45.7 cm). Black cap contrasts with brown back. Above, sooty brown; underparts, half collar, and conspicuous white V at base of tail; • *black cap contrasts sharply with white cheek.*
Similarities
Bill of Greater is thin, black, and shorter than Cory's.
Remarks
This species reverses the normal pattern of migration. During our winter it breeds in the Tristan da Cunha Islands in the Atlantic, then spends its winters in the North Atlantic during our summer.
Range
Breeds in S. Hemisphere; appears offshore in N. Atlantic waters during late spring to summer.

SOOTY SHEARWATER
Puffinus griseus **2:4**

Description
Size, 16–18 in. (40.6–45.7 cm). All-dark, sooty brown, appearing black at a distance; underwings sometimes lighter, but difficult to see at a distance. Looks like a small, all-black gull that glides (not flaps) on stiff wings over the waves.
Voice
"Low, guttural *wok-wok-wok* when much excited" (Rich).
Range
Breeds in subantarctic; appears offshore in N. Atlantic waters during summer.

MANX SHEARWATER
Puffinus puffinus **2:10**

Description
Size, 12½–15 in. (31.8–38.1 cm). Bill, cap, and • *upperparts black; underparts white;* white undertail coverts; feet pinkish.
Similarities
Audubon's is slightly smaller, and has gray undertail coverts.
Habits
Flight resembles that of Greater. Manx glides more, flaps less than Audubon's; wingbeats slower, flight less fluttery.
Voice
"Cuck-cuck-oo, repeated several times" (Saunders).
Range
Breeds in e. Atlantic; appears rarely offshore along Atlantic Coast.

AUDUBON'S SHEARWATER
Puffinus lherminieri

Description
Size, 12 in. (30.5 cm). Feet slaty; white-bellied, with short, gray undertail coverts.
Habits
Flaps more, glides less than Manx.
Voice
Weird cacklings on breeding grounds: *"whither did he go"* (hence the local name, Wedrigo).
Range
Breeds in tropical seas; appears along coastline in summer, n. to N.Y.

STORM-PETRELS

Family Hydrobatidae

These small, dark seabirds have white rumps and fly close to the surface of the sea, with their feet hanging down. Their habitat is the open ocean, occasionally large bays, and they feed on small invertebrates.

LEACH'S STORM-PETREL

Oceanodroma leucorhoa **2:6**

Description
Size, 7½–9 in. (19.1–22.9 cm). Dusky black; • *white rump;* wings long, top of wing with pale band; legs short, dark, and in flight do not show beyond end of tail; tail forked.
Habits
Flight like that of Common Nighthawk; wingbeats slower, more irregular than Wilson's, which has rounded tail; glides with wings curved down; swims in water with wings uplifted, then springs into air; not a ship follower.
Voice
On breeding grounds, 7 to 8 low cooing notes; also twitterings, screams, trills.
Eggs
1; white; 1.3 x 0.9 in. (3.3 x 2.4 cm).
Range
Breeds along Atlantic Coast, from Labr., s. to Maine; winters at sea.

WILSON'S STORM-PETREL

Oceanites oceanicus **2:9**

Description
Size, 7 in. (17.8 cm). Dusky black; wings with narrow white line; legs long, in flight show beyond end of tail; yellow feet; square tail.
Similarities
Leach's appears lighter with longer wings and rounded tail.
Habitat
Open ocean. More common than Leach's.
Habits
Flight is swallowlike. A ship follower.
Voice
A peeping, when feeding at sea.
Eggs
1; 1.3 x 0.9 in. (3.3 x 2.4 cm).
Range
Breeds in Antarctic; appears offshore along Atlantic and Gulf coasts in summer.

Pelicans and Allies

Order Pelecaniformes

PELICANS

Family Pelecanidae

These are large, fish-eating water birds, with oversized pouched beaks and long wings. They fly, with necks drawn in, in follow-the-leader or V-shaped lines, alternately flapping and gliding in unison.

AMERICAN WHITE PELICAN

Pelecanus erythrorhynchos *Fig. 2*

Description
Size, 54–70 in. (137.2–177.8 cm). • *White;* wing tips black, bill yellow to orange. The only white pelican in the east.
Habitat
Breeds on western freshwater lakes, migrates along river valleys, winters on saltwater bays and bayous.
Habits
Frequently soars.
Voice
Low croaks on breeding ground.
Eggs
2; white; 3.5 x 2.2 in. (8.9 x 5.6 cm). Nest is of vegetation, stones, and debris with a raised rim; on ground or flattened vegetation, in colonies on islands in shallow water
Range
Breeds from nw. Canada, s. to w. Ont.; winters from Fla. along Gulf Coast, southward.

Fig. 2

BROWN PELICAN

3:4

Pelecanus occidentalis *Fig. 2*

Description
Size, 44–55 in. (111.8–139.7 cm). • *Brown;* the only brown pelican in east. Strictly coastal.
Habitat
Resident with some movement on south Atlantic and Gulf coasts, in small numbers.
Habits
Fishes from surface or dives from low height.
Voice
Usually silent, rarely croaks.
Eggs
2–3; dusky white; 2.8 x 1.8 in. (7.3 x 4.6 cm). Nest is variably bare ground to stick nest in low trees; on islands offshore.
Range
Resident along coastline, from N.C. to Fla., southward.

GANNETS

Family Sulidae

These are large, big-billed seabirds that fly on stiff, long, pointed wings. They eat fish, which they capture by diving, often deeply and from a considerable height.

GANNET

Morus bassanus **3:3**

Description
Size, 35–40 in. (88.9–101.6 cm). Larger than any gull. Adult: white; head sometimes yellowish; • *wing tips black*; bill long, sharp, carried point down; tail pointed; brilliant white of adults visible and recognizable from a great distance; silhouette when soaring makes an almost-perfect cross. Immature: brown above, flecked with white; light below. Birds in changing plumage often mixed black and white, especially on mantle.
Habitat
Open coastal sea, sometimes large bays.
Habits
Often seen along eastern coasts in spring and fall, diving offshore; flies, soars, dives well; alternately flaps and sails low over waves; wingbeats relatively rapid. Breeds in great colonies.
Voice
Guttural croaks and grunts on breeding grounds, *kur-r-r-ruk, kur-r-ruk.*
Eggs
1; bluish-white; 3.0 x 1.8 in. (7.6 x 4.6 cm). Nest is large mound of vegetation on top or on the ledge of an island cliff.
Remarks
The cliffs of Bonaventure Island off the Gaspé Peninsula, in Quebec, contain one of the largest Gannet colonies.
Range
Breeds on islands in Gulf of St. Lawrence; winters along coast from New England to Fla.

CORMORANTS AND ANHINGAS

Family Phalacrocoracidae

These are black, long-necked water birds about the size of geese. They have long, hook-tipped bills and are darker and longer-tailed than any goose or loon. They swim well, sometimes with only the head, neck, and upturned bill showing. They fly with heronlike flaps and intermittent glides, with their necks slightly raised. Their diet is fish. They nest in colonies on the ground on rocky islands or ledges, sometimes in trees. The nest is of seaweed, sticks, twigs, and grasses. The three to four pale blue eggs have a chalky covering. Although they swim and dive, cormorants and anhingas seem incompletely adapted to the water, for their feathers are not waterproof like a duck's and, when perched, they habitually spreadeagle their wings to dry them.

GREAT CORMORANT

Phalacrocorax carbo **3:6**

Description
Size, 30–40 in. (76.2–101.6 cm). One quarter larger than Double-crested. In breeding plumage (March–June), only cormorant in the east with white flank patches. Adult: black; throat pouch white, chin pouch yellow; except in breeding plumage, flanks dark.

Immature: dark brown, underparts whitish (generally), as far back as undertail coverts; bill heavy; neck stocky.

Similarities

Can be confused with Double-crested, but area of regular overlap is small: Long Island and New Jersey in winter, Gulf of St. Lawrence, and Newfoundland in breeding season.

Habitat

Ocean and large bays; prefers rocky shores.

Habits

As for family; wingbeats slower than Double-crested's; flies in lines, wedges, or irregular bunches; can soar.

Voice

Deep croaks.

Eggs

3–4; 2.5 x 1.6 in. (6.4 x 4.1 cm).

Range

Breeds in Canadian Maritime Provinces, s. to N.S.; winters along coast s. to N.J.

DOUBLE-CRESTED CORMORANT

Phalacrocorax auritus **3:5**

Description

Size, 30–36 in. (76.2–91.4 cm). Smaller, slimmer than Great, especially in the neck. Adult: black; • *throat pouch orange*; never any white on flanks or chin. Immature: brown; breast and forebelly white with brownish cast; hindbelly black.

Similarities

Care is necessary to distinguish this from Great; size is a poor characteristic, unless direct comparison is possible. Some adult Double-cresteds have bleached throat pouches, which look yellowish; some immatures may have unusually light bellies.

Habitat

Lakes, rivers, bays, ocean.

Habits

Wingbeats more rapid than Great.

Voice

Normally silent; a rare croak when alarmed.

Eggs

3–4; pale blue; 2.4 x 1.5 in. (6.1 x 3.8 cm).

Range

Breeds in U.S. and Canada; winters s. from N.Y. and n.–cen. U.S.

NEOTROPICAL (OLIVACEOUS) CORMORANT

Phalacrocorax olivaceus **3:7**

Description

Size, 23–29 in. (58.4–73.7 cm). Adult: black; • *orange throat pouch has narrow white border*, visible at close range.

Similarities

Double-crested is larger, heavier.

Habitat

Salt, fresh, and brackish water.

Voice

Grunts or croaks.

Eggs

2–6; chalky-bluish; 2.2 x 1.4 in. (5.6 x 3.6 cm). Nest is in bush, tree, or on ground; of small twigs.

Range

Resident in s. U.S., along coast of La. and Tex., southward.

ANHINGA
Anhinga anhinga **3:2**

Description
Size, 32–36 in. (81.3–91.4 cm). Bill long, pointed; thin, snakelike neck. Mainly black, female with brownish foreparts; long, fanlike, white-tipped tail.
Habitat
Quiet southern marshes and swamps in fresh or salt water.
Habits
Frequently swims with only head and bill above water; often spreads wings to dry.
Voice
Long, chattering call.
Eggs
2–5; 2.0 x 1.4 in. (5.1 x 3.6 cm). Nest is singly or in colonies in trees.
Range
Breeds along Atlantic and Gulf coasts, from N.C., s., and in Mississippi Valley from Ark., southward; winters from Fla. and Gulf Coast southward.

FRIGATEBIRDS
Family Fregatidae

These are long-winged, gliding and soaring birds of the tropical seas. They feed at the water's surface on jellyfish, fish, squid, and sewage items, or by robbing other fish-eating birds. None breed in east.

MAGNIFICENT FRIGATEBIRD
Fregata magnificens **3:1**

Description
Size, 37–41 in. (94.0–104.1 cm); wingspread to 8 ft. (2.4 m). Male: glossy black. Female: browner with breast and sides of abdomen white. Immature: white-headed are most often seen in eastern waters. Long, forked tail and narrow, crooked wings with a long, hooked bill are diagnostic. Feeds over sea, but in sight of land.
Range
Breeds s. of U.S.; appears along coasts of Fla., La., and Tex.

Long-Legged Waders
Order Ciconiiformes

These usually large wading birds of marshes, mud flats, and shores have long bills, necks, and legs and short tails.

HERONS AND BITTERNS
Family Ardeidae

Members of this narrow-bodied family have sharp, pointed bills; slender necks; and large, rounded wings. Many flock to large roosts in evening. They fly with necks drawn in and legs extended; their wingbeats are slow and deliberate. Their food consists of crayfish, small fishes, and other invertebrates.

GREAT BLUE HERON

Ardea herodias **5:1**

Description
Size, 48 in. (121.9 cm). Adult: head white with 2 black plumes, neck brownish-gray, shoulder patches black, rest of plumage grayish-blue; largest heron in eastern range. Immature: duller, with dark cap and no plumes.
Similarities
Sandhill Crane is larger, more uniform gray in color, and flies with neck extended. In far southern Florida, an all-white form, the "Great White Heron," is larger than Great Egret and has yellow, not black, legs.
Habitat
Shallow water anywhere, but rarely oceanfront. Most common heron in east.
Habits
Fishes by standing motionless in water for protracted periods; flies with slow, steady, heavy wingbeats; can soar well.
Voice
Guttural squawks; a honk reminiscent of Canada Goose but flatter, harsher, longer.
Eggs
4; greenish-white; 2.5 x 1.8 in. (6.4 x 4.6 cm). Nest is a platform of sticks in swamp trees, on rocky islets, in marshes; loosely colonial.
Range
Breeds throughout s. Canada and U.S., from N.S. and Que. s. to Mexico; winters from New England and Great Lakes southward.

GREEN HERON

Butorides striatus **4:9**

Description
Size, 16–22 in. (40.6–55.9 cm). A small, dark heron. Adult: neck maroon, front white, upperparts blue-green (often looks more blue than green, and may appear dark at a distance), • *legs orange.* Immature: browner; throat streaked, legs greenish.
Similarities
Least Bittern is only heron that is smaller.
Habitat
Any shallow water, even tiny ditches and pools, usually near trees.
Habits
Stretches neck and raises crest when alarmed; often alights in trees. In flight, shows downward bend to wing tips.
Voice
Various squawks and grunts; a loud *skyau.*
Eggs
4–5; greenish-white; 1.5 x 1.2 in. (3.8 x 3.0 cm). Nest is of sticks; in tree, shrub, or grass clump, not necessarily near water; nests singly or in small colonies.
Range
Breeds from N.B. s. to Minn. and throughout U.S.; winters from S.C. and Gulf Coast southward.

GREAT EGRET

Casmerodius albus **4:2, 5:2**

Description
Size, 37–41 in. (94.0–104.1 cm). Largest white heron in the east; only one with a • *yellow bill.* • *Legs, feet, and extreme tip of bill*

black, bill largely yellow; up to 50 long plumes (aigrettes) on its back in breeding season.

Similarities
See Habits, below.

Habitat
Streams, ponds, marshes; fresh or salt water.

Habits
Flocks to evening roosts, waits motionless for prey (does not rush around like Snowy); wingbeats slower than smaller Snowy's.

Voice
Low, heavy croak.

Eggs
3–4; pale bluish-green; 2.2 x 1.6 in. (5.6 x 4.1 cm). Nest is of sticks; in trees near water; in colonies.

Former name
Common or American Egret.

Range
Breeds from n. U.S. s. throughout rest of U.S.; winters from S.C. and Gulf Coast southward.

LITTLE BLUE HERON

Egretta caerulea **4:6**

Description
Size, 20–24 in. (50.8–61.0 cm). • *Bill bluish with black tip;* • *feet and legs greenish.* Adult: completely dark; head and neck deep maroon (no white in front); rest of plumage slaty-blue; looks black at a distance. Immature: all-white. Changing birds show dark and white patches.

Similarities
Snowy Egret has black bill and black legs, yellow feet. Green Heron is smaller, short-necked.

Habitat
Fresh- and saltwater marshes.

Habits
Gathers at communal roosts at sunset; moves about actively when feeding (but motions slower than Snowy's); in flight, wingbeats more rapid than Great Blue or Great Egret.

Voice
Croaks, and screams, *"tell you what, tell you what"* (Chapman).

Eggs
4–5; bluish-green; 1.7 x 1.3 in. (4.3 x 3.3 cm). Nest is of sticks, in bushes and trees near water.

Range
Breeds mainly along Atlantic Coast, from New England s. and from Ark. and Tenn. s.; winters in s. states, southward.

SNOWY EGRET

Egretta thula **4:5**

Description
Size, 20–27 in. (50.8–68.6 cm). Adult: all-white with many beautiful plumes (aigrettes) on its back in the breeding season; bill narrow and black with yellow or red skin near base; • *legs black, feet yellow.* Immature: has yellow line up back of legs.

Similarities
Little Blue Heron's bill is shorter, heavier, bluish with black tip.

Habitat
Marshes and shallow water, fresh or salt.

Habits
Shuffles legs and dashes about erratically when feeding (Little Blue Heron and other egrets do not); wingbeats more rapid than Great Egret's.
Voice
Harsh hiss.
Eggs.
4–5; pale bluish-green; 1.7 x 1.3 in. (4.3 x 3.3 cm). Nest is of sticks; in trees, shrubs, or marsh grass; in colonies.
Range
Breeds in Atlantic coastal states from Maine s. and along Gulf Coast; winters from S.C. s. and along Gulf Coast.

LOUISIANA HERON

Egretta tricolor **4:3**

Description
Size, 26 in. (66.0 cm). Dark, very slim, white rump and belly. Adult: upperparts slaty-blue with some maroon. Immatures: head, neck, and part of wings brown; rest of upperparts grayish.
Similarities
Little Blue; changing from white to dark plumage, sometimes shows white belly with dark upperparts, but there would also be some white feathers above.
Habitat
Coastal and larger inland marshes.
Habits
Fairly active feeder; flocks to roost in evening; engages in aerial games over breeding pond.
Voice
Low groans and grunts.
Eggs
4–5; pale bluish-green; 1.7 x 1.3 in (4.3 x 3.3 cm). Nest is of sticks (with a soft lining); in low trees and bushes; in colonies.
Range
Breeds along Atlantic and Gulf coasts from Fla. w.; after breeding, some wander n. to Mass.

REDDISH EGRET

Egretta rufescens **4:4**

Description
Size, 27–32 in. (68.6-81.3 cm). More common dark phase: gray-backed with rusty brown head and neck. White phase: all-white. • *Bill pink with black tip.* Legs bluish in both phases.
Habitat
Feeds along coast and in adjacent brackish and fresh water.
Habits
Usually dashes about, actively hunting.
Voice
Croaks, honking noise.
Eggs
3–4; bluish-gray; 2.0 x 1.5 in. (5.1 x 3.8 cm). Nest is on ground or in bush.
Range
Resident along Tex. Gulf Coast and in s. Fla.

CATTLE EGRET

Bubulcus ibis **4:1**

Description
Size, 20 in. (50 cm). Adult: white, with buff patches on head, breast, and back (buff paler in nonbreeding season); reddish bill and legs. Immature: all-white; bill yellow; legs dark.
Similarities
Stockier than other white-colored herons; bill shorter, eye red.
Habitat
Wet fields.
Habits
Associates with cattle, flocks to roost in evening.
Voice
Various croaks when breeding.
Eggs
4–5; bluish-white; 1.8 x 1.3 in. (4.6 x 3.3 cm). Nest is of sticks; in trees; in colonies.
Range
Breeds from s. New England s. and w. to Great Lakes; winters throughout se. states from Va. and w. to Ark.

BLACK-CROWNED NIGHT HERON

Nycticorax nycticorax **4:8**

Description
Size, 23–28 in. (58.4–71.1 cm). The stockiest heron; neck and legs short, legs yellow (red in breeding season). Adult: • *crown and back black*, wings gray, underparts and 2 long plumes from nape white. Immature: brown above with white spots, white below with brown streaks.
Similarities
See American Bittern and Yellow-crowned Night Heron.
Habitat
Ponds, lakes, rivers, bays, and marshes, both fresh and salt water; also built-up areas.
Habits
Most active at night, often roosts in small groups by day; stands with neck pulled in, with a hunched-up appearance. In flight, looks short-necked; flies with slow flaps; sometimes glides with wings bent down.
Voice
Loud *quawk*.
Eggs
3–6; pale bluish-green; 2.0 x 1.5 in. (5.1 x 3.8 cm). Nest is of sticks; in trees and shrubs; in colonies, not necessarily near water.
Range
Breeds from N.B. to Sask. s. throughout U.S.; winters along Atlantic Coast from N.J. s. and in s. states to Gulf Coast.

YELLOW-CROWNED NIGHT HERON

Nycticorax violacea **4:7**

Description
Size, 24 in. (61.0 cm). Adult: blue-gray with black markings on back and wings, • *head black*, crown pale yellow or white, cheek stripe white. Immature: gray-brown above with white dots, white with brown streaks below.

Similarities
Immature Yellow-crowned is grayer than Black-crowned, dots smaller. Stocky, but slimmer than Black-crowned; bill blunter, neck and legs longer, legs yellower.
Habitat
Swamps and marshes, fresh or salt water.
Habits
Not as nocturnal as Black-crowned. Feet and part of legs extend beyond tail in flight.
Voice
Similar to, but higher pitched and more pleasant than, Black-crowned's.
Eggs
3–4; greenish-blue; 2.0 x 1.5 in. (5.1 x 3.8 cm). Nest is of sticks (with lining); in trees, bushes, or on the ground; in small colonies.
Range
Breeds along Atlantic Coast, from Mass. to Fla. w. to Mississippi R. and Tex. with some to Kans., Ky.; winters mainly along Gulf Coast.

LEAST BITTERN
Ixobrychus exilis **4:10**

Description
Size, 11–14 in. (27.9–35.6 cm). A tiny heron. Male: crown, back, primaries, and tail greenish-black; cheeks and neck bright chestnut-buff; white line down either side of back; underparts buffy; • *buff wing patches*. Female: black replaced by dark brown, neck streaked. Young: like female, but paler. Very rare and local dark phase: buff replaced by dark red-brown.
Habitat
Marshes, especially with cattails.
Habits
Climbs along plant stems; slips through reeds like a rail; "freezes" like American Bittern when alarmed; flushes with fluttering wings, legs dangling.
Voice
Dovelike or cuckoolike *coo-coo-coo-coo*, somewhat similar to Pied-billed Grebe.
Eggs
4–5; bluish-white; 1.2 x 0.9 in. (3.0 x 2.3 cm). Nest is of twigs and grasses; in thick vegetation over water.
Range
Breeds from s. Canada and throughout U.S. s. to W. Indies; winters from Fla. and Gulf Coast southward.

AMERICAN BITTERN
Botaurus lentiginosus **4:11**

Description
Size, 27 in. (68.6 cm). Above rich brown; below tan, heavily marked with brown; bill yellow; throat white; legs green; primaries black.
Similarities
Immature Black-crowned Night Heron is grayer, has yellow legs, black bill, and no black on wings or neck. Immature Yellow-crowned is grayer still.
Habitat
Marshes, fresh and salt water.

Habits
When alarmed, stands still, bill pointing straight up. During flight, wings are less curved and wingbeat is faster than other herons; wingbeat has "snap" to downstroke.
Voice
"Pumping sound, *plum pudd'n*" (Collins); a noise as of driving a stake; when flushed, a croaking, *kok-kok-kok*.
Eggs
4–5; buffy-brown; 1.9 x 1.4 in. (4.8 x 3.6 cm). Nest is of vegetation; just above water.
Range
Breeds from Newfoundland, s. to Fla. and w. from Man. to Gulf Coast; winters from Del. and Ohio, southward.

STORKS
Family Ciconiidae

WOOD STORK (WOOD IBIS)
Mycteria americana **5:6**

Description
Size, 34–47 in. (86.4–119.4 cm). Long-legged white bird with dark, naked head. Bill decurved, head bare, tail black; rear of white wings black.
Similarities
See White Pelican, White Ibis. Note also that white herons lack black in wings and retract neck in flight. Rare Whooping Crane is very similar, with straight bill.
Habitat
Marshes, swamps of Deep South.
Habits
Feeds quietly from water on shore. Flies with legs and neck extended.
Voice
Usually silent, but occasionally a hoarse croak.
Eggs
3–4; white; 2.7 x 1.7 in. (7.0 x 4.4 cm). Nest is of sticks in trees; in colonies.
Range
Resident from S.C. to Fla. and w. to Tex., southward; some wander n. in summer.

IBISES AND SPOONBILLS
Family Threskiornithidae

These are long-legged marsh birds with decurved or spatulate bills; they fly with legs and bills extended. They feed on crustaceans and insects.

GLOSSY IBIS
Plegadis falcinellus **5:8** *Fig. 3*

Description
Size, 22 in. (55.9 cm). A dark ibis, very common to the East Coast. Adult: dark, glossy chestnut with some iridescent green; looks black at a distance; narrow white border on skin between bill and eye in breeding condition; legs grayish. Immature: grayer, neck streaked with white.

Similarities
In flight, silhouette resembles Whimbrel, but is bigger, heavier-billed.
Habitat
Ponds, wet fields, mud flats, and marshes, both fresh and salt.
Habits
Flies in long, often undulating lines or ranks; flocks to roost in evening.
Voice
Combination grunt and bleat.
Eggs
3–4; bluish-white; 2.0 x 1.5 in. (5.1 x 3.8 cm). Nest is of sticks; in trees, bushes, or reeds over water; in colonies.
Range
Breeds along Atlantic Coast, from Maine to Fla. and along Gulf Coast to Tex.; also inland to Great Lakes; winters in Fla. and along Gulf Coast.

Note: The **WHITE-FACED IBIS,** *Plegadis chihi (Fig. 3)*, size, 23 in. (58.4 cm), on the western Gulf Coast and east to Louisiana. A close relative of the Glossy Ibis, it is exceedingly similar. Adults in breeding condition may be differentiated from Glossy with care under favorable circumstances by a broader area of white feathering extending back behind eye.

Fig. 3

WHITE IBIS

Eudocimus albus **5:5**

Description
Size, 22–28 in. (55.9–71.1 cm). Adult: all-white except for some black on wing tips; • *decurved bill* and • *red legs*. Immature: streaky brown with white belly, dark legs, and bill.
Similarities
Wood Stork has dark, naked head.
Habitat
Coastal fresh- and saltwater areas.
Habits
Gregarious.
Voice
Squawking notes.
Range
Resident from Ga. s. along Atlantic Coast and w. along Gulf Coast to Tex.

ROSEATE SPOONBILL

Ajaia ajaja **5:9**

Description
Size, 30–34 in. (76.2–86.4 cm). • *Pinkish* with bare head (adults) or white (immature) with long, spatulate bill.
Habitat
Coastal salt- and freshwater marshes and lagoons.
Habits
In small groups, moves rapidly through water sweeping head and bill from side to side.
Eggs
1–4; white with brown spots; 2.5 x 1.75 in. (6.4 x 4.4 cm). Bulky nest of sticks in island trees.
Range
Breeds locally from Fla., southward, and along Gulf Coast; winters s. from s. Fla. and s. Tex.

Swans, Geese, and Ducks

Order Anseriformes

SWANS, GEESE, AND DUCKS

Family Anatidae

These are medium to large birds of fresh or salt water, with webbed feet and bills that have tiny, toothlike projections along the edges. Their flight is swift and sustained; but, when the flight feathers are molting, most species for a time are unable to fly. Except when specifically noted otherwise, all make ample-sized nests of vegetation on the ground near water. This family includes some of our choicest game birds. The major groups of waterfowl are:

Swans: Very large, long-necked birds, usually all-white when adult. They feed on aquatic vegetation from the bottom by extending their necks deep under water.

Geese: Smaller, shorter-necked than swans and larger than ducks, geese have blunt, triangular bills. The sexes are alike. They feed in stubble and grain fields or grassy marshes, consuming seeds, aquatic plants, and grasses.

Whistling-Ducks: Also called Tree Ducks, these are small, gooselike ducks that perch upright, may feed away from water on seeds and grass, and they nest in trees.

Surface-feeding or Pond Ducks (Genus *Anas* and Allies): These largely freshwater ducks spring almost vertically into the air when taking flight. They feed on aquatic plants and seeds by "tipping up," with heads below the surface and tails pointing skyward. Males are often highly patterned about the head. Females are predominantly brown.

Diving Ducks (Pochards [Genus *Aythya*], Goldeneyes, Eiders, Scoters): The legs of diving ducks are set farther back than those of the surface-feeders, and the hind toe is free and lobed. Confusingly similar females are often best identified by their accompanying males. They dive for aquatic plants, snails, and insects, and patter along the water's surface before they fly.

Mergansers: These fish-eating ducks have crests and long, saw-toothed, slightly hooked bills, with some white in the wing and on the secondaries. They feed on fish, crayfish, and some amphibians.

Stiff-tailed Ducks: These are small diving ducks that often erect the tail as they swim. They feed on water plants, insects, and other small aquatic creatures.

MUTE SWAN

Cygnus olor *Fig. 4*

Description
Size, 60 in. (152.4 cm). The only swan with a • *black knob at the base of the orange bill.* Knob is less conspicuous in female and almost absent in young. Immature: bill pink or dusky, blackish at base.
Similarities
Bill of immature Mute is darker than young Whistling's.
Habitat
Freshwater marshes, ponds; coastal waters in winter.
Habits
Sits higher in water than Whistling; carries neck gently arched, bill pointed down; wing secondaries often raised over back aggressively (no other swan does this); wingbeats make a singing, throbbing sound audible from afar; very aggressive in defense of nest.
Voice
Usually silent, but can hiss.
Eggs
4–8; bluish-white; 4.5 x 2.9 in. (11.4 x 7.4 cm).
Range
Resident s. New England to Md.; introduced in U.S. from Europe and established along middle Atlantic Coast.

Fig. 4

WHISTLING SWAN

Cygnus columbianus **6:4** *Fig. 4*

Description
Size, 52 in. (132.1 cm). Adult: bill black with yellow spot at base. Immature: bill pinkish, tip dusky, lacks yellow spot at base.
Similarities
Smaller than Mute Swan. Immature is difficult to distinguish from immature Mute.
Habitat
Lakes and ponds.
Habits
Holds neck erect, bill horizontal, wings not arched above back; spends more time feeding with head under water than does Mute.

Voice
Musical whooping whistle and soft trumpeting; *wow-wow-ou.*
Eggs
4–5; creamy-white; 4.2 x 2.7 in. (10.7 x 6.9 cm).
Range
Breeds in far n. Canada from Baffin Is. w.; winters coastally from Md. w. to Tex., occasionally w. to Great Lakes.

CANADA GOOSE
Branta canadensis **6:5**

Description
Size, 22–43 in. (55.9–109.2 cm). Quite variable, with several different and disputed races. Above grayish-brown; • *head, neck, and tail black;* • *cheek patches white*; base of black "stocking" neck clearly defined against whitish underparts; bill, legs black. Smaller races very stubby-billed; western races often darker-breasted than eastern races.
Habitat
Ponds, rivers, bays, fields.
Habits
Usually flies in V formation. In flight, black neck stretched out and slightly downcurved; usually very vocal. (Cormorants are silent and darker.)
Voice
Loud, from 2-syllabled honking to high-pitched yelping, depending on race.
Eggs
4–6; white; 2.3–3.0 x 1.5–2.5 in. (5.8–7.6 x 3.8–6.4 cm). Nest usually on ground near water, well guarded by adults.
Range
Breeds s. from n. Canada to N.C.; winters along Atlantic Coast s. from N.J. and interior from Nebr. to Gulf Coast.

BRANT
Branta bernicla **6:3**

Description
Size, 24 in. (61.0 cm). A small goose with a black breast clearly set off from the pale underparts and an all-black head. Above dark brown; head, neck, breast, and tail black; white patch on each side of throat; upper- and undertail coverts white, making white V over tail in flight; • *underparts light brownish-gray;* on water, • *sides look white.* Immature lacks throat patch.
Similarities
Canada Goose has white on breast above waterline and white cheek patches.
Habitat
Sheltered salt water.
Habits
Sits high in water like a gull; flight somewhat undulating, hence hunters' name of "wavy"; flight rapid for a goose, in irregular flocks, rarely V-shaped; wings pointed.
Voice
Deep, loud, grunting honks.
Eggs
3–5; creamy; 2.8 x 1.9 in. (7.1 x 4.8 cm).
Range
Breeds in Arctic; winters along Atlantic Coast to S.C.

WHITE-FRONTED GOOSE

Anser albifrons **6:2**

Description
Size, 29 in. (73.7 cm). Adult: gray-brown; • *area around bill* and upper- and undertail coverts *white;* black speckles on belly (hence hunters' name, Specklebelly); bill pink; • *legs yellow or orange.* Immature: similar, but bill yellow; no white around bill or black speckles on belly.
Similarities
Immature White-fronted resembles immature blue form of Snow, but has yellow bill and legs.
Habitat
Ponds, lakes, rivers, bays; likes shallow water.
Habits
Flies rapidly in V formation.
Voice
High-pitched, double-noted *wah-wah, wah-wah,* like human laughter.
Eggs
5–7; white; 3.1 x 2.1 in. (7.9 x 5.3 cm).
Range
Breeds in Greenland and nw. Canada; winters s. from Ill. to Gulf Coast; uncommon on Atlantic Coast.

SNOW GOOSE

Chen caerulescens **6:1**

Description
Size, 23–31 in. (58.4–78.1 cm). Adult: • *body white,* variegated, or entirely gray-brown; legs pink; head white (may be rust-stained); • *wings always with black tips;* bill pink with dark basal streak. Immature: similar but duskier; darker bill and feet.
Similarities
Swans have longer necks, no black in wings.
Habitat
Ponds, lakes, rivers, bays, marshes, and fields.
Habits
Flies in loose V's.
Voice
Single nasal honk.
Eggs
5–7; white; 3.2 x 2.1 in. (8.1 x 5.3 cm).
Remarks
"Blue" and white phases now considered same species. Birds wintering in Atlantic Coast are mainly white, while both blue and white phases occur on the Gulf Coast.
Range
Breeds in Arctic; winters along Atlantic Coast from N.J. to Fla. and w. along Gulf Coast to Tex.; casual to interior.

FULVOUS WHISTLING-DUCK

Dendrocygna bicolor **7:2**

Description
Size, 18–21 in. (45.7–53.3 cm). Upright posture, buff and rusty color, and • *white rump.*
Similarities
Cinnamon Teal is smaller, deeper color.

Habitat
Pastures and rice fields, and adjacent waters.
Habits
Active by night and day, feeds in fields as well as in water, dives regularly.
Voice
High-pitched double whistle.
Eggs
12–17; white; 2.2 x 1.5 in. (5.6 x 3.8 cm). Nest is in dense ground vegetation in wet areas.
Range
Resident along Gulf Coast from La. to s. Tex.

MALLARD

Anas platyrhynchos **7:10, 8:2**

Description
Size, 23 in. (58.4 cm). • *Speculum bluish-purple with white borders,* wing linings white. Male: breast chestnut; body gray; tail white; some black tail coverts curl forward; • *head green, narrow white neckband,* bill yellow. Female: brown, bill mottled orangish, tail with only some white. Southeast coastal (Florida and Gulf Coast) Mottled Duck **(7:10)** is a subspecies in which males are colored as females, but both sexes are darker than female Mallards, with a clear orange or yellow bill; wing speculum is bordered with black (very little white is evident).
Similarities
Northern Shoveler has uncrested green head, but breast is white, sides chestnut, and bill huge. Black Duck is darker than female Mallard and lacks white borders to speculum. Female Pintail is longer and neck more slender than female Mallard, has only 1 white rear border on speculum, wing linings not white, bill gray, tail more pointed. Female American Wigeon has more white in wing.
Habitat
Almost any water.
Habits
Flight is like Black Duck, wingbeat not very deep.
Voice
Male, a low, reedy *kwek-kwek-kwek;* female, the familiar barnyard quack.
Eggs
8–10; olive-green; 2.3 x 1.6 in. (5.8 x 4.1 cm). Nest is not always near water.
Remarks
This species is the ancestor of most of our domestic ducks. It frequently hybridizes with the Black Duck in the northeast.
Range
Breeds throughout Canada and n. U.S., from Greenland to Alaska, s. to Va. and Tex.; winters s. from s. Canada, throughout cen. and s. states.

BLACK DUCK

Anas rubripes **7:9, 8:1**

Description
Size, 23 in. (58.4 cm). Sexes similar. • *Very dark brown,* throat and cheeks paler; speculum purple bordered with black; paler head contrasting with very dark body; bill yellow to greenish; legs dusky

or red. Female Mallard is lighter, shows less head and body contrast, and has an orangish bill.

Similarities

Scoters have bigger heads, shorter necks, and are found along the coast.

Habitat

Ponds, lakes, often in woodland.

Habits

Similar to Mallard. In flight, shows all-silver wing linings, flashing against a nearly black body.

Voice

Similar to Mallard.

Eggs

6–12; creamy to greenish; 2.3 x 1.7 in. (5.8 x 4.3 cm). Nest is usually on ground, occasionally in an abandoned crow or hawk nest in a tree.

Range

Breeds throughout e. and cen. North America, from Labr. and Man. s. to Va., Ill., and Wis.; winters from coastal Newfoundland and s.-cen. Canada to Fla. and Tex.

GADWALL

Anas strepera **7:8, 8:3**

Description

Size, 20 in. (50.8 cm). The only surface-feeding duck with a • *white speculum,* conspicuous in flight, not easily seen on the water. Male: • *gray;* head and neck light brown, shoulder red-brown; • *rump and undertail coverts black* (conspicuous in flight), bill gray. Female: light brown, bill yellow-brown. In both sexes, belly white, feet yellow.

Similarities

White wing patch of American Wigeon more conspicuous and on forepart, not hindpart, of wings; on water, Wigeon shows white on flanks. Female Wigeon is ruddy-flanked, gray-headed. Female Mallard is larger. Female Pintail has blue-gray bill.

Habitat

Ponds, lakes, streams, marshes; fresh water and brackish.

Habits

Wingbeat faster than Wigeon's; flight swift, direct, in small compact flocks; wings long, pointed.

Voice

Male whistles and trills, female quacks.

Eggs

7–13; creamy; 2.2 x 1.6 in. (5.6 x 4.1 cm). Nest is not necessarily near water.

Remarks

No other duck has so wide a distribution. Of the larger regions of the world, it is missing only from South America and Australia.

Range

Breeds from s. Canada, s. to N.C. along Atlantic Coast and interior to Gulf Coast; winters from s. New England southward.

NORTHERN PINTAIL

Anas acuta **7:11, 8:4**

Description

Size, 22–27 in. (55.9–68.6 cm). Male: A noticeably long, pointed tail; distinctive white line up side of neck. Female: light brown, somewhat darker above; lacks very long central tail feathers. Both sexes, white below, speculum brown; bill and feet blue-gray. In flight, long-necked and long-tailed, slender, white underparts and white line on trailing edge of wing visible from afar.

Similarities
Female Mallard is larger, with shorter neck and tail, orangish bill, and speculum with 2 white borders. Female Gadwall has white speculum. The only other long-tailed duck, the Oldsquaw, is a diving species.

Habitat
Ponds, lakes, marshes, rivers; in winter, saltwater bays.

Habits
A fast flier, sits high on water.

Voice
Male, in flight, a loud *kwa, kwa;* a Teal-like whistle; female, a hoarse quack.

Eggs
6–12; olive-buff; 2.1 x 1.5 in. (5.3 x. 3.8 cm). Nest can be as much as a mile from water.

Range
Breeds from n. Canada, s. to w. Pa. and w. to Nebr. and casually farther e. in U.S.; winters s. from s. New England and s. Ill.

COMMON TEAL

Anas crecca **7:5, 8:5**

Description
Size, 14 in. (35.6 cm). Male: spotted tan breast, gray sides, • *chestnut-colored head;* • *green ear patch.* American form (Green-winged Teal) has • *white vertical crescent behind breast;* European form, which is rare in our area, lacks vertical bar but has horizontal white stripe on back above wing. Female: grayish-brown, speckled below; European form indistinguishable from American form.

Similarities
Shorter-necked, shorter-bodied than Blue-winged, and without blue on wing. From below, in flight, male Common shows light belly; male Blue-winged shows dark. Female resembles a pint-sized female Mallard. If a small duck without conspicuous wing colors springs up from a marsh, it is probably this species.

Habitat
Marshes, ponds, lakes, streams; brackish and salt water.

Habits
Sometimes walks about in mud feeding like shorebird; flight fast, buzzy, and erratic, in compact flocks wheeling like pigeons; wings whistle in flight.

Voice
Male, piping whistles; female, high-pitched quack.

Eggs
10–12; pale buff; 1.8 x 1.4 in. (4.6 x 3.6 cm). Nest is occasionally found some distance from water.

Range
Breeds from Que. to N.Y. and from Man. to Nebr.; winters s. from New England and from Mo. southward.

BLUE-WINGED TEAL

Anas discors **7:7, 8:6**

Description
Size, 15 in. (38.1 cm). Bill relatively large. A very small duck, with • *chalky-blue shoulder patches.* Male: grayish above, tan spotted

with dark below; white patch on rear of flanks; large • *white face crescent.* Female: brownish-gray above, pale gray marked with darker below.

Similarities

Northern Shoveler bill is shaped similar to Blue-winged. Female Northern Shoveler is larger and much heavier-billed; female Common Teal has no blue on wing, bill is smaller.

Habitat

Freshwater marshes, ponds; rarely salt water.

Habits

Flight erratic, blue wing patch may look white in poor light; dabbles in mud when feeding.

Voice

Male, whistling peep; female, quack similar to Common Teal's.

Eggs

6–12; white; 1.8 x 1.3 in. (4.6 x 3.3 cm).

Range

Breeds from Newfoundland, Que., and Man. s. to N.C. and Gulf Coast; winters from Md. and s. Ill. southward.

EURASIAN WIGEON

Anas penelope **7:3**

Description

Size, 19 in. (48.3 cm). Dusky axillars. Male: gray upperparts and flanks; pinkish breast; white wing coverts, rump, rear flanks, and belly; black primaries and tail coverts; green speculum, • *red-brown head, creamy-colored crown.* General impression is of a gray duck with a red-brown head.

Similarities

Male American Wigeon lacks brown on the head; female American looks grayer, has gray head, and white axillars. Redhead has black breast.

Habitat, Habits, Eggs

Similar to American Wigeon.

Voice

Shrill whistling *whee-you.*

Remarks

Common in Europe, this bird is regular, but quite rare, in this country.

Range

Breeds in Eurasia; regularly appears in small numbers along East Coast.

AMERICAN WIGEON

Anas americana **7:4, 8:8**

Description

Size, 20 in. (50.8 cm). Head and neck gray; speculum green; wing coverts and belly white; bill blue, black-tipped. Male: • *shiny white crown,* ear patch green (looks black at a distance); back brown; breast and sides pink; rump, rear of flanks, and large patch on front of wing white; primaries and undertail coverts black. In flight, white front wing patches and black undertail coverts conspicuous. Female: head gray, wing patches grayish-white, upper parts ruddy-brown, breast and sides tan, undertail coverts white.

Similarities

Gadwall lacks white crown and flank patches, sits lower on water; blue wing patches of Northern Shoveler and Blue-winged Teal may sometimes look white. See Eurasian Wigeon. Female Mallard is larger, darker, and has less white on wing.

Habitat
Lakes, streams, rivers, marshes, bays; prefers fresh water but also seen on salt.
Habits
Sits high on water, pivots as it feeds; flies in compact, irregular flocks.
Voice
Male: "a wild and musical note, *whew, whew, whew*." Female: "a *qua-awk, qua-awk*" (Kortright).
Eggs
6–12; creamy; 2.1 x 1.5 in. (5.3 x 3.8 cm). Nest is often at a distance from water.
Other name
Baldpate.
Range
Breeds in Nw., s. to Nebr. and Minn.; rare farther e.; winters along coast from N.J. southward, and in s. states, southward.

NORTHERN SHOVELER

Anas clypeata **7:6, 8:7**

Description
Size, to 19 in. (48.3 cm). Has a large, flat, spoon-shaped bill. Male: head green (looks black at a distance); body and tail white; • *sides rufous*; on water shows most white of any surface-feeding duck. In flight, from head to tail shows alternating pattern of dark-light-dark-light-dark. Female: brownish above, paler below. Both sexes have chalky-blue wing coverts, green speculum.
Similarities
Shoveler is larger and bigger-billed than Blue-winged Teal.
Habitat
Marshes and shallow water, fresh or brackish.
Habits
Flies and swims with bill pointing down at an angle, wings seem set far back on body; flight slower, more hesitating than teal's; sits low on water, feeds from surface, using bill as a strainer.
Voice
Male: "*woh, woh, woh*" (Collins); female: a weak quack.
Eggs
6–14; pale olive-green or buffy; 2.1 x 1.5 in. (5.3 x 3.8 cm).
Range
Breeds from N.S. and Man. s. to Del., N.C. and w. to Nebr.; winters s. from Ga.; some s. from New England.

WOOD DUCK

Aix sponsa **7:1, 8:9**

Description
Size, 18 in. (45.7 cm). Has a conspicuous crest. Male: boldly patterned with iridescent maroon, green, purple, and white; eclipse plumage much duller and with little or no crest. Female: head gray, crested; eye-ring, throat, and underparts white; back gray-brown, speculum blue. In flight, head held above level of body, bill pointed down at an angle; short neck and long square tail conspicuous; also dark breast, wings, and tail; white belly and white trailing edge of wing.
Similarities
See American Wigeon.
Habitat
Freshwater marshes, swamps, and creeks.

Habits
Often perches in trees within swamp; sits high on water; flight swift and direct.
Voice
Distinctive whistle in flight.
Eggs
10–15; dull white; 2.0 x 1.6 in. (5.1 x 4.1 cm). Nest is in hole in tree, often some distance from water.
Range
Breeds from N.S. and s.-cen. Canada s. to Fla. and Gulf Coast, w. to Tex.; winters from N.C. southward.

REDHEAD

Aythya americana

9:1, 11:1

Description
Size, 20 in. (50.8 cm). Male: • *rounded, brownish-red head,* gray back; blue bill with black tip. Female: brown; belly and area around bill white; head rounded.
Similarities
Canvasback has blackish bill, long sloping forehead, rustier head, whiter body. Female Common Goldeneye has brown head, white collar. Female scaup have broad white circle around base of bill. Female Ring-necked Duck is smaller, darker, and has white eye-ring and ring around bill. In flight, Redhead is shorter and darker than Canvasback, wingbeats more rapid, and flight more erratic; wing stripe of Redhead is long and gray, not white as in scaup.
Habitat
Ponds, lakes, rivers, bays.
Habits
Gathers in big rafts on large lakes and bays in winter.
Voice
Male: a catlike *me-ow.* Female: "a growl *r-r-r-rwha, r-r-r-r-wha*" (Griscom).
Eggs
10–15; pale olive-buff; 2.4 x 1.7 in. (6.1 x 4.3 cm).
Range
Breeds in nw. Canada, s. and e. to Wis. and Nebr., and occasionally farther e.; winters from Great Lakes s., and from s. New England southward.

RING-NECKED DUCK

Aythya collaris

11:3

Description
Size, 17 in. (43.2 cm). Head rather triangular in shape; bill with white ring, speculum bluish-gray. Male: head, breast, and back black; head with purple iridescence, sides gray (sometimes seem white), chestnut ring on neck seldom visible; on the water, shows a • *vertical white crescent* on its side behind the breast. Female: brown with white eye-ring and indistinct white area near bill. In flight, black back and forewing in male (dark brown in female) and gray wing stripe.
Similarities
Male scaups have lighter backs; female Redhead is larger and has less white on cheeks. Female scaup have more distinct white area around base of bill. In flight, male scaup show broad white wing stripe.
Habitat
Ponds, wooded lakes, streams; in South, saltwater bays.

Habits
Travels in small groups, alights without circling.
Voice
Similar to Lesser Scaup, seldom heard.
Eggs
6–12; dark olive-buff; 2.3 x 1.6 in. (5.8 x 4.1 cm).
Range
Breeds from Newfoundland and Man. s. to Maine and Iowa; winters from New England and Ill. southward.

GREATER SCAUP

Aythya marila **9:3, 11:5**

Description
Size, 18 in. (45.7 cm). Black with a white saddle, this is only eastern duck with a broad white wing stripe extending almost to wing tip; bill bluish. Male: black head, breast, primaries, and tail; gray back; white sides; • *head glossed with green* and quite rounded. Female: dark brown with sharply defined white patch around bill.
Similarities
Lesser Scaup has a thinner neck, shorter wing stripe, and purple gloss on the somewhat angular head of the male.
Habitat
Lakes, ponds, rivers, bays, often ocean.
Habits
Large rafts collect on bays in winter; in east, more inclined to salt water and larger bodies of fresh water than Lesser Scaup.
Voice
"A loud, discordant *scaup, scaup*" (Kortright).
Eggs
8–9; buffy-olive; 2.5 x 1.7 in. (6.4 x 4.3 cm).
Remarks
The extreme similarity of the 2 scaup is noteworthy. Presumably they represent formerly separated populations that evolved into separate species, the Greater in Eurasia, the Lesser in North America. The Greater is the more northerly and westerly, and probably has recently entered North America from Asia.
Range
Breeds in nw. Canada, s. and e. to Mich.; winters along Atlantic and Gulf coasts from N.B., southward, and in Great Lakes.

LESSER SCAUP

Aythya affinis **9:4, 11:4**

Description
Size, 17 in. (43.2 cm). Male: • *head iridescent purple* (but iridescence may vary with angle of light), flanks grayish, head somewhat angular. Female: dark brown, bill clearly "masked" with white at base.
Similarities
Smaller than Greater, especially in apparent bulk, and white stripe on wing noticeably shorter. Ring-necked Duck has a black back; female Redhead and Canvasback are larger; female Ring-necked has less distinct white around bill but has white eye-ring.
Habitat
In the east, in smaller bodies of fresh water than the Greater and less often in salt water.
Habits
Flight swift, erratic; often in closely bunched large flocks.

Voice
Male: a coarse *scaup;* in flight, a repeated *pppr-pppr.*
Eggs
6–15; buffy-olive; 2.3 x 1.6 in. (5.8 x 4.1 cm).
Other names
Bluebill, Broadhull.
Remarks
If a male scaup has a purple head sheen, it is a Lesser, but this is usually difficult to determine. On the water, watch for one to flap its wings, showing the length of the white wing stripe.
Range
Breeds in nw. Canada, se. to Nebr. and Iowa; occasionally farther e.; winters from Mass. and Ill. s. to Gulf Coast.

CANVASBACK
Aythya valisineria **9:2, 11:2**

Description
Size, 21 in. (53.3 cm). • *Bill long, blackish, merges into forehead* in a long continuous slope. Male: • *rusty head,* white back. Female: head and breast light brown, forehead sloping, back grayish.
Similarities
Male Redhead is gray above, bill bluish, head rounded. Similar females all have rounded foreheads.
Habitat
Ponds, lakes, rivers, bays.
Habits
In flight, long bill, head, and neck carried slightly down; pointed wings appear set far back on body. Flies in lines or V's; collects in great rafts on the water.
Voice
Male grunts, female quacks.
Eggs
7–9; grayish-olive; 2.5 x 1.8 in. (6.4 x 4.6 cm).
Range
Breeds in nw. Canada, s. and e. to Minn. and Nebr.; winters from Mass. and Great Lakes s. to Gulf Coast.

COMMON GOLDENEYE
Bucephala clangula **9:5, 10:3**

Description
Size, 18 in. (45.7 cm). Chunky body, high-domed head, short neck, golden eye, white wing patch; in flight, looks large-headed, short-necked. Male: white-sided duck with black head and • *round white jowl spot*; head, upperparts, and bill black (head with greenish gloss), rest of body white; great white squares in wings and blackish lining to wings from below. Female: brown head; gray back and sides; collar, breast, belly, and divided wing patch white; outer third of bill yellow in spring.
Similarities
Barrow's has shorter bill; winter females indistinguishable in field. Scaups have a black breast; mergansers are long-necked and have a rakish, not a stocky, look.
Habitat
Lakes, rivers, bays, and ocean.
Habits
Wings whistle in flight; dives frequently when feeding; flock rises all at once, does not string out in flight; female sits lower in water than male.

Voice
Male: "a penetrating *spear, spear*" (Collins). Female: a low quack.
Eggs
8–15; pale green; 2.4 x 1.7 in. (6.1 x 4.3 cm).
Range
Breeds from Labr., s. to n. New England, w. to Minn.; winters in n. waters, s. to Fla. and Gulf Coast.

BARROW'S GOLDENEYE
Bucephala islandica **10:1**

Description
Size, 18 in. (45.7 cm). Male: • *head has purplish gloss* and bulges fore and aft; crown has a snoodlike droop at rear (not high-domed, as in Common); has a white • *crescent-shaped jowl spot.* On water, sides and wings show much more black than does Common; black wedge points down at shoulder, separating sides from breast; row of white spots on black scapulars; black line divides white wing patch. Female: in spring breeding plumage, bill all-yellow, head darker than female Common; forehead more abrupt; in winter, indistinguishable in field.
Similarities
Similar to the more abundant Common Goldeneye, but bill shorter and thicker at base, head shaped differently.
Habitat
Lakes, bays, ocean.
Habits
In winter, remains farther north than Common.
Voice
Croakings.
Eggs
6–14; pale greenish; 2.4 x 1.7 in. (6.1 x 4.3 cm).
Range
Breeds along Labr. coast; winters from Maritime Provinces s. to N.Y.

BUFFLEHEAD
Bucephala albeola **9:6, 10:2**

Description
Size, 14 in. (35.6 cm). Small, puffy head on chunky body, short-necked; bill blue-gray, stubby; white wing patches are conspicuous in flight. In flight, Bufflehead resembles goldeneye, but flies faster and nearer water. Male: on water, whitest eastern duck; suggests a small goldeneye, with a • *large, white head patch* extending over the rear of crown. Female: upperparts dusky, head with a slanting white cheek patch; at a distance, suggests female goldeneye.
Similarities
Male Hooded Merganser looks dark, has long head and body, long, thin black bill, black neck; its white head patch has a black border.
Habitat
Lakes, rivers, bays, ocean.
Habits
Occurs in small groups, can dive from wing, and can fly directly into air from under water.
Voice
Seldom heard; male whistles, female quacks.
Eggs
6–14; buffy; 1.9 x 1.4 in. (4.8 x 3.6 cm).

Other names
Buffleheads are locally known as Butterball, Cock-dipper, Didapper, Dopper, Marionette, Scotchman, Shotbag, Spirit Duck, and Woolhead.
Range
Breeds from Ont. to Man.; winters along Atlantic Coast from N.B. to Fla. and from Great Lakes to Gulf Coast.

OLDSQUAW

Clangula hyemalis **9:9, 10:5**

Description
Size, male, 21 in. (53.3 cm); female, 16 in. (40.0 cm). Only duck in this area with • *all-dark wings* and white on the body. On water looks chunky, bill small. Male: in summer, dusky with • *white on face and belly;* in winter, white with dark cheeks, back, and breast. From below, in flight, white with conspicuous black wings and breast band. In winter has an all-white crown and at all seasons has long, pointed tail. Female: largely brown, including crown and ear/cheek patch, but plumage lighter in winter than in summer; tail short. Immature: in fall often quite dark.
Similarities
Head pattern of young female resembles changing Harlequin.
Habitat
Large lakes, bays, ocean.
Habits
In flight, wings low, curved, pointing to rear; wingbeats rapid; flight erratic, buzzy, low over water, in small flocks veering (like shorebirds) and flashing black and white; often alights with great splash. Dives for shellfish.
Voice
A "musical, gabbling *south, south-southerly* or *how doodle do*" (Collins).
Eggs
6–10; light grayish-olive; 2.1 x 1.6 in. (5.3 x 4.1 cm).
Range
Breeds in Arctic, s. to Hudson Bay; winters along Atlantic Coast s. to S.C.

HARLEQUIN DUCK

Histrionicus histrionicus **10:4**

Description
Size, 17 in. (43.2 cm). On water, small and chunky; in flight, suggests goldeneye but is dark; bill small. Male: slate-blue body, white markings, and rusty flanks are unique, but looks black at a distance; conspicuous white marks on head, neck, sides, wings; changing male presents bizarre pattern. Female: brown with 3 white spots on head, no wing patch.
Similarities
Female has shape of Bufflehead and pattern of female scoters. Female Surf and White-winged scoters are bigger, heavier, and have 2 white head spots. Female Bufflehead has 1, and a white wing patch; female Oldsquaw has whiter head and neck.
Habitat
In winter, the ocean, especially near rocky shores; in summer, mountain streams.
Habits
Floats high in water in close formation, often with tail cocked;

feeds around rocks; flies fast, low to water, in compact flocks; can dive from wing.

Voice
Whistle or squeak.

Eggs
5–10; buffy; 2.3 x 1.6 in. (5.8 x 4.1 cm). Nest is in rock crevices, logs, or holes in trees.

Range
Breeds in Arctic, s. to Labr.; winters along Atlantic Coast s. to N.Y.

COMMON EIDER
Somateria mollissima *Fig. 5*

Description
Size, 24 in. (61.0 cm). Head wedge-shaped, bill and forehead form a continuous slope. Male: • *black underparts*, crown, and tail, white head and breast; black wings with white coverts. Female: reddish-brown barred with black. First-year male: often with a white collar, is intermediate between female and adult male, or all-dark with whitish breast and wing patch (like female goldeneye with a slanting forehead).

Similarities
Slope of forehead resembles Canvasback. Female Common Eider is similar to female King Eider, but top of bare upper bill slopes almost to eye.

Habitat
Rocky seacoasts, islands; the ocean over shoals.

Habits
Flies low, sluggishly with neck drooping; heavy wingbeats alternate with gliding; looks very large in flight; swims with neck in and bill down.

Voice
Male gives a low, plaintive *he-ho-ha-ho*, female quacks.

Eggs
5–10; dull green; 3.1 x 2.1 in. (7.9 x 5.3 cm).

Range
Breeds from Greenland and Arctic to Maine; winters along Atlantic Coast s. to N.Y.

Fig. 5

KING EIDER
Somateria spectabilis **9:10, 10:9** *Fig. 5*

Description
Size, 22 in. (55.9 cm). Male: distinguished by his large orange forehead shield; pearly crown; black back, wings, and belly; white

breast, wing patch, and flank spot near tail; face white with greenish cast. Forehead slopes up abruptly from bill, back appears shorter than in Common; in flight, shows large white wing patches; at a distance looks white in front, black to rear. Female: difficult to distinguish from female Common Eider, but forehead more abrupt, top of bare upper bill much shorter, plumage lighter, more rusty. Immature male: dusky; abrupt forehead; dark brown head with indication of adult facial pattern; light breast.

Similarities
Female Common Eider very similar; see above.

Habitat
Large lakes, rocky ocean shorelines, offshore reefs.

Habits
In flight, similar to Common, but beak droops less.

Voice
Male moans, female quacks.

Eggs
4–7; dull green; 2.0 x 2.0 in. (5.1 x 5.1 cm).

Range
Breeds in Arctic from Greenland to Hudson Bay; winters along Atlantic Coast s. to Mass.; casually to great Lakes.

WHITE-WINGED SCOTER

Melanitta fusca **9:11, 10:8**

Description
Size, 21 in. (53.3 cm). Both male and female have • *white wing patches.* These patches are not always evident, when the bird is in the water, until it flaps its wings. Male: white streak under eye; bill orange with a black knob. Female: 2 whitish cheek patches prominent in younger birds, sometimes obscure in adult females.

Similarities
See Black Guillemot.

Habitat
Ocean, bays, large lakes.

Habits
Travels in large flocks.

Voice
In flight, a low, bell-like whistle; also a croak.

Eggs
9–14; pinkish-buff; 2.6 x 1.8 in. (6.6 x 4.6 cm).

Range
Breeds from Man. s. to N.Dak.; winters s. to S.C. and La.

SURF SCOTER

Melanitta perspicillata **9:12, 10:7**

Description
Size, 19 in. (48.3 cm). No white wing patches; bill forms an even slanting line with forehead. Male: white forehead and nape; bill varicolored red, white, orange, and black. Female: face has 2 light patches in younger birds, more obscure in older females (similar to female White-winged, but without wing patch); sometimes shows whitish patch on nape.

Similarities
In flight, resembles Black Scoter, but lacks silvery sheen under flight feathers.

Habitat
Oceans, bays, large lakes.

Habits
Similar to White-winged, but alights with wings held upward; wings hum in flight.
Voice
Croak.
Eggs
5–9; creamy; 2.4 x 1.7 in. (6.1 x 4.3 cm).
Range
Breeds in n. Canada s. to n. Que.; winters on Atlantic and Gulf coasts.

BLACK SCOTER

Melanitta nigra **10:6**

Description
Size, 19 in. (48.3 cm). Smallest-looking, most ducklike scoter, primaries with silvery sheen beneath. Male: all-black with an orange bill knob. Female: brown with black crown and whitish cheeks.
Similarities
Female White-winged and Surf Scoters have 2 light patches on cheeks; winter Ruddy is smaller, paler, and with white chest.
Habitat
Coast and tundra, rare inland.
Habits
Travels in large flocks, flight more clean-cut than other scoters; on water, often cocks pointed tail, rides high with head high and bill horizontal or uptilted (other scoters carry bill pointed down).
Voice
Musical *cour-loo.*
Eggs
6–10; pale ivory-yellow; 2.5 x 1.7 in. (6.4 x 4.3 cm).
Former name
Common or American Scoter.
Range
Breeds in Labr. and Newfoundland; winters along Atlantic Coast s. to S.C.; rarely in interior.

HOODED MERGANSER

Mergus cucullatus **9:7, 11:6**

Description
Size, 18 in. (45.7 cm). The smallest, slimmest, merganser, male unlike others in pattern; bill small, thin. Male: • *black-bordered white crest* is unique among eastern water birds; head, upperparts, and 2 vertical lower breast stripes black; underparts and wing patch white, sides brown; in flight, crest shows as a white streak on head, which is held lowered. Female: crest buffy; smooth-edged • *brown head*; gray-brown upperparts, flanks, and breast; white wing patch and belly.
Similarities
See Bufflehead.
Habitat
Ponds, streams, lakes, rivers.
Habits
On water, sometimes cocks tail, can rise into air with great speed; usually found in pairs or small groups; male frequently raises and lowers crest.
Voice
Grunting *crew, crew.*

Eggs
6–12; white; 2.1 x 1.8 in. (5.3 x 4.6 cm). Nest is in hole in tree or stump near water.
Range
Breeds from N.S. and Man. s. to Tenn. and Nebr.; occasionally seen in se.; winters from New England and Nebr. s. to Gulf Coast.

COMMON MERGANSER
Mergus merganser **11:8**

Description
Size, 24 in. (61.0 cm). Large, loonlike, long bill. Male: crest not noticeable; back, primaries, and tail black; • *head green-black*, bill red; rest of bird is white, breast with rosy blush. In flight, shows more • *white on body* and wings than any other duck. Female: moderate crest; head bright red-brown, sharply contrasting with white throat and neck; upperparts gray, wing-patch white.
Similarities
Red-breasted male has red breast, conspicuous crest; female is more crested, and red-brown of head blends into white of throat and breast; female goldeneyes are shorter and stockier, with shorter bill.
Habitat
Lakes, ponds, rivers.
Habits
Submerges by jumping, then diving, or by gradually sinking; flies in string formation, low, loonlike, horizontal; flight shape rakish.
Voice
Usually silent; occasionally "an unmelodious squawk" (Swarth).
Eggs
6–17; creamy; 2.5 x 1.8 in. (6.4 x 4.6 cm). Nest is in hollow tree or stump, sometimes on ground.
Former name
American Merganser.
Range
Breeds from Newfoundland and Man. s. to New England and Mich.; winters s. to Gulf Coast.

RED-BREASTED MERGANSER
Mergus serrator **9:8, 11:7**

Description
Size, 22 in. (55.9 cm). Often has a • *double-pointed crest* and red bill. Male: black head and back, head with a greenish gloss and an untidy crest; sides gray; collar, wing patch, and • *underparts white*; breast red-brown, dark patch with white spots near shoulder. On water, looks slim and dark; in flight, red breastband conspicuous between white neck and belly; white wing patch appears framed. Female: see Common Merganser.
Similarities
Uncrested male Common Merganser is mostly white, and has more white on wing patch; female Common has a brighter head, more of a crest, and sharp line of demarcation between throat and breast. See also Red-necked Grebe.
Habitat
Lakes, ponds, rivers; in winter, bays and ocean. (Common prefers fresh water.)
Habits
Flight is swift, noiseless, and direct, with head, neck, and body horizontal; flattens crest before diving; flies in string formation.

Voice
Guttural croaks when alarmed.
Eggs
6–12; creamy buff; 2.5 x 1.8 in. (6.4 x 4.6 cm).
Range
Breeds in Arctic from Labr. s. to Maine, Mich.; winters along Arctic Coast s. to Gulf Coast.

RUDDY DUCK
Oxyura jamaicensis **11:9**

Description
Size, 15 in. (38.1 cm). Thick-necked, chunky; bill broad, and upturned; • *cheeks white*; wings entirely brown. Male: in summer, mostly rich red-brown, cap black, bill bright blue, cheeks white; in winter, red is replaced by gray and brown, cap becomes dark brown, and bill much duller. Female: similar to winter male, but with black streak on cheek and bill dusky.
Similarities
Winter male resembles female Black Scoter but is smaller, whiter-cheeked, and seldom seen on ocean.
Habitat
Ponds, lakes, streams, rivers, sheltered bays.
Habits
Swims buoyantly with tail often cocked; can sink slowly under water or dive abruptly; needs long run for takeoff into air; flight buzzy.
Voice
Silent, except for weak cluck when courting.
Eggs
5–15; pale buff; 2.5 x 1.8 in. (6.4 x 4.6 cm).
Range
Breeds in cen. Canada from Que. s. to Gulf Coast, southward; winters s. from Mass. and Ill.

Vultures, Hawks, and Falcons
Order Falconiformes

These birds have hooked beaks and, except for the vultures, strong toes and long claws. Females are generally larger than males. (Where two sizes are given, the smaller is the average size for the male, the larger the average for the female.) These birds hunt by day. Vultures feed on carrion, the others usually on freshly killed animal life. Many are powerful fliers and soarers.

AMERICAN VULTURES
Family Cathartidae

The vultures in east are black and have naked heads. They have excellent sight and are magnificent soarers. Recent studies show that they may locate their food—carrion—by smell. Their voice is a hiss. They nest in hollow logs, crevices in rocks, or other hideaways on the ground, and they feed their young by regurgitation.

TURKEY VULTURE

Cathartes aura **12:1**

Description
Size, 29 in. (73.7 cm); wingspread, 6 ft. (1.8 m). All-black bird that soars on long wings held at an angle. Head small, red (dark in immatures); flight feathers lighter than coverts, giving wings a 2-toned look from below; outer flight feathers spread like fingers; tail long, narrow, rounded.
Similarities
Black Vulture has a shorter, squared tail and a white spot under end of wings; it is clumsier, and flaps more. Northern Harrier also holds wings at an angle, but is smaller, slimmer, and white-rumped. Eagles soar on horizontal wings.
Habitat
Usually seen soaring over mountains, woods, farm country, but rare over unbroken forest.
Habits
Perches with shoulders hunched, roosts in groups; soars effortlessly, flaps infrequently.
Eggs
1–3; white, marked with brown and purple; 2.8 x 1.9 in. (7.1 x 4.8 cm).
Range
Breeds throughout most of e. U.S., s. to Gulf Coast; winters from s. New England s. and w. to Nebr.

BLACK VULTURE

Coragyps atratus **12:2**

Description
Size, 25 in. (63.5 cm); wingspread, 5 ft. (1.5 m). An all-black bird with whitish patches near the ends of its wings. Head small, black; tail short, square (legs often protrude in flight); underwings silvery.
Similarities
Black's wings are shorter, broader than Turkey Vulture's. Eagles and dark phases of large hawks have large heads.
Habitat
Open sky, beaches, garbage dumps, slaughterhouses.
Habits
Gregarious; in air, appearance more compactly built, heavier than Turkey Vulture; wings held almost horizontal, tail often fanned; flaps wings more often, flight more labored than Turkey's.
Eggs
1–3; bluish-white, marked with brown; 3.1 x 2.0 in. (7.9 x 5.1 cm).
Range
Resident from Pa., Ind., and Kans. southward.

KITES, HAWKS, EAGLES, AND HARRIERS

Family Accipitridae

In the past, the predatory habits and seeming love of independence of these birds made them suitable symbols for coats of arms. Today hawks are recognized as of great value to agriculture because of the millions of rodents, grasshoppers, and other insects they annually destroy.

Kites are graceful, falcon-shaped hawks with pointed wings. They often have forked tails, weak bills, and feet adapted for capturing insects and small reptiles and mammals.

The accipiters, or bird hawks (Genus *Accipiter*), have short, rounded wings, are finely barred below, and have long tails barred with narrow black bands. Adults are slate-gray above with dark caps and barred below. They frequent woods.

The buteos (Genus *Buteo*) are medium-sized to large hawks with broad wings and generally short, fan-shaped tails. All are dark above. They hunt by soaring and circling, then dropping down to seize prey, mainly rodents.

The eagles are essentially very large hawks. They have fully feathered heads, large bills, and long tails and wings that are held more or less horizontally in flight. They make huge nests on trees or ledges.

Harriers are slender birds of prey that hunt small birds or rodents in open rangeland. They have slim wings and long tails and fly lazily at low altitude, on dihedral wings.

WHITE-TAILED KITE

Elanus leucurus **16:4**

Description
Size, 15–16 in. (38.1–40.6 cm); wingspread, 3⅓ ft. (1.0 m). A gull-like gray-and-white hawk with black shoulders. Adult: • *square tail* and • *whitish head, black* of *wing bend* visible in flight. Immature: similar but with rusty streaking above and below.

Habitat
Meadows and marshes.

Habits
Circles low over meadows, hovering in search of prey, then pounces.

Voice
Whistled call.

Food
Small mammals.

Eggs
3–5; blotchy; 1.6 x 1.2 in. (4.2 x 3.2 cm). Nest is of twigs in a tree near open country.

Range
Resident s. Tex., southward.

SNAIL KITE

Rostrhamus sociabilis **15:3, 16:7**

Description
Size, 15–17 in. (38.1–43.2 cm). Black or brown, white-rumped hawk with a long needle-hook on the bill and square tail. Male: black with white base of tail and rump, reddish legs and base of bill. Females heavily blotched, immatures streaked with brown on head and underparts; whitish in tail and wings, orange legs.

Similarities
Northern Harrier is distinguished from female Snail Kite by white on its tail.

Habitat
Freshwater lakes and marshes.

Habits
Glides low over marshes, dropping to the ground for food, then perching to eat.

Voice
Shrieking note.

Food
Freshwater snails.
Eggs
3–4; brown, marked white; 1.7 x 1.4 in. (4.3 x 3.6 cm).
Former name
Everglade Kite.
Remarks
This bird is endangered in the United States and is limited to a small area of Florida, where at most a few score remain.
Range
Resident in s. Fla., chiefly Lakes Okeechobee and Loxahatchee.

MISSISSIPPI KITE

Ictinia mississippiensis **16:3**

Description
Size, 14 in. (35.6 cm); wingspread, 3 ft. (0.9 m). A • *black-tailed*, falcon-shaped hawk. Adult: bluish-gray above; • *pale head, with unmarked gray below;* long, pointed wings; dark primaries, light secondaries; tail notched or squared; eyes and legs reddish. Immature: streaked above, spotted below; tail banded.
Similarities
Graceful flight and gray color of adult distinguish this from hawks of similar size.
Habitat
Open country and woods.
Habits
Flight buoyant and gull-like; often tilts, showing 2-toned upper surface of gray wings; gathers in flocks in migration.
Voice
"Phee-phew" (Sutton).
Food
Insects.
Eggs
2–3; blue-green; 1.6 x 1.3 in. (4.1 x 3.3 cm). Lined nest of twigs in treetop.
Range
Breeds in se. and s.-cen. U.S.; winters from Fla. and Tex. southward.

Note: In southern Florida observers may see an occasional **SWALLOW-TAILED KITE,** *Elanoides forficatus* **(15:4),** size, 22–24 in. (55.9–61.0 cm). A beautiful, fork-tailed bird of prey, this is often seen in small groups over swampy forests and adjacent marshes. It is considered an occasional visitor to the Gulf Coast. White head and underparts, with blackish wings and tail characterize this distinctive bird.

NORTHERN GOSHAWK

Accipiter gentilis **13:1**

Description
Size, 20–26 in. (53.3–63.5 cm); wingspread, 4 ft. (1.2 m). A large gray hawk with short, rounded wings and a long, nearly squared tail. Adult: • *white line over eye*, dark cap and ear patch, and finely gray-barred pearly breast; paler above than Cooper's. Immature: brown above, streaked below, distinct white eye line.
Similarities
In flight, proportionately heavier about head and neck than smaller

Cooper's, and tail barring less conspicuous. Gyrfalcon has long, pointed wings.

Habitat

Forests, especially coniferous.

Habits

Silently glides and swoops through woods, dropping onto prey.

Voice

Silent except when breeding, then various cackles.

Food

Large birds, rabbits, other mammals.

Eggs

2–5; bluish-white; 2.3 x 1.8 in. (5.8 x 4.6 cm). Nest of sticks is in tree.

Range

Breeds throughout Canada from Newfoundland and n. Que. s. to New England and Mich.; seen along Appalachians s. to Md.; winters s. to Va., Tenn., and Tex.

COOPER'S HAWK

Accipiter cooperii **13:2**

Description

Size, 14–20 in. (35.6–50.8 cm); wingspread, 3 ft. (0.9 m). A medium-sized hawk with short, rounded wings and a • *long, rounded tail*; size about that of a crow. Adult: breast barred with red brown. Immature: usually more finely streaked below than young Sharp-shinned.

Similarities

Broad-winged Hawk has longer wings, which are light and unbarred below, and a fan-shaped tail; wings of Merlin are long and pointed; difference between tails of Cooper's and Sharp-shinned is best seen when tail is folded.

Habitat

Deciduous forest.

Habits

Wingbeats slower than Sharp-shinned's; circles and soars more.

Voice

Noisier than Sharp-shinned; "a shrill *quick, quick, quick*" (May), sometimes suggestive of a Common Flicker.

Food

Birds, small mammals.

Eggs

3–5; bluish-white, often with brown spots; 1.9 x 1.5 in. (4.8 x 3.8 cm).

Range

Breeds in s. Canada, s. to Fla.; winters from Maine and s. Ont. southward.

SHARP-SHINNED HAWK

Accipiter striatus **13:3**

Description

Size, 10–14 in. (25.4–35.6 cm); wingspread, 2 ft. (0.6 m). A small hawk with • *short, rounded wings* and a • *long, square tail*, sometimes slightly notched. Adult: slate-gray above, breast barred with red-brown. Immature: above brown; below white with brown streaks; more reddish in female, darker in male. Tail may look slightly rounded if spread like fan. When perched, Sharp-shinned's wings reach to lower one-third or one-half of tail.

Similarities
Merlin's wings when perched reach almost to tip of tail. Large female Sharp-shinned may be as large as a small male Cooper's, but Cooper's tail is rounded; American Kestrel and Merlin have long, pointed wings.
Habitat
Woodland and edges.
Habits
Rarely soars, except in migration, but occasionally ascends in tight circles with much flapping.
Voice
Cack, cack, cack.
Food
Mainly small birds.
Eggs
3–5; bluish-white, beautifully spotted with brown and lilac; 1.5 x 1.2 in. (3.8 x 3.0 cm).
Remarks
Hawk Mountain Sanctuary, Pa., is famous for fall migrations of this and other hawks.
Range
Breeds in cen. Canada, from Newfoundland s. to Fla. and Tex.; winters s. from New England southward.

RED-TAILED HAWK

Buteo jamaicensis **13:7, 14:4**

Description
Size, 19–25 in. (48.3–63.5 cm); wingspread, 4½ ft. (1.4 m). A large hawk with a red tail; chunkier than Red-shouldered and with wider wings, shorter tail. Adult: from below, pale, white area on breast set off by dark area of throat and dark band on lower breast; often with black wrist mark. Individuals vary to almost wholly dark below, but adults normally have a red tail. Immature: underparts streaked with brown; no wrist mark; tail brown above, barred low. Red tail is not always obvious from below; wait until the soaring hawk wheels, showing upper surface of tail. Adult of northern Plains race is very pale, with whitish head and pinkish tail. Western race typically rufescent below, but dark phase occurs. Far northwestern race is very dark, tail mottled without red; winters in small numbers in southern Plains to Louisiana and Arkansas. All races may mix on Southern Plains in winter.
Similarities
Hard to distinguish Red-tailed from young Red-shouldered, although latter is usually uniformly streaked below, without tendency to zoning, as in Red-tailed; immature Red-taileds are sometimes whitish at base of tail; compare with Swainson's, Rough-legged, Ferruginous.
Habitat
Common in wooded or open areas, marshes.
Habits
Soars and circles for long periods, often twisting tail at angle to body (wings shorter than Turkey Vulture and held horizontally), occasionally hovers.
Voice
Squealing *kee-a-a-a-r-r-r*, suggesting escaping steam; longer and higher than "*blue-kay*" note of Red-shouldered.
Food
Rodents and other small mammals.

Eggs
2–4; white, sparingly spotted with brown; 2.3 x 1.9 in. (5.8 x 4.8 cm).
Former name
The dark northwestern form of this hawk was known as Harlan's Hawk.
Range
Breeds throughout s. Canada and U.S.; winters s. from Maritime Provinces southward.

RED-SHOULDERED HAWK

Buteo lineatus **13:5, 14:1**

Description
Size, 17–24 in. (43.2–61.0 cm); wingspread, 4 ft. (1.2 m). Noted for the narrow white bands on its black tail; intermediate in size between larger Red-tailed and smaller Broad-winged. Wings and tail longer and slimmer than other buteos; white "translucent" spots near wing tip are set off by dark reddish-brown underwing coverts; upperwings barred with black and white. Adult: underparts cross-barred with bright rufous. Immature: duller, often with pale band on upper wing near tip; underparts uniformly streaked, tail bands usually brown and reddish. Races from southern Florida and the Keys are fairly to very pale about the head and chest.
Similarities
Immature Red-tailed's streaks tend to be in bands; Broad-winged has broad white bands on tail.
Habitat
Wooded or open areas, often found in moister terrain and smaller woodlands than Red-tailed.
Habits
Perches less conspicuously, soars less, and has less buoyant flight than Red-tailed.
Voice
Clear whistled *kee-yer*; never wheezy, like Red-tailed; often imitated by Blue Jay.
Food
Rodents and other small mammals, reptiles, and amphibians.
Eggs
2–4; white, marked with brown and buff; 2.2 x 1.7 in. (5.6 x 4.3 cm).
Range
Breeds in s. Canada s. to Gulf Coast; winters from s. New England and s. Minn. southward.

BROAD-WINGED HAWK

Buteo platypterus **13:4, 14:3**

Description
Size, 15–17 in. (38.1–43.2 cm); wingspread, 3¾ ft. (.8 m). A buteo with • *2 or 3 broad, white bands on its black tail;* chunky, like a little Red-tailed; the broadest-winged, broadest-tailed, and smallest buteo. Throat and eyebrow white; wings from below mainly white with dark tips and black wrist marks, without the translucent spots of the Red-shouldered. Adult: cross-barred with rusty below. Immature: underparts buff, streaked with brown; dark tail bands more numerous, white bands narrower and less distinct than in adult (more like Red-shouldered, but wings relatively broader, tail shorter—sometimes a difficult identification).

Similarities
Broad-winged is duller than Red-shouldered, and without zoning of Red-tailed.
Habitat
Deciduous woods.
Habits
Sluggish, tame; much given to soaring; when hunting, sometimes hovers.
Voice
Drawn-out, high-pitched whistle.
Food
Insects, mammals, reptiles, amphibians.
Eggs
2–4; whitish, marked with reddish-brown; 1.9 x 1.6 in. (4.8 x 4.1 cm).
Range
Breeds from s. Canada s. throughout e. U.S.; winters s. of U.S.

SWAINSON'S HAWK

Buteo swainsoni **13:8, 14:2**

Description
Size, 19–22 in. (48.3–55.9 cm); wingspread, 4½ ft. (1.4 m). In light plumage, with • *dark breast* and • *2-toned wing linings*; wings long and somewhat pointed for a buteo; tail long. Adult: light phase shows throat and belly white; tail dark gray above, lighter at base, with 9–12 indistinct dark bands below; dark phase shows nearly entire plumage sooty brown, sometimes with some white underwing or ashy bars on tail. Immature: buff, streaked with dark brown below (usually darker than other young buteos); breast usually darker; wings and tail indistinctly barred with brown. Many variations among these 3 plumages occur.
Similarities
Red-tailed has white breast, streaks at rear; dark phase of Red-tailed has reddish tail above. Dark phase of Rough-legged has clear white underwing flight feathers.
Habitat
Open areas.
Habits
Sluggish, tame; wingbeats faster than Red-tailed's; hunts by cruising low over prairie with wings in open V, as does Northern Harrier or Rough-legged, not by soaring high as does Red-tailed; gregarious, especially in migration.
Voice
Long "plaintive whistle, *kree-e-e-e*" (Bent), suggesting Broad-winged.
Food
Grasshoppers, rodents.
Eggs
2–4; dull white, with umber spots; 2.3 x 1.8 in. (5.8 x 4.6 cm).
Range
Breeds chiefly in w. North America; occasionally e. to Minn., Mo., Tex.; winters in S. America, small numbers in Fla.

Note: The **SHORT-TAILED HAWK,** *Buteo brachyurus* **(14:5),** size, 17 in. (43.2 cm), is a buteo the size of a crow. It is found in swampy forests of South Florida and the Keys. It has 2 phases: the adult white phase has a white belly and underwing coverts, with a conspicuous white patch on its forehead; the adult black phase is black on the belly and has underwing coverts; tails of both show indistinct barrings.

ROUGH-LEGGED HAWK

Buteo lagopus **13:9, 14:6**

Description

Size, 20–23 in. (50.8–58.4 cm); wingspread, 4½ ft. (1.4 m). Large and dark, with rather long and pointed wings for a buteo and longish tail; • *rump and base of tail white*; feathered legs sometimes visible as feet drop before swoop. Adult: light phase shows head and upper breast buffy streaked with brown, with black patch at bend of light underwing; • *belly, broad end of white tail*, wing tips, and rest of plumage *blackish*; in dark phase shows all-dark except for white on underwing and base of tail, occasionally even lacks white base of tail. Immature: similar to light adult, but dark belly band more pronounced. Birds with intermediate plumages not uncommon. In most of our area only in winter.

Similarities

Dark phase of Rough-legged may resemble young Golden Eagle. Dark phase of Ferruginous usually has some rufous mixed with the black, and has more white at base of primaries. Dark phase of Red-tailed usually has some rufous in tail.

Habitat

Open areas.

Habits

Sluggish, perches on observation post by meadow, hovers in air like Kestrel, or hangs motionless on updraft. Quarters low over meadows like a Northern Harrier, but is larger, with broader wings, and may light on a tree, which Harrier does not. Soars with wings "fingered," and sometimes spirals upward; at other times alternately flaps heavily and sails.

Voice

Silent in winter; a squealing *hurry-up* when breeding.

Food

Mice, lemmings.

Eggs

2–5; whitish, blotched with brown; 2.2 x 1.8 in. (5.6 x 4.6 cm).

Range

Breeds in Arctic s. to Newfoundland and Man.; winters s. to Va.

FERRUGINOUS HAWK

Buteo regalis

Description

Size, 22½–25 in. (57.2–63.5 cm); wingspread, 4¾ ft. (1.5 m). The largest buteo; the reddish-brown legs make a distinctive dark V against light underparts. Light area in extended primaries in all plumages; more slender than Rough-legged and with longer tail and larger bill. Adult: light phase shows shoulders, rump, and thighs rusty; • *head and tail whitish;* underwings and underparts also whitish, lightly streaked and barred. Dark phase (rare) shows all-dark with some rufous mixed in, and with tail light or with several narrow white bands. Immature: "underparts and tail whiter than in other hawks" (Bent).

Similarities

Western race of Red-tailed lacks reddish-brown upperparts of Ferruginous. The darker Rough-legged has a broad black band on its tail. Ferruginous is smaller than young Golden Eagle, with wings more pointed and tail proportionately longer. Swainson's is smaller, usually with finely barred tail and dark rear edge of wing.

Habitat

Open areas, prairies, badlands.

Habits
Perches on 1 leg on observation post; takeoff is slow and heavy, but flight is swifter than Rough-legged's; quarters like a Northern Harrier alternately flapping and sailing; also soars like a Red-tailed.
Voice
Various squeals; also a gull-like *kaah*.
Food
Rodents, especially ground squirrels.
Eggs
3–5; white marked with brown; 2.4 x 1.9 in. (6.1 x 4.8 cm).
Range
Breeds primarily in W., from Canadian prairies, s. to Ariz. and Okla.; winters in sw. U.S.

GOLDEN EAGLE
Aquila chrysaetos **12:6**

Description
Size, 31–40 in. (78.7–101.6 cm); wingspread, 7½ ft. (2.3 m). The only eagle that is all-dark below, including the wing linings. Adult: above dark brown, with golden-brown nape (visible only at close range); base of undertail lighter. Immature: base of tail white both above and below, giving • *"ring-tailed"* effect; base of primaries white, rest of body dark brown; amount of white diminishes with age.
Similarities
Wings and tail narrower in Bald Eagle; immature Bald has white on coverts, not flight feathers, and never has sharply ringed tail.
Habitat
Woods, mountains, badlands (rare).
Habits
In flight, tips of flight feathers are outspread and upcurved; soars high, then dives for prey; in straight flight, alternately flaps and glides; wings beat faster than Bald Eagle's; occasionally hovers.
Voice
"A shrill whistled *kee-kee-kee*" (Bendire).
Food
Mainly smaller mammals, a few birds.
Eggs
1–3; white, blotched with red-brown; 3.0 x 2.3 in. (7.6 x 5.8 cm).
Range
Breeds in n. Canada, s. to N.Y. and Nebr.; also occurs in Appalachians s. to N.C.; winters casually s. to Gulf states.

Note: The Golden Eagle of the mountains also nests near the sea abroad. The **WHITE-TAILED** or **GRAY SEA EAGLE,** *Haliaeetus albicilla*, size, 27–36 in. (69–91 cm), resembles the Golden Eagle. It breeds in western Greenland and possibly winters into the eastern Canadian Arctic. It is a coastal eagle, like the Bald Eagle, but with a brown head and fully white tail.

BALD EAGLE
Haliaeetus leucocephalus **12:3**

Description
Size, 32–40 in. (81.3–101.6 cm); wingspread, 7½ ft. (2.3 m). Adult: large, brown; • *head and tail white*; bill and legs yellow. Immature: dark brown with mottled white on wing linings; bill dusky; tail and head gradually assume full white plumage with age.

Similarities
Dark phase of Rough-legged and Ferruginous hawks have shorter wings. See also Golden Eagle.
Habitat
Near water; rare.
Habits
Wings held horizontal in flight; soars like a Red-tailed; steals fish from Ospreys.
Voice
Loud cackles.
Food
Dead fish, small animals, rarely birds.
Eggs
2–3; white; 2.8 x 2.3 in. (7.1 x 58.4 cm). Nest of sticks is in trees.
Range
Breeds in n. and e. Canada, n. U.S., and Fla.; winters coastally and along large rivers in interior.

NORTHERN HARRIER

Circus cyaneus **13:6, 15:5**

Description
Size, 18–22 in. (45.7–55.9 cm); wingspread, 4½ ft. (1.4 m). A hawk with long, slim wings and tail and a white rump. Adult male: • *pale gray* above, whitish below; black wing tips. Female: dark brown above, lighter below. Immature: like female but underparts rufescent with less streaking.
Similarities
Rough-legged is larger; has broader wings and tail; and has white mainly on upper tail, not on rump.
Habitat
Marshes, fields.
Habits
Flight buoyant, gull-like; flies low over grass with wings angled upward; flaps and glides low over grass; in courtship or migration, may soar high and circle on level wings like a buteo.
Voice
Low *chu-chu-chu*; a weak nasal *pee-pee.*
Food
Small mammals, birds.
Eggs
4–6; whitish; 1.8 x 1.4 in (4.6 x 3.6 cm). Nest is of grass and reeds, usually in marsh.
Former name
Marsh Hawk.
Range
Breeds in cen. Canada, from Newfoundland to Va.; winters from s. Ont. and n. U.S. southward.

OSPREYS

Family Pandionidae

OSPREY

Pandion haliaetus **12:5, 15:2**

Description
Size, 21–24½ in. (53.3–62.2 cm); wingspread, 6 ft. (1.8 m). The only hawk that dives into the water. Wings long, narrow; dark

brown above, with narrow, blackish bands on tail; • *white crown, throat, and underparts;* • *black eye patch* and wrist mark on underwings; immature has dark crown. In flight, wings characteristically angled, not flat or upturned.

Habitat
Near water.

Habits
Flaps slowly; sometimes sails, often hovers before plunging; dives from up to 100 ft. (30.5 m) in the air.

Voice
Whistled *you-you-you*; a "complaining *shriek, shriek, shriek*" (Cruickshank); also chickenlike peeps.

Food
Fish.

Eggs
2–4; white to rusty, blotched with deep brown; 2.4 x 1.8 in. (6.1 x 4.6 cm). Nest is of sticks on tree or other bare pole.

Range
Breeds in cen. Canada, s. to Fla. and Gulf Coast; winters from Gulf Coast, southward.

CARACARAS AND FALCONS

Family Falconidae

This family contains fast-flying, long-legged birds of prey having large heads; long, narrow, pointed wings; and long tails. The sexes are alike in color, but females are larger. The various species primarily eat small birds and animals or carrion. The major group of these species is the true falcons (Genus *Falco*).

CRESTED CARACARA

Polyborus plancus **12:4, 15:1**

Description
Size, 22–24 in. (55.9–61.0 cm); wingspread, 4 ft. (1.2 m). Adult: dark above; long-legged, long-necked; black crest and red face; throat and breast white; belly black; tail white, tipped with black, presenting an alternating flight pattern of light and dark viewed from below; conspicuous whitish patches near wing tips. Immature: duskier; breast streaked, not barred.

Habitat
Open rangeland, prairies.

Habits
Often observed on fence posts or feeding with vultures.

Voice
Rattling cackle.

Eggs
2–3; whitish, blotched; 2.3 x 1.8 in. (5.9 x 4.6 cm). Nest is reeds or sticks; atop tree, yucca, or cactus.

Range
Resident in s. Ariz., s. Tex.; disjunct population in prairies of cen. Fla.

GYRFALCON

Falco rusticolus **16:5**

Description
Size, 20–25 in. (50.8–63.5 cm); wingspread, 4 ft. (1.2 m). Gull-sized; white phase is all-white with some black markings; gray

phase is gray above, paler below; dark phase is almost solid black with some white markings; various difficult-to-identify intermediates occur—dark above, lighter below, with underparts barred or spotted in adult, streaked in immature.

Similarities

Peregrine is smaller, slimmer; darker above, lighter below; with black crown and mustache; relatively faster wingbeats. See Goshawk, Rough-legged.

Habitat

Northern coasts and tundra.

Habits

Flight swift but heavier than Peregrine's; rapid wingbeats alternate with short sails; may hover before stooping for prey.

Voice

Chattering and screaming.

Eggs

3–4; white to rusty, marked with brown; 2.3 x 1.8 in. (5.8 x 4.6 cm).

Range

Breeds from Greenland and n. Canada s. to cen. Canada; winters in breeding range; occasionally wandering s. to n. U.S.

PEREGRINE FALCON

Falco peregrinus **16:6**

Description

Size, 15–21 in. (38.1–53.3 cm); wingspread, 3¾ ft. (1.4 m). Sturdy-bodied; dark above with a black cap, light below. Adult: slaty above, tail lightly banded, throat and upper breast white, underparts barred. Immature: brown above, streaked below; throat buffy.

Similarities

Color of Gyrfalcon is more uniform, less contrasting; Merlin is smaller, more streaked below, tail more banded, less conspicuous mustache.

Habitat

Cliffs, coasts, high buildings; many migrate along seashore. Rare.

Habits

Wingbeat rapid, flight swift, resembling that of a pigeon; captures prey by pursuit or by stooping (plunging); observation perch a crag or dead branch. In pursuit one of fastest of all birds.

Voice

Loud cackles, *kak, kak, kak*, and other cries.

Eggs

2–4; creamy, blotched with chocolate; 2.1 x 1.7 in. (5.3 x 4.3 cm). Nest is on ledge.

Range

Breeds in Greenland; winters from Mass. southward.

MERLIN

Falco columbarius **16:2**

Description

Size, 10–13½ in. (25.4–34.3 cm); wingspread, 2 ft. (0.6 m). A medium-sized falcon with a heavily banded tail. Male: slaty above, buffy steaked with brown below. Female, immature: gray upperparts replaced by dark brown.

Similarities

Sharp-shinned has short, rounded, not long pointed, wings; see American Kestrel.

Habitat
Woodlands; also frequents seashore in migration and open country in winter.
Habits
Flies low and fast, somewhat like a pigeon, or alternates beats with glides; perching outlook is a post, knoll, or dead branch; like Kestrel, sometimes hovers and pumps tail.
Voice
Various cries and cackles.
Eggs
3–6; white to maroon, with dark blotches of brown; 1.6 x 1.2 in. (4.1 x 3.0 cm). Nest is in evergreens or on ledge.
Former name
Pigeon Hawk.
Range
Breeds from Labr. s. to N.S., and Mich.; winters from Newfoundland southward.

AMERICAN KESTREL
Falco sparverius **16:1**

Description
Size, 9–12 in. (22.9–30.5 cm); wingspread, 2 ft. (0.6 m). Back and tail rufous; head multicolored with black head markings; tail with black band near tip; underparts buffy. Male: wings blue-gray, tail unbarred. Female: duller; wings rusty, rufous tail barred.
Similarities
Merlin seems heavier and with shorter wings. Sharp-shinned has short, rounded, not long pointed, wings; very dark immature Kestrel sometimes suggests Merlin.
Habitat
Open country, roadsides, cities.
Habits
Flight usually unhurried, frequently hovers, pumps tail when perched; keeps lookout from telephone wires, poles, dead branches.
Eggs
3–5; white to reddish-white, spotted with brown; 1.4 x 1.2 in. (3.6 x 3.0 cm). Nest is in holes, boles, nooks.
Former name
Sparrow Hawk.
Range
Breeds from Newfoundland s. to S. America; winters from New England and Ill. southward.

Fowl-like Birds
Order Galliformes

These are small-headed, full-bodied birds with short, rounded wings and strong legs and feet. They usually live on the ground, migrate little, and are good scratchers and runners. Their wings make a loud whirring sound in flight. Their food is largely vegetation. The young run about almost as soon as hatched; all are considered game birds.

GROUSE, PARTRIDGES, PHEASANTS, AND TURKEYS

Family Phasianidae

Grouse are ground-dwelling birds with feathered legs that feed on seeds, buds, berries, and some insects. Ptarmigan differ from other grouse in having feathered toes as well as feathered legs, an adaptation to their often snowy environment.

Partridges and pheasants are similar in many ways to grouse, differing mainly in that their legs are not fully feathered. They nest in concealed, lined hollows on the ground, and eat insects, grains, and berries.

The Turkey is the largest upland game bird. It is the species from which all domestic turkeys are descended. Its food is acorns and other nuts, berries, plants, seeds, and insects.

SPRUCE GROUSE

Canachites canadensis **16:8**

Description
Size, 16 in. (40.6 cm). Male: grayish; breast, throat, and tail black; red over eye. Female and immature: brown barred with black above, lighter below.

Similarities
Ruffed Grouse is larger and lighter brown; spotted, not barred, above; with black band at end of longer tail.

Habitat
Northern coniferous forest, especially spruce.

Habits
Solitary, often remarkably tame, flushes to small tree and sits.

Voice
Cluckings; drums with wings.

Eggs
6–15; brown spots on buff; 1.6 x 1.2 in. (4.3 x 3.1 cm).

Range
Resident in cen. Canada from N.S. to Man., s. to New England, n. N.Y., and Mich.

RUFFED GROUSE

Bonasa umbellus **16:9**

Description
Size, 18 in. (45.7 cm). Has a conspicuous black tail band. Sexes similar; with a slight crest; above • *red-brown* or • *gray-brown* (2 phases); tail rufous or gray, white-tipped, fan-shaped, bare red area over eye; black ruff on neck; underparts white barred with dark brown. Female: duller, lacks crest and ruff.

Similarities
Female Ring-necked Pheasant has longer, pointed tail; slower wingbeats; less noisy flight; and is found in more open country.

Habitat
Woodlands with openings, second growth.

Habits
Tame in wilderness, wary near man; seeks safety by lying close, flying behind tree, or running away and flying at a distance; flushes with an explosive whir.

Voice
Nervous *quit-quit*; various clucks and coos; male makes a characteristic drumming with its wings in the breeding season.
Eggs
6–15; buffy to olive, spotted with brown; 1.4 x 1.1 in. (3.8 x 2.9 cm). Nest is on the ground.
Range
Resident in n. Canada s. to S.C. and S.Dak.; also in Appalachians s. to Ga.

WILLOW PTARMIGAN
Lagopus lagopus **17:2**

Description
Size, 16 in. (40.6 cm). Tail black. Spring male: white; head, neck, breast, and front of back red-brown. Summer male: brownish, darkest above; wings white. Spring and summer female: black and buff; scaly above, barred below; usually found with mate. Much variation; individuals in molt are mottled brown and white. Winter: pure white; bill and tail black.
Similarities
Bill heavier and more arched than in Rock Ptarmigan; female Rock in summer almost indistinguishable; male and some female Rocks in winter have black patch between bill and eye.
Habitat
In summer, low tundra, upland valleys; in winter, willow bottoms.
Habits
Gregarious after breeding season; roosts in snow; numbers subject to periodic fluctuations.
Voice
Crows, cackles.
Eggs
5–10; blotchy-red; 1.7 x 1.2 in. (4.4 x 3.1 cm). Nest is tundra depression lined with grass and feathers.
Remarks
Ptarmigans differ from other grouse in having feathered toes as well as feathered legs, an adaptation to their snowy environment.
Range
Resident in n. Canada from Baffin I. and Greenland s. to Newfoundland, cen. Que., and Man.

ROCK PTARMIGAN
Lagopus mutus **17:1**

Description
Size, 13 in. (33.0 cm). Similar to Willow Ptarmigan but somewhat smaller; female only pigeon-sized; bill smaller and all-black. Male: in summer gray-brown, not red-brown, below. Female: in fall, looks more like male Rock. Winter: all males and 60% of females have black between bill and eye; the other females, except for the bill, look like small Willow Ptarmigan.
Similarities
Willow is more reddish in summer than male Rock. Female Willow is more buffy than female Rock.
Habitat
In summer, the most barren rocky areas; in winter, more sheltered situations.
Habits, Voice
Similar to Willow Ptarmigan.

Eggs
6–15; buffy-olive, spotted with brown; 1.7 x 1.2 in. (4.4 x 3.2 cm). Nest is on ground.
Range
Resident in Arctic from Greenland and Ellesmere I. s. to Newfoundland and n. Que.

GREATER PRAIRIE CHICKEN
Tympanuchus cupido **17:3**

Description
Size, 18 in. (45.7 cm). Brown and buff with light scales above, • *dark bars* below; • *tail square, black* in male, barred in female.
Similarities
Female pheasant has long, pointed tail; Sharp-tailed Grouse has white on sides of pointed tail; Ruffed Grouse has fan-shaped tail.
Habitat
Prairie grasslands.
Habits
In flight, alternately flaps and sails; in courtship male raises elongate neck feathers and tail.
Voice
Clucks and cackles; male in courtship utters hollow, booming notes.
Eggs
7–17; olive, spotted; 1.7 x 1.2 in. (4.4 x 3.2 cm). Nest is grassy depression in prairie.
Range
Resident locally from Man. s. through Great Plains to Okla., and in Ill., Mich., and Wis.; also along e. Tex. coastal prairies.

Note: The **LESSER PRAIRIE CHICKEN,** *Tympanuchus pallidicinctus,* size, 16 in. (40.6 cm), occurs in southeastern Colorado and southwestern Kansas. It is smaller and paler than the Greater. Some consider them a single species.

SHARP-TAILED GROUSE
Tympanuchus phasianellus **17:4**

Description
Size, 15–20 in. (38.1–50.8 cm). Brown above, marked with white; below white with dark V-shaped marks on breast; • *tail with white sides, short and pointed*; neck sacs of male purple; in flight, white spots on wings conspicuous; in winter, more rufous and buffier.
Similarities
Tail of female Ring-necked Pheasant is long and pointed; of prairie chickens, short, dark, and rounded; of Ruffed Grouse, fan-shaped and banded.
Habitat
Open woodlands and brush.
Habits
Flight speedy, straight; rapid wingbeats alternate with glides on downcurved wings; in winter, perches in trees, roosts in snow; has elaborate courtship similar to that of Greater Prairie Chicken.
Voice
Clucking "*whucker, whucker, whucker*" (Bent).
Eggs
7–13; brownish-green, spotted; 1.7 x 1.2 in. (4.4 x 3.2 cm). Nest is grassy depression under bush or in tall grass.
Range
Resident from Que. and n. Man. s. to Mich. and Nebr.

BOBWHITE

Colinus virginianus **16:10**

Description
Size, 10 in. (25.4 cm). Body plump, brownish, and with much white below; throat and line over eye white in male, buffy in female; tail short and dark.
Similarities
Much smaller than Ruffed Grouse; lacks white outer tail feathers of meadowlarks.
Habitat
Pastures, brush.
Habits
When flushed, explodes into flight, then scales quickly down into cover; group or covey roosts on ground in circle facing out.
Voice
Clear whistled *bob-white* or *poor-bob-white*; also covey call, *quoi-lee* (suggests Sora).
Eggs
10–20; white; 1.2 x 0.9 in. (3.0 x 2.3 cm).
Range
Resident from Mass. and s. Ont., s. to Fla. and Gulf Coast.

GRAY PARTRIDGE

Perdix perdix **16:11**

Description
Size, 13 in. (33.0 cm). Brownish-gray above, tan face, gray breast with chestnut V, barred flanks, • *red-brown tail* (conspicuous in flight).
Similarities
Bobwhite is smaller and not as gray, with whitish face marks.
Habitat
Fields, brushy edges.
Habits
Flight swift, direct, noisy, followed by scaling down to cover; covey sleeps on ground in circle facing out.
Voice
Loud *kee-hah*.
Eggs
15–17; olive; 1.4 x 1.0 in. (3.6 x 2.5 cm).
Range
Locally in s. Canada and n. U.S., from N.S., N.B., and Ont., s. to Ohio, Ind., s. Mich., Iowa, and Minn., w.; introduced from Eurasia, but generally unsuccessful in becoming established.

RING-NECKED PHEASANT

Phasianus colchicus **17:5**

Description
Size, male, 30–36 in. (76.2–91.4 cm); female, 21–25 in. (53.3–63.5 cm). Usually with a white neck ring, this pheasant is easily distinguished by brilliant plumage. Male: • *tail very long*, tan barred with black. Female: brownish, tail similar but shorter.
Similarities
Ruffed Grouse has shorter, fan-shaped tail; Sharp-tailed Grouse has white on sides of shorter tail.
Habitat
Brushy edges, pasture.

Habits
Flushes with loud whir, flies fast, scales quickly into cover; enjoys dust baths, roosts on ground or in trees, flocks in winter.
Voice
Harsh *c-a-a-a-a*, like antique automobile horn.
Eggs
6–14; greenish-brown; 1.6 x 1.3 in. (4.1 x 3.3 cm).
Range
Resident from Maritime Provinces to Ont., s. to N.J., Md., and Okla.; introduced from Asia and Europe.

TURKEY

Meleagris gallopavo *Fig. 6*

Description
Size, male, 48 in. (121.9 cm); female, 36 in. (91.4 cm); wingspread, 5 ft. (1.5 m). Identical to domestic bird but • *tail tip brown.*
Similarities
None.
Habitat
Extensive open woodlands.
Habits
Good flier, but prefers to escape by running; roosts in trees.
Voice
Gobbles like barnyard bird.
Eggs
8–15; buff spotted with gray; 2.7 x 1.8 in. (6.9 x 4.6 cm).
Range
Resident locally from n. New England to Ill., s. to Gulf Coast.

Fig. 6

Cranes and Rails

Order Gruiformes

These are a diverse group of marsh and open-country birds that nest on the ground. The young can run about almost immediately after they leave the egg.

CRANES

Family Gruidae

The long-necked cranes, our tallest birds, are stockier than herons but their straight bills are shorter and less pointed. The long, curved tertials create a "tufted" effect over the tail. They fly with their necks and legs stretched straight out. They are waders and walkers, and never perch in trees. Their food is reptiles, amphibians, insects, grains, aquatic plants.

WHOOPING CRANE

Grus americana **5:4**

Description
Size, 50 in. (127.0 cm); wingspread, 7½ ft. (2.3 m). A tall, long-necked white bird with black primaries. Adult: head with naked red area, bill dark with yellow base, legs black. Immature: tan above, no naked red area, bill all-dark.
Similarities
Snow Goose has much shorter bill, neck, and legs; White Pelican is heavy, not tall and slender; egrets lack black wing tips; White Ibis is much smaller and has curved bill.
Habitat
Marshes; very rare.
Habits
Very wary; walks about marsh; in flight, wingbeats slow, fly in single file; sometimes spirals to great height and performs aerial revolutions.
Voice
Loud vibrating trumpeting.
Eggs
2; buff, blotched with brown; 3.9 x 2.5 in. (9.9 x 6.4 cm).
Range
Breeds in nw. Canada (Wood Buffalo Park); winters on Tex. Gulf Coast (Aransas National Wildlife Refuge).

SANDHILL CRANE

Grus canadensis **5:3**

Description
Size, 34–48 in. (86.4–121.9 cm); wingspread, 7 ft. (2.1 m). Uniformly gray or brown. Adult: gray; bill and legs dark; forehead red. Immature: all-brown, no red.
Similarities
Great Blue Heron is varicolored and flies with neck drawn in.
Habitat
Marshes and open country.
Habits
Gregarious; walks much; in flight, neck extended, wingbeats slow and with flick on upstroke; at times spirals high in air.
Voice
Ringing trumpetings, "*garoo-oo-oo-oo, garoo-oo-oo*" (Laing).
Eggs
2; drab, sparingly spotted with brown; 3.6 x 2.3 in. (9.1 x 5.8 cm).
Range
Breeds in Arctic and nw. Canada, s. to Mich. and Minn.; also along Gulf Coast from Fla. to Tex.; winters s. of U.S. and in Tex.

LIMPKINS

Family Aramidae

LIMPKIN

Aramus guarauna **5:7**

Description
Size, 24–28 in. (61.0–71.1 cm). A long-legged brown marsh bird with • *white spots on head, neck, and back,* and a long, slightly decurved bill; often flies with dangling legs.
Similarities
Immature night herons have straight bills, shorter, heavier necks.
Habitat
Marshes.
Habits
Feeds mainly at night.
Voice
Wailing *rr-ow,* piercing the night air.
Food
Snails.
Eggs
5–6; buff; 2.2 x 1.7 in. (5.7 x 4.4 cm). Nest of marsh plants is fastened to vegetation or platform in bush or tree.
Range
Resident locally in S., from s. Ga. to Fla., from Okefenokee Swamp to Everglades, and s. to S. America.

RAILS, GALLINULES, AND COOTS

Family Rallidae

Members of this family are somewhat chickenlike birds of wet areas. They seldom fly far, except on migration. They are often heard, especially at dawn, but less often seen. Their nest is a platform in the marsh. The young chicks are downy black. Rails and crakes are thin skulkers amid marsh reeds and grasses. The more ducklike gallinules and coots often swim about in open fresh water. Rails (Genus *Rallus*) have slightly curved bills and conspicuously barred flanks; crakes are small and short-billed with barred flanks. Rails prefer running to swimming, and prefer swimming to flying; if flushed, they usually flutter only a few yards with feet dangling.

KING RAIL

Rallus elegans **18:5**

Description
Size, 17 in. (43.2 cm). A red-brown rail of the freshwater marsh, this is the largest. Above brown streaked with black, cheeks and underparts cinnamon.
Similarities
Clapper replaces cinnamon with buff and gray. Virginia Rail is similar to King but much smaller, and with gray cheeks.
Habitat
Freshwater marshes, sometimes salt marshes in winter.
Habits
Skulks.
Voice
Grunting *umph-umph-umph-umph* on same pitch (call of Virginia Rail descends scale).

Food
Insects, aquatic animal life, seeds.
Eggs
6–12; buffy, blotched with red-brown; 1.6 x 1.1 in. (4.1 x 2.8 cm).
Range
Breeds from Mass. to Minn., s. to Fla. and Tex.; winters from Gulf Coast southward.

CLAPPER RAIL

Rallus longirostris **18:4**

Description
Size, 15 in. (38.1 cm). A large, gray-brown rail of salt marshes. Some buff below, throat and undertail coverts white.
Similarities
Freshwater King Rail (by some recognized as part of same species) has cinnamon instead of gray, blacker streaks above, and rusty on wings; the browner young Clapper resembles young King.
Habitat
Salt marshes.
Habits
Swims well and dives; often seen swimming and walking about salt marsh during high or flood tides; head bobs in swimming; tail often raised to reveal white undertail coverts.
Voice
Frequently heard and oft-repeated staccato *kek-kek-kek-kek*.
Food
Crustaceans, mollusks, worms.
Eggs
5–14; buffy, marked with red-brown; 1.7 x 1.2 in. (4.3 x 3.1 cm).
Range
Breeds along Atlantic Coast, from Mass. to S. America.; also along Gulf Coast to Tex.; winters from N.J., southward.

VIRGINIA RAIL

Rallus limicola **18:6**

Description
Size, 10 in. (25.4 cm). A small, red-brown rail. Above olive streaked with dusky; eye red, cheeks gray; below; red-brown, forewing reddish (conspicuous in flight); immatures have much black but with long bill.
Similarities
King Rail is considerably larger and has red-brown cheeks. Sora has short, yellow bill.
Habitat
Freshwater marshes, rarely brackish.
Habits
Can climb reeds and vines.
Voice
Harsh *kid-ik, kid-ik, kid-ik*; other "kicking," squealing, clucking, and vaguely chickenlike noises.
Food
Aquatic animal life, seeds, berries, insects.
Eggs
5–12; similar to King's; 1.3 x 0.9 in. (3.3 x 2.3 cm).
Range
Breeds s. Canada, from Newfoundland to Minn. s. to C. America; winters from Va. southward, occasionally farther north.

SORA
Porzana carolina **18:7**

Description
Size, 9 in. (22.9 cm). Adult: chunky shape, dark brownish above, black face and throat, gray below, • *short yellow bill.* Immature: buffy below, bill duller.
Similarities
Buffy Yellow Rail is smaller, rarer, and has white wing patches.
Habitat
Freshwater marshes; salt marshes in migration.
Habits
Swims, dives well; migrates at night, often striking obstructions.
Voice
High-pitched, horselike whinny; a whistled *cur-wee* (like covey call of Bobwhite); a single *keek*; other rail-like noises.
Food
Insects; wild rice and other seeds.
Eggs
6–15; buffy, marked with dull brown; 1.2 x 0.9 in. (3.0 x 2.2 cm).
Range
Breeds in cen. Canada, from Newfoundland to Ont., s. to Pa. and Okla.; winters from S.C., southward.

YELLOW RAIL
Coturnicops noveboracensis **18:9**

Description
Size, 7 in. (17.8 cm). Buffy yellow; back streaking has "checkerboard" effect; • *white wing patch*; bill very short, yellow.
Similarities
Immature Sora is larger, with no white in wing.
Habitat
Wet meadows, higher parts of fresh and salt marshes; extremely rare.
Habits
Most secretive, almost never flies except at night on migration; can conceal itself in extremely short, sparse grass.
Voice
High-pitched clicking notes in series. "*kik-kik-kik-kik-queeah*" (Ames).
Food
Little known; includes snails.
Eggs
7–10; buff with small, red-brown dots; 1.1 x 0.8 in. (2.8 x 2.0 cm).
Range
Breeds in cen. Canada and n. U.S., from N.B. and Que. s. to Maine and N.Dak.; winters from S.C., Fla., and Gulf Coast southward.

BLACK RAIL
Laterallus jamaicensis **18:8**

Description
Size, 6 in. (15.2 cm). A tiny black rail, the size of a sparrow, with a maroon-brown nape. Above black with white dots; below dark gray; • *bill black.*
Similarities
Downy young chicks of all rails are black, but they lack the Black's nape patch and white dots; they cannot fly.

Habitat
Fine grass salt marshes; grassy edges of freshwater marshes.
Habits
Very hard to flush; runs like a mouse, with head down and neck out.
Voice
Female: "*croo-croo-croo-o* . . . like the commencement of the song of the Yellow-billed cuckoo" (Wayne). Male: "*kik, kik, kik, kik* or even *kuk, kuk, kuk, kuk*" (Wayne). "*Did-ee-dunk, did-ee-dunk, did-ee-dunk*" (McMullen).
Food
Isopods (small crustaceans).
Eggs
4–9; white, marked with red-brown; 1.0 x 0.8 in. (2.5 x 2.0 cm).
Range
Breeds along Atlantic and Gulf coasts from N.Y. to S. America and locally inland; winters from Gulf Coast southward.

PURPLE GALLINULE
Porphyrula martinica **18:1**

Description
Size, 13 in. (33.0 cm). • *Blue forehead shield* and a purple breast. Adult: back and wings glossy green; head, neck, and underparts glowing purple and blue (looks black in poor light); undertail coverts white; bill red with a yellow tip; • *legs bright yellow.* Immature: above dull brown with some greenish reflections; below, dingy buff and white; bill dull.
Similarities
Common Gallinule has a red frontal shield and white stripe along sides; immature much grayer below.
Habitat
Freshwater marshes, especially those with small ponds.
Habits
Runs or walks on lily pads with dovelike nods, flirting white undertail; clambers about marsh reeds, climbs vines; perches in shrubs and trees; bobs head as it swims; but less aquatic than Common; flight weak and with legs dangling.
Voice
Loud, chickenlike; in flight, *kek, kek, kek, kek.*
Food
Frogs, shellfish, aquatic insects, seeds.
Eggs
6–10; white, lightly spotted with brown; 1.6 x 1.1 in. (4.1 x 2.8 cm).
Range
Breeds from S.C. and Tenn. s. to Fla. and Tex.; winters from Fla. and Tex. s. to S. America

COMMON GALLINULE
Gallinula chloropus **18:3**

Description
Size, 13 in. (33.0 cm). Head, neck, and underparts gray; back olive; divided undertail coverts white. • *White stripe along flanks;* • *red bill* and frontal shield with yellow tip; green legs. Immature: paler below; bill dusky.

Similarities
American Coot is plumper, shorter-necked, uniform slate gray, larger headed with a white bill and shield. Immature Purple is buffier-white below.
Habitat
Freshwater marshes, occasionally salt.
Habits
Runs and walks over lily pads jerking tail; tips up like a pond duck to feed in shallow water; also feeds on land; short flights weak with legs dangling; long flights with neck and legs extended; swims stern high, pumping head (but Coot swims more); can dive.
Voice
Loud, chickenlike, extremely varied; *cac, cac, cac,* etc.
Food
Vegetation, snails, insects.
Eggs
6–12; buffy, marked with brown; 1.7 x 1.2 in. (4.3 x 3.0 cm).
Other name
Moorhen.
Range
Breeds throughout U.S., from s. Canada to Fla. and Gulf Coast; winters from Ga. and Gulf Coast s. to S. America.

AMERICAN COOT

Fulica americana **18:2**

Description
Size, 15 in. (38.1 cm). A • *slaty-black* ducklike water bird with a conspicuous, short • *white bill.* Adult: head black, bill short, undertail coverts white, legs greenish, toes with gray lobes. Immature: paler below, and with a duller bill.
Habitat
Freshwater marshes, ponds, rivers; at times, salt bays.
Habits
Gregarious; swims, dives, and tips up; floats high on water with back level (not stern high, as do gallinules); pumps head when swimming; patters along surface before taking to air. In flight, shows white on trailing edge of wing; neck and legs are extended; feet protrude behind tail. Where protected, becomes quite tame, often approaching people to be fed.
Voice
"Coughing sounds, froglike plunks, and a rough sawing or filing *kuk-kawk-kuk, kuk-kawk-kuk*, as if tree saw were dull and stuck . . . [and] a grating *kuk kuk kuk kuk kuk*" (Bailey); and many other noises.
Food
Aquatic plants, grass, grain.
Eggs
8–12; whitish, with pinhead-sized dark brown dots; 1.9 x 1.3 in. (4.8 x 3.3 cm).
Range
Breeds from s. Canada s. to Gulf Coast and Fla.; winters s. from Md. and Ohio Valley to n. S. America.

Shorebirds, Gulls, Auks, and Allies

Order Charadriiformes

OYSTERCATCHERS

Family Haematopodidae

COMMON OYSTERCATCHER

Haematopus palliatus

19:16
Fig. 7

Description
Size, 19 in. (48.3 cm). A shorebird with a long, • *heavy red bill.* Head and neck black; back, wings, and tail dark brown; underparts, rump, and wing stripe white; eye red, feet pink; bill brownish in young; conspicuous bold black-and-white appearance in flight.
Habitat
Sand beaches, rocky coasts.
Habits
Wary; walks sedately, wades to belly; can swim and dive, prods in wet sand with bill; flight strong, swift.
Voice
Penetrating *wheep, wheep.*
Food
Shellfish, crustaceans, marine worms.
Eggs
2–3; buff, blotched with dark brown; 2.2 x 1.5 in. (5.6 x 3.8 cm). Nest is in slight hollow in beach.
Range
Breeds along Atlantic and Gulf coasts, from Mass., s. to Fla. and S. America; winters from N.C., southward.

Large Shorebirds in Flight

PLOVERS

Family Charadriidae

These small to medium-sized plump, round-headed, short-necked shorebirds are runners rather than waders. They have pigeonlike bills that are shorter than their heads, and shorter and thicker than those of sandpipers. They have a stop-and-go manner of feeding, rather than the continuous probing of sandpipers. When standing still, many bob the body forward and back. The bulk of their food is shellfish, other marine invertebrates, and worms. Plovers all nest in a depression in the sand or on the ground; the eggs are pear-shaped. The young can run about almost as soon as they are hatched. The parent simulates a broken wing to lure intruders away from the nest or young. The Ringed Plovers (Genus *Charadrius*) are usually small, dark brown or whitish above and white below, with one or two black breastbands and some white on the tail; females and young are paler.

SEMIPALMATED PLOVER

Charadrius semipalmatus **19:2, 23:1**

Description
Size, 7 in. (17.8 cm). Spring: above color of wet sand; • *black band across upper breast* and black line through eye to bill; white forehead, throat, underparts, and sides of tail; yellow-orange bill with black tip, yellow-orange legs. Fall and immature: black areas browner, more white on forehead, legs paler, bill more dusky.

Similarities
Piping Plover is lighter, with fewer face markings. Killdeer has 2 breastbands.

Habitat
Mud flats, beaches, shorelines, wet fields.

Habits
Often flies in compact flocks wheeling in unison; on alighting, birds spread out to feed, stoop intermittently to pick up food.

Voice
Plaintive rising *cheer-wee.*

Eggs
3–4; buff, marked with blackish-brown; 1.3 x 0.9 in. (3.3 x 2.3 cm).

Range
Breeds in Arctic, s. to N.S., Que., and n. Man.; winters from Gulf Coast southward.

PIPING PLOVER

Charadrius melodus **19:1, 23:2**

Description
Size, 7 in. (17.8 cm). The only pale-backed plover on East Coast and Great Lakes. Above color of dry sand. Spring: tip of bill, line over forehead, and end of tail black; single black band across upper breast, sometimes broken; base of bill and • *legs yellow-orange*; underparts, rump, and wing stripe white. Fall and immature: black on head and breast lost or replaced by brown, bill darker.

Habitat
Sandy beaches, mud flats, and fill.

Habits
Similar to Semipalmated but prefers drier sand of higher beach.

Voice
Soft, whistled descending *peep-lo.*

Eggs
3–4; creamy, finely dotted with chocolate; 1.2 x 1.0 in. (3.0 x 2.5 cm). Nests amid dunes.

Range
Breeds along Atlantic Coast from Newfoundland, s. to Va. and along shores of Great Lakes; winters along Atlantic and Gulf coasts, from S.C., southward.

Note: The **SNOWY PLOVER,** *Charadrius alexandrinus* **(19:4)**, size, 6 in. (15.2 cm), is a Piping Plover-like bird with a paler back, broken breastband, thinner black bill and slaty legs; it occurs in eastern Colorado and southwest Kansas, and along entire Gulf Coast.

WILSON'S PLOVER

Charadrius wilsonia **19:5**

Description
Size, 8 in. (20.3 cm). • *Bill thick and long, black* at all seasons. Spring male: above color of wet sand (head and nape of male sometimes rusty); very broad black breastband; legs pinkish. Female and fall plumages: black areas are brown.
Similarities
Semipalmated Plover is smaller, slightly darker, with a much smaller bill, narrower breastband, and yellow legs.
Habitat
Sandy beaches, mud flats, shores of salt marshes.
Habits
Movements quite deliberate; not as shy or nimble as Piping Plover; often active after dark.
Voice
Loud whistled *wheat*; on wing, a *tut-tut* (like alarm note of Wood Thrush, but lower); near nest, a high-pitched *"queet, queet, quit it, quit it"* (Pennock).
Eggs
3; olive or drab, marked with dark brown; 1.3 x 1.0 in. (3.3 x 2.5 cm).
Range
Breeds along Atlantic and Gulf coasts from s. N.J. to Fla. and w. to Tex.; winters along Gulf Coast s. to n. S. America.

KILLDEER

Charadrius vociferus **19:3, 23:3**

Description
Size, 10 in. (25.4 cm). Distinguished by • *2 black breastbands*; earth-brown above, with white forehead, throat, collar (continuing around neck), underparts, wing stripes, and edges of tail; • *rump and upper part of rather long tail orange-brown*; black bill; pinkish legs; young, paler, and with grayish breastband.
Similarities
Other ringed plovers are smaller with only 1 band.
Habitat
Mud flats, fields, parks, open areas, and fill; often near water.
Habits
Swift runner and flier, often active after dark and on moonlit nights; has spectacular nuptial flight; if nest is approached, circles in air screaming above intruder, or gives "broken-wing" display on ground.
Voice
Loud *kill-dee, kill-dee;* a noisy and persistent trilling.
Eggs
4; buff, blotched with brown; 1.4 x 1.1 in. (3.6 x 2.8 cm).
Range
Breeds throughout most of U.S., from Newfoundland s. to West Indies; winters from N.J. and Ohio, southward.

AMERICAN GOLDEN PLOVER

Pluvialis dominica **19:7, 21:11, 23:6**

Description
Size, 9½–11 in. (24.1–27.9 cm). Spring: speckled gold above, all-black below; forehead and sides of neck white; underparts black to tail. Fall: brownish or yellowish-brown, darker above, tail always dark, bill black, legs slate, no hind toe. In flight, wings gray below.

Similarities
Black-bellied is slightly larger, less slender, and has shorter wings; underparts black only to thigh. In winter, Black-bellied is grayer, has thicker bill and neck; in flight, has white wing stripe and rump, some contrast between gray belly and white undertail coverts, and black wing pits.

Habitat
Marshes, fields, mud flats, beaches, open areas.

Habits
Gregarious, flight swifter, more buoyant than Black-bellied; on ground, more aggressive; raises wings on alighting, often bobs head.

Voice
Quavering *quee-i-i-a* or a harsh *queedle.*

Eggs
4; buffy-olive, marked with brown and black; 2.0 x 1.3 in. (5.1 x 3.3 cm).

Range
Breeds in Arctic and nw. Canada; winters in S. America. Fall migration is from Ne. coast across Atlantic; spring migration is through cen. states and Canadian provinces to tundra. The Golden Plover's migration covers 2400 miles from N.S. to S. America.

BLACK-BELLIED PLOVER

Pluvialis squatarola **19:6, 21:12, 23:8**

Description
Size, 11 in. (27.9 cm). Stout head and bill; hind bill and legs black; toe present; wing stripe, • *black wing pits*, rump, black-barred tail, and undertail coverts white. Spring: pale speckled gray above, black (to thighs) below; forehead and sides of neck white. Fall: grayish, lighter below.

Similarities
White area of Golden is smaller; for other comparisons, see American Golden Plover.

Habitat
Beaches, mud flats, salt-marsh meadows.

Habits
Sedate, somewhat stolid shoreline figure of erect carriage; white forehead is conspicuous in birds on ground amid grass.

Voice
Wild, plaintive, somewhat Bluebird-like *toor-a-wee.*

Eggs
4; buffy-olive, marked with brown and black; 2.1 x 1.4 in. (5.3 x 3.6 cm).

Range
Breeds in Arctic and s. Baffin Is.; winters along coast from Va. to S. America.

WOODCOCK, SNIPE, SANDPIPERS, AND ALLIES

Family Scolopacidae

These shorebirds have thin, relatively long bills; long, pointed wings (except Woodcock); rather long legs; and short tails. Most are waders and prefer moist areas or shallow water and shorelines; most can swim and dive. Almost all have a distinctive courtship song, occasionally heard in migration. With one exception (Solitary Sandpiper), all nest in a depression upon the ground, often lined with grasses, and usually near water. Parents may simulate a broken wing to lure intruders from the nest. The eggs are usually four and pear-shaped; the young can run about almost immediately after leaving the egg. Their diet consists primarily of shellfish, crustaceans, insects, and seeds.

AMERICAN WOODCOCK

Scolopax minor **20:17, 23:9**

Description

Size, 11 in. (27.9 cm). Chunky; • *long bill*, short neck, big eyes; upperparts mottled like dead leaves, underparts orange-tan; crown black with narrow, tan crossbands; legs pinkish, • *wings rounded*; no seasonal change.

Similarities

Common Snipe is thinner, darker above, rises in a zigzag on thin, pointed, noiseless wings.

Habitat

Swamps, wet woods, and thickets.

Habits

Solitary, secretive, largely nocturnal; probes deep into mud when feeding, leaving visible borings; rises straight, with round body, long bill pointing down, and on short whistling wings.

Voice

Peent somewhat like Common Nighthawk.

Eggs

3–4; buff, marked with brown; 1.5 x 1.1 in. (3.8 x 2.8 cm).

Range

Breeds in e. N. America from s. Canada to Gulf Coast; winters from Va. and Mo. southward.

COMMON SNIPE

Gallinago gallinago **20:14, 23:7**

Description

Size, 11 in. (27.9 cm). Wings rather pointed; • *bill long*; head striped; streaked brown above; breast spotted, belly white; • *tail orange*, tail corners pale; legs greenish; no seasonal change.

Similarities

Dowitchers have conspicuous white stripe on lower back and frequent seashore.

Habitat

Meadows and marshes, fresh or salt.

Habits

Solitary, secretive; hides by squatting; most active at dawn, dusk, and on cloudy days; usually rises in zigzags, uttering nasal escape note; can swim and dive.

Voice

A "rasping *escape, escape*; on breeding grounds, a melodious *wheat wheat*" (Collins).

Eggs
3–4; dull olive, boldly blotched with brown; 1.5 x 1.1 in. (3.8 x 2.8 cm).
Range
Breeds from Labr. s. to Mass., Ind. and some farther s.; winters from Va. southward.

RUDDY TURNSTONE
Arenaria interpres **19:8, 21:5, 23:10**

Description
Size, 9 in. (22.9 cm). Ruddy upperparts of spring replaced by brown in fall; in flight at all seasons, calico pattern with 5 prominent white stripes on back and wings; black bill may be slightly upturned; legs short, orange.
Habitat
Primarily coastal; beaches, shorelines, mud flats, rocks, jetties.
Habits
Pokes bill under pebbles, shells; digs shallow holes in sand; pugnacious; sometimes perches off ground; can swim; large migrating flocks separate into small groups on ground.
Voice
Harsh *chut-chut*; a melodious *quit-tock.*
Eggs
4; cream, splashed with brown; 1.5 x 1.1 in. (3.8 x 2.8 cm).
Range
Breeds in Arctic; winters from S.C. s. to S. America; some wander farther n.

WHIMBREL
Numenius phaeopus **19:13** *Fig. 7*

Description
Size, 15–18 in. (38.1–45.7 cm); bill long and decurved, 2¾–4 in. (7.0–10.2 cm). A common curlew, the only one with • *prominent stripes on the head.* Grayish brown above, barred pinkish-buff underwings, grayish-white below, blue-gray legs.
Similarities
Much smaller, grayish not buffy, and bill shorter than Long-billed.
Habitat
Salt marshes, tidal flats, shorelines, river bars, tundra.
Habits
Flight steady; high over land in flocks like ducks, or in long lines low over water; often scales on set wings.
Voice
Soft musical *cur-lew*; a series of harsh *ku-ku-ku* notes on same pitch.
Eggs
4; buff, marked with brown; 2.4 x 1.6 in. (6.1 x 4.1 cm).
Former name
Hudsonian Curlew.
Range
Breeds in Arctic and nw. Canada; winters from Gulf Coast s. to S. America.

LONG-BILLED CURLEW

Numenius americanus **19:14**

Description
Size, 20–26 in. (50.8–66.0 cm); bill, 5–7 in. (12.7–17.8 cm). An impressive shorebird distinguished by its very long, downcurved bill. Buffy; head unstriped; underwings and • *wing pits bright cinnamon.*
Similarities
Whimbrel has center crown stripe, no cinnamon underwings.
Habitat
Plains, prairies, open areas near water; lagoons and mud flats along coast in winter.
Habits
Wary, flies in V-shaped flocks; can swim.
Voice
Loud whistled *curlew-curlew*; a rapid "*wheety, wheety, wheety*" and a loud, rattling "*que-he-he-he-he-he*" (Bent).
Eggs
4; buff, marked with brown and lavender; 2.6 x 1.9 in. (6.6 x 4.8 cm).
Range
Breeds in w. N. America; winters from La. to C. America and from S.C. to Fla.

UPLAND SANDPIPER

Bartramia longicauda **19:9, 23:5**

Description
Size, 12 in. (30.5 cm). A medium-sized upland species with a straight bill and white outer tail feathers. Streaked brownish above, lighter below; small head, slender bill, rather long neck and tail; line over eye and outer tips of tail white; underwing black and white; dark rump, yellowish legs.
Similarities
Buff-breasted Sandpiper is smaller and with an unmarked breast.
Habitat
Grassy inland fields, prairies.
Habits
Flies brief distances with short strokes of downcurved wings like Spotted Sandpiper; extended flight is swift, buoyant; often perches on fence posts.
Voice
Song is a mournful, mellow whistle, "*wh-e-e-e-e-e-e-e-e-o-o-o-o-o-o-o-o-o-o* . . . [alarm note] *quitty-quit-it-it*" (Knight).
Eggs
4; buff, finely spotted with brown; 1.8 x 1.3 in. (4.6 x 3.3 cm).
Former name
Upland Plover.
Range
Breeds in cen. Canada, s. to cen. U.S.; winters in S. America.

SPOTTED SANDPIPER

Actitis macularia **21:18, 22:6**

Description
Size, 8 in. (20.3 cm). Olive-brown above; line over eye, wing stripe, and edges and tip of tail white. Spring: distinctly • *spotted with dark below.* Fall: spots lacking; dusky sides of neck separated from wing by white mark.

Similarities
Fall Solitary is darker above, has streaked breast, lacks white wing stripe; bobs, does not teeter.
Habitat
Shorelines.
Habits
Usually solitary, constantly teeters tail up and down as it walks; flight highly characteristic, somewhat like a meadowlark, short rapid beats of downcurved wings alternate with glides low over water; also has a seldom-seen full, free flight like a yellowlegs; can swim and dive from water or wing; perches on posts, wires, and branches.
Voice
Low, distinct *peet-weet.*
Eggs
4; cream, spotted with chocolate and gray; 1.3 x 0.9 in. (3.3 x 2.3 cm). Nest is grassy scrape near water or in brush.
Range
Breeds from n. Canada to s. U.S.; winters from S.C. s. to S. America.

SOLITARY SANDPIPER

Tringa solitaria **21:16, 22:9**

Description
Size, 7½–9 in. (19.1–22.9 cm). No seasonal change. Upperparts and underwing dark brown; eye-ring white; • *tail appears white* with dark center; breast streaked, underparts white; bill dark, slender; legs olive, no wing stripe. Young more white-dotted above.
Similarities
Spotted is paler above, has black spots below in spring, and white wing stripe all seasons. Yellowlegs are larger, lighter, taller, and with white rump.
Habitat
Freshwater edges, streams, pools in marshes and woods, ditches.
Habits
Rather solitary; flight light, airy, often zigzagging; wings have good upstroke (unlike Spotted); engages in extensive aerial revolutions; however, short-distance flight is jerky with wings only partially spread; drops abruptly to a landing, raises wings on alighting; bobs head, as if hiccuping (does not teeter); can swim, dive.
Voice
One or more *peet* notes, higher-pitched than Spotted's.
Eggs
4–5; pale greenish, with red-brown spots and blotches; 1.4 x 1.0 in. (3.6 x 2.5 cm). Nests in the deserted tree nests of other birds.
Range
Breeds in n. forest fringe at edge of tundra in Canada; winters from cen. U.S. and Gulf Coast s. to S. America.

GREATER YELLOWLEGS

Tringa melanoleuca **21:21**

Description
Size, 12½–15 in. (31.8–38.1 cm). • *Bright yellow legs* and • *white rump* distinguish the 2 yellowlegs. Slender, above grayish, no wing stripe; • *tail whitish;* below, white with streaks on breast; long legs. No seasonal change.
Similarities
Lesser is smaller, with smaller, straighter bill and different call.

Habitat
Marshes, mud flats, rain pools, streams, muskeg.
Habits
Not so gregarious as Lesser; bobs up and down, wades deeply, occasionally swims; very noisy; its loud cries warn of intruders; responds to imitations of its note.
Voice
Clear whistled *yew,* usually repeated 3 or 4 times; "a rolling *toowhee, toowhee*" (Nichols).
Eggs
4; grayish, splashed with brown and lilac; 1.8 x 1.3 in. (4.6 x 3.3 cm). Nest is a muskeg hollow.
Range
Breeds in s. and cen. Canada from Newfoundland to James Bay; winters from S.C. and cen. U.S. s. to S. America.

LESSER YELLOWLEGS

Tringa flavipes **21:20, 22:15**

Description
Size, 11 in. (27.9 cm). • *Bright yellow legs,* a white rump, a • *straight, thin bill* and its call identify this smaller yellowlegs. Similar to Greater, but smaller, and with a shorter, more slender bill. Other differences slight; size alone deceptive, but direct comparison in field often possible.
Similarities
Stilt Sandpiper in fall is smaller, with greenish legs and bill slightly drooped at tip. Fall Wilson's Phalarope is smaller and has more needlelike bill, lacks yellow legs, and usually swims in groups.
Habitat
Same as Greater.
Habits
More gregarious than Greater, less suspicious; flight "more buoyant and hence not so suggestive of momentum" (Brewster); goes north later than Greater and south earlier.
Voice
Usually only 1 or 2 *yew* notes, similar to Greater, somewhat flatter and less forceful; rolling *toowhee* and several other notes similar to those of larger bird.
Eggs
4; 1.7 x 1.1 in. (4.3 x 2.8 cm); in other respects similar to Greater. Nest is ground hollow in open, often far from water.
Range
Breeds from n. Man. to cen. Que.; winters from S.C. and cen. U.S. s. to Gulf Coast and S. America.

Note: Fall-plumaged **RUFFS (20:11)** (males) and **REEVES** (females), *Philomachus pugnax,* size, 10–12½ in. (25.4–31.8 cm), occasionally occur in company with Lesser Yellowlegs, which they somewhat resemble. Fall: distinguished by generally browner plumage; narrow wing stripe; shorter, stouter bill; shorter legs, often yellow; and 2 long, oval white patches on sides of dark rump and tail. Very rare in spring when male has large, distinctive ruff varying greatly in color and neck and ear tufts. Voice: *To-wit.* This is an Old World species, in which the sexes bear different names (smaller female called Reeve).

WILLET

19:12, 21:17

Catoptrophorus semipalmatus *Fig. 7*

Description
Size, 14–17 in. (35.6–43.2 cm). A plain-colored shorebird with a broad • *white stripe* on its • *black wings;* one of the larger species. On ground, rather nondescript; grayish above, white below with numerous black markings in spring, unmarked in fall; wing pattern occasionally visible; bill thick, black, and with a bluish base; legs bluish; white rump and pale tail conspicuous in flight.
Similarities
Greater Yellowlegs is less robust, has plain wings, and bright yellow legs.
Habitat
Marshes, beaches, mud flats, sloughs.
Habits
Bobs less than yellowlegs; flight is strong, direct, ducklike; wingbeat is flat; occasionally scales, may hold wings aloft for several seconds after alighting; often perches on trees, fences, or posts; wades up to belly, swims well.
Voice
Whistling *pill-will-willet;* "a loud vehement *wek, wek, wek* or *kerwek, kerwek, kerwek*" (Bent).
Eggs
Variable; often grayish-olive spotted with brown; 1.9 x 1.5 in. (4.8 x 3.8 cm). Nest is grassy cup in green.
Range
Breeds locally in s. Canada, N.S., and N.Y., s. along coast to Fla. and w. along Gulf Coast; winters from Va. and Gulf Coast s. to S. America.

PURPLE SANDPIPER

Calidris maritima **20:10, 21:7, 22:4**

Description
Size, 9 in. (22.9 cm). Plump. Fall and winter; • *slaty above;* gray breast, white underparts and wing stripe, bill orange with a black tip, • *legs yellow.* Spring: browner and with a more heavily streaked breast.
Habitat
Wave-washed rocks and jetties; with Sanderling and Dunlin the only regularly wintering sandpipers from New York north.
Habits
Usually seen in groups; stands low; can swim.
Voice
Like *kip* of Sanderling; also a double-noted *twit-twit.*
Eggs
3-4; olive-drab, boldly marked with brown; 1.5 x 1.0 in. (3.8 x 2.5 cm).
Range
Breeds in Arctic; winters along Atlantic Coast from Maritime Provinces s. to N.J.

RED KNOT

Calidris canutus **20:12, 21:10, 22:11**

Description
Size, 11 in. (27.9 cm). Legs and bill short. Spring: reddish-brown and gray with heavy black streaks above; breast is robin-red. Fall: gray above, white below, giving a washed-out appearance; wing stripe faint. Rump and tail whitish; legs greenish.

Similarities
Dowitcher has a long bill; Sanderling and Dunlin are smaller, less chunky, and dark rumped.
Habitat
Shorelines, flats, beaches, salt marshes, tundra.
Habits
Gregarious on ground, sluggish; in flight, occurs in large, tight, twisting flocks.
Voice
Rather quiet, a low harsh chut; a soft *wah-quoit;* "a soft *whit whit,* like a man whistling for a dog" (Hoffmann).
Eggs
3–4; light-greenish, marked with brown; 1.8 x 1.2 in. (4.6 x 3.0 cm).
Range
Breeds in Arctic; winters from Mass. and Gulf Coast s. to S. America.

PECTORAL SANDPIPER
Calidris melanotos **20:8, 22:3**

Description
Size, 8–9½ in. (20.3–24.1 cm). Noted for its • *streaked, buffy breast* abruptly divided from a white belly; male larger than female. Above brownish striped with black; rump pied; legs greenish; wing stripe very thin and invisible at any distance; greenish bill sometimes droops at tip.
Similarities
All peeps are similar; all but Baird's have white wing stripe. Baird's has a scaly back pattern.
Habitat
Grassy edges of mud flats and marshes, both fresh and salt.
Habits
Lies close; if flushed, jumps suddenly with a harsh *kriek* and hurriedly zigzags away; then may circle high before pitching abruptly down; suggests Common Snipe in flight behavior; occurs in scattered flocks; can swim.
Voice
Reedy *kriek-kriek.*
Eggs
3–4; pale olive, spotted with umber; 1.5 x 1.0 in. (3.8 x 2.5 cm). Nest is grassy tundra hollow.
Range
Breeds in Arctic, s. to Hudson Bay; winters in S. America; migrates coastally and in interior.

WHITE-RUMPED SANDPIPER
Calidris fuscicollis **20:4, 21:6, 22:1**

Description
Size, 7½ in. (19.1 cm). Above rusty in spring, gray in fall; • *breast streaked* and dark in front of wing; wing stripe white; prominent white line over eye in fall; bill straight, heavy for a peep, occasionally slightly drooped at tip; legs dark greenish; • *rump white.*
Similarities
Semipalmated has white in front of wing; when feeding in mixed flocks, it can be distinguished as larger than Least or Semipalmated. This species and next 6 are termed "peep."
Habitat
Mud flats, rocky beaches, shorelines, salt marshes; fresh marshes inland.

Habits
Tame; actions deliberate; feeds in water up to its belly.
Voice
"A squeaky mouselike *jeet*" (Nichols); like 2 marbles being knocked together, uttered on wing.
Eggs
3–4; olive, marked with dark brown; 1.3 x 0.9 in. (3.3 x 2.3 cm).
Range
Breeds in Arctic s. to n. Hudson Bay; winters in S. America. Spring migration mainly through interior; fall migration more commonly along Atlantic Coast.

BAIRD'S SANDPIPER

Calidris bairdii **20:5**

Description
Size, 7½ in. (19.1 cm). A buffy-brown, long-winged peep with a scaly pattern on its back; no seasonal change. • *Head and breast buffy;* dark in front of wing, white stripe weak or missing; white throat; black legs and bill.
Similarities
Least and Semipalmated are smaller, slightly darker above; Semipalmated shows white in front of wing, and its breast is paler. Pectoral is browner with bill shorter and more slender; Pectoral and White-rumped are more streaked on crown and back. Sanderling has a prominent wing stripe. Western has a longer bill. Dunlin has a curved bill. Buff-breasted is a trifle larger and has the crown paler, breast and throat buffy. This identification requires special care.
Habitat
Grassy areas inland, shorelines, sometimes tidal flats, high-altitude lakes in migration. Rarest peep on coast.
Habits
Allows close approach, less active than other peeps; flies with them, but feeds by itself.
Voice
Similar to that of Semipalmated; a distinctive *creep* in flight.
Eggs
4; clay-colored, spotted with umber; 1.3 x 0.9 in. (3.3 and 2.3 cm).
Remarks
Like many of its relatives, this sandpiper migrates long distances. Many individuals that summer in the Arctic winter in Chile, sometimes in areas up to 13,000 feet (3962 m) in the Andes.
Range
Breeds in Greenland and Canadian Arctic; winters in S. America; migrates mainly through interior.

LEAST SANDPIPER

Calidris minutilla **20:2, 21:2**

Description
Size, 6 in. (15.2 cm). The smallest sandpiper, with • *yellowish or greenish legs,* thin bill.
Similarities
Resembles Semipalmated, but browner above, with neck and breast more streaked in spring and darker in fall; bill thinner, legs dusky yellow. Suggests a diminutive Pectoral.
Habitat
Shorelines, mud flats, wet fields; most common away from beach.

Habits
Perhaps the most common shorebird; quite tame; when flushed, zigzags off like Common Snipe.
Voice
A *peep*; also a short *kreep,* higher-pitched and more squeaky than Semipalmated's.
Eggs
3–4; buff, with rich, dark markings; 1.1 x 0.8 in. (2.8 x 2.0 cm). Nest is mossy depression in bog or tundra marsh.
Range
Breeds in Arctic, s. to James Bay and Newfoundland; winters from N.C. and Gulf Coast s. to S. America.

CURLEW SANDPIPER
Calidris ferruginea **20:9; 21:8**

Description
Size, 8½ in. (21.6 cm). Slender; • *bill noticeably downcurved;* rufous in spring; gray above, white below in fall; white wing stripe, • *rump white.*
Similarities
Dunlin has bill drooped at tip only and lacks white rump in fall. White-rumped has shorter, straight bill.
Habitat
Coastal. Very rare.
Habits
Usually seen as a solitary bird with Dunlins.
Voice
Soft, musical double-note *chirrup,* quite different from harsh ternlike call of Dunlin.
Range
Breeds in Asian Arctic; winters in Africa, Asia, and Australia; in migration, seen along Atlantic and Gulf coasts.

DUNLIN
Calidris alpina **20:7, 21:9, 22:5**

Description
Size, 8½ in. (21.6 cm). Short-legged, rather stocky; • *bill long, heavy, and with droop at tip;* legs black. Spring: bright reddish above; wing stripe, sides of rump, and underparts white; • *belly black.* Fall and winter: gray above, white below with grayish band across breast.
Similarities
Purple Sandpiper is darker, Sanderling lighter with straight bill.
Habitat
Beaches, shorelines, flats.
Habits
On ground, tame, sluggish; often looks hunched up.
Voice
Harsh, rather loud *cheer-ur;* "flushing note . . . a fine *chit-l-it*" (Nichols).
Eggs
4; buff, marked with chestnut brown; 1.5 x 1.0 in. (3.8 x 2.5 cm).
Former name
Red-backed Sandpiper.
Range
Breeds in Arctic s. to Hudson Bay; winters along Atlantic Coast from New England s. to Fla.

SEMIPALMATED SANDPIPER

Calidris pusilla **20:3, 21:1**

Description
Size, 5½–6¾ in. (14.0–17.1 cm). A small, common peep. Streaked gray-brown above, slightly browner in spring; breast ashy with light streaks; paler or unmarked in fall; black bill stout and straight. • *Slaty legs* sometimes appear greenish when slimy with mud.
Similarities
Least very similar, but browner and less chubby; breast more streaked in fall; legs yellow. Bill of Western is longer, thicker at base, and with slight droop at tip.
Habitat
Beaches, shorelines, flats; more common than Least near beach.
Habits
Often stands or hops on 1 leg, dashes about feeding with head down (unlike plovers); retreats and advances before waves, often in groups or flocks; at high tide, rests higher up on beach in groups behind some shelter or facing wind.
Voice
Flight call a grating *churp* or *check,* shorter than Least's; flushing note *ki-i-ip.*
Eggs
3–4; variable, buffy marked with brown; 1.2 x 0.8 in. (3.0 x 2.0 cm).
Range
Breeds in n. Canada s. to Hudson Bay; winters s. of U.S.

WESTERN SANDPIPER

Calidris mauri **20:1, 21:3, 22:2**

Description
Size, 6½ in. (16.5 cm). Similar to Semipalmated, but • *bill longer, thicker* at base, and with slight • *droop at tip.* Fall: almost indistinguishable, but "somewhat larger, rangier, paler, grayer" than Semipalmated and with "better developed white stripes over the eyes" (Nichols). Spring: more reddish above, giving a 2-toned rusty-and-gray effect; breast more streaked, often with dark V's running down the sides.
Habitat
Similar to Semipalmated's.
Habits
Carries bill pointed down more than does Semipalmated; usually feeds in deeper water.
Voice
Many notes like those of Semipalmated, but flight note *chee-rp (ee* as in *kreep* of Least Sandpiper).
Eggs
Similar to Semipalmated's.
Remarks
Only typical birds should be identified in the field.
Range
Breeds in Asia and Alaska; winters along coasts, from N.C. to S. America.

SANDERLING

Calidris alba **20:6, 21:4, 22:8**

Description
Size, 8 in. (20.3 cm). Somewhat larger and stockier than other peep. Conspicuous • *white wing stripe;* underparts and sides of

tail white; bill and legs black. Spring: head, upperparts, and breast bright red-brown. Fall: upperparts light gray, dark at bend of wing, forehead and underparts foam-white.

Habitat

Sea beaches, tidal flats, sandy edges of bays.

Habits

Follows edge of advancing and receding waves, often in small flocks; when flushed, flock semicircles out over breakers and lands farther up the beach; often stands on 1 leg; makes lines of little holes probing in sand.

Voice

Sharp and distinct *kit,* singly or in series.

Eggs

4; dull olive, marked with brown; 1.4 x 1.0 in. (3.6 x 2.5 cm).

Range

Breeds along Arctic coast s. to Hudson Bay; winters from s. New England s. along coast to Fla., Gulf Coast, and S. America.

STILT SANDPIPER

Micropalama himantopus **21:13, 22:14**

Description

Size, 7½–9 in. (19.1–22.9 cm). Legs long, greenish; no wing stripe; white on uppertail coverts • *horseshoe-shaped,* not straight, across back; whitish tail; long bill, slightly drooped at tip. Spring: • *rusty ear streak,* dark back, and evenly barred underparts. Fall: gray above, line over eye and underparts white.

Similarities

Lesser Yellowlegs is larger, bill does not droop, and legs are longer and yellow. Dowitchers are heavier billed, chunkier, have a white wing stripe and back stripe. Stilt "looks and acts more like a short-billed, long-legged, pale-breasted dowitcher" (Pough).

Habitat

Shallow pools, fresh or salt.

Habits

Flight similar to Lesser Yellowlegs; quiet, sedate, frequently feeds in compact groups in pools up to its belly in water, often with head and neck under water and bill thrusting down perpendicularly, or swinging widely from side to side; does not bob head, as do yellowlegs.

Voice

Low *thu,* resembling note of Lesser Yellowlegs.

Eggs

3–4; dull white, marked with brown; 1.4 x 1.0 in. (3.6 x 2.5 cm).

Range

Breeds in n. Canada; winters in S. America; migrates w. of Mississippi except for a small number along Atlantic Coast.

BUFF-BREASTED SANDPIPER

Tryngites subruficollis **19:10, 23:4**

Description

Size, 7½–8¾ in. (19.1–22.2 cm). No seasonal change. Bill short, head rounded; appears big-eyed because of conspicuous white eye-ring; head, neck, and • *underparts plain buff,* a few dark spots on the side; upperparts blackish-brown tinged with buff; wing stripe obscure, • *wing linings whitish;* tail dark; wings long, pointed, reaching at rest beyond end of tail; legs yellow.

Similarities

Remotely resembles a diminutive Upland Sandpiper, which is larger.

Habitat
Grassy open areas; rare along coast.
Habits
Tame; while on ground often raises wing and extends head; seeks safety by hiding; behavior often similar to that of Upland Sandpiper; flies in compact flocks.
Voice
Rather silent; occasionally a sharp, thin, clicking *tik;* also a Robin-like *chwup,* or a *pr-r-r-reet.*
Eggs
4; clay-colored; boldly marked with umber and slate; 1.5 x 1.0 in. (3.8 x 2.5 cm).
Range
Breeds in Arctic and nw. Canada; winters in S. America; migrates through Great Plains, rarely to Atlantic coast.

SHORT-BILLED DOWITCHER

Limnodromus griseus **20:15, 21:14, 22:12**

Description
Size, 10½–12 in. (26.7–30.5 cm); bill, male, 2¼ in. (5.7 cm); female, 2½ in. (6.4 cm). Chunky, wing stripe on trailing edge, not in middle as in some other shorebirds; barred tail; greenish legs. Spring: brown above, rusty below. Fall: gray instead of red-brown and with a white eye line.
Similarities
See Long-billed Dowitcher, Red Knot, Common Snipe, yellowlegs. Fall Stilt Sandpiper is more slender, shorter-billed, and longer-legged.
Habitat
Marshes, mud flats, shorelines, inner beaches, still waters.
Habits
Flies in compact flocks with bill pointed partly down; feeds in deeper water than most shorebirds, probing like a sewing machine in the mud with its long bill; swims readily.
Voice
Flight call resembles that of Lesser Yellowlegs, but is faster, softer; it is a whistled, *too, too, too* or *dow-itch,* or *dow-itch-er.*
Eggs
4; grayish, with dark brown spots; 1.6 x 1.1 in. (4.1 x 2.8 cm).
Remarks
Dowitchers are the only shorebirds with long straight bills and white rumps, continuing as a wedge up the back. The bill of the Short-billed averages shorter than that of the Long-billed, but is not a reliable factor.
Range
Breeds in cen. to s. Canada, from Que, to ne. Man.; winters from S.C. to Gulf Coast and S. America.

LONG-BILLED DOWITCHER

Limnodromus scolopaceus

Description
Size, 12 in. (30.5 cm); bill, male, 2½ in. (6.4 cm); female, 3 in. (7.6 cm). Spring: dark above, underparts red-brown to undertail coverts, which are barred, not spotted. Winter: very like Short-billed; bill longer (see Voice).
Similarities
Very difficult to distinguish from Short-billed except by voice in winter plumage; tends to be longer billed, in spring has rusty belly with bars.

Habitat, Habits
Similar to Short-billed's.
Voice
High *keek* or *keek, keek, keek.*
Eggs
4; olive, marked with brown; 1.7 x 1.2 in. (4.3 x 3.0 cm).
Range
Breeds in nw. Canada; winters from Fla. and Gulf Coast to C. America.

MARBLED GODWIT

19:11
Limosa fedoa
Fig. 7

Description
Size, 16–20 in. (40.6–50.8 cm); bill, 4¼ in. (10.8 cm). One of the largest shorebirds. • *Upturned bill,* sometimes almost straight; buffy brown with black markings; no white; legs blue-gray. In flight, patch of cinnamon under wings; wings also show black outer primaries.
Similarities
Hudsonian Godwit is smaller and has a white rump; Long-billed Curlew has downcurved bill.
Habitat
Shorelines, flats, salt marshes, wet grasslands. Rare.
Habits
Flight swift, strong, direct, head somewhat drawn in; bill straight forward, legs stretched out behind.
Voice
Noisy on breeding grounds, *god-WIT, god-WIT, god-WIT* or *you're-crazy-crazy-crazy* and *cor-RECT, cor-RECT;* flight call is *queep, queep, queep.*
Eggs
4; buff, spotted with brown; 2.2 x 1.5 in. (5.6 x 3.8 cm).
Range
Breeds in Prairie provinces, from s.-cen. Canada to Minn. and S.Dak.; winters from S.C. to Fla. and Gulf Coast to C. America.

HUDSONIAN GODWIT

20:13, 21:15
Limosa haemastica
Fig. 7

Description
Size, 14–16½ in. (35.6–41.9 cm). • *Bill slightly pink, upturned, black at tip,* sometimes virtually straight; wings blackish with narrow white wing stripe; underwing and wing pit black (conspicuous in flight); tail black, with base and tip white; legs blue-gray. Spring: below rich chestnut marked with black. Fall: above dark gray; below whitish.
Similarities
Willet in flight has much more white on wing and tail is whitish, not ringed; standing, Willet is sturdier and has a thicker, shorter bill. Greater Yellowlegs is smaller, has yellow legs, and bobs. Neither has an upcurved bill.
Habitat
Similar to Marbled Godwit, but breeds on tundra, not prairie.
Habits
Probes in mud and deep water, often up to belly and with head submerged; flies in dark, undulating lines.

Voice
Low *qua-qua* when flushed; lower, less harsh than Marbled's; a low *ta-it* in flight.
Eggs
2–4; dull olive, marked with brown; 2.2 x 1.4 in. (5.6 x 3.6 cm).
Range
Breeds on Canadian tundra; winters in S. America; migrates through interior in spring and reaches Atlantic Coast in fall.

PHALAROPES

Family Phalaropodidae

Phalaropes, a group (often merged into sandpiper family) of only three species, are small, long-necked, swimming sandpipers with pointed wings and lobed toes. The females are larger and more brilliantly plumaged than the males, and they take the lead in courtship. Phalaropes fly swiftly and swim buoyantly, often whirling. Their food is insects and small aquatic animals. The male builds the nest on the ground, incubates the eggs, and cares for the young.

RED PHALAROPE

Phalaropus fulicarius **20:18, 21:23, 22:10**

Description
Size, 7½–9 in. (19.1–22.9 cm). Bill thick, wholly or partly yellow. Spring female: • *reddish below* and on neck, black crown, and white cheeks and wing stripe; dark-tipped yellow bill. Spring male much duller. Fall: pale, unstreaked gray above; head, neck, and underparts white; black patches on head; notably curved "phalarope mark" behind eye. Bill short, blunt, thicker at base.
Similarities
Rather than needle bill of other phalaropes, Red has shorter, stouter bill; see Wilson's and Northern Phalaropes. Whiter-backed Sanderling in fall seldom swims, and it lacks mark on head.
Habitat
Most maritime of the phalaropes; breeds on tundra, winters on open sea.
Habits
Active, unsuspicious; flies like a Sanderling, swims like a tiny gull, bobs head and dabs bill for food, or tips up and feeds; in feeding, makes twice as many dabs per second as Northern.
Voice
Similar to Sanderling, but thinner and higher-pitched.
Food
Jellyfish, fish.
Eggs
4; buffy or greenish, marked with brown; 1.2 x 0.8 in. (3.1 x 2.2 cm).
Range
Breeds in Arctic; winters in S. Hemisphere oceans; migrates along Atlantic Coast.

WILSON'S PHALAROPE

Phalaropus tricolor **20:16, 21:19, 22:13**

Description
Size, 8½–10 in. (21.6–25.4 cm). Has the thinnest, most needlelike bill of any shorebird. White rump and tail, no white wing stripe.

Spring female: underparts and cheeks white; black band from eye turns • *chestnut* on back. Spring male: much duller. Fall: plain brownish-gray above, plain white below; legs turn greenish-yellow.

Similarities

Other phalaropes have wing stripes and dark marks through mid-rump and tail. Winter Sanderling has thicker, shorter bill and conspicuous wing stripe. Stilt Sandpiper has similar flight pattern, but is darker, bill is much heavier, forehead is dark. Lesser Yellowlegs has streaks on breast and lemon-yellow legs.

Habitat

Pools, shallow lakes inland; not oceanic.

Habits

Spins and runs nervously in shallow water, making abrupt twists and turns; probes with bill in mud and swings bill from side to side; feeds on foot in wet meadows and mud flats.

Voice

Soft nasal *oit-oit*, unlike other phalaropes.

Food

Crustaceans, insects.

Eggs

4; buffy or greenish, marked with brown; 1.2 x 0.9 in. (3.3 x 2.3 cm). Nest is grassy depression in either wet or dry meadow.

Range

Breeds from s.-cen. Canada, s. to Ind. and Mich.; winters in S. America; migrates through Mississippi Valley region and occasionally to Atlantic Coast.

NORTHERN PHALAROPE

Phalaropus lobatus **20:19, 21:22, 22:7**

Description

Size, 6½–9 in. (16.5–20.3 cm). Small head, thin neck; needlelike black bill, white wing stripe, dark tail and legs. Spring female: gray above, darkest on head; • *sides of neck chestnut*; throat and belly white; back streaked; Spring male: much duller. Fall: above blackish streaked with gray; below white; black eye stripe.

Similarities

Fall Red Phalarope is larger, more heavily built, darker above, and has shorter, heavier bill. Fall Wilson's has white rump, no white wing stripe, and no face markings. Sanderling has a longer wing stripe, shallower wingbeat.

Habitat

Breeds on tundra, winters on ocean, sometimes seen on inland ponds.

Habits

Gregarious, tame; flight swift, erratic; often alights on floating seaweed and runs about; bathes with characteristic upjerk and ducking motion.

Voice

Similar to Sanderling's.

Food

Insects, crustaceans, plankton.

Eggs

4; buffy or greenish, marked with brown; 1.1 x 0.7 in. (2.9 x 2.0 cm).

Range

Breeds in Arctic, s. to Labr. and St. James Bay; winters in S. Hemisphere oceans; migrates off Atlantic Coast.

AVOCETS AND STILTS

Family Recurvirostridae

These are very long-legged, thin-billed wading birds spectacularly patterned with black and white. They frequent shallow freshwater, alkaline or brackish ponds and coastal flats, feeding on insects, shellfish, and seeds.

AMERICAN AVOCET

Recurvirostra americana **19:17**

Description
Size, 15½–20 in. (39.4–50.8 cm). • *Bill upturned*; black-and-white pattern unique, both at rest and in flight; head and neck buffy in spring.
Similarities
Black-necked Stilt is more slender, has black on head and neck, and has red legs.
Habitat
Shallow lakes and marshy ponds inland.
Habits
Flight direct, strong, and fairly swift, with neck and legs extended. Feeds by walking quickly through water, working bill from side to side in the mud; or while swimming, by tipping up.
Voice
Loud yelp.
Eggs
3–4; olive, marked with brown; 2.0 x 1.3 in. (5.1 x 3.3 cm).
Range
Breeds in nw. Canada, s. to Tex.; winters from s. Tex. s. to C. America; casual to Atlantic Coast in migration.

BLACK-NECKED STILT

Himantopus mexicanus **19:15**

Description
Size, 13–17 in. (33.0–43.2 cm). A tall shorebird, black above and white below. Adult male: long, slender black bill; • *very long, thin red legs;* white rump and tail. In flight, appears to have a white body, black wings. Immature: browner, legs paler.
Habitat
Shallow water, primarily fresh.
Habits
Aggressive; feeds while wading, steps high in walking and often tosses head. In flight, neck somewhat drawn in, legs extended.
Voice
Noisy; loud, high-pitched, insistent yipping.
Eggs
3–5; buff, marked with dark brown; 1.7 x 1.2 in. (4.3 x 3.0 cm).
Range
Breeds along Atlantic Coast from Del. s. to Fla. and S. America; winters s. of U.S.; casual in migration along Atlantic Coast.

SKUAS AND JAEGERS

Family Stercorariidae

This is a family of strong-flying, hook-billed, hawklike, predominantly darkish seabirds flashing white on their wings. Adult jaegers are generally dark above, with black caps and whitish collars, and light below. The amount of wing flash varies with the species and each has its characteristic lengthened tail feathers. The young have dark, barred breasts and lighter bellies barred on the sides. Their tails are not lengthened, thus making identification difficult and sometimes impossible. Members of this family breed in the Arctic, winter at sea, and only occasionally are seen from shore in migration. These predators feed by harassing other seabirds into dropping their food, which they then seize. Problems of identification of jaegers are enhanced by the occurrence of dark and light color phases.

POMARINE JAEGER

Stercorarius pomarinus **2:2** *Fig. 8*

Description
Size, 20–23 in. (50.8–58.4 cm). Center • *tail feathers blunt and twisted* in adults. Typically dark-capped, light below, but also all-dark phase.

Similarities
• *Broader-winged*, larger than other jaegers. Young larger, bill heavier, and buffier below than other jaegers; more white in wings.

Habitat
Arctic coasts when breeding; offshore winter.

Habits
Aggressive.

Voice
"A sharp *which-yew*" (Rich).

Food
Mainly fish.

Eggs
2; olive-brown with dark brown spots; 2.2 x 1.6 in. (5.7 x 4.1 cm). Nest is in grass-lined depression on ground.

Range
Breeds in Greenland and Arctic Canada; winters at sea, s. from N.C.

PARASITIC JAEGER

Stercorarius parasiticus **2:1**

Description
Size, 16–21 in. (40.6–53.3 cm). Black legs, pale breastband, bill smaller than Pomarine's; center tail feathers pointed, project beyond others.

Similarities
Less contrast between cap and back in light phase than in Long-tailed; also yellow on head is paler, and white collar on hind neck is narrower.

Habitat
See Pomarine.

Voice
Various wails, shrieks, and cries.

Food
Small birds, lemmings in summer, fish at other times.

Eggs
2; olive-brown with dark brown spots; 2.2 x 1.6 in. (5.7 x 4.1 cm). Nest is tundra hollow.

Range
Breeds in Greenland and n. Canada, s. to cen. Canada; winters at sea, s. to S. America; migrates along coasts.

LONG-TAILED JAEGER

Stercorarius longicaudus **2:3**

Description
Size, 20–23 in. (50.8–58.4 cm), including tail. • *Bluish legs;* breast white shading into gray on belly; tail pointed, over 4 in. (10.2 cm). Light phase: pale, ashy back contrasts sharply with black cap, cheeks bright yellow, white collar on hindneck broader than Parasitic's and almost surrounds neck.
Similarities
Other jaegers are larger-bodied with shorter central tail feathers.
Habitat
Open sea when not breeding.
Habits
Less aggressive, harasses gulls less frequently than larger-bodied jaegers.
Voice
"A shrill *pheu-pheu-pheu-pheo* . . . often followed by a harsh *qua*" (Nelson).
Food
Lemmings, young birds in summer, fish, marine invertebrates otherwise.
Eggs
2–3; olive, spotted with sepia; 2.1 x 1.5 in. (5.5 x 3.9 cm).
Range
Breeds in Arctic; winters at sea, s. of Va.; migrates along Atlantic Coast and Great Lakes area.

NORTHERN SKUA

Catharacta skua *Fig. 8*

Description
Size, 21 in. (53.3 cm). Large, robust; • *wings broad with conspicuous white patch.* Buteolike; sooty brown above, rusty below; • *tail short,* slightly uptilted, faintly forked in young.
Similarities
Herring Gull lacks white wing patches.
Habits
Flight hawklike or eaglelike, and very rapid.
Voice
"Skua" (McGillivray).
Food
As jaegers; carrion.
Eggs
2–3; white blotched with brown; 2.7 x 1.9 in. (7.0 x 4.9 cm).
Range
Breeds in Iceland and n. Europe; winters at sea off Newfoundland s. to New England.

Fig. 8

GULLS AND TERNS
Family Laridae

GULLS
Subfamily Larinae

Gulls are fairly large, long-winged, generally white water birds, many with black wing tips, often with a pearly "mantle" (back and top of wings), sometimes with a black head or back. Juveniles and immatures frequently are of varying degrees of dusky brown, as it sometimes takes several years to reach the adult plumage. Gulls fall into three plumage groups: (1) the white-wings: Glaucous, Iceland, and Ivory; (2) the hooded gulls: Laughing, Franklin's, Bonaparte's, Black-headed, Little, and Sabine's; (3) the pearly-backs: all the others except the Great Black-backed.

Gulls fly with the bill straight forward, soar for protracted periods, often follow boats, swim buoyantly, pick food from the surface, and, being scavengers, often collect in great numbers at sewer outlets, fish-processing areas, and garbage dumps. They perch on buoys, pilings, and roofs, and walk easily. Gulls sometimes frequent croplands and follow the farmer as he plows in order to eat the grubs in the furrow.

GLAUCOUS GULL
Larus hyperboreus **24:1**

Description
Size, 28 in. (71.1 cm). Almost size of Great Black-backed, the largest gull. Folded, white-tipped wing does not reach tip of tail. Adult: white with a pale gray mantle, bill heavy and yellow with red spot near tip, eye-ring and eye yellow in breeding season, feet pinkish, • *white wing tips*. First-winter immature: pale buffy throughout, primaries equally pale. Second-winter immature: mantle gradually changes to creamy white, then to all-white in third year; bill white, bill spot black.
Similarities
Adult Iceland is smaller, has a less robust head and bill, eye-ring sometimes red, and longer wings; immature has smaller, darker bill. Iceland is more buoyant; at rest, wings reach beyond tail.
Habitat
Coast, offshore.
Habits
Predatory; flight steady, soaring, somewhat hawklike.
Voice
Hoarse *ku-ku-ku, ku-lee-oo*.
Food
Carrion, refuse, birds, marine invertebrates.
Eggs
2–3; olive-buff, marked with dark brown; 3.0 x 2.1 in. (7.6 x 5.3 cm). Nest of seaweed, grasses, etc., up to 3 ft. (0.9 m) high, on cliffs and ledges; in colonies.
Range
Breeds in Arctic; winters in N. America to N.Y. and the Great Lakes.

ICELAND GULL
Larus glaucoides **24:3**

Description
Size, 24 in. (61.0 cm). Folded wing reaches beyond tip of tail. Adult: white with a pale gray mantle, yellow bill with red spot near tip, eye ring purple, iris mottled to dark in breeding season, wings white-tipped like Glaucous or white with gray spots. Feet pinkish. Immature: plumages similar to immature Glaucous, but smaller bill more extensively black-tipped and turns greenish in second year. The subspecies called "Kumliens" **(24:3c)** has white wing tips with gray spots.
Similarities
Glaucous is larger and has a proportionately heavier head and bill, whereas Iceland's wings are longer, proportionately more pointed, its flight more buoyant. However, size alone is confusing; large Iceland may be as large as small Glaucous and vice versa. In breeding season Glaucous and Herring Gulls have yellow eyes and eye-rings, the Iceland and Thayer's Gulls have purple eye-rings and mottled or dark eyes; see Thayer's Gull.
Habitat
Coasts, inland lakes, and rivers.
Habits
Less aggressive and predatory than Glaucous.
Voice
Similar to Herring's, but shriller.
Food
Fish, refuse, carrion, offal, crowberries.
Eggs
2–3; buff, often tinged with green and marked with brown; 2.8 x 1.7 in. (7.1 x 4.3 cm). Nest of vegetation on cliffs; in colonies.
Range
Breeds in Greenland, s. to Baffin Is.; winters in ne. U.S. waters s. to N.J. and w. to Great Lakes.

THAYER'S GULL
Larus thayeri

Description
Size, 22 in. (55.9 cm). Resembles Herring Gull and Iceland Gull closely, but dark-eyed. Adult: pearly mantle; bill and legs as Herring Gull; iris dark, surrounded by dark (violet) eye-ring; wing tip variable, gray with white spots to gray-black with white spots (between Herring and Iceland in color). Immature: indistinguishable from Herring Gull.
Similarities
Herring Gull has yellow eyes and black-and-white wing tips. Usually distinguishable from Iceland Gull by darker mantle; darker wing tips; Kumlien's race of Iceland has paler gray spots in wing tips. A very difficult identification.
Habitat, Habits, Voice, Food
Similar to Herring Gull's.
Eggs
Similar to Herring Gull's but nest is of vegetation on Arctic cliffs.
Range
Breeds in Greenland and n. Canada; winters along north Pacific Coast; rarely to Atlantic Coast and Great Lakes.

HERRING GULL

Larus argentatus **24:4, 26:11**

Description
Size, 22½–26 in. (57.2–66.0 cm). Adult: white with pearly-gray mantle, black wing tips with white spots; iris and eye-ring yellowish in breeding season, bill yellow with a red spot near tip, • *legs pinkish;* in winter, head and neck streaked with brown. First-winter immature: almost uniform sooty. Second-winter immature: dusky, broad diffuse dark band near end of tail; body becomes lighter and uppertail coverts whiter with age; bill dark, sometimes with black ring near end.
Similarities
Ring-billed Gull is smaller, has black ring around middle of bill and yellowish legs, and has more buoyant flight. Immature Great Black-backed is larger, darker above, lighter below. Immature Ring-billed is lighter above and below, and has narrow blackish band near end of white tail. See also Iceland and Thayer's Gulls.
Habitat
The commonest gull in most areas including shorelines, bodies of water, rivers, garbage dumps.
Habits
Gregarious; a great ship-follower and soarer, often high in air; "trades" back and forth between bodies of water morning and late afternoon, at middle height; swims, drinks sea water; drops mollusks from height on a hard surface to break shells.
Voice
"Queeeeeeah-ah, kak, kak, kak" (Collins).
Food
Carrion, garbage, refuse, marine animals, eggs, young birds.
Eggs
2–5; variable, often olive-buff spotted and blotched with dark brown; 2.8 x 1.9 in. (7.1 x 4.8 cm).
Range
Breeds from Greenland to N.C.; winters from s. Canada, s. to W. Indies and Mississippi Valley.

RING-BILLED GULL

Larus delawarensis **24:5, 26:10**

Description
Size, 18–21 in. (45.7–53.3 cm). Adult: white with pearly mantle slightly paler than Herring's; black wing tips with white spots, wing tips all-black below; • *yellowish legs;* black ring around bill. First-winter immature: dusky. Second-winter immature: light dusky above and whitish below, narrow blackish band near end of white tail, bill and legs usually pinkish, bill with a black tip.
Similarities
Immature Herring Gull tail mainly brown; Ring-Billed has dark-tipped white tail, body whiter.
Habitat
Coast and inland waters; garbage dumps.
Habits
Gregarious; carriage more sprightly than Herring's; follows plow and ship; flight said to be lighter, more buoyant, more dovelike than Herring's, but not easy to distinguish.
Voice
Notes higher-pitched than Herring's; alarm note a hawklike *cree-cree.*
Food
Garbage, refuse, carrion, aquatic animals, rodents, insects.

Eggs
2-4; similar to Herring Gull but smaller; 2.3 x 1.6 in. (5.8 x 4.1 cm).
Range
Breeds from Labr. to Great Lakes; winters from s. New England s. to Cuba.

Note: The **MEW GULL**, *Larus canus*, size, 17 in. (43.2 cm), occurs in East only as a breeder east to W. Mackenzie, northeastern Alberta, northern Saskatchewan, and recently Wisconsin. It resembles a Ring-billed Gull but has a smaller unmarked yellow bill.

GREAT BLACK-BACKED GULL

Larus marinus **24:2**

Description
Size, 29 in. (73.7 cm); white-headed, • *slate-colored back and wings.* Adult: bill yellow with red spot near tip, legs pinkish. First-winter immature: dusky streaked with dark band near end of tail; back darker, head and underparts lighter than Herring, older young approach adult appearance; bill blackish, feet whitish to pinkish. In flight, bigger head and neck protrude more than Herring.
Similarities
Immature Glaucous Gull is lighter and lacks blackish primaries and dark band near tip of tail. Immature Herring has streakier head, less dark on mantle.
Habitat
Coastal waters.
Habits
Predatory, flight strong, wingbeats slow, impressive; soaring flight.
Voice
Keeaw, lower than Herring's; also a quick *gaw-gaw-gaw;* a high *key-key-key;* and a hoarse laugh, *ha-ha-ha.*
Food
Carrion, refuse, marine invertebrates, birds, eggs, rodents.
Eggs
2–3; olive or buffy gray, blotched with brown; 3.2 x 2.1 in. (8.1 x 5.3 cm).
Range
Breeds along Atlantic Coast from Greenland to N.J.; winters s. to Ga. and w. to Great Lakes.

Note: The accidental **LESSER BLACK-BACKED GULL**, *L. fuscus*, size, 23 in. (58 cm), of Europe, is smaller (size of Herring Gull), slimmer, lighter above, and has yellow legs.

BLACK-HEADED GULL

Larus ridibundus **25:5**

Description
Size, 15 in. (38.1 cm). Summer adult: white with • *brown head, gray mantle, bill red to yellowish.* Winter adult: like winter Bonaparte's, but bill and underwing differ as above. Immature: As Bonaparte's, but younger birds browner above, bill and legs yellow to pinkish, yellow bill with black tip.
Similarities
Bonaparte's is smaller, with smaller black bill.
Habitat
Sewer outlets, estuaries, harbors, usually in winter.

Habits
Similar to Bonaparte's.
Voice
"A harsh *kwarr,* a short *kwuh*" (Mountfort).
Range
Breeds in Eurasia; winters s. of Europe to Africa and in s. Asia; casual to Atlantic Coast in winter.

LAUGHING GULL
Larus atricilla **25:1**

Description
Size, 17 in. (43.2 cm). Summer adult: wing tips black, • *trailing edge of wing white;* neck, tail, and underparts white; bill and legs dark red; bill relatively long. Winter adult: head white with dark smudge; bill and legs duller. First-winter immature: head and breast dark, rump and belly white, tail with black band near white tip, bill and legs blackish.
Similarities
The only black-headed gull lacking a distinctive black or white wing-tip pattern.
Habitat
Shorelines, rivers; breeds in marshes.
Habits
Flight light, strong, but heavier than other black-headed gulls; often soars, catches insects in air; follows plow.
Voice
Noisy; a characteristic laughter; *"hah-ha-ha-ha-ha-hah-hah-hah"* (Langille).
Food
Aquatic animals, insects, carrion, offal, garbage, refuse.
Eggs
3; variable, often buff blotched with umber; 2.1 x 1.5 in. (5.3 x 3.8 cm).
Range
Breeds along Atlantic Coast from Maine to W. Indies and along Gulf Coast to Tex.; winters s. from N.C.

FRANKLIN'S GULL
Larus pipixcan **25:2**

Description
Size, 14 in. (35.6 cm). Summer adult: head black, • *white bar* or "window" between gray primary base and white-tipped black ends, bill and legs dark red, underparts with faint rosy bloom, bill relatively long. Winter adult: white head with dark band behind head from eye to eye. First-winter immature: forehead white, nape black, wings dark all over, rump and underparts white, legs blackish.
Similarities
Somewhat similar, Laughing Gull is strictly coastal, and in east never occurs in the summer range of Franklin's.
Habitat
Prairies, marshes, fields, and bodies of water inland.
Habits
Gregarious; follows plow, sometimes soars upward in spirals, captures insects in air, swims buoyantly, often flies in V-shaped flocks.

Voice
Unlike Laughing; "a soft *krruk* . . . a louder and more plaintive . . . *pway* or *pwa-ay*" (Bent); when soaring, a "*weeh-a-weeh-a weeh-a po-lee po-lee po-lee po-lee*" (Miller).
Food
Grasshoppers, other insects, fish, frogs, mollusks.
Eggs
2–4; buffy, marked with brown; 2.0 x 1.4 in. (5.1 x 3.6 cm).
Range
Breeds in prairie marshes, s. from s. Canada to Minn. and S.Dak.; winters from La. and Tex., s. to S. America.

BONAPARTE'S GULL
Larus philadelphia **25:3**

Description
Size, 13 in. (33.0 cm). Summer adult: gray mantle, black-headed; long, white wedge down outer side of black-tipped primaries; wing tips white from below; bill is small, black; legs orange-red. Winter adult: white head with black spot behind ear coverts. First-winter immature: similar to winter adult but with narrow black band near tip of tail; upper wing with grayish band from wrist to back; bill dark with light base; legs pinkish.
Similarities
Laughing adult has much darker mantle; immature, neck and sides of breast dark and broader band on tail. Franklin's has much darker mantle in adult. Kittiwake young is larger; black bar on back of neck, yellowish legs.
Habitat
Coastal and interior waterways; breeds in coniferous forests.
Habits
Gregarious; flight somewhat ternlike, but seldom dives; points bill down in flight, which seems desultory but is rapid; swims buoyantly.
Voice
Shrill, nasal *peer;* "sparrowlike conversational notes" (Jones).
Food
Largely insects, some fish, crustaceans, worms.
Eggs
2–4; olive-buff, spotted with chocolate; 1.9 x 1.3 in. (4.8 x 3.3 cm). Nest of sticks in conifers, to height of 20 ft. (6.1 m).
Range
Breeds in nw. Canada; winters along Atlantic Coast from s. New England to Fla. and Gulf Coast.

LITTLE GULL
Larus minutus **25:6**

Description
Size, 11 in. (27.9 cm). This rare European visitor is the only gull with • *all-black wing linings.* Wings and tail somewhat rounded. Summer adult: black head; gray mantle and wings above; wings blackish below with white trailing edges; bill and legs red. Winter adult: head white with black ear patch, bill blackish. First-winter immature: wide, black open V on gray wing; black band near tip of tail; wing linings white; perched bird shows dark stripe on wing.
Similarities
Immature resembles larger small Kittiwake, but has black on back and distinct crown patch. Like immature Bonaparte's but more black on back and crown; primaries solid black without white bases.

Habitat
Ocean coasts and Great Lakes harbors; sewer outlets in winter.
Habits
Behavior in pursuit of insects suggests Black Tern; tame, fearless; feeds from surface of water in flight; often found with Bonaparte's.
Voice
"A rather low *kek-kek-kek*" (Mountfort).
Food
Fish, aquatic life, insects, refuse.
Eggs
3; olive-brown with dark spots; 1.6 x 1.2 in. (4.1 x 3.1 cm).
Range
Breeds in Europe and recently in Ont. and Wis.; winters in E., particular along Atlantic Coast from N.B. to N.J., and around Great Lakes.

IVORY GULL
Pagophila eburnea

Description
Size, 17 in. (43.2 cm). Adult: white, plump; wings long, wedge-shaped; bill dark with yellow tip; • *legs black.* Immature: blackish spots above bill dusky.
Habitat
On Arctic coasts in summer, edge of pack ice in winter.
Habits
Not highly gregarious; restless, runs on ice like large plover; flight ternlike; perches on ice pinnacles, seldom swims.
Voice
"Shrill and not unlike that of the Arctic Tern" (Yarrell).
Food
Carrion, refuse, lemmings, marine organisms.
Eggs
2; buff-olive with dark blotches; 2.4 x 1.7 in. (6.1 x 4.4 cm).
Range
Breeds primarily in E. Hemisphere, some on Somerset Is. in Arctic; winters at sea in Arctic, sometimes s. to N.Y. and N.J.

BLACK-LEGGED KITTIWAKE
Rissa tridactyla

25:7, 26:9

Description
Size, 17 in. (43.2 cm). Only eastern gull with a solid • *black triangular tip on a gray wing.* Wings long; tail long, broad, slightly forked. Adult: white head, gray mantle, yellow bill, black legs, nape gray in winter. First-winter immature: dark open V on each wing; bill, outer primaries, and end of slightly notched tail black; spot behind eyes; dusky bar on nape.
Similarities
Young Bonaparte's is smaller, lacks dark bar on back of neck. See also Little and Sabine's Gulls.
Habitat
Coasts, ocean.
Habits
Gregarious; drinks salt water, sleeps on waves; follows ships; flight buoyant, graceful, swallowlike, and distinctive at great distance; wingbeats rapid; only gull that dives from wing and swims under water.

Voice
"*Kitti-wake, ka-ake;* sharp and piercing *ki, ki, ki* . . . harsh rattling *kaa, kaa, kae, kae* and *kaak-kaak*" (Townsend).
Food
Fish, crustaceans, mollusks, refuse (but less of a scavenger than larger gulls).
Eggs
1–2; olive-buff, marked with brown; 2.2 x 1.6 in. (5.6 x 4.1 cm). Nest of seaweed, etc., on cliffs; in colonies.
Range
Breeds from Arctic Ocean to Gulf of St. Lawrence; winters off Atlantic Coast, s. to Bermuda.

Note: The **ROSS' GULL,** *Rhodostethia rosea,* size, 14 in. (35.6 cm), is a small, wedge-tailed gull with rosy-tinged underparts. It breeds in Siberia and winters in the Arctic seas. It is an accidental visitor to eastern coastal or Great Lakes areas.

SABINE'S GULL
Xema sabini **25:4**

Description
Size, 13–14 in. (33.0–35.6 cm). Summer adult: • *tail forked;* wings with long, black triangular area at tip; long, white triangular area at rear of wings. Winter adult: head white, brownish smudge around back of head. First-winter immature: head gray above, forked tail black-tipped, wing pattern like adult.
Habitat
Tundra, coast, ocean.
Habits
Feeds ploverlike over flats at low tide; flight ternlike, seldom dives.
Voice
Cry similar to Arctic Tern's, but shorter, harsher.
Food
Aquatic life, insects.
Eggs
2–3; olive-buff, faintly spotted with brown; 1.7 x 1.3 in. (4.3 x 3.3 cm).
Range
Breeds along Arctic shoreline; winters at sea in N. Atlantic; seen in migration along Atlantic Coast from Labr. to N.Y.

TERNS
Subfamily Sterninae

Terns are generally smaller, lighter, more streamlined than gulls; and have sharp, pointed bills, long pointed wings, and forked tails. They start changing into fall plumage in midsummer. These "sea swallows" often hover, usually fly with the bill pointing down, and dive readily for food (or at intruders around the nest), but swim and walk little. The young can run about soon after hatching, but, unlike shorebirds, they are dependent for a long time upon their parents for food. Members of the large Genus *Sterna* have thin bills, black caps, pearl-gray or black mantles, and forked tails.

GULL-BILLED TERN

Gelochelidon nilotica

26:5, 27:5

Description

Size, 14 in. (35.6 cm). • *Body white, stocky*; feet black; • *bill black*. Spring adult: black cap. Fall adult: head white with black ear patch and mottled crown. Immature: similar but with brown band on end of tail, mottling on back, and dark-tipped, lighter bill.

Similarities

Tail less forked than in Common or Forster's; immature particularly gull-like, but has notched tail.

Habitat

Coastal islands, marshes.

Habits

Somewhat like Laughing or Franklin's Gull; flight characteristic, wingbeats slower than *Sterna* terns, dives less, picks food off surface more; hawks back and forth over marshes for insects; follows plow.

Voice

"*Katydid, katydid*; a gull-like *ka-ka-ka*" (Collins).

Food

Insects, some crustaceans.

Eggs

2–3; variable, often buff marked with brown; 1.9 x 1.3 in. (4.8 x 3.3 cm). Nest is in depression in sand or of vegetation among grass.

Range

Breeds along Atlantic Coast from N.Y., s. to Gulf of Mexico and W. Indies; winters from Gulf Coast to S. America.

FORSTER'S TERN

Sterna forsteri

26:4, 27:7

Description

Size, 14–16½ in. (35.6–41.9 cm). • *Whitish primaries* in breeding plumage. Spring adult: bill orange-red with a black tip; cap black; tail gray, moderately forked, dark inner edges, light outer edges; feet orange-red. Fall adult: black ear patch on white head, pale nape, dusky bill, yellowish feet.

Similarities

Bill occasionally without black tip; Common is darker, with shorter tail, paler, and tail has reverse color—dark outer and light inner edges. Fall Common has a larger black area on head going across nape and crown from eye to eye.

Habitat

Coast, marshes, inland waters.

Habits

Wingbeats have quicker, sharper snap than Common's.

Voice

Rasping *tsa-a-ap, zreep,* suggesting a Nighthawk; also a rapid peeping *pip, pip, pip*.

Food

Insects, fish, frogs.

Eggs

3–4; similar to Common Tern's; 1.7 x 1.2 in. (4.3 x 3.0 cm). Nest is usually of vegetation on ground or marsh; in colonies.

Range

Breeds along Atlantic Coast, from Md. to Gulf Coast, w. to Tex., and inland w. of Ill.; winters along coasts from S.C. to C. America.

COMMON TERN

Sterna hirundo **26:3, 27:9**

Description
Size, 13–15 in. (33.0–38.1 cm). A common tern, distinguished in breeding plumage by its black-tipped red bill and dusky primaries. Spring adult: black cap, pearl-gray mantle, moderately forked tail white except for gray outer edge of outermost feathers, orange-red legs. Fall adult: head white with black band around nape from eye to eye, bill blackish, legs paler. At rest, wings extend beyond tip of tail. Immature young: resembles fall adult; similar in color to young Arctic or Roseate.

Similarities
Bill occasionally without black tip in late summer; see Roseate. Bill and legs of Common are longer and tail shorter than in Arctic. See also Forster's.

Habitat
Coast, ocean, shorelines, bodies of water inland.

Habits
Gregarious; flocks collect over schools of fish, thus guiding fishermen to good fishing areas.

Voice
Descending *tee-arr*, a rapid *kik-kik-kik*.

Food
Fish, aquatic life, insects.

Eggs
2–4; variable, often pale brown blotched with gray or lilac; 1.6 x 1.2 in. (4.1 x 3.0 cm). Nest is depression in sand, of vegetation.

Range
Breeds throughout e. N. America from Labr. s. to W. Indies and w. to Wis.; winters from Fla. s. to S. America.

ARCTIC TERN

Sterna paradisaea **26:6, 27:8**

Description
Size, 13–16 in. (33.0–40.6 cm). • *Bill blood-red to the tip*, legs short. Spring adult: black cap with white border above grayish face. Fall adult: bill black, legs blackish. At rest, body is closer to ground than Common, wings and tail about equal in length. Bill does, rarely, have a black tip; identifying this species requires care.

Similarities
Face of Common is all-white; Arctic is grayer above and below than Common and tail is longer and more deeply forked. Young Common as Roseate is similar in color to fall Arctic.

Habitat
Coast, ocean, interior lakes.

Habits, Food
Similar to Common.

Voice
Rising *key, key, key*; a rising *tee-arr* shriller than Common.

Eggs
2; similar to Common's; 1.6 x 1.2 in. (4.1 x 3.0 cm). Nest is depression in sand in colonies.

Range
Breeds in n. Canada from Ellesmere Is. s. to Newfoundland, Mass., and w. to Que. and Man.; winters at sea s. of Equator. The 22,000 mile-(35,405 km) a-year flight of Arctic Terns from the Arctic to the Antarctic is the longest migration of any bird.

ROSEATE TERN

Sterna dougallii **26:7, 27:6**

Description

Size, 14–17 in. (35.6–43.2 cm). This has the longest, most deeply forked tail of any tern. Pale mantle evenly colored throughout; at rest, tail projects beyond wings. Spring adult: • *bill black*, sometimes with red at base; black cap; faint rosy tinge to creamy underparts; primaries light; tail pure white; feet red. Fall adult and immature: crown and forehead white, nape dark.

Similarities

Similar to young Common or Arctic, but forward edge of inner wing not dark. Bill dark and slimmer compared with mainly red bill of Common; tail longer than wings at rest, while it is shorter than wings in Common; and about the same as wings in Arctic.

Habitat

Coast.

Habits

Flight most graceful of all terns, "the greyhound of its tribe" (Bent). Wingbeats a little slower, deeper than Common's; flies lower over water than most terns.

Voice

Distinctive rasping *krack*, louder, lower than other terns, like tearing a piece of strong cloth; a musical whistlelike *chee-bee*.

Food

Fish.

Eggs

2–3; buff, spotted with brown; 1.7 x 1.2 in. (4.3 x 3.0 cm). Nest is depression in sand or crevice in rocks; in colonies.

Range

Breeds along Atlantic Coast from Gulf of St. Lawrence s. to S. America; winters from Gulf Coast to S. America.

Note: The **SOOTY TERN,** *Sterna fuscata (Fig. 9),* size, 16 in. (40.6 cm), of the Gulf of Mexico is occasionally blown north into our area from West Indian seas by hurricanes. Adults are black above, white below with a black cap; black bill and feet; and a forked, mainly black tail. Immatures are almost all brown, flecked above with white. The accidental **BRIDLED TERN,** *S. anaethetus,* a rarer hurricane waif from the West Indies, is similar but paler gray, not black, with a white collar between black cap and gray back.

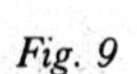

Fig. 9 Sooty Tern

LEAST TERN

Sterna albifrons **26:1, 27:10**

Description

Size, 8½–9 in. (21.6–22.9 cm). A • *small tern*. Spring adult: white forehead, underparts, and tail; black cap, eye line, and outer primaries; yellow feet; tiny black tip to bill, • *bill yellow*. Fall adult: head white, only nape and line to eye black, bill dusky, feet fuller. Immature: similar, but with primaries and front edge of wing dusky.

Similarities
Immature Black has front edge of wing light, flanks grayish (not white), and tail dark.

Habitat
Sandy coasts, interior islands, and shorelines.

Habits
Very rapid and distinctive wingbeat.

Voice
A "rasping *zreeeep*; a rapid, high *kik kik kik*" (Collins).

Food
Small fish and aquatic life.

Eggs
2–3; variable, often olive-buff marked with drab; 1.2 x 1.0 in. (3.0 x 2.5 cm). Nest in depression in sand; in colonies.

Range
Breeds along Atlantic Coast from Maine s. to S. America and also in Mississippi Valley; winters from Gulf Coast southward.

ROYAL TERN

Sterna maxima **26:12, 27:1**

Description
Size, 18–21 in. (45.7–53.3 cm). A big, crested tern with a • *thick orange-red bill.* At rest, wing tips do not extend beyond end of tail. Spring adult: black cap, well-forked tail, underwings light, feet black. Fall adult: more white on forehead. Immature: sometimes with yellow legs.

Similarities
Bill thinner than Caspian's.

Habitat
Coast.

Habits
Flight heavier than Common's, lighter than Caspian's.

Voice
Shrill *keer*, a squawking *kowk*; a rolling, liquid whistle, *tourrreee*, suggesting Upland Sandpiper.

Food
Fish.

Eggs
1–2; whitish, evenly spotted with dark brown; 2.5 x 1.8 in. (6.4 x 4.6 cm). Nest is in depression in sand; in colonies.

Range
Breeds along Atlantic Coast from Va. to Fla. and along Gulf Coast to Tex.; winters from Gulf Coast, southward.

Note: The uncommon **SANDWICH TERN,** *Sterna sandvicensis* **(26:8, 27:4),** size, 15 in. (38.1 cm), breeds with Royal Terns along the Virginia coast, wintering to the south. It is small, gray-backed with a white, forked tail, and has a black cap with a slight crest and a black, yellow-tipped bill. Voice: *"Kirhitt, kirhitt"* (Yarrell). Nests with Royal Terns but the eggs are smaller, more variable—often buff with brown and lavender marks.

CASPIAN TERN

Sterna caspia **26:13, 27:2**

Description
Size, 19–23 in. (48.3–58.4 cm). A big, • *crested tern* with a • *thick red bill.* Heavy body, broad wings; at rest, wing tips extend beyond end of tail. Spring adult: black cap; primaries dusky above, darker beneath; feet black. Fall adult: crown streaked, head appears dusky above to below eyes. Immature: similar to fall adult, but marked with black above; bill more orange.
Similarities
Bill is thicker, tail less forked than Royal's; build is stockier, flight heavier, and has less crest than Royal.
Habitat
Coasts, large interior lakes, rivers.
Habits
Makes powerful Gannet-like dives; flight heavy; gull-like; flies low when fishing, at a great height and with bill forward when traveling; soars occasionally; often seen with Ring-billed Gulls.
Voice
"A hoarse, croaking *kraaa*" (Bent).
Food
Fish.
Eggs
1–3; buff, sparingly spotted with brown; 2.7 x 1.9 in. (6.9 x 4.8 cm). Nest is depression in sand; in colonies.
Range
Breeds from Newfoundland s. to Gulf Coast and w. to Great Lakes; winters from N.C., southward.

BLACK TERN

Chlidonias nigra **26:2, 27:3**

Description
Size, 9–10¼ in. (22.9–26.0 cm). Mantle and tail slate-gray, tail slightly forked, bill black. Spring adult: • *black head* and body, dark red feet. Fall adult: nape black with point extending down behind eye; forehead, collar, and underparts white; presents a pied appearance when molting. Immature: mottled brown or mantle.
Similarities
In fall, gray color and slightly forked tail are helpful; immature Least is paler, but with darker, more contrasting wings.
Habitat
Breeds in marshes and lakes, migrates along coast and marshes; seen on North Atlantic Coast only in fall.
Habits
Flight in pursuit of insects like a Nighthawk's; usually picks food from surface rather than by diving.
Voice
Short, shrill *crik* (longer, not as high-pitched as Least's) or longer screaming *creek.*
Food
Insects, fish, small aquatic life.
Eggs
2–3; variable, often buffy-olive blotched with brown; 1.3 x 1.0 in. (3.3 x 2.5 cm). Nests on ground or in marsh; in colonies.
Range
Breeds from N.S. s. to Pa. and w. to Mo.; winters in S. America.

SKIMMERS
Family Rynchopidae

BLACK SKIMMER
Rynchops niger *Fig.10*

Description
Size, 18 in. (45.7 cm). Distinctive for lower bill, which is longer than upper, and very long wings. Adult: black above, white below; bill and legs red, bill with black tip. Immature: streaked brownish above, bill and legs paler.
Habitat
Sheltered bays on sandy coasts.
Habits
Flies low in flocks, cutting water surface with lower mandible, often flying back over same route; often active at night.
Voice
Like yelping of hounds.
Food
Fish, small aquatic organisms.
Eggs
4; pale buff, blotched with dark brown; 1.7 x 1.3 in. (4.3 x 3.3 cm). Nest is depression in sand; in colonies.
Range
Breeds along Atlantic and Gulf coasts from Mass. to Fla. and w. to Tex.; winters from S.C., southward.

Fig. 10

Black Skimmer

AUKS, MURRES, AND PUFFINS
Family Alcidae

Alcids, as these birds are called, have short necks, short pointed wings, short tails, and webbed feet. Their legs are placed far back on the body. The plumage is generally black above and white below. They swim well, dive from the surface, and patter along the surface of the water to get aloft. They have a buzzy flight with rapid wingbeats. Their wings are used in swimming under water. These birds frequent rocky coasts, northern coastal cliffs, or rocky islands. The larger species feed mainly on fish; smaller alcids feed on crustaceans, mollusks, and other invertebrates. Dovekies are notable for their small size. They are compact seabirds with extremely short necks and white throats. Alcids are the Arctic ecological counterparts of the Antarctic penguins, which they superficially resemble.

RAZORBILL

Alca torda **28:5**

Description
Size, 17 in. (43.2 cm). Summer: heavy head and neck; black head and upperparts, white below; • *white line from eye down side of bill.* Winter: throat, lower face, and sides of neck white. Immature: bill much smaller, but curved and blunt.
Similarities
Murres are less robust, have thinner, pointed bills.
Habitat
Coastal cliffs, offshore in water.
Habits
Swims lightly with head drawn in, bill tilted upward, and tail cocked; looks heavy-headed, hump-backed in flight; migrates in lines low over water; inquisitive.
Voice
Guttural croaks.
Eggs
1; often bluish-white spotted with brown; 3.0 x 2.0 in. (7.6 x 5.1 cm).
Range
Breeds along coast from Greenland to Maine; winters s. to N.J.

COMMON MURRE

Uria aalge **28:1**

Description
Size, 16–17 in. (40.6–43.2 cm). Bill thin, pointed; • *no white patch;* white line on rear of wing. Summer: black head, throat, and upperparts; white below. Ringed phase: eye-ring and streak behind eye white. Winter: throat and side of head white, black line back from eye.
Similarities
Head smaller, bill thinner, neck longer than Razorbill's; see also Thick-billed Murre. Winter loons are larger and have spotted backs and no white line on wing.
Habits
Flies with head and neck extended and drooping; looks long and slender in flight (Razorbill looks short and compact; Thick-billed Murre is intermediate); feet project beyond short tail.
Voice
Purring *mur-r-r-r-e.*
Eggs
1; light green, marked with brown and lilac; 3.2 x 2.0 in. (8.1 x 5.1 cm).
Former name
Atlantic Murre.
Range
Breeds in Arctic s. to Gulf of St. Lawrence; winters along Atlantic Coast s. to N.J.

THICK-BILLED MURRE

Uria lomvia **28:6**

Description
Size, 18 in. (45.7 cm). Bill stout, pointed. Summer adult: black above, white below, as Common Murre, but pale line along nape. Winter: lower cheek only and throat white.
Similarities
Thicker, shorter bill than Common Murre; winter Common

Murre has more white on sides of head; Razorbill has thicker bill; winter loons lack white line on wing.
Habits
Flies with head and neck extended; less extended and less drooping than Common, less "heavy-headed" than Razorbill; can swim high or low in water.
Voice
Adults, a soft purr; young, an emphatic *beat-it, beat-it.*
Eggs
1–2; pear-shaped, variable, but usually indistinguishable from Common Murre's; 3.2 x 2.0 in. (8.1 x 5.1 cm).
Former name
Brünnich's Murre.
Range
Breeds in Arctic s. to Gulf of St. Lawrence; winters along Atlantic Coast s. to N.J.

DOVEKIE
Alle alle **28:4**

Description
Size, 8 in. (20.3 cm). Unmistakable by its small size, only that of a Starling. Bill stubby, neck short. Summer: head and upperparts black; streaks on back, line on wing, and underparts white. Winter: similar, but throat and sides of nape white.
Habits
Inquisitive; can swim high or low in water; flight suggests that of Chimney Swift, not as fast as guillemot or murre; drops off cliffs rapidly.
Voice
Al-le.
Eggs
1; pale greenish-blue; 1.7 x 1.2 in. (4.3 x 3.0 cm).
Range
Breeds in n. Arctic s. to Greenland; winters off Atlantic Coast from N.S. s. to N.J.

BLACK GUILLEMOT
Cepphus grylle **28:2**

Description
Size, 13 in. (33.0 cm). Summer adult: black with • *white patch on wing* (this patch is rarely missing), • *inside of mouth and feet red.* Winter adult: foreback mottled gray, end of wing black, much of plumage white. Immature: dark above with white specks and much white on wing, barred with brown on sides, white below.
Similarities
The much larger, longer-billed White-winged Scoter is remotely similar in summer.
Habits
Can swim high or low in water, often dabbles nervously in water with bill, usually flies close to surface, takes flight readily.
Voice
"A high-pitched, complaining whine" (Snyder).
Eggs
2; whitish, sometimes spotted with brown; 2.4 x 1.5 in. (6.1 x 3.8 cm).
Range
Breeds in Arctic s. to Maine; winters off Atlantic Coast s. to N.Y.

COMMON PUFFIN

Fratercula arctica **28:3**

Description
Size, 13 in. (33.0 cm). Bill high and brightly colored. Short-necked, chunky. Summer adult: black upperparts and collar, white underparts and cheek, • *large and varicolored bill,* red feet. Winter adult: cheek grayish, but head pattern similar; bill smaller. Immature: similar to winter adult, but bill small, yellow.

Similarities
Immature is smaller, floats higher in water, and looks chunkier than Razorbill or Thick-billed Murre.

Habits
Inquisitive, unsuspicious; stands high; flocks wheel in unison.

Voice
"A low, purring note, *purr-la-la-la*" (Townsend); a "deep nasal *hey-al*" (Cruickshank).

Eggs
1–2; dull white, occasionally freckled, usually earth-stained; 2.5 x 1.8 in. (6.4 x 4.6 cm). Nest is in burrows.

Range
Breeds along coast from Greenland s. to Maine; winters offshore in breeding range.

Land Birds

Pigeons and Doves

Order Columbiformes

PIGEONS AND DOVES

Family Columbidae

This is a family of rapid-flying, small-headed birds with slender bills swollen at the base, long pointed wings, and short legs. They spend much time on the ground; bob their heads as they walk; eat grain, seeds, and nuts; and drink, unlike other birds, without raising their heads.

ROCK PIGEON

Columba livia **30:1**

Description
Size, 13 in. (33.0 cm). This is the common street pigeon introduced from Europe and often gone wild. Tail squared; most wild birds bluish-gray with a white rump, 2 black bars on wing, and 1 on tail; eyes and legs red; feral birds may be highly variegated with shades of gray, brown, white, and black, but populations tend to revert to original wild plumage in time.
Similarities
Mourning Dove has pointed tail, tan color.
Habitat
Originally cliffs; now also adapted to cities, parks, bridges, freight yards, farmyards, beaches.
Habits
Gregarious; tame, may take food from the hand; flight strong.
Voice
Characteristic cooing.
Eggs
2; white; 1.5 x 1.1 in. (3.8 x 2.8 cm). Nest is of trash, on building ledge, cliff.
Other names
Domestic Pigeon, Rock Dove.
Range
Resident throughout much of N. America, from s. Canada, s. to Mexico.

Note: The **WHITE-CROWNED PIGEON,** *Columba leucocephala* **(30:2),** size, 13 in. (33.0 cm), occurs in southernmost Florida and the Keys. Looking much like an all-dark Rock Pigeon, it has a snowy white crown. It is a gregarious, colonial species feeding on fruits.

MOURNING DOVE

30:3

Zenaida macroura *Fig. 11*

Description
Size, 11–13 in. (27.9–33.0 cm). One of the most abundant and widespread North American birds. Wings long, pointed; • *tail long,* white-edged, • *pointed;* head brown; body buffy-gray with a • *bluish cast on wings.* Slimmer than Rock Pigeon and with pointed tail.

Similarities
American Kestrel in flight lacks sharp downstroke of dove, has heavier head and shoulders, and squared tail.
Habitat
Open woodlands, suburban roadsides.
Habits
Feeds on ground, often takes dust baths or picks gravel from roadside; flight direct, wings whir as it rises; in dry areas flies daily to water; forms loose flocks in winter; frequently seen on wire lines; may nest in cities on fire escapes.
Voice
Mournful *coo-ah, coo, coo.*
Eggs
2; white; 1.1 x 0.9 in. (2.8 x 2.3 cm); 2 or more broods in South. Nest is usually in tree, loosely built of twigs, or on ground.
Range
Breeds from N.S. w. across Canada and s. through U.S. to C. America; winters from n. U.S. southward.

Fig. 11

COMMON GROUND DOVE

30:4
Columbina passerina *Fig. 11*

Description
Size, 6–7 in. (15.2–17.8 cm). A tiny dove; showing • *rusty patches in wings* during flight. Spots on wings, tail short and black.
Similarities
Mourning Dove larger with pointed tail.
Habitat
Open areas, usually near coast.
Habits
Aptly named, found usually feeding on ground.
Voice
Repetitive *woo-oo, woo-oo,* often with rising intonation.
Eggs
2; white; 0.7 x 0.6 in. (1.9 x 1.7 cm). Nest is on ground, in tangle, or tree, sometimes old nest of other bird, of twigs.
Range
Resident from s. U.S., s. to S. America.

Parrots

Order Psittaciformes

PARROTS

Family Psittacidae

This family includes the various parrots and parakeets. None of these are illustrated here; the only native eastern parrot, the Carolina Parakeet, *Conuropsis carolinensis,* is extinct. The only parrot likely to be seen away from larger cities (where zoo and aviary parrots of diverse species often escape and puzzle bird-watchers) and southern Florida (where several "escaped" parrots are feral and breeding) is the Monk Parakeet, *Myiopsitta monachus,* a long-tailed, gray-headed, greenish parrot that is highly social. It was introduced accidentally near New York City, and now occurs in scattered eastern areas.

Cuckoos and Allies

Order Cuculiformes

CUCKOOS

Family Cuculidae

Members of this family are slender birds with long, slightly curved bills; long, narrow, rounded tails; and zygodactyl feet (two toes forward, two toes behind). North American cuckoos usually do not lay their eggs in other birds' nests as do their European counterparts. The family feeds on seeds, grasshoppers, and other insects and small fruits.

YELLOW-BILLED CUCKOO

Coccyzus americanus **32:8**

Description
Size, 11–13½ in. (27.9–34.3 cm). Eyelids and • *lower bill yellow;* olive-brown above, • *rufous on wing,* white underparts, • *large white spots on tail.*
Similarities
Black-billed has small tail spots, no rufous on wings.
Habitat
Second growth, orchards, thicketed streamsides.
Habits
Secretive, moves noiselessly about upper foliage of small trees, sits motionless; flight in open is direct, horizontal.
Voice
Wooden *kuk-kuk-kuk-kow-kow-kow* (or *kiaow-kiaow-kiaow* or *kowlp-kowlp),* the *kow, kiaow,* or last notes retarded; other *kek, kow,* and *koo* notes, some indistinguishable from Black-billed's.
Eggs
2–6; bluish-green; 1.2 x 0.8 in. (3.0 x 2.0 cm). Nest is frail platform of twigs 4–10 ft. (1.2–3.0 m) up in shrub or low tree.
Range
Breeds from Maine and Minn. s. to Mexico; winters in S. America.

BLACK-BILLED CUCKOO

Coccyzus erythropthalmus **32:9**

Description
Size, 11–12½ in. (27.9–31.8 cm). Uniform olive-brown above, white below; • *bill black;* • *eye-ring red.*
Similarities
Yellow-billed is not as slim. Distinguish from Yellow-billed by absence of rufous on wing, large white spots on tail.
Habitat
Second growth.
Habits
Secretive, similar to Yellow-billed, but more active at night.
Voice
Notes evenly spaced, not retarded at end; a series of 2-syllable notes, *kuk* or *coo* in groups. Yellow-billed may give similar calls.
Eggs
2–6; bluish-green; 1.3 x 0.8 in. (3.3 x 2.0 cm). Nest is a platform of twigs, firmer than Yellow-billed's; 2–10 ft. (0.6–3.0 m) up in shrub.
Range
Breeds from s. Canada s. to S.C. and w. to Kans.; winters in S. America.

MANGROVE CUCKOO

Coccyzus minor **32:6**

Description
Size, 11–12 in. (27.9–30.5 cm). Resembles Yellow-billed, but has buffy underparts, blackish ear coverts, and no rusty patch on the wings.
Similarities
Resembles other cuckoos in form and movements but has black mask and buffy underparts.
Habitat
Mangrove swamps.
Voice
Call resembles Yellow-billed's, but is slower, huskier.
Range
Resident only in s. Fla., s. to W. Indies and from Mexico to S. America.

Note: In south Colorado, and central Kansas south to Texas, the **ROADRUNNER,** *Geococcyx californianus (Fig. 12),* size, 23 in. (58.4 cm), is a very large, ground-dwelling, long-tailed, streaky cuckoo of open country. It is unmistakable as it runs about, occasionally erecting its crest and eating lizards and insects. The **SMOOTH-BILLED ANI,** *Crotophaga ani,* size, 12 in. (29.4 cm), is an all-black, social cuckoo with a deeply arched bill, occurring locally in open areas of southern Florida and the Keys.

Fig. 12 Roadrunner

Owls

Order Strigiformes

Owls have big eyes that face forward, acute hearing, a broad head with a facial disk, a short neck, and soft, fluffy plumage that covers the base of the hooked bill and legs. They have sharp talons and are silent-flying, largely nocturnal birds of prey, although some are abroad by day. Owls feed extensively on rodents and are highly beneficial to the farmer. They usually perch upright, and periodically disgorge pellets of indigestible fur and bone. Owls are usually silent except in the mating season. Their eggs are white, almost round.

BARN OWLS

Family Tytonidae

These owls are "monkey-faced" and have long legs extending beyond the tail in flight.

BARN OWL

Tyto alba **29:8**

Description

Size, 14–20 in. (35.6–50.8 cm). Among owls, this species is unique for its "monkey-face" and pale breast. Long wings, dark eyes; buffy-brown above, pale below, relatively unmarked; •*face white, heart-shaped.* In flight, looks big-headed, slender-bodied, white below.

Habitat

Wood edges, farmland, haunts of man.

Habits

Quite nocturnal; frequents old belfries, water towers, barns, deserted buildings; flies with deep wingbeats, buoyantly, often reeling, legs trailing behind.

Voice

"(1) a discordant scream . . . (2) a snapping of the bill . . . (3) a flight call, resembling *ick-ick-ick-ick-ick*" (Potter and Gillespie); young in nest hiss.

Food

Rodents, insects.

Eggs

5–8; more oval than most owls; 1.8 x 1.3 in. (4.6 x 3.3 cm). Nest is in hollow trees, buildings, or cavities elsewhere.

Remarks

One of the few bird species of almost worldwide distribution, this owl in its various races is found on every continent, and even in islands as remote from each other as Curaçao and Madagascar.

Range

Resident throughout most of U.S. from s. Canada s. to S. America.

TYPICAL OWLS

Family Strigidae

These owls have round or squarish faces and short legs that usually do not reach beyond the tail in flight.

SCREECH OWL

Otus asio **29:3**

Description
Size, 7–10 in. (17.8–25.4 cm). Breast streaked, eyes yellow; • *ear tufts.* Red phase: rufous; the reddest owl. Gray phase: brownish-gray; our grayest owl.
Similarities
Long-eared is taller, thinner, browner; ear tufts are set closer together.
Habitat
Orchards, open woods, suburbs, small towns.
Habits
Strictly nocturnal; often lives in hollow tree and sits in entrance hole by day; enjoys bathing.
Voice
Tremulous wail, descending, or all on 1 key; *"oh-o-o-o-o that I had never been bor-or-or-or-orn"* (Thoreau).
Food
Rodents, insects, other animal food.
Eggs.
3–5; white; 1.4 x 1.3 in. (3.6 x 3.3 cm). Nest is in hole in tree, nook, or nesting box.
Range
Resident throughout most of U.S. from New England, w. across s. Canada and s. to Gulf Coast and Mexico.

SNOWY OWL

Nyctea scandiaca **29:10**

Description
Size, 20–27 in. (50.8–68.6 cm). Eyes yellow. Adult male • *white* with some dark scaly barring, heavier on female and young.
Similarities
White Gyrfalcon is slenderer, has smaller head, longer neck, and pointed wings. Many young owls are whitish in downy plumage.
Habitat
Rolling tundra; in U.S., coastal marshes, farmland.
Habits
Forages mainly during the day; perches on dune, post, or other elevation for a lookout; flight strong, direct, but jerky, upbeat faster than down; often sails.
Voice
"A deep angry *krohgogogok,* almost like a raven" (Hantzsch).
Food
Small mammals (especially lemmings), ptarmigan.
Eggs
4–10; white; 2.3 x 1.8 in. (5.8 x 4.6 cm). Nest is depression in tundra.
Range
Breeds in Arctic and n. Canada; winters s. rarely to S.C.; populations migrate further s. when food is scarce in n.

GREAT HORNED OWL

Bubo virginianus **29:11**

Description
Size, 18–25 in. (45.7–63.5 cm). • *A very large owl with ear tufts;* female larger. Wings long, broad; eyes yellow. Dark brown, heavily barred, and streaked with black; throat white; northern races paler. In flight, ear tufts flattened; dark, looks neckless.

Similarities
Long-eared is smaller; streaked, not barred below; has ear tufts closer together. Red-tailed Hawk is smaller, with proportionally smaller head and wings.
Habitat
Woodland edges, woods.
Habits
Most powerful of owls; sometimes hunts by day as well as by night, often sails on fixed wings; sometimes soars like a Red-tailed.
Voice
Bass; a deep *hoo-hoo-hoo-hoo, hoo-hoooo* (deeper than Barred); a *waugh-HOO;* less commonly, a blood-curdling shriek.
Food
Small mammals, birds.
Eggs
2; white; 2.2 x 1.8 in. (5.6 x 4.6 cm), often laid in February. Nest is often in deserted hawk's or crow's nest, usually in heavy timber.
Range
Resident throughout most of N. America from Arctic to S. America, excluding W. Indies.

HAWK OWL
Surnia ulula **29:5**

Description
Size, 14½–17½ in. (36.8–44.5 cm). Appearance dark, plump; wings short, pointed; • *tail long,* graduated, banded. Gray-brown above, head with white spots and black sideburns.
Similarities
Boreal and Saw-whet lack dark bars and long tail.
Habitat
Natural openings in northern coniferous forest.
Habits
Diurnal; often perches hawklike on tops of trees with body bent forward; pumps, sometimes cocks tail; flight falconlike, direct, swift, close to ground, and with glides; hovers, drops on prey like a shrike; tame.
Voice
"A trilling whistle *tu-wita-wit, tuwita-tu-wita, wita, wita*" (Henderson).
Food
Rodents, other small mammals, grouse.
Eggs
3–7; white; 1.5 x 1.3 in. (3.8 x 3.3 cm). Nest is in hollow tree, top of stump, or deserted bird's nest.
Range
Breeds in n. Canada from Newfoundland w. to Que.; winters from breeding range s. to n. U.S.

BURROWING OWL
Athene cunicularia **29:1**

Description
Size, 9–11 in. (22.9–28.9 cm). Lives on the ground. Legs long; tail short; eyes yellow; brown above, spotted with white; white below, barred with brown.
Habitat
Open places, even vacant lots in cities.
Habits
Lives in rodent or other holes, often standing near entrance or

perching in eminence on ground, post, wire, or shrub; bobs head and tail up and down in comical fashion on its long legs; often follows moving animals; flies little.

Voice
In flight, a chattering note; alarm note, *tsip-tsip*.

Food
Insects, small vertebrates.

Eggs
5–9; white; 1.3 x 1.1 in. (3.3 x 2.8 cm). Nest is grass-lined chamber at end of manure-bordered burrow in ground.

Range
Resident in Fla., W. Indies, and s. to S. America.

BARRED OWL

Strix varia **29:9**

Description
Size, 17–24 in. (43.2–61.0 cm). A big, round-headed brownish owl without ear tufts. Bill yellow, eyes dark; gray-brown above; pale below, with • *dark crossbars on breast* and collar and dark streaks on belly.

Similarities
Great Horned and Long-eared have ear tufts and reddish facial disks; all other owls, except Barn, have yellow eyes.

Habitat
Moist woods.

Habits
Wingbeat slow, flight buoyant; inquisitive.

Voice
Hoots higher-pitched than Great Horned, commonly 8 hoots in 2 groups of 4; at distance may be confused with dog barking.

Food
Mice, other small animals, birds.

Eggs
2–3; white; 2.0 x 1.6 in. (5.1 x 4.1 cm). Nest is in hollow tree or deserted nest of hawk, crow, or squirrel.

Range
Resident from cen. Canada s. to Gulf of Mexico and w. to Rockies.

GREAT GRAY OWL

Strix nebulosa **29:12**

Description
Size, 24–33 in. (61.0–83.8 cm). A big, round-headed gray owl with no ear tufts. • *Eyes yellow;* plumage loose, fluffy; gray, streaked below with black; wings and tail long and broad; face disks very large, black chin spot; no bars on breast.

Similarities
Barred Owl is smaller and browner, with bars on breast, smaller face disk, larger and darker eyes, and shorter tail. Great Horned Owl has ear tufts, is browner.

Habitat
Northern coniferous forest, rarely wandering south in winter.

Habits
Flies with slow flaps of broad, rounded wings; hunts by daylight in Arctic summer; tame.

Voice
"Several deep-pitched *whoos*" (Grinnell and Storer); also a wavering cry suggesting Screech Owl.

Food
Small mammals.

Eggs
2–5; white; 2.2 x 1.8 in (5.6 x 4.6 cm). Nest is of sticks from 20 ft. (6.1 m) up in evergreens.
Range
Resident throughout much of Canada and most of U.S., from Arctic to S. America.

LONG-EARED OWL
Asio otus **29:7**

Description
Size, 13–16 in. (33.0–40.6 cm). Long wings and tail; ear tufts close together; blackish-brown above; rusty face; • *streaked and lighter below;* yellow eyes. In flight, ear tufts not visible; brown spot shows on buff underwing.
Similarities
Great Horned is larger, bulkier; has ear tufts wider apart, a white throat, and more crossbars below; in flight, Short-eared looks lighter and occurs in more open country.
Habitat
Mixed woodlands or, preferably, evergreens.
Habits
Flight wavering; perches close to trunk of tree, stretches body very thin, can successfully hide in slight cover; may give distraction displays at nest, spreading wings and body, looking fierce, or feigning injury, leading one away from nest. Flocks in migration, and in winter often in groves of dense evergreens.
Voice
A "dove-like *hoo hoo hoo* . . . a slurred whistle, *WHEE-you*" (Griscom); snaps bill and makes great variety of noises at nest, when disturbed.
Food
Small mammals.
Eggs
3–8; white; 1.6 x 1.3 in. (4.1 x 3.3 cm). Nest is in conifer, of sticks.
Range
Breeds from s. Canada s. to s. U.S.; winters in s. parts of breeding range s. to Gulf states and Mexico.

SHORT-EARED OWL
Asio flammeus **29:6**

Description
Size, 13–17 in. (33.0–43.2 cm). A day-flying owl with its streaked breast, wavering flight, but no ear tufts. Bright buffy-brown, darker on back; head and pale breast streaked with black; facial disk dark, bordered with white; eyes yellow; wings with light patch above and black spot near bend below; appears neckless in flight.
Similarities
Buffier, and with longer wings than Long-eared. Owl head, short tail, and absence of white rump distinguish it in flight from Northern Harrier. Barn Owl is paler, lacks wing patches. Rough-legged Hawk also has black mark under wing, but has smaller head, more neck, white rump, and less wavering flight.
Habitat
Marshes, meadows, prairies, open country.
Habits
Often hunts by day, quarters low over ground like a Northern Harrier or sits on observation post; roosts on ground; has impressive aerial courtship flight.

Voice
Toot-toot-toot (to 20 times), higher than Great Horned; a sharp, catlike *ski-ow.*
Food
Mice.
Eggs
4–9; white; 1.5 x 1.3 in. (3.8 x 3.3 cm). Nest is on ground near clump of vegetation in marsh or meadow.
Remarks
This cosmopolitan species hunts by night the same marshes hunted by the Northern Harrier by day, but at times both may be abroad in daylight.
Range
Breeds locally from n. Canada s. to N.J.; winters from s. parts of breeding range s. to C. America.

BOREAL OWL

Aegolius funereus **29:2**

Description
Size, 8½–12 in. (21.6–30.5 cm). • *No ear tufts;* • *yellow bill and eyes;* brown above, with white spots on forehead and back, and • *black rim around face disk;* underparts pale, streaked with brown.
Similarities
Darker brown and with longer wings and tail than red phase of ear-tufted Screech-Owl. Saw-whet is smaller, streaked instead of spotted on forehead, with no black ring around face disk, and with a black bill.
Habitat
Coniferous forests.
Habits
Very tame; hunts by daylight in Arctic summer.
Voice
Notes like dripping water.
Food
Mice, birds, insects.
Eggs
4–7; white; 1.2 x 1.0 in. (3.0 x 2.5 cm). Nest is in holes in trees or deserted nests of other birds.
Former name
Richardson's Owl.
Range
Breeds from Labr. w. across Canada and s. to N.B., Ont., and s. Man.; winters from s. parts of breeding range s. as far as N.J.; Ill. (rare).

SAW-WHET OWL

Aegolius acadicus **29:4**

Description
Size, 7–8½ in. (17.8–21.6 cm). A very small owl with a black bill and • *no ear tufts.* Wings long and broad; eyes yellow; brown above, with white streaks on head; underparts white, heavily streaked with rufous. Young: chocolate brown with • *dusky face* and white V on forehead.
Similarities
Boreal is larger and has yellow bill, black rim around face, and white spots on forehead.
Habitat
Evergreen forests, swamps.

Habits
Strictly nocturnal, inquisitive; in taking off, bird often drops a little before flying forward; flight suggests that of a woodcock.
Voice
In spring, like the filing or whetting of a saw with notes tapering off at end; also a ventriloquial "soft *co-co-co-co-co-co*" (Eckstrom).
Food
Insects, small mammals.
Eggs
3–7; white; 1.0 x 0.9 in. (2.5 x 2.3 cm). Nest is in woods or swamps in old woodpecker hole.
Range
Breeds from N.S. w. to Man. and s. to Md. and Kans.; winters from s. parts of breeding range s. to Gulf Coast and C. America.

Goatsuckers

Order Caprimulgiformes

GOATSUCKERS

Family Caprimulgidae

Goatsuckers have huge, often bristle-bordered mouths; huge, flat heads; big, dark eyes; long wings, and often long tails; and small, weak feet. Their fluffy plumage is associated with their silent, wavering flight. Their mottled colors afford protection when they rest motionless by day on leaves or gravel, or lengthwise on the branch of a tree. These birds are most active by night, have highly distinctive voices, feed on the wing on insects, and lay their protectively colored eggs, with no nest, on bare leaves or gravel. The name goatsucker comes from ancient times and the erroneous belief that with their wide, gaping mouths they sucked milk from goats.

CHUCK-WILL'S-WIDOW

Caprimulgus carolinensis **30:7**

Description
Size, 12 in. (30.5 cm). Has bristles around its very wide mouth; • *tail rounded,* longer than the rounded, owl-like wings; mottled buffy-brown, no white spot on wing. Male: narrow white band across lower throat, white tail corners. Female: white on throat and tail replaced by buff. Whip-poor-will has more white on throat and outer tail feathers.
Habitat
Deciduous woods, common in southern woodlands.
Habits
Starts calling at dusk; its eyes often shine in headlights along country roadsides.
Voice
Emphatic *chuk-wills-WID-ow;* also a *quak* when flushed.
Food
Large insects such as moths, rarely small birds.
Eggs
2; white, mottled with gray and brown; 1.4 x 1.0 in. (3.6 x 2.5 cm). Laid on leaves or bare ground.
Range
Breeds from N.Y. w. to Kans. and s. to Gulf states; winters in W Indies and from Mexico, s. to S. America.

WHIP-POOR-WILL

Caprimulgus vociferus **30:9**

Description
Size, 10 in. (25.4 cm). Bristles about mouth; • *tail rounded,* longer than wings; mottled brown, sides unbarred, no white wing spot. Male: fairly narrow white band low on black throat and much white on outer tail feathers. Female: buff instead of white on throat and dark outer tail feathers.
Similarities
Common Nighthawk is gray-brown, has white spot on wing, and has forked tail.
Habitat
Woods, especially near fields.
Habits
Calls most near dusk, dawn, and on moonlit nights; feigns broken wing to lure intruder from young; flight low, erratic; sails on horizontal wings, tilting them sometimes almost vertically in turning. Red eyes sometimes shine in headlights along roadside.
Voice
Cluck often precedes the *whip-poor-WILL,* more accurately written *pur-ple RIB.*
Food
Various moths, beetles, flying insects.
Eggs
2; white, mottled with brown and gray; 1.2 x 0.8 in. (3.0 x 2.0 cm). Laid on leaves or bare ground.
Range
Breeds from s. Canada s. to s. U.S., and in mountains s. to Mexico; winters from Gulf of Mexico s. to C. America.

POOR-WILL

Phalaenoptilus nuttallii **30:8**

Description
Size, 7–8½ in. (17.8–21.6 cm). Tail short, rounded; no white spot on rounded wing; mottled with gray-brown; throat white, buffy in young; breast very dark; extreme corners of tail white, narrower in female.
Similarities
Suggests a small Whip-poor-will, has less white in tail.
Habitat
Dry, open country.
Habits
Pink eyes often shine in headlights along roadsides, bird flutters up as car passes; rests by day on ground, sometimes in tree; flight low, mothlike, in short sallies.
Voice
At a distance sounds like *poor-will* or, farther off, *p'will;* also a clucking note in flight.
Food
Nocturnal flying insects.
Eggs
2; plain white; 1.1 x 0.8 in. (2.8 x 2.0 cm). Laid on bare ground.
Remarks
This is one of the few birds that ever become dormant. In winter, torpid specimens with distinctly lowered body temperatures have been found in rock crevices.
Range
Breeds in W., but e. to Nebr. and Iowa, s. to cen. Tex.; winters from s. Tex. southward.

COMMON NIGHTHAWK

Chordeiles minor **30:6**

Description
Size, 8½–10 in. (21.6–25.4 cm). No bristles about mouth; wings long, thin, extend beyond end of forked tail; mottled gray-brown; sides pale, barred; male has • *broad white band on throat* and near end of tail, throat band in female is buffy.
Similarities
Wings of Whip-poor-will are not as pointed.
Habitat
Open country, cities.
Habits
Often seen by day; flies high in bouncing, erratic flight, hawking insects in spectacular aerial revolutions; also swoops low over cities and near street lights; often seen in migration, especially along river valleys, in loose flocks of 20–100 birds. Defends young by attacking intruder, hissing, with mouth agape or by feigning injury or death; sometimes roosts on fence posts.
Voice
Harsh *peenk,* given in flight with 3 double-speed flips of wings; suggests ground call of courting male Woodcock; in courtship, a booming sound at end of dive.
Food
Diverse insects, including mosquitoes.
Eggs
2; grayish-white, blotched with brown; 1.2 x 0.9 in. (3.0 x 2.3 cm). On bare ground or flat roofs in towns and cities.
Range
Breeds throughout most of Canada and U.S. from cen. Canada s. to C. America; winters in S. America.

Swifts and Hummingbirds

Order Apodiformes

SWIFTS

Family Apodidae

Swifts have short bills; long, narrow curved wings; short heads and tails; and small, weak feet. The ends of the tail feathers are stiff. They are among the swiftest fliers. Their habitat is the sky, where they feed on flying insects, drink, mate, and collect twigs (without perching), which they cement into a nest with their saliva.

CHIMNEY SWIFT

Chaetura pelagica **30:5**

Description
Size, 4¾–5½ in. (12.1–14.0 cm). An all-dark bird that looks like a flying cigar; the only swift in most of the East. Wings sicklelike, tail squared, plumage sooty throughout.
Habitat
The sky, usually near habitations.
Habits
Alternately flies and sails, often high up, on set wings that seem not to bend; 2 or 3 birds frequently fly together; roosts clinging upright in chimney, well, cave, or hollow tree, supported by its stiff tail; many collect for migration into a great flock, which spirals funnellike at evening into a selected chimney to roost.

Voice
Characteristic repeated twittering in flight.
Eggs
4–6; white; 0.7 x 0.5 in. (1.8 x 1.3 cm). Nest is inside chimney or hollow tree.
Range
Breeds throughout much of e. U.S. from N.S. w. to s. Man. and s. to Gulf of Mexico; winters in S. America.

Note: The **WHITE-THROATED SWIFT,** *Aeronautes saxatilis,* size, 7 in. (17.8 cm), may be seen in the mountains and valleys of the western fringes. It is larger than the Chimney Swift and is easily distinguished by its white throat.

HUMMINGBIRDS

Family Trochilidae

This family includes the smallest birds—those with the fastest wingbeat and the only ones that can fly backward. "Hummers" have long, needlelike bills; extensible tongues; partly iridescent plumage; and small, weak feet.

RUBY-THROATED HUMMINGBIRD

Archilochus colubris **34:15**

Description
Size, 3½ in. (8.9 cm). This is the only hummingbird in most of the East. Male: tail forked; bronze-green above, whitish below; with • *ruby throat* (which appears black in some lights); green back. Female: tail rounded, throat and tips of outer tail feathers white. Note absence of rufous edgings in tail feathers, and relatively silent flight; do not confuse with large hawkmoths.
Habitat
Gardens, woodlands, areas with flowers.
Habits
Flies 500 miles across Gulf of Mexico in migration; readily frequents feeding stations.
Voice
Shrill squeals, chirps, and chippering.
Food
Small insects, some nectar.
Eggs
2; white; 0.5 x 0.3 in. (1.3 x 0.9 cm).
Range
Breeds from s. Canada s. to Gulf Coast; winters from Mexico s. to C. America, with some n. to Gulf Coast.

Note: The **BROAD-TAILED HUMMINGBIRD,** *Selasphorus platycercus,* size, 4–4½ in. (10.2–11.4 cm), is a Rocky Mountain species frequenting the western fringe of the East, especially in migration. Males are very like the Ruby-throated, but have a paler, rosier throat, no strong fork in the tail, and shrill sound of whistling wings; females have rusty sides and rufous in the tail.

Kingfishers and Allies

Order Coraciiformes

KINGFISHERS

Family Alcedinidae

Kingfishers have strong, pointed bills, longer than the head; big heads; big eyes; short tails; and small legs and feet. The third and fourth toes are joined together. They fish from a perch above water or by hovering and diving headfirst, feeding on fish, some insects, and lizards. The scales and bones are ejected in pellets.

BELTED KINGFISHER

Ceryle alcyon **35:12**

Description
Size, 13 in. (33.0 cm). Crested, looks top-heavy. Male: upperparts and breastband gray-blue, underparts and collar white. Female: with an additional band of chestnut on lower chest and flanks.
Habitat
Shorelines.
Habits
Solitary; each has own territory along watercourse with a series of observation posts; hovers before diving; flight straight, 2 regular wingbeats alternate with several faster ones.
Voice
Loud, wild rattle, often heard in flight.
Eggs
5–8; white; 1.3 x 1.1 in. (3.3 x 2.8 cm). Nest is chamber at end of 4–8 ft. (1.2–2.4 m) tunnel excavated in earth or gravel bank, often away from water.
Range
Breeds from n. Canada s. through most of U.S.; winters in C. America.

Woodpeckers and Allies

Order Piciformes

WOODPECKERS

Family Picidae

Woodpeckers are arboreal, medium-sized birds with strong skulls; strong, pointed bills; extensible tongues; stiff tails; and zygodactyl feet (two toes pointing forward and two pointing back). Many species are resident, but the Common Flicker, Red-headed Woodpecker and Yellow-bellied Sapsucker are partly migratory. Most feed largely by clinging to the bark of a tree with their feet, bracing with the tail, and chiseling with the beak in search of wood-boring insects. A woodpecker often signals its presence by the tattoo it plays on a dead limb. Woodpeckers have an undulating flight. They nest, and usually roost, in holes, which they excavate with their bills in dead timber.

COMMON FLICKER

Colaptes auratus **31:2**

Description
Size, 13 in. (33.0 cm). Upperparts brown with barring; • *white rump* (conspicuous in flight), • *yellow* (usually) *or reddish underwings*, black crescent on breast, black spots below. Males have black (or red) mustaches, lacking in female.
Habitat
Suburbs, farms, wood edges, woods.
Habits
Spends much time on ground, may loudly drum on dead limb or tin roof; glides upward as it alights on tree; has noteworthy displays in breeding season.
Voice
Long series of *wik* notes; alarm *pee-ah;* a repetitive *wick-a, wick-a* or *wick-up, wick-up* during displays; many others.
Food
Ants, other insects, fruits, berries.
Eggs
6–8; white; 1.0 x 0.8 in. (2.5 x 2.0 cm).
Former names
Yellow-shafted Flicker, Red-shafted Flicker.
Range
Breeds from Newfoundland w. to Man. and s. to s. Fla. and Gulf Coast; winters from N.Y. and s. Ont. s. to Gulf Coast and Mexico.

PILEATED WOODPECKER

Dryocopus pileatus **31:1**

Description
Size, 16–19 in. (40.6–49.5 cm). A crow-sized woodpecker, with a crest. Male: black, with • *red crest* and *mustache* and big, heavy bill; white stripe on thin neck; big, white wing patch conspicuous in flight. Female: less red on crest and no red mustache.
Habitat
Woods with large trees.
Habits
Strips off quantities of bark and chisels big rectangular holes in dead or diseased trees; shy; flight vigorous, somewhat slow, either straight and crowlike or in long undulations.
Voice
A *yuk-yuk-yuk* sounding hollow, louder, heavier, slower than Common Flicker's; in flight, a slow *puck, puck;* "call, a loud *cack-cack-cack*" (Bendire); drumming signals are higher, faster, and louder in the middle, dying away at end.
Food
Wood-boring beetles and ants, berries, nuts.
Eggs
3–6; white; 1.4 x 1.0 in. (3.6 x 2.5 cm).
Remarks
This big woodpecker seems gradually adapting itself to well-wooded suburbs, even nesting within sight of New York City.
Range
Resident from s. Canada to Gulf Coast and s. almost to Mexico.

Note: The probably extinct (except for a few in Cuba) **IVORY-BILLED WOODPECKER,** *Campehilus principalis*, size, 11 in. (48.2 cm), has been reported but not proven to exist in certain southeastern states (Florida, South Carolina, Louisiana, Texas).

Similar to the Pileated Woodpecker, it has a black throat, white patches in front and back of wings in flight, and conspicuous, large, white patch at rest.

RED-BELLIED WOODPECKER

Melanerpes carolinus **31:5**

Description
Size, 10 in. (25.4 cm). Distinctive • *red nape;* • *barred back and wings,* pale, unmarked face. Adult: white rump and wing patches conspicuous in flight; gray underparts; reddish wash on belly inconspicuous; entire crown red in male, gray in female. Immature: duller, head all-gray.
Habitat
Woods, edges, farms, suburban areas, swamps.
Habits
Comes to feeders, will also feed on ground; stores acorns, nuts in crevices for future use.
Voice
Noisy; *cha-cha-cha;* a *churr, churr,* like Red-headed's, but lower; a wide variety of other notes.
Food
Ants, other insects, fruits, acorns.
Eggs
4–5; white; 1.0 x 0.8 in. (2.5 x 2.0 cm).
Range
Resident throughout most of se. U.S. from Atlantic Coast w. to Tex., and from Conn. s. to Fla.

RED-HEADED WOODPECKER

Melanerpes erythrocephalus **31:10**

Description
Size, 9 in. (22.9 cm). Adult: above dark blue-black; • *head and neck all-red;* white rump and underparts; conspicuous white wing patch; in some light appears red, white, and blue. Immature: mottled brown head and body, but has white rump and wing patch like adult.
Habitat
Orchards, roadsides, farming country, open woods.
Habits
Social, quarrelsome; catches insects in air like a flycatcher; stores nuts in cracks and cavities; tends to drop from tree almost to ground and then fly low; many killed by automobiles; distribution spotty.
Voice
Churr, churr, higher-pitched than Red-bellied's; a trill like the Gray Tree Frog: *"ker-r-r-ruck, ker-ruck-ruck-ruck"* (Merriam).
Food
Beechnuts, acorns; insects, fruit; occasionally eats young birds.
Eggs
4–6; white; 1.0 x 0.8 in. (2.5 x 2.0 cm).
Range
Breeds from Que. and Man. s. to Fla., Gulf Coast, and La.; winters in S. America; scarce in ne. U.S.

Note: The **LEWIS' WOODPECKER,** *Melanerpes lewis,* size, 11 in. (27.9 cm), is found on the western fringe of the East. This dark woodpecker has a distinctive gray-white collar, pinkish underparts, and a red face. Its flight is direct, unwoodpeckerlike. Often found in rather open country.

YELLOW-BELLIED SAPSUCKER

Sphyrapicus varius **31:8**

Description
Size, 8 in. (20.3 cm). Conspicuous • *white wing stripe;* • *red forehead* and black upper breast; belly yellowish. Male: red throat. Female: white throat. Immature: very different, brownish.
Similarities
Immature Red-headed has differently placed wing patch, shows separation of brown upper breast and throat from whitish rest of underparts. (No such contrast in sapsuckers.)
Habitat
Forests, wood lots, orchards.
Habits
Perforates bark of trees with even rows of small holes, mainly to get sap, but also insects attracted to it; secretive in migration, perching quietly, often on side of tree away from the observer.
Voice
Noisy in spring, a querulous *wha-ee;* a ringing clear, 5–6 times; drums in a disconnected roll, individual taps often being well separated.
Food
Various insects in and on trees, also regularly drills holes and feeds on sap flowing from them.
Eggs
5–7; white; 0.9 x 0.7 in. (2.3 x 1.8 cm).
Range
Breeds from cen. Canada, s. to Va.; winters from N.J. and Ohio Valley s. to C. America W. Indies.

HAIRY WOODPECKER

Picoides villosus **31:4**

Description
Size, 8½–10½ in. (21.6–26.7 cm). Size of Robin. Black-and-white pattern above and on wings, white down center of back and below; 3 outer tail feathers white. Male has • *red on back of head.*
Similarities
Bill much longer than Downy's, length twice the distance from base of bill to eye.
Habitat
Woods.
Habits
Woodland counterpart of the dooryard Downy; sometimes travels in mixed groups with Downies, nuthatches, and chickadees; comes closer to habitations in winter; shier, noisier, and more restless than smaller Downy.
Voice
Loud *peek,* louder and sharper than Downy's; a rattle, loud and on same pitch; drumming usually louder than Downy's, but often indistinguishable at a distance.
Food
Borers, caterpillars, other tree-bark insects.
Eggs
3–6; white; 0.9 x 0.7 in. (2.3 x 1.8 cm).
Range
Resident from n. Canada s. to s. U.S. and C. America.

DOWNY WOODPECKER
Picoides pubescens **31:3**

Description
Size, 6–7 in. (15.2–17.8 cm). Size of House Sparrow; commonest small woodpecker. Similar to Hairy but smaller; 3 outer tail feathers have black bars; bill shorter, length equals distance from base of bill to eye.
Similarities
Hairy is larger, has no black bars in outer tail feathers.
Habitat
Open woodland, wood edges, trees about habitations, orchards.
Habits
Easily attracted to feeders, tame; often travels in mixed flocks with chickadees, nuthatches, kinglets, and Brown Creepers.
Voice
Short, sharp *pik;* a rattle of 15 or so rapid staccato notes, like Hairy's, but not so loud, normally descending in pitch; tattoo a long roll.
Food
Borers, surface insects, some vegetation.
Eggs
4–7; white; 0.7 x 0.6 in. (1.8 x 1.5 cm).
Range
Resident from n. Canada, s. to s. U.S.

Note: The **LADDER-BACKED WOODPECKER,** *Picoides scalaris,* size, 6–7½ in. (15.2–19.1 cm), barely reaches the southwestern fringe of the East. It has a striped face and barred back, and is a black-and-white woodpecker of the desert and riverside timber.

RED-COCKADED WOODPECKER
Picoides borealis **31:6**

Description
Size, 8 in. (20.3 cm). • *Ladder back, black crown;* large white cheek patch. Male: has a small, often hidden red "cockade" at side of crown behind eye. Immature: buffy below, instead of white, and with red crown patch.
Habitat
Mature southern pine forests with old trees.
Habits
Somewhat social, noisy, active, quarrelsome, wary.
Voice
Call note suggests small woodwind instrument; also a nasal, querulous, scolding note.
Food
Mainly wood-boring insects.
Eggs
3; white, 1.0 x 0.7 in. (2.5 x 1.8 cm).
Remarks
Its requirement for old pines with rotted centers has made it vulnerable through pinewoods management and timber practices. On the decrease, it is classed as an Endangered Species.
Range
Resident of se. U.S. from Md. and Ky. s. to Fla. and w. to e. Tex.

BLACK-BACKED WOODPECKER
Picoides arcticus **31:7**

Description
Size, 9–10 in. (22.9–25.4 cm). • *Black-backed* with barred sides. Male has a • *yellow crown patch.*
Similarities
Three-toed is smaller, back barred or mainly white.
Habitat
Coniferous forests, often near water.
Habits
Solitary, unsuspicious, movements deliberate; makes periodic winter incursions south of normal range; freshly barked tree trunks are often a sign of this species.
Voice
Sharp, metallic *pik;* distinctive rattle call, a screaming, *wreeoh;* noisy and aggressive about nest.
Food
Mostly subsurface arboreal insects, few nuts and fruits.
Eggs
4–5; white; 0.9 x 0.7 in. (2.3 x 1.8 cm).
Former names
Arctic or Black-backed Three-toed Woodpecker.
Remarks
Look for this species in boggy areas amid dead conifers, and about clearings or burns.
Range
Resident from cen. Canada s. to n. U.S.

THREE-TOED WOODPECKER
Picoides tridactylus **31:9**

Description
Size, 8–9½ in. (20.3–24.1 cm). Noticeably smaller, browner, less contrastingly black-and-white than Black-backed. Male has a • *yellow crown patch;* in flight, • *barred back* is conspicuous and tail flashes white.
Similarities
Some eastern birds show little white on back and may be mistaken for Black-backeds—look for more broadly white face marks, black marks in white of outer tail, and signs of white on back to distinguish from Black-backed.
Habitat
Coniferous forest, especially spruce.
Habits
Solitary, unsuspicious, sedentary; often rarer, more local and is less noisy, and flight less vigorous than Black-backed.
Voice
Call differs from Black-backed, a weak *pik* or *peet;* a rattle, like Hairy's, but softer.
Food
Beetle larvae and other bark insects.
Eggs
4; white; 0.9 x 0.7 in. (2.3 x 1.8 cm).
Former names
American or Northern Three-toed Woodpecker.
Range
Resident from Arctic s. to n. U.S.

Perching Birds

Order Passeriformes

This order includes far more species than any other, and has most of the land birds. The feet have three toes forward and one behind, making them well designed for grasping a perch. The young are hatched naked, blind, and helpless and are cared for in the nest until fledged.

TYRANT FLYCATCHERS

Family Tyrannidae

Flycatchers have flattened bills with a little hook at the tip and bristles about the broad base. They make their headquarters on a perch, from which they sally forth to snap up passing flying insects, their main food. Some species are quite pugnacious, whence the family name. All, except the Eastern Phoebe, spend the winter in southern regions.

The *Empidonaces* (Genus *Empidonax*) are flycatchers 5½ to 6 inches (14–15.2 cm) in size that look alike, so most species cannot safely be identified by plumage alone. All have olive-grayish, or brownish, backs and whitish or yellowish breasts. Most have conspicuous light eye-rings, two wing bars, and pale lower mandibles. However, their notes in the breeding season are distinctive. The somewhat smaller wood pewees (Genus *Contopus*) and Eastern Phoebe are larger and lack white eye-rings.

EASTERN KINGBIRD

Tyrannus tyrannus **33:7**

Description
Size, 8–9 in. (20.3–22.9 cm). Black above, white below; 2 very narrow white wing bars; prominent white band at end of tail; male with a concealed crimson crown patch.
Similarities
Western lacks white band at tail tip and has yellow in plumage.
Habitat
Roadsides, farms, orchards; woods, meadows, streams; shorelines.
Habits
Very pugnacious; attacks crows, hawks, vultures, sometimes alighting on their backs; flies horizontally on quivering wings.
Voice
Incisive *tzee,* alone or rapidly repeated; an excited *kipper, kipper* with descending inflection; a shrill twitter; also a dawn song.
Eggs
3–5; creamy, with brown spots; 0.9 x 0.7 in. (2.3 x 1.8 cm). Nest of twigs and grasses; bulky; ragged without, neat within; 3–20 ft. (0.9–6.1 m) up in bush, tree, stump, or structure.
Range
Breeds from cen. Canada s. to s. U.S.; winters in S. America.

WESTERN KINGBIRD

Tyrannus verticalis **33:6**

Description
Size, 8–9½ in. (20.3–24.1 cm). Olive above, gray head and upper breast, • *black tail with white edges on outer feathers,* yellow belly; black ear patch.

Similarities
Young sometimes lack white edges of outer tail feathers; Eastern Kingbird is much darker above. Great Crested Flycatcher, a woodland bird, has a rufous tail.
Habitat
Open country, ranches, towns, roadsides.
Habits
Flight less fluttery than Eastern's.
Voice
Noisier than Eastern, many similar notes; also "a single *kip* . . . a *quer-ick*" (Stevens); and a *"ker-er-ich-ker-er-ich"* (Barnes).
Eggs
3–5; creamy, with brown spots; 0.9 x 0.7 in. (2.3 x 1.8 cm). Nest is similar to Eastern's, but more often on man-made structures.
Former name
Arkansas Kingbird.
Range
Breeds from Man. s. to Kans., Okla., and Mexico; winters in C. America.

Note: The **CASSIN'S KINGBIRD,** *Tyrannus vociferans* **(33:5),** size, 8–9 in. (20.3–22.9 cm), occurs on the westernmost fringe of eastern region. It resembles the Western Kingbird, but is darker, with a white throat and a white-tipped tail, and it lacks the white edges of the outer tail feathers of the western. The **GRAY KINGBIRD,** *Tyrannus dominicensis* **(33:8),** inhabits Florida, and the Florida Keys especially, but sometimes wanders north to the Carolinas. It is 7 in. (17.8 cm), very like the Eastern, but lacks white on the tail, is pale gray above, and has a larger bill.

GREAT CRESTED FLYCATCHER
Myiarchus crinitus **33:1**

Description
Size, 8–9 in. (20.3–22.9 cm). Crested; olive-brown above; gray throat and breast, yellow belly; rufous on wings, 2 white wing bars; tail rufous.
Habitat
Woods, orchards, farms.
Habits
Aggressive; glides from tree to tree on outspread wings; often feeds from tops of tall forest trees.
Voice
Raucous *wheap;* a whistled *whit-whit, whit-whit;* a "rattling cry, *creep* or *cr-r-r-reep*" (Allen).
Eggs
3–8; creamy, with black and brown lines and penlike scratches; 0.9 x 0.7 in. (2.3 x 1.8 cm). Nest is in cavity, 5–60 ft. (1.5–18.3 m) high, of trash, often with discarded snakeskin.
Range
Breeds throughout much of e. U.S., from s. Canada to Gulf of Mexico; winters from Mexico to S. America.

SCISSOR-TAILED FLYCATCHER
Muscivora forficata **33:4**

Description
Size, 11–15 in. (27.9–38.1 cm). Conspicuous long streamer tail. Gray above, flanks and underwings pink, underparts white, scarlet patch on underwing and crown (concealed).

Similarities
Western Kingbird resembles immature with short tail, but is yellowish below instead of pinkish.
Habitat
Plains, prairies; roadsides, farms, ranches.
Habits
Often seen on wire, tail not spread when perched; in flight it opens and shuts; flight swift, graceful, low over ground.
Voice
"Twittering *psee, psee, psee;* a harsh *thish-thish*" (Collins); also a twilight song.
Eggs
3–6; creamy, spotted with brown; 0.9 x 0.7 in. (2.3 x 1.8 cm). Nest of twigs, 4–20 ft. (1.2–6.1 m) up on limb of a deciduous tree.
Range
Breeds in s.-cen. U.S. from Mo. and Nebr., s. to Mexico; winters from Mexico s. to C. America; also winters in s. Fla.

EASTERN PHOEBE
Sayornis phoebe **33:3**

Description
Size, 6¼–7¼ in (15.9–18.4 cm). • *Bill black,* tail notched. Olive-gray above, darker on head and tail; grayish on throat and breast, whitish below; • *no eye-ring or wing bars.* In fall, yellowish below, young with cinnamon wing bars.
Similarities
Eastern Wood Pewee has wing bars and does not pump its shorter tail.
Habitat
Farms, orchards, gardens, streamsides; about buildings.
Habits
Tame; an early migrant; gets some food from ground; head looks blackish in flight; tail pumping diagnostic.
Voice
Burred, often repeated *PHE-bee,* sometimes uttered in flight; a chip like a Swamp Sparrow's.
Eggs
4–5; white, rarely dotted; 0.7 x 0.5 in. (1.8 x 1.4 cm); 2 broods. Nest is of mud and moss, often under a bridge, eaves, cave, or man-made structure.
Range
Breeds from s. Canada s. to n. Gulf states; winters from Va. and Gulf Coast s. to Mexico.

SAY'S PHOEBE
Sayornis saya **33:2**

Description
Size, 7–8 in. (17.8–20.3 cm). Pale, gray-brown back; black tail; • *rusty below.*
Habitat
Open country, ranches, ravines.
Habits
Early migrant, very active, restless; flight is low, strong, zigzaggy; wingbeats deeper, slower than Eastern's; occasionally pumps tail.
Voice
Plaintive *phee-ur* (given with twitch of tail); "a swift *pit-tsee-ar*" (Hoffmann).

Eggs
4–5; white, rarely dotted; 0.7 x 0.5 in. (1.9 x 1.5 cm). Nest is similar to Eastern, with less mud (often wool-lined).
Range
Breeds chiefly in W., but e. to N.Dak., e. Nebr., and cen. Tex.; winters from s. Tex. southward.

YELLOW-BELLIED FLYCATCHER
Empidonax flaviventris **33:15**

Description
Size, 5½ in. (14.0 cm). This *Empidonax* says *pu-ee.* Olive-green above, eye-ring yellow. This species in spring may be identified by its color, yellow below including throat, but in fall other *Empidonaces* may be yellowish below.
Habitat
Near streams in coniferous forests; cold bogs; alder swamps in migration.
Habits
Secretive; jerks head and tail and flutters wings as it sings.
Voice
Rising *pu-EE;* a sneezelike *pse-ek;* less musical than Acadian, which does not breed within its range; in migration, "a plaintive *pea*" (Maynard); flight song *pu-ee, killik, pu-ee, killik.*
Eggs
4; creamy, finely dotted with brown; 0.6 x 0.5 in. (1.7 x 1.3 cm).
Range
Breeds from Newfoundland and Man. s. to Maine, Wis., and N.Dak. and Pa. mountains; winters from Mexico to C. America.

ACADIAN FLYCATCHER
Empidonax virescens **33:14**

Description
Size, 6 in. (15.2 cm). This *Empidonax* says *ka-ZEEK.* Somewhat greener than other *Empidonaces.* Throat white, pale yellowish below, brighter in fall.
Habitat
Woods and swamps near streams in southern woodlands; likes beeches.
Habits
Sings through heat of day, twitches tail as it sings, frequents the middle layer of branches.
Voice
Explosive *ka-ZEEK,* second syllable higher; call a sharp *peet.*
Eggs
2–4; pale cream, sparingly spotted with brown; 0.7 x 0.5 in. (1.8 x 1.4 cm). Nest of rootlets, etc., swung like a hammock between twigs near end of branch, 4–20 ft. (1.2–6.1 m) up, often over water.
Range
Breeds from s. New England, Pa., and Great Lakes s. to se. U.S.; winters from C. to S. America.

WILLOW FLYCATCHER
Empidonax traillii **33:12**

Description
Size, 5½–6¾ in. (13.3–17.1 cm). This *Empidonax* says *fitz-BEW.* Greenish-brown above with whiter throat than most *Empidonaces.*

Similarities
Only by its song can this species be distinguished with certainty from the Alder.
Habitat
Usually near water, low bushes along streams, edges of marshes and meadows; occasionally in dry fields; usually in more open, grassy, less wooded areas than Alder.
Habits
Late migrant; active, restless; sings from tops of shrubs.
Voice
A *fitz-BEW,* buzzy, an even buzzier *fizz-BEW;* also "a buzzy *creet*" (R. C. Stein).
Eggs
3–4; whitish, with fine brown spots; about 0.7 x 0.5 in. (1.8 x 1.3 cm). Nest is bulky, of shreds, grasses, low in shrub, to 3 ft. (0.9 m) up.
Remarks
This and the Alder Flycatcher were formerly considered one species, called the Alder or Traill's Flycatcher.
Range
Breeds from Maine, N.Y., and Alta. s. to Va., Ark., and Okla.; winters in S. America.

ALDER FLYCATCHER
Empidonax alnorum **33:12**

Description
Size, 6 in. (15.2 cm). This *Empidonax* says *fe-BE-O.* Tends to be grayer, less green above than Willow, with same whitish throat.
Similarities
Alder cannot be distinguished from Willow except by voice.
Habitat
Low bushes in wet areas, swamps, around marshes, streamsides, near woods.
Habits
As Willow Flycatcher.
Voice
A *fe-BE-O,* the *fe* rasping, the *BE-O* whistled.
Eggs
3–4; buffy, with large purplish-brown spots; about 0.7 x 0.5 in. (1.8 x 1.3 cm). Nest is compact, of cottony material and grasses; resembles that of Yellow Warbler; whitish, in shrubs 3 to 8 ft. (0.9–2.4 m) up.
Range
Breeds from cen. Que. and Man. s. to s. New England and Minn.; winters in C. and S. America.

LEAST FLYCATCHER
Empidonax minimus **33:13**

Description
Size, 5¼ in. (13.3 cm). This *Empidonax* says *che-bec.* Grayest above, whitest below of *Empidonaces;* lower mandible fairly dark.
Similarities
Light lower mandible in other eastern *Empidonaces,* noticeable at close range.
Habitat
Orchards, woods, streamsides, rural countryside, haunts of man.
Habits
Jerks head and tail as it calls.

Voice
Noisy; an emphatic *che-BEC,* much repeated; call, a short *whit;* also a flight song, *"che-BEC-treee-treo, che-BEC-treee-cheu"* (Samuels).
Eggs
3–6; creamy; 0.6 x 0.4 in. (1.6 x 1.2 cm). Nest is of plant fibers and grasses, in upright fork of tree, 8–40 ft. (2.4–12.2 cm) up.
Range
Breeds from s. Canada s. to cen. U.S., and in n. Ga. mountains; winters in C. America.

Note: The western fringes of our region are reached in summer by 2 western species. The **DUSKY FLYCATCHER,** *Empidonax oberholseri,* size, 5½ in. (14.0 cm), is a gray-olive species that frequents low trees in the wooded foothills of the Rocky Mountains. Its song is a *clip-whee-ZEE.* In the same region the more yellow-greenish **WESTERN FLYCATCHER,** *Empidonax difficilis,* size, 6 in. (15.2 cm), favors open places, especially near streams within the forests. Its vocalizations include a wheezy *pee-IST,* and its song is *ps-SEET, ptsick, sst.*

EASTERN WOOD PEWEE
Contopus virens **33:9**

Description
Size, 6–6¾ in. (15.2–17.1 cm). Wings longer than Eastern Phoebe's. • *Lower mandible partly yellow;* dark olive-gray above, with • *2 white wing bars* (buffy in young); whitish below, mid-breast with light line dividing darker sides.
Similarities
Eastern Phoebe has all-black bill, no wing bars, pumps longer tail and usually feeds near ground. *Empidonaces* are smaller, have white eye-ring, and lack whitish line down breast. See also Western Wood Pewee.
Habitat
Pinewoods, orchards, deciduous woods, shade trees.
Habits
Sits motionless, does not pump tail; not pugnacious, frequents middle and upper layers of branches.
Voice
Sweet, plaintive *PEE-wee, PEE-a-WEE,* or *pee-AAA* with a rising inflection, repeated; also a more elaborate song at dawn and twilight; sings well into August.
Eggs
2–4; with ring of brown or purple spots; 0.7 x 0.5 in. (1.8 x 1.3 cm). Nest is shallow, constructed of fibers covered with lichens 6–50 ft. (1.8–15.2 m) up on a horizontal branch.
Range
Breeds from se. Canada s. to Gulf of Mexico; winters from C. to S. America.

WESTERN WOOD PEWEE
Contopus sordidulus **33:10**

Description
Size, 6–6½ in. (15.2–16.5 cm). Plumage similar to Eastern Wood Pewee's; slightly grayer and less greenish above, darker below, but a difficult identification without voice.
Habitat
Open woodlands.

Habits
Similar to Eastern.
Voice
Nasal *peeer,* suggesting note of Common Nighthawk, a sad *dear-me; "Tswee-tee-teet, TSWEE-tee-teet, bzew (bzew,* a downward slur)" (Miller); often sings at night.
Eggs
3–4; similar to Eastern's. Nest is like Eastern's; tends to be larger, deeper, lined with yellow grasses, and usually without lichens.
Range
Breeds chiefly in W., but from Man. s. through Nebr. and cen. Tex. to C. America; winters in C. and S. America.

OLIVE-SIDED FLYCATCHER
Contopus borealis **33:11**

Description
Size, 7–8 in. (17.8–20.3 cm). Large bill and head, sturdy body. Olive-gray above; • *gray sides and white stripe down mid-breast* suggest gray vest half open over white shirt; white tufts in side of rump visible at times.
Similarities
Wood pewees are smaller, with light wing bars, and lack contrast between sides and center of breast.
Habitat
Dead timber in coniferous forest, often near water.
Habits
Selects a particularly exposed perch.
Voice
Loud whistle, quick *THREE BEERS*; usually silent in migration.
Eggs
3; creamy, blotched with chestnut; 0.9 x 0.7 in. (2.3 x 1.8 cm). Nest is of twigs, 10–50 ft. (3.0–15.2 m) up, in crotch of horizontal branch on conifer.
Range
Breeds from s.-cen. Canada s. to Mich., Wis., and n. Minn.; winters in S. America.

LARKS
Family Alaudidae

Largely terrestrial birds, larks are mostly brown streaked, have musical voices, are gregarious; the sexes are virtually alike. The hind claw is elongated, and nearly straight.

HORNED LARK
Eremophila alpestris **48:7**

Description
Size, 7–8 in. (17.8–20.3 cm). Long hind claw. Male: brownish above and on sides, unique face pattern with • *tiny black "horns"* and black sideburns, • *tail black*, outer tail feathers and underparts white, line over eye yellow or white; the "yellow-faced" northern form occurs in winter; the "white-faced" form breeds in our area. Female and immature: duller, except the juvenal, but all show suggestion of face pattern.
Similarities
Juvenal might be confused with molting longspur, but no longspur has Lark's tail pattern, and Lark's bill is thinner.

Habitat
Grasslands, golf courses, fields, parkway shoulders, beaches, dunes, marshes.
Habits
In winter, gregarious; walks (does not hop) or runs mouselike on ground; or "freezes" into invisibility. Flight is light, bounding, showing contrast of black tail and white underparts.
Voice
Common flight note, *"p-seet"* (Collins); song, thin, jumbled tinkling and bubbling often given in flight, or as it parachutes earthward from a height.
Food
Seeds, insects.
Eggs
3–5; grayish, marked with brown; 0.9 x 0.6 in. (2.3 x 1.5 cm). Nest is on ground, of grasses.
Range
Breeds from Arctic s. to N.C., Mo., and Tex. Gulf Coast; winters from s. Canada s. to Gulf Coast and southward.

SWALLOWS
Family Hirundinidae

Swallows have short, flat bills and wide mouths; long, pointed wings; weak, small feet; and usually notched tails. They frequent open country and bodies of water, and spend much time catching insects on the wing. Mixed flocks often hawk for insects low over ponds prior to bad weather. Swallows are gregarious and migrate in flocks by day, often along the coasts. None, except some Tree Swallows, winter in our area.

TREE SWALLOW
Iridoprocne bicolor **32:3**

Description
Size, 5–6½ in. (12.7–15.9 cm). Adult: glossy greenish-blue above, including area around eyes; • *white below.* Immature: dusky-brown above, sometimes with faint, broken dark band across upper breast.
Similarities
Brown Rough-winged has dingy throat; resembles young Tree in late summer. Bank has complete breastband.
Habitat
Open areas, especially near water; often breeds in dead trees in open swamps.
Habits
Flight slightly flickering, wingbeats faster than Barn's; when sailing, seems to arch back and lower wing tips; wings look triangular in air, somewhat like Purple Martin's. An early migrant; collects in vast numbers on coastal marshes in fall migration, swarming over trees, wires, bushes, roads, beaches.
Voice
Chirrups, twitters, *silip, silip, silip* or *"tsip-prrup, tsip-prrup-prrup"* (Allen), much repeated.
Eggs
3–6; white; 0.7 x 0.5 in. (1.9 x 1.3 cm). Nest is single or in colonies, in woodpecker holes or boxes, often near water.
Range
Breeds from Newfoundland w. to n. Man. and s. to Md. and Nebr.; winters from N.C. southward.

Note: The **VIOLET-GREEN SWALLOW,** *Tachycineta thalassina,* size, 5¼ in. (13.3 cm), is found in the mountains and valleys from the Black Hills and Pine Ridge of Nebraska westward. It is colored somewhat like the Tree Swallow but with white around the eyes and a white patch on either side of the rump.

BANK SWALLOW

Riparia riparia **32:4**

Description
Size, 4¾–5¼ in. (12.1–13.3 cm). • *Brown above;* head, wings, and tail darker; • *white below;* distinctive • *brown breastband.*
Similarities
Immature Tree is larger and whiter below, darker above.
Habitat
Meadows, ponds, other open areas, banks.
Habits
Flight low, erratic, somewhat fluttering and mothlike; keeps wings close to body when sailing; rows of holes in bank mark site of nesting colony.
Voice
Notes more gritty than other swallows; "soft abrupt *ffrrutt*" (Cruickshank); song, a twitter, *"speedz-sweet, speedz-sweet"* (Dickey), much repeated; "when alarmed . . . *te-a-rr* [reminiscent] of the Common Tern" (Stoner).
Eggs
3–7; white; 0.7 x 0.5 in. (1.8 x 1.3 cm). Nest is in colonies, in hole excavated in bank along river, shoreline, right-of-way, or gravel pit.
Range
Breeds from Labr. and n. Man. s. to Va. and Tex.; winters in S. America.

ROUGH-WINGED SWALLOW

Stelgidopteryx ruficollis **32:5**

Description
Size, 5–5¾ in. (12.7–14.6 cm). Uniform • *grayish-brown above* and dusky on throat, white below.
Similarities
Smaller Bank Swallow has white throat, distinct breastband. Young Tree has white throat.
Habitat
Open areas, creeks, ponds, rivers.
Habits
Flight more direct, less erratic than Bank's; wingbeats deeper and slower, more glides and sails; more solitary, less colonial than Bank.
Voice
Twitter similar to, but lower-pitched than, Bank's; call, "a *trit, trit-trit,* or *tri-ri-ri-rit*" (Saunders); a "rasping squeak . . . *quiz-z-z-zeep, quiz-z-z-zeep*" (Dickey); unmusical *"burp-burp"* (Cruickshank).
Eggs
4–8; white; 0.7 x 0.5 in. (1.8 x 1.3 cm). Nest is singly or in small groups, in holes in banks, caves, or crevices, usually near water.
Range
Breeds throughout most of U.S. from N.B. w. to Mich. and s. to Fla. and S. America; winters from Mexico southward.

BARN SWALLOW

Hirundo rustica **32:2**

Description
Size, 7 in. (17.8 cm). Adult: iridescent steel-blue above, chestnut forehead and throat, thin necklace, white on inner tail feathers, buffy below, • *long, forked tail.* Immature: paler, with buffy-brown throat; lacks tail streamers.
Habitat
Meadows, marshes, ponds, coast, often near water or buildings.
Habits
Flight strong, swift, graceful; drives through air, wings close to and parallel with body at end of stroke; gregarious; large flocks migrate along coast.
Voice
Various cheerful twitterings, *"sweeter-sweet, sweeter-sweet"* (Cruickshank); a *kittik, kittik* or *kvik, kvik,* repeated.
Eggs
3–6; white, spotted with red-brown; 0.8 x 0.6 in. (2.0 x 1.5 cm). Nest is of mud, often inside barns, boathouses, or other structures, on rafter or against wall, usually in small colonies.
Range:
Breeds from Maritime Provinces s. to S.C. and w. to Ark.; winters in S. America.

CLIFF SWALLOW

Petrochelidon pyrrhonota **32:7**

Description
Size, 5–6 in. (12.7–15.2 cm). Adult: creamy forehead, chestnut face, gray collar, • *buffy rump;* rest of upperparts blue, underparts white, • *tail square.* Immature: more brownish, rump paler.
Similarities
Barn has long forked tail.
Habitat
Meadows, marshes, open area.
Habits
Colonial, often quite local.
Voice
Squeaky, creaking, grinding but pleasant series of chirrupings, huskier than Barn; a nasal *y-e-a-a.*
Eggs
4–6; white, thickly marked with reddish-brown, 0.7 x 0.6 in. (1.9 x 1.7 cm). Nest is of mud, gourd-shaped, on outside of buildings under eaves, cliffs, bridges.
Range
Breeds from N.S. w. to Ont. and s. to Va., Mo., and C. America; winters in S. America.

PURPLE MARTIN

Progne subis **32:1**

Description
Size, 7½–8½ in. (18.4–21.6 cm). The largest swallow. Male: • *all-purple* (looks black at a distance). Female and immature: grayish throat and breast, whitish underparts, some gray on dark upperparts.
Similarities
Immature Tree Swallow is smaller and with white below.
Habitat
Seashore, meadows, about houses, even common in some midwestern towns.

Habits
Wings triangular in flight; rapid wingbeats alternate with sails, often in a great arc; flight vaguely suggests Starling, but wings much longer; often heard at night; prior to migration collects in big flocks in groves, where their united voices sound like escaping steam.
Voice
Rich, liquid, loud chirruping, a somewhat guttural *too-too* and *too-too-too-weadle;* call, "a harsh *zhupe, zhupe*" (Stone); alarm, a *kerp*.
Eggs
3–5; white; 0.9 x 0.7 in. (2.3 x 1.8 cm). Nesting is colonial, originally in holes in trees; then also in hollow gourds put up by the Indians; now largely in artificial birdhouses.
Range
Breeds from N.S. and Man. s. to Fla., the Gulf Coast, and the W. Indies; winters in S. America.

JAYS, MAGPIES, AND CROWS

Family Corvidae

These medium to large birds, collectively called corvids, have a stout, pointed bill; rounded wings; and a rounded or wedge-shaped tail. Crows and ravens are black, jays usually colorful in blue and green, magpies black and white with long tails. These birds are omnivorous, eating insects, berries, nuts, and seeds.

GRAY JAY

Perisoreus canadensis **35:9**

Description
Size, 10–13 in. (25.4–33.0 cm). Adult: white forehead, throat, and collar; • *black cap* on back of head; • *rest of plumage gray,* darker above. When fluffed up, suggests a giant chickadee. Immature: dark slate, head blackish.
Similarities
Northern Shrike has black wings and tail.
Habitat
Coniferous forest and clearings.
Habits
Tame, inquisitive, bold; takes food about camps, robs traps, stores food. In flight, seems to float in air; glides more than most jays, wingbeats deeper; sails from top of one tree to bottom of another, then hops up branches in a spiral around tree, and repeats; on ground, hops.
Voice
"A loud, hawklike whistle, very like that of the Red-shouldered" (Brewster); a querulous *quee-ah, kuoo* or *whah* (Knight); other varied notes, some suggesting Blue Jay or Red Squirrel; a Catbird-like song in spring; a raucous rattle; somewhat of a mimic.
Eggs
3–5; gray, wreathed with brown markings at large end; 1.1 x 0.8 in. (2.8 x 2.0 cm), laid while snow is still on ground. Nest is of twigs lined with moss; large, 4–30 ft. (1.2–9.1 m) up, on branch near trunk of conifer.
Former name
Canada Jay.
Range
Resident from Labr. s. to New England and N.Y. and w. along n. U.S.; some populations move south in winter as far as Pa. and Great Plains.

BLUE JAY

Cyanocitta cristata **35:6**

Description
Size, 12 in. (30.5 cm). • *Crested;* tail long, wedge-shaped; • *violet-blue above, azure on wings and tail;* white face, throat, wing bars, and outer tail tips; pale below and under tail.
Habitat
Woods, edges, suburbs.
Habits
Active, aggressive, drives other birds away from feeder; its loud cries warn of intruders, but it is furtive about its own nest; flight steady, slow, on level plane, and with primaries spread; group flies across open space one at a time; collects in flocks in migration; stores food by burying acorns, etc.; hops, does not walk.
Voice
Noisy, varied; a shriek, *jay, jay;* a scream, *teearr,* like Red-shouldered Hawk; a creaky *tea-cup, tea-cup;* a *chillak, chillak;* also notes like a toy trumpet; song, a low mixture of whistles and sweet notes, often imitating various small birds.
Eggs
3–6; variable, often olive-blotched with brown; 1.1 x 0.9 in. (2.8 x 2.3 cm). Nest is of twigs, ragged, bulky, usually 8–20 ft. (2.4–6.1 m) up, near trunk.
Range
Mainly resident e. of Rockies from s. Canada s. to Gulf of Mexico.

Note: The **STELLER'S JAY,** *Cyanocitta stelleri,* size, 12 in. (30.5 cm), occurs along the western fringe of the eastern range, from Colorado to the Black Hills, and casually farther east. It is a blue-black, crested jay lacking the Blue Jay's head patterning and is white on the wings, tail, and underparts. In Florida a form of the **SCRUB JAY,** *Aphelocoma coerulescens* **(35:8),** size, 11½ in. (29.2 cm), is found, far isolated from the mainly western range of this species. It is blue and gray, lacks a crest, and has a blue-bordered white throat. This jay is social, often nesting in small groups.

BLACK-BILLED MAGPIE

Pica pica **35:11**

Description
Size, 17½–22 in. (44.5–55.9 cm); tail 9½–12 in. (24.1–30.5 cm). A large • *black-and-white bird.* Wings short, rounded; • *tail glossy green, long.*
Habitat
Foothills, roadsides, streamsides, thickets, fields, pastures, ranches.
Habits
Often found in small groups; conspicuous from car or train as one approaches Rockies; suspicious but inquisitive; spends much time on ground, twitches tail constantly in its jerky walk, hops when in a hurry; flight short, wavering, jaylike rather than crowlike.
Voice
Noisy; a harsh *ca-ca-ca,* higher than Common Crow; "a nasal querulous *maag?* or *aag-aag?*" (Peterson); other chattering and whistling notes.
Eggs
6–9; grayish, blotched with brown; 1.3 x 0.9 in. (3.3 x 2.3 cm). Nest is sometimes in colonies; huge round mass of sticks with hole in one side, lined with grass, often in willow thickets.
Range
Resident chiefly in W., but e. to Alta. and Man. and s. to n. Ariz.; in winter, casual to Iowa.

COMMON RAVEN

Corvus corax *Fig. 13*

Description
Size, 21½–27 in. (54.6–68.6 cm). Largest species in Order Passeriformes. Larger and with longer, thicker neck than Common Crow; all-black, with heavier bill; throat "bearded." In flight, wings angled at shoulder, outspread primaries, horizontal wings, • *wedge-shaped tail* (longer than Crow's).
Habitat
Seashore, mountains.
Habits
Aggressive but wary; flight with heavy slow wingbeats, often hawklike aerial revolutions in courtship; hovers, drops shellfish from height to break it open; a scavenger; sometimes mobbed by crows; seems to pair for life.
Voice
Various croaks, such as *c-r-r-r-u-u-k,* and others (young sometimes *caw* like crow); also a high-pitched *tok-tok-tok.*
Eggs
4–7; greenish, spotted with brown; 1.8 x 1.4 in. (4.6 x 3.6 cm). Nest is very large, of sticks, near top of conifer or on ledge.
Range
Resident throughout much of Canada and in w. U.S. from n. Canada and Greenland s. to n. New England, w. to Daks.

Fig. 13

COMMON CROW

Corvus brachyrhynchos *Fig. 13*

Description
Size, 17–21 in. (43.2–53.3 cm). All-black, tail gently rounded.
Similarities
Tail of Common Raven is longer; see also Fish Crow.
Habitat
Fields, woods, coasts, parks.
Habits
Wary, intelligent; hops and walks; frequents parkways feeding on animal life killed by cars; often seen flying overhead with steady, deep wingbeat; soars with wings in shallow V; congregates in great roosts in winter.
Voice
Caw, caw and variations thereof. Young: *car, car* (somewhat like adult Fish Crow but louder).

Eggs
3–6; greenish, blotched with brown; 1.7 x 1.2 in. (4.3 x 3.0 cm). Nest is bulky, of sticks, 10–50 ft. (3.0–15.2 m) up in a tree.
Range
Breeds from Newfoundland, s. to Fla. and Gulf Coast and w. across Canada and U.S.; winters from n. U.S. s. to Mexico and southward.

FISH CROW

Corvus ossifragus

Fig. 13

Description
Size, 17 in. (43.2 cm). All-black; tail gently rounded.
Similarities
Common Crow is virtually indistinguishable except by voice and direct size comparison, although Common is slimmer and wings are more pointed at tips and broader at base.
Habitat
Seashores, river valleys, occasionally vicinity of inland waters.
Habits
Feeds on beach, hovers over food, engages in aerial revolutions, soars more and flies with more alternate sailing and flapping than Common Crow; somewhat more gregarious at all seasons.
Voice
Nasal *car, car* much like squawk call of young Common Crow, actually "closer to a Black-crowned Night Heron's *quowk*" (Pough); "hoarse *craaa-craaa* and a staccato *ca-a*" (Cruickshank).
Eggs
3–5; greenish, blotched with brown; 1.5 x 1.1 in. (3.8 x 2.8 cm). Nest is of sticks, often in small colonies, in swamps or coastal pines.
Range
Resident along Atlantic and Gulf coasts, from Mass., s. to Fla. and w. to Tex.; also found inland along large rivers from Va. to Tenn. and Ark.

Note: Two other crowlike birds of western North America reach the western fringe of our region. The **PINYON JAY,** *Gymnorhinus cyanocephalus,* size, 10 in. (25.4 cm), is a dull bluish, short-tailed bird with a thinner bill than other jays. It is a flocking species with crowlike habits, sporadically reaching the western Plains. The **CLARK'S NUTCRACKER,** *Nucifraga columbiana* **(35:10)**, size, 12 in. (30.5 cm), of western mountains resembles the Pinyon Jay in shape, but is pale gray with black and white wings and tail. It wanders eastward in some numbers after breeding.

CHICKADEES AND TITMICE

Family Paridae

These are small, plump, big-headed birds with short, straight bills; beady black eyes; rounded wings; and soft, fluffy plumage, predominantly gray, black, and white in color. They are very active and often hang head down from a branch as they feed. They are confiding and inquisitive, and can be attracted by squeaking or by imitations of their notes. Their natural foods are insects, insect eggs or larvae, seeds and berries, but they readily frequent feeding trays.

BLACK-CAPPED CHICKADEE

Parus atricapillus **34:10**

Description
Size, 4¾–5¾ in. (12.1–14.6 cm). • *Black cap and bib*; gray above with white edges on wing feathers; white below, pale chestnut wash on flanks; • *white cheeks* noticeable at a distance.
Similarities
Narrow line of white feather edges on folded wing and somewhat larger size distinguish this species from the similar Carolina Chickadee when together, but their voices provide the best identifying feature. Eastern Phoebe's *fee-bee* note is hoarse, not whistled, as in this species. Spring male Blackpoll Warbler does not have black bib.
Habitat
Woods, edges, gardens, towns.
Habits
Tame, may be taught to eat from hand; somewhat migratory; tends to wander after breeding season in mixed flocks with Downy Woodpeckers, nuthatches, kinglets.
Voice
Call, a *chick-a-dee-dee-dee;* song, a sweet whistled *fee-bee* or *SPRING's-come,* or *FEE-be-be,* the first note higher; hisses if disturbed on nest.
Eggs
4–9; white, finely spotted with reddish-brown; 0.6 x 0.5 in. (1.5 x 1.3 cm).
Range
Breeds from Newfoundland and cen. Man. s. to s. N.J. and Mo., mts. to Tenn.; winters from breeding range s. to Md. and Tex.

CAROLINA CHICKADEE

Parus carolinensis **34:7**

Description
Size, 4½ in. (15.7 cm).
Similarities
Resembles Black-capped Chickadee but shows • *little or no white in wings;* best distinguished by voice.
Habitat
Coastal-plain swamps; woods, especially pinewoods; residential areas.
Habits
Similar to Black-capped Chickadee; roams in mixed species flocks when not breeding.
Voice
Chick-a-dee-dee-dee, higher pitched than Black-capped and faster; song is a short, whispered sibilant *su* preceding each syllable of *fee-bee* song, thus *sufee-subee;* "the pumphandle strain . . . *spee-deedle-dee-deedle-dee*" (Dickey).
Eggs
Similar to Black-capped's.
Remarks
There is virtually no overlap in breeding season, but Black-cappeds move into the adjacent range of the Carolinas in winter.
Range
Resident from cen. N.J. and Mo. s. to Fla., Tex., and Okla.

BOREAL CHICKADEE

Parus hudsonicus **34:6**

Description
Size, 5–5½ in. (12.7–14.0 cm). Brownish above, • *brown cap,* whitish below, bib black, flanks chestnut.
Similarities
Black bib is smaller, chestnut flanks brighter than in Black-capped.
Habitat
Northern evergreen forest.
Habits
Similar to Black-capped Chickadee's, but not as lively and works closer to tree trunk.
Voice
Somewhat like a Black-capped Chickadee, but notes slower, wheezier, and blurred; *chick-a-day-day;* song, a warble.
Eggs
5–9; white, finely spottted with reddish-brown; 0.5 x 0.4 in. (1.4 x 1.1 cm).
Former name
Brown-capped Chickadee.
Range
Resident in much of Canada, from Labr. w. to Man. and s. to n. New England; some wander s. to Md., Ohio, and Ill. in winter.

TUFTED TITMOUSE

Parus bicolor **34:4**

Description
Size, 5–6 in. (12.7–15.2 cm). A small • *gray bird with a crest.* Gray above, cheeks and underparts whitish, chestnut wash on flanks.
Similarities
Cedar Waxwing, also crested, is larger, brownish, and with bright markings on wings and tail.
Habitat
Swamps, deciduous woodlands, residential areas.
Habits
Somewhat like Black-capped Chickadee; visits feeders; after breeding season roams in mixed flocks with chickadees, Downy Woodpeckers, nuthatches; not migratory.
Voice
Loud whistled *peter, peter, peter*, often confused at a distance with Carolina Wren or Northern Oriole; a chickadeelike *dee-dee-dee;* and whisperings; a scolding *ya-ya-ya.*
Eggs
4–7; white, thickly dotted with reddish-brown; 0.7 x 0.5 in. (1.8 x 1.3 cm).
Range
Resident in much of e. U.S. from Maine and Wis. s. to Fla., Gulf Coast, and Tex., s. to Mexico.

NUTHATCHES

Family Sittidae

Nuthatches are small, stout birds with thin, sharp bills; short, square tails; and strong legs and feet. They clamber about tree trunks and branches, often head down. They nest in cavities, which they frequently excavate themselves in the soft wood of dead stumps, stubs, or branches, from 4–120 feet (1.2–36.6 m) up. Their food consists of insects, seeds, nuts, berries, and fruit.

WHITE-BREASTED NUTHATCH

Sitta carolinensis **34:2**

Description
Size, 5–6 in. (12.7–15.2 cm). Black cap and nape, beady black eye. White completely encircling eyes; blue-gray back, white underparts, chestnut undertail coverts, white markings on outer tail feathers. Female often has grayer crown.
Similarities
Red-breasted has black eye stripe.
Habitat
Large trees in deciduous woods and about habitations.
Habits
Tame, comes to feeders, likes suet; stores food in crevices; except in breeding season, often travels in mixed flocks; flight slightly undulating.
Voice
Nasal *yank-yank-yank;* a short, high, conversational *hit-hit;* song, a low-pitched *"to what what what what"* (Thoreau).
Eggs
4–8; white spotted with reddish-brown; 0.6–0.8 x 0.5–0.6 in (1.5–2.0 x 1.3–1.5 cm).
Range
Resident from N.S. and Ont. s. to Fla., Gulf Coast, and Mexico.

RED-BREASTED NUTHATCH

Sitta canadensis **34:5**

Description
Size, 4½ in. (11.4 cm). Both a black and a white face stripe. Male: black cap, white line over eye, • *black line through eye,* dark blue-gray upperparts, reddish below. Female: paler.
Similarities
White-breasted is larger, has no eye stripe.
Habitat
Northern evergreen forest, also deciduous woods. In winter, when food supplies fail in north, given to irregular southward incursions, and it then appears in all sorts of places.
Habits
More active than White-breasted, forages to end of branches, captures insects in air; tame, frequents feeders, stores food; flight undulating; smears entrance to nesting hole with pitch; except in breeding season, often travels in small mixed species flocks.
Voice
High-pitched nasal *ink-ink-ink,* like a little tin horn; much higher than White-breasted's *yank;* a *hit-hit* similar to White-breasted; song somewhat similar to that of White-breasted but faster, higher, reedier.
Eggs
4–8; white, spotted with reddish-brown; 0.6–0.8 x 0.5–0.6 in. (1.5–2.0 x 1.3–1.5 cm).
Range
Breeds from Newfoundland and Man. s. to N.J., N.C., and w. across U.S.; winters from breeding range s. to Gulf Coast and Mexico.

BROWN-HEADED NUTHATCH
Sitta pusilla **34:3**

Description
Size, 4½ in. (11.4 cm). A southern pinewoods species lacking any nasal calls. • *Brown cap coming down to eye,* blue-gray back, underparts and spot on nape white.
Similarities
Red- and White-breasteds are larger.
Habitat
Southern pinewoods, especially loblolly; also mixed woods, cypress swamps.
Habits
Forages to end of twigs; flight gently undulating; gregarious and sociable, often traveling with chickadees, titmice, and woodpeckers.
Voice
Fast, high *yit, yit, yit;* a chickadeelike *dee-dee-dee.*
Eggs
4–8; white, spotted with reddish-brown, 0.6–0.8 x 0.5–0.6 in. (1.5–2.0 x 1.3–1.5 cm).
Range
Resident from Del., Mo., and e. Tex. s. to Fla. and Gulf Coast.

Note: The western mountain pine forest **PYGMY NUTHATCH,** *Sitta pygmaea,* size, 4 in. (11 cm), is possibly only a subspecies of the Brown-headed Nuthatch. It reaches the western fringe of the eastern region in the Black Hills. It is somewhat grayer-capped but otherwise closely resembles the Brown-headed—their ranges do not meet.

CREEPERS
Family Certhiidae

Creepers are small, stiff-tailed, slender birds. They creep up and around tree trunks, probing for bark insects with slightly curved bills.

BROWN CREEPER
Certhia familiaris **34:1**

Description
Size, 5½ in. (14.0 cm). Creeps up tree trunks. Streaked brownish above, white below; pale band across relatively long wing; • *bill thin, curved.*
Habitat
Woodlands, swamps, shade trees.
Habits
Using long tail as prop, it creeps up trunk of tree often in a spiral, or along underside of horizontal branch, feeding as it goes; then drops to base of neighboring tree and repeats; solitary, but often travels with groups of chickadees, nuthatches, woodpeckers, kinglets; by means of its protective coloration, it often escapes detection; comes to feeders for suet, chopped peanuts, peanut butter.
Voice
Long, thin, high *tseee,* longer than Golden-crowned Kinglet's and not quickly repeated in series; song, *seee-see-see, swee-swee.*
Eggs
4–9; white, with reddish-brown spots; 0.6 x 0.5 in. (1.5 x 1.3 cm). Nest is under loose-hanging slab of bark low on a tree trunk.

Range
Breeds from Newfoundland and Ont. s. to C. America; winters from n. U.S., s. to Gulf Coast and Fla.

Note: The **DIPPER,** *Cinclus mexicanus,* size, 8 in. (20.3 cm), is the only member of its family, the Cinclidae, found in North America. It is an unmistakably plump, short-tailed, large gray wrenlike bird, and is a resident along mountain streams in which it feeds. It is found in the Black Hills, and from the Rocky Mountains west.

WRENS

Family Troglodytidae

Wrens are plump, vivacious brown birds, generally smaller than sparrows, that often carry the tail cocked over the back. They have thin, curved bills and short, rounded wings. The sexes look alike. They spend most of their time on or near the ground, feeding on insects and spiders.

HOUSE WREN

Troglodytes aedon **34:14**

Description
Size, 4½–5¼ in. (11.4–13.3 cm). Gray-brown above, grayish below, brownish flanks, faint barring on wings and tail, longish tail, • *no eye stripe.*
Similarities
Winter Wren is smaller, has shorter tail, and dark barring below; see also Bewick's Wren.
Habitat
Woods, thickets, gardens, farming country. Common.
Habits
Aggressive, scolds intruders; frequents bird bath and feeder; will nest in wren house; male builds extra dummy nests.
Voice
Song, a long gushing, bubbling melody; alarm note, a grating chatter like a Common Yellowthroat's.
Eggs
5–12; pinkish, heavily dotted with reddish-brown; 0.6 x 0.5 in. (1.5 x 1.3 cm). Nest is of twigs, etc., in virtually any cavity or bird box.
Range
Breeds from Maine and Ont. s. to N.C.; winters from S.C. s. to Gulf Coast and Fla.

WINTER WREN

Troglodytes troglodytes **34:11**

Description
Size, 4 in. (10.2 cm). Dark brown above and on belly, breast lighter; • *belly barred with black;* • *tail stubby;* faint eye stripe.
Similarities
See House Wren.
Habitat
In summer, northern or mountain forest; in migration, thickets, brush piles.
Habits
Secretive, mouselike; bobs up and down and cocks tail; feeds along stone walls, woodpiles, and banks of streams; keeps near cover.
Voice
Song is one of the most beautiful in North America, a long, wild,

ringing, tinkling melody, first part often ending in a trill, second part ending in a trill an octave higher; call, a *tik, tik* like chip of Song Sparrow; also a *chirr.*

Eggs

4–10; white, dotted with red-brown and purple; 0.7 x 0.5 in. (1.8 x 1.3 cm). Nest is of sticks and moss in root mass of fallen tree or other tangle near ground.

Remarks

This is the only species of wren in the Old World and the Wren of English literature.

Range

Breeds from Newfoundland and s. Man. s. to Conn. and Mich.; also in Ga. mountains; winters from s. parts of breeding range, s. to Gulf Coast.

BEWICK'S WREN

Thryomanes bewickii **34:13**

Description

Size, 5½ in. (14.0 cm). Slim; • *fanlike tail longer than wings, with white corners;* brown above, whitish below.

Similarities

House Wren lacks white eye stripe, white underparts, and white tail corners. Rock Wren is grayer, with a streaked breast.

Habitat

Gardens, farms, woodpiles, thickets, woodlands.

Habits

Less aggressive than House Wren, movements more deliberate than House Wren or Carolina, comes to feeders, often fans and flirts tail.

Voice

Song is loud, beautiful; often confused with that of Song Sparrow, but longer, more varied, and ends on a trill; alarm, a chatter somewhat like Carolina's; a low *pit.*

Eggs

4–7; white or pinkish, heavily dotted with red-brown; 0.7 x 0.5 in. (1.8 x 1.3 cm). Nest is of twigs, etc., in cavities virtually anywhere; also in boxes.

Range

Resident throughout much of s. U.S. from cen. Pa. and Nebr. s. to Gulf of Mexico and Mexico, excluding Atlantic Coast.

CAROLINA WREN

Thryothorus ludovicianus **34:17**

Description

Size, 6 in. (15.2 cm). Largest eastern wren. • *White eye stripe,* buffy belly, and no white in the tail; • *reddish-brown above,* buffy below; chin white; undertail coverts barred; narrow black barring on wings and tail; bill rather long and curved.

Similarities

Smaller Bewick's Wren is similar but is whitish below and has white edges of the tail.

Habitat

Woods, swamps, thickets, brushy slopes, gardens, orchards.

Habits

Teeters body, jerks tail, often escapes by running mouselike through cover; inquisitive but quick to hide; likes brush piles; does not migrate, cold winters kill off many in northern portions of their range; comes to feeders.

Voice
Song is a clear whistled *tea-kettle, tea-kettle, tea-kettle,* or *sweet-william, sweet-william, sweet-william* with many variations; mimics other birds; sings throughout year. Alarm is a buzzy chatter; many other notes.
Eggs
4–8; creamy white, heavily spotted with reddish-brown; 0.8 x 0.6 in. (2.0 x 1.5 cm). Nest is of vegetation in any cavity or nesting box, up to 40 ft. (12.2 m) high.
Range
Resident throughout se. U.S. from N.Y., Mich., and Wis. s. to Fla. and Gulf Coast.

LONG-BILLED MARSH WREN

Cistothorus palustris **34:9**

Description
Size, 4¾–5½ in. (12.1–14.0 cm). Dark brown above, with a blackish, unstreaked crown; white line over eye; whitish below, including undertail coverts; buffy flanks; • *blackish back with white lines.*
Similarities
Bewick's has longer white-cornered tail and no white lines on back. Short-billed has shorter bill, streaked crown, buffy breastband, and brownish undertail coverts.
Habitat
Freshwater cattail or salt marshes.
Habits
Sings from stalks in marsh, often rises 6–10 ft. (1.8–3.0 m) in air and flutters down singing, climbs nimbly about reeds; males often polygamous and given to building extra dummy nests.
Voice
Short, rich, bubbling, somewhat guttural song; alarm, a chatter, often suggesting a Red-winged Blackbird; sometimes sings at night.
Eggs
3–7; pale to dark brown, often with darker dots; 0.6 x 0.5 in. (1.5 x 1.3 cm); 2 broods. Nest is a ball of grasses with a side entrance, attached to reeds in marsh.
Range
Breeds from N.B. and Man. s. to Fla., the Gulf Coast, and Mexico; winters from N.J. along Atlantic and Gulf coasts to Mexico.

SHORT-BILLED MARSH WREN

Cistothorus platensis **34:8**

Description
Size, 4–4½ in. (10.2–11.4 cm). Tiny. Streaked crown and back, bill and tail short, brownish above, buff band across breast, throat and belly white, undertail coverts and flanks brownish.
Similarities
Long-billed has eye stripe.
Habitat
Wet meadows and grassy marshes, fresh or salt; unevenly distributed, common in some areas, absent in others.
Habits
Secretive, stays near ground, sings from weed stalks; if flushed, flight is short, bird soon drops into marsh; male builds extra dummy nests.

Voice
Short series of clinking notes, somewhat like knocking pebbles together; *chap, chap, chapperm-chapperm-chapper,* running down scale and increasing in tempo.
Eggs
4–10; white; 0.6 x 0.5 in. (1.5 x 1.3 cm). Nest is round ball of grass with side entrance, attached to growing grass stalks.
Range
Breeds from N.B. and Man. s. to Del., Mo., and Kans. and s. to S. America; winters from N.J. and Ill. s. to Fla. and Mexico.

Note: The **CANYON WREN,** *Catherpes mexicanus,* size, 5½–5¾ in. (14.0–14.6 cm), can be seen in rocky canyons and ravines in the western fringe of the eastern region. It is a rusty-brown wren with a clear, white throat and upper breast.

ROCK WREN
Salpinctes obsoletus **34:16**

Description
Size, 5–6¼ in. (12.7–15.9 cm). Upperparts gray-brown; white line over eye; • *underparts white with dusky streaked breast;* tail rounded, and with black band just before buffy tip.
Similarities
Bewick's Wren lacks breast streaks.
Habitat
Open rocky places, including deserts, and mountains above timberline.
Habits
Bobs up and down on rock, stays close to ground. In flight, quick, jerky.
Voice
Song sometimes suggests a Brown Thrasher; a "harsh, grating . . . *kerEE kerEE kerEE . . . chair chair chair chair, deedle deedle deedle deedle, tur tur tur tur, kerEE kerEE kerEE trrrrrrrrrr*" (Nice); call, *tick-ear,* often given with a bob; also a loud, purring trill.
Eggs
5–8; white, lightly dotted with red-brown; 0.7 x 0.5 in. (1.8 x 1.3 cm). Nest is of grass, moss; cup-shaped, often in rocky cranny.
Range
Breeds chiefly in W., but e. to N. Dak., southward; winters from Calif. and Tex. southward.

MOCKINGBIRDS AND THRASHERS
Family Mimidae

Members of this family are slim, medium-sized birds with short, rounded wings and long, rounded tails, which they pump when excited. They feed on wild berries, insects, and fruit on the ground, in thickets, or in the open.

NORTHERN MOCKINGBIRD
Mimus polyglottos **35:2**

Description
Size, 9–11 in. (22.9–27.9 cm). Slender; gray above, whitish below; no black on face; outer tail feathers with much white; • *white wing patches* conspicuous in flight; long tail.

Similarities
Shrikes are stouter, with hooked bills, shorter tails; black on face, wings, and tail; and less white in wings.
Habitat
Gardens, farms, roadsides, towns.
Habits
Active, aggressive; runs, hops, and feeds on ground with tail raised; but sings from fence post, TV antenna, top of building, or roadside wire; frequently raises wings, flies with rapid wingbeats, long tail noticeable; frequents feeders (likes raisins, nutmeats, suet, crumbs); chases other birds.
Voice
Song is varied, mellifluous; notes often repeated 3 or more times. This species is one of the world's great songsters and an accomplished mimic; often sings in flight and by moonlight; call, a harsh *chat.*
Eggs
3–6; greenish-blue, spotted with red-brown; 1.0 x 0.8 in. (2.5 x 2.0 cm). Nest is bulky, of twigs, in shrub or thicket, 2–20 ft. (0.6 x 6.1 m) up.
Range
Resident from N.S. throughout much of U.S. s. to W. Indies and Mexico.

GRAY CATBIRD
Dumetella carolinensis **35:3**

Description
Size, 8–9½ in. (20.3–24.1 cm). • *Slate-gray,* with a black cap. • *Undertail coverts chestnut.*
Habitat
Thickets, wet or dry; hedges.
Habits
Friendly, inquisitive; comes to feeders; often nests near houses; usually sings from concealed perch; common.
Voice
Song is squeaky, simple version of Northern Mockingbird's or Brown Thrasher's, but always with some catlike mews and phrases not repeated; somewhat of a mimic; often sings at night, one of earliest songs heard at dawn; note, a catlike mew.
Eggs
3–6; greenish-blue; 1.0 x 0.7 in. (2.5 x 1.8 cm); 2 broods. Nest is of twigs, in thickets, 2–8 ft. (0.6–2.4 m) up.
Range
Breeds from N.S. and Man. s. to Fla., Tex.; winters from Gulf Coast to C. America.

BROWN THRASHER
Toxostoma rufum **35:1**

Description
Size, 10½–12 in. (26.7–30.5 cm). • *Reddish-brown above;* whitish below with • *brown streaks on breast;* eye yellow; bill long, curved; 2 white wing bars.
Similarities
Wood Thrush is smaller; is less reddish and more brown; has shorter tail, and has straight bill, brown eye, and no wing bars.
Habitat
Dry thickets.

Habits
Aggressive; will not brook intruders at nest; sings from lofty perch; scratches for food on ground in thicket; walks, runs, or hops, likes dust baths and water baths; less common about habitations than Gray Catbird.
Voice
Long, somewhat like Northern Mockingbird's and Gray Catbird's, but few imitations and each phrase repeated twice, as *"drop it, drop it, cover it, cover it, I'll pull it up, I'll pull it up"* (Thoreau); call, a distinctive kissing sound.
Eggs
3–6; whitish, heavily dotted with red-brown; 1.1 x 0.8 in. (2.8 x 2.0 cm); often 2 broods. Nest is bulky, of twigs, on or near ground in thicket.
Range
Breeds throughout much of e. U.S. from n. New England, Ont., and Man. s. to Fla. and Gulf Coast; winters in s. parts of breeding range.

Note: The sagebrush-dwelling **SAGE THRASHER,** *Oreoscoptes montanus* **(35:4)**, size, 9 in. (22.9 cm), occurs from southern Wyoming and eastern Colorado west, with an isolated population in southern Saskatchewan (rarely wanders eastward). It is a gray, open-country bird with white wing bars and a streaked breast.

THRUSHES, BLUEBIRDS, AND ALLIES

Family Turdidae

Most thrushes are woodland birds that forage on the ground. The young all have spotted breasts. This family is known for its songsters. They feed on insects, spiders, worms, grubs, wild fruits, berries, and seeds.

AMERICAN ROBIN

Turdus migratorius **36:2**

Description
Size, 9–11 in. (22.9–27.9 cm). Slaty back, • *reddish breast,* bill yellow, head black, belly and tips of outer tail feathers white; female paler than male; juvenal with black spots on buffy breast.
Habitat
Lawns, gardens, suburbs, fields, swamps, clearings.
Habits
Adapts well to civilization, hops or runs on lawn, cocks head, nests near houses, likes to bathe; flocks spend winter in woods and swamps, sometimes gather in huge roosts; one of first species to sing at dawn. Flight is direct, with deep, fairly rapid wingbeats, straight back, and white rear flanks noticeable.
Voice
Song is a carol, *cheerily cheer-up, cheerily cheer-up*; notes include a querulous *cuk, cuk;* a *ssssp;* a nervous *bup, bup;* a rattling *chi, il, il, il, il;* and others.
Eggs
3–4; Robin's-egg blue; 1.1 x 0.8 in. (2.8 x 2.0 cm); 2 broods. Nest is cup-shaped, of grass and mud, in crotch or branch of tree, or in man-made nook, usually 5–20 ft. (1.5–6.2 m) up.
Range
Breeds throughout most of Canada and U.S., from Newfoundland and Man. s. to S.C., Ark., and C. America; occasional breeder along Gulf Coast; winters from Newfoundland, s. Ont., s. to Gulf Coast and southward.

Note: The **VARIED THRUSH (36:1)**, size, 9–10 in. *Zoothera naevia* (22.9–25.4 cm), is a western bird that sporadically wanders into the eastern region during migration. It resembles an American Robin, but both sexes of the Varied Thrush have a buffy or rusty eye stripe and wing bars; the male has a black band across its orange-brown breast.

WOOD THRUSH

Hylocichla mustelina **36:11**

Description
Size, 8 in. (20.3 cm). Plump; bright brown above, becoming • *reddish on head, neck, and upper back;* breast creamy with large, round, dark spots.
Similarities
Brown Thrasher is longer tailed and yellow eyed.
Habitat
Shade trees, moist woods.
Habits
Originally a woodland bird, now common in suburbs and seen on lawn like American Robin; aggressive, less furtive than other brown thrushes.
Voice
Song is an unhurried, liquid, flutelike melody, *"eeohlay-ayolee-ahleelee-ayleahlolah-iloliee"* (Saunders); alarm note, *quirt-quirt*; a rattling *putt, putt, putt.*
Eggs
3–4; Robin's-egg blue; 1.1 x 0.7 in. (2.8 x 1.8 cm). Nest is of mud, grass, and paper, 3–12 ft. (0.9–3.7 m) up in tree or shrub.
Range
Breeds throughout much of e. U.S. from N.S. and Ont. s. to Fla. and Gulf of Mexico; winters from Mexico s. to C. America, but some winter from s. Fla. and Tex.

HERMIT THRUSH

Catharus guttatus **36:5**

Description
Size, 6½–7¾ in. (16.5–19.7 cm). Russet-brown above; • *rump and tail rufous;* hint of rufous in primaries; below whitish with small dark wedge-shaped spots on breast.
Similarities
Fox Sparrow has thick bill, entire upperparts reddish-brown.
Habitat
Damp, mixed woods; in winter, edges, thickets.
Habits
Raises and slowly lowers tail on alighting; usually seen near ground, but sings from treetop; hardy, a few linger in north in winter; the only brown thrush to winter in U.S.; visits feeders.
Voice
Song is a clear, flutelike introductory note, then various 5–12 note phrases, a pause, then another introductory note on a different pitch, and so on; notes are sweet, clear, bell-like; also a soft *chuck,* a catlike *pay,* a scolding *tuk-tuk.*
Eggs
3–5; greenish-blue; 0.9 x 0.7 in. (2.3 x 1.8 cm). Nest is of moss, twigs, grasses, on or near ground.
Range
Breeds from Labr. and Sask. s. to N.Y. and Mich.; also in mountains to Va.; winters from s. New England s. to Fla. and C. America.

SWAINSON'S THRUSH

Catharus ustulatus **36:9**

Description
Size, 6½–7¾ in. (16.5–19.7 cm). Above, uniform olive-brown; breast light buff with small dark spots; belly whitish; • *buffy eye-ring* and cheek contrasting with rest of head.
Similarities
Gray-checked Thrush has gray cheeks, approximately same color as rest of head, breast whitish; it is sometimes hard to distinguish these 2 species in fall.
Habitat
Breeds in spruce and fir of Canadian zone; in migration, woods, shrubbery, even suburban lawns.
Habits
If flushed, flies to low branch and remains motionless, then flits on farther; sometimes pumps tail, but not habitually as Hermit does; sings from treetop.
Voice
Song is simpler than Wood or Hermit Thrush; seems to ascend in spirals; *"whao-whayo-whiyo-wheya-wheeya"* (Saunders). Notes, a *whit;* a catlike *twee-ur;* in migration at night overhead, a short *heap.*
Eggs
3–5; greenish-blue, spotted with reddish-brown; 0.9 x 0.7 in. (2.3 x 1.8 cm). Nest is of grass, moss, twigs, usually in spruce or fir, 4–15 ft. (1.2–4.6 m) up.
Range
Breeds in Canada from Newfoundland and Man. s. to n. New England and Mich.; also in mountains to W. Va.; winters in S. America.

GRAY-CHEEKED THRUSH

Catharus minimus **36:10**

Description
Size, 7 in. (17.8 cm). Above, olive-brown; below, whitish with small dark spots on breast; eye-ring absent or faint; • *cheeks gray,* approximately as dark as rest of head.
Similarities
See Swainson's Thrush.
Habitat
Willows or alders in tundra; Hudsonian zone of northern coniferous forest; in migration, woods and shrubbery.
Habits
Similar to Swainson's, but shyer.
Voice
Least impressive of eastern thrushes; fainter than Swainson's and without upward spiral; a subdued, faraway-sounding *"wee-oh, chee-chee-wee-oh, wee-oh"* (Gillespie); also, *"chook-chook, wee-o, wee-o, wee-i-ti-t-ter-ee"* (Wallace); also a similar flight song. Call is a melancholy *pheu,* or penetrating *queep,* often heard from migrants overhead at night.
Eggs
3–5; greenish-blue, spotted with red-brown; 0.9 x 0.7 in. (2.3 x 1.8 cm). Nest is of grass, twigs, on low limb or bush.
Range
Breeds in n. Canada from Newfoundland and n. Ont. s. to N.Y.; winters in W. Indies, C. and S. America.

VEERY

Catharus fuscescens **36:8**

Description
Size, 7 in. (17.8 cm). • *Above bright cinnamon-brown;* below light buff with • *inconspicuous spots on throat;* throat appears white at a distance.
Similarities
Swainson's is same size, but has broad, buffy eye-ring and browner, less rufous upperparts. Hermit has reddish tail. Gray-cheeked has more gray-brown, no buff on chest.
Habitat
Undergrowth in swamps, wet woodlands, ravines, and dry hillsides.
Habits
Keeps close to ground, even when singing; progresses by jumping, stands upright.
Voice
Song is a series of hollow *whree-u*'s descending the scale, or *"ta-weel'ah, ta-weel'ah, twil'uh, twil'ah"* (Ridgeway); dawn chorus in June on breeding grounds is hauntingly beautiful; call, a distinctive *view.*
Eggs
3–5; greenish-blue; 0.9 x 0.7 in. (2.3 x 1.8 cm). Nest is of grass and twigs on ground.
Range
Breeds from Newfoundland and s. Man. s. to N.J. and Ind.; also some in mountains to Ga.; winters in S. America.

EASTERN BLUEBIRD

Sialia sialis **36:6**

Description
Size, 6½–7½ in. (16.5–17.8 cm). Spring male: • *bright blue above, breast reddish,* belly white. Spring female: grayish, with bright blue on wings and tail. Juvenal: spotted above and below. Fall adult: duller.
Habitat
Orchards, wood edges, roadsides, farmlands.
Habits
Often seen on roadside wire, looks hunched when perched; catches some insects in air, drops from perch to pounce on others.
Voice
Song is a soft mellow warble, *purity, purity;* call, *oola* or *aloola.*
Eggs
3–7; light blue; 0.8 x 0.6 in. (2.0 x 1.5 cm). Nest is hole in stump or tree, often abandoned woodpecker's hole 3–30 ft. (0.9–9.1 m) up, or in bird box 3–6 ft. (0.9–1.8 m) up, with entrance hole 1½ in. (3.8 cm) in dia.
Range
Breeds locally throughout e. U.S. from s. Canada s. to Gulf of Mexico and w. to Rockies; winters from s. New England and Mo. s. to Mexico.

MOUNTAIN BLUEBIRD

Sialia currucoides **36:7**

Description
Size, 6½–7¾ in. (16.5–19.7 cm). Male: paler below, belly white. Female: brownish gray, gray below, with blue on wings and tail. Both duller in winter.

Similarities
Eastern Bluebird is not as slim; Mountain sits more erect. Juvenal has much less spotting than Eastern.
Habitat
Open woods in mountains; ranches, open country.
Habits
Seizes insects in air, or hovers and then drops on them on ground.
Voice
Usually silent; near dawn "a sweet, clear *trually tru-al-ly* like that of the Eastern species, and a mellow warble" (Wheelock); note, a low *turr.*
Eggs
4–6; pale blue; 0.8 x 0.6 in. (2.0 x 1.5 cm). Nest is of grass, in hole in stump or tree, or beneath eaves of building; also in nesting box.
Range
Breeds chiefly in W. but e. to s. Man. and s. to w. S.Dak. and Nebr.; winters from Nebr. s. to Tex. and Mexico.

Note: The **COMMON WHEATEAR,** *Oenanthe oenanthe,* size, 6 in. (15.2 cm), is a small gray and white bird with black wings. It breeds in the Arctic and winters in Africa, but is seen in migration occasionally in Canada. The **TOWNSEND'S SOLITAIRE,** *Myadestes townsendi,* size, 9 in. (22.9 cm), is found in mountain canyons and woods from the Black Hills west. It is an erect-perching, grayish thrush with a white eye-ring that forages for insects by flycatching. It somewhat resembles the Northern Mockingbird, but has white edges on its outer tail feathers, a shorter bill, and buff rather than large white wing patches.

GNATCATCHERS, KINGLETS, AND OLD-WORLD WARBLERS

Family Sylviidae

This is a family of tiny, active woodland birds with small bills adapted to a diet of insects. Gnatcatchers are distinguishable by their comparatively long tails, kinglets by their brightly colored crowns.

BLUE-GRAY GNATCATCHER

Polioptila caerulea **34:12**

Description
Size, 4–5 in. (10.2–12.7 cm). Slender; blue-gray above, eye-ring and underparts white; male has black forecrown; • *long, floppy tail* conspicuous in flight; • *tail white-bordered.*
Similarities
Resembles a tiny Northern Mockingbird.
Habitat
Thickets, swamps, tall trees.
Habits
Active, graceful, tail always in motion; often hovers in front of a twig, frequently catches insects in air.
Voice
Call is a shrill, high-pitched, scolding *spee, spee, spee;* song, "rarely heard, a low, pleasant, exquisite, warbling ditty . . . *zee-u, zee-u, ksee-ksee-ksee-ksee-ksee-ksee-ksu*" (Simmons).

Eggs
4–5; bluish-white, spotted with red-brown and slate; 0.6 x 0.4 in. (1.5 x 1.1 cm). Nest is tiny felted cup in desert shrub.
Range
Breeds from N.H., s. Ont., and cen. Minn. s. to Bahamas, Gulf Coast, and Ark.; winters from N.C. and the Gulf Coast southward.

GOLDEN-CROWNED KINGLET

Regulus satrapa **41:15**

Description
Size, 3½–4 in. (8.9–10.2 cm). Smaller than any wood warbler. Black-bordered orange or yellow crown. Bill tiny; tail short, notched; olive-gray above, underparts, 2 wing bars, and line over eye white; • *crown yellow* in female, and with • *orange center* in male (sometimes concealed).
Similarities
Ruby-crowned has broken white eye-ring, no white line over eye, male crown patch scarlet.
Habitat
Coniferous and mixed woods; deciduous trees, bushes, as well as evergreens in migration and winter.
Habits
Tame, confiding; very active, clambers about or flutters at ends of twigs, often flips out wing tips over back; catches insects in air as well as on twigs; in winter, frequently travels in mixed species flocks.
Voice
Call is a high, thin *see-see-see,* shorter than note of Brown Creeper and rapidly repeated; song, similar but longer *"zee, zee, zee, zee, zee, why do you shilly-shally?"* (Stanwood).
Eggs
5–10; creamy, marked with brown and lavender; 0.5 x 0.4 in. (1.3 x 1.0 cm). Nest is of moss, lichens; round, suspended from twig of conifer, 4–50 ft. (1.2–15.2 m) up.
Range
Breeds from Newfoundland and Man. s. to Mass. and Mich.; also into mountains of N.C.; winters from s. parts of breeding range to Fla., Gulf Coast, and C. America.

RUBY-CROWNED KINGLET

Regulus calendula **41:13**

Description
Size, 3¾–4⅓ in. (9.5–11.4 cm). A very small, green, short-tailed bird with a • *white eye-ring.* Bill tiny; tail notched; olive-gray above, whitish below; 2 white wing bars, rear one with black border; male has concealed patch of ruby on crown, visible when bird erects it in display; eye-ring gives black beady eye a staring look.
Similarities
Any kinglet without a visible crown patch is this species; *Empidonax* flycatchers have white eye-rings, but they sit still, do not flit about.
Habitat
Breeds in northern bogs and forests; in migration, shrubbery, orchards, second growth, thickets.

Habits
Tame, active, restless, flips wing tips out over back; often associates with warblers and nuthatches in migration.
Voice
Ringing song, loud for so small a bird, seems to go up and down, *"see-see-see-you-you-you-just-look-at-me just-look-at-me"* (Cruickshank); also a harsh, wrenlike note.
Eggs
5–11; creamy white, usually finely dotted with red-brown; 0.5 x 0.4 in. (1.4 x 1.1 cm). Nest is of moss and lichens; suspended, or on limb of conifer, 2–50 ft. (0.6–15.2 m) up.
Range
Breeds from Newfoundland and Man. s. to n. New England and Ont.; winters from s. New England s. to Fla., Gulf Coast, and C. America.

PIPITS

Family Motacillidae

Pipits are slender, larklike open country birds with thin bills and long hind claws. They are brownish above, and have streaked breasts and white outer feathers on their tails, which they frequently pump. They eat insects, grubs, small mollusks, and crustaceans.

WATER PIPIT

Anthus spinoletta **48:5**

Description
Size, 6–7 in. (15.2–17.8 cm). Unstreaked dark back and dark legs. Spring: gray-brown above, • *buffy below* with streaks on upper breast, white outer tail feathers, bill slender. Fall: darker above, buffier below, breast more heavily streaked.
Similarities
Vesper Sparrow has short, thick bill and hops, does not walk or wag tail. Horned Lark has pipitlike flight, but is heavier, has white belly, different pattern, and narrower white edges on black tail.
Habitat
Breeds on mountains above timberline, tundra; plowed fields, dunes, flats.
Habits
Terrestrial, restless, nods head like dove as it walks, usually seen in flocks; perches on trees, fences, wires; wags tail more than Sprague's.
Voice
Note, heard in migration, "a sharp *tsip-tsip, tsip-it* . . . [the lark's is lower . . . more rolling . . . *sleek-slik-seezik*]" (Cogswell); song on breeding grounds simple, pleasant, often with trills; also has a flight song.
Eggs
4–7; whitish, heavily marked with chocolate; 0.8 x 0.6 in. (2.0 x 1.5 cm). Nest is of grass on ground in shelter of rock or hummock.
Range
Breeds in Arctic regions of Greenland, Newfoundland, and Hudson Bay, s. to n. Maine; winters from s. Pa., w. Va., s. to Fla., Gulf Coast, and C. America.

SPRAGUE'S PIPIT

Anthus spragueii **48:4**

Description
Size, 6½–7 in. (16.5–17.8 cm). Paler than Water; head, neck, and • *back streaked with black and buff;* white below, washed with buff, finely streaked; more white in tail than Water Pipit; bill lighter; legs yellowish.
Similarities
Vesper Sparrow is heavier, has short, thick bill; hops, does not walk. Horned Lark is heavier, has less white in tail in flight. Water Pipit is darker, more grayish, has unstreaked back, more streaks on breast, pumps tail more. Longspurs have thicker bills.
Habitat
Prairie, plains.
Habits
Male spends much time in air ascending in circles; often occurs singly or in pairs; flight more bouncy than Water Pipit's.
Voice
Famous flight song delivered at average height of 300 ft. (91 m), "As it flies around, its flight rises and falls. Each time it rises the bird sings; when it falls, he is silent . . . . The song is clear, sweet, and musical" (Saunders); call, single notes, harsher than paired notes of Water Pipit.
Eggs
4–5; grayish-white, thickly blotched with purplish-brown; 0.8 x 0.6 in. (2.0 x 1.5 cm). Nest is of grass, on ground.
Range
Breeds from cen. Man. s. to Minn. and Mont.; winters from Miss. and Tex. s. to Mexico.

WAXWINGS

Family Bombycillidae

Waxwings are slim, crested brown birds, with sleek plumage; black chins and foreheads; red, waxlike tips to their secondaries; and a yellow band across the end of their squared tails. They eat berries, especially cedar and mountain ash, and seeds and insects.

CEDAR WAXWING

Bombycilla cedrorum **36:4**

Description
Size, 6½–8 in. (16.5–20.3 cm). Above, reddish-brown; flanks yellow, undertail coverts white; • *crested.* Juvenal: duller, grayer with blurry streaking below.
Similarities
Throat black more restricted toward chin than in Bohemian; redder brown above.
Habitat
Open woods, edges, orchards.
Habits
Social; occurs in small flocks here today, gone tomorrow; flock alights on tree in compact body, birds remain upright, motionless, before feeding; sometimes catches insects in air, occasionally gets drunk on overripe berries; wing shape similar to Starling's; a late nester.
Voice
Thin, high hiss; once learned it is the best means of detecting their presence.

Eggs
3–6; often pale-bluish, spotted with blackish; 0.9 x 0.7 in. (2.3 x 1.8 cm). Nest is bulky cup of twigs, grass, moss on horizontal branch 4–40 ft. (1.2–12.2 m) up.
Range
Breeds from Cape Breton Is. and cen. Man. s. to Ga. and Ark.; winters from New England s. to Fla., Gulf Coast, and C. America.

BOHEMIAN WAXWING
Bombycilla garrulus **36:3**

Description
Size, 7½–8¾ in. (19.1–22.2 cm). Throat black; flanks gray; • *undertail coverts chestnut; yellow markings* on tail and wings, white streak in wings.
Similarities
Grayer above than Cedar Waxwing.
Habitat
Breeds in northwestern coniferous forests, flocks visit northern United States occasionally in winter.
Habits
Gregarious at all seasons, flight undulating in tight flocks; catches insects in air; may come to feeder for raisins.
Voice
"Alarm note *tzee-tzee*" (Forbush); no song.
Eggs
3–6; grayish, spotted with dark brown; 0.9 x 0.7 in. (2.3 x 1.8 cm). Nest of twigs, grass, in conifer, 8–50 ft. (2.4–15.2 m) up.
Range
Breeds chiefly in nw. N. America, but e. to ne. Man. and s. to s. Alta.; winters s. and e. of breeding range, to s. Ont.

SHRIKES
Family Laniidae

American shrikes are gray, black, and white predators with heavy heads, stout hooked beaks, black masks, and slim tails. They are birds of the open country, where there are scattered trees. They prey upon insects, small mammals, and birds, and impale their prey on thorns.

NORTHERN SHRIKE
Lanius excubitor **35:7**

Description
Size, 9–10¾ in. (22.9–27.3 cm). Less extensive than Loggerhead. • *Black mask;* feathers just over the bill are whitish; bill light below at base except in spring and summer. Adults: gray above, rump light, wings and tail black with some white on wings and outer tail feathers; underparts whitish, faint barring on breast in fall and winter. Immature: browner, plainly barred below, bill brown; this plumage worn through first winter.
Similarities
Loggerhead Shrike is darker above, whiter below, with black feathers over bill.
Habitat
Breeds in openings in northern forest; in winter, open country, swamps, orchards.

Habits
Bold, aggressive; perches on exposed observation post or wire like a Kestrel and watches for prey; sometimes hovers, occasionally pumps tail; flight slightly undulating; usually catches birds on the wing; this species more apt to be seen in northern part of our area and in winter; range may barely overlap with next only in winter.
Voice
Call is "a harsh shrieking *jo-ree*" (Knight); song prolonged, a Robin-like or Catbird-like carol with some harsh notes; also mimicking notes of other birds; both sexes sing.
Eggs
4–7; whitish, with darker markings; 1.1 x 0.8 in. (2.8 x 2.0 cm). Nest is of twigs and leaves in thorny bush or on limb of tree, 5–20 ft. (1.5–6.1 m) up.
Range
Breeds in n. Canada from Labr. w. to n. Man.; winters from s. parts of breeding range s. to Va. and Tex.

LOGGERHEAD SHRIKE
Lanius ludovicianus **35:5**

Description
Size, 8–10 in. (20.3–25.4 cm). • *Black mask* is complete; feathers just over the bill are black. Dark above, unbarred below; bill all-black; immature faintly barred on pale gray underparts.
Similarities
Slimmer Northern Mockingbird lacks face mask; Northern Shrike is larger.
Habitat
Farming country with scattered trees.
Habits
Similar to Northern Shrike; much more apt to be seen in southern part of eastern area; range may barely overlap with the Northern only in winter.
Voice
Harsh *chack-chack; tsirp-see, tsirp-see;* song somewhat resembles Gray Catbird's or Brown Thrasher's, but lower, harsher.
Eggs
4–8; white to greenish-gray, variously marked; 1.0 x 0.7 in. (2.5 x 1.8 cm). Nest is of rootlets, twigs, in bush or tree 5–20 ft. (1.5–6.1 m) up.
Range
Breeds locally throughout most of U.S. from Maritime Provinces and s. Man. s. to s. Fla., Gulf Coast, and Mexico; winters from Va. southward.

STARLINGS
Family Sturnidae

This is a large and varied Old World family of gregarious, sharp-billed, short-tailed birds, many looking like blackbirds. Only one introduced species occurs in East. They are omnivores, eating insects, worms, grubs, seeds, fruits.

STARLING
Sturnus vulgaris **42:3**

Description
Size, 7½–8½ in. (19.1–21.6 cm). • *Short-tailed,* chubby, black bird. Spring adult: iridescent black, white dots on back; strong, pointed

• *yellow bill*; squared tail. Fall adult: similar, but liberally covered with white spots, some brown spots; black bill. Juvenal: grayish, blurry whitish streaking on throat and belly, black bill.

Similarities

Young have longer bills, shorter tails than female cowbirds.

Habitat

Cities, suburbs, farmlands, edges, lawns, pastures; even deserts.

Habits

Gregarious, persistent, wary, aggressive; comes to feeders; walks or waddles energetically and erratically on ground; collects in great flocks in cities, suburbs, and marshes; dense flocks in air perform intricate revolutions in concert; looks triangular or spindle-shaped in flight with short, pointed, fast-beating wings, alternately flapping and sailing; often flocks with Red-winged Blackbirds and other American blackbirds; usually feeds on ground.

Voice

Varied, including whistles, squeaking, squealing, harsh and grating notes; mimics many other species.

Eggs

4–7; pale blue; 1.2 x 0.9 in. (3.0 x 2.3 cm); 2 broods. Nest is of sticks, grass, in hole in tree or cactus, or other natural or man-made cavity, 10–25 ft. (3.0–7.6 m) up.

Range

Resident throughout much of Canada and most of U.S., from Que., s. to Gulf Coast and Mexico; introduced from Europe in 1890, now well established.

VIREOS

Family Vireonidae

Vireos are sparrow-sized arboreal birds, olive-gray above, white or yellowish below, with thick, slightly hooked bills. They are larger than warblers and less active. Most are quite deliberate in their movements as they creep about the outer branches gleaning insects; hence, they are more often heard than seen. Vireos all build dainty cuplike nests suspended between forked twigs; the male helps in incubation and sings on the nest. Their diet consists of insects.

VIREO COMPARISON CHART

Species	*Wing Bars*	*Breast*	*Eye Markings*
Red-eyed	no	white	black and white line
Warbling	no	white	light line
Philadelphia	no	yellowish	light line
Bell's	faint	white	white spectacles
Solitary	yes	white	white spectacles
White-eyed	yes	white	yellow spectacles
Yellow-throated	yes	yellow	yellow spectacles

WHITE-EYED VIREO

Vireo griseus **40:18**

Description

Size, 6 in. (15.2 cm). Only vireo with 2 white wing bars, • *white breast*, and • *yellow spectacles*. Greenish-olive below, eyes white, flanks pale yellow; immature has dark eyes.

Similarities

Bell's is smaller, has faint wing bars, dark eyes, and faint white spectacles.

Habitat
Thickets, often near water.
Habits
Aggressive, restless, noisy, inquisitive, fearless; but also shy, hard to see, retiring; pumps tail; often victimized by cowbird.
Voice
Emphatic *"Chick! ticha WHEEyo chick!"* (Saunders); and many variations of these, some ventriloquial; also notes mimicking other birds; a harsh mew, loud whistle, and a short *tik*.
Eggs
4; white, faintly dotted with purple and brown; 0.8 x 0.6 in. (2.0 x 1.5 cm). Nest is in shrub, 2–8 ft. (0.6–2.4 m) up.
Range
Breeds throughout much of se. U.S. from cen. New England and Nebr. s. to s. Fla. and Mexico; winters from Gulf States s. to W. Indies and C. America.

BELL'S VIREO
Vireo bellii **40:15**

Description
Size, 4½–5 in. (11.4–12.7 cm). Nondescript. Olive above, rump brighter, white below, faint white eye-ring, 1 or 2 pale wing bars, pale yellowish on sides.
Similarities
See White-eyed Vireo. Warbling lacks any eye-ring and wing bars. Ruby-crowned Kinglet has shorter bill, distinct eye-ring.
Habitat
Bottomland shrubbery, mesquite thickets, hedgerows.
Habits
Similar to White-eye; active, keeps out of sight, but inquisitive, fearless about nest.
Voice
Song is *"whillowhee, whillowhee, WHEE"* (Nice); this phrase is repeated, first rising as if asking a question, then falling as if answering it; also a hoarse scolding note; call, not as harsh as White-eyed's.
Eggs
3–5; white, sparingly spotted with brown; 0.7 x 0.5 in. (1.9 x 1.4 cm). Nest is low in briar patch.
Range
Breeds from Ind. and S.Dak. s. to Tex. and C. America; winters in C. America.

YELLOW-THROATED VIREO
Vireo flavifrons **40:16**

Description
Size, 6 in. (15.2 cm). Greenish above, grayer on rump and tail; • *yellow throat* and spectacles, white belly.
Similarities
Pine Warbler is more slender, has no yellow spectacles, and frequents pines.
Habitat
Shade trees, orchards.
Habits
Sluggish; feeds slowly, deliberately, climbing about upper branches.
Voice
Song is faintly reminiscent of American Robin, includes as its main

phrase a distinctive nasal *three-eight;* much deeper and less continuous than Red-eyed's; also a scolding note.

Eggs

3–4; white, spotted with purplish-brown; 0.9 x 0.6 in. (2.3 x 1.5 cm). Nest is in tree, 3–50 ft. (0.9–15.2 m) up.

Range

Breeds throughout much of e. U.S. from s. New England and Minn. s. to Fla. and Gulf states; winters from Mexico to S. America.

SOLITARY VIREO

Vireo solitarius

40:13

Description

Size, 5–6 in. (12.7–15.2 cm). Olive-green back contrasts sharply with • *blue-gray head;* • *white throat, spectacles,* wing bars, and underparts; sides yellowish; western races lack bluish cast on head.

Habitat

Coniferous and mixed open forests and edges.

Habits

Tame, sedate, an early migrant; wanders about when singing, unlike Red-eyed, which moves little; commonly victimized by cowbird.

Voice

Song is in character between that of Red-eyed and Yellow-throated, sweeter than Red-eyed's, with longer pauses between phrases and including a rising *tu-wee-tu;* and a scolding note.

Eggs

3–5; white, dotted with red-brown and umber; 0.8 x 0.6 in. (2.0 x 1.5 cm). Nest is in tree, 5–12 ft. (1.5–3.7 m) up.

Former name

Blue-headed Vireo.

Range

Breeds from Newfoundland and Man. s. to Conn. and Mich.; also into mountains of Ga.; winters from N.C. and Gulf Coast s. to C. America.

RED-EYED VIREO

Vireo olivaceus

40:14

Description

Size, 5½–6½ in. (14.0–16.5 cm). A "whiskerless" vireo with • *black and white eye stripe.* Upperparts olive, grayer on crown; white below; • *no wing bars;* red eye inconspicuous at a distance.

Similarities

Warbling is smaller, grayer above and lacks black eye stripe. Philadelphia is yellowish below, dark eyed. Tennessee Warbler has thinner bill, lacks conspicuous black eye stripe.

Habitat

Virtually any woodland or shade trees.

Habits

Movements deliberate, but less sluggish than Yellow-throated and Solitary; sings through heat of day; often parasitized by cowbird; very common and widespread.

Voice

Song is a continuing repetition of little phrases, remotely Robin-like, *now you see me, now you don't, here I am, there I go;* or *"Hello, hello, Are you there? Can you hear me? This is Vireo. Yes, yes, Vireo, Vireo"* (Taylor). Alarm, a descending, catlike *kway.*

Eggs
3–4; white, sparingly spotted with blackish; 0.9 x 0.6 in. (2.3 x 1.5 cm). Nest is in sapling or tree, 4–50 ft. (1.2–15.2 m) up.
Range
Breeds throughout much of e. U.S. from Gulf of St. Lawrence and Ont. s. to Fla., Gulf Coast, and Tex.; winters in S. America.

Note: The **BLACK-WHISKERED VIREO,** *Vireo altiloquus* **(40:17)**, size, 5 in. (12.7 cm), has extended its range from southern to northern Florida recently. It very closely resembles the Red-eyed, but it has a fine, black whisker mark; its song is like that of Red-eyed, but phrases paired.

PHILADELPHIA VIREO

Vireo philadelphicus **40:12**

Description
Size, 4½–5½ in. (11.4–12.7 cm). • *Yellowish breast* and • *no wing bars.* Olive above, light white line over eye, dark spot between bill and eye, chin and belly whitish.
Similarities
See Red-eyed Vireo. Warbling Vireo is whitish below, but sometimes has yellowish wash on sides; also, Warbling has light, not dark, spot between bill and eye. Tennessee and Orange-crowned Warblers sometimes similar in color, but more slender, with thin bill and warblerlike build and actions. Female Black-throated Blue Warbler has small, white square on wing.
Habitat
Edges of woods, low trees, and shrubs, often near water.
Habits
Like Warbling Vireo; tame, when feeding, rather active for a vireo, sometimes flutters in front of leaves, hangs upside down in twig, or catches insects in air; uncommon, but frequently overlooked, in migration in East.
Voice
Similar to Red-eyed's, but a trifle higher, less often repeated, and containing frequently a characteristic abrupt, rising 2-syllable note.
Eggs
4; white, sparingly spotted with umber; 0.8 x 0.5 in. (2.0 x 1.3 cm). Nest is in shrub or tree, 9–40 ft. (2.7–12.2 m) up.
Range
Breeds in s. Canada, s. to n. U.S.; winters in C. America.

WARBLING VIREO

Vireo gilvus **40:10**

Description
Size, 4½–5½ in. (11.4–12.7 cm). Olive-gray above, white spot between bill and eye, sometimes a yellowish wash on sides, whitish breast, • *no wing bars.*
Similarities
See Red-eyed and Philadelphia Vireos.
Habitat
Shade trees and large forest-edge trees, often in upper branches.
Habits
Sings much, but not as consistently as Red-eyed; often parasitized by cowbird.
Voice
Pleasant, continuing warble, somewhat reminiscent of Purple Finch, but softer, more languid; call, a snarl, *myee.*

Eggs
3–5; white, often spotted with red-brown; 0.7 x 0.5 in. (1.9 x 1.4 cm). Nest is in tree, 20–70 ft. (6.1–21.3 m) up.
Range
Breeds throughout much of U.S., from N.E. and Man. s. to N.C., La., and Mexico; winters in C. America.

WOOD WARBLERS

Family Parulidae

Wood Warblers are smaller and more slender than sparrows, with thin, straight, unhooked bills and wings usually longer than the tail. Most are active and restless, seeking insects in diverse habitats. They do not warble, but have distinctive songs. All wood warblers, except the Yellow-rumped and Palm, are normally transients or summer visitors to the eastern area, although an occasional Yellow-breasted Chat or Common Yellowthroat may linger into the winter. Warblers are sometimes confused with vireos, which average larger and have heavier bills and sluggish movements; or with kinglets, which are smaller, shorter-tailed, and rounder and given to flirting their wings.

Many warblers in fall have a plumage quite different from that of spring; immatures may also differ. Illustrated and described are the principal plumages. For greatest help in identification, check both text and illustration. Occasionally, a bird in molt fails to match either and may offer real problems in identification.

BLACK-AND-WHITE WARBLER

Mniotilta varia **38:11**

Description
Size, 4½–5½ in. (11.4–12.7 cm). • *Black and white,* with a striped crown. Male: • *striped black and white;* 2 white wing bars. Female and immature: paler, more white below, lacking black throat.
Similarities
See Blackpoll Warbler.
Habitat
Woodlands, preferably deciduous, moist or dry.
Habits
Creeps along trunks and branches like a nuthatch, head up or down; sometimes catches insects in air.
Voice
Thin, wiry, repeated *zee, zee, zee,* or *we-see, we-see, we-see,* spiraling upward, other similar songs; call, a short *pip,* a louder *jink.*
Eggs
4–5; creamy white, well spotted with dark brown; 0.7 x 0.5 in. (1.8 x 1.3 cm). Nest is of grasses and rootlets, on ground at foot of tree, stump, or rock.
Range
Breeds throughout most of e. U.S. from s. Canada s. to s. U.S., e. of Rockies; winters from s. Gulf states s. to S. America.

PROTHONOTARY WARBLER

Protonotaria citrea **39:9**

Description
Size, 5½ in. (14.0 cm). Bill long (for a warbler) and slightly curved. Male: olive back; • *blue-gray wings,* rump, and tail, latter

with white on outer feathers; no wing bars. Unmarked • *golden head and breast.* Female and immature: duller.

Similarities
Smaller Yellow Warbler is lighter above, tail corners yellow.

Habitat
Swamps, moist bottomlands, edges of watercourses.

Habits
Active, tame, low-ranging; creeps nuthatchlike about trunks; flight swift, direct, like a waterthrush.

Voice
Song is a loud, clear *tweet-tweet-tweet* of 7–12 notes, all on same key, reminiscent of Solitary Sandpiper; also a low, sweet flight song, *"che-wee—che-wee chee-chee, chee-chee-che-wee—che-wee"* (Walkinshaw), given while hovering; call, a soft chip, alarm note sharp, like Louisiana Waterthrush.

Eggs
4–6; creamy white, often liberally marked with brown and purple; 0.7 x 0.5 in. (1.9 x 1.4 cm). Nest is in woodpecker hole, cavity, or bird box, 1–30 ft. (0.3-9.1 m) up.

Range
Breeds throughout much of e. U.S., from N.Y. and Minn. s. to Fla. and Gulf Coast; winters from Mexico to S. America.

SWAINSON'S WARBLER

Limnothlypis swainsonii **37:10**

Description
Size, 6 in. (15.2 cm). Bill rather long. Adult: underparts and • *line over eye whitish;* brownish-olive, unstreaked above; red-brown crown; no wing bars. Immature: olive above, yellowish eye stripe and underparts, red-brown wings.

Similarities
Worm-eating has black stripes on crown. Waterthrushes are streaked below. Palm Warbler has red-brown cap, but has streaks on sides, some yellow below, and wags tail.

Habitat
Coastal swamps; rhododendron-clad Appalachian hillsides, bottomlands, and swamps, up to 3000 ft. (914 m).

Habits
Low-ranging; walks like an Ovenbird on ground, sings from perch 1–20 ft. (0.3–6.1 m) up; elusive, more often heard than seen.

Voice
Song is loud, ringing, suggesting Louisiana Waterthrush but shorter, *"whee, whee, whee, WHIP-poor-WILL"* (Brooks); call, a clear *churp.*

Eggs
3; bluish-white; 0.8 x 0.6 in. (2.0 x 1.5 cm). Nest is bulky, of leaves, pine needles, in cane or bush, 2–10 ft. (0.6–3.0 m) up.

Range
Breeds in se. U.S. from Md. and Ill. s. to Fla. and Gulf Coast; winters in Caribbean, southward.

WORM-EATING WARBLER

Helmitheros vermivorus **37:9**

Description
Size, 5 in. (12.7 cm). • *Head buff,* with 4 bold • *black stripes;* rest of upperparts brownish-olive; underparts buffy white; no wing bars; sexes similar.

Similarities
Lacks golden crown and spotted breast of Ovenbird.
Habitat
Dry, wooded hillsides; damp woods; ravines.
Habits
Low-ranging, shy; bobs head as it walks on ground with tail high, often sings from stub or sapling, sometimes clambers about trunk of tree like Black-and-White.
Voice
Song is dry trill somewhat like a Chipping Sparrow's but faster, buzzier; also a goldfinchlike flight song; call, a sharp *pip,* like a Black-and-White's; alarm, a *chip-chip-chip.*
Eggs
4–5; white, spotted with brown and lavender; 0.7 x 0.5 in. (1.8 x 1.3 cm). Nest is of leaves, on steep hillside in dense underbrush.
Range:
In se. U.S. from Mass. and Iowa s. to Fla. and Gulf Coast; winters in W. Indies and from Mexico to C. America.

GOLDEN-WINGED WARBLER

Vermivora chrysoptera **39:7**

Description
Size, 4½–5 in. (11.4–12.7 cm). Spring male: • *black bib* and ear patch; • *gold on crown and wings;* rest of upperparts blue-gray; underparts white. Spring female: like male, but bib and ear patch dull gray. Fall adults and immature: greener, duller; black or gray areas obscured.
Habitat
Bogs, wet woodland edges; also wet or dry hillside thickets.
Habits
Tame; feeds from low shrubbery to tall treetops, often hanging upside down from end of twig.
Voice
Song is *see, buzz-buzz-buzz,* sometimes 3 or 5 notes; similar song of Blue-winged normally has 2 notes (but these 2 species occasionally interchange songs); also a *"th-th-th-th-th-th-three"* (Tyler); alarm note, a sharp *chip.*
Eggs
3–5; white, speckled with brown; 0.7 x 0.5 in. (1.8 x 1.3 cm). Nest is bulky, of leaves, bark strips, in tussock of grass, goldenrod, bush.
Range
Breeds from N.H. and Minn., s. to N.J. and Iowa; also to mountains of Ga.; winters from Mexico. s. to S. America.

Note: The Golden-winged and Blue-winged Warblers overlap and hybridize frequently in a very complex fashion. The hybrids, two of the more typical forms, "Lawrence's" and "Brewster's" **(39:8, 39:10),** are variable, mixing colors of the parental forms, except that they either have the face and throat patch of the Golden-winged, or lack it completely as the Blue-winged. Blue-winged-like hybrids with a face and throat patch, usually found where Blue-wingeds are commoner, are known as **LAWRENCE'S WARBLER**, originally described as a distinct species, *Vermivora lawrencei.* The more common hybrids lacking the face and throat patch but tending more or less toward the Golden-winged in other features are called **BREWSTER'S WARBLER,** also originally described as a species, *Vermivora leucobronchialis.* The hybrid males may sing like Blue-winged or Golden-winged, or sing both songs. Hybrids have habits like the two parental forms, and may be expected wherever Golden-wingeds or Blue-wingeds occur.

BLUE-WINGED WARBLER

Vermivora pinus **39:12**

Description
Size, 4½–5 in. (11.4–12.7 cm). Spring male: bright yellow head and underparts, olive-green back, blue-gray wings with 2 usually • *white wing bars*, gray tail with some white, white undertail coverts, • *black line through eye.* Spring female, fall adults, and immature: duller, crown more olive, wing bars white to yellowish.
Habitat, Eggs
Same as Golden-winged.
Habits
Forages near ground, sings from exposed branch, 12–30 ft. (3.7–9.7 m) up, motions somewhat deliberate.
Voice
Song, a buzzy *seee-buzzzz,* second note lower (similar song of Golden-winged normally has 4 notes); also a "longer song, *wee-chi-chi-chi-chi, chur, chee-chur*" (Chapman).
Range
Breeds from cen. New England and Nebr. s. along Atlantic Coast states to S.C. and inland to s. Ill.; winters from Mexico to C. America.

BACHMAN'S WARBLER

Vermivora bachmanii **39:15**

Description
Size, 4¼ in. (10.8 cm). Male: • *black throat and crown patch;* gray nape, wings and tail; back olive, tips of outer tail feathers white; eye-ring, breast and belly yellow; no wing bars. Female: similar, but black replaced by gray on crown, yellow on throat; eye-ring and bend of wing yellow. Immature: male, less black on throat; female, less black, and back grayer.
Similarities
In the larger male Hooded a complete black hood joins black crown and throat. Female Nashville has more gray on head and white eye-ring. Female Wilson's, Yellow-throated and Hooded lack yellow eye-ring of female Bachman's and have olive, not gray, crown; also female Hooded is larger and has white tail corners.
Habitat
Heavily wooded swamps.
Habits
Very active; forages somewhat like a Parula from 2 to 40 ft. (0.6–12.2 m) up; sometimes hangs upside down like a chickadee; often flies hundreds of yards between feeding spots; sings from tops of tall trees, seen there also in migration; elusive, very hard to find. Very rare.
Voice
Song is "*zee zee zee zee zee see zee zee chew*" (Borror), *zee*'s somewhat similar to those of Golden-winged; "wiry and insect-like, strongly suggesting a 'listless' Parula Warbler . . . ventriloquial." "*Zrr, zrr, zrr, zrr, zrr, zrr* [sometimes ending in a] *chwit*" (Barnes).
Eggs
3–4; white; 0.6 x 0.5 in. (1.5 x 1.3 cm). Nest is of grass and leaves, in bush or cane, 2–5 ft. (0.6–12.2 m) up.
Range
Breeds locally in se. U.S., in S.C., Ala., Ark., and Mo.; winters in Cuba and Isle of Pines.

ORANGE-CROWNED WARBLER

Vermivora celata **37:7, 40:9**

Description
Size, 4½–5½ in. (11.4–14.0 cm). Spring adult: perhaps the most nondescript spring warbler; olive-green above, yellowish-green below; pale eye stripe; • *faint stripes on flank;* hidden patch of orange in crown; yellow undertail coverts; no wing bars. Fall adult and immature: similar but duskier, often quite grayish, and hardly any paler below; crown patch absent or not visible.
Similarities
Nashville is yellow below, unstreaked. See fall Tennessee. Philadelphia Vireo is larger, sluggish, has thicker bill, light stripe over eye, no side streaks, and is an early fall migrant.
Habitat
Open woodlands with heavy brush, especially on slopes or near water.
Habits
Forages low. Rare in East in spring, less uncommon in late fall (October).
Voice
Song is somewhat like a Chipping Sparrow's, a series of 18–22 *si* notes, dropping slightly in middle, rising and fading toward end; note, a distinctive *chip.*
Eggs
4–6; white, dotted with reddish-brown; 0.7 x 0.5 in. (1.8 x 1.3 cm). Nest is of grass, rootlets, on ground in shrubbery on brushy (or other) hillside.
Range
Breeds from Labr. and Man. s. to Tex.; winters from S.C. and Ark. s. to C. America.

NASHVILLE WARBLER

Vermivora ruficapilla **39:11, 40:5**

Description
Size, 4–5 in. (10.2–12.7 cm). Spring male: olive back, chestnut crown patch (often concealed); gray head, undertail coverts whitish, • *yellow underparts, white eye-ring;* no wing bars, no white in tail; female duller. Fall adults and immature: duller; yellow below but with color of head and back blending; immature quite brownish above.
Similarities
Connecticut, Mourning, and MacGillivray's are somewhat similar in color but are larger and throats not yellow.
Habitat
Breeds in northern bogs and second growth, up to 15 ft. (4.6 m) high; in migration, edges of woods, orchards, shade trees.
Habits
Active, forages at all levels; relatively early spring and late fall migrant.
Voice
Song is *"see it see it see it see it, ti-ti-ti"* (Gunn); note, a *chip.*
Eggs
3–5; white, dotted with red-brown; 0.6 x 0.5 in. (1.5 x 1.3 cm). Nest is of grass, leaves, concealed in or under tussock on ground.
Range
Breeds from N.S. and Man. s. to Conn. and Minn.; winters from Fla. and Mexico, s. to C. America.

TENNESSEE WARBLER

Vermivora peregrina **37:8, 40:11**

Description
Size, 4½–5 in. (11.4–12.7 cm). Spring adult male has gray crown, • *white line over eye,* and dark line through it, rest of upperparts greenish, white below; female has crown more olive, underparts somewhat yellowish. Fall adult and immature: bright greenish above, unstreaked yellowish below, yellowish line over eye (grayer above, whiter below in male), pale wing bars, undertail coverts white (sometimes with a trace of yellow).

Similarities
Fall Nashville has white eye-ring, yellow undertail coverts. Fall Orange-crowned is darker, dingier, with faint streaks on sides, yellow undertail coverts, no line over eye; occurs in late (not early) fall. Vireos are larger, have heavier bill, and are sluggish. Red-eyed has more conspicuous eye lines.

Habitat
Breeds in openings in northern coniferous forest; in migration, woodlands, edges, shade trees, brush.

Habits
Restless; prefers upper branches in spring, any height in fall; on breeding grounds sings from outer branch of top of tree 5–30 ft. (1.5–9.1 m) up.

Voice
Song is loud, in 3 parts, last somewhat like a Chipping Sparrow's; *tenne-tenne-tenne-tenne, chip-chip, ssee-ssee-ssee-ssee-ssee;* or *"tsip-pit-tsip-PIT, tsip-PIT, tsip-PIT, tsip-pit-tsee, tsee, tsee, tsee,* more rapid and higher pitched toward end" (Harrison); reminiscent of Nashville's song but latter's is only 2-parted.

Eggs
4–7; creamy white, with red-brown spots; 0.6 x 0.5 in. (1.5 x 1.3 cm). Nest is cup-shaped, of vegetable fibers and grass, close to ground in conifer.

Range
Breeds from Labr. and Man. s. to Maine and Wis.; winters from Mexico to S. America.

NORTHERN PARULA

Parula americana **37:5, 41:12**

Description
Size, 4½ in. (11.4 cm). Smallest warbler; 2 white wing bars. Spring male: • *Yellow throat,* breast, and eye-ring; greenish-yellow patch on back, • *blue above with a chestnut breastband.* Spring female: blue above, yellow below, breastband narrower or missing. All plumages are duller in fall.

Habitat
Breeds in swamps, wood edges, orchards, and parks where trees bear *Usnea* lichen (gray moss) or Spanish moss; in migration, any trees or shrubs.

Habits
Tame; sometimes hangs head down at end of limb, forages more often at medium and upper levels, often sings from treetops.

Voice
Song is a rising, sizzling trill, *zzzzzzzz-ZIP;* or a *bz-bz-bz-bz-zzzzzz-ZIP;* call, a *chip.*

Eggs
4–5; white, wreathed at large end with brown spots; 0.7 x 0.5 in. (1.8 x 1.3 cm). Nest is gourd-shaped, of *Usnea* or Spanish moss, near end of branch, 5–30 ft. (1.5–9.1 m) up.

Range
Breeds throughout much of e. U.S. from se. Canada s. to Gulf of Mexico; winters from s. Fla. s. to W. Indies and C. America.

YELLOW WARBLER

Dendroica petechia **39:6, 40:1**

Description
Size, 4½–5¼ in. (11.4–13.3 cm). • *All-yellow.* Male: darker yellow above, lighter below, chestnut streaks along sides, • *yellow wing bars*, slightly duller in fall. Female and immature: similar but darker above, paler below, and fewer streaks on sides. Alaskan race in fall is duller, grayer above.
Similarities
Fall immature Yellow is almost as dusky as Orange-crowned, but has yellow tail corners. American Goldfinch has black wings and tail. Female Wilson's has all-dark tail.
Habitat
Swamps, shrubbery, willows or alders near water; orchards, gardens.
Habits
Tame, a low forager; an early fall migrant; frequently victimized by Brown-headed Cowbird, but often builds new floor in nest over egg of Cowbird so as to incubate only its own eggs.
Voice
Song is *sweet, sweet, sweet, I'm so sweet,* with a goldfinchlike ending; call, a loud *chip.*
Eggs
3–5; greenish-white, marked with brown and purple; 0.7 x 0.5 in. (1.8 x 1.3 cm). Nest is cup-shaped, of silver-gray plant fibers in fork of low bush or sapling.
Range
Breeds from Newfoundland and n. Que. s. to S.C. and Tex. and s. to S. America; winters from Mexico to S. America.

MAGNOLIA WARBLER

Dendroica magnolia **38:2, 41:10**

Description
Size, 5 in. (12.7 cm). From below, • *tail looks white with black band at end.* Spring male: gray crown and wings; yellow rump and underparts; black ear patch, back, and streaks on breast and sides; 2 white wing bars. Spring female and fall adults: similar but duller, back grayer, side streaks less pronounced. Immature: even duller; olive above, side streaks faint but has white eye-ring and tail pattern of adult.
Similarities
Fall Nashville lacks yellow rump and white tail band.
Habitat
Breeds principally in coniferous forest; in migration, woods, edges, gardens.
Habits
Active, a low forager; often droops wings, spreads tail.
Voice
Song is short, variable; W. W. G. Gunn records 8 different varieties on his warbler song record; one common form he renders as *chew-chew-chew-chew wit-see,* another as *wit wit witty swit;* call, a sharp *chip.*

Eggs
4–5; white, spotted with brown and lavender; 0.7 x 0.5 in. (1.8 x 1.3 cm). Nest is cup-shaped, of twigs, grass, on limb of conifer, 3–35 ft. (0.9–10.7m) up.
Range
Breeds from cen. Canada s. to ne. U.S. and into mountains of Va.; winters in W. Indies and C. America.

CAPE MAY WARBLER
Dendroica tigrina **38:3, 41:7**

Description
Size, 5–5½ in. (12.7–14.0 cm). Spring male: olive with black streaks above, • *chestnut cheek patch,* yellow with black streaks below, side of neck and rump yellow, wing patch and tail corners white. Spring female and fall adults: much duller; cheek patch gray, light eye stripe, • *dull yellowish patch behind ear,* 2 narrow white wing bars, streaks fainter, underparts white. Immature: similar to female, back not streaked, wing bars buffy, rump duller yellow, sparsely streaked below with whitish belly.
Similarities
Pine has fainter side streaks, lacks yellow rump. Fall Yellow-rumped is browner above with a streaked back, yellow undertail coverts, and virtually no wing bar.
Habitat
Breeds in northern coniferous forests; in migration seeks edges, shrubbery.
Habits
Sings from treetops, often at extreme pinnacle of spruce or fir; catches some insects in air; found in migration at all levels.
Voice
Song is a thin, high *see-see-see-see-see;* also, less commonly, a series of 6–8 *to-be*'s preceding a *see-see-see;* call, a faint *chip.*
Eggs
4–9; greenish-white, dotted with browns and lilac; 0.7 x 0.5 in. (1.8–1.3 cm). Nest is large for warbler; of sphagnum moss, in end clump of vegetation, near top of spruce or fir, 30–60 ft. (9.1–18.3 m) up.
Range
Breeds from Que. and Man. s. to N.S., Maine, N.Y., Mich., and N.Dak.; winters from W. Indies s. to C. America.

BLACK-THROATED BLUE WARBLER
Dendroica caerulescens **37:4, 40:7**

Description
Size, 5 in. (12.7 cm). Small • *white wing spot* in all adult plumages. Male: • *throat and sides black,* contrasting sharply with • *white underparts;* duller, but essentially similar in fall. Female: very different; brownish above, buffy below with dark cheek and light line over eye; immature fall females lack white wing spot.
Similarities
Fall Tennessee is much greener above, yellower below; has yellowish line over eye; lacks dark cheek and wing spot.
Habitat
Breeds in deciduous or mixed woodlands with underbrush, especially on slopes; in migration, shrubs, trees, gardens, orchards.
Habits
Tame, movements deliberate; while perched often partly opens its wings; forages at all levels, sometimes catches insects in air.

Voice
Song is a hoarse, buzzy, deliberate *bee-tee-bee* (BTB, its initials), or *zero, zero, zeee,* last note in each case rising.
Eggs
4–5; whitish, marked with browns and lavender; 0.7 x 0.5 in. (1.8 x 1.3 cm). Nest is cup-shaped, of bark shreds, rootlets in bushes.
Range
Breeds from se. Canada to ne. U.S. and into mountains of n. Ga.; winters s. of U.S.

YELLOW-RUMPED WARBLER
Dendroica coronata **38:1, 41:8**

Description
Size, 5–6 in. (12.7–15.2 cm). Spring male: rump, sides, crown, and in West, throat, bright yellow; blue-gray above, white below; heavily streaked with black above; black cheek patch, breastband, and heavy streaking on sides; 2 white wing bars east, or white wing patch in west; white tail corners. Spring female, immature and fall adults: gray replaced by brown, less yellow on sides, markings below fainter.
Similarities
Magnolia and spring Cape May have yellow rumps, but are yellow below.
Habitat
Breeds in northern coniferous forest; in migration, almost anywhere, in trees, shrubs, even on ground; common in winter in coastal bayberry bushes.
Habits
Visits feeders, catches insects in air; flight zigzaggy, jerky; an early migrant, winters in southern part of our area.
Voice
Song is a thin, colorless jumble, weaker in spring, a stronger rising or falling trill in summer; call, a highly characteristic *chip.*
Eggs
3–5; white, spotted brown and purple; 0.7 x 0.5 in. (1.8 x 1.3 cm). Nest is cup-shaped, rather bulky, of bark shreds, twigs, in conifer, 3–40 ft. (9.1–12.2 m) up.
Remarks
Formerly Myrtle and Audubon Warblers, now recognized as single species.
Range
Breeds from cen. Que. and n. Man. s. to Maine, Mass., n. N.Y., and Mich.; winters from s. parts of breeding range s. to C. America.

BLACK-THROATED GREEN WARBLER
Dendroica virens **38:6, 41:4**

Description
Size, 4½–5 in. (11.4–12.7 cm). Contrasting • *yellow cheeks* in all plumages; olive above; 2 white wing bars. Spring male: • *black throat* and streaks on sides, white below and on tail. Spring female and fall male: similar, but duller and no black throat, back unstreaked.
Similarities
Fall Blackburnian has a streaked back, darker cheek patch, yellowish breast.
Habitat
Breeds in northern coniferous or mixed forest, or in cypress swamp in south; in migration, prefers conifers.

Habits
Forages at all levels, but usually well up; migrates across Gulf of Mexico.
Voice
Song is a dreamy *zee-zee-zee-zoo-zee;* or *please, please, pass the cheese,* second and third notes from end lower.
Eggs
4–5; white, marked with various shades of brown; 0.6 x 0.5 in. (1.5 x 1.3 cm). Nest is cup-shaped, of twigs, birchbark, bound with spider webs, on limb of conifer, 15–70 ft. (4.6–21.3 m) up.
Range
Breeds from Newfoundland and Ont. s. to N.J., Ohio, and Minn., and into mountains in Ga.; winters from Fla. and Tex. s. to S. America.

Note: The **BLACK-THROATED GRAY WARBLER,** *Dendroica nigrescens* **(38:9),** size, 4½–5 in. (11.4–12.7 cm), regularly wanders eastward somewhat from its western breeding range. It is black, white, and gray; the head and throat pattern of the male closely resembles that of the Black-and-white Warbler, but its back is gray with fine black streaks, not black-and-white streaked.

CERULEAN WARBLER
Dendroica cerulea **37:6**

Description
Size, 5 in. (12.7 cm). Male: sky-blue above, white below; black streaks on back and sides; • *blue-black breastband* (incomplete in fall); dusky cheek patch; 2 wing bars and spots on tip of tail white. Female: greenish above with bluish tinge especially on crown; yellowish wash below; pale streak over eye, dark line through it; no black streaks on back or sides; no breastband. Immature: similar, but duller, more olive-green above.
Similarities
Female resembles female Black-throated Blue, but has 2 wing bars, not single white wing spot. Immature resembles fall Blackpoll, but is brighter green above and with more distinct eye stripe. Immature Northern Parula is deeper yellow below, no line over eye.
Habitat
Large trees in relatively open situations, as along river valleys.
Habits
Keeps to treetops, sometimes catches insects in air, or creeps along branches like Black-and-White; migrates across Gulf of Mexico.
Voice
"The songs of the Parula and Cerulean are very similar but . . . the pattern is reversed. . . . the Parula's song is *bzz, bzz, bzz, trill,* while that of the Cerulean is *trill, trill, trill, bzz.* The Cerulean's song can be expressed by the phrase *Just a little sneeze!*" (Murray); note, a sharp *chip.*
Eggs
3–4; white, spotted with red-brown; 0.7 x 0.5 in. (1.8 x 1.3 cm). Nest is shallow, cup-shaped, of grasses, bound with spider webs, decorated with lichens, suspended in fork of branch, 20–90 ft. (6.1–27.4 m) up.
Range
Breeds from N.Y. and Great Lakes s. to se. U.S. and cen. states; winters in S. America.

BLACKBURNIAN WARBLER

Dendroica fusca **38:5, 41:6**

Description
Size, 4½–5½ in. (11.4–14.0 cm). Spring male: • *flaming orange throat* and breast with black streaks on sides, orange on crown and sides of head, back black with white streaks, large white wing patch, white tail corners. Spring female and fall adults: similar but duller and paler, 2 white wing bars. Immature: even duller; similar pattern in brownish and dull yellow, 2 white wing bars.
Similarities
Immature resembles immature Black-throated Green, but has dark ear patch, light stripes on back, and yellow throat.
Habitat
Breeds in northern and mountain forest, especially in conifers.
Habits
Active, forages well up in trees, sings from treetops; but found at all levels in migration.
Voice
Song is a high, thin, wiry *tsee-tsee-tsee-tsee-tsee-zi-zi-zi-zi;* call, a faint *tseeck.*
Eggs
4–5; blue-green to whitish, spotted with lilac and brown; 0.7 x 0.5 in. (1.8 x 1.3 cm). Nest is of twigs in conifers near end of limb, 10–80 ft. (3.0–24.4 m) up.
Range
Breeds from se. Canada s. to ne. U.S., and s. in mountains to Ga.; winters in S. America.

YELLOW-THROATED WARBLER

Dendroica dominica **38:7**

Description
Size, 5½ in. (14.0 cm). Bill rather long. Spring adult: • *Yellow bib;* unstreaked • *gray back;* white wing bars; forehead, cheeks, and stripes along sides black; line over eye, underparts, and tail corners white. Fall adult: similar, but washed with brownish above. Immature: still browner above, throat paler yellow.
Similarities
Female Blackburnian has yellow stripes on head.
Habitat
Bottomlands with Spanish moss; pines.
Habits
Creeps along branches like Black-and-White, forages in middle and upper levels, movements somewhat deliberate; likes to bathe.
Voice
Song is quite suggestive of Indigo Bunting, especially in opening phrases, "*tee-ew, tew tew tew tew—tew wi* or [western race] *see-wee, see-wee, see-wee, swee swee swee swee* . . . [or] *sweetie sweetie, sweetie*" (Sprunt).
Eggs
3–4; whitish, with dark dots and spots; 0.7 x 0.5 in. (1.9 x 1.4 cm). Nest is of twigs, bark strips, on branch of pine, oak, or sycamore, 10–120 ft. (3.0–36.6 m) up.
Range
Breeds from s. N.J. and Ill. s. to cen. Fla., Gulf states, and Tex.; winters from Gulf states s. to C. America.

CHESTNUT-SIDED WARBLER

Dendroica pensylvanica **37:3, 41:14**

Description
Size, 4½–5¼ in. (11.4–13.3 cm). Spring male: • *yellow crown* and black mark below eye, back olive with black stripes; • *sides chestnut*; 2 yellow wing bars; cheeks, underparts, and tail corners white. Spring female and fall adults: similar but duller, with less chestnut and black. Immature: bright lemon-green above, white below; white eye-ring; gray cheeks; buffy wing bars.
Similarities
Spring Bay-breasted has chestnut crown and throat.
Habitat
Second growth, shrubbery, dry thickety hillsides.
Habits
Tame; forages low where breeding, but at all levels in migration; often victimized by Brown-headed Cowbird; may migrate across Gulf of Mexico.
Voice
Song resembles Yellow's but with emphatic final double note, *pleased, pleased, pleased t'MEETcha* or *I want to see Miss BEECHer;* many variations around this theme; also a quite different late-season song; call is a soft *tsip* or louder *chip.*
Eggs
Usually 4; white, spotted chiefly at larger end with brown and lavender; 0.6 x 0.5 in. (1.7 x 1.3 cm). Nest is loose, of bark strips and plant fibers, in low bush.
Range
Breeds from s. Canada s. to s.-cen. U.S. and into Appalachians; winters from C. to S. America.

BAY-BREASTED WARBLER

Dendroica castanea **37:2, 41:1**

Description
Size, 5½ in. (14.0 cm). Spring male: • *chestnut (or bay) on back of head,* throat, and sides; black face; dark crown; • *buffy neck patch;* above gray with black stripes; wing bars, underparts, and outer tail tips white. Spring female: similar but much less chestnut, face grayish. Fall adults and immature: olive-green above, buffy below, dusky cheek patch, black streaks on back, 2 white wing bars, legs normally black (but on occasion light), often a trace of chestnut on flanks, undertail coverts buffy.
Similarities
The similar fall Blackpoll has a greenish-yellow tinge below, usually appears streaked; and has light legs, white undertail coverts, stronger streaking above, and no chestnut. Fall Pine is unstreaked above.
Habitat
Breeds in bogs and openings in spruce forests; relatively uncommon.
Habits
In breeding season, forages at all levels; movements somewhat deliberate; in migration, keeps lower, often occurs in woods, particularly pine; a late spring and early fall migrant (August, September), before most Blackpolls.
Voice
Song is a high, thin *"seetzy-seetzy-seetzy-seetzy-see"* (Gunn); reminiscent of Blackburnian, Cape May, and Golden-crowned Kinglet; call, a fine *tsip.*

Eggs
4–7; often bluish-green, spotted with brown; 0.7 x 0.5 in. (1.8 x 1.4 cm). Nest is of twigs and moss on branch of conifer, 3–20 ft. (0.9–6.1 m) up.
Range
Breeds in se. Canada and ne. U.S.; winters in C. and S. America; migrates along Atlantic Coast.

BLACKPOLL WARBLER

Dendroica striata **38:10, 41:2**

Description
Size, 5½ in. (14.0 cm). Black with white wing bars in all plumages. Spring male: olive above, white below, black stripes on back and sides, cheek and throat white, • *black cap*. Spring female: cap olive streaked with black, streaks less conspicuous, below white, cheeks gray. Fall adults and immatures: olive-green above, white below with tinge of greenish-yellow, black streaks on back and faintly on sides, legs pale, undertail coverts white.
Similarities
See Bay-breasted Warbler for difficult distinction between these 2 species in fall; use caution. Black-and-White Warbler has a black-and-white striped crown.
Habitat
Breeds in northern and mountain coniferous forest, especially by bogs and in areas of stunted trees.
Habits
Movements rather deliberate, a good flycatcher; forages and sings chiefly at middle and upper levels in migration; a late-spring night migrant, many hit lighthouses, beacons, and TV towers.
Voice
Song is easy to learn and remember, a high, penetrating *zee-zee-Zee-Zee-Zee-Zee-ZEE-ZEE-Zee-Zee-zee-zee,* of even pitch, becoming louder in middle, fading away at end; call, a sharp *chip.*
Eggs
3–5; whitish, boldly spotted with red-brown; 0.7 x 0.5 in. (1.8 x 1.3 cm). Nest is of twigs, moss, low in spruce.
Range
Breeds from n. Canada s. to s. Canada and ne. U.S.; winters in S. America; migrates through W. Indies.

PINE WARBLER

Dendroica pinus **38:12, 41:3**

Description
Size, 5½ in. (14.0 cm). All plumages have white wing bars and tail corners, dark legs. Spring male: unstreaked yellow-green above, • *yellow breast* with obscure streaking, gray wings and tail, yellow line over eye, white undertail coverts. Spring female and fall adults: much duller, less yellow on breast. Immature female: duller, more brownish above; whitish below, often with little trace of yellow.
Similarities
Similar Yellow-throated Vireo has yellow spectacles and is more deliberate in its movements. Palm Warbler has red-brown cap. Orange-crowned lacks wing bars. In fall, similar Bay-breasted and Blackpoll have streaked backs; Bay-breasted has buffy undertail coverts; Blackpoll has yellowish legs.
Habitat
Pines, but any trees or shrubs in migration.

Habits
Rather tame; often creeps along branches like Brown Creeper; occasionally feeds on ground; catches insects in air; an early spring and late fall migrant, often seen with Palm Warblers.
Voice
Song is a Chipping Sparrow-like trill but slower, softer, with intervals between notes longer than in Chipping, much longer than in Worm-eating; call, a *chirp*.
Eggs
4; whitish, boldly marked with lilac and purple; 0.7 x 0.5 in. (1.8 x 1.3 cm). Nest is of bark strips, pine needles, on pine limb, 10–50 ft. (3.0–15.2 m) up.
Range
Breeds from s. Canada s. to Gulf of Mexico; winters from s. parts of breeding range s. to W. Indies.

KIRTLAND'S WARBLER
Dendroica kirtlandii **38:8, 41:5**

Description
Size, 5¾ in. (14.6 cm). Very local and declining. Rarely seen except where breeding. Spring male: blue-gray above, yellow below; black streaks on back, breast, and sides; incomplete white eye-ring; blackish mask; 2 obscure white wing bars; dark legs. Spring female: duller and without mask. Fall adults: head, back, and sides brownish; all markings duller.
Similarities
Canada has unmarked back, black necklace, light legs.
Habitat
Breeds in open stands of jack pines, 3–18 ft. (0.9–5.5 m) high. Extremely rare.
Habits
Tame; jerks tail, often forages on ground by flying or hopping along, sometimes catches insects on wing.
Voice
Song is a loud, clear *"chip chip chip-chip-chip-teoo-teoo-weet-weet"* (Borror); call, a *tsip*.
Eggs
4; white, finely spotted with brown; 0.7 x 0.5 in. (1.8 x 1.4 cm). Brown-headed Cowbirds severely parasitize the nests of this warbler. Nest is of bark and vegetable fiber, on ground, in groves of jack pines; usually in loose colonies.
Range
Breeds exclusively in n.-cen. Mich.; winters in Bahamas.

PRAIRIE WARBLER
Dendroica discolor **38:14, 41:11**

Description
Size, 5 in. (12.7 cm). Spring male: olive-green above, with patch of • *chestnut streaks on back*; 2 yellowish wing bars; white tail corners and undertail coverts; all-yellow below with bold • *black streaks down sides*. Spring female, fall adults, and immature: similar pattern but duller, grayer above; chestnut spots concealed or missing; wing bars and side streaks inconspicuous or absent.
Similarities
Fall Pine is larger, has more conspicuous wing bars.
Habitat
Second-growth low pine and scrub oak.

Habits
Forages low, wags tail (but not as much as Palm), catches insects in air, also hovers before leaf, sometimes sings from top of medium-sized tree.
Voice
Song is a distinctive, easy-to-learn *zee-zee-zee-zee-zee-zee,* rising in pitch throughout; call, a soft *chip.*
Eggs
3–5; white, spotted with brown; 0.6 x 0.5 in. (1.5 x 1.3 cm). Nest is of bark, plant stems, in fork near top of low shrub.
Range
Breeds from cen. New England and S.Dak. s. to Fla., Gulf Coast, and Okla.; winters from s. Fla. s. to C. America.

PALM WARBLER
Dendroica palmarum **38:13, 41:9**

Description
Size, 4½–5½ in. (11.4–14.0 cm). Spring adult: olive-brown above with chestnut cap; rump greenish-yellow; line over eye; throat, upper breast, and coverts yellow; belly yellow to white; gray cheek patch; light chestnut streaks on breast; white tail corners. Fall adult and immature: browner above, concealing chestnut cap; line over eye, throat, and breast white; obscure streaks below. Belly always white in western race; chestnut brighter and belly always yellow in eastern race.
Similarities
Young Cape May has white undertail coverts.
Habitat
Breeds in northern bogs and muskeg; seen everywhere in migration, especially lawns, fields, swamps, and in undergrowth of open woods.
Habits
Pumps tail, has gliding walk, hops, runs, spends much time on ground or in low bushes; often seen with Yellow-rumped or Pine Warblers, kinglets and chickadees, or sparrows in field; hardy, a few winter north to New England.
Voice
Song is "a short rapid series of thin, lisping notes. . . . Very similar to a Chipping Sparrow; *thi, thi, thi, thi, thi*" (Griscom); call, a *chip.*
Eggs
4–5; creamy white, spotted with brown and lilac; 0.7 x 0.5 in. (1.8 x 1.3 cm). Nest is of grass, in moss at foot of bush or tree.
Range
Breeds in cen. Canada s. to n. U.S.; winters from s. U.S. s. to C. America.

OVENBIRD
Seiurus aurocapillus **37:11**

Description
Size, 5½–6½ in. (14.0–16.5 cm). Olive above, with a black-bordered • *rufous-buff crown;* eye-ring and underparts white; breast and sides heavily streaked with black; legs pinkish; no wing bars.
Similarities
Waterthrushes lack rusty crown.
Habitat
On or near ground in open deciduous or mixed woodlands; in migration, shrubbery in gardens, parks.

Habits
Inquisitive; suggests a small thrush, feeds and walks daintily on forest floor, sings from branch.
Voice
Song is a loud, ringing crescendo often described as *tea-cher, tea-cher,* plus a varied, melodious, and gushing evening flight song sometimes also given at night; alarm note, a sharp hard *chik.*
Eggs
3–6; creamy white, finely spotted with red-brown and lilac; 0.8 x 0.6 in. (2.0 x 1.5 cm). Nest is arched, and with side entrance; bulky, of grass and leaves, concealed, on forest floor.
Range
Breeds from cen. Canada s. to n. Gulf states; winters from Gulf of Mexico s. to S. America.

NORTHERN WATERTHRUSH
Seiurus noveboracensis **37:12**

Description
Size, 5½–6½ in. (14.0–16.5 cm). Dark olive-brown above, greenish-yellow below; heavy, • *dark streaks on breast and sides;* fine spots on throat; legs pinkish, no wing bars.
Similarities
Louisiana lacks spots on throat, is whiter below, has white eye line, but, in fall, western subspecies of Northern also has white eye line.
Habitat
Breeds in northern bogs and swamps and places with quiet water; in migration, found along streams, in shrubbery anywhere.
Habits
Feeds on ground, sings from branch; constantly teeters like Spotted Sandpiper as it walks or runs, often along edge of pond or watercourse; often sings at middle level, birds chase each other through woods in fast zigzag flights; an early fall migrant.
Voice
Song is a loud, carrying *"sweet sweet sweet swee-wee-wee-chew-chew-chew-chew"* (Gunn), last *chew* notes lower, diagnostic; or *"hurry, hurry, hurry, pretty, pretty, pretty"* (Sutton); call, a sharp distinctive *clink.*
Eggs
4–5; pinkish-white, spotted and scrawled with red-brown and lavender; 0.8 x 0.6 in. (2.0 x 1.5 cm). Nest is of moss, in moss, at base of stump or tree, near water.
Range
Breeds from n. Canada s. to n. U.S.; winters from Mexico s. to S. America.

LOUISIANA WATERTHRUSH
Seiurus motacilla **37:13**

Description
Size, 5¾–6¾ in. (11.4–17.1 cm). Similar to Northern, but averages larger, whiter below, often with buffy flanks; • *white eye line;* • *unmarked throat* diagnostic (at close range).
Similarities
See Northern Waterthrush.
Habitat
Breeds in ravines, gorges, near flowing streams; frequents swamps and other wet areas in southern portion of range and on migration.
Habits
Similar to Northern Waterthrush, sometimes sings in flight.

Voice
Song is wild, ringing; 3 loud introductory notes, then a sweet, jumbled, descending warble; more penetrating than song of Northern; also an evening flight song; call, like Northern's.
Eggs
4–6; white, marked with brown and lilac; 0.8 x 0.6 in. (2.0 x 1.5 cm). Nest is bulky, of leaves, grass, hidden in cavity in stream bank.
Range
Breeds from cen. New England and Minn. s. to Ga. and Tex.; winters from Mexico s. to S. America.

KENTUCKY WARBLER
Oporornis formosus **39:1**

Description
Size, 5½ in. (14.0 cm). Olive-green above, black forehead, • *yellow spectacles,* pale legs; male has more • *black in mustache* and on forehead than female or immature.
Similarities
Common Yellowthroat is smaller, has white belly, and no yellow spectacles.
Habitat
Swampy, wet woods.
Habits
Walks, occasionally teeters like waterthrush.
Voice
Song is a loud, ringing *turtle, turtle, turtle* somewhat like a Carolina Wren's; alarm note, a metallic *chip.*
Eggs
4–5; white, spotted with red-brown and lilac, chiefly at larger end; 0.8 x 0.6 in. (2.0 x 1.5 cm). Nest is bulky, of leaves, grass, well-hidden on ground in woods.
Range
Breeds from N.J., Pa., Ind., and Iowa s. to se. U.S.; winters from Mexico s. to S. America.

CONNECTICUT WARBLER
Oporornis agilis **39:14, 40:4**

Description
Size, 5¼–6 in. (13.3–15.2 cm). Olive above, yellow below; yellow undertail coverts are ⅔ the length of tail; • *white eye-ring complete.* Male: • *gray hood.* Female: duller. Immature: even duller; crown, throat, upper breast, and flanks brownish; eye-ring slightly buffy.
Similarities
Nashville has yellow throat. Male Mourning has black on breast. Fall Mourning has incomplete eye-ring.
Habitat
Forest bogs, brush; in migration, thickets, overgrown fields.
Habits
If flushed, flits to low branch and sits motionless.
Voice
Song is a loud ringing *"chuckety chuckety chuckety chuck"* (Allen); call, a metallic *pink.*
Eggs
3–5; creamy white, with a few dark spots wreathed around larger end; 0.8 x 0.6 in. (2.0 x 1.5 cm). Nest is of grasses, in moss, on ground in tamarack swamps.
Range
Breeds from s.-cen. Canada s. to Mich., Wis., and n. Minn.; winters in S. America.

MOURNING WARBLER

Oporornis philadelphia **39:13, 40:6**

Description
Size, 5–5¾ in. (12.7–14.6 cm). Olive above, yellow below; • *gray hood, black apron;* undertail coverts less than half length of tail; legs pinkish; no eye-ring. Male: gray hood with black apron, no eye-ring. Female and immature: similar to larger female and immature Connecticut, but usually no eye-ring in spring and eye-ring broken in fall.
Similarities
Spring male Connecticut has gray, not black, apron.
Habitat
Wet woods, thickets.
Habits
Very late migrant.
Voice
Song is a somewhat low "*wee surree surree surree surree* with a falling inflection" (Allen); also an emphatic "*cheese cheese cheese, cheese sweet*" (Gunn); call, "a strong almost bi-syllabic *tchip,* a bit sharper than a Yellowthroat's . . . quite distinctive" (Peterson).
Eggs
Indistinguishable from Kentucky's. Nest is of plant stalks, bark shreds, on ground in tussock in woods.
Range
Breeds from N.Dak. and Alta. s. to s. Canada and n. U.S.; winters from C. to S. America.

Note: The **MacGILLIVRAY'S WARBLER,** *Oporornis tolmiei,* size, 5½ in. (14.0 cm), replaces the Mourning Warbler in the Black Hills, the Cypress Hills, and Rocky Mountains. The two are indistinguishable in fall, but in spring the MacGillivray's has a white eye-ring, broken in front and at the rear. These two very similar forms interbreed in Alberta, and may belong to the same species.

COMMON YELLOWTHROAT

Geothlypis trichas **39:2, 40:8**

Description
Size, 4½–5¾ in. (11.4–14.6 cm). Olive upperparts, yellow throat, white belly, yellow undertail coverts, no white wing bars or tail corners. Male: distinctive black mask. Female: similar but with whitish eye-ring and no black mask. Immature: similar but throat often buffy and sides brownish (males may show trace of black mask). Fall females and immatures among the most nondescript of warblers.
Similarities
Nashville has gray crown and all-yellow underparts. Connecticut and Mourning have dark, not yellow, throat.
Habitat
Damp undergrowth, marshes, thickets.
Habits
Active, inquisitive, wrenlike; skulks near ground and in underbrush.
Voice
Song is a distinctive, easy to learn, but quite variable *WITCH-i-ty, WITCH-i-ty, WITCH,* or *which IS it, which IS it, which IS it?* also a flight song; alarm note, a chatter like House Wren; call, a *check.*

Eggs
Usually 4; creamy white, marked with red-brown and lilac; 0.7 x 0.5 in. (1.8 x 1.3 cm). Nest is large, cup-shaped, of grass, leaves, underbrush in marsh.
Range
Breeds from Newfoundland and Ont. s. to Fla., Gulf Coast, and Mexico; winters from Gulf states southward.

YELLOW-BREASTED CHAT
Icteria virens **39:3**

Description
Size, 6½–7½ in. (16.5–19.1 cm). The largest and most unwarblerlike warbler; bill stout, tail long. Olive above; bright- • *yellow throat and breast;* • *white spectacles,* mustache, belly, and undertail coverts.
Similarities
Yellow-throated Vireo is smaller, has 2 white wing bars, frequents upper level of trees.
Habitat
Thickets, especially dry hillsides, briar patches.
Habits
Rises in air from thicket and sings with wings flopping and feet and tail dangling; secretive and hard to see, but is attracted by hand-kissing; actions and behavior seem clownish, eccentric.
Voice
Song is a highly variable medley of calls, clucks, mews, whistles, and gurgles, often with long pauses, remotely suggesting a Northern Mockingbird; a long bugle note; often sings at night.
Eggs
3–5; white, marked with brown and lilac; 0.9 x 0.7 in. (2.3 x 1.8 cm). Nest is large, cup-shaped, of grass, leaves; well hidden in thicket 1–5 ft. (0.3-1.5 m) up.
Range
Breeds locally from Mass. and Ont., s. to Fla., Gulf Coast, and Mexico; winters from Mexico, s. to C. America.

HOODED WARBLER
Wilsonia citrina **39:5, 40:3**

Description
Size, 5½ in. (14.0 cm). Olive above, • *yellow face* patch about beady dark eye, forehead, and underparts, including undertail coverts; white in outer tail feathers; • *no wing bars.* Male has black hood and throat; female and immature with less or no black on hood, ear region dusky.
Similarities
Very similar female Wilson's has no white in tail. Female Yellow has yellow tail corners, less conspicuous eye. Immature Canada has dark forehead and white undertail coverts; see also Bachman's Warbler.
Habitat
Lower story of wet woods.
Habits
Often opens and shuts tail, revealing white spots on outer tail feathers; flycatches.
Voice
Song is a loud, sweet ringing *"Monte Monte video"* (Sprunt); or *"wit a wit TEOO* or *toway toway toway TEE-TOO"* (Saunders); call, a *chick.*

Eggs
3–5; white, spotted and blotched with chestnut and lilac; 0.7 x 0.5 in. (1.8 x 1.3 cm). Nest is cup-shaped, of leaves, bark shreds, in bush, 1–5 ft. (0.3–1.5 m) up, in upland or marsh.
Range
Breeds from n. U.S. s. to Gulf Coast, winters from Mexico s. to C. America.

WILSON'S WARBLER

Wilsonia pusilla **39:4, 40:2**

Description
Size, 4¼–5 in. (10.8–12.7 cm). Male: olive-green above, bright yellow below; • *black cap; no wing bars* or white in tail. Female and immature: yellow line over beady black eye; olive about ears; may or may not show trace of black cap.
Similarities
See Hooded, Yellow.
Habitat
Low trees and shrubs, underbrush.
Habits
Catches insects, twitches tail, flips wings.
Voice
Song is a quick, dry chatter, last half faster, lower, *chee-chee-chee-chee-chee, chi-chi-chi-chi-ch-ch;* call, a *chip.*
Eggs
3–6; white, with brown spots, wreathed around large end; 0.6 x 0.5 in. (1.5 x 1.3 cm). Nest is of grass, at base of tree in shrubbery.
Range
Breeds from Newfoundland and n. Man. s. to n. New England and Minn.; winters from Mexico s. to S. America.

CANADA WARBLER

Wilsonia canadensis **38:4**

Description
Size, 5–5¾ in. (12.7–14.6 cm). Slate-blue above, bright yellow below; white undertail coverts; no white on wings or tail. Male: black forehead, trim • *black necklace of spots, yellow spectacles.* Female and immature: duller with only faint necklace and spectacles.
Habitat
Shrubbery, understory of woods; but at all levels in migration.
Habits
Active, flycatches.
Voice
Song is a short jumble, often containing a *witchity* note like yellowthroat's introduced by a *chip,* same as call note.
Eggs
3–5; white; speckled with red-brown, mainly around large end; 0.7 x 0.5 in. (1.8 x 1.3 cm). Nest is of grass, leaves, well hidden on ground in woods.
Range
Breeds from s. Canada s. to n. U.S., e. of Rocky Mountains, and into mountains of n. Ga.; winters in S. America.

AMERICAN REDSTART

Setophaga ruticilla **37:1**

Description
Size, 4½–5¾ in. (11.3–14.6 cm). In any plumage, an interrupted broad orange or yellow band across mid-tail. Adult male: black with • *salmon-orange patches on shoulder,* wings, and tail; white belly and undertail coverts. Adult female: salmon-orange replaced by yellow, head gray, upperparts olive, underparts white. First-year male: like female, but orange where female is yellow.
Habitat
Deciduous woodlands, swamps, shrubbery, clearings.
Habits
Most restless warbler, frequently spreads patterned tail, droops wings; good flycatcher.
Voice
Song is a thin, incisive series of single notes with emphasized note near end, *zee-zee-zee-zee-Zee zee-oo (zee-oo* descending); or of double notes, *teetza-teetza-teetza-teetza, teetza;* sometimes alternates these songs; alarm note, a sharp *chik.*
Eggs
3–5; whitish, marked with lilac and brown, wreathed at larger end; 0.6 x 0.5 in. (1.7 x 1.3 cm). Nest is cup-shaped, of bark, shreds, leaf stalks, in fork of bush or tree, 3–30 ft. (0.9–9.1 m) up.
Range
Breeds from Newfoundland and cen. Man. s. to Ga. and se. Okla.; winters from Mexico s. to S. America.

WEAVER FINCHES

Family Ploceidae

This is a widely distributed, sparrowlike family of the Old World, two species of which were introduced during the nineteenth century into North America. They feed on grain, fruit, seeds, garbage, and insects.

HOUSE SPARROW

Passer domesticus **45:2**

Description
Size, 6 in. (15.2 cm). Bill stout; tail short, slightly notched. Male: white cheek and wing bars; chestnut nape; red-brown and gray above, grayish white below; • *black throat* mixed with gray in winter and in immature. Female: streaked, buffy; gray-and-brown above, pale brownish-gray below; • *buffy line over eye.* Albinos of all degrees are not uncommon, giving rise to reports from beginners of unidentifiable birds.
Habitat
Cities, suburbs, farms, ranches, and other haunts of man. Common.
Habits
Tame, gregarious, aggressive, hardy, prolific; hops, does not walk; frequents feeders, likes dust baths; does not "kick" feet in foraging like many native sparrows; gathers in large roosts except when breeding.
Voice
Calls, a *chissik, chissik;* a chirp; alarm note, *tell, tell;* and a noisy chatter.
Eggs
5–6; gray-white marked with brown and gray; 0.9 x 0.6 in. (2.3 x 1.5 cm). Nest is bulky, of straw, debris, in ivy, tree, or cranny in habitations.

Range
Resident throughout much of Canada and most of U.S., from Newfoundland and n. Man. s. to Mexico.

Note: Unlike the related House Sparrow, the **EUROPEAN TREE SPARROW**, *Passer montanus (Fig. 14)*, size, 6 in. (15.2 cm), has not spread from the vicinity of St. Louis, where it was introduced in 1870. Found only about that city in Missouri and immediately adjacent Illinois, it can be told from the House Sparrow by its black throat in all plumages, its chestnut crown, and its black ear spot.

Fig. 14 European Tree Sparrow

MEADOWLARKS, BLACKBIRDS, AND ORIOLES

Family Icteridae

This family is quite varied in character; but its members all have more or less conical, pointed bills and rounded, not notched, tails. They feed on seeds and insects.

BOBOLINK

Dolichonyx oryzivorus **48:1**

Description
Size, 6–8 in. (15.2–20.3 cm). Bill rather short, tips of tail feathers look worn. Spring male: head and • *underparts black;* nape buffy; large • *white areas on wings, back,* and rump. Females and fall male: rich buff, darker above; crown black with buff stripe down middle; buff and brown stripes on back.

Similarities
Female Dickcissel has slightly notched tail and rusty shoulders. Female Red-winged Blackbird has longer bill and heavy streaks on breast. Stripes on Bobolink's back suggest giant Grasshopper Sparrow.

Habitat
Breeds in wet grassy fields, meadows, and along river valleys; in migration, marshes, rice fields.

Habits
Sings as it flies low or hovers over meadow; perches on grass stalk, shrub, tree, or fence post; gathers in flocks in migration, flies high.

Voice
Song, a rich, bubbling medley *"bob-o-link, bob-o-link, spink, spank, spink"* (William Cullen Bryant) given in flight; call, a metallic *pink,* often heard overhead in migration.

Eggs
4–7; very variable, gray to red-brown, spotted and blotched with browns and purples; 0.8 x 0.6 in. (2.0 x 1.5 cm). Nest is of grass, weed stems, in tall grass in meadows, hidden.

Range
Breeds from Newfoundland and Man. s. to Md., Ohio, and Kans.; winters in S. America.

EASTERN MEADOWLARK
Sturnella magna **48:2**

Description
Size, 8½–11 in. (21.6–27.9 cm). Plump; bill rather long; tail short, wide, rounded. Black V on yellow breast; wide black stripes on crown, white cheeks, brown upperparts, streaked sides; • *white outer tail feathers* conspicuous in flight.
Similarities
Dickcissel is much smaller, slimmer, and has short bill. See Western Meadowlark.
Habitat
Prairies, hayfields, pastures, coastal marshes.
Habits
Flicks tail as it walks on ground; sings from tree or fence post, early spring to late fall. Flight quaillike (rapid wingbeats, then a glide), usually low over a field, frequently sailing with wings outstretched and pointed slightly down; silhouette in flight looks like a Starling. Gathers in small flocks in winter, migrates by night; males are polygamous.
Voice
Song is a sweet, whistled *"ah-tick-seel-yah"* (Thoreau); sometimes rendered as *spring-o'the-YE-ar;* call, an emphatic *dzhert;* alarm note, a rapid, guttural chatter (often as it flies away).
Eggs
3–7; white, speckled with brown and lavender; 1.0 x 0.8 in. (2.5 x 2.0 cm). Nest is arched saucer of grasses and weeds, under tuft of grass in field.
Range
Breeds from N.S. and se. Ont. s. to Fla., Gulf Coast and Mexico; winters from New England s. to C. America.

WESTERN MEADOWLARK
Sturnella neglecta

Description
Size, 8½–11 in. (21.6–27.9 cm).
Similarities
Closely resembles Eastern Meadowlark; a little paler, yellow of chin carried on to mustache area near base of bill, but cannot be safely identified in field except by voice.
Habitat
Prefers drier grasslands than Eastern.
Habits
Same as Eastern.
Voice
Song is varied, different from Eastern's; louder, warbled, variously rendered as *"Hip! Hip! Hurrah! boys; three cheers!"* (Dawson and Bowles); or *"tung-tung-tung-ah, tillah'-tillah'-tung"* (Ridgeway); call, a metallic *tuk,* sharper, harsher than Eastern's; alarm note, a chatter.
Eggs
Same as Eastern's.
Range
Breeds from Man. s. to La. and cen. Tex., spreading eastward; winters from Nebr. and Utah southward.

YELLOW-HEADED BLACKBIRD

Xanthocephalus xanthocephalus **42:1**

Description
Size, 8½–11 in. (21.6–27.9 cm). Male: black, base of bill black, spot on wing white, • *yellow head* and neck. Female and immature: dark brown; yellowish cheek and line over eye; yellow throat and breast; female smaller than male, lacks white in wing.
Similarities
Grackles have longer tails.
Habitat
Fields, marshes, farmyards.
Habits
Gregarious; flight slow, deliberate, undulating like the Red-winged; flocks long, loose (not wide like Red-winged); often flocks with Red-wingeds, grackles, cowbirds; in fall, gathers in great roosts; follows plow, climbs about reeds.
Voice
"Oka WEE wee" (Bent), *oka* guttural, *WEE wee,* loud whistles; an unmusical *"klick-kluck-klee-klo-klu-klel-kriz-kri-zzzzzzz-zeeeeee"* (Taverner), delivered with much effort; alarm, a vehement *"klookoloy, klookoloy, klook ooooo"* (Dawson); call, a low *kek.*
Eggs
3–7; white to greenish, heavily marked with red-brown and gray, and with some hairlines; 1.0 x 0.7 in. (2.5 x 1.8 cm). Nest is in colony; of blades of sedge and grass attached to cattails, reeds; 1–3 ft. (.3–.9 m) up, in marshes.
Range
Breeds chiefly in W., but e. to cen. Wis. and s. to Ark.; winters from Tex. southward.

ORCHARD ORIOLE

Icterus spurius **43:3**

Description
Size, 6–7 in. (15.2–17.8 cm). Slender. Adult male: has 1 white wing bar; black with • *chestnut breast,* belly, rump. Female: olive-green above, 2 white wing bars, rump and underparts yellow. Immature male: as female, but with black face and chin. Later intermediate plumages show scattered chestnut.
Similarities
Female Northern is brownish-olive or grayer above, dull orange below; some have black throats.
Habitat
Orchards, tall hedgerows, and farmlands with scattered trees.
Habits
Lively; pumps tail, may hang head downward; migrates by day; males precede females; sometimes nests in a loose colony.
Voice
Song is a lively, pleasant warble, suggestive of Purple Finch or Blue Grosbeak, quite variable, often ending in a tanagerlike downward-slurred note, *"Look here, what cheer, what cheer, whip yo, what cheer, wee yo"* (Townsend); not as loud, but longer than Northern's; call, a *chak* like Red-winged's, a longer rattle.
Eggs
4–6; bluish-white, marked with browns and purples; 0.8 x 0.6 in. (2.0 x 1.5 cm). Nest is of grass, 3 in. (7.6 cm) deep, hanging from small branch usually of fruit tree.
Range
Breeds from cen. New England and s., Minn. s. to Gulf Coast, and w. to Nebr. and Tex.; winters from Mexico s. to S. America.

NORTHERN ORIOLE

Icterus galbula **43:1**

Description
Size, 7–8 in. (17.8–20.3 cm). Adult male: bright • *orange; black head and throat;* patterned black and yellow tail, white wing bar; colors conspicuous in flight. Female: variable; brownish-olive or grayish above, brighter on tail; dull orange or yellowish rump; underparts yellowish-orange to whitish; 2 white bars on blackish wings; sometimes some black on head. Immature male (to second summer, may breed): as female but duskier above and shows black on throat.

Similarities
Female Scarlet Tanager has no wing bars. Female and immature Northerns always have some orange, lacking in female and young Orchards.

Habitat
Shade trees and edges.

Habits
Sings from treetop or upper branch; migrates by day, flying high.

Voice
Song is loud, robust whistles; highly varied, usually a series of 1- and 2-note phrases; sometimes sings on wing; call note, a whistled *pe-ter*; alarm, a chatter.

Eggs
4–6; grayish-white, streaked and scrawled with browns and black; 0.9 x 0.6 in. (2.3 x 1.5 cm). Nest is of grasses and string, gourd-shaped, carefully woven, hanging from branch.

Remarks
Former (eastern) Baltimore and (western) Bullock's Orioles are now regarded as 1 species.

Range
Breeds from N.S. and Sask. s. to Ga., La., and Mexico; winters from Mexico southward.

Note: In the central Great Plains, from Saskatchewan to Oklahoma, there is a zone of hybridization in which variable hybrids of the eastern (Baltimore) and western (Bullock's) subspecies are abundant. The free interbreeding and the existence of the hybrid population indicates that these two birds belong to the same species. The hybrids show diverse mixtures of the patterns of the two forms, typical individuals of which are illustrated on Plate 43. From the western plains westward, orioles of this species are of the Bullock's type.

In the Miami, Florida, area is found the **SPOT-BREASTED ORIOLE,** *Icterus pectoralis* **(43:2)**, size, 8 in. (20.3 cm), introduced from Central America. Both sexes are orange with an all-black tail, black and white wings, a black bib and area in front of the eye, and a sprinkling of black spots on the sides of the breast; immature much duller.

RED-WINGED BLACKBIRD

Agelaius phoeniceus **42:2**

Description
Size, 7–9 in. (17.8–24.1 cm). Adult male: black with • *red-and-yellow shoulder patch.* Female: like a large, dark sparrow, with a sharp, pointed bill; heavily streaked below, light stripe over eye. Immature: like a dark female with a reddish shoulder patch. When perched, especially in winter, red on shoulder sometimes concealed by the yellow.

Similarities
See Bobolink.
Habitat
Fields, marshes, edges.
Habits
Noisy, gregarious; sings from reeds, tree, or fence post, spreading wings and tail; comes to feeders; on the ground, walks, runs, or hops. Flight undulating, a gentle rise on beating wings, a short sharp drop on a glide; flocks advance on a wide front and wheel in unison; often occurs in mixed flocks with cowbirds, grackles, and Starlings; in spring, males arrive before females.
Voice
Song is a "pleasing *conk-er-EEE* or *oolong TEA*" (Collins); call, a loud *chak*; also a *tee-urr*; and others.
Eggs
3–5; pale, bluish-green variously marked with brown, purple, and black; 1.0 x 0.7 in. (2.3 x 1.8 cm). Nest is bulky, bowl-shaped, of grasses attached to reeds in marsh or in bush or grass.
Range
Breeds from Newfoundland and cen. Man. s. to Fla., Gulf Coast, and Mexico; winters from Pa. s. to C. America.

RUSTY BLACKBIRD
Euphagus carolinus **42:5**

Description
Size, 8½–9¾ in. (21.6–24.8 cm). Eye yellow, tail of medium length. Spring male: black head with dull green sheen. Spring female: gray. Fall and winter adults: barred with rusty on head and body, black wings and tail. Immature: similar but even rustier.
Similarities
Brewer's female has dark eye, head of male has purple sheen. Grackles have longer tails.
Habitat
Swamps, marshes, woods.
Habits
Gregarious, sometimes in mixed flocks; on ground, walks and runs, nodding head; flirts tail, often gives voice with bodily contortions.
Voice
Song is an even, rhythmic "*tolalee-eek-tolalee-eek*; [also a rapid] *kawicklee kawicklee*" (Saunders); flock in song utters a creaky chorus; call, a *kik* and a rattling, a *chuck* like call of wood frog.
Eggs
3–6; light bluish-green, blotched with chocolate and gray; 1.0 x 0.8 in. (2.5 x 2.0 cm). Nest is bulky; of leaves, grass, in alder or willow, 1–2 ft. (0.3–0.6 m) above water in swamp.
Range
Breeds from n. Canada s. to ne. U.S.; winters from cen. New England and s. Man. s. to Fla., Gulf Coast, and Tex.

BREWER'S BLACKBIRD
Euphagus cyanocephalus **42:6**

Description
Size, 8–10 in. (20.3–25.4 cm). Tail of medium length. Male: black with an • *iridescent purplish head*; body with greenish gloss, but looks black at distance; eye yellow. Female: gray, paler over eye and on throat, eye dark. Immature: male may show slight grayish edgings.

Similarities
Rusty Blackbird's plumage is rusty in fall and winter. Grackles are iridescent, have long tails. Starling's tail is short. Female Brown-headed Cowbird is smaller, lighter, and has short, conical bill.
Habitat
Fields, ranches, farmyards.
Habits
Wanders in flocks over countryside; gathers in roosts at night; follows plow, nods head as it walks.
Voice
Laidlaw Williams recognized a scolding *tschup*, a hoarse, whistled *squee*, a clear, whistled *tee-uuu*; also a *kit-tit-tit-tit*; and, when elevating tail, a *chug-chug-chug, tucker-tucker-tucker*, or *tit-tit-tit*.
Eggs
4–6; grayish, often heavily blotched with brown; 1.0 x 0.7 in. (2.5 x 1.9 cm). Nest is in small colonies; bulky; of twigs, bark, and mud, in bush or tree, 1–30 ft. (0.3–9.1 m) up, often in colonies.
Range
Breeds chiefly in w. U.S., but e. to Great Lakes and s. to Ind. and Tex. winters from Wis. s. to Mexico.

BOAT-TAILED GRACKLE
Quiscalus major **42:8**

Description
Size, male, 16 in. (40.6 cm), female, 12 in. (30 cm), tail, 7 in. (17.8 cm). Male: • *Long, wide tail;* tan, gray, or pale yellow eyes. Female: paler and smaller.
Similarities
Much larger than Common Grackle.
Habitat
Coastal salt marshes and adjacent towns.
Habits
Singing males have wing-flutter display as they sing.
Voice
Male has 3-part song with harsh *tireet* notes, rapid series of ascending notes, then another *tireet* series; warning note in colonies, a high-pitched *kle-teet* or just *teet;* no whistled call of males to other males (Selander and Giller).
Eggs
3–4; pale blue, with umber and vinaceous clouding at smaller end and with fine black and brown blotches and marks, most prominent at larger end; 1.3 x 0.9 in. (3.3 x 2.3 cm). Nest is invariably in salt marsh, of marsh grasses, reeds, etc., with some paper, rags, feathers, and mud or dung wrapped in it.
Range
Resident along Atlantic and Gulf coasts from N.J., s. to La.; also inland in Fla.

Note: The **GREAT-TAILED GRACKLE,** *Quiscalus mexicanus,* male, size, 17 in. (43.2 cm), is a sibling (i.e., virtually identical) species that occurs from Mexico to southwest Louisiana. Although very similar, males differ in being larger, darker, more uniform-colored; with more glossy purple and less blue-green; and with longer, wider tail than Boat-tailed. However, the main field difference in the area of overlap is its bright yellow eyes. Females also have bright yellow eyes and are darker, especially below, than Boat-tailed. Less strictly coastal than Boat-tailed, it inhabits marshes, ponds, lakes, irrigated desert, farms, and towns. The male's song is a 4-part crackling, hissing, ending usually in 2 (1–5) long, piercing *cha-we* notes, the end phrase often being given alone; the warning note in the colonies is a loud, low *clack*. Its nest is usually in trees or man-made structures, similar in structure to Boat-tailed's, but of sticks with papers, moss, rags, feathers, mud or dung, when in trees. The eggs have a darker blue tone and with greater clouding of vinaceous and umber and more numerous black and brown blotches and lines at smaller end.

COMMON GRACKLE

Quiscalus quiscula **42:7**

Description
Size, 11–13½ in. (27.9–34.3 cm). Eye yellow, • *tail wedge-shaped.* Male: black with iridescent hues above of purple, blue, green, and bronze. Female: smaller, duller, iridescent only on front of body. Immature: dull brown, with a brown eye.
Similarities
Rusty Blackbird and Brewer's Blackbird have shorter tail. Various races have either a purplish or bronze back, with many intermediates. See Boat-tailed.
Habitat
Lawns, shade trees.
Habits
Walks; often gives voice with bodily contortions. Flight is direct, steady (not rising and falling).
Voice
Medley of harsh squeaking and guttural noises, some suggesting a rusty hinge, but not altogether unmusical; call, a *chack,* louder, lower than Red-winged's.
Eggs
4–6; greenish-white to reddish-brown, variously marked with brown; 1.2 x 0.9 in. (3.0 x 2.3 cm). Nest is bulky; of twigs, grass, and mud; on branch or in fork of conifer; 5–40 ft. (1.5–12.2 m) up.
Former name
Purple Grackle, Bronzed Grackle.
Range
Breeds from Newfoundland and cen. Man. s. to Fla., the Gulf states, and Tex.; winters from n. U.S. s. to Mexico border.

BROWN-HEADED COWBIRD

Molothrus ater **42:4**

Description
Size, 6–8 in. (15.2–20.3 cm). Eye dark; bill short, almost sparrowlike. Male: black with iridescent reflections of purple and green on upper back, • *head and neck coffee-brown.* Female and immature: gray; rather streaked below in juvenal plumage.
Similarities
Starling has longer bill, shorter tail. In mixed flocks can be recognized by smaller size; female Rusty and Brewer's are larger and with longer, thinner bills.
Habitat
Fields, open woods, edges.
Habits
Gregarious, often flocks with Starlings, Red-wingeds, grackles; walks with tail up; associates with cattle; spreads wings and tail as it squeaks; flight like Red-winged's.
Voice
Male, a characteristic squeak, like the swinging of an unoiled gate; also a liquid *glub, glub, glee;* female, a rattling chatter; call, a *chuk.*

Eggs
4–5; white, speckled with brown; 0.9 x 0.7 in. (2.3 x 1.8 cm). No nest; lays eggs in other birds' nests—only parasitic species in east.
Range
Breeds from Newfoundland and cen. Man. s. to Va., La., and Mexico; winters from Md. southward.

TANAGERS

Family Thraupidae

Tanagers are arboreal birds of tropical origin. They have notched tails and are somewhat like finches in appearance, but have longer, often notched bills. The males are brilliantly plumaged, the females modestly so. Tanagers are deliberate, even sluggish in their movements. They feed on insects, fruits, and berries.

WESTERN TANAGER

Piranga ludoviciana **43:5**

Description
Size, 6¼–7½ in. (15.9–19.1 cm). Adult male: yellow with • *red head* and black back. Black also on wings and tail, 2 pale yellow wing bars. Yellow rump, face largely yellow in fall. Female and immature plumages: as those of Scarlet Tanager, olive above, yellow below, but with 2 pale yellow wing bars.
Similarities
Female orioles have sharp, pointed bills and lighter cheeks.
Habitat
Upper canopy of open pine or mixed forests and wood edges.
Habits
Often catches insects on wing; flight straight, wingbeats fairly rapid; tame, visits feeders.
Voice
Song is similar to Scarlet Tanager's; call, *pit-ik* or *pit-itik*.
Eggs
3–5; similar to Scarlet's. Nest is saucer-shaped, of bark shreds and grass, on low branch.
Range
Breeds chiefly in w. U.S., but e. to Nebr. and w.-cen. Tex.; winters from Mexico southward; occasional birds farther e. in migration.

SCARLET TANAGER

Piranga olivacea **43:4**

Description
Size, 7 in. (17.8 cm). Spring male: in spring, scarlet with • *black wings* and tail; in fall, has scarlet replaced by olive-green above, yellow-green below; wings blackish; summer birds in molt show patches of red and green. Female: like fall male, but wings dark olive-gray (not black); undertail coverts canary yellow.
Similarities
Male Summer Tanager has red wings. Male Cardinal has crest and red wings. Female orioles and Western Tanager have wing bars.
Habitat
Canopy of deciduous woodlands, especially where oaks are present.
Habits
In migration, sometimes seen low or on ground.
Voice
Song is like a hoarse Robin's; call, a burry *KIP-purr*.

Eggs
3–5; pale greenish-blue, spotted with brown; 1.0 x 0.7 in. (2.5 x 1.8 cm). Nest is cup-shaped, thin; of bark shreds, twigs; near end of branch, 10–30 ft. (3.0 –9.1 m) up.
Range
Breeds from se. Canada s. to e.-cen. U.S.; winters in S. America.

SUMMER TANAGER
Piranga rubra **43:6**

Description
Size, 7½ in. (19.1 cm). Adult male: entirely • *rose-red* with • *no black*. Female: above olive-green washed with yellow; below yellow with orange tinge, overall more orange-yellow, less yellow-green than female Scarlet. Immature male: as female, or with varying patches of rosy-red (but lacks black wings of similarly changing Scarlet Tanager).
Similarities
Female Scarlet is darker, greener above, more yellow below, and has dark gray wings. Orioles have wing bars.
Habitat
Shade trees; deciduous, pine, and mixed woodlands.
Habits
Catches insects in air.
Voice
Song is a sweet, long carol suggesting both a Robin and a Rose-breasted Grosbeak; call, a *chicky-chucky-tuck*.
Eggs
3–4; bluish-green, spotted with purplish-brown; 0.9 x 0.7 in. (2.3 x 1.8 cm). Nest is cup-shaped, thin; of bark shreds and grass; near end of limb, 5–30 ft. (1.5–9.1 m) up.
Range
Breeds from Del. and Wis. s. to Gulf Coast and Mexico; winters from Mexico southward.

GROSBEAKS, FINCHES, SPARROWS, LONGSPURS, AND BUNTINGS
Family Fringillidae

Members of this large family have strong, short, conical bills adapted for eating seeds, a major item in their diet, which often includes berries, fruits, buds, and insects. Grosbeaks, finches, and buntings tend to be arboreal and brightly colored; sparrows, terrestrial and brownish. The term "winter finches" includes the Pine and Evening Grosbeaks, Pine Siskin, crossbills, and redpolls. Most of these northern birds are seen only in the colder months. Most sparrows are streaked brown above, pale and often streaked below. They are mainly ground-dwelling, but frequently sing from an elevated perch. Longspurs are sparrowlike birds of the open country. They walk and run rather than hop.

CARDINAL
Cardinalis cardinalis **44:1**

Description
Size, 7½–9 in. (19.1–22.9 cm). Bill red-orange. Male: crested; • *red* with a black face and throat. Female: reddish hue confined to crest, wings, and tail; back brownish-gray; face blackish; underparts buffy.

Similarities
Male Scarlet Tanager has black wings and tail. Male Summer Tanager has no crest.
Habitat
Thickets, gardens.
Habits
Tame, frequents feeders (likes sunflower seeds); sings from high perch.
Voice
Song is a loud, clear whistle, *wheat-wheat-wheat, what-cheer, what-cheer, what-cheer*; call, a sharp *tik*.
Eggs
3–4; whitish, blotched with brown and lilac; 1.1 x 0.8 in. (2.8 x 2.0 cm). Nest is ragged; of twigs, bark shreds; in low bush.
Range
Resident from N.S. and S.Dak. s. to Fla., Gulf Coast and s. Tex., s. to Mexico.

ROSE-BREASTED GROSBEAK

Pheucticus ludovicianus **43:12**

Description
Size, 7–8½ in. (17.8–21.6 cm). Spring male: heavy whitish bill, black head and upperparts, white on wings and lower back conspicuous in flight, • *rose-red triangle on breast*. Fall male: black of head and back replaced by streaked dark brown, rose and white of underparts spotted with dusky. Female: like a large brown sparrow; white line over eye, 2 white wing bars, heavy buff bill. Immature male: like female, but with some rose on breast.
Similarities
Female Purple Finch is much smaller, with smaller bill.
Habitat
Wood edges, open woods, shade trees, thickets.
Habits
Sluggish; male sings fron high perch (also on nest), feeds from near ground to treetops.
Voice
Song is a beautiful rolling warble, like a Robin that has had singing lessons! Call, a distinctive sharp *ink*.
Eggs
3–5; bluish-green, blotched with brown and lilac; 1.0 x 0.7 in. (2.5 x 1.8 cm). Nest is saucer-shaped; of grass, twigs; in bush or tree, 5–20 ft. (1.5–6.1 m) up.
Range
Breeds from S. Canada s. to cen. U.S., and in mountains to n. Ga.; winters from Mexico s. to S. America.

BLACK-HEADED GROSBEAK

Pheucticus melanocephalus **43:11**

Description
Size, 6½–7¾ in. (16.5–19.7 cm). A Western species. Male: black head; orange-brown underparts, nape, and rump; • *white wing bars,* conspicuous in flight. Female: streaked brown above; buffy brown below, especially across breast, virtually unstreaked; white line over eye and under cheek; pale heavy bill. Both sexes: yellowish belly, yellow wing linings.
Similarities
Female Rose-breasted is whiter with more streaked breast.

Habitat
Deciduous and mixed woods, thickets, edges.
Habits
Comes to feeders and picnic tables.
Voice
Song similar to Rose-breasted's; call, a sharp *tik,* somewhat like Rose-breasted's.
Eggs
Same as Rose-breasted's.
Range
Breeds chiefly in W., but e. to S.Dak. and cen. Kans. where hybrids occur with Rose-breasted; winters from La. s. to C. America. Rarely reaches e. Coast in fall.

BLUE GROSBEAK

Guiraca caerulea **43:7**

Description
Size, 6–7½ in. (15.2–19.1 cm). Bill heavy. Male: Purplish-blue (looks black in poor light) with • *2 rufous wing bars.* Female and immature: brown above, often with some bluish; buffy brown below; bill tan-colored; • *2 buffy wing bars;* immature male in spring is like female, but with considerable blue.
Similarities
Indigo Bunting is smaller, small-billed, lacks wing bars, and male is all blue.
Habitat
Scattered shrubs in dry fields, thickets near water, farms.
Habits
Sluggish, shy; sings from bush top.
Voice
Song is a finchlike warble suggesting Indigo Bunting and also Orchard Oriole; call, a loud *chuk.*
Eggs
3–5; light blue; 0.9 x 0.6 in. (2.3 x 1.7 cm). Nest is of grass, rootlets, snakeskin; in shrub or on low branch.
Range
Breeds from N.J. and Mo. s. to Fla., Gulf Coast, and Mexico; winters in C. America.

DICKCISSEL

Spiza americana **48:6**

Description
Size, 6–7 in. (15.2–17.8 cm). Spring male: black crescent on yellow breast; streaked brown above; head gray; bill bluish; throat white; yellow line over eye; dark, thin mustache streak; chestnut on shoulder; belly whitish. Fall male: often lacks black bib. Female and immature: paler, no black V, breast with some streaks; suggestion of male head pattern; chestnut shoulder, yellow wash on breast.
Similarities
Female fall Bobolink, male are buffier, have head stripes, and lack chestnut on shoulder. Female House Sparrow lacks chestnut shoulder, blackish mustache streak, yellow on breast.
Habitat
Fields, especially alfalfa; prairies.
Habits
Breeds in loose colonies; sings from wire, bush, or stalk, persistently and into late summer; gathers in flocks in migration.

Voice
Song is *dick, dick, dick-cissel,* much repeated, with varying numbers of *dick*'s and *cissel*'s.
Eggs
3–5; greenish-blue; 0.8 x 0.6 in. (2.0 x 1.5 cm). Nest is of grass; on ground by tussock of grass.
Range
Breeds from n.-cen. U.S. s. to Gulf states; winters from Mexico s. to S. America.

INDIGO BUNTING

Passerina cyanea **43:9**

Description
Size, 5½–5¾ in. (14.0–14.6 cm). Bill small. Spring male: • *rich blue;* looks black in poor light; in fall, strong mixture brown on back and head, and whitish below (molting birds blue and brown). Females and immature: sparrowlike; warm brown above, paler below, faintly streaked; often some blue in tail and wings in adult female. First-year males: often have white belly, otherwise blue.
Similarities
Blue Grosbeak is larger, bill heavier, male with buffy brown wing bars.
Habitat
Dense brush in bottomlands, clearings, edges, roadsides.
Habits
Sings from exposed perch on wire or near top of tree, persistently and through heat of day.
Voice
Loud song, notes paired, seems to "change gears" in middle; a "high, strident whistle, *sweea sweea sit sit seet seet sayo*" (Saunders); alarm note, a sharp *chit.*
Eggs
3–4; pale blue; 0.8 x 0.6 in. (2.0 x 1.5 cm); often 2 broods. Nest is cup-shaped; of grass, weeds; in bush.
Range
Breeds from New England s. to Gulf states and w. to Utah and Tex.; winters from Mexico s. to C. America.

PAINTED BUNTING

Passerina ciris **43:8**

Description
Size, 5¼ in. (13.3 cm). No wing bars. Adult male: • *violet head* and nape, • *yellow-green back, red rump* and underparts. Female: dull green, paler below than above. Immature male: in spring, like female but with some blue on head.
Habitat
Dry thickets, hedgerows, abandoned fields, streamsides, gardens.
Habits
Shy; sings from elevated perch.
Voice
Song suggests that of Indigo, but weaker; call, a sharp *chip.*
Eggs
3–5; white, marked with red-brown; 0.7 x 0.5 in. (1.9 x 1.4 cm). Nest is cup-shaped; of leaves, grass; in bush or tree in river flood plains
Range
Breeds from N.C. and Mo. s. to se. states and w. to Okla.; winters from Gulf states s. to Cuba, and from Mexico to C. America.

LAZULI BUNTING

Passerina amoena **43:10**

Description
Size, 5–5½ in. (12.7–14.0 cm). A western species. Adult male: • *azure-blue* head and rump, • *orange-brown breastband,* white belly and wing bars. Immature male: often has blue on head only. Female: head and back unstreaked gray-brown, some blue on wings, rump, and tail, light wing bars, underparts pale with buffy wash on breast, no streaking.
Similarities
Female Blue Grosbeak is larger, paler, has a much bigger bill, and brownish, not light, wing bars.
Habitat, Habits, Voice, Eggs
Like those of Indigo Bunting.
Remarks
Observers in the plains should be aware that Indigo and Lazuli hybridize; mixed traits of the two may be seen in hybrids on breeding grounds or migration in eastern Plains states.
Range
Breeds chiefly in W., but e. to e. S.Dak. and e. Nebr.; winters in Mexico.

EVENING GROSBEAK

Coccothraustes vespertina **44:3**

Description
Size, 7–8½ in. (17.8–21.6 cm). Chunky body, stout yellowish-white bill, short notched tail. Male: golden-yellow body, forehead; dusky brown head; black wings and tail; large white wing patches. Female: much of brown and yellow replaced by gray; dingier white on wings, white tail.
Similarities
Female Pine Grosbeak longer, bill is stubby, black; no white in tail.
Habitat
Breeds in northern coniferous forests; in winter, found among trees or by edges.
Habits
Gregarious; wanders widely in winter; tame, frequents feeders (likes sunflower and hemp seeds). Flight is swift, undulating; attracted by salt.
Voice
Song is a sweet warble; also a metallic cry; a call like an oversized House Sparrow's and a chattering.
Eggs
3–4; bluish-green, marked with gray and brown; 0.9 x 0.6 in. (2.3 x 1.7 cm). Nest is saucer-shaped; of twigs, grass; in top of conifer, 15–20 ft. (4.6–6.1 m) up.
Range
Breeds from N.S. and s. Man. s. to n. New England and Minn.; winters from breeding range s. to S.C. and Tex.

HOUSE FINCH

Carpodacus mexicanus **44:7**

Description
Size, 5½ in. (14.0 cm). Male: more orange-red than similar Purple Finch and streaked below. Female and immature: like small, dingy, less-patterned Purple Finch.

Similarities
See Purple Finch.
Habitat
Cities, towns, open country.
Habits
Gregarious; comes to feeders.
Voice
Song is a continuous, variable warble, higher, longer than that of Purple Finch; notes various, some like House Sparrow's; also a chatter.
Eggs
4–5; pale blue, with some black spots; 0.7 x 0.5 in. (1.8 x 1.3 cm). Nest is of grass, paper, rags; in vines, conifers, often near houses.
Remarks
A number of House Finches were apparently surreptitiously released on western Long Island in 1940. The first bird in the wild, presumably one of those released, was reported on April 11, 1941, and the first nest reported in 1943. They spread rapidly, and by 1970 were established from Connecticut and Long Island, through southern New York, New Jersey, to parts of Delaware, Maryland, and Virginia. They may be partly replacing the Purple Finch, whose numbers have diminished during this period.
Range
Resident originally in W., but introduced to e. U.S. and now established from Maine s. to S.C.

PURPLE FINCH

Carpodacus purpureus **44:5**

Description
Size, 5½–6¼ in. (14.0–15.9 cm). Stout bill, notched tail; Male: • *raspberry-colored,* no streaks on side or belly. Female and immature: sparrowlike; heavily streaked all over; gray-brown above, white below; pale line over eye; dark stripe below dark cheek.
Similarities
Pine Grosbeak is much larger. Male House Finch has dark streaks on sides and belly and much less red overall. Female House Finch is less distinctly marked; face unpatterned; ventral streaks paler, narrower, and less contrasting, hence appearing dingy.
Habitat
Woods, especially conifers; wood edges, shade trees.
Habits
Sings from treetops; wanders in fall, frequents feeders (likes seeds of sunflower, hemp, millet).
Voice
Song is a mellifluous warble like Warbling Vireo, but louder; call, a *tik, tik;* in flight, a sharp *pink.*
Eggs
4–5; blue-green, with dark spots chiefly around larger end; 0.8 x 0.6 in. (2.0 x 1.5 cm). Nest is cup-shaped, frail; of dark shreds, rootlets; in conifer.
Remarks
The Purple Finch is easily confused with the House Finch. In addition to the above characters, it is larger, chunkier, and less trim than the House Finch.
Range
Breeds from Newfoundland and Que. s. to N.J. and Minn.; winters from N.S. s. to Fla., Gulf Coast, and Tex.

PINE GROSBEAK

Pinicola enucleator **44:2**

Description
Size, 8–10 in. (20.3–25.4 cm). Big and chunky. Tail long, notched; bill, wings, and tail black; 2 white wing bars. Adult male: rose-red and gray. Female: rosy-red replaced by olive-yellow and gray. Immature male: like female but often with some red or orange-brown.

Similarities
White-winged Crossbill is smaller, has crossed bill. Purple Finch lacks white wing bars.

Habitat
Coniferous woods, edges, shrubbery.

Habits
Tame; in winter occurs in flocks, varying widely in numbers from year to year; flight undulating; likes to bathe in soft snow.

Voice
Song is warble and whistle reminiscent of Robin and of Purple Finch; a whistled *yew, yew, yew* somewhat like the Greater Yellowlegs's; flight call, *pee-ah.*

Eggs
3–5; greenish, spotted with purple; 0.9 x 0.6 in. (2.3 x 1.5 cm). Nest is of twigs, rootlets, grass; in conifer.

Range
Breeds from Newfoundland and n. Que. s. to n. New England and Man.; winters from s. parts of breeding range s. to Pa. and Kans.

GRAY-CROWNED ROSY FINCH

Leucosticte tephrocotis **48:13**

Description
Size, 5¾–6¾ in. (14.6–17.1 cm). Tail notched; bill and forehead black; crown gray; foreparts deep chestnut brown; wings, belly, and rump pink; bill yellow in winter; females paler than males.

Similarities
Sparrows and finches with which females might be confused lack dark, unmarked breast.

Habitat
In summer, in mountains above timberline; in winter, open mountain areas, foothills, plains.

Habits
Active; ground feeder, flycatches at times; wanders in flocks after breeding.

Voice
Song is a high-pitched series of *chips;* flight note, a *"chee-chee-chi-chi-chi"* (Hoffmann).

Eggs
4–5; white; 0.9 x 0.6 in. (2.5 x 1.5 cm). Nest is cup-shaped; of grasses; in rocky crevice above timberline.

Range
Breeds chiefly in nw. North America; winters e. sporadically to w. fringe of Great Plains.

COMMON REDPOLL

Carduelis flammea **44:8**

Description
Size, 5¼ in. (13.3 cm). Tail notched. Winter male: • *red cap on forecrown;* streaked gray brown above, paler on rump; 2 white

wing bars, whitish below; • *chin blackish;* breast pink. Female and immature: similar, but no pink on breast. Browner in spring plumage.

Similarities

Male Purple Finch has entire head reddish. Pine Siskin is more heavily streaked. See Hoary Redpoll.

Habitat

Tundra, openings in northern forest; fields, swamps.

Habits

Gregarious, active; clings to weed stems; sings from perch or in air; flocks twitter as they feed; frequents feeders (likes rolled oats and seeds of hemp, millet, and sunflowers). Flight is goldfinchlike; flocks wheel in unison.

Voice

Call, a dry *ch-ch*, repeated; song, same alternated with a trill.

Eggs

3–6; pale blue, spottted with brown and lavender; 0.7 x 0.5 in. (1.8 x 1.4 cm). Nest is of grass, plant down; in low bush or tree.

Range

Breeds from n. Que. s. to Newfoundland and n. Man.; winters from s. parts of breeding range s. to N.C. and Okla.

AMERICAN GOLDFINCH

Carduelis tristis **44:10**

Description

Size, 4½–5½ in. (11.4–14.0 cm). Notched tail, no streaks. Summer male: • *yellow*, with • *black wings, tail, forehead;* wings and tail with white markings; bill pink. Female and immature: yellow replaced by a brownish olive-yellow, no black forehead; female somewhat browner, duller in winter. Winter male: like female, but with yellow shoulder.

Similarities

Pine Siskin and redpolls are streaked. Warblers have thin bills. Yellow Warbler is yellow all over.

Habitat

Open country, weedy fields, edges, gardens.

Habits

Lively, gregarious; characteristic roller-coaster flight, utters flight note on top of rise; nests late, often into August; in winter often flocks with redpolls, siskins, crossbills.

Voice

Song is long, pleasing, somewhat canarylike; also a medley of trills and other notes, often including a *swee;* or a *dee-ar;* flight note, *per-CHIK-o-ree.*

Eggs

3–7; bluish-white; 0.6 x 0.5 in. (1.7 x 1.3 cm). Nest is cup-shaped; of grass, bark shreds, thistledown; in fork of bush or sapling.

Range

Breeds from Newfoundland and s. Man. s. to Ga., Ark., cen. Okla.; winters from n. U.S. s. to Gulf Coast and Mexico.

Note: The southwestern **LESSER GOLDFINCH,** *Carduelis psaltria* **(44:11)**, size, 4 in. (10.2 cm), is found in open areas in Colorado to eastern Texas. Males are black above, yellow below, and retain a black cap even in immature plumage. Females are like the American Goldfinch, but the rump is colored like back, not paler as in American.

HOARY REDPOLL
Carduelis hornemanni

Description
Size, 5 in. (12.7 cm). Virtually identical to Common Redpoll; distinguished by white, unstreaked rump under optimum conditions only.
Habitat, Habits, Voice, Eggs
Same as Common Redpoll.
Range
Breeds in Arctic; winters from Arctic s. to n. U.S.

PINE SISKIN
Carduelis pinus **44:9**

Description
Size, 4½–5¼ in. (11.4–13.3 cm). A • *heavily streaked* finch with a flash of • *yellow in wings and tail;* tail notched, wings with 2 white bars and yellow notch at base of primaries; base of outer tail yellow.
Similarities
American Goldfinch is never streaked. Redpolls are paler, streaks lighter. Female Purple and House finches are larger, have larger bills, and lack yellow in tail.
Habitat
Breeds in conifers; at other times found almost anywhere.
Habits
Active, gregarious; often occurs in flocks with goldfinches, crossbills, redpolls; flight goldfinchlike, frequently recognized by *shree* note as it flies overhead; frequents feeders (likes millet).
Voice
Song is like American Goldfinch's, but lower, longer, rougher; calls, a rasping, rising *shre-e-e-e-e-e;* a soft *it-it-it;* a louder *klee-up*; "2 or 3 very rapid, twangy syllables that suggest striking a tightly stretched wire" (Harrison).
Eggs
3–6; pale green-blue, spotted with brown; 0.7 x 0.5 in. (1.8 x 1.3 cm). Nest is saucer-shaped; of grass and twigs; in conifer, 8–30 ft. (2.4–9.1 m) up.
Range
Breeds from cen. Que. and s. Man. s. to N.S. and Nebr., and in mountains to N.C.; winters from s. parts of breeding range, s. to Fla., Tex., and Mexico.

RED CROSSBILL
Loxia curvirostra **44:4**

Description
Size, 5¼–6½ in. (13.3–16.5 cm). Crossed mandibles; bill at a distance looks slender; tail short, notched. Adult male: • *brick-red*, brightest on rump; wings and tail dusky. Female: olive-gray, rump and breast olive-yellow, wings and tail dusky. Immature male: intermediate in color between female and adult male.
Similarities
Crossed bill, lack of wing bars distinguish it from all other reddish birds.
Habitat
Conifers, especially pine and spruce.
Habits
Tame, actions parrotlike; extracts seeds from evergreen cones with its crossed mandibles; sings from treetop. Flight is undulating;

travels in small flocks, occurrence irregular both in summer and winter; nests from January to July.

Voice

Song is *"too-tee too-tee, too-tee, tee, tee"* (Hoffmann); also a trill; call, a *pip*, or *pip-pip*.

Eggs

4–5; pale greenish, spotted with brown and lavender; 0.8 x 0.6 in. (2.0 x 1.5 cm). Nest is of evergreen twigs and moss; usually in conifer, 5–20 ft. (1.5–6.1 m) up.

Range

Breeds from Newfoundland and Man. s. to N.C. and Wis.; winters from breeding range s. to Gulf Coast.

WHITE-WINGED CROSSBILL

Loxia leucoptera **44:6**

Description

Size, 6–6¾ in. (15.2–17.1 cm). Crossed mandibles, • *2 white wing bars* conspicuous in flight. Adult male: • *rose-red*, brightest on rump; wings and tail black with bold, white wing bars. Female: olive-gray, yellowish on rump; more streaked than Red Crossbill; wings as male. Immature male: intermediate in color between female and adult male.

Similarities

Pine Grosbeak is much larger.

Habitat

Conifers, especially spruce.

Habits

Travels in small flocks, occurrence irregular winter or summer.

Voice

Song is loud, long, varied, finchlike, sometimes given in air; calls, a clear *cheap*, a dry *chif-chif*.

Eggs

2–4; pale bluish-green, marked with brown and lavender; 0.8 x 0.6 in. (2.0 x 1.5 cm). Nest is of twigs, birch bark, moss; in tip of conifer branch.

Range

Breeds from n. Que. and cen. Man. s. to Newfoundland and s. Ont.; winters in breeding range and s. to New England and N.Y.; occasional to N.C.

RUFOUS-SIDED TOWHEE

Pipilo erythrophthalmus **45:6**

Description

Size, 7–8½ in. (17.8–21.6 cm). Tail long, rounded; eyes red, white in southeast Atlantic Coast race. Male: black upperparts, hood, and breast; rufous flanks; white in wing and on belly; white tips of outer tail feathers conspicuous in flight. Female: black replaced by brown. Juvenal (summer): like a large, slender sparrow with white tail corners, 2 buffy wing bars, streaked below. Western "spotted" form (found in Great Plains river valleys) has much white spotting on back and wings.

Habitat

Dry woods, especially second growth; edges; thickets.

Habits

Forages on ground, scratching noisily among dead leaves; sings from bush or low branch; opens and shuts tail.

Voice

Song is *drink your tea-e-e-e, see towhee*, or sometimes just *teeeee;* call, a *che-WINK!*

Eggs
3–6; white, finely dotted with red-brown; 1.0 x 0.8 in. (2.5 x 2.0 cm). Nest is of leaves, bark shreds; well hidden on ground in woods.
Range
Breeds from Maine and s. Man. s. to Fla., La., and C. America; winters from Md. and Nebr. southward.

Note: The **GREEN-TAILED TOWHEE,** *Pipilo chlorurus* **(45:5)**, size, 7 in. (17.8 cm), occurs from southern Wyoming and central Colorado west. It is a bird of brushy mountain slopes, rusty-capped, olive above, gray below and on face, and with a distinct white throat. It winters to central Texas and occasional strays are recorded eastward.

LARK BUNTING
Calamospiza melanocorys **48:3**

Description
Size, 6 in. (15.2 cm). Spring male: black with large • *white wing patch* conspicuous in flight. Female: streaked brown above, white below; brown on cheeks; white wing patch smaller than in male. Fall male: similar to female, but chin, wings, and tail black.
Similarities
Female chunkier than most sparrows. Male Bobolink has white on base of wing and back, and yellow hindneck.
Habitat
Grasslands.
Habits
Sings from elevated perch or from air; gregarious when not nesting; flocks wheel in unison.
Voice
Song is a series of warbled trills; flock often sings in chorus; flight call, a sweet *whoo-ee.*
Eggs
4–6; pale blue; 0.8 x 0.6 in. (2.2 x 1.7 cm). Nest is of grass, plant down; in tussock of grass.
Range
Breeds from Minn. and Man. s. to Tex.; winters from s. Tex. and Ariz. s. to Mexico.

GRASSHOPPER SPARROW
Ammodramus savannarum **46:11**

Description
Size, 4½–5¼ in. (11.4–13.3 cm). Head large, flat; short tail bristly tipped. Adult: buff stripe through dark crown; yellow spot before eye and on bend of wing; back heavily streaked, throat and • *breast unstreaked buff.* Immature: breast buffier than adult.
Similarities
Savannah has notched tail, streaked breast.
Habitat
Meadows, dry fields, prairies.
Habits
Runs mouselike through grass, sings from top of weed stalk, often quite persistently. Distribution uneven. In flight, close to ground, fluttering, wrenlike.
Voice
Song is an insectlike buzz, *tit-zeeeeeeeeee;* also a long, low, varied song later in season, like elfin music; call, a *tlik.*

Eggs
3–5; white, sparingly spotted with brown; 0.7 x 0.5 in. (1.8 x 1.4 cm). Nest is cup of grass in grass.
Range
Breeds from N.H. and Man. s. to Fla., W. Indies, and Mexico; winters from N.C. and Tex. southward.

BAIRD'S SPARROW
Ammodramus bairdii **46:7**

Description
Size, 5–5½ in. (12.7–14.0 cm). Appearance somewhat like commoner Grasshopper Sparrow, but with necklace of sharp black streaks on buffy breast; head yellow-brown streaked with black and prominent ochre crown stripe; tail notched.
Similarities
Savannah has narrower, lighter crown stripe and more extensive breast streakings.
Habitat
Dry long-grass prairies.
Habits
Sings from weed stalk.
Voice
Song is somewhat like Savannah's, but more musical and with more body—several *chip*'s, then a trill.
Eggs
3–5; whitish, marked with red-brown and black; 0.8 x 0.6 in. (2.0 x 1.5 cm).
Range
Breeds from Man. s. to Minn. and Mont.; winters in Tex. and Mexico.

HENSLOW'S SPARROW
Ammodramus henslowii **46:6**

Description
Size, 5 in. (12.7 cm). Large bill and head; short, spiky tail; pale stripe on buffy-olive crown, no yellow before eye; hindneck greenish; back and wings tinged with chestnut, streaked with white; upper breast pale buff with fine black streaks.
Similarities
Adult Grasshopper Sparrow has unstreaked, buffy breast and less prominent white streaks on back.
Habitat
Fields and meadows, often near water. Distribution irregular; only locally common.
Habits
Scurries mouselike through grass; flies low and jerkily, twisting tail; sings from top of weed stalk or fence post, often quite persistently.
Voice
Song is a short *s'LICK,* often heard at night.
Eggs
Similar to Grasshopper's.
Range
Breeds from ne. U.S. s. to e.-cen. states; winters from Ga. and Gulf states, southward.

SAVANNAH SPARROW

Passerculus sandwichensis **46:5**

Description
Size, 4½–5¾ in. (11.4–14.6 cm). • *Yellow in front of eye;* light stripe through crown; breast streaked with black, and with dark central spot; underparts whiter than most sparrows; legs pinkish; • *tail short,* notched. The Sable Island, Nova Scotia race, called the "Ipswich Sparrow" and recognized by some as a full species, is recognizably larger, and paler above; it winters on Atlantic Coast beaches.
Similarities
Song Sparrow has larger breast spot; wider and browner streaks; longer, rounded tail, lacks yellow on face. Vesper has white outer tail feathers. Other similar sparrows have rounded tails.
Habitat
Fields, fresh or salt meadows, beaches, dunes.
Habits
Hops, rarely walks; runs through cover with head low; if flushed, makes short zigzag, undulating flight revealing notched tail, then drops back into meadow; sings from low perch or wire; feigns injury to lure intruder from nest.
Voice
Song is a buzzy *tsip-tsip-tsip, seeee-saaay* or *tsip-tsip-tsip, saaay-seee;* call, a *thlip.*
Eggs
4–6; pale greenish-white spotted with red-brown and purple-brown; 0.8 x 0.6 in. (2.0 x 1.5 cm).
Range
Breeds from Labr. s. to N.J. and Mo.; winters from Mass. along coast and from cen. Ohio s. to Gulf Coast.

SEASIDE SPARROW

Ammospiza maritima **46:10**

Description
Size, 6 in. (15.2 cm). Dark olive-gray above, faintly streaked with gray below; large head and bill; • *yellow mark before eye;* black jawline separates white mustache from white throat. Distinctive races (formerly considered species) occur locally in parts of coastal eastern and southern Florida.
Similarities
Female Dickcissel has heavier, conical bill. See Sharp-tailed.
Habitat
Salt marshes.
Habits
Skulks mouselike in grass; if flushed, soon drops again into marsh; wades like a sandpiper; sometimes sings in flight, rising in air; inquisitive, can be attracted by hand-kissing.
Voice
Song is a "buzzy *tip-tip-ZEE-reeeeeeeee* with a *z* sound in gradually fading trill" (Saunders); *ZEE* note distinguishes it from Sharp-tailed; call, a harsh, Red-winged-like *check.*
Eggs
4–6; grayish-white, coarsely spotted with red-brown; 0.8 x 0.6 in. (2.0 x 1.5 cm).
Remarks
Dusky Seaside Sparrow and Cape Sable Seaside Sparrow now considered part of same species.
Range
Breeds along Atlantic and Gulf coasts from s. New England s. to Fla. and Tex.; winters in s. parts of breeding range.

LE CONTE'S SPARROW

Ammospiza leconteii **46:9**

Description
Size, 4½–5½ in. (11.4–14.0 cm). Tail short, spiky-tipped; crown stripe white; hindneck and nape with purple-chestnut (in young, buffy and unstreaked); • *eye line buffy-ocher,* throat and breast yellow-brown, sides streaked.

Similarities
Henslow's breast much less buffy. Henslow's and Grasshopper have buffy crown stripe. Grasshopper lacks streaked sides. Sharp-tailed lacks white crown stripe (but young Sharp-taileds have bright, buffy breasts).

Habitat
Prairie marshes, fields with matted cover, wet meadows.

Habits
Secretive, prefers running to flight; if flushed, flies jerkily for short distance, then drops back into cover; sings from top of weed or bush.

Voice
Song is thin, high, grasshopperlike, with a *chip* at end.

Eggs
4–5; whitish, heavily spotted with red-brown; 0.7 x 0.5 in. (1.8 x 1.3 cm). Nest is grassy cup in grass.

Range
Breeds from s.-cen. Canada s. to n.-cen. U.S.; winters in cen. and s. states.

SHARP-TAILED SPARROW

Ammospiza caudacuta **46:8**

Description
Size, 5–6 in. (12.7–15.2 cm). • *Gray ear patch bordered with buff;* short, spiky-tipped tail. Adult: dark cap (no obvious central stripe), gray nape; light streaks on dark back; more or less streaked on buffy breast. Young: much buffier and with dark streaking above and below. Subspecies (interior and coastal races) are variable in brightness of facial lines, amount of breast streaking, and overall dark or light coloration.

Similarities
Young Bobolink is similar but larger than young Sharp-tailed. When sparrows are flushed in salt marsh, a Sharp-tailed looks brownish above with a bristly rounded tail; a Savannah looks brownish above with a notched tail; a Seaside looks dark grayish above.

Habitat
Salt- or freshwater marshes.

Habits
Scurries mouselike through grass; if flushed, soon drops again into marsh, tail down; has flight song in spring.

Voice
Song is a "wheezy *tsup tsup shreeeeeeeee* with a *sh* sound in gradually fading trill" (Saunders).

Eggs
4–5; grayish-white, finely dotted with brown; 0.7 x 0.6 in. (1.9 x 1.5 cm). Nest is loosely woven grass cup in marsh.

Range
Breeds locally from cen. Canada to Minn. and along Atlantic Coast from lower St. Lawrence R., s. to N.C.; winters along Atlantic and Gulf coasts, from S.C. s. to Fla. and w. to Tex.

VESPER SPARROW

Pooecetes gramineus **46:2**

Description
Size, 5–6½ in. (12.7–16.5 cm). • *White outer tail feathers,* conspicuous in flight; streaked grayish-brown above, whitish below, with dark streaks ending sharply on lower breast; no breast spot; • *bend of wing chestnut; pale eye-ring* and rather distinct ear patch.
Similarities
Song Sparrow is browner above, has dark breast spot, lacks white outer tail feathers. Water Pipit is more slender; walks, does not hop. Lark Sparrow has white tail corners.
Habitat
Upland fields, pastures, fencerows.
Habits
Hops, does not walk; likes dust baths; sings more toward evening and at dawn; has courtship flight song; seen in small flocks in migration.
Voice
Song is *"ah ah ay ay tetetetetetetatatatata toto tu"* (Saunders). The introductory notes are long and low (those of Song Sparrow are short and high); call, a short *tsi.*
Eggs
4–6; grayish-white dotted with red-brown; 0.8 x 0.6 in. (2.0 x 1.5 cm). Nest is grass cup in grass.
Range
Breeds from N.S. and Ont. s. to N.C. and Tex.; winters from New England s. to Gulf Coast and Mexico.

RUFOUS-CROWNED SPARROW

Aimophila ruficeps **47:8**

Description
Size, 5–6 in. (12.7–15.2 cm). Adult: reddish-brown and gray streaks on back (appears dark); crown solid rufous; eye line gray; black whiskerlike streak bordering sides of throat; breast unstreaked grayish-white; tail rounded. Juvenal: crown brown, streaked; thin, dark streaks on breast.
Similarities
Chipping Sparrow has white line over eye, notched tail, black stripes on back. Swamp Sparrow is more rufous and with black stripes above, grayer below.
Habitat
Dry, grassy hillsides with low scrub, open pine-oak woods.
Habits
Sometimes occurs in small, loose colonies; skulks mouselike on ground, sings from low perch, looks hunched when sitting.
Voice
Song is several *mew*'s, then a gurgling warble (first part rising, last part falling), somewhat like House Wren's; call, a musical *deer, deer* or a nasal *chur, chur, chur.*
Eggs
3–5; bluish-white; 0.8 x 0.6 in. (2.0 x 1.5 cm). Nest is grassy cup on ground.
Range
Resident in sw. U.S., but e. rarely to w.-cen. Okla. and Tex., s. to Mexico.

BACHMAN'S SPARROW
Aimophila aestivalis **47:4**

Description
Size, 6 in. (15.2 cm). Brownish above, streaked with some gray and black; dull white below, unstreaked, with • *tan wash across breast;* pale bill.
Similarities
Field Sparrow has pink, not horn-colored, bill. Grasshopper has much shorter tail.
Habitat
Open pine woods, old brushy fields, cut-over areas, old orchards.
Habits
Shy, secretive, but sings from low branch or top of bush, February to August.
Voice
Song is clear, tuneful, of 5–12 phrases of different pitches and with pauses between, each phrase consisting of a long note and a trill; somewhat suggests pattern of Hermit Thrush; alarm note, a snakelike hiss.
Eggs
4; white; 0.7 x 0.6 in. (1.9 x 1.5 cm).
Former name
Pinewoods Sparrow.
Range
Breeds locally in se. U.S. from Pa., Ohio, and Ill. s. to cen. Fla., Gulf Coast, and Tex.; winters along Atlantic Coast from N.C., s. to Fla. and along Gulf Coast to s. Tex.

Note: **CASSIN'S SPARROW,** *Aimophila cassinii,* size, 5¼–5¾ in. (13.3–14.6 cm), breeds in the southwestern area from west-central Kansas and central Colorado, south and west. A bird of arid grasslands, it is plain-breasted; streaked, gray-brown above, with a streaked crown and white eye stripe. The **BLACK-THROATED SPARROW,** *Amphispiza bilineata* **(45:1),** size, 5 in. (12.7 cm), is a characteristic desert bird of the Southwest that reaches arid southeastern Colorado and the southern Texas coast. It is distinguished by its striking face pattern of stripes plus a black throat.

DARK-EYED JUNCO
Junco hyemalis **45:4**

Description
Size, 5½–6½ in. (14.0–16.5 cm). Solid gray, brown-gray, or blackish head, pinkish-white bill. Females, and especially immatures, tend to be duller and browner. Juvenals are streaked below. Dark tail with white outer feathers, white belly. Variously gray above and on breast and sides (eastern "slate-colored" form); gray with 2 white wing bars ("white-winged" Black Hills form); pinkish or brown sides with red-brown back, and in males gray (n. Rocky Mt. "pink-sided" form) to black (western "Oregon" form) hood.
Habitat
Diverse conifers, mixed, deciduous woods, edges, brushy fields, suburban gardens.
Habits
Tame; winters in flocks, often with other sparrows; sings from tree.
Voice
Song is a Chipping Sparrow-like trill, but more musical; also a warbled song; call, a *click*.

Eggs
3–6; grayish-white, thickly spotted with lilac and brown; 0.8 x 0.6 in. (2.0 x 1.5 cm).
Former names
White-winged Junco, Oregon Junco, Slate-colored Junco, Gray-headed Junco.
Remarks
Diverse forms of this variable species hybridize freely wherever they are in contact, much as in the Common Flicker. Intermediate patterns of color among hybrids are common. Although typical individuals of the different forms can be distinguished by the observer, the birds appear to disregard these differences in mating; hence, we consider them as one species. Individuals of western races wander east in winter, even to the Atlantic Coast. This is one of the commonest wintering species and a common visitor at bird feeders.
Range
Breeds from Newfoundland and Man. s. to Ga. and Mexico; winters from s. parts of breeding range s to Gulf Coast and Mexico.

LARK SPARROW
Chondestes grammacus **45:3**

Description
Size, 5½–6¾ in. (14.0–17.1 cm). Adult: • *chestnut crown stripes* and ear patch; white eye line and mustache; black line on either side of chin; dark spot in middle of unmarked white breast; tail black, • *fan-shaped,* and with • *white sides and corners.* Juvenal: fine, dark streaks replace dark breast spot.
Similarities
Vesper has white outer tail feathers.
Habitat
Grassy areas near trees; farms, ranches.
Habits
When sitting, raises crown at intervals; sings from elevated perch, or hovering in air.
Voice
Song is "somewhat like that of Indigo Bunting, but louder, clearer, much finer" (Forbush).
Eggs
3–5; white, spotted and scrawled with black; 0.8 x 0.6 in. (2.0 x 1.5 cm). Nest is grassy cup on ground or in low bush.
Range
Breeds from s. Ont. and n. Minn. s. to s. S.C., cen. Ala., and Tex.; winters from Gulf Coast s. to C. America.

TREE SPARROW
Spizella arborea **47:6**

Description
Size, 5½–6½ in. (14.0–16.5 cm). Rufous cap; • *black spot on plain breast;* bill dark above, light below; tail slightly notched; streaked brownish above, breast whitish; gray on sides of face; 2 white wing bars; a little white on tail.
Similarities
Lark Sparrow has a striped crown, much white in tail.
Habitat
Low trees, brush, weedy fields, gardens.

Habits
Gregarious in winter; visits feeders, often sings in concert.
Voice
Song is a sweet *"eee eee tay tititite tay"* (Saunders); call, a *tee-lo* and a *tseet.*
Eggs
4–5; light green, dotted with light brown; 0.7 x 0.5 in. (1.9 x 1.4 cm).
Range
Breeds from n. Que. and n. Man. s. to Newfoundland, cen. Que., and cen. Man.; winters from n. U.S. s. to N.C. and Ark.

CHIPPING SPARROW
Spizella passerina **46:13**

Description
Size, 5–5¾ in. (12.7–14.6 cm). Trim and slim; tail notched. Spring adult: streaked brown above; breast grayish-white, unmarked; chestnut cap; • *black line through eye,* white line over eye; 2 thin white wing bars. Immature: brown cap with pale central stripe and black streaks; buffy eye line; brown ear patch (not bordered with black), gray rump.
Similarities
See Clay-colored.
Habitat
Pinewoods, wood edges, lawns, gardens, country roadsides.
Habits
Tame, feeds on ground, sings from wire or high branch; often victimized by Brown-headed Cowbird.
Voice
Song, a simple unmusical trill, all on one pitch, suggesting songs of Worm-eating and Pine Warblers, junco, Swamp Sparrow, and others, but is the most mechanical; call, a *tsip.*
Eggs
3–5; greenish-blue with wreath of blackish dots about larger end; 0.7 x 0.5 in. (1.8 x 1.3 cm). Nest is hair-lined cup in tree or bush.
Range
Breeds from Newfoundland and Man. s. to Fla. and C. America; winters from Md. and Tenn., southward.

CLAY-COLORED SPARROW
Spizella pallida **46:12**

Description
Size, 5½ in. (14.0 cm). Adult: head mainly white with black molar streak, • *whitish center crown stripe* separating brown areas, black-bordered • *brown ear patch,* white stripe over eye; nape unmarked, pale gray; back gray-brown striped with black; 2 fine, white wing bars; entire underparts white. Immature: much browner.
Similarities
Immature Chipping has brown cheek patch, more bordered with black, and buffier rump. Larger Lark Sparrow has somewhat similar head pattern, but with dark spot on breast.
Habitat
Fields, open brush and underbrush, often near water.
Habits
Feeds low or on ground; in breeding season sings persistently from bush top.
Voice
Song is 3–4 slow insectlike buzzes; call, a *chip.*

Eggs
3–5; light green-blue, with brown spots about larger end; 0.6 x 0.5 in. (1.7 x 1.3 cm). Nest like Chipping Sparrow's.
Range
Breeds from n.-cen. Canada and Ill. s. to Nebr. and Mont.; winters from s. Tex., southward; occasional e. to w. N.Y., Conn.

Note: The westernmost fringe of our area in summer is visited by the **BREWER'S SPARROW,** *Spizella breweri,* size, 5 in. (12.7 cm), which resembles the Clay-colored, but has a finely streaked rather than broadly striped crown.

FIELD SPARROW
Spizella pusilla **47:5**

Description
Size, 5¼–6 in. (13.3–15.2 cm). • *Pink bill;* tail slightly notched; reddish ear patch; dull gray border to narrow white eye-ring; 2 white wing bars; underparts unmarked grayish-white, with brown wash on breast and sides.
Similarities
Tree Sparrow has dark spot on breast, stronger eye stripe, mustache stripe, grayer face and nape. Chipping Sparrow has white line over eye.
Habitat
Brushy hillsides, edges of fields, pastures, near bushes, trees.
Habits
Sings from low perch, sometimes by moonlight; gregarious in winter.
Voice
Song is a sweet, musical *twee-twee-twee, te-te-te-te-te-te,* first notes slow, trill becoming faster, and sometimes rising or dropping in pitch; call, a *tsip.*
Eggs
4–5; grayish-white dotted with red-brown and lilac; 0.7 x 0.5 in. (1.8 x 1.3 cm).
Range
Breeds from n. New England, s. Wis., and N.Dak. s. to Ga., Miss., La., and cen. Tex.; winters from Mass., Ohio and Kans. s. to Fla. and Gulf Coast.

HARRIS' SPARROW
Zonotrichia querula **47:1**

Description
Size, 7½ in. (19.1 cm). Spring adult: • *black cap,* face, throat and upper breast; bill pinkish. Brownish above, whitish belly, sides streaked; cheeks gray; 2 white wing bars. Fall adult: black crown, partly veiled with gray. Immature: spotted brown crown; buffy face, flanks and undertail coverts; white throat; necklace of brown streaks.
Similarities
Spring male Lapland Longspur is smaller, has chestnut nape.
Habitat
Breeds in dwarf timber near tundra; in migration and winter, thickets.
Habits
Similar to White-crowned and White-throated.
Voice
Song is 1–5 clear whistles, usually on same pitch, then a pause, followed by several notes on another pitch; alarm note, a loud *winck.*

Eggs
3–5; greenish-white, spotted and blotched with brown; 0.8 x 0.6 in. (2.4 x 1.7 cm).
Range
Breeds from n. Canada s. to n. Man.; winters from Iowa s. to Tex.

WHITE-CROWNED SPARROW
Zonotrichia leucophrys **47:3**

Description
Size, 5½–7 in. (14.0–17.8 cm). Black-and-white striped crown, pearly-gray breast, • *pinkish bill.* Adult: streaked brownish above, 2 white wing bars. Immature: similar, but crown striped with red-brown and buff, underparts washed with brown.
Similarities
White-throated is darker billed, adult has whiter throat sharply set off by black lines at sides, yellow spot before eye; immature much dingier on breast.
Habitat
In summer, shrubby mountainsides, dwarf willows; in migration, brush, edges, thickets, roadsides.
Habits
Has dignified bearing, can erect crown; forages on ground; sings from elevated perch, sometimes at night; often found with other species of sparrows.
Voice
Song is a clear, whistled *aaaa-ee-aay;* a husky, descending *see-say-so;* call, a *chink.*
Eggs
3–5; greenish-white, heavily spotted with red-brown; 0.8 x 0.6 in. (2.2 x 1.7 cm).
Range
Breeds from Labr. and Newfoundland w. across n. Canada and s. in far West; winters from N.J. s. to Ga., Gulf Coast, and Mexico.

WHITE-THROATED SPARROW
Zonotrichia albicollis **47:2**

Description
Size, 6–7 in. (15.2–17.8 cm). • *Black-and-white crown,* dark bill. Adult: black-and-white striped crown with yellow spot before eye, back brownish streaked with black, 2 white wing bars, cheeks and underparts gray; white throat sharply set off by black malar line. Immature: similar, but usually with some blurry streaking below, crown stripes brown and buff, white on throat duller. There is considerable variation with sharply marked, brighter, and duller, less white-faced phases.
Similarities
White-crowned may have grayish-white throat, but less distinct, not bordered by dark line as in White-throated. Adult Swamp has rusty crown; immature Swamp is smaller, very faint wing bars.
Habitat
Various woodlands, slash piles, brushy pastures, thickets.
Habits
Active; common; visits feeders, forages noisily on ground using feet to kick aside debris; sings from ground or low perch, sometimes at night and on spring migration; gathers in small flocks in fall.

Voice
Song is a set of clear whistles, often written as *Old Sam PEAbody, PEAbody, PEAbody;* notes, a distinctive, longish *ssst;* also a *chip.*
Eggs
4–5; grayish-white, dotted with red-brown; 0.8 x 0.6 in. (2.0 x 1.5 cm).
Range
Breeds from Newfoundland and cen. Que., s. to Pa., Wis., and N.Dak.; winters from s. New England and Kans., s. to Gulf Coast and Mexico.

FOX SPARROW
Passerella iliaca **46:1**

Description
Size, 6¼–7¼ in. (15.9–18.4 cm). Large, stocky; • *streaked red-brown and gray above,* rusty brightest on wings, rump, and tail; white below, heavily streaked with red-brown, forming a more or less conspicuous spot on the breast.
Similarities
Hermit Thrush has thinner, longer bill, no streaks on back.
Habitat
Open woods, edges, thickets.
Habits
Scratches noisily on ground with both feet at once, sings from elevated perch, visits feeders; gregarious except in breeding season.
Voice
Song is "loud, beautiful whistled melody, *hear hear I sing-sweet sweeter most-sweetly*" (Cruickshank); alarm note, a *smack;* call, a *sssp.*
Eggs
4–5; pale green, heavily spotted with rusty brown; 0.9 x 0.7 in. (2.3 x 1.8 cm). Nest is grass cup, feather lined, in bush or on ground.
Range
Breeds from n. Que. and n. Man. s. to N.B. and n. U.S.; winters from cen. states s. to Fla. and Gulf Coast.

LINCOLN'S SPARROW
Melospiza lincolnii **46:3**

Description
Size, 5–6 in. (12.7–15.2 cm). Appearance like a slender, elegant Song Sparrow; streaked grayish-brown above, white below; • *fine dark streaks largely confined to buffy band across breast,* seldom merge into dark spot; brown crown with light-gray central stripe; narrow white eye-ring.
Similarities
Song Sparrow is heavier, browner above, with broader, more extensive streaks below forming a spot on the breast; juvenal Song is finely streaked below but has no eye-ring or buffy wash on breast. Swamp is browner above, grayer below with rusty cap; immature Swamp has less distinct breast streaks than Lincoln's. Baird's is shorter-tailed, has buffy crown stripe, no buffy on breast. Savannah has shorter, notched tail; yellow before eye; no buff on breast.
Habitat
Breeds near bogs and water in northern forest; in migration, wet areas; dense thickets, edges; stone walls, bushy fences.

Habits
Secretive; does not sing in migration; skulks or runs mouselike along ground and under cover; hard to see, but responds to squeaking.
Voice
Song "suggests the bubbling, guttural notes of the House Wren, combined with the sweet, rippling music of the Purple Finch" (Dwight); call, a low *tsip*.
Eggs
4–5; white, heavily spotted with brown; 0.8 x 0.6 in. (2.0 x 1.5 cm).
Range
Breeds from Labr. and Newfoundland w. to n. Que. and s. to New England and n. Wis.; winters from Gulf Coast s. to C. America.

SWAMP SPARROW
Melospiza georgiana **47:7**

Description
Size, 5–5¾ in. (12.7–14.6 cm). • *Rusty crown;* • *gray-bordered, white throat,* tail rounded; eye stripe, grayish-white line over eye; mustache stripe; gray wash on breast; rusty wings with no wing bars. Spring adult dark rufous above, very faint bars; rusty crown; cheeks, nape, and underparts gray; white throat clearly outlined with gray. Fall adult: often buffy on sides and underparts, crown streaked with black, light center stripe. Immature: crown brown with light center stripe; belly, breast, sides buffy; faint streaks on breast.
Similarities
See Lincoln's Sparrow. Song Sparrow has longer, more rounded tail; is lighter above; and has streaked breast. Chipping is more slender, has notched tail, and black-and-white eye stripes. Tree and Field have white wing bars.
Habitat
Swamps, freshwater marshes; in migration, also weedy fields.
Habits
Sings from reed or bush in marsh.
Voice
Song is a 1-pitch trill, slower, louder, sweeter than Chipping Sparrow's; note, a *chip,* similar to White-throated's.
Eggs
4–5; variable, often bluish-white, heavily spotted with brown; 0.8 x 0.6 in. (2.0 x 1.5 cm).
Range
Breeds from e.-cen. Canada s. to e.-cen. U.S.; winters from N.Y. and Nebr. s. to Fla. and Gulf Coast.

SONG SPARROW
Melospiza melodia **46:4**

Description
Size, 5–6 in. (12.7–15.2 cm). One of the commonest sparrows. Brown streaked below with streaks converging into spot on breast. Adult: tail long, rounded; variable in tone; light line across top of brown crown and over each eye; blackish mustache stripe; white below; occasionally without central black spot. Juvenal: streaked buffy band across breast; hard to distinguish from Lincoln's or young Swamp Sparrow.
Similarities
Savannah has shorter, notched tail.

Habitat
Thickets, shrubbery, especially in damp or wet areas.
Habits
Visits feeder, likes to bathe; pumps tail up and down in flight, often hunches back, lowers tail before landing; forages on ground, sings from elevated perch.
Voice
Song is variable, but easy to learn: *"Maids! maids! maids! hang up your teakettle-ettle-ettle"* (Thoreau); call, a *tsak.*
Eggs
3–7; variable, often whitish, spotted with red-brown; 0.8 x 0.6 in. (2.2 x 1.5 cm); 2 broods. Nest is grassy cup on ground or in low bush.
Range
Breeds from Newfoundland and n. Man., s. to S.C. and N.Dak.; winters from s. Canada, s. to Gulf Coast and Mexico.

McCOWN'S LONGSPUR
Calcarius mccownii **48:12**

Description
Size, 6 in. (15.2 cm). Black inverted T on rather short, white tail. Spring male: black crown, whisker, and small breast patch; white eye line, throat, and belly; • *gray collar;* streaked brown-gray above, chestnut shoulder patch. Fall male: more tawny above, black largely replaced by gray. Female and immature: streaked brownish and buffy, similar to female and young Chestnut-collared, but tail pattern different.
Similarities
Spring male Chestnut-collared has chestnut collar. Female Smith's is buffier below, has white on outer rectrices only. Lapland has more distinct breast streaks, white outer rectrices only. Horned Lark has thin bill, black sideburns.
Habitat
Short-grass plains.
Habits
Flight undulating; after breeding season, gregarious; often found with Horned Larks and Chestnut-collared Longspurs.
Voice
Pleasant warble, with twitters, uttered hovering in air with wings seemingly straight up; call, a *chirrup-chirrup.*
Eggs
3–4; pale greenish, dotted with brown; 0.8 x 0.6 in. (2.0 x 1.5 cm).
Range
Breeds from se. Man. s. to N.Dak. winters from Nebr. s. to Mexico.

LAPLAND LONGSPUR
Calcarius lapponicus **48:9**

Description
Size, 6–7 in. (15.2–17.8 cm). Legs dark; white outer tail feathers in all plumages. Spring male: crown, face, throat, sides, and tail black; collar chestnut; upperparts brownish, streaked; wing bars, stripe from eye down sides, underparts, and outer tail feathers white. Fall male and female: black replaced by dark smudge on sides of lower throat and ear; legs black. In flight, wings look dark.
Similarities
Other longspurs have more white on tail, different tail patterns, except Smith's, which is buffy, not white, below. Snow Bunting is much lighter, has light, not dark, wings. Horned Lark has thin

bill, yellow and black throat markings, longer tail, less undulating flight.

Habitat
Tundra, plains, fields, beaches.

Habits
Gregarious, except in breeding season; often clings to weed stalks; hard to detect when it remains motionless on a field; flies like Snow Bunting, often seen with it.

Voice
Song is short *"tee-tooree, tee-tooree, teereeoo"* (Snyder), uttered on wing; in winter, a hoarse *churrr;* and a rattle followed by a whistled *dicky-dick-do.*

Eggs
6; variable, often greenish-gray, heavily blotched with red-brown; 0.8 x 0.6 in. (2.2 x 1.5 cm).

Range
Breeds from Greenland and Canadian Arctic s. to n. Que.; winters from n. U.S. s. to N.Y. and n.-cen. states.

SMITH'S LONGSPUR

Calcarius pictus **48:11**

Description
Size, 6½ in. (16.5 cm). Underparts buffy, bill thinner and more pointed than in other longspurs, legs paler; white outer tail feathers. Spring male: crown and cheek triangle black; unmistakable pattern of black-and-white face, buff nape, and underparts. Female, fall male, and immature: similar, but head streaked like rest of upperparts; underparts paler, but still completely buffy; throat and breast often somewhat streaked; fall male often shows small, white shoulder patch of spring plumage.

Similarities
See McCown's Longspur.

Habitat
Tundra, plains.

Habits
Gregarious in fall and winter, often occurs with other longspurs and Horned Lark; sings from ground.

Voice
Sharp clicking notes when flushed; song longer, but includes similar notes.

Eggs
4–6; whitish, spotted and blotched with purple-brown; 0.8 x 0.6 in. (2.0 x 1.7 cm).

Range
Breeds from n. Ont. and n. Man. s. to Hudson Bay; winters from Nebr. s. to Tex.

CHESTNUT-COLLARED LONGSPUR

Calcarius ornatus **48:10**

Description
Size, 6 in. (15.2 cm). Black triangle on white tail. Spring male: • *black cap* and underparts; white line over eye; chestnut collar; streaked brown above, whitish cheek and throat. Fall male: black and chestnut replaced by brown. Female and immature: streaked brown above, buffy below, with faint streaks on sides and breast.

Similarities
McCown's lacks breast streaks.

Habitat
Plains, prairies.
Habits
Similar to Lapland and McCown's; gregarious, often found with other longspurs; flight undulating, showing much white on tail.
Voice
Song is short, high, weak, and twittery, uttered on wing; remotely suggests Western Meadowlark; flight call, a twitter.
Eggs
4; pale greenish, dotted with brown; 0.8 x 0.6 in. (2.0 x 1.5 cm).
Range
Breeds from Man. s. to Minn.; winters from Iowa s. to Tex. and Mexico.

SNOW BUNTING
Plectrophenax nivalis **48:8**

Description
Size, 6–7¼ in. (15.2–18.4 cm). Summer male: • *white;* with back, bend of wing, wing tips, and middle tail feathers black. Summer female: similar, but black replaced by dusky. Winter male: white; cap, ear patch, and spots on rump rusty; bend of wing, wing tips, midtail, and spots on back black; winter female rustier; Immature: the rustiest, least white of all. In flight (in winter) looks white below including tail, dark above; wings always flash some white.
Similarities
In flight, Water Pipit and Horned Lark show dark wings and black tails.
Habitat
Mountain slopes, tundra; fields, salt marshes, flats, beaches in winter.
Habits
Walks, runs, hops; gregarious except in breeding season; compact flocks wheel in unison, and settle on field like snowflakes; often seen with Lapland Longspurs; in north, frequents camps, villages; called "House Sparrow of the Arctic."
Voice
Song is *"turee-turee-turee-turiwee"* (Snyder); notes include a high whistled *teer;* also a purring note.
Eggs
4–7; variable, whitish spotted with brown; 0.8 x 0.6 in. (2.0 x 1.5 cm).
Range
Breeds in Arctic, from Greenland w. across Canada and s. to n. Que.; winters from n. Canada s. to Pa., Ind., and Kans.

Mammals

Consulting Editor
Sydney Anderson
Curator of Mammalogy
American Museum of Natural History

Illustrations
Plates 49–52, 54, 56–58, 60, 61 Nina L. Williams
Plates 53, 55, 59, 63 John Hamberger
Plate 62 Jennifer Emry-Perrott
Text Illustrations by Nina L. Williams and Jennifer Emry-Perrott

Mammals

Class Mammalia

Mammals are warm-blooded, air-breathing, milk-producing vertebrates with hair. They give their young a period of parental care, and all but the platypus and echidna, which lay eggs, bring forth their young alive. Most have several different kinds of teeth.

Warm-blooded means the animal is capable of maintaining a relatively constant internal body temperature irrespective of the outside air or water temperature. The normal internal temperature for most active mammals lies in the range of from 90° to 104° Fahrenheit. In hibernating individuals, however, the temperature is reduced and may be only a few degrees above freezing. All mammals have hair, although with some mammals the hair may be in the form of bristles. Even whales, at least embryonic whales, have some bristles. In contrast, the reptiles, from which mammals descended, are cold-blooded vertebrates with scales or plates instead of hair, and they do not produce milk or give their young parental care. Reptilian teeth are less differentiated, as well.

Most mammals also have four feet and a tail. In the seals, whales, and manatees, the forefeet are transformed into flippers; and in the whales and manatees, the hind limbs have totally disappeared externally. In a few species the tail is greatly reduced or missing, as in bears and human beings.

Early mammals had five digits on each foot. This number was gradually reduced in many evolutionary lines; in the horse only one digit, bearing the hoof, remains. Human beings retain the original five, which of course, has been of crucial importance in their evolution.

Evolution

Mammals arose from therapsid reptiles in the late Triassic period and early Jurassic. In addition to the differences already mentioned, mammals represent an advance over reptiles in various other ways. Being warm-blooded means that, unlike reptiles or amphibians, mammals can occupy cold climates and deal with freezing winters by continuing activity or hibernation. Having differentiated teeth, they can chew their food and prepare it for quick digestion, increasing its availability as a source of energy. This makes mammals more active than reptiles, which, after swallowing their prey unchewed, may be sluggish for a week or two as the meal is slowly digested.

Increased activity means greater ease in catching prey and escaping from predators, both desirable in the struggle for existence. This activity is aided also by the development of a diaphragm, which promotes deep breathing, hence more rapid aeration of the blood and quicker metabolism. A reptile's rib-controlled breathing can never provide oxygen as quickly as the bellows action of the mammalian diaphragm. Activity is also furthered by a four-chambered heart, which separates arterial blood from venous blood, helping in the more rapid and efficient utilization of oxygen. The red blood cells of mammals are nonnucleated, thus improving oxygen-carrying capacity.

Evolution has given rise to three quite different groups of living mammals. It is presumed that the subclass Prototheria arose in the Jurassic period. The present representatives of this subclass are the primitive egg-laying monotremes, the platypus and echidna, which are confined to Australia and New Guinea.

In the succeeding period, the Cretaceous, two new groups arose. The less advanced were the Metatheria, the marsupials, or pouched mammals, which bring forth living young, but in an embryonic condition. In most species, these are kept and nourished in an external pouch for a protracted period. The more advanced were the Eutheria, the placental mammals, which retain the young for a longer period within the body of the mother and which lack any outer pouch. At the end of the Cretaceous and in the Paleocene, great development took place among the placentals. This group then proved superior to the marsupials, which have survived to the present almost exclusively in Australia and South America, continents long isolated from the world of the more aggressive placentals.

Adaptations
Mammals have dominated the globe for 70 million years. In that time they have become adapted to every major type of environment. From a central, walking terrestrial stock, various running, or cursorial, forms progressively lost digits as they increased their speed, until they evolved into the fleet antelope with two hoofs and the horse with one. From the central stock some forms, such as the pocket gophers and moles, developed adaptations for a life underground.

Others, such as the squirrels and monkeys, took to the trees, where they now lead semiarboreal or arboreal lives. Flying squirrels suggest an attempt at the conquest of the air by rodents, a triumph achieved by bats in the Paleocene or earliest Eocene. Notable as well were the adaptations required for successful reinvasion by mammals of the sea, the ancestral home of all life. Seals and walruses have not yet freed themselves from dependence on the land, on which they still must bring forth their young; but the whales and sirenians live entirely in the water, to which they are almost as well adapted as fish, save for their dependence on the air above for oxygen. Along with their adaptations to all environments, mammals have successfully adapted to all altitudes, latitudes, and climates, and have grown to various sizes.

It is in intelligence, however, that mammals are supreme. Compared with most other mammal genera, *Homo* is a newcomer, perhaps two or three million years old, compared, for example, with the eight-million-year-old lineage of his companion, *Felis,* the house cat.

Conservation
With mammals, as with almost all the wildlife covered in this book, the greatest enemy today is people. Once shooting and trapping were the main dangers; they may still be for the bighorn and grizzly, the mountain lion and fisher. More important now, however, is loss of habitat through the expansion of civilization. This loss continues and the adverse pressure increases, particularly on many of the larger mammals. A great deal still needs to be known about the habits and ecologic relationships of most mammals, including even some of the more common species.

Field Study
Field study of mammals is quite different from that of birds. A few common species, such as rabbits, squirrels, skunks, woodchucks, and some mice and rats, are frequently encountered. But most are not usually seen. Many are nocturnal; most are secretive; the larger ones are quite wary; sea mammals are usually seen by accident.

Therefore mammal study is usually part of a general field trip on which some mammals and the tracks, scats, and signs of others may be seen. A flashlight will surprise some mammals at night. Taking a "stand" (that is, sitting quietly at a likely place in the woods) will often afford a view of some of the shyer species.

Habitat
The best places to see mammals on land include woods, wood edges, stone walls, the unkempt corners of suburbs and countryside (small mammals), the edges of lakes and streams (moose, beaver, otter, muskrat), deep woods and swamps (bear, bobcat), caves and old mines (bats), and old houses and outbuildings (mice, squirrels).

Mammals in general may be seen throughout the year; many, however, appear to be most active in spring and fall. The hibernators, of course, are hidden away in winter, although if their roosts are known, they may be seen readily.

Teeth and Food
The primitive placental mammalian tooth count was forty-four, consisting, in each jaw, of three incisors, one canine, four premolars, and three molars on either side. In this chapter the dental formula will be written as 3143/3143. The digits of the first number refer to the number of incisors, canines, premolars, and molars, respectively, on each side of the upper jaw; the second number, to those on each side of the lower. Incisors are for cutting, canines for tearing, premolars and molars for grinding or shearing. Mammals with different diets require different kinds of teeth, and their tooth structure has developed along with their diet. Omnivorous human beings have the formula 2123/2123. The rodents, which gnaw but do not tear, have lost their canines; armadillos, which neither gnaw nor tear, have lost both incisors and canines; many mice have no premolars; some whales have no teeth at all. The food of an animal determines its habits; its teeth reflect its food. Thus, given in the chapter are the dental formulae, as a matter of interest, and as an aid in the identification of skulls that the naturalist sometimes encounters in the field, in caves, or in the pellet of a raptor.

Reproduction
Placental mammals are so named from the placenta through which the young are nourished inside the body of the mother. This provides a safe first home in which young mammals can develop for a longer time and hence reach a more advanced state of development. This characteristic has enabled these mammals to concentrate on producing a few or only one young, instead of the dozens or hundreds of young or eggs that reptiles require in order to assure the perpetuation of their kind.

Despite the prolonged internal development, newborn mammal young are often quite helpless. Many require parental, or at least maternal, care for a period after birth. This affords the parent an opportunity to teach the young, and the species thus acquires a more varied behavioral pattern than the stereotyped responses characteristic of the reptiles. The increased emphasis on learned responses provides greater opportunity for the use and development of intelligence.

Taxonomy
The names and organization of the Class Mammalia in this chapter follow the most recent checklist of J. Knox Jones, Jr., Dilfold C. Carter, and Hugh H. Genoways of Texas Tech University, entitled *Revised Checklist of North American Mammals North of Mexico, 1979.*

Range and Scope
There are about 4000 mammal species in the world, representing 122 families; of these, there are in North America, north of Mexico, about 410 species in 40 families, of which 183 nondomesticated species occur in the region covered in this chapter.

The area of coverage includes that region of North America east of the 100th meridian and north of the 25th parallel. This encompasses the United States and Canada from the Atlantic Ocean west to the Dakotas, Kansas, Nebraska, Oklahoma, and east Texas; and from northern Canada to southern Florida. Distributions apply to the species as a whole only within the areas of geographic coverage of the chapter.

Species may vary widely in color over different parts of their range, or vary as a result of age, sex, or season. Those illustrated are of a typical color, and the text describes the range of variation. Longevities given are mostly for captive species; maximum age in the wild is usually much less.

Nomenclature

The scientific names used in the chapter are from Hall and Kelson's *The Mammals of North America,* Ronald Press, 1959, as revised in scientific literature through 1976. Common English names follow *Revised Checklist of North American Mammals North of Mexico,* 1979, by J. Knox Jones, Jr., D. C. Carter, and H. H. Genoways, of the Museum of Texas Tech University, Lubbock, Texas.

USEFUL REFERENCES

Anderson, S., and Jones, J. K., Jr., eds. 1967. *Recent Mammals of the World.* New York: Ronald Press. (453 pp., 70 figs.)

Blair, W. F., et al. 1968. *Vertebrates of the United States.* New York: McGraw-Hill. (616 pp., illustrated.)

Burt, W. H. 1972. *Mammals of the Great Lakes Region.* Ann Arbor: Univ. of Michigan. (246 pp., 54 figs., many maps.)

Burt, W. H., and Grossenheider, R. P. 1976. *A Field Guide to the Mammals.* 3rd ed. Boston: Houghton Mifflin. (319 pp., 24 col. pl. and other illustrations.)

Cahalane, V. H. 1947. *Mammals of North America.* New York: Macmillan. (682 pp., illustrated.)

Hamilton, W. J., Jr. 1943. *The Mammals of Eastern United States.* Ithaca, N.Y.: Comstock. (Cornell Univ.) (432 pp., illustrated.)

Jackson, H. H. 1961. *Mammals of Wisconsin.* Madison: Univ. Wisconsin Press. (504 pp., 81 maps, illustrated.)

Lowery, G. H. 1974. *The Mammals of Louisiana and Its Adjacent Waters.* Baton Rouge: Louisiana State Univ. Press. (565 pp., illustrated.)

Murie, O. J. 1954. *A Field Guide to Animal Tracks.* Boston: Houghton Mifflin. (374 pp., illustrated.)

Palmer, E. L. 1957. *Fieldbook of Mammals.* New York: Dutton. (321 pp., illustrated.)

Palmer, R. S. 1954. *The Mammal Guide.* New York: Doubleday. (384 pp., 30 col. pl. and other illustrations.)

Peterson, R. L. 1966. *The Mammals of Eastern Canada.* Toronto: Oxford Univ. Press. (465 pp., 233 figs., 107 maps.)

Schwartz, C. W., and Schwartz, E. R., 1972. *The Wild Mammals of Missouri.* Columbia: Univ. Missouri Press and Missouri Conserv. Comm. (341 pp., illustrated.)

Marsupials
Order Marsupialia

NEW WORLD OPOSSUMS
Family Didelphidae

VIRGINIA OPOSSUM
Didelphis virginiana

51:6
Fig. 15

Description
Size: head and body, 15–20 in. (38.1–50.8 cm); tail, 9–13 in. (22.9–33 cm); weight, 6–12 lb. (0.27–0.54 kg); 50 teeth, 5134/4134; mammae, usually 13. Fur coarse, usually gray above, but varies from white to near-black, only slightly paler below; face white, ears and inner part of tail black. Snout long, legs short; 5 toes on each foot; first toe on hind foot opposable (thumblike) and lacks claw; tail is long, scaly, prehensile.

Habitat
Variable, but prefers woodlands.

Habits
Nocturnal; climbs well; feigns death or "plays possum" if frightened.

Reproduction
2 litters per year; nest located in a burrow, hollow log, hollow tree; gestation 12½ days; 8–18 (occasionally more) young that weigh $\frac{3}{100}$–$\frac{7}{100}$ oz. (1–2 g), born at an early stage of development; young travel immediately to pouch, or *marsupium,* where they attach firmly to a teat and remain for 55–70 days; pouch young usually number 6–9.

Remarks
The Virginia Opossum, which is the only marsupial that occurs north of Mexico, has extended its range northward in both prehistoric and recent times. The family Didelphidae dates back to Cretaceous times, making it one of the oldest of living mammalian families.

Range
Throughout e. U.S. and southernmost Ont.

Other name
Common Opossum.

Fig. 15

Insectivores

Order Insectivora

The insectivores of the eastern United States are small terrestrial or fossorial mammals with small, sharp teeth; long snouts; and flat, claw-bearing feet. Of all placental mammals living today, these are regarded as the most primitive, and therefore closest to the ancestral mammalian stock.

SHREWS

Family Soricidae

At first glance a shrew looks like something halfway between a mouse and a mole. It is small, mouse-colored, and has a long, pointed nose. Its inconspicuous eyes and ears distinguish it from a mouse, and it has five toes on each foot. Most mice have only four on the front foot. Shrews lack the strong digging forelimbs of moles, and are swift runners above ground. They are small, three and one-half to six inches (8.9–15.2 cm) long, usually grayish or brownish, generally lighter beneath; the sexes look alike. Females have six mammae, usually two abdominal, four inguinal. The teeth, brown-tipped, number thirty-two, formula 3133/1113, except in the Least and Desert Shrews.

Favorite habitats include moist forest floors, swamps, marshes, bogs, rocks, tundra, and mountains, but a few species inhabit dry areas. Shrews make small, round holes in leaf litter about rocks or logs and use the surface runways of mice and voles. They do not hibernate. In disposition they are aggressive, irascible, and nervous.

The diet includes insects, earthworms, grubs, snails, other invertebrates, and mice, as well as berries and soft vegetation. They are voracious feeders and many apparently eat their own weight in food every twenty-four hours just to sustain life. Shrews breed early in the year and many species have more than one litter; young are brought forth in a little round nest of shredded vegetation concealed in leaves, rocks, logs, or a burrow. Their gestation averages eighteen to twenty-two days, and litters range from four to ten. The young are often independent within three weeks. Maximum age rarely exceeds two years. Many utter tiny, high-pitched squeaks. Young shrews have a pencil of hairs at the tip of the tail. These wear off with age, and the tail tip of an old shrew is nearly naked.

It is difficult to distinguish the individual species of shrews.

LONG-TAILED SHREWS

Genus *Sorex*

Shrews of the Genus *Sorex* are widespread in North America, occurring discontinuously from the Bering Strait to Guatemala. These shrews are also widely distributed in northern Eurasia.

MASKED SHREW

Sorex cinereus **52:14**

Description

Size: head and body, 2½ in. (6.4 cm); tail, 1¼ in. (3.2 cm); weight, $\frac{1}{10}$–$\frac{1}{5}$ oz. (2.8–5.6 g). Grayish-brown above, paler below; tail dark above, buffy below. Pelage in winter darker above than in

summer, paler below. Five unicuspids on each side of upper jaw, third and fourth slightly smaller than first and second.
Habitat
Moist areas from salt marshes to alpine meadows; may often be found in drier areas.
Habits
Can eat 3 or 4 times its own weight in 24 hours.
Range
Most widely distributed in the n. part of the E., s. to Ill. and the Appalachian Mts.
Other name
Cinereus or Common Shrew.

SOUTHEASTERN SHREW

Sorex longirostris *Fig. 16*

Description
Size: head and body, 2½ in. (6.4 cm); tail, 1½ in. (3.8 cm); weight, ⅛ oz. (3.5 g). Reddish-brown, paler below; tail relatively short, slightly bicolored.
Habitat
Cool, open fields and woods, grassy edges of ponds and bogs.
Range
In s. states, from Ill., southward, and from Coast, w. to Mo.

Long-tailed Shrew, p. 231
Southeastern Shrew

WATER SHREW

Sorex palustris *Fig. 17*

Description
Size: head and body, 3⅓–3⅗ in. (8.5–8.7 cm); tail, 2½–3 in. (6.4–7.6 cm); weight, ⅓–½ oz. (9.5–14.2 g). Largest long-tailed shrew, with stiff hairs on the sides of its large hind feet. Glossy black or blackish-gray above; gray, grayish-brown, or silver below; tail bicolored. Middle toes partly webbed.

Pygmy Shrew, p. 231
Smoky Shrew, p. 230
Fig. 17
Water Shrew

Habitat
Bogs, edges of streams, ponds, and lakes in northern and mountainous areas.
Habits
Swims, dives, and can run on surface of water.
Range
Throughout e. Canada, s. through Appalachians and w. to e. Mont. and Wyo.
Other name
Northern Water Shrew.

SMOKY SHREW
Sorex fumeus *Fig. 17*

Description
Size: head and body, 2½–3 in. (6.4–7.6 cm); tail, 1¾–2 in. (4.4–5.1 cm); weight ⅕–⅖ oz. (5.7–11.3 g); all 6 mammae inguinal. Dull brown, grayish in winter, with bicolored tail and pale feet.
Habitat
Damp, shady forests, especially in deep moss, leaf litter, and rotting logs.
Habits
Feeds primarily on earthworms and insects; seems to experience more fluctuations in numbers than do most shrews.
Range
Throughout e. Canada, s. through Appalachians and w. into cen. Midwest.

ARCTIC SHREW
Sorex arcticus **52:13**

Description
Size: head and body, 3 in. (7.6 cm); tail, 1¼–1⅔ in. (3.2–4.1 cm); weight, ¼ oz. (7.1 g). Brightly tricolored; back dark brown, sides light brown, underparts pale; brighter in winter.
Habitat
Spruce and tamarack swamps in southern parts of range and tundra farther north.
Range
Throughout ne. U.S. and Canada.

GASPÉ SHREW
Sorex gaspensis

Description
Size: head and body, 2 in. (5.1 cm); tail, 2 in. (5.1 cm). Similar to Long-tailed, but smaller.
Habitat
Moist streamsides.
Range
Shickshock Mts. of Gaspé Peninsula, Que., and nearby Mt. Carleton, N.B.

LONG-TAILED SHREW

Sorex dispar *Fig. 16*

Description
Size: head and body, 2¾ in. (7 cm); tail, 2–2½ in. (5.1–6.4 cm); weight, ⅕ oz. (5.7 g). Tail proportionally long for the genus. In summer, dark gray above, only slightly paler below; in winter, slate-gray above and below.
Habitat
Damp, rocky situations in coniferous forests.
Range
Throughout Appalachians, from s. Que. to e. Tenn.
Other name
Rock Shrew.

OTHER SHREWS

Genera *Microsorex, Blarina, Cryptotis,* and *Notiosorex*

PYGMY SHREW

Microsorex hoyi *Fig. 17*

Description
Size: head and body, 2–2½ in. (5.1–6.4 cm); tail, 1–1⅖ in. (2.5–3.6 cm); weight, ¹⁄₁₂ oz. (2.4 g); 32 teeth, 3133/1113. Smallest shrew in the East. Gray or grayish-brown above, paler below; 5 unicuspids, but only 3 are easily visible in lateral view.
Habitat
Wet and dry woods and adjacent grass clearings.
Remarks
Regarded as America's smallest living mammal, 6 of these tiny shrews do not equal ½ ounce (14.2 g).
Range
Throughout most of e. Canada, southward into e. U.S., to Pa. and Ohio, w. to Wis. and e. N. and S. Dakota.

SHORT-TAILED SHREW

Blarina brevicauda **52:11**

Description
Size: head and body, 3–4 in. (7.6–10.2 cm); tail, ¾–1 in. (1.9–2.5 cm); weight, ⅖–1 oz. (11.3 -28.4 g). Slate-gray; eyes barely apparent; ears small and hidden in pelage; short tail; stocky build; muzzle not as pointed as long-tailed shrews; skull ridged, other shrews have smoother skulls. Teeth dark chestnut; first and second unicuspids larger than third and fourth.
Habitat
Forests and moist grassy areas having copious vegetation and litter.
Habits
Makes flattened runways, 1 x ¾ in. (2.5 x 1.9 cm), through leaf mold.
Remarks
One of the few poisonous mammals; the saliva in its bite can poison and paralyze its prey, and it is painful but not otherwise dangerous to humans.
Range
Throughout entire ne. U.S. and adjacent areas of s. Canada.

SOUTHERN SHORT-TAILED SHREW
Blarina carolinensis

Description
Size: head and body, 2½–3½ in. (6.4 –8.9 cm); tail, ½–¾ in. (1.3–1.9 cm); weight, ¾ oz. (21.3 g). Like Short-tailed except smaller. The short-tailed shrews of the Dismal Swamp, Virginia, may be of a separate species.
Habitat and Habits
Like those of Short-tailed.
Range
S. of the range of Short-tailed, throughout se. U.S.

LEAST SHREW
Cryptotis parva **52:12**

Description
Size: head and body, 2⅕–2½ in. (5.6–6.4 cm); tail, ½–¾ in. (1.3–1.9 cm); weight, ⅐–⅕ oz. (4.1–5.7 g); 30 teeth, 3123/1113, dark-tipped. Small, short-tailed. Ashy below; young are dark slaty-gray.
Habitat
Brushy and grassy situations, dried marshes, damp woods.
Habits
In captivity, when ample food is available, are docile and nest in groups; also known in wild; have been observed to hunt in packs and jointly attack animals many times their own size.
Range
Throughout cen. and s. parts of e. U.S., n. from N.Y. and S.Dak.

DESERT SHREW
Notiosorex crawfordi

Description
Size: head and body, 2½–2¾ in. (6.4–7 cm); tail, 1–1½ in. (2.5–3.8 cm), weight, ⅙–¼ oz. (4.7–7.1 g); 28 teeth, 3113/2013. Pale ashy. Three unicuspids: only shrew having less than 30 teeth.
Habitat
Desert areas and short grass plains.
Habits
Poorly known; often found in association with wood-rat dens or apiaries.
Range
In e. Okla., e. Tex., and w. Ark.

MOLES
Family Talpidae

Moles are small, burrowing animals with dark, soft, velvety fur that rubs both ways. They weigh one and one-half to four ounces (43–115 g) and have small eyes and external ears. Their outwardly turned forefeet make them strong, powerful diggers. They can utter a high-pitched squeak. Moles occupy all humid habitats from sea level to above timberline and are active at all hours and seasons; they do not hibernate. The tunnels of the Eastern and Hairy-tailed Moles make ridges on top of the ground.

The mole diet is largely earthworms and other invertebrates, but may include some plant material. The nest is a six-inch (15.2 cm) agglomeration of dry vegetation in a sheltered or underground earth mound. The female brings forth her two to six naked young from April to June. Moles check wasteful runoff by churning up the forest or meadow floor, thus helping rain soak into the ground. The family dates back to the Eocene.

HAIRY-TAILED MOLE

Parascalops breweri **52:15**

Description
Size: head and body, 4¼ in. (10.8 cm); tail, 1–1½ in. (2.5–3.8 cm); 44 teeth, 3143/3143. Fur blackish, sometimes yellowish on belly or snout. Nose naked, pointed; forefeet equally broad as long. Distinguished from other moles in East by its hairy tail.
Habitat
Well-drained soil with wooded cover, lawns, golf courses, to 3000 ft. (915 m).
Habits
Ridges less conspicuous than those of Eastern Mole; also makes low mounds; reputed to be capable of eating 3 times its own weight in earthworms in 24 hours.
Range
New England, throughout Appalachians, and adjacent parts of Canada.

EASTERN MOLE

Scalopus aquaticus **52:16**

Description
Size: head and body, 4¼–6½ in. (10.8–16.5 cm); tail, 1–1½ in. (2.5–3.8 cm); 36 teeth, 3133/2033. Most widespread and commonest eastern mole. Color ranges from dark gray-brown to pale whitish. Nose naked; nostrils open upward; forefeet broader than long, and much larger than hind feet; tail naked.
Habitat
Well-drained soil in open areas.
Habits
Makes circular mounds; Pocket gophers' mounds are crescent-shaped. Most active in spring and fall and after summer rains.
Range
Entire e. U.S., from coast to 100th meridian.

STAR-NOSED MOLE

Condylura cristata **52:17**

Description
Size: head and body, 4¼–5 in. (10.8–12.7 cm); tail, 3–3⅓ in. (7.6–8.5 cm); 44 teeth, 3143/3143. A "star nose," or ring of 22 fleshy pink tentacles at end of nose which move as the animal searches for food. Fur dark brown or black. Eyes visible; forefeet as long as broad. Tail scaly with hairs thickest in middle, narrower near body; thicker in winter.
Habitat
Moist ground near water.
Habits
Makes mounds in muck 1 ft. (0.3 m) wide, can swim and dive.
Range
In ne. U.S. and se. Canada.

Bats

Order Chiroptera

Bats are the only mammals that fly. They do this by means of thin membranes (double-layered extensions of the body skin) that connect all fingers with the arm, sides, legs, and tail. The middle finger usually is the longest; the thumb is free. The sexes generally look alike, but young frequently are darker in color.

Bats have certain peculiar structures. The interfemoral membrane joins leg and tail. The calcar is a cartilage that extends from the foot along the outer edge of this membrane and acts as a brace. If some membrane extends beyond the calcar, the calcar is said to be keeled. The tragus is a leaflike formation inside the ear.

Bats utter high-pitched squeaks, sometimes fifty or more per second, which are inaudible to humans. Aided in part by the tragus, they hear the echoes of these sounds as they bounce back off insects and nearby obstacles; thus, bats can find their prey in the dark and avoid hitting obstructions. Bats are crepuscular (twilight-flying) and nocturnal. They roost by day, hanging upside down by their feet in caves, mine shafts, rocky crevices, hollow trees, thick foliage, or buildings, or behind loose bark or shutters. Some species are solitary; others are colonial. In winter bats either hibernate, often in colonies or caves, or, like birds, migrate south. Some species both migrate and hibernate.

The diet of bats in eastern North America is exclusively insects, which they capture on the wing, either in their mouths or in the "dragnet" of their tail membrane when cupped and used as a scoop.

In the females of some hibernating species the sperm or embryo can lie dormant for several months; thus an early fall mating may give rise to a late spring young.

Females often assemble in "maternity roosts" away from the males before giving birth to the young, usually one or two, which are born from May to July. Young babies often cling to their mother when she flies; later they hang by themselves at their roost until they are able to fly, which is usually in five to six weeks.

Bats have few natural enemies. When not hibernating, bats seem unable to endure long fasts; protracted cold and windy, rainy weather, which keeps insects from flying, may result in some mortality. Bats are complementary to birds in providing a natural check on insects. Birds feed on day-flying insects, bats on nightfliers.

Bats should be encouraged, and not feared, because they are valuable insect eaters.

COMMON OR PLAIN-NOSED BATS

Family Vespertilionidae

In this family, the tail protrudes only slightly beyond the edge of the interfemoral membrane. There are two mammae, except in the Red, Yellow, and Hoary Bats, which have four. In most, the interfemoral membrane is not furred or only slightly so. The number of teeth ranges from thirty to thirty-eight. In winter, vespertilionids either hibernate or migrate southward. The family dates back to the middle Eocene. In the Genus *Myotis* the thirty-eight teeth are all 2133/3133. There is usually one offspring born in late spring or early summer, and it weans in about six weeks.

LITTLE BROWN MYOTIS

Myotis lucifugus **49:5**

Description
Size: forearm, 1½ in. (3.8 cm). Small. Dark brownish above, buffy-gray below; calcar unkeeled.
Habitat
Hangs in caves, hollow trees, buldings, under loose bark or shingles, and behind shutters.
Habits
Colonial; flits on rounded wings over fields, forests, wood edges or water; flight is erratic, feeble; sometimes flies by day in forest clearings; hibernates.
Range
Throughout most of E., from tree limit s. to cen. Ga.
Other name
Little Brown Bat.

SOUTHEASTERN MYOTIS

Myotis austroriparius

Description
Size: forearm, 1½ in. (3.8 cm). Small, fur thick. Medium orange-brown to buff-brown above, dull buff below.
Habitat
Principally caves and old mines.
Range
In se. U.S., s. from Ind. and Ill. to Gulf Coast.

GRAY MYOTIS

Myotis grisescens

Description
Size: forearm, 1¾ in. (4.4 cm). Medium-sized, fur velvety. Color is russet smoky-brown. Hairs not banded from base to tip, same color throughout; wing membrane attached on ankle, not to base of toes, as in other small eastern bats.
Habitat
Roosts in caves, sometimes sewers or buildings.
Habits
Colonial.
Range
In se. U.S., from Ky., w. to Mo. and Kans., s. through Ala.

CAVE MYOTIS

Myotis velifer

Description
Size: forearm, 1¾ in. (4.4 cm). Medium-sized. Dull brown above, paler below; base of hairs dark. Hair has "woolly" appearance; wing membrane starts at base of toes.
Habitat
Roosts in caves and, in summer, buildings.
Habits
Colonial.
Range
In e. range, se. Kans.

KEEN'S MYOTIS

Myotis keenii *Fig. 18*

Description
Size: forearm, 1½ in. (3.8 cm). Small. Fur dark yellowish-brown. Ears, if laid forward, extend slightly beyond end of nose.
Habitat
Wooded regions.
Habits
Hangs in caves, trees, buildings; flight direct, strong.
Range
Throughout e. U.S., from s. Canada to n. Ala.

Fig. 18

Keen's Myotis

INDIANA MYOTIS

Myotis sodalis

Description
Size: forearm, 1½ in. (3.8 cm). Small, fur glossy. Hairs 3-banded from base to tip, giving "purple" hue to pelage.
Similarities
Little Brown is similar but calcar not keeled and feet are larger.
Habitat
Limestone caves and nearby areas; rare.
Range
In n.-cen. U.S. from New England s. through w. N.C. to se. Ala. and w. to s. Mich. and s. Wis., s. to Mo.
Other name
Social Myotis.

SMALL-FOOTED MYOTIS

Myotis leibii **49:3**

Description
Size: forearm, 1½ in. (3.8 cm). Smallest myotis; fur long, silky. Color is golden brown in contrast to blackish ears and membranes; blackish mask across face.
Habitat
Caves, especially in hemlock forests in mountains or hilly country, particularly in rocky areas.
Range
From w. Maine and w. N.C., westward; also in extreme s. Canada.
Other name
Sometimes called *Myotis subulatus.*

SILVER-HAIRED BAT

Lasionycteris noctivagans **49:6**

Description
Size: forearm, 1⅔ in. (4.2 cm); 36 teeth, 2123/3133. Medium-sized. Brownish to blackish front frosted with silvery-white; females somewhat paler than males. Ears, membrane black; tragus blunt; basal half of tail membrane furred.
Similarities
Hoary Bat is larger and has buffy throat.
Habitat
Lakes, streams in wooded regions, up to evergreens in mountains.
Habits
Hangs in hollow trees, under loose bark, amid thick leaves, in buildings. Flight distinctive, fluttery with short sideways turns, sallies, and glides. In summer, males solitary, females gregarious; in winter, many presumably migrate south, while some hibernate in buildings.
Range
Throughout most of e. U.S., from Hudson Bay, s. to Ga. and La.

EASTERN PIPISTRELLE

Pipistrellus subflavus **49:8**

Description
Size: forearm, 1⅕ in. (3 cm); 34 teeth, 2123/3123. Smallest eastern bat. Yellowish-brown to pale brownish, somewhat grizzled; ears and forearms pinkish-brown contrasting with blackish wing membranes. Fur tricolored; darkish brown at base, then lighter yellowish-brown with tips darker; can be seen by blowing on fur to separate the hairs. Thumb large, ⅕ length of forearm; tragus blunt.
Habitat
Arid lowlands to evergreen mountain forests; woods, canyons, cliffs, rocky hillsides; waterways; buildings.
Habits
Comes out early, often before sundown; flight erratic, weak, mothlike, often undulating. Hibernates in caves, never in clusters.
Reproduction
Females congregate as do those of other bat species to bring forth single young in late spring; mating takes place in fall and occasionally in winter.
Range
Throughout much of ne. U.S., from s. Canada to n. Va. and w. to Nebr. and Kans.

BIG BROWN BAT

Eptesicus fuscus **49:2**

Description
Size: forearm, 2 in. (5.1 cm); wingspread, 12 in. (30.5 cm); 32 teeth, 2113/3123. Largest brown bat. Uniformly dark brown above in the East to pale brown in the West, paler below; ears, membranes black. Tragus blunt.
Habitat
Varied, cities to wilderness, sea level to mountains; caves, mines, crevices, buildings.
Habits
Flight strong with sudden, frequent changes in direction; often seen about street lights; enters hibernation late, emerges early; may fly

on mild, sunny middays in winter. Commonest large bat about buildings, in which it often hibernates.

Voice
Like escaping steam; also a click.

Reproduction
Mates in fall; single young (East) or twins (West) born in late May or June.

Age
To 9 years.

Range
S. Canada, s. throughout U.S. to S.A.

RED BAT
Lasiurus borealis **49:10**

Description
Size: forearm, 1⅔ in. (4.1 cm); 32 teeth, 1123/3123; mammae, 4. Medium-sized; only reddish bat. Male is light reddish-orange; female much duller and with frosted appearance. Fur soft, fluffy; ears short, tragus blunt; wings narrow; tail membrane fully furred above.

Habitat
Forested regions.

Habits
Roosts in vegetation, often low; flies early, often high and descending in spirals. Migrates April–May, September–late November, sometimes flying in numbers like birds. Common; migrates in winter.

Reproduction
Mates in flight in August; 1–4 young, born late May–mid-June.

Range
Throughout s. Canada, s. through e. U.S. to Gulf Coast.

SEMINOLE BAT
Lasiurus seminolus **49:11**

Description
Size: forearm, 1½ in. (3.8 cm); 32 teeth, 1123/3123. Resembles Red Bat, but color darker, rich mahogany-brown, lightly frosted with grayish-white.

Habitat
Forested regions.

Habits
Like Red Bat.

Reproduction
Mates in flight in August; 1–4 young, born late May–mid-June.

Range
Along Atlantic Coast states, from s. N.Y. to cen. Fla. and Gulf Coast; w. to cen. Pa., n. Ga., and e. Tex.

HOARY BAT
Lasiurus cinereus **49:4**

Description
Size: forearm, 2+ in. (5.1+ cm); wingspread, 14 in. (35.6 cm) or more; 32 teeth, 1123/3123; mammae, 4. Largest eastern bat, recognized by large size, swift flight, and narrow, pointed wings. Fur frosted, general appearance grayish-white; hairs are dark brown or blackish at base, followed by band of golden yellow, band

of dark brown or blackish subterminally, and a white tip. Tragus blunt; tail membrane fully furred above.

Similarities

Silver-haired is colored similarly but lacks buffy throat.

Habitat

Forested regions.

Habits

Hangs alone in trees, bushes; emerges late in evening; frequents meadows, watercourses; flight direct, purposeful; seldom seen.

Reproduction

1–2 young (usually twins) born in late May or June, weigh nearly ⅙ ounce (5.0 g) each at birth; carried by female in flight until old enough to be left hanging in tree when she forages.

Range

Throughout s. Canada and e. U.S. in warm months; migrates southward in winter.

NORTHERN YELLOW BAT

Lasiurus intermedius **49:9**

Description

Size: forearm, 1⅞ in. (4.8 cm); 30 teeth, 1113/3123; mammae, 4. Yellowish to yellowish-brown; tail membrane fully furred.

Similarities

Red and Hoary are similar but Yellow is larger than former, somewhat smaller than latter, and differs in color and absence of small upper premolar.

Habitat

Forested regions.

Habits

Much the same as other members of genus.

Reproduction

1–2 young (usually twins) born in late May or June.

Range

Principally in se. U.S. and Mexico, but extends from N.J., southward through coastal states.

EVENING BAT

Nycticeius humeralis **49:1**

Description

Size: forearm, 1½ in. (3.8 cm); 32 teeth, 1123/3123. Medium-sized. Fur sparse; dark brown above, washed with buff below; ears and membranes black. Ears thick; tragus curved, blunt.

Similarities

Little Brown Myotis is confusingly similar, but Evening is slightly larger and has fewer teeth.

Habitat

Buildings and tree hollows.

Habits

Hangs in hollow trees, under bark, in buildings; flight steady, straight, slow; often flies high to travel, low to feed; migratory east of 100th meridian.

Reproduction

Usually 2 young, born May–July.

Range

Throughout s. and cen. parts of e. U.S., from s. Canada s. to Fla. and w. to Iowa and Nebr.

TOWNSEND'S BIG-EARED BAT

Plecotus townsendii **49:7**

Description
Size: forearm, 1¾ in. (4.4 cm); 36 teeth, 2123/3133. Medium-sized. Fur long, woolly. Warm brown to cinnamon above, pinkish-buff to pale brownish below; 2 conspicuous lumps on nose. Tragus pointed; ears extremely long, joined at base.
Habitat
Limestone caves, forested regions.
Habits
Hangs, with long ears spiraled alongside neck like a ram's horn, in hollow trees, mine shafts, buildings, and semidark areas of dry caves; emerges after dark; flight agile, swift; can hover; wary. Hibernates.
Reproduction
Single young born in spring and early summer; able to fly at 3 weeks of age.
Range
From w. Va. and W.Va., westward in band through Ky., Ind., s. Ill., to Mo. and s. Kans.; also in Black Hills of S.Dak. and adjacent areas.
Other name
Lump-nosed Bat.

RAFINESQUE'S BIG-EARED BAT

Plecotus rafinesquii *Fig. 19*

Description
Resembles Townsend's Big-eared Bat but tips of ventral hairs whitish instead of brownish.
Habits
Much the same as for Townsend's.
Range
In se. U.S., from cen. Va. and W. Va. s. to Fla. and w. to Ind. and Mo.

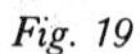

Fig. 19

Rafinesque's Big-eared Bat

FREE-TAILED BATS

Family Molossidae

Bats in this family have tails that extend beyond the interfemoral membrane, hair that is short, ears that are blunt, and tragi that are very small. The family dates back to the late Oligocene.

BRAZILIAN FREE-TAILED BAT

Tadarida brasiliensis *Fig. 20*

Description
Size: forearm, 1⅘ in. (4.6 cm); 32 teeth, 1123/3123, upper incisors converge at tips. Medium sized. Fur is dark brown, short, not glossy. Upper lip grooved; ears not united at base and, when laid forward, do not extend beyond end of nose; wings long, narrow.

Habitat
Caves.
Habits
Hangs in caves, crevices, buildings; highly colonial. Flies over open areas; take-off from cave mouth is straight. Flight rapid with abrupt changes; perhaps the most rapid flier among bats. Migrates southward from range.
Range
In se. U.S., w. along Gulf Coast and to e. Kans. and se. Nebr., with some n. to Ohio.

Fig. 20

Brazilian Free-tailed Bat

BIG FREE-TAILED BAT
Tadarida macrotis

Description
Size: forearm, 2½ in. (6.4 cm); 30 teeth 1123/2123. Large; similar to Brazilian Free-tailed, but larger, base of hairs whitish; ears united at base, and, if laid forward, extend well beyond end of nose. Upper incisors do not converge at tips.
Habitat
Caves, crevices in cliffs, buildings.
Habits
Nocturnal, colonial.
Range
In e. range, Iowa and e. Kans.

Edentates
Order Edentata

ARMADILLOS
Family Dasypodidae

NINE-BANDED ARMADILLO
Dasypus novemcinctus

51:8
Fig. 15

Description
Size: head and body, 15–17 in. (38.1–43.2 cm); tail, to 16 in. (40.6 cm); weight (adults), to 17 lb. (7.7 kg); 28–36 teeth, 007–9/007–9. Teeth degenerate, premolars indistinguishable from molars. Upper parts tan or yellowish. Covered with horny armor, with 9 flexible bands in the center. Snout long; feet strong, well-clawed, with 4 toes in front, 5 behind; upperparts and ears nearly naked.

Habitat
Brush, open woodlands.
Habits
Most active by night; a burrower.
Voice
Grunt.
Food
Insects, other small invertebrates, some vegetable matter.
Reproduction
Young, usually 4, identical quadruplets, born March–April in a burrow.
Range
From Kans., s. to Tex., extending its range in some se. states where it has been introduced.

Hares, Rabbits, and Pikas

Order Lagomorpha

HARES AND RABBITS

Family Leporidae

Hares and rabbits are distinguished by their soft fur, long legs and ears, and short tails, usually white below. The larger species are called hares or jackrabbits, the smaller ones rabbits or cottontails. The classical distinction between hares and rabbits is that young of the former are precocial, whereas those of the latter are altricial. The fur is generally brownish or grayish above, sometimes white in the north or in winter, and paler or white below. They have twenty-eight teeth, formula 2033/1023. One pair of upper incisors is directly behind the other.

These mammals occupy all habitats from desert to moist forest, from sea level to beyond tree line, and from the northern limit of land southward to the Gulf. They do not hibernate. Hares generally like open country; rabbits prefer shrubby cover. Both run rapidly and jump well with their powerful hind legs. They rest in their "forms" or protected hiding places, read the wind with their noses, and thump on the ground with their hind feet to warn others of danger. All are timid, inoffensive animals that rely for safety on hiding or flight.

Their diet is vegetarian and includes a wide variety of plant food, from grass and garden greens to buds, twigs, bark, and leaves of trees. By breeding early and often they withstand the heavy tolls levied by fox and owl, gun and auto. Many species have periodic fluctuations in population numbers. The family dates back to the Eocene and has endured well.

COTTONTAILS

Genus *Sylvilagus*

Cottontails are smaller and have shorter ears and hind legs than hares; they usually seek safety by hiding. After a gestation period of about twenty-eight days, usually four or five naked, blind, and helpless young are born in a fur-lined nest on the ground; they are weaned in less than three weeks. Females often breed after nine months of age and produce several litters a year.

EASTERN COTTONTAIL

Sylvilagus floridanus **50:12, 51:1**

Description
Size: head and body, 14–17 in. (35.6–43.2 cm); ear, to 3 in. (7.6 cm). Grayish-brown, underside of tail cotton-white; rusty nape. Ears short.
Habitat
From swampy wood to upland thickets and farmlands.
Habits
Most active at dawn and dusk; timid, restricts movements to area of only a few acres.
Range
Entire e. U.S., except n. New England.

NEW ENGLAND COTTONTAIL

Sylvilagus transitionalis **50:14, 51:1**

Description
Size: head and body, 15–17 in. (38.1–43.2 cm); ear, to 2½ in. (6.4 cm). Small. Summer: reddish-brown. Winter: reddish-gray, with only a pale rusty nape patch. Ear tips and patch between ears black, feet whitish.
Similarities
Eastern is larger, with longer ears. Snowshoe Hare has longer ears.
Habitat
Primarily forests.
Habits
More nocturnal than Eastern; prefers thicker cover.
Range
New England and through Appalachians to ne. Ala.

SWAMP RABBIT

Sylvilagus aquaticus **50:15**

Description
Size: head and body, 14–17 in. (35.6–43.2 cm); ear, to 4 in. (10.2 cm). Fur short, sleek. Rich brownish-gray, with small nape patch; hind feet rusty above. Tail slender; toes large, splayed, and with sharp nails.
Similarities
Eastern Cottontail is smaller; has longer, less sleek fur, with more pronounced rusty nape patch, whitish hind feet, and thicker tail.
Habitat
Moist bottomland.
Habits
Often walks rather than hops, stays very close to cover, swims well.
Range
In s.-cen. states, n. Ga., w. to e. Tex. and Gulf of Mexico; also from Ind. and Ill., w. to s. Kans.

MARSH RABBIT

Sylvilagus palustris *Fig. 21*

Description
Size: head and body, 14–16 in. (35.6–40.6 cm); ear, to 3 in. (7.6 cm). Dark brown. Small hind feet, ears, and tail. Differs from all other cottontails in the East by having brown or grayish underside of the tail, rather than white.

Habitat
Swamps and marshes, possibly not occurring above 500 ft. (152.4 m).
Habits
Nocturnal, secretive; often walks instead of hopping while feeding; uses own trails through dense marshy vegetation; swims readily.
Range
From the Dismal Swamp of Va., s. through Fla.

Fig. 21

Marsh Rabbit

HARES AND JACKRABBITS
Genus *Lepus*

Hares and jackrabbits are larger and have longer ears and hind legs than do cottontails; they usually seek safety in flight. The female makes no nest. After a gestation period of thirty to forty-three days, three to eight young are born. They are furred, with eyes open, and are able to move about within a few minutes. There are usually two or more litters a year.

SNOWSHOE HARE
Lepus americanus **50:16, 51:3**

Description
Size: head and body, 13–18 in. (33–45.7 cm); ear, to 4 in. (10.2 cm). Smaller than jackrabbits, larger than cottontails. Summer: brown above, white below; tail dark above; feet brownish, not whitish as in cottontails. Winter: only tips of hairs white, bases dark; ear tips dark; tail all white. Large, long-furred feet permit travel on heavy snow, whence its name.
Habitat
Northern and mountainous swamps, forests and brush; laurel and rhododendron thickets.
Habits
Spends day in cover, feeds at night in open; uses own trails.
Age
5 years in wild, 8 in captivity.
Remarks
This is a classic North American example of a species with widely fluctuating numbers. During 1 peak year in Ontario a density of 3400 individuals to a square mile was reached. At the 10-year low, however, there may be only 1 or 2 per square mile.
Range
Entire Canada, s. through U.S. to New England, n. Mich., and n. Minn.; also s. throughout Appalachians.
Other name
Varying Hare.

ARCTIC HARE

Lepus arcticus *Fig. 22*

Description
Size: head and body, 17–24 in. (43.2–61 cm); ear, to 4½ in. (11.4 cm). Large. Winter: white; also white all year in Greenland, Ellesmere Island, and northern Baffin Island, hairs white to base. Winter: gray or brown farther south. Ear tips black, tail always all-white.

Habitat
Tundra and rocky slopes from tree limit north.

Habits
Usually in groups; often unwary; sometimes stands on hind feet only and hops on them. Migratory, moving south in winter and returning to north in summer.

Range
Arctic regions, especially around Hudson Bay and Baffin Is.; also in Greenland.

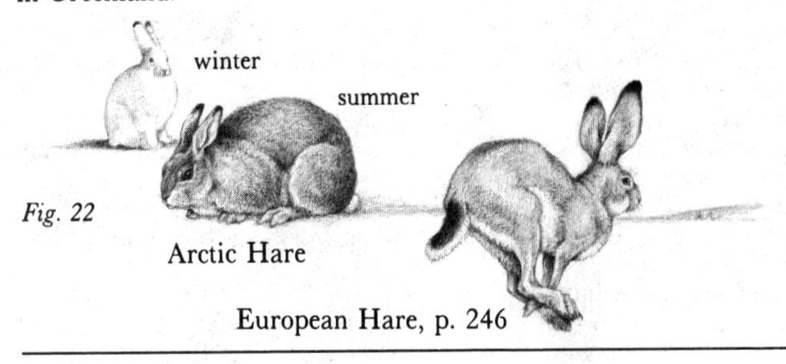

Fig. 22

Arctic Hare

European Hare, p. 246

WHITE-TAILED JACKRABBIT

51:2

Lepus townsendii *Fig. 23*

Description
Size: head and body, 18–22 in. (45.7–55.9 cm); ear, to 6 in. (15.2 cm). Large. Summer: brownish-gray. Winter: paler in south, almost white in north. Tail all-white below and above or with dusky or buffy middorsal stripe that does not extend onto back. Ear tips black.

Similarities
Snowshoe Hare is smaller, with shorter ears, and has preference for brush and thickets. Black-tailed Jackrabbit is smaller and with only partially white tail.

Habitat
Open country and exposed mountain slopes.

Habits
Most active at dawn and dusk, but may be seen infrequently during the day and may be active at any time of night.

Range
In n. parts of Kans., Iowa, and Minn., westward.

Fig. 23

White-tailed Jackrabbit

BLACK-TAILED JACKRABBIT
Lepus californicus **50:13, 51:2**

Description
Size: head and body, 17–21 in. (43.2–53.3 cm); ear, to 7 in. (17.8 cm). Large. Grayish-brown above, pure white below; ear tips, rump, and top of tail black.
Habitat
Open country.
Habits
Relies on speed and rapid changes of direction to elude predators.
Range
In e. range, from w. Mo. and e. Tex., w. to s. S.Dak.

EUROPEAN HARE
Lepus capensis *Fig. 22*

Description
Size: head and body, 25–27 in. (63.5–68.6 cm); ear, to 5 in. (12.7 cm). Large. Brownish-gray above, pure white below; tail black above, white below.
Habitat
Prefers areas where open fields and pastures are bordered by hedgerows and woodlots.
Habits
Does not make burrow or den but hides in dense vegetation by day.
Age
To 12 years in wild.
Range
In Great Lakes region, from s. Canada, s. to New England and e. Pa.; introduced from Europe.
Other name
Cape Hare.

Rodents
Order Rodentia

Rodents are small- to medium-sized gnawing mammals distinguished by possessing two, and only two, incisors in each jaw, which continue to grow throughout life. They lack canine teeth and instead have a conspicuous space between the incisors and the grinding cheek teeth. Most have four toes on each front foot, five behind. Members of the only other order of gnawers, the Lagomorpha, which includes the hares and rabbits, have two pairs of incisors in the upper jaw, one small pair just behind the big front pair. Considered as a whole, rodents are one of the most successful orders of mammals. They are widely and abundantly distributed from the tropics to the tundra, from the sea beach to the alpine meadow.

SQUIRRELS
Family Sciuridae

Members of this family include woodchucks; prairie dogs; ground, tree, and flying squirrels; and chipmunks. All have hairy, sometimes bushy, tails. The tooth formula is 1023/1013, except for the chipmunks and the Fox and Red Squirrels, in which the formula is 1013/1013. Incisors are yellow except for the Woodchuck, in which they are white.

These mammals live either on the ground or in trees, and they range throughout the East, north to the Arctic Ocean and up to the tops of high mountains. In winter, to the north and in high altitudes the ground squirrels hibernate. All but the flying squirrels are diurnal. All nest in the ground in burrows or under logs or rocks except the tree-nesting flying squirrels and tree squirrels. Most members of the family are capable of vocalization, such as a high-pitched whistle, barking, or chattering. The family dates back to the Miocene.

EASTERN CHIPMUNK

Tamias striatus **50:11, 54:8**

Description
Size: head and body, 5–6 in. (12.7–15.2 cm); tail, to 4 in. (10.2 cm). Small. Reddish-brown above, white below; stripes on back and sides end at reddish rump; stripes extend onto face. Has internal cheek pouches.
Similarities
Thirteen-lined Ground Squirrel has no face stripes.
Habitat
Hardwood forests and edges, rocks, stone walls, outbuildings.
Habits
Unsocial; a ground dweller that burrows and climbs, often sits upright on stumps when frightened or, in a hurry, runs with tail straight up or at raised angle; hibernates in North.
Food
Nuts, berries, seeds, insects, and small animals; stores food.
Reproduction
Usually 3–5 young, born in an underground chamber; usually 2 litters yearly.
Age
To 8 years.
Range
Throughout most of E. from s. Canada, s. to cen. Miss. and w. to e. Daks., Nebr., and Kans.

LEAST CHIPMUNK

Eutamias minimus **50:10**

Description
Size: head and body, 3½–4½ in. (8.9–11.4 cm); tail, to 4½ in. (11.4 cm). Yellowish, reddish, or grayish with stripes on face, back, and sides; stripes on back and sides extend to base of tail. Has internal cheek pouches.
Habitat
Rocks and openings in evergreen or mixed woodland; brush, edges of woods, lakes; outbuildings, sawmills.
Habits, Food, Reproduction
Similar to Eastern Chipmunk.
Age
To 8 years.
Range
In Canada, from s. of Hudson Bay to n. Mich. and Wis.; w. to w. Daks. and w. Nebr.

WOODCHUCK

Marmota monax **50:6, 54:2**

Description
Size: head and body, 16–20 in. (40.6–50.8 cm); tail, to 7 in. (17.8 cm). Large. Above dark brown to yellowish-brown; grizzled; paler, sometimes rusty below. No white between eyes; sides of neck same color as back; feet black to dark brown. Heavy-bodied, ears small, legs and tail short.
Habitat
Dry woods and adjacent open spaces; brushy ravines, rocky slopes; fields, mowed parkway borders.
Habits
Most active in early morning or late afternoon; hibernates in winter, not always too deeply; can climb and swim; frequently hunted; main den entrance frequently has fresh dirt about it, especially in spring.
Food
Clover, alfalfa, other perennial plants.
Reproduction
Grass-lined nest in side tunnel; young, usually 4, born naked, blind, and weighing about 1 ounce.
Age
To 5 years.
Remarks
This species has been made famous by folklore relating to Ground Hog Day, February 2 each year.
Range
In most of e. U.S., from Hudson Bay to n. Ala. and w. to edge of prairies.

GROUND SQUIRRELS

Genus *Spermophilus*

Ground squirrels have unstriped faces and internal cheek pouches. They live near burrows, which they dig; in winter they hibernate. Their food includes seeds, plants, and insects. The young, as many as thirteen in some species, are born blind and naked, but they become independent within a month. By churning up the soil, ground squirrels check run-off, but they also damage crops and pastures.

RICHARDSON'S GROUND SQUIRREL

Spermophilus richardsonii *Fig. 24*

Description
Size: head and body, 8–9½ in. (20.3–24.1 cm); tail, to 4½ in. (11.4 cm). Large. Dull; buffy gray, paler below; no spots or stripes. Tail light brown, bordered with buff or white.
Similarities
Franklin's is larger, darker, and has longer tail.

Fig. 24

Richardson's Ground Squirrel — Plains Pocket Gopher, p. 253

Habitat
Meadows, prairies, sagebrush.
Habits
Colonial; adults stay underground from July to late winter; young go below in September.
Reproduction
Usually 1 litter per year, normally consisting of 6–7 young that are born in mid-May.
Range
From Minn., w. to 100th meridian in Daks. and w. Nebr.
Other name
Picket Pin.

THIRTEEN-LINED GROUND SQUIRREL

Spermophilus tridecemlineatus **50:8**

Description
Size: head and body, 4½–6½ in. (11.4–16.5 cm); tail, to 5¼ in. (13.3 cm). Only striped ground squirrel. Back with a series of alternating dark (brownish or blackish) and pale longitudinal stripes; a row of squared whitish spots in each dark dorsal stripe; stripes on sides less well defined than those on back.
Habitat
Prairies, pastures, brushy edges of woods, golf courses, roadsides.
Habits
Most active during warmer hours of the day; often seen scurrying across highways; runs with tail straight out behind. Burrow is a small, round hole without telltale dirt, hidden in vegetation.
Reproduction
May have 2 litters per year, the first in late April or May; 4–12 young per litter.
Range
From cen. Man., s. to Ohio and Mo., w. to 100th meridian.

SPOTTED GROUND SQUIRREL

Spermophilus spilosoma **50:9**

Description
Size: head and body, 5–6 in. (12.7–15.2 cm); tail, to 3½ in. (8.9 cm). Small; only spotted ground squirrel. Pale reddish-brown or grayish-brown with squarish light spots on back; white below. Tail not bushy.
Habitat
Semi-arid plains; prefers sandy soils.
Reproduction
2 litters per year, 3–8 young per litter.
Range
Plains of w.-cen. Nebr. and Kans.; adjacent areas in S.Dak., Wyo., Colo.

FRANKLIN'S GROUND SQUIRREL

Spermophilus franklinii

Description
Size: head and body, 9–10 in. (22.9–25.4 cm); tail, to 6 in. (15.2 cm). Gray with brownish wash, only slightly lighter below; no stripes or spots.
Similarities
Richardson's is smaller, has shorter tail. Gray Squirrel has longer, bushier tail, longer ears, and sides not buffy.

Habitat
Tall-grass prairies, edges of fields, open woodland, pastures; frequently in tall and dense vegetation.
Habits
Most active on sunny days; can climb trees; forms small colonies.
Reproduction
1 litter of 4–10 young, born in May or June.
Range
From Man., s. to w. Ind. and w. to w. Nebr.

ROCK SQUIRREL
Spermophilus variegatus **50:5**

Description
Size: head and body, 10–11 in. (25.4–27.9 cm); tail, to 10 in. (25.4 cm). Large. Varying shades of gray washed with brownish, giving mottled effect; light gray or buffy below; sometimes almost black on head, back, or entire upper parts. Tail bushy and about as long as head and body.
Habitat
Rocky areas.
Habits
Often sits on top of boulder; climbs bushes, trees.
Reproduction
About 6 young, born in spring and summer.
Age
To 9 years.
Range
In e. range, only se. Colo.

PRAIRIE DOGS
Genus *Cynomys*

BLACK-TAILED PRAIRIE DOG
Cynomys ludovicianus **50:7**

Description
Size: head and body, 11–13 in. (27.9–33 cm); tail, to 4 in. (10.2 cm). Yellowish above, whitish below; outer third of tail black. Stout; short-legged; ears small. Has internal cheek pouches. Females slightly smaller than males; young paler than adults.
Habitat
Short-grass prairie of Great Plains.
Habits
Highly colonial; sits erect by entrance to burrow, a 1–2-ft. (0.3–0.6 m)-high mound; hibernates in cold weather.
Reproduction
Usually 5 young, naked, blind, born in spring in grass-lined nest inside tunnel of burrow.
Age
To 10 years.
Range
In w.-cen. parts of Daks., Nebr., and Kans.

TREE SQUIRRELS

Genera *Sciurus, Tamiasciurus,* and *Glaucomys*

Tree squirrels have no internal cheek pouches. They are primarily arboreal, but are also seen on the ground. They do not hibernate. Their food, some of which they store, includes nuts, berries, fruits, seeds, buds, twigs, bark, eggs, fungi, and insects. They usually nest in tree cavities or leaf nests. Gestation takes forty to forty-five days. The one to seven young are born naked and blind; there are one to two litters a year.

GRAY SQUIRREL

Sciurus carolinensis **50:2, 54:11**

Description
Size: head and body, 8–10 in. (20.3–25.4 cm); tail, 8–10 in. (20.3–25.4 cm). Fur gray above, often washed with fulvous, whitish below; light eye-ring; in winter, no fulvous, back of ears white; long, bushy tail of same general tone as body; also has an all-black phase. More graceful than the larger Fox Squirrel; head not squarish.

Habitat
Usually found in heavily forested areas, especially bottomlands; also in suburbia and city parks.

Habits
Most active early morning and late afternoon; highly arboreal; more secretive than Fox Squirrel.

Age
To 15 years.

Range
Along s. border of Canada and throughout e. U.S. w. to e. parts of Daks., Nebr., and Kans.

FOX SQUIRREL

Sciurus niger **50:4**

Description
Size: head and body, 10–15 in. (25.4–38.1 cm); tail, 9–14 in. (22.9–35.6 cm). Color varies: in North, generally yellowish-rust above, pale yellow to orange below, tail with rusty or buffy border; in South, grayish or grizzled; dark head with white nose and ears; in mid-Atlantic states, an all-steel-gray phase occurs. Intermediate variations also exist; a black phase also known. Heavy-bodied, squarish face.

Habitat
Groves of oak, longleaf pine; borders of cypress swamps, thickets; in western part of range restricted to trees along streams, in woodlots, and in urban areas.

Habits
Walks or climbs without speed or grace of Gray; tolerates more open situations than Gray; active throughout day.

Age
To 10 years.

Range
Throughout entire e. U.S., except New England, N.Y., and n. parts of Wis., Minn., N.Dak.

RED SQUIRREL

Tamiasciurus hudsonicus **50:1**

Description
Size: head and body, 7–8 in. (17.8–20.3 cm); tail, 4–6 in. (10.2–15.2 cm). Reddish, sometimes yellowish; white below; tawny border to tail; black line along sides; in winter, paler above and with ear tufts.
Similarities
Gray and Fox are larger.
Habitat
Evergreen forests, less common in hardwoods.
Habits
Leaves piles of pine-cone cuttings on rocks, logs; sometimes migrates.
Age
To 9 years.
Remarks
Most commonly seen mammal of the Great North Woods.
Range
From tree limit n., s. throughout Appalachian region, and w. to Ind. and Iowa.

NORTHERN FLYING SQUIRREL

Glaucomys sabrinus *Fig. 25*

Description
Size: head and body, 5½–6 in. (14–15.2 cm); tail, to 5½ in. (14 cm). Similar to Southern Flying Squirrel, but is slightly larger and heavier. Belly hairs lead-colored at base, white only at tips.
Habitat
Mixed woods and conifers.
Habits
As Southern.
Range
From tree limit, s. to Appalachian region and s. Wis. and N.Dak.

Fig. 25 Northern Flying Squirrel

SOUTHERN FLYING SQUIRREL

Glaucomys volans **50:3**

Description
Size: head and body, 6 in. (15.2 cm); tail, 4 in. (10.2 cm). Small. Fur thick, soft, glossy. Olive-brown above, white below; belly hairs white to base. Eyes large, tail flattened.
Habitat
Woodlands with tall trees.
Habits
Nocturnal; gregarious; sleeps by day in hole in tree (or attic); can glide 125 ft. (38.1 m) given a high enough takeoff point. Lands with an audible thump, as on a camp roof.
Age
To 13 years.

Remarks
Flying squirrels have a fold of loose, furred skin along their sides from wrist to ankle. When extended this enables them to glide (not fly) from tree trunk to tree trunk.
Range
Throughout most of e. U.S., w. to e. Kans. and e. Nebr.

POCKET GOPHERS

Family Geomyidae

Chunky, toothy, big-headed, these brownish, rat-sized burrowers are noted for their underslung jaws and external, fur-lined cheek pouches, which they use for carrying food to storage. There are twenty teeth, formula 1013/1013. The long-clawed, sturdy forefeet and the front teeth are adapted for fast and protracted digging. Eyes and ears are small; the mouth closes with the upper incisors outside. The tail is sensitive to touch, nearly naked, and shorter than the body.

Even where gophers are common and their mounds abundant, they themselves are rarely seen. When they are seen, it will probably be the front end only as the animal hurriedly peers from its burrow. Most of their lives are spent underground. Their mound is fan-shaped, the mole's is round. The hole beneath leads off at an angle, whereas the mole's smaller hole leads straight down.

Gophers eat roots, bulbs, and other vegetation. Extensive surface tunnels are used in foraging; a smaller, deeper system contains the nest, food-storage chambers, and toilet chamber. These mammals are unsocial. If they meet—except to mate—they fight. Gestation is presumed to be about twenty-eight days; the litter size is usually one to five, with one to three litters a year. Species are sometimes hard to identify when ranges overlap: if the front teeth have two distinct grooves, it is a Plains Pocket Gopher or Southeastern Pocket Gopher; if no grooves (or very faint), it is a Northern Pocket Gopher. The family dates back to the Oligocene.

NORTHERN POCKET GOPHER

Thomomys talpoides

Description
Size: head and body, 5–6½ in. (12.7–16.5 cm); tail, to 3 in. (7.6 cm); mammae, usually 10. Males larger. Brownish, lightly washed with black or gray; nose black or brown; patches on back of ears black; 1 faint groove on front of each incisor; claws relatively short.
Habitat
Uplands and mountains with thin soil.
Range
From s. Canada, southward and from e. Daks., westward.

PLAINS POCKET GOPHER

Geomys bursarius *Fig. 24*

Description
Size: head and body, 5½–9 in. (14–22.9 cm); tail, to 4½ in. (11.4 cm); mammae, 6. Tawny in West to nearly black in Illinois; albinos and white-spotted individuals are known. Claws long; 2 grooves on incisors.

Habitat
Deep soil, frequently sandy, in treeless regions, hay and alfalfa fields, pastures and edges of highways, and railroad rights-of-way.
Range
Throughout plains from cen. Ind. and Ill., w. to Daks., and from Minn. and N.Dak., s. to Kans.

SOUTHEASTERN POCKET GOPHER
Geomys pinetis *Fig. 26*

Description
Similar to Plains Pocket Gopher. Three other closely related species with very small ranges occur in the same general area and can be distinguished only by specialists.
Habitat
Dry sandy soils.
Range
In s. Ga., n. and cen. Fl., and se. Ala.

Fig. 26 Southeastern Pocket Gopher

POCKET MICE AND KANGAROO RATS
Family Heteromyidae

This family is found only west of the Mississippi. It consists of small-eared rodents with hind limbs much larger than forelimbs. These small, long-tailed, nocturnal, burrowing rodents have external fur-lined cheek pouches which they use to carry food to storage. They have twenty teeth, 1013/1013. Their habitat is arid or semiarid plains and prairies. They make underground tunnel systems with sleeping, nesting, and food-storage chambers and inconspicuous surface holes that are plugged up during the day.

Their principal foods are seeds and greens, which they store in underground chambers for future use. They are inactive in cold weather. Populations are subject to periodic fluctuations. The one to eight, usually four, young are born in spring or summer; there are one or more litters a year. The family dates back to the Oligocene.

POCKET MICE
Genus *Perognathus*

Pocket mice are the smallest family members, with moderately long and untufted tails. Their faces are unmarked; a buffy lateral stripe separates the darker back from the white belly. These mice inhabit sandy soil; they are poor jumpers. Their voice is a thin, high squeak. The soles of the hind feet are naked.

OLIVE-BACKED POCKET MOUSE
Perognathus fasciatus

Description
Size: head and body, 3 in. (7.6 cm); tail, 2½ in. (6.4 cm). Fur soft, silky. Olivaceous above; yellow spots behind ears; yellowish lateral line.
Range
From Canadian prairies, s. through Daks.

PLAINS POCKET MOUSE
Perognathus flavescens

Description
Size: head and body, to 3¾ in. (9.5 cm); tail, to 2⅗ in. (6.6 cm). Back pale buff, variously washed with blackish; spot behind ear and lateral line clear buff; belly white.
Range
In plains, from Minn. and e. N.Dak., s. to nw. Mo. and Kans.

HISPID POCKET MOUSE
Perognathus hispidus **52:8**

Description
Size: head and body, 4½–5 in. (11.4–12.7 cm); tail, to 4½ in. (11.4 cm). Largest pocket mouse in East. Hair coarse. Back ochreous, mixed with blackish; sides only faintly paler than back. Lateral line distinct; uncrested tail equal to, or shorter than, head and body.
Range
From cen. N.Dak., S.Dak., Neb., and Kans., westward and southward through Tex.

KANGAROO RATS
Genus *Dipodomys*

Kangaroo rats all have long tails tufted at the tip, very long hind legs, and distinct facial markings. They prefer arid or semiarid country and easily worked soil. The belly is white and the soles of the hind feet are moderately haired.

ORD'S KANGAROO RAT
Dipodomys ordii **52:5, 54:3**

Description
Size: head and body, 4–4½ in. (10.2–11.4 cm); tail, to 6 in. (15.2 cm). Fur silky; eyes large. Pale to bright orange-brown above, white below. Face with white crescent through eye and under ear; lower incisors rounded, not flat across front. White lateral stripe crosses thigh and extends out along side of long, tufted tail. Dark tail stripes wider than white ones. Hind legs long, strong; 5 toes on each hind foot; soles have short, stiff fur. Young darker than adults.
Habitat
Arid areas, sandy, soft, or hard soils.
Habits
Nocturnal; runs by leaping with its hind feet. Seldom seen aboveground in very cold (or very hot) weather.
Range
Extreme s.-cen. N.Dak. southward and westward.

BEAVER
Family Castoridae

The beaver is the largest rodent in the United States of America and the only land mammal with a broad, flat tail. The body is thickset and compact, the legs short, ears small, hind feet large, toes webbed.

BEAVER
51:5
Castor canadensis
Fig. 27

Description
Size: head and body, 25–30 in. (63.4–76.2 cm); tail, to 10 in. (25.4 cm); weight, to 60 lb. (27.2 kg); 20 teeth 1013/1013. Fur dense, brown, waterproof; head massive; incisors large, chestnut-colored; grinding teeth high-crowned; hind feet webbed, tail scaly.
Habitat
Streams and lakes.
Habits
Builds relatively watertight dam of sticks and mud across a stream, a cone-shaped house in a pond; trees gnawed a foot from the ground are beaver signs. Most active at dawn, dusk, and by night.
Voice
Various sounds made only within the lodge; outside, slaps water with its tail as a sign of warning.
Food
Bark of aspen, alder, birch, maple, willow, and other vegetation.
Reproduction
Young, called kits, usually 2–4, are born from April-July in the lodge.
Range
Throughout entire e. U.S., absent from the Gulf Coastal Plain.

Fig. 27
Porcupine, p. 269
Beaver

NEW WORLD RATS AND MICE
Family Cricetidae

These small rodents have sixteen teeth, 1003/1003. The rear foot has five toes. Mice and rats generally have large eyes and ears, long tails, and four toes on the front foot. Voles and lemmings have small eyes and ears, short tails, and four or five toes on the front foot. All groups have one, sometimes two or more, litters a year of three to four young, usually in a nest of vegetation on the ground; none hibernate. The family dates back to the Oligocene.

MARSH RICE RAT

Oryzomys palustris **53:13**

Description
Size: head and body, 5 in. (12.7 cm); tail, to 7 in. (17.8 cm). Fur short, coarse. Above grayish-brown, underparts and feet whitish. Tail long, scaly, paler beneath. Two rows of tubercles on molars.
Similarities
Norway and Black Rats have 3 rows of tubercles on molars; tails are longer and not lighter below. Wood rats are larger and have smoother fur.
Habitat
Saltwater and freshwater marshes and meadows, and vicinity; upland clearings.
Habits
Nocturnal, social; swims, dives, and makes runways in salt meadows.
Food
Plants, seeds, crustaceans, snails.
Range
In se. U.S., from cen. N.J., s. to Fla. and from Va., w. to s. Ill. and s. Mo.

HARVEST MICE

Genus *Reithrodontomys*

These are brownish mice with a longitudinal groove on each upper incisor, conspicuous ears, no external cheek pouches, and medium-sized, thinly haired tails. The young are darker than the adults. Preferred habitats are overgrown fields, marshes, and woods edges. They are nocturnal and good climbers. Their voice is a high ventriloquistic bugling. The diet consists of plant cuttings, seeds, and insects.

PLAINS HARVEST MOUSE

Reithrodontomys montanus **52:10**

Description
Size: head and body, 2–3 in. (5.1–7.6 cm); tail to 2⅗ in. (6.6 cm). Fur pale gray above with a faint tawny cast, middle of back often darker; pale gray to whitish below; narrow blackish dorsal stripe on tail.
Similarities
Tail of Western usually more than 2½ in. (6.4 cm); that of Fulvous more than 3⅓ in. (8.5 cm).
Habitat
Principally an upland species, especially where grasses are relatively short and sparse.
Range
From e. Nebr. and e. Kans., westward.

EASTERN HARVEST MOUSE

Reithrodontomys humulis **53:6**

Description
Size: head and body, 3 in. (7.6 cm); tail, to 2½ in. (6.4 cm). Fur rich brown, grayish below, often with pinkish hues; tail bicolored, feet light.
Habitat
Brier patches, roadside ditches, pasture, bogs.
Range
From Md., s. to Fla. and w. to s. Ohio and e. Tex.

WESTERN HARVEST MOUSE
Reithrodontomys megalotis *Fig. 28*

Description
Size: head and body, 3 in. (7.6 cm); tail, 2⅓–3⅕ in. (5.8–8.1 cm). Color varies geographically from pale gray to warm brown above, often with darker stripe down back, less pronounced than in Plains; buff to white below. Dorsal stripe on tail broad, blackish.
Similarities
Plains has thin-stripe tail, usually under 2½ in. (6.4 cm). Fulvous has bright tawny sides and tail more than 3⅓ in. (8.5 cm).
Habitat
Grassy areas, also dense grasses, arid regions in riparian situations.
Range
From Canada, s. to Minn., w. Wis., Mo. and westward.

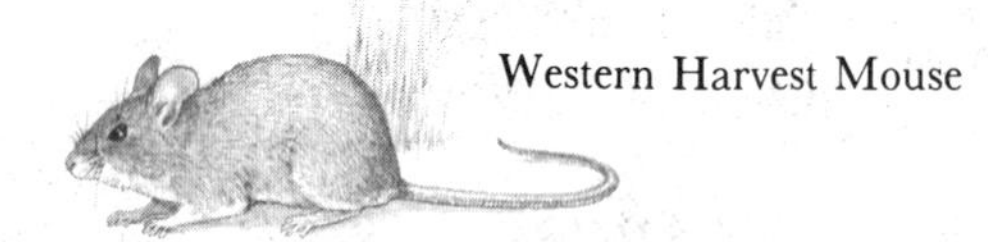
Fig. 28 Western Harvest Mouse

FULVOUS HARVEST MOUSE
Reithrodontomys fulvescens **53:5**

Description
Size: head and body, 3⅕ in. (8.1 cm); tail, 3⅓–4 in. (8.5–10.2 cm). Largest harvest mouse in the East. Above grayish-brown, sides bright orange-buff, below whitish, tail darker above.
Habitat
Grassy and brushy lowlands along streams.
Range
From se. Kans. and s. Mo., s. to the Gulf.

WHITE-FOOTED MICE
Genus *Peromyscus*

These common mice have no grooves on their upper incisors. They are usually brownish above, sometimes grayish, with white feet and underparts. The base of the hairs is slaty. Tails are long and hairy. Those mice that live in woodlands are dark; those that live in the open are pale. The young are grayish, difficult to tell apart. Most are terrestrial; all are nocturnal. Food consists of seeds, berries, fruit, insects, carcasses.

DEER MOUSE
Peromyscus maniculatus **52:3, 54:4**

Description
Size: head and body, 2⅘–4 in. (7.1–10.2 cm); tail, 2–5 in. (5.1–12.7 cm). Color varies geographically from gray to brown to pale yellowish and in between, sometimes with a dark, dorsal stripe; tail sharply bicolored. Tufts in front of ears often whitish. Hind foot relatively short in comparison with that of White-footed Mouse.
Habitat
Woods, prairies, dry uplands, sand beaches, farmland.

Habits
Ground-dwelling, but can climb.
Range
Areas e. of 100th meridian except coastal areas from R.I. southward and most of the s. coastal states.

FLORIDA MOUSE

Peromyscus floridanus **53:3**

Description
Size: head and body, 4½–5 in. (11.4–12.7 cm); tail, about 3½ in. (8.9 cm). Relatively large. Color brownish-gray.
Habitat
Sandy, upland ridges.
Habits
Often uses burrows of other animals.
Range
Most of peninsular Fla.

OLDFIELD MOUSE

Peromyscus polionotus **53:7**

Description:
Size: head and body, about 3½ in. (8.9 cm); tail, about 1½–2½ in. (3.8-6.4 cm). Small, often pale.
Habitat
Sand beaches and old fields.
Habits
Nocturnal; makes burrows in reasonably dry areas.
Range
In se. states, from S.C., s. to parts of Fla. and w. to Ala.

COTTON MOUSE

Peromyscus gossypinus

Description
Size: head and body, 3½–4½ in. (8.9–11.4 cm); tail, 2¾–3½ in. (7–8.9 cm). Dark grayish-brown to tawny above.
Similarities
White-footed is smaller, with brighter whitish underparts.
Habitat
Lowland woods, swamps; invades camps, houses.
Range
From Dismal Swamp of Va., s. along coast, and w. to Ky., Ill., s. Mo.

TEXAS MOUSE

Peromyscus attwateri **53:4**

Description
Size: head and body, 3⅗–4⅕ in. (9.1–10.1 cm); tail, 3⅗–4⅖ in. (9.1–11.2 cm). Brown or grayish-brown above, tawny sides. Well-haired tail dark above, light below; tail usually longer than head and body.
Habitat
Brushy areas and rocky situations.
Habits
Nests under rocks, in crevices, or in piles of brush or sticks.
Range
In s. Mo. and se. Kans.

WHITE-FOOTED MOUSE
Peromyscus leucopus **52:1**

Description
Size: head and body, 3½–4⅕ in. (8.9–10.7 cm); tail, 2–4 in. (5.1–10.2 cm). Reddish-brown above, white below; occasionally brown or gray; often with darker dorsal stripe; upper lip dusky. Tail frequently not sharply bicolored, shorter than head and body, less hairy than in Deer Mouse.
Similarities
Cotton is darker, heavier, without white beneath.
Habitat
Mixed woods, hardwoods; thickets, streamsides.
Age
To 5 years.
Range
Throughout e. U.S. except the coastal plain from S.C. to La.

OTHER MICE
Genera *Ochrotomys, Onychomys, Baiomys,* and *Sigmodon*

GOLDEN MOUSE
Ochrotomys nuttalli **53:9**

Description
Size: head and body, 3⅖–3⅘ in. (8.6–9.7 cm); tail, 3–3½ in. (7.6–8.9 cm). Thick, soft fur. Golden above, white below.
Habitat
Evergreen forests, hardwoods, thickets, swamps.
Habits
Largely arboreal; builds grapefruit-sized nests of Spanish moss, grasses, and other vegetation, up to 10 ft. (3 m) aboveground in shrubs, vines, or trees.
Range
In se. U.S., along coast from Va. to cen. Fla., and w. to Ky., s. Ill.

NORTHERN GRASSHOPPER MOUSE
Onychomys leucogaster **52:7, 54:5**

Description
Size: head and body, 4–5 in. (10.2–12.7 cm); tail, to 2½ in. (6.4 cm). Two phases, with varying shades of gray or pinkish-cinnamon above, white below, giving bicolored pattern. Fur short, dense; body stocky.
Habitat
Prairies.
Habits
Nocturnal.
Range
From s. Man., s. to w. Iowa and e. Kans. and westward.

NORTHERN PYGMY MOUSE
Baiomys taylori **53:1**

Description
Size: head and body, 2–2½ in. (5.1–6.4 cm); tail, about 1½ in. (3.8 cm). A tiny gray mouse. Length of hind foot ⅝ in. (1.5 cm). Color

grayish, like a House Mouse or young *Peromyscus,* but has tail paler beneath than House Mouse and more delicate feet than young *Peromyscus.*

Habitat

Grassy areas.

Habits

Uses small runways in the grass, eats seeds.

Range

Mexican-U.S. border areas, well into s. Tex.

HISPID COTTON RAT

Sigmodon hispidus **53:10**

Description

Size: head and body, 5–8 in. (12.7–20.3 cm); tail, 3–5 in. (7.6–12.7 cm). Fur long, coarse. Grizzled brown and black above, whitish beneath; darker in the East, paler in the West; feet gray. Thinly haired tail dark above, light below, and shorter than head and body. Hind foot longer than in Prairie Vole or Southern Bog Lemming; molars somewhat rounded, not angular.

Habitat

Moist grassy areas, meadows, ditches; also dry places in areas with ample ground cover.

Habits

Active at all hours; strews little piles of cut grasses along its runways.

Range

Along Atlantic and Gulf Coasts, and w. to Tenn., Nebr., and Kans.

WOODRATS

Genus *Neotoma*

These rats have conspicuous ears and eyes, soft fur, white feet, hairy tails, and flat molars. In comparison, Old World rats have smaller ears, dusky feet, scaly tails, and cusped molars. When alarmed, woodrats thump with their hind feet. Scats are frequently deposited in large piles. Their food is seeds, fruits, leaves, berries, cactus pulp, grass, and insects. These nocturnal rats collect unusual objects, such as cans, silver, belt buckles, and so forth at their nest sites and sometimes replace objects they take with other items. Other names include pack rat and trade rat.

EASTERN WOODRAT

Neotoma floridana **53:11**

Description

Size: head and body, 8–9 in. (20.3–22.9 cm); tail, 6–8 in. (15.2–20.3 cm). Above brown; underparts, throat, and chest white; midline of abdomen pure white; sides of belly white, with hairs gray at the roots; tail shorter than head and body, hairy but not bushy, dark above, light below. Norway and Black Rats have longer, scaly tails and their underparts are never white.

Habitat

Caves, cliffs, rocky outcrops, hedgerows; nests on ground in West, on ground or in trees in East.

Range

From N.Y., s. to Va. and w. to Nebr. and Kans.; also s. to Gulf Coast.

SOUTHERN PLAINS WOODRAT
Neotoma micropus **53:12**

Description
Size: head and body, 7½–8½ in. (19–21.6 cm); tail, 5½–6½ in. (14–16.5 cm). Above all-gray; below whitish; throat, chest, and feet white; tail darker above than below.
Habitat
Dry plains, usually in areas where cactus abounds.
Range
In s. Kans.

VOLES
Genera *Clethrionomys* and *Phenacomys*

SOUTHERN RED-BACKED VOLE
Clethrionomys gapperi **52:9, 55:6**

Description
Size: head and body, 3⅔–4⅔ in. (9.3–11.8 cm); tail, to 2 in. (5.1 cm). Three phases: red, gray, and black. Red has reddish back and gray sides. Gray lacks reddish back and is hard to distinguish externally from other voles. All phases are silvery below. Tail thin; black above, gray below, lacking bristles except at tip.
Habitat
Northern forests, clearings, and swamps.
Habits
Terrestrial; a good climber; active at all hours.
Voice
High, musical squeak.
Food
Plants, seeds, nuts, berries, lichens.
Range
Throughout Canada and s. to Appalachians, S.Dak., and Iowa.
Other name
Boreal Red-backed Mouse.

HEATHER VOLE
Phenacomys intermedius **55:2**

Description
Size: head and body, 3½–4¾ in. (8.9–12.1 cm); tail, to 1⅔ in. (4.2 cm). Nose and (sometimes) face yellowish, tail bicolored.
Similarities
Rock Vole has longer tail. Red-backed and Meadow and Prairie Voles lack yellow nose.
Habitat
Grassy areas in spruce forests; by streams.
Habits
Terrestrial; rare.
Food
Twigs and needles of conifers.
Range
Most of Canada, and s. to n. Minn.

MEADOW MICE AND OTHER VOLES

Genus *Microtus*

These voles are best known by the popular name of "meadow mice." They have long, grayish-brown fur; short ears and tails; and small eyes. They have ungrooved teeth, whereas the bog lemmings have grooved teeth. Their tails usually are more than an inch long, and are not brightly colored. They live on the ground usually in grassy terrain, where they make runways and leave cut grass stems and deposit brown droppings. They are active at all hours and can swim and dive. In the winter they make round holes to the surface through the snow. Except for the Woodland Vole, their voice is a high-pitched squeak. They eat grass, roots, bark, and seeds.

MEADOW VOLE

Microtus pennsylvanicus **52:4, 54:7, 55:4**

Description
Size: head and body, 3½–5 in. (8.9–12.7 cm); tail, 1⅖–2⅗ in. (3.6–6.6 cm). Commonest and most widespread vole. Grizzled fur varies from gray in the West to a rich brown in the East. Underparts vary from silvery to buff to dark gray; tail bicolored. Ears almost hidden in fur.
Habits
Population of 15 to an acre may increase to 250 in 2 years.
Range
Throughout Canada and n. U.S., from Hudson Bay, s. to s. Va. and n. Kans.

Note: The **BEACH VOLE,** *Microtus breweri* (**55:1**), is similar to the Meadow Vole, but larger, longer-tailed, and paler. It is the only vole found on Muskeget Island, Massachusetts, and is the sole mammal species peculiar to that state.

ROCK VOLE

Microtus chrotorrhinus **55:5**

Description
Size: head and body, 4¾ in. (12.1 cm); tail, 2 in. (5.1 cm). Fur grayish-brown; nose yellow to reddish-orange, but color not always conspicuous; grayish below.
Habitat
Lives in small colonies in rocky places; rather uncommon.
Range
From coastal Labr., n. Minn. and s. to N.C.
Other name
Yellow-nosed Vole.

PRAIRIE VOLE

Microtus ochrogaster **55:7**

Description
Size: head and body, 3½–5 in. (8.9–12.7 cm); tail, 1⅗ in. (4.1 cm); mammae, 6. Common vole of the prairies. Fur grayish to blackish-brown with yellowish grizzle, paler to the Northwest, darker to the South and East; tawny below; yellowish-rusty at base of short tail.
Similarities
Meadow usually has a longer tail, darker back, and silver-gray belly. Woodland has long, lax fur.

Habitat
Prairies, fencerows, fields.
Habits
Active at all hours, burrows where cover is scant.
Range
From W.Va., w. to 100th meridian and from s. Alta. to Kans.

WOODLAND VOLE
Microtus pinetorum **55:3**

Description
Size: head and body, 3–4 in. (7.6–10.2 cm); tail, ⅔–1 in. (1.7–2.5 cm); mammae, 4. Fur soft, thick; lacks guard hairs, has no set, can be rubbed forward or backward without ruffling. Above auburn, with buffy sides in Northeast; gray or buff below; ears and tail short. Not grizzled like the longer-tailed Meadow or Prairie Voles.
Habitat
Deciduous forests, orchards, gardens, fields.
Habits
Fossorial; tunnels through leaf mold, pushing up the mold into little ridges. Frequently causes damage to orchards by girdling trees beneath snow in winter.
Voice
Chitter's and *chirr*'s; alarm is a thrushlike *cheer, cheer.*
Range
From Maine to n. Fla. and from Wis. to e. Kans.
Other name
Pine Vole

WATER RATS
Genera *Neofiber* and *Ondatra*

ROUND-TAILED MUSKRAT
Neofiber alleni *Fig. 29*

Description
Size: head and body, 7–8 in. (17.8–20.3 cm); tail, 4–6 in. (10.2–15.2 cm). Legs and ears short; fur dense. Color is a rich brown.
Habitat
Shallow, grassy marshes.
Habits
Builds small, dome-shaped houses of grass about 7–25 in. (17.8–63.5 cm) across, also feeding platforms of packed grass fragments.

Fig. 29

Food
Plants.
Range
Most of Fla., except the w. peninsula and Okefenokee vicinity in southern Ga.
Other name
Florida Water Rat.

MUSKRAT
Ondatra zibethicus

51:4
Fig. 29

Description
Size: head and body, 10–14 in. (25.4–35.6 cm); tail, to 11 in. (27.9 cm); mammae, 8–10. Above thick, dark brownish or blackish fur with coarser guard hairs; silvery below. Tail black, naked, laterally compressed; hind feet partly webbed.
Habitat
Fresh water and saltwater marshes; lakes, ponds, watercourses.
Habits
Aquatic, active at any hour; often seen swimming with head appearing wedge-shaped, while Beaver's head appears squarish. Builds a cone-shaped lodge, 5 ft. (1.5 m) in diameter at base, up to 3 ft. (0.9 m) above the water, of mud and sticks in marsh; has an underwater entrance; burrows in banks, has feeding platform on a mat of cattails.
Food
Stems of cattails, grasses; mussels.
Voice
Moans, squeals, chatters from within lodge.
Reproduction
5–7 born in lodge.
Range
Throughout the East; absent from Fla. and adjacent areas.

LEMMINGS
Genera *Lemmus* and *Synaptomys*

These lemmings have curved upper incisors with shallow grooves near the outer edges and very short tails. Their long, grizzled, grayish-brown fur almost hides their ears. The thumb has a large digging claw. They are colonial and active day or night.

BROWN LEMMING
Lemmus sibiricus

55:9

Description
Size: head and body, 4½–5½ in. (11.4–14 cm); tail, to 1+ in. (2.5+ cm). Thick, long hair. Upper incisors not grooved. Grayish head, reddish body, brown rump; cream to medium brown below, no dark dorsal stripe.
Habitat
Tundra and adjacent forests.
Habits
Leaves piles of dung pellets in runways; active at all hours.
Range
N.W.T. and n. Man.

SOUTHERN BOG LEMMING

Synaptomys cooperi **55:8**

Description
Size: head and body, 3⅖–4⅖ in. (8.6–11.2 cm); tail, ⅜–⅞ in. (0.95–2.2 cm). Grizzled brown above, underparts grayish and lighter.
Habitat
Bogs, swamps, and meadows with thick ground vegetation.
Range
From s. Canada throughout ne. U.S. to Tenn. and sw. Kans.

NORTHERN BOG LEMMING

Synaptomys borealis **52:6**

Description
Size: head and body, to 4¾ in. (12.1 cm); tail, to 1 in. (2.5 cm). Pale to dark brown above, lighter below. Long thick claws on front toes in winter.
Similarities
Southern has shorter tail.
Habitat
Locally distributed in open or wooded, moist or dry areas.
Range
N.W.T. and Alta., s. to Maine, N.H., and n. Minn.

COLLARED LEMMINGS

Genus *Dicrostonyx*

The members of this circumpolar genus are the only rodents that turn white in winter. Another distinctive feature of the group is the bulbous central extensions of the third and fourth claws of the forefeet in winter, probably an adaptation for grooming the long winter pelage and digging in snow; these structures are lost in summer. Collared lemmings are not known to occur south of the treeless parts of the Arctic. One to eight (usually three to four) young are born in each of two litters annually (June and July). The upper incisors are not grooved and the ears are entirely concealed in fur. The voice has been reported as a squeal or chuckling note.

COLLARED LEMMING

Dicrostonyx torquatus **55:10**

Description
Size: head and body, 4–5½ in. (10.2–14 cm); tail, to ⅘ in. (2 cm). Summer: above brownish-black with some buff; tawny color across throat; creamy-buff below.
Similarities
Brown Lemming is less stout, lacks dark dorsal stripe, and stays brown in winter.
Habitat
Dry gravelly or sandy tundra.
Range
Greenland, N.W.T., and n. Man.
Other name
Greenland Collared Lemming.

LABRADOR COLLARED LEMMING
Dicrostonyx hudsonius

Description
Size: head and body, 5–6 in. (12.7–15.2 cm); tail, ⅔–1 in. (1.7–2.5 cm). Summer: buffy-gray above, dark stripe down middle of back; flanks, throat, and underparts buffy or rusty.
Habitat
Open tundra and rocky hillsides.
Range
Labr. and n. Que. (Ungava Bay).

OLD WORLD RATS AND MICE
Family Muridae

Old World rats and mice are grayish-brown to black above, usually grayish below. They have even-colored, nearly naked, long scaly tails; sixteen teeth, 1003/1003; and molars with three longitudinal rows of tubercles. They frequent buildings, dumps, ships, or fields; are active at all hours; swim; and do not hibernate. Their varied diet includes grain, groceries, garbage, and meat.

These animals are our unwanted companions, having entered the Western Hemisphere on ships from the Old World. For survival they rely on fecundity. Their nests are of anything soft in almost any kind of hole. Gestation takes from nineteen to twenty-three days; there are five to nine young, and several litters a year.

BLACK RAT
Rattus rattus *Fig. 30*

Description
Size: head and body, 7 in. (17.8 cm); tail, 9 in. (22.9 cm). Black to tawny above, slate-colored to white below; tail naked, scaly, usually longer than head and body.
Similarities
Norway Rat has less slender body and shorter ears and tail.
Habitat
Includes ships, trees.
Habits
Climbs; commonest in the South.
Remarks
Fleas that infest this rat are the carriers of the plague, or "black death."
Range
S. and e. temperate climates, especially large coastal cities.

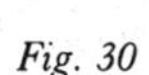

Fig. 30

Norway Rat, p. 268 Black Rat

NORWAY RAT
Rattus norvegicus

54:10
Fig. 30

Description
Size: head and body, 7–10 in. (17.8–25.4 cm); tail, to 8 in. (20.3 cm). Fur coarse, ears short. Underparts usually grayish but sometimes white; tail naked and scaly, usually shorter than head and body.
Similarities
Woodrats have hairy tails and white underparts and feet. Black Rat has a longer tail.
Habitat
Cities and farmyards.
Habits
Destructive; yearly damage runs into millions.
Voice
A squeal.
Range
All areas inhabited by humans from Hudson Bay, southward.

HOUSE MOUSE
Mus musculus

53:2

Description
Size: head and body, 3⅖ in. (8.6 cm); tail, 2¾–3⅘ in. (7–9.7 cm). Dull gray to grayish-brown above, underparts grayish or buffy; fur short; ears nearly naked; notch on posterior tip of upper incisors best seen in side view. Notch separates the House Mouse from all the native mice in the East.
Habitat
In or near buildings.
Habits
Prolific; can breed when 40 days old.
Voice
A squeak.
Remarks
The white strain is kept as a pet and for experimentation.
Range
Throughout most of e. U.S. in and around habitations of humans; feral populations known in many places as well.

JUMPING MICE
Family Zapodidae

These are small, delicate mice with very long tails and hind legs by which they can leap more than six feet. They differ from pocket mice and kangaroo rats by the absence of external cheek pouches; their tails are scantily haired and are never tufted at the tip. They are yellow- or orange-brown above with darker, somewhat bristly hairs in a band along the back; the underparts are whitish. The tail is dark above and white underneath. Each upper incisor has a groove down the front surface.

Swamps, bogs, meadows, and woodlands are their preferred habitat. They do not occur in arid regions. Jumping mice hibernate from October or November to April or May, depending on the latitude, in winter nests located in burrows at least a foot below the surface in well-drained ground. Insects, seeds, berries, and other fleshy fruits make up the diet. Usually two litters of four to six young are produced each summer. The summer nests, rounded in shape, are made of grasses and leaves, and are lined with finer materials; they are placed on the ground at the end of short burrows, under logs and other debris, in tufts of grass or in clumps of shrubs.

Jumping mice are secretive, nocturnal, and uncommon; they apparently make no runways. There is no mound of earth at the burrow entrance, but the entrance may be plugged with earth.

MEADOW JUMPING MOUSE

Zapus hudsonius **52:2**

Description
Size: head and body, 3–3½ in. (7.6–8.9 cm); tail, 4–5¾ in. (10.2–14.6 cm); teeth, 1013/1003. Sides grayish-yellow, dorsal band dark; tail long, bicolored, lacks white tip.
Habitat
Mesic areas, especially grassy lowlands along streams or around lakes.
Habits
Swims readily.
Age
To 2 years.
Range
Throughout most of e. N.A. with the exception of extreme n. parts and s. of N.C. and Mo.

WOODLAND JUMPING MOUSE

Napaeozapus insignis **53:8**

Description
Size: head and body, 3–4 in. (7.6–10.2 cm); tail, 4½–6 in. (11.4–15.2 cm); teeth, 1003/1003. Sides bright orange, dorsal band dark; tail with white tip.
Habitat
Moist, cool, deep forests and forest clearings, frequently along lake and stream beds.
Habits
Dives into water when frightened; climbs vines and bushes for berries.
Range
From New England and Canada s. through Pa. and Appalachians.

PORCUPINE

Family Erethizontidae

The spiny-quilled porcupines are large, blackish rodents about the size of a small dog. Most of the body, especially the rump and tail, is thickly set with long, sharp, needle-tipped spines.

PORCUPINE

Erethizon dorsatum **51:7**
Fig. 27

Description
Size: head and body, 18–22 in. (45.7–55.9 cm); tail, 7–9 in. (17.8–22.9 cm); weight, to 30 lb. (13.6 kg); 20 teeth, 1013/1013. Underfur short, soft, and covered by longer, coarse guard hairs, amid which grow quills. Underfur blackish, guard hairs and quills often light-tipped.

Habitat
Forests, preferably with conifers or poplars.
Habits
Usually nocturnal, but sometimes seen by day shuffling over forest floor or hunched into a large ball in a tree. If bothered, the animal erects its spines, arches its back, and tucks its head between its front legs; keeping its back toward the tormentor, it strikes out with its tail, heavily armed with spines; the spines are not thrown or "shot."
Voice
Snort and bark; groaning or crying sounds.
Food
Twigs, leaves, and buds in summer; inner bark of conifers and hardwoods in winter.
Age
To 10 years.
Reproduction
1 young; well-developed, can climb trees and eat leaves when 2 days old. Nest is in a rocky den, burrow, or hollow log.
Range
Throughout N., except for treeless regions; most abundant where conifers abound; s. to Pa., Mich., Wis.

NUTRIAS

Family Capromyidae

NUTRIA

Myocastor coypus *Fig. 29*

Description
Size: head and body, 22–25 in. (55.9–63.5 cm); tail, 12–17 in. (30.5–43.2 cm); weight, 15–18 lb. (6.8–8.2 kg). Underfur brownish, fine, covered by longer coat of coarse overhair. Somewhat resembles a Muskrat, but is larger and has a long, scantily haired, rounded tail.
Habitat
Swamps, marshes, and rivers.
Habits
Semiaquatic, swims well; makes shallow burrows in banks with an enlarged nesting chamber at the rear.
Food
Aquatic vegetation.
Reproduction
2–8 young.
Range
Watercourses in s. U.S.; introduced.
Other name
Coypu.

Cetaceans

Orders Odontoceti and Mysticeti

Whales and their allies, collectively known as cetaceans and comprising two orders, Odontoceti (toothed) and Mysticeti (baleen), are the most completely aquatic mammals. They are distinguished by nostrils (blowholes) set high on the head; fins instead of forelimbs; a laterally flattened tail bearing horizontal flukes; the absence of external hind limbs; a thick, subcutaneous layer of blubber; and a body nearly devoid of hair. The largest animal that ever lived, the Blue, or Sulphur-bottomed, Whale, reaches a length of more than 100 feet (30.5 m) and a weight in excess of 100 tons (90.7 t).

The larger species can be located by their characteristic "blow." Cetaceans frequently occur together in groups; groups of whales are called gams or pods; groups of porpoises or dolphins are called schools. The latter are often seen leaping through the waves in coastal areas and playing about or following ships.

The habitat of many whales is the open sea; smaller species are sometimes also seen in bays or near shore. They concentrate where their food is most abundant, which may be the plankton-rich edge of the polar ice or more temperate latitudes rich in squid and fish. Certain species have special calving grounds, often in shallow tropical waters where they repair in the breeding season. Some migrate regularly, following their sources of nourishment.

Some cetaceans indulge in extensive courtship before mating. The newborn must immediately learn to breathe, hold its breath under water, suckle without drowning, and swim to keep up with the mother. Cetaceans are quite intelligent, show parental affection, and can be taught to perform. Many, perhaps all, species employ sonarlike echolocation for navigation and finding food.

The toothed whales, as their name implies, are equipped with teeth. Additionally, they have a single blowhole, an asymmetrical skull, and nasal bones that form no part of the roof of the narial passage; the skull is highly telescoped. The suborder ranges in geologic time from the late Eocene. It contains five living families and more than thirty living genera.

In contrast, the baleen whales lack teeth, being equipped instead with dense sheets of baleen (whalebone) that grow downward from the upper jaws and form an effective sieve by which microorganisms are strained from the sea. The narial openings, or blowholes, are paired. The suborder dates back certainly to the middle Oligocene.

Mystery still surrounds the immediate ancestors of the whales and the steps these ancestors took in abandoning the land, for scientists are unanimous in assigning them a terrestrial origin. Probably they split from primitive eutherian stock in the Cretaceous period. Ancestral cetaceans may have been inhabitants of freshwater lakes and streams. Perhaps adaptation to the deep by the proto-cetaceans was relatively rapid. Upon the disappearance of the marine reptiles of the Mesozoic, there must have been a vast and attractive food supply there and a relative scarcity of enemies. But little is known, and it makes a fascinating field of inquiry for the paleontologist.

Whales today are threatened with extinction. Within the last 100 years whalers have virtually extirpated them, in commercial quantities, from the Northern Hemisphere, and in the past few years have greatly reduced populations in the Southern Hemisphere.

BEAKED WHALES

Family Ziphiidae

Beaked whales have tapering heads and "beaks," and a dorsal fin placed behind the midpoint of the back. They are medium-sized to moderately large cetaceans and are characteristically brownish or grayish but sometimes dark above and paler on the sides and below. The functional teeth are reduced in number; the remaining are vestigial. Little is known of the biology of beaked whales. Evidently preferring the cold waters of open oceans, they are seldom seen near shore. Their food probably consists of squid and other cephalopods, but some fish also are eaten. The family dates back to the early Miocene.

RARE BEAKED WHALES

Genus *Mesoplodon*

This is a genus of rare whales that are encountered usually only when occasionally one is cast ashore. Adults range in overall length from fifteen to twenty-two feet and are generally blackish in color. Little is known of the natural history of these whales since they are of no commercial value. Females are thought to bear a single calf, as is the rule in most cetaceans. Members have only two teeth and two grooves on the throat that meet in front. The size and positioning of the teeth are useful in distinguishing between the four species which occur in eastern waters, but are not a sure means of identifying every individual of each species.

NORTH ATLANTIC BEAKED WHALE

Mesoplodon bidens *Fig. 31*

Range

Principally North Sea and adjacent areas in the eastern Atlantic; occasional records along Atlantic Coast of N.A., from Newfoundland to Mass.

Fig. 31

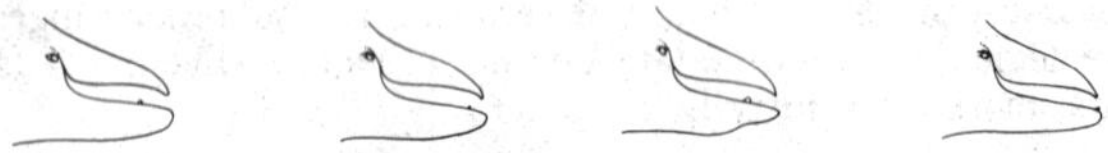

North Atlantic Tropical Gervais' True's

TROPICAL BEAKED WHALE

Mesoplodon densirostris *Fig. 31*

Range

Along Atlantic Coast, from N.S., s. to Carolinas.

GERVAIS' BEAKED WHALE

Mesoplodon europaeus *Fig. 31*

Range

Along Atlantic Coast, from Long Is., N.Y., southward.

TRUE'S BEAKED WHALE

Mesoplodon mirus **56:6** *Fig. 31*

Range

North Atlantic Ocean, from N.S., occasionally to Fla.

OTHER BEAKED WHALES

Genera *Ziphius* and *Hyperoodon*

GOOSE-BEAKED WHALE

Ziphius cavirostris **56:7**

Description
Size: length, to 28 ft. (8.5 m). Body thick; beak short, forehead fairly prominent; 2 teeth, rounded or conical, at tip of lower jaw. Color varies from gray to black, sometimes with white on head or back; sides brownish or spotted; belly whitish; pattern occasionally reversed. No other beaked whale has a projecting lower jaw and a distinct ridge from the dorsal fin to the tail.
Habits
Travels in gams of 30 or more, but often solitary; some are taken commercially.
Range
Along Atlantic Coast, from R.I., s. to Fla.

NORTH ATLANTIC BOTTLE-NOSED WHALE

Hyperoodon ampullatus **56:3**

Description
Size: length, 20–30 ft. (6.1–9.1 m); 1 large pair of teeth at front of lower jaw. Snout sharp, short. Only beaked whale with a bulging forehead. Color varies from black to yellowish, becoming lighter with age; head and belly whitish. Males have high forehead with a white patch and a white dorsal fin.
Habits
Highly sociable, travels in gams; migratory, often stranded; individuals can remain submerged for 30–45 minutes.
Range
Arctic and Atlantic oceans, s. at least to R.I.

GIANT AND PYGMY SPERM WHALES

Families Physeteridae and Kogiidae

These families contain three species: the sperm whale, and two pygmy sperm whales. The latter, regarded in the past as representing a distinct family, Kogiidae, are now considered a subfamily of Physeteridae. Physeterids have a fossil history dating back to the early Miocene.

SPERM WHALE

Physeter macrocephalus **56:5**

Description
Size: length, to 60 ft. (18.3 m); weight, to over 110,000 lb. (50,000 kg). Males much larger. Head enormous, squared, containing huge, barrel-shaped spermaceti organ filled with high-grade translucent oil. Upper jaw toothless, lower jaw slender and containing about 24 functional teeth. Single narial passage on left and directed forward, old right narial passage modified as sound-producing organ. Dorsal fin obtuse or rounded. Uniform slate-gray.
Habits
Migratory, gregarious, polygamous; males may move to edge of pack ice in summer but females generally remain in tropical or temperate waters.

Food
Principally squid.
Reproduction
Sexual maturity reached in about 8 years; females bear single calf every 4 or 5 years after gestation of about 15½ months; young nurse up to 2 years.
Range
All oceans, except polar ice fields; most common between 50°N and 50°S.

PYGMY SPERM WHALE
Kogia breviceps **56:4**

Description
Size: length, to 13 ft. (4 m); weight, to 881 lb. (400 kg). Head blunt, rounded; lower jaw narrow and with 8–11 teeth on each side. Dorsal fin small, curved, and posterior to center of back. Blowhole as in Sperm Whale. Blackish above, light below.
Habits
Probably solitary, rarely seen, occasionally stranded.
Food
Squid, octopi.
Reproduction
Sexual maturity reached in about 2 years; mating occurs in late summer and females bring forth a single calf in spring.
Range
Worldwide in tropical and temperate waters; north along Atlantic Coast to N. S.

Note: Little is known of the biology of Pygmy Sperm Whales. A second species, the **DWARF SPERM WHALE**, *Kogia simus,* also may occur in coastal areas in the southern part of the United States. *K. simus* is smaller than *K. breviceps:* length, to 9 feet (2.7 m), weight, to 595 pounds (270 kg). It has a high dorsal fin, and usually 12 to 16 teeth (as opposed to 8 to 11) in each lower jaw.

WHITE WHALE AND NARWHAL
Family Monodontidae

These are small whales without a dorsal fin and with a single blowhole. The family contains but two species, each of which occurs only in Arctic and north temperate waters. These are gregarious animals that move southward in winter beyond the ice cap or congregate in areas kept free of ice by tides, currents, or winds. The family is known only from the Pleistocene.

WHITE WHALE
Delphinapterus leucas **56:1**

Description
Size: length, to 18 ft. (5.5 m); weight, to 4000 lb. (1814.4 kg). Both jaws with 8–10 teeth on each side. The only white whale. Young are dark gray, then mottled, then yellowish or bluish before becoming white as adults.
Habits
Travels in gams; migrates, often ascends rivers.
Food
Squid, fish, crustaceans.
Reproduction
Young born April–June; nurse 8 months, mature in 2 to 3 years.

Remarks
This is one of the few cetaceans that enter rivers. Schools of them live in the lower St. Lawrence, where the alert boat passenger may occasionally glimpse them from the bow. White Whales are hunted by Eskimos for their meat, oil, and hides.
Range
Arctic and subarctic seas of Atlantic and Pacific, rarely s. to Mass.
Other name
Beluga Whale.

NARWHAL
Monodon monoceros **56:2**

Description
Size: length, to 18 ft. (5.5 m); weight, to 3000 lb. (1360.8 kg). Snout blunt, eye small; ridge along mid-back. Above mottled gray, below white; young are gray, may be confused with young White Whale. Male has single, forward-projecting, spiraled tusk, which is the left, rarely the right, canine tooth, up to 9 ft. (2.7 m) long. Females occasionally have small tusks, but there are no other teeth in either sex.
Habits
May dive to 1200 ft. (365.8 m); use of tusk is unknown.
Voice
Females may roar in calling young.
Food
Shrimp, cuttlefish, fish.
Range
Arctic seas; virtually limited to n. parts of Hudson Bay and coast of Labr.

PORPOISES AND DOLPHINS
Family Delphinidae

Members of this family are small whales, usually with teeth in both jaws. They have a single blowhole far back from the snout, and the dorsal fin is near the middle of the back. They are generally dark above and paler below. Adults range in size from less than six feet in the Atlantic Harbor Porpoise to more than thirty feet in the Killer Whale. The family dates back to the Miocene.

STRIPED PORPOISE
Stenella caeruleoalba

Description
Size: length, to 8 ft. (2.4 m); teeth, 44–50 in each half of each jaw. Dark above, white below; black bands from eye to base of flipper and from eye to vent.
Range
Widely distributed in temperate and tropical waters the world around.

ATLANTIC SPOTTED DOLPHIN
Stenella plagiodon *Fig. 32*

Description
Size: length 5–7 ft. (1.5–2.1 m). The only really spotted dolphin, although other species may show various light marks. Small and

numerous white spots on back, paler below. Young are not spotted. Moderately long snout separated from forehead, as in Bottle-nosed Dolphin.

Range
Coastal waters, from N.C., southward; fairly common in Gulf of Mexico.

Fig. 32

Atlantic Spotted Dolphin

ROUGH-TOOTHED PORPOISE

Steno bredanensis

Description
Size: length, to 8 ft. (2.4 m); 20–27 rugose teeth in each half jaw. Beak not distinctly set off from forehead. Purplish-black above, yellowish or whitish spots on sides, whitish to purplish-pink below.

Range
Tropical and warm temperate seas; s. from coastal Va.

COMMON DOLPHIN

Delphinus delphis **57:6**

Description
Size: length, to 8½ ft. (2.6 m); weight, to about 180 lb. (81.6 kg); teeth small, conical, approximately 40–50 in each half of both jaws. Only delphinid that is black above, white below, and with yellowish flanks. Jaws slender; eye-rings white and connected by 2 white lines. Flippers black; flanks with yellowish, grayish, or whitish bands.

Habits
Gregarious; frequently travels in schools of up to 100 individuals; follows ships, makes graceful leaps out of water; normally submerges for 1–3 minutes.

Age
To about 15 years.

Reproduction
Single calf born in spring after gestation of 9–10 months.

Range
Atlantic Coast, from N.S., southward.

Other name
Saddle-backed Dolphin.

BOTTLE-NOSED DOLPHIN

Tursiops truncatus **57:2**

Description
Size: length, to 12 ft. (3.7 m); weight, to 400 lb. (181.4 kg); teeth large, smooth, 19–26 in each half of both jaws. The most common dolphin. Transverse groove between beak and rounded forehead; lower jaw longer than upper; dorsal fin prominent; belly paler than back. Color, grayish; short snout, and "bottle nosed."

Habits
Migratory, travels in schools, leaps in air at surface.

Food
Fish.
Reproduction
Single calf, 3–4 ft. (0.9–1.2 m) long and weighing 25 lb. (11.3 kg); born every other year after gestation of about 12 months.
Range
Atlantic Coast from N.B. to Mexico.

WHITE-BEAKED DOLPHIN

Lagenorhynchus albirostris **57:1**

Description
Size: length, to 10 ft. (3 m); weight, to 600 lb. (272.2 kg); teeth, 22–25 in each half of both jaws. Only dolphin with white beak. Above black, pale stripe along side, belly whitish.
Habits
Travels in schools, sometimes of 1000 or more individuals.
Range
North Atlantic, rarely s. to Gulf of Mexico.

ATLANTIC WHITE-SIDED DOLPHIN

Lagenorhynchus acutus **57:3**

Description
Size: length, to 9 ft. (2.7 m); weight, to 500 lb. (226.8 kg); teeth, 30–34 in each half of both jaws. Snout short, blunt; dorsal fin prominent. Back black, belly white, bright whitish stripe on sides.
Habits
Gregarious.
Food
Fish and squid.
Range
North Atlantic, s. to Mass.; most common off Newfoundland.

GRAMPUS

Grampus griseus **57:7**

Description
Size: length, to 13 ft. (4 m); functional teeth, 2–7 in each half of lower jaw only. Head bulging, yellowish; no beak. Blunt, but not projecting, nose and slender flippers. Males bluish-white above, females brownish; flippers mottled grayish to blackish, belly grayish-white.
Habits
Solitary or in small schools.
Food
Fish, cuttlefish.
Reproduction
A single calf, about 5 ft. (1.5 m) long, born in winter.
Range
All temperate and tropical seas from Mass., southward.
Other name
Risso's Dolphin.

KILLER WHALE
Orcinus orca **57:5**

Description
Size: length, 15–30 ft. (4.6–9.1 m); weight of males, to about 10,000 lb. (4535.9 kg), of females, to about 5000 lb. (2268 kg); teeth long, narrow, 10–14 in each half of both jaws. Body stout, rounded; nose blunt; eye large; belly and spot behind eye white; recognized by its tall dorsal fin, glossy black back, and white rear flank.
Habits
Travels in schools of up to 40 or more, a fast swimmer.
Reproduction
Length of a single calf at birth 7–9 ft. (2.1–2.7 m).
Range
Along Atlantic Coast.

COMMON PILOT WHALE
Globicephala melaena **57:8**

Description
Size: length, to 28 ft. (8.5 m); weight, to about 6000 lb. (2721.6 kg); teeth, 10 in each half of both jaws. Only dolphin that is almost uniformly black and has a bulging forehead. Body cylindrical; dorsal fin large, curved, set forward of midback; flippers 1/5 length of body. Belly with a light stripe that enlarges to a heart-shaped area on throat.
Habits
Spouts to 5 ft. (1.5 m); migratory, travels in large schools, which are occasionally stranded.
Food
Squid and some fish.
Reproduction
Breeds in spring; single (usually) calf born after gestation of 16 months; nurses up to 2 years.
Range
Coastal waters from Greenland to Va.

SHORT-FINNED PILOT WHALE
Globicephala macrorhynchus

Description
Size: length, 15–20 ft. (4.6–6.1 m); teeth 7–9 in each half of both jaws. Similar to larger Common Blackfish, but flippers only 1/6 length of body.
Habits, Food, Reproduction
Same as Common Pilot Whale
Range
Principally tropical, from N.J. along Atlantic Coast.

HARBOR PORPOISE
Phocoena phocoena **57:4**

Description
Size: length, to 6 ft. (1.8 m) weight, to about 50 lb. (22. 7 kg); teeth, 22–27 in each half of both jaws, have spadelike crowns. Only small dolphin with a triangular dorsal fin. Body small, thick; mouth round, eye small. Back black, sides pinkish, belly white; dark line from corner of mouth to flipper.

Habits
Spout short, vertical; makes puffing sound; active before squalls; travels in schools; frequents shorelines, harbors.
Food
Fish, squid.
Range
In North Atlantic from Greenland to N.J.

FALSE KILLER WHALE
Pseudorca crassidens *Fig. 33*

Description
Size: length, 13–18 ft. (4–5.5 m); teeth, relatively large, 8–10 on each side above and below. Black, slender, small, without bulging forehead. Dorsal fin relatively small, projecting slightly backward and slightly ahead of the middle of the body; snout bluntly rounded, longer than lower jaw, without beak or bulging forehead.
Food
Fish and squid.
Remarks
Mostly known from mass strandings.
Range
Atlantic Coast from N.C., southward.

False Killer Whale

Fig. 33

FINBACK WHALES
Family Balaenopteridae

Members of this family are commonly called fin whales or rorquals. They generally are characterized by short, broad plates of whalebone, or baleen, hanging in rows from each side of the upper jaw, and they possess no teeth as adults. They have a dorsal fin, many furrows in the throat, and a double row of blowholes. These whales have been the mainstay of the recent whaling industry and, after being ruthlessly exploited in northern waters, are now being systematically hunted in the Southern Hemisphere. The family dates back to the early Miocene.

FIN WHALE
Balaenoptera physalus *Fig. 34*

Description
Size: length, to over 80 ft. (24.4 m); weight, to 70 tons (63.5 t); baleen, to 3 ft. (0.9 m), purple and white, sometimes grayish; 70–80 throat furrows. Up to about 400 plates of baleen on either side of upper jaw. Head flat; dorsal fin small, with slightly concave rear edge, just in front of flukes. Above gray, below white.
Habits
Spout 15–20 ft. (4.6–6.1 m) high, inclined forward, narrow at start, then elliptical, made with loud whistling sound; fastest of all whales; sometimes comes near ships. Gregarious, usually occurring

in pods of 10–15 but there are reports of many hundreds flocking together.

Food

Plankton, small crustaceans.

Reproduction

Gestation period 1 year; a single calf, occasionally 2, weighing up to 4000 lb. (1814.4 kg) and 20–22 ft. (6.1-6.7 m) long at birth.

Range

All oceans from Greenland to the Caribbean.

SEI WHALE

Balaenoptera borealis

Description

Size: length, to 60 ft. (18.3 m); weight, rarely exceeds 20 tons (18.1 t); baleen, to 3 ft. (0.9 m), blackish; 40–60 throat furrows. Somewhat similar to Fin Whale, but body smaller, dorsal fin relatively larger and set farther forward. Back bluish-black, sides gray, belly white.

Habits

Spout to 14 ft. (4.3 m), vertical; migratory; uncommon.

Food

Small marine invertebrates.

Reproduction

Single calf born about November after gestation of 1 year; weaned after 5 months when 26–30 ft. (7.9–9.1 m) long.

Range

All oceans; from coast of Labr., southward.

MINKE WHALE

Balaenoptera acutorostrata *Fig. 34*

Description

Size: length, to 33 ft. (10.1 m); weight, to about 3 tons (2.7 t); baleen, to 8 in. (20.3 cm), whitish; 50–70 throat furrows. The only cetacean with a broad white band across the flipper. Snout triangular from above; dorsal fin far back, tip curved. Varying shades of dark above, white below.

Habits

Spout faint, vertical; migratory, travels singly or in small groups; frequents coastal waters.

Food

Mostly small school fish.

Reproduction

Single calf.

Range

Widely distributed in all oceans; from Baffin Bay, s. to Fla.

Other name

Little Piked Whale.

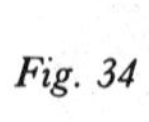

Fig. 34

BLUE WHALE

Balaenoptera musculus

Description
Size: length, to 106 ft. (32.3 m); weight, to 140 tons (127 t); baleen, to 40 in. (101.6 cm), bluish-black; 80–100 throat furrows. Identified by its U-shaped snout (from above) and great size. Dorsal fin small, set far back. Above slaty- to bluish-gray; below yellowish or whitish.
Habits
Spout to 20 ft. (6.1 m), vertical; travels singly or in gams; migrates to warm water to breed; at other times frequents water near pack ice.
Food
Mainly small, shrimplike invertebrates (krill).
Reproduction
Probably monogamous, mating under water from May–August; gestation period probably 10–11 months; calves (25 ft. [7.6 m] long and 2–4 tons [1.8–3.6 t] at birth) are weaned at 5–7 months when about 50 ft. (15.2 m) long.
Remarks
This is the largest animal that ever lived on land or sea. Weighing a ton or more a foot, a 106-ft. (32.3-m) specimen would be twice or thrice the presumed weight of *Brontosaurus*, the largest dinosaur.
Range
All oceans; occurs along entire Atlantic Coast.

HUMP-BACKED WHALE

Megaptera novaeangliae *Fig. 35*

Description
Size: length, to 50 ft. (15.2 m); weight, 25–45 tons (22.7–40.8 t). Baleen varies in length; shorter at front of mouth, less than 3 ft. (0.9 m) long; all-black, with black or olive-black bristles. No other finback has a humped back and scalloped flippers and flukes. Flipper ⅓ length of body; dorsal fin small.
Habits
Spout to 20 ft. (6.1 m) in an expanding column; travels singly or in small schools; can leap clear of the water.
Food
Small crustaceans.
Reproduction
Mating takes place in late winter or spring; gestation period about 10 months; calves 9–15 ft. (2.7–4.6 m) long at birth, weigh up to 4000 lb. (1814.4 kg).
Range
All oceans; occurs along entire Atlantic Coast.

Fig. 35

RIGHT WHALES
Family Balaenidae

These are the only whales with flexible plates of whalebone, or baleen, smooth throats, double blowholes, and no dorsal fin. When diving deeply (sounding), the flukes are thrown clear of the water. The family dates back to the early Miocene.

BLACK RIGHT WHALE
Balaena glacialis *Fig. 35*

Description
Size: length, to 60 ft. (18.3 m), weight, to 60 tons (54.4 t); baleen, to 9 ft. (2.7 m), black. Only baleen whale with high arched mouth and head about ⅕ the total length of the body. Body stout, blackish, sometimes with patches of white below.
Habits
Spout 10–15 ft. (3–4.6 m) high, in 2 columns which diverge to form a V; slow-moving, easy to capture; rare.
Food
Plankton and other small invertebrates.
Reproduction
Little known; gestation thought to last about 10 months; 1 calf 10–20 ft. (3–6.1 m) long at birth and nurses for about 1 year.
Range
Temperate waters of both hemispheres, from at least Gulf of St. Lawrence southward; said once to have occurred as far n. as Greenland.

BOWHEAD WHALE
Balaena mysticetus

Description
Size: length, to 65 ft. (19.8 m); weight, to over 50 tons (45.4 t); baleen, to 14 ft. (4.3 m), black. Only whale whose head is more than ⅓ of body length. Body very stout; lower jaw bowed gently upward, not arched, as in Right Whale. Above dark grayish-brown, belly may be spotted with white.
Habits
Spout V-shaped; does not migrate; not gregarious, normal submersion 20–30 minutes.
Food
Small crustaceans.
Reproduction
Little known; 1 calf, 10–18 ft. (3–5.5 m) long, born in late winter or early spring; bluish-gray in color, later blackish.
Range
Polar seas, s. to St. Lawrence R.; now rare.

Carnivores

Order Carnivora

Terrestrial carnivores are distinguished by their large canine teeth (although the Virginia Opossum, a marsupial, also has large canines) and their strong jaws, legs, and claws. All have five toes on the front foot and four or five on the hind foot. Their sight and sense of smell are often keen, and many are fleet of foot. The smallest, the Least Weasel, weighs less than two ounces; the largest, the Grizzly Bear (a western species), weighs up to 850 pounds (385.6 kg). Females are smaller than males, sometimes by as much as one-third. Carnivores occur in all latitudes and altitudes and from low, hot deserts to cool, damp woods. As a recognizable order, carnivores go back to the early Paleocene. By the Oligocene most of the principal modern families had arisen.

A few carnivores, such as the bears, become inactive during the winter, but most are active at all seasons. Many have voices which, like the cry of the wolf, can be heard from afar. All are flesh- or fish-eaters, but a number vary their diet with berries, fruits, and nuts. Females breed once or twice a year, bringing forth a litter of blind and usually furred young (called cubs, pups, or kits) in a secluded den, hollow log, or burrow. These usually stay with the mother into the summer or fall or, in the case of some species, the next year. Some mustelids have a prolonged gestation period (up to one year), but most embryo growth takes place in final few weeks.

COYOTES, WOLVES, AND FOXES

Family Canidae

All members of this family are doglike. They have four toes on each hind foot. The dental formula of all eastern species is 3142/3143. The young are born in a hollow log, den, or burrow, and the gestation period is seven to ten weeks. The family dates back to the Eocene. The mammae always occur in pairs.

COYOTE

Canis latrans **59:1, 60:5, 60:8**

Description
Size: head and body, to 54 in. (137.2 cm); tail, to 16 in. (40.6 cm); weight, to 50 lb. (22.7 kg). Buffy-gray with rusty legs, feet, and ears; underparts whitish. Nose and ears pointed, nose pad narrow; legs long; long, bushy tail with black tip.
Similarities
Gray Wolf is larger, grayer, and has broader nose pad.
Habitat
Brush country, grasslands, farmlands; prefers open spaces.
Habits
Largely nocturnal, but frequently active by day; carries tail drooping when on run.
Voice
High-pitched bark and howl, often in evening chorus.
Food
Small mammals and ground-nesting birds; in summer, also fruits and berries.
Reproduction
Usually 6–7 pups.
Range
In much of U.S., from New England westward, in se. states.

RED WOLF

Canis rufus **60:3**

Description
Size: head and body, to 49 in. (124.5 cm); tail, to 17 in. (43.2 cm); weight, to 80 lb. (36.3 kg). Similar to Gray Wolf, but smaller, with color varying from grayish-brown to reddish and black. Light-colored individuals can be distinguished from Coyote by their broader nose patch (over 1 in. [2.5 cm] wide).
Habitat
Plains, forests.
Habits
Runs with tail horizontal.
Voice
Like Gray Wolf.
Food
Small or weak deer, small mammals.
Reproduction
Usually 4–6.
Range
Formerly most se. states, now reduced to remnants in Tex. and La.

GRAY WOLF

Canis lupus **60:3, 60:9** *Fig. 36*

Description
Size: head and body, to 43–48 in. (109.2–121.9 cm); tail, to 19 in. (48.2 cm); weight, to 170 lb. (77.1 kg). Usually gray but varies from white to gray to black; pure white on Arctic islands and Greenland; all above colors in Alaska. Fur long.
Similarities
Coyote has less rounded ears and narrower nose pad; is less doglike in appearance.
Habitat
Wilderness only; tundra, plains, forests.
Habits
Travels singly or in pairs or packs; often hunts cooperatively; carries tail level or upraised when on the run.
Voice
Loud howl; various barks and growls.
Food
Young, sick, and aged deer; mountain sheep, bison, and musk oxen; small mammals; also berries and fruits.
Reproduction
First breeds at 2 years of age; 1–13 pups born in spring (usually April), after gestation of 60–63 days.
Range
Formerly occurred throughout eastern U.S., now restricted to wilderness areas in extreme n. U.S. and Canada.
Other name
Timber Wolf.

Fig. 36 Gray Wolf

ARCTIC FOX

Alopex lagopus

60:4
Fig. 37

Description
Size: head and body, to 31 in. (78.7 cm); tail, to 15 in. (38.1 cm); weight, 7–15 lb. (3.2–6.8 kg). Two color phases: blue phase in summer is brownish or gray above with yellowish-white sides and belly; in winter, slate-blue, sometimes with brown on head and feet, no white tip to tail. White phase in summer is same as blue phase; in winter, all-white. Feet heavily furred; ears short and rounded.

Habitat
Tundra and Arctic coast.

Habits
Unsuspicious; burrows in snow for a temporary den; solitary or travels in pairs, but never in packs.

Voice
High-pitched yapping bark.

Food
In winter, dead seals, walruses, whales; in summer, rodents, birds, eggs, berries.

Reproduction
Reaches sexual maturity in 10 months; mates in March to April; litter of 6–7 kits born following gestation of 52–57 days.

Range
N.W.T. and adjacent areas in Canada; coastal areas of Greenland.

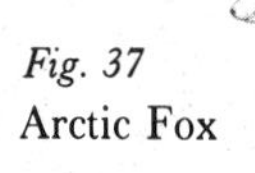

Fig. 37
Arctic Fox

summer winter

SWIFT FOX

Vulpes velox

59:3, 60:4

Description
Size: head and body, to 31 in. (78.7 cm); tail, to 12 in. (30.5 cm); weight, 4–6 lb. (1.8–2.7 kg). Gray to pale yellowish-brown. Body slender, ears large, tip of tail black; blackish spot on each side of snout.

Habitat
Plains, dry foothills, deserts.

Habits
Nocturnal, shy.

Voice
Weak bark.

Food
Mice and other small mammals, insects, some fruit.

Reproduction
Mates in December and January; gestation about 50 days; litter size usually 4–7.

Remarks
Once nearly extirpated from much of its range, now seems to be increasing substantially in number.

Range
Principally Great Plains; reaches Prairie Provinces of Canada.

RED FOX

Vulpes vulpes **59:2, 60:4, 60:7**

Description
Size: head and body, to 42 in. (106.7 cm); tail, to 16 in. (40.6 cm); weight, 8–15 lb. (3.6–6.8 kg). Two color phases: red phase is reddish above, white below, feet black; black phase is all-black except white tip on tail. Cross phase is reddish-brown, with heavy black markings on shoulder. Ears large, triangular; tail large, bushy, always having a white tip. Somewhat resembles small collie or Coyote, but both lack white-tipped tail.
Habitat
Extremely adaptable, including the ability to thrive in proximity to humans; probably prefers sparsely inhabited areas combining forests and open country.
Habits
Primarily nocturnal but occasionally active by day; males assist in rearing young.
Voice
Male, "a short yelp, ending in a *yurr,* as if gargling"; female, "a yapping scream" (Palmer).
Food
Rabbits, rodents, snakes, berries, and fruit.
Reproduction
Reaches sexual maturity in about 1 year; mates in January or February; gestation period is about 52 days; single litter each year containing 4–5 kits.
Range
Throughout most of e. U.S. except the Gulf Coastal plain.

GRAY FOX

Urocyon cinereoargenteus **59:5, 60:4**

Description
Size: head and body, to 44 in. (111.8 cm); tail, to 16 in. (40.6 cm); weight, 7–13 lb. (3.2–5.9 kg). Above salt-and-pepper gray; sides, legs, and feet rusty; underparts gray; tail black-tipped and bushy, with a prominent blackish area down middorsal line.
Habitat
Brush, wooded lowlands; swamps.
Habits
Largely nocturnal, solitary; can climb trees.
Voice
Harsh, rarely heard.
Food
Rabbits, rodents, reptiles, berries, and fruits.
Reproduction
Most individuals breed in January or February in their first year of life; the 1 annual litter (3–5 kits) born after 63-day gestation.
Range
From Canadian border southward, w. to S.Dak. and w. Nebr.

BEARS

Family Ursidae

Bears are the largest living carnivores. They have five toes on each foot, and their short tails are almost hidden in long fur. The tooth formula is 3142/3143 and the gestation period is seven to nine months. Bears walk on their entire foot, like human beings. The family dates back to the Miocene.

BLACK BEAR

Ursus americanus **60:1, 61:3**

Description
Size: length, to 6 ft. (1.8 m); height at shoulder, to 3 ft. (0.9 m); weight, usually to 350 lb. (158.8 kg). The smallest and most common bear. Black phase most common in East, cinnamon (all-brown) phase most common in West; face always brown, breast often has patch of white. Snout long, face profile straight; front claws short, rounded; no hump on shoulders.
Habitat
Forests, swamps, mountains.
Habits
Solitary, quarrelsome; inactive in winter but does not hibernate; marks home range by clawing boundary trees.
Voice
Various whines, grunts, and huffs.
Food
Omnivorous; flesh, fish, carrion, fruits, and berries.
Reproduction
2 cubs, born late January or February; weight 8 oz. (226.8 g) or less; breeds every other year.
Range
Formerly occurred throughout entire E. except in Greenland and extreme n. Canada; no longer found in heavily populated areas.

POLAR BEAR

Ursus maritimus *Fig. 38*

Description
Size: length, to 8 ft. (2.4 m); height at shoulder, to 4 ft. (1.2 m); weight of males, usually to 800 lb. (362.9 kg). Coat white, often with a yellow tinge; young whiter than adults. Eye, nose pad, and foot pad black. Head small, neck long.
Habitat
Arctic coast and ice floes.
Habits
Roams widely, digs winter dens in snow, sand, or dirt; a powerful fighter.
Voice
A roar.
Food
Seals, young walruses, stranded whales.
Reproduction
Cubs (usually 2) born in November or December in winter den; females probably breed every other year, may have prolonged gestation.
Range
Arctic Coast of Canada and Greenland.

Fig. 38 Polar Bear

WALRUS

Family Odobenidae

Walruses are pinnipeds, aquatic carnivorous mammals that come to shore to breed and sometimes to rest. Their limbs have evolved into flippers, with which they can haul themselves out on the beach or ice floe; otherwise they are poorly adapted for locomotion on land because the limbs are enclosed within the body to the level of the elbow and knee; the digits of the paddle-shaped hand and foot usually retain claws. They have short tails. The males are usually much larger than females.

Walruses evolved directly from terrestrial carnivores, and are first reported from the Miocene. Being still tied to the land to breed, they are less adapted to marine life than the whales, whose ancestors may have entered salt water 30 million years earlier.

WALRUS

Odobenus rosmarus **62:2**

Description
Size: length, to 12 ft. (3.7 m); weight, to 3000 lb. (1360.8 kg); 18 teeth, 1130/0130. Only marine mammal with 2 large white tusks, smaller in cows than bulls. Body tan to gray, almost hairless; hide thick, in folds; no external ears; hind limbs bend forward when not in water.

Habitat
Ice floes, usually over shallow water; seldom goes ashore.

Habits
Gregarious, feeds standing on head in water, may inflate neck and float on water when asleep.

Voice
Bellow like voice of St. Bernard dog; elephantlike trumpeting.

Food
Mollusks.

Reproduction
Gray calf, born in spring, lacks tusks; travels on cow's neck.

Range
Arctic coasts, s. in Atlantic rarely as far as Maine.

RACCOONS AND ALLIES

Family Procyonidae

With the exception of the Lesser Panda, all recent procyonids are restricted to the New World. The dental formula is 3142/3142 for the species in the East; the cheek teeth are less shearing than those of canids but not so flattened as the teeth of bears. Procyonids walk on the entire flattened surface of the foot, have five toes on each foot, and nonretractile or semiretractile claws. The family is first known from the Oligocene.

COATI

Nasua narica *Fig. 39*

Description
Size: head and body, 20–25 in. (50.8–63.5 cm); tail, 20–25 in. (50.8–63.5 cm); weight, to about 20 lb. (9.1 kg). Long-nosed, long-tailed, with indistinctly ringed tail tapering toward end. White spots near eyes; general color grizzled brown.

Habitat
Shrubby hillsides and ravines.
Habits
Tough nose pad and movable snout used in rooting out food; active day or night; forages in groups of various sizes; tail often carried aloft.
Voice
Loud grunts when alarmed.
Food
Invertebrates, lizards, small rodents.
Reproduction
Gestates 72 days; 2–7 young; weaned at 4 months, mature at 2 yrs.
Range
Chiefly Mexican, but into s. Tex. and border areas farther w.

Fig. 39

Coati

RACCOON

Procyon lotor **54:6, 61:7**

Description
Size: head and body, to 33 in. (83.8 cm); tail, to 12 in. (30.5 cm); weight, 6–35 lb. (2.7–15.4 kg). Fur is gray, brown, and black, giving a grizzled appearance; face with black mask, tail ringed. Body stout, fur long; snout and ears pointed.
Habitat
Woods, swamps.
Habits
Nocturnal, sometimes washes food; a good fighter, climber, and swimmer.
Voice
Varied; includes barks, growls, a throaty cry, a whine, and an owllike quaver.
Food
Fish, crayfish, birds, eggs, corn, vegetables, fruit.
Reproduction
Females mature sexually in first year, males in second year; litter of 3–4 young weighing less than ½ lb. (226.8 g) born in den in ground or hollow tree in April or May after 63-day gestation period; some females have second litter in August, especially if first is lost; weaned at 7–16 weeks.
Range
Throughout e. U.S. and adjacent Canada.

RINGTAIL CAT

Bassariscus astutus *Fig. 40*

Description
Size: head and body, to 17 in. (43.2 cm); tail, to 19 in. (48.2 cm); weight, 30–45 oz. (870–1300 g). Buff above, overcast with blackish or brownish; white or buffy-white below; eye ringed with black or brown.

Habitat
In northern part of range, rocky outcrops, frequently along streams.

Habits
Agile, good climber, nocturnal.

Food
Omnivorous; eats small mammals, birds, eggs, berries, and other fruit.

Voice
"Spits" or "growls" defensively.

Reproduction
3–4 young born in den in rocks or a hollow tree in May or June; mature at one year, weaned at 4–8 weeks.

Range
In s. Kans.

Fig. 40 Ringtail Cat

WEASELS, SKUNKS, AND ALLIES

Family Mustelidae

These mammals are often predominantly brown, but vary considerably in color and size. They usually have long bodies, short legs, rounded ears, and large, paired scent glands. The family dates back to the Oligocene.

MARTEN

Martes americana **58:1, 59:7**

Description
Size: head and body, to 17 in. (43.2 cm); tail, to 9 in. (22.9 cm); 38 teeth, 3141/3142. Above yellowish-brown to dark brown (in Labrador almost black); head, belly paler; has a pale buff patch on the throat and chest; legs, tail, and ears dark. Head small, feet partly furred.

Similarities
Mink is darker, and with shorter tail and small white patch on chin. Fisher is darker and larger.

Habitat
Coniferous forests.

Habits
Largely arboreal, but also hunts on the ground and occasionally in water; does not hibernate.

Voice
High-pitched screams and squeals.

Food
Variable, including squirrels, mice, birds, eggs, and fish.
Reproduction
3–5 young, born in leaf-lined nest in hollow tree, in March or April, after protracted gestation of 220–265 days.
Range
Formerly throughout boreal forest of Canada and n. U.S.; now rare or extirpated over much of s. part of former range.
Other name
Sable.

FISHER

Martes pennanti **58:6, 59:9**

Description
Size: head and body, 25 in. (63.5 cm); tail, to 15 in. (38.1 cm); 38 teeth, 3141/3142. Blackish-brown, head and shoulders grizzled; legs and tip of tail black; tail bushy, tapering.
Similarities
Marten is smaller, lighter, and has light patch on chest.
Habitat
Dense mixed forests.
Habits
Mainly nocturnal; somewhat arboreal, solitary.
Voice
Scream and hiss.
Food
Squirrels, mice, raccoons, rabbits, some vegetable matter, carrion.
Reproduction
1–5 (usually 3) young, born in late March or April; females mate shortly after birth of young, delayed implantation accounting for the prolonged period of gestation.
Range
Same as Marten.

WEASELS

Genus *Mustela*

Weasels have small heads, long necks, and a dental formula of 3131/3132. They are terrestrial, largely nocturnal, and ferocious for their size. They are carnivorous, feeding chiefly on small animals, both warm- and cold-blooded.

ERMINE

Mustela erminea **59:6**

Description
Size: head and body, to 9 in. (22.9 cm); tail, to 4 in. (10.2 cm). In summer brown above; feet and underparts white. In winter, white (in the East); end of tail always black. Spring and fall molts show transition between brown and white upperparts.
Habitat
Field borders, open woodlands, brushy and rocky places.
Habits
Primarily terrestrial, but good swimmer and climber; sometimes hunts in pairs.
Voice
Varied squeals, barks, hisses, and chatters.

Reproduction
Usually 4–7 young, born in fur-lined nest on ground; weight at birth, 1/14 oz. (2 g).
Range
Throughout Canada, coastal Greenland, s. into U.S. to Md. and w. to Iowa.
Other name
Short-tailed Weasel.

LEAST WEASEL
Mustela nivalis **58:9, 59:4**

Description
Size: head and body, 6½ in. (16.5 cm); tail, to 1½ in. (3.8 cm). In summer, brown above, white below. In winter, white in northern part of range, some individuals brown above in southern areas. Tip of tail not black, but may have few black hairs.
Habitat
Open woods, brush piles, grassy areas.
Habits
Home range about 2 acres; can swim.
Voice
A weak bark or shriek.
Reproduction
3–10 (usually 5) young, evidently born in any season, in underground nest lined with mouse fur or grass.
Range
Hudson Bay, southward, excepting New England states; s. to N.C. and w. to n. Kans.

LONG-TAILED WEASEL
Mustela frenata **58:8, 61:4**

Description
Size: head and body, 10½ in. (26.7 cm); tail, to 6 in. (15.2 cm). In summer, brown above, yellowish-white below; tail black-tipped. In winter, in northern part of range, white, with black-tipped tail; in southern part of range, same as summer.
Similarities
Summer Ermine has white, not brown, feet; winter Ermine is similar but smaller. Mink is dark below.
Habitat
Variable, including farmlands, prairies, woodlands, swamps.
Habits
Primarily nocturnal, but often seen by day, usually solitary; climbs well.
Voice
Hisses, screams, purrs.
Reproduction
4–9 young, born in nest in stump, rock pile, or borrow.
Range
Throughout e. U.S., from N.B., Que. and s. parts of Prairie Provinces of Canada to s. U.S.

BLACK-FOOTED FERRET

Mustela nigripes **59:8**

Description
Size: head and body, to 18 in. (45.7 cm); tail, to 6 in. (15.2 cm). Only eastern weasel with a black mask and black feet. Above, buffy-yellow; underparts paler, tip of tail black.
Habitat
Plains, prairies coinciding with range of prairie dogs, on which it feeds.
Habits
More active at night than by day; wary.
Voice
Hiss and chatter.
Reproduction
Mates in March and April; gestation 42–45 days; litter of 1–5 (usually 3–4).
Remarks
Formerly abundant, now extremely rare.
Range
Canada, s. to e. Nebr. and cen. Kans.

MINK

Mustela vison **58:7, 61:5**

Description
Size: head and body, to 17 in. (43.2 cm); tail, to 9 in. (22.9 cm). Only uniformly dark brown weasellike animal in east with white patch on its chin. No seasonal color change.
Habitat
Near water; streams, marshes; in winter, woods.
Habits
Nocturnal, solitary, wary; spends much time in water; emits strong odor when cornered; stores food in den.
Voice
Various hisses, screams, barks, purrs.
Reproduction
Single litter of 5–8 young, born in nest in rocks or other natural cavity, often in bank of stream, river, or lake.
Range
Throughout most of e. U.S.

OTHER CARNIVORES

Genera *Gulo, Taxidea, Spilogale, Mephitis,* and *Lutra*

WOLVERINE

Gulo gulo **58:2, 59:10**

Description
Size: head and body, to 32 in. (81.3 cm); tail, to 9 in. (22.9 cm); weight, to 40 lb. (18.1 kg); 38 teeth, 3141/3142. Color varies from yellowish-brown to almost black, paler on head; has a broad light stripe on each side. Fur long, ears small, back arched, feet large, tail bushy.
Habitat
Wilderness, chiefly brushlands, forests, and mountains.
Habits
Active at all times of day and year; mainly terrestrial, but can climb.

Voice
Snarl and growl.
Food
Any animal it can kill; also carrion.
Reproduction
Usually 2–3 young, born in March or April after a long gestation period involving delayed implantation.
Range
Formerly in Canada and n. U.S., s. to Pa., Ind., and Nebr., but now limited to n. wilderness areas; may be increasing in abundance in some areas.

BADGER

Taxidea taxus **54:1, 61:8**

Description
Size: head and body, to 22 in. (55.9 cm); tail, to 12 in. (30.5 cm); 34 teeth, 3131/3132. No other mammal has white stripe from its nose back over the top of its head. Above grayish-yellow, giving grizzled appearance; underparts paler; head black and white; feet black; tail short, yellowish. Fur long; body flattened, stout; front claws very long.
Habitat
Dry, open prairie and farmlands.
Habits
Nocturnal, but also abroad by day; a strong digger, lives in an underground den.
Voice
Grunts and growls, but usually silent.
Food
Ground squirrels, other small mammals, birds, eggs, reptiles.
Reproduction
1–5 (usually 2) young, born in April or May in grass-lined, subterranean nest.
Range
Throughout prairies, from w. N.Y. and Alta., southward; absent in e. Canada and se. part of the region.

EASTERN SPOTTED SKUNK

Spilogale putorius **58:3, 61:2**

Description
Size: head and body, 13½ in. (34.3 cm); tail, to 9 in. (22.9 cm); 34 teeth, 3131/3132. Stripes on neck, back, and sides; spots on head; and tip of tail all variable, including white.
Habitat
Brushy areas, edges of woods, rocky outcrops.
Habits
Nocturnal; terrestrial, but also a good swimmer, climber, and burrower; very playful.
Food
Rodents, birds, eggs, insects, fruits.
Reproduction
2–6 (usually 4) young, born in a dry, vegetation-lined den in early spring.
Range
In s. states, from n. Minn., Ind., and s. Pa., southward.

STRIPED SKUNK

Mephitis mephitis **58:4, 61:1**

Description
Size: head and body, to 18 in. (45.7 cm); tail, to 10 in. (25.4 cm); 34 teeth, 3131/3132. Narrow white stripe up forehead; tail bushy, may have white tip; scent glands well developed; can often be identified by its characteristic pungent odor.
Habitat
Woods, plains, meadows, suburbs.
Habits
Hunts at night; protects itself by using scent glands if molested.
Voice
Low *churr*'s and growls.
Food
Rats, mice, chipmunks, insects, fruits, berries.
Reproduction
4–7 (usually 5) young, born in vegetation-lined den or sometimes under a building.
Range
Throughout e. U.S. and adjacent Canada.

Note: The **HOG-NOSED SKUNK,** *Conepatus mesoleucus,* has been reported east from extreme southwestern Texas. It resembles the Striped Skunk in size, but differs in having a single, broad, white stripe on the back and a tail that is often entirely white.

RIVER OTTER

Lutra canadensis **58:5, 61:6**

Description
Size: head and body, to 30 in. (76.2 cm); tail, to 19 in. (48.2 cm); weight, to 30 lb. (13.6 kg); 37 teeth, 3141/3132. Above brown to grayish; chin and throat grayish-white; underparts with silvery sheen. Snout broad, rounded; tail round, stout, tapering; hind feet webbed.
Habitat
Watercourses and edges of bodies of water.
Habits
Makes earthen slides into the water, on which it slides apparently in sport; active at all times of day and year.
Voice
Varied, including grunts, chatters, chuckles.
Food
Crayfish, frogs, and fish.
Reproduction
1–4 (usually 4) tiny young, born in vegetation-lined den near water in March or April; animals mature sexually in 2 years, but males frequently do not breed successfully until 5–6 years old.
Range
Formerly from n. Que. and n. Mackenzie, southward; extirpated over much of s. part of former range.

HAIR SEALS

Family Phocidae

The hind limbs of these mammals extend behind the tail and cannot be rotated forward. They have short, thick necks; claws on all their digits; and no external ears. Their habitat is coastal and offshore marine waters and ice floes. They feed on fish, squid, crustaceans, and mollusks. The family dates back to the Miocene.

RINGED SEAL

Phoca hispida **62:1**

Description
Size: length, to 6½ ft. (2 m); weight, to at least 200 lb. (90.7 kg); teeth, as in Harbor Seal. Color variable; above usually grayish-brown with ringed spots and streaks, often continuous along back; below, white or yellow, sometimes spotted; area around eyes black; ring marks on sides white to yellowish-brown.

Similarities
Harbor has more pronounced pattern and lacks streaks and rings around spots.

Habits
Inhabits open, marine waters and pack ice; congregates in winter in sheltered bays and inlets.

Reproduction
Pup is white, born February-April in cavity in snow.

Remarks
Of all mammals, this species regularly ranges the farthest north.

Range
Arctic Ocean, s. to Labrador.

HARP SEAL

Phoca groenlandica **62:4**

Description
Size: length, to 6 ft. (1.8 m); weight, to about 400 lb. (181.4 kg); teeth, as in Harbor Seal. No other light-colored seal has a dark saddle across its shoulders and along its sides, and sometimes dark spots on its feet and neck. Adult male has upperparts yellowish to grayish, head and saddle brownish-black, underparts white. Adult female has upperparts brownish to dull white, flippers and saddle brownish-black; face lighter than in male, underparts white to pale brownish. Bedlamer (2–3 years) is above gray or brown, spotted; head and flippers darker. Rusty (under 1 year) is pale gray above with light spots; lighter below. Pup is whitish; eyes and muzzle dark.

Habits
Gregarious, migrates south in winter; on way north in April adults and bedlamer gather on ice to fast and molt.

Reproduction
Usually 1 pup, born in February or March on ice.

Voice
Males emit a long "hoh" and also roar.

Remarks
This seal migrates up to 1000 miles (1609.3 km) each way each year from Greenland to Newfoundland. There, on the pack ice offshore and in the Gulf of St. Lawrence, the young are born in March, and thousands are taken each year for pelts by the sealers. For sustained yield this controversial harvest may need to be curtailed. The Canadian population has been estimated to total more than 3 million.

Range
Arctic, s. to Gulf of St. Lawrence, occasionally farther.

HARBOR SEAL

Phoca vitulina **62:5**

Description
Size: length, to about 5 ft. (1.5 m); weight, to 255 lb. (115.7 kg); 34 teeth, 3141/2141; mammae, 2. Yellowish with brown spots, or brownish with yellow spots. Face doglike.
Similarities
Ringed has streaks as well as spots; its spots are often ringed; and its pattern is generally softer. Gray is larger, grayish, and with squarer, heavier snout.
Habits
Hunts alone along bottom of shallow water, frequently in quiet harbors and bays; congregates in loosely organized colonies when on land.
Voice
Unimpressive grunts and barks.
Reproduction
Pup is usually grayish above, white below, born in spring or early summer.
Remarks
Only seal found regularly south of Nova Scotia, and the only spotted seal south of Labrador.
Range
Arctic and Atlantic coasts, s. to N.C.

BEARDED SEAL

Erignathus barbatus **62:6**

Description
Size: length, to 12 ft. (3.7 m); weight, to 1000 lb. (453.6 kg); teeth, as in Harbor Seal; mammae, 4. Only eastern seal with thick tufts of bristles on each side of its muzzle. Above gray to yellow and brown; below silvery to yellowish; spots absent or obscure. Third finger on forelimb longest.
Habitat
Solitary, sluggish, nonmigratory; feeds on bottom of ocean shallows.
Reproduction
Pup to almost 5 ft. (1.5 m), gray, born April–May on ice floes.
Range
Arctic, s. occasionally to Newfoundland.

GRAY SEAL

Halichoerus grypus **62:3**

Description
Size: length, to 12 ft. (3.6 m); weight, to 800 lb. (362.9 kg); teeth, as in Harbor Seal. Adults: above light gray to almost black, often with obscure spots; lighter below. Yearlings: yellowish with gray blotches on sides. Body stout; snout long, squarish; upper teeth hooked; skin has odor of tar.
Similarities
Harbor is smaller and not as heavy, with less square head. Hooded has hood.
Habits
Gregarious, but rare in w. Atlantic; more deliberate in movements than Harbor Seal; slapping of water if alarmed can be heard from afar. Frequents rough waters with strong currents near rocks, cliffs or reefs.
Voice
"Gloomy . . . between a moo and a howl" (Palmer).

Reproduction
1 pup (rarely twins), woolly-white, born on a ledge, sand bar, or ice, in January–February; molts into yellowish coat after 6 weeks.
Range
Greenland to Labr.

HOODED SEAL
Cystophora cristata **62:7**

Description
Size: length, to 10 ft. (3 m); weight, to 850 lb. (385.6 kg); 30 teeth, 2141/1141. The commonest seal in the North Atlantic. Only seal with an inflatable sac or hood on top of its snout. Adults: blue-gray to slaty-black above, lighter below; muzzle black; sides with white spots on dark blotches. Female paler, hood smaller. Young: called "graybacks," bluish above, flippers blackish-white below; sometimes confused with Harbor or Gray Seal.
Habits
Gregarious, migratory; bull, when excited, inflates hood and puffs out mucous membrane into crimson bladders; can keep farther out of water than any other species of seal.
Voice
Growl.
Reproduction
1 pup, born on ice in February or March, weighs 25–30 lb. (11.3–13.6 kg), nurses for 2–3 weeks.
Range
Greenland to N.S.

CATS
Family Felidae

Members of this family, which includes the domestic cat, have short faces and small, somewhat rounded ears. Five toes are present on each front foot, four on each hind foot; claws are retractile. The dental formula is 3131/3121. The family dates back to the Pliocene.

MOUNTAIN LION
60:11
Felis concolor *Fig. 41*

Description
Size: head and body, 42–54 in. (106.7–137.2 cm); tail, to 36 in. (91.4 cm); height at shoulder, to 31 in. (78.7 cm); weight, to 260 lb. (117.9 kg). Females, ⅓ smaller. The largest North American cat, and only one that is unspotted when adult and that has a long tail. Above yellowish-brown; underparts whitish. Fur short, head small and round, whiskers prominent.
Habitat
Remote mountains, plains, woodlands, and swamps.
Habits
Active at all hours, ranges widely for food, climbs trees when chased.
Voice
Prolonged scream.
Food
Mainly deer, also smaller animals.
Reproduction
1–6 (usually 2–3) buffy, spotted kits, with ringed tails, are born in

a cave or den in dense woods; litters usually born in late spring, but summer births not uncommon; gestation period approximately 70 days.

Range

Formerly throughout U.S., as well as adjacent regions of Canada, from N.B., Ont., and Alta., southward; presently exists only in remote areas of former range.

Other names

Cougar, Puma, Panther.

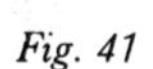

Fig. 41

Mountain Lion

LYNX

Felis lynx — *Fig. 42*

Description

Size: head and body, to 40 in. (101.6 cm); tail, to 4 in. (10.2 cm). Color grayish-buff, lightly spotted. Ear tufts prominent; feet large. Somewhat similar to Bobcat, but the Lynx's tail is black only on the tip.

Habitat

Forests, thickets, swamps, tundras.

Habits

Furred feet give indistinct footprints, larger than those of the Bobcat; climbs and swims well, usually hunts alone.

Voice

Catlike, seldom heard.

Food

Small mammals, mainly Snowshoe Hare.

Reproduction

1–4 (usually 3) kits, which are more strongly marked than adults, born in well-hidden den in some natural cavity in early spring.

Range

Formerly throughout most of Canada and n. U.S., s. to Ind. and Nebr.; extirpated over much of s. part of former range.

Other name

Canada Lynx.

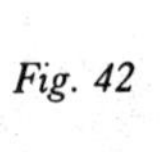

Fig. 42

Bobcat

Lynx

BOBCAT

60:10

Felis rufus — *Fig. 42*

Description

Size: head and body, to 30 in. (76.2 cm); tail, to 5 in. (12.7 cm). Above pale brown to reddish-brown with black streaks and spots; below whitish with dark spots; ear tufts small. Tail black only above, white at extreme tip.

Habitat
Forests, swamps, deserts, mountains.
Habits
Nocturnal; swims, climbs well.
Voice
Catlike.
Food
Hares, rabbits, and other small mammals; ground birds.
Reproduction
Breeds in first year; usually 4 (1–6) kits, born in March or April, in den in ledge, log, or underbrush; usually 1 but occasionally 2 litters per year.
Range
Throughout e. U.S. and also in s. parts of adjacent provinces of Canada.

Dugong and Manatee

Order Sirenia

MANATEES

Family Trichechidae

These are aquatic animals not related to whales, considered to date back to the Pleistocene. There are three species known, but only the Manatee is found in eastern U.S. waters.

MANATEE

Trichechus manatus *Fig. 43*

Description
Size: head and body, 10–13 ft. (3–4 m). Body bulky, head broad; lips thick and bristled. No hind limbs; front limbs are flippers and tail is broadly rounded and horizontally flattened. Skin wrinkled; hairs inconspicuous. Color is gray to brown.
Habitat
Both fresh and salt water, especially shallow waters near coast.
Habits
Relatively slow-moving, hence sometimes injured by motor boats.
Food
Aquatic vegetation.
Range
Warm waters of Fla. and s. states.
Other name
Sea Cow.

Fig. 43

Manatee

Even-Toed Ungulates

Order Artiodactyla

Mammals of this order usually have either two or four toes on each foot. Each toe ends in a hoof. This group is virtually worldwide in distribution. Many artiodactyls have been domesticated.

OLD WORLD SWINE

Family Suidae

WILD BOAR

Sus scrofa *Fig. 44*

Description
Size: length, to 5 ft. (1.5 m); height at shoulder, to 3 ft. (0.9 m); 44 teeth, 3143/3143 (usually not all teeth are present, however); mammae, normally 12. A wild pig. Tusks to 1 foot, upcurved. Color pale gray to blackish. Legs long.
Habitat
Mountains, forests.
Habits
Good swimmer and fierce fighter; will breed with domestic swine.
Voice
Various grunts.
Food
Roots, tubers.
Reproduction
3–12 piglets, dark-spotted and striped lengthwise; active very soon after birth.
Range
Introduced from Europe into parts of N.H., N.C., and elsewhere as game.

Fig. 44

Wild Boar

DEER

Family Cervidae

These are the only hoofed animals that have true antlers, which they shed every year. Females are appreciably smaller than males and lack antlers, except in the Caribou. Deer hear well and have an excellent sense of smell. The family dates back to the Oligocene.

ELK

Cervus elaphus **63:2**

Description
Size: length of bull, to 9½ ft. (2.9 m); height at shoulder, to 5 ft. (1.5 m); weight, to 1100 lb. (499 kg); 34 teeth, 0133/3133. Antlers unpalmated, branched. In summer, light brown, head and limbs darker, rump buffy, hair short, mane slight. In winter, grayish-brown, head and limbs dark, rump buffy, mane longer and darker. Calf primarily brown with light spots till early fall, rump buffy.
Habitat
Semiopen woodlands.
Habits
Summers in mountains, winters in valleys; alert, curious.
Voice
Far-carrying bugling; a loud bark of alarm.
Food
Grass, leaves, twigs.
Reproduction
Usually 1 calf, born May–June.
Remarks
The elk was extirpated in the East long ago. It has since been reintroduced from the West. The male is the second largest deer, moose is larger.
Range
Once common over much of the temperate part of E.; now occurs only in semicaptivity in small, reintroduced populations.
Other name
Wapiti.

MULE DEER

Odocoileus hemionus **54:9, 63:5**

Description
Size: length, to 6½ ft. (2 m); height at shoulder, to 3½ ft. (1.1 m); weight, to 400 lb. (181.4 kg); 32 teeth, 0033/3133; mammae, 4. Brownish in summer, brown-gray in winter; belly, throat patch, and rump patch white. Ears large, tail ropelike and with blackish tip, scent glands on legs large. Principal tines of antlers branch equally. Fawns have white spots.
Similarities
Antlers of White-tailed are prongs from common base.
Habitat
Open forests, brushy areas, rock uplands.
Habits
Has a jumping gait; carries tail down at all times; more gregarious than White-tailed; summers in mountains or hilly areas, winters in valleys.
Voice
A *baaaa*.
Food
Grass, forbs, moss, leaves, twigs.

Reproduction
Usually 2 (1–3) fawns, born late spring, bred in November–December; lie hidden to 6–8 weeks.
Range
From Canada, southward, and from Minn. and Iowa, w. to 100th meridian.

WHITE-TAILED DEER

Odocoileus virginianus **54:9, 63:6**

Description
Size: length, to 6 ft. (1.8 m); height at shoulder, to 3¾ ft. (1.1 m); weight, to 400 lb. (181.4 kg); teeth, same as Mule Deer. Only eastern deer with tail white below and the same color as the back above. Adults tawny above in summer, blue-gray in winter; white below. Antlers have erect, unbranched tines, arising from a main base. Fawn, white spots on reddish coat for 3½ months.
Similarities
Mule Deer has the first tines branched; color duller in summer.
Habitat
Low mixed woodlands, forest edges, second growth.
Habits
Secretive, alert; in flight, raises tail to show white flag of underside; in winter of heavy snow, congregates and keeps packed-down "yards" at feeding grounds.
Voice
Whistling snort when startled.
Food
Grass, leaves, twigs.
Reproduction
Usually 2 (1–4) fawns, born in hideaway in brush, in May or June.
Range
Throughout E., w. to 100th meridian and s. from edge of tundra.

MOOSE

Alces alces **63:1**

Description
Size: length, to 10 ft. (3 m); height at shoulder, to 7½ ft. (2.3 m); weight, to 1400 lb. (635 kg); 32 teeth, 0033/3133; mammae, 4. This huge deer is distinguished by its overhanging snout and, in the male, by the huge broad antlers. Coat blackish or brownish, legs lighter. Antlers palmate and with small prongs at border, spread to 6 feet; shed annually. Bull has mane and also bell of skin and hair hanging from throat. Cow lacks antlers, mane, and bell; smaller than males. Calf is dull reddish-brown.
Habitat
Wilderness forests near shallow lakes.
Habits
Feeds in shallow water at dawn, dusk; a good swimmer; hunters call bull on moose horn by imitating voice of cow.
Voice
Bull, a rising *moo*; cow, softer, more like domestic cow.
Food
Water plants, leaves, and twigs.
Reproduction
Usually 2 (1–3) calves, born mid-May–early June.
Range
Formerly edge of tundra, s. to Pa., Minn., and N.Dak.; now rare or extirpated over parts of former range.

CARIBOU

Rangifer tarandus **63:4**

Description
Size: length, to 8 ft. (2.4 m); height at shoulder, to 4 ft. (1.2 m); weight, to about 600 lb. (272.2 kg); 34 teeth, 0133/4033. Unique member of the deer family in that both males and females possess antlers. Only deer in which part of the antlers project toward the nose. Color varies from almost pure white to blackish-brown; white band above hooves.
Habitat
Tundra, muskegs, coniferous woodlands; mountains to above timberline.
Habits
Migratory, travels in herds; when in motion, feet make clicking sound; gait, loping and running to bounding.
Voice
Various snorts and grunts; when in rut, bucks roar.
Food
Lichens, reindeer moss, grasses, browse of willows and birches.
Reproduction
Usually 1 calf, born in May or June, unspotted, weight 10 to 15 lb. (4.5–6.8 kg).
Range
From Greenland, s. to n. border of U.S.; now extirpated in some parts of former range, particularly in s.

PRONGHORNS

Family Antilocapridae

PRONGHORN

Antilocapra americana **63:3**

Description
Size: length, to 4½ ft. (1.4 m); height at shoulder, to 3½ ft. (1.1 m); weight, to 140 lb. (63.5 kg); 32 teeth, 0033/3133; mammae, 4. Only deerlike animal with white rump and 2 white throat bands. Above tan, grayish in winter; below white. Both sexes horned, horn sheath (composed of fused hairs) shed each year. Horn sheath branched about two-thirds of the distance from base to tip; bony horn core unbranched, laterally flattened. The horn core is never shed. Hoofs pointed, lacking dewclaws. Buck with face and patch on side of neck black; horns longer than ears. Doe with black mask and patch almost absent, horns seldom longer than ears. Kid (to 3 months) with coat gray, wavy.
Habitat
Plains, prairies, sagebrush flats, deserts.
Habits
Active day and night; travels in bands; erects white hairs of rump patch when disturbed; said to run up to 50 mph (80 kph).
Food
Browse plants, sagebrush, weeds, grasses.
Reproduction
1–2 kids, born in June.
Range
From s. Canada, s. to high plains of w. Kans., w. Nebr., and the Daks.; formerly found as far east as Iowa and Minn.
Other name
Antelope.

WILD CATTLE

Family Bovidae

Members of this family have, in both sexes, true, unbranched horns that are not shed. The tooth formula for the wild cattle is 0033/3133. The family dates back to the Miocene.

BISON

54:12

Bison bison *Fig. 45*

Description
Size: length, to 11½ ft. (3.5 m); height at shoulder, to 6 ft. (1.8 m); weight, to 2000 lb. (907.2 kg). Head huge, neck short, high shoulder hump. Hair on shoulders and front legs shaggy, forming a beard on chin, tail tufted. Bull has brownish-black on head, neck, tail, and shoulder; remainder of body brown to light brown. Cow smaller, more evenly colored, less bearded, smaller hump, horns more slender and curved. Calf reddish-yellow, legs and belly lighter. Adult in winter, yellow-brown.
Habitat
Grasslands and open woodlands.
Habits
Highly gregarious, formerly migratory; eyesight fairly poor, hearing and sense of smell keen; likes to wallow in mud or dust.
Voice
A bellow.
Food
Grass.
Reproduction
Usually 1 calf, born in May; hump not noticeable until fall; stays up to 3 years with cow.
Range
Formerly throughout e. U.S., except for Atlantic Coast, New England, and ne. Canada; extirpated over most of former range and at present restricted to small herds on parks and game preserves, notably Wind Cave National Park and Custer State Park in the Black Hills of S.Dak.
Other name
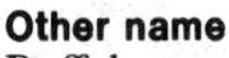
Buffalo.

Bison

Fig. 45

MUSKOX

Ovibos moschatus *Fig. 46*

Description
Size: length, to 6 ft. (1.8 m); height at shoulder, to 5½ ft. (1.7 m); weight, to 900 lb. (408.2 kg). Deep brown or blackish, nose and patch behind shoulders pale, long fur hangs down almost to its feet. Horns flat, broad, close to skull, and with curved tips pointing

forward; shoulders with slight hump; legs and neck short; tail very short. Cow frequently has paler face and more slender horns.

Habitat

Arctic and subarctic coast, tundra, and foothills.

Habits

Gregarious; habit of forming circle with calves inside, if attacked, has made it an easy prey to the rifle; species now dangerously reduced in numbers.

Voice

A bellow; also various snorts.

Food

Sedges, grasses, leaves, twigs.

Reproduction

Usually 1 calf, born April–June on open tundra.

Range

From Greenland and n. Canada, s. to 60° latitude.

Fig. 46

Muskox

Reptiles

Consulting Editor
Patricia Gardner Haneline
Andrew Mellon Fellow
Department of Biological Sciences, University of Pittsburgh

Illustrations
Plates 64–69 John Cameron Yrizarry
Text Illustrations John Cameron Yrizarry and
Jennifer Emry-Perrott

Reptiles

Class Reptilia

A reptile is a cold-blooded vertebrate that lays amniotic eggs covered with a hard shell on land, or that bears living young on land or in the sea. But it is not dependent, as are amphibians, on the return to the water to breed. Fertilization is internal, there is no free larval stage, and the young, as well as the adult, has a dry skin bearing scales or horny plates. The term cold-blooded, in the strict sense, is a popular misnomer. It means not that the blood necessarily is "cold," but that the body temperature varies with that of the surrounding atmosphere, and is not maintained internally at a constant level, as it is in birds and mammals. Ectothermic in a better term. A controversy is currently raging over the physiology of dinosaurs, and whether they were warm-blooded like birds and mammals. If so, they were probably more active than has been popularly portrayed. Eastern American reptiles certainly cannot survive freezing temperatures and must hibernate in holes or burrows below the frost line in climates with cold winters. Reptiles are more abundant both as to species and individuals in the tropics and subtropics than in the temperate zones.

Of all land vertebrates, reptiles have enjoyed the longest sway. During the entire Mesozoic era, the age of dinosaurs, they ruled the land and sea and air. Brontosaurus in the swamp, Tyrannosaurus on land, Pteranodon in the air, Ichthyosaurus and Plesiosaurus in the sea were the largest and most powerful creatures earth had ever seen. Compared with this seemingly endless era of 130 million years, the subsequent reign of the birds and mammals has been short-lived, and that of man is but as minutes out of a twenty-four-hour day.

The group of reptiles known as dinosaurs has now long been extinct. But other groups still live, namely the orders of turtles, crocodilians, lizards and snakes, and rhynchocephalians. Of this last, only one species, the Tuatara, survives, a primitive lizardlike form virtually unchanged since the Jurassic. It is now confined to a few small islands off New Zealand.

Size

The sizes given for identification are maximums. When an unusually large individual has been found, that record is listed separately.

Range and Scope

In this chapter are described forty-two species of turtles, one crocodilian, eighteen lizards, and sixty snakes. These all occur within the range, which covers the United States from the Canadian border (50th parallel) to the south of Florida (26th parallel); and from the 100th meridian east to the Atlantic Coast.

Nomenclature

The common and scientific names in this chapter follow *Standard Common and Current Scientific Names for North American Amphibians and Reptiles.*

USEFUL REFERENCES

Anderson, P. 1965. *The Reptiles of Missouri.* Columbia: Univ. of Missouri Press.

Barbour, R. W. 1971. *Amphibians and Reptiles of Kentucky.* Lexington: University Press of Kentucky.

Bellairs, A. 1969. *The Life of Reptiles.* 2 vols. London: Weidenfeld and Nicolson.

Cagle, F. R. 1968. *Vertebrates of the United States.* 2d ed. pp. 213–68. New York: McGraw-Hill.

Carr, A. 1952. *Handbook of Turtles.* Ithaca, N.Y.: Comstock.

———. 1963. *The Reptiles.* New York: Time, Inc.

———. 1956. *The Windward Road.* New York: Knopf.

Collins, J. T. 1974. *Amphibians and Reptiles in Kansas.* Lawrence: University of Kansas Mus. of Nat. Hist.

Conant, R. 1975. *A Field Guide to Reptiles and Amphibians of Eastern and Central North America.* 2d ed. Boston: Houghton Mifflin.

Ernst, C. H., and Barbour, R. W. 1973. *Turtles of the United States.* Lexington: University Press of Kentucky.

Goin, C. J., and Goin, O. B. 1978. *Introduction to Herpetology.* 3d ed. San Francisco: Freeman.

Kauffeld, C. 1957. *Snakes and Snake Hunting.* New York: Hanover House.

———. 1969. *Snakes: The Keeper and the Kept.* New York: Doubleday.

Lazell, J. D., Jr. 1976. *This Broken Archipelago: Cape Cod and the Islands, Amphibians and Reptiles.* New York: Quadrangle.

Leviton, A. E. 1972. *Reptiles and Amphibians of North America.* New York: Doubleday.

Minton, S. A., and Minton, M. R. 1973. *Giant Reptiles.* New York: Scribner's.

Mount, R. H. 1975. *The Reptiles and Amphibians of Alabama.* Auburn: Auburn Univ. Agricultural Experiment Station.

Oliver, J. A. 1955. *Natural History of North American Amphibians and Reptiles.* Princeton, N.J.: Van Nostrand.

Pope, C. H. 1955. *The Reptile World.* New York: Knopf.

———. 1937. *Snakes Alive and How They Live.* New York: Viking.

———. 1946. *Turtles of the United States and Canada.* New York: Knopf.

Porter, K. R. 1972. *Herpetology.* Philadelphia: W. B. Saunders.

Pritchard, P. C. H. 1967. *Living Turtles of the World.* New York: T.F.H. Publications.

Schmidt, K. P. 1953. *A Check List of North American Amphibians and Reptiles.* 6th ed. Chicago: American Soc. of Ichthyologists and Herpetologists.

Schmidt, K. P., and Davis, D. D. 1944. *Field Book of Snakes.* New York: G. P. Putnam's Sons.

Schmidt, K. P., and Inger, R. F. 1957. *Living Reptiles of the World.* Garden City, N.Y.: Doubleday.

Smith, H. M. 1946. *Handbook of Lizards.* Ithaca, N.Y.: Comstock.

SSAR Committee on Common and Scientific Names. 1978. *Standard Common and Current Scientific Names for North American Amphibians and Reptiles.* Available from D. H. Taylor, Dept. of Zoology, Miami University, Oxford, Ohio.

Stebbins, R. C. 1954. *Amphibians and Reptiles of Western North America.* New York: McGraw-Hill.

Wright, A. H., and Wright, A. A. 1957. *Handbook of Snakes.* 2 vols. Ithaca, N.Y.: Comstock.

GLOSSARY

Anterior Toward the front (of the body).

Carapace Upper shell of a turtle or tortoise, including bony plates and horny shields.

Caudal scale Straplike scale extending across ventral surface of tail.

Cloaca Area into which internal wastes discharge.

Dewlap Skin fold hanging from neck region.

Diurnal Daytime.

Dorsolateral Pertaining to the upper sides of the animal.

Exfoliation Scaling off in flakes.

Frontals Bony membranes that form the forehead.

Interspace Area of merging of two dorsal color patterns in lizards and snakes.

Lateral Of or pertaining to the side of the body.

Maxillary bone Bone on each side of head, forming side border of upper jaw and bearing most of the upper teeth.

Occipitals Bony membranes that form the posterior part of the skull.

Parietal bones A pair of membrane bones in the roof of the skull between the frontals and occipitals.

Plastron Underpart of the shell of a turtle or tortoise.

Posterior Toward the rear (of the body).

Postocular Behind the eye.

Postorbital Behind the eye.

Reticulate Having the form or appearance of a net.

Riparian Relating to the bank of a river, lake, or pond.

Rugose Rough, wrinkled.

Shield In turtles, any one of the horny plates that cover the shell.

Temporal horns In horned lizards, horns toward the sides of the crown.

Tubercle Any of various small knoblike prominences.

Vent Opening on the surface of the cloaca.

Venter Belly.

Vertebral stripe Stripe down the midline of the back.

Vertical pupil Eye in which pupil is elliptical; long axis is vertical.

Crocodilians

Order Crocodilia

Crocodilians are large, lizardlike reptiles with an elongate skull, four sturdy limbs, and a long, powerful tail. They flourish in fresh- and saltwater habitats in tropical and subtropical regions, and alligators are found in certain temperate climates. Crocodilians are familiar to us through animal parks, museums, and books. In the South, the American Alligator was once struggling for its survival, but fortunately, enforcement of conservation laws during this last century has enabled it to regain healthy populations. To encounter one of these gigantic beasts in the field is a memorable experience.

Features of Crocodilians

Crocodilians are usually divided into three groups. The crocodiles occur in Africa, Asia, Australia, widely in the Pacific, and in the Americas. The alligators are found in North America and in the Yangtze drainage of China (if they are not already extinct), and their close relatives, the caimans, are widely distributed in Central and South America. The strangest species of all is the Gavial, or Gharial, which inhabits rivers in India and feeds largely on fish by lashing its long and extremely thin jaws back and forth in the water.

Beneath the thick, horny scales are bony plates called osteoderms on neck, back, and tail and, in some species, on the belly. Crocodiles are unusual among reptiles in that they have a four-chambered heart to prevent mixing of oxygenated and deoxygenated blood and a good secondary palate that allows them to eat and breathe simultaneously. Their teeth are implanted in sockets in the jaw and are generally similar in shape. Crocodilians are carnivorous, devouring prey whole or tearing it to pieces with their strong jaws. Their nostrils are located at the tip of the long snout, and the eyes are located at the top of the skull, enabling them to expose as little of the body as possible while cruising in the water looking for food. Crocodilians are also commonly seen basking or building nests on river banks and coastal shores.

Identification

The American Alligator is discussed below because it is widely distributed in the South. Problems of identification will occur only in Florida: The American Crocodile breeds in the Florida Keys and may travel up the coast, and the Spectacled Caiman has been introduced through the pet trade. In the American Crocodile the snout tapers to a point, and the fourth lower tooth is greatly enlarged and visible from the side when jaws are closed. In the American Alligator the snout is broad and is rounded in front, and the fourth tooth, while enlarged, fits into a pit in the upper jaw and is not exposed at the side when jaws are closed. The Spectacled Caiman is small and has a bony ridge between the eyes.

Reproduction

Crocodilians are unusually active in the care of their young. The mating and nesting behavior of the American Alligator is well known and will be described below, but all species exhibit similar behaviors. During courtship the male strokes the female with his forelimbs and head and finally embraces her and bends his hindquarters under her body. The female builds a nest on land of soil and vegetation. First she beats down an area about four feet by six feet (1.2 m by 1.8 m) in size and then gathers twigs, leaves, and other plants, which are piled and compacted to a height of three feet (0.9 m). In a depression that she makes in the top of the mound the eggs are laid, and then they are covered over. The female remains near the nest and periodically moistens it. When, with the aid of an egg caruncle on the snout of the embryo, the young hatch, they usually make noises in concert and this alerts the female to come and open the nest. Some species have been seen carrying young in the mouth to water. In any case, the female may stay with them for as long as a year, sheltering them in the kind of burrow that all adult alligators make near river banks, consisting of a hole four to six feet (1.2–1.8 m) deep connected to an underground passage some fifty feet (15.2 m) in length and about ten feet (3.0 m) in width. Predation is heaviest on the eggs and young, common predators in the United States being the peccary and bears, skunks, and raccoons. When an alligator has reached adult size it has few enemies other than man.

ALLIGATORS

Family Alligatoridae

AMERICAN ALLIGATOR

Alligator mississippiensis *Fig. 47*

Description

Length, to 19 ft. 2 in. (5.8 m); young 9 in. (22.9 cm) at hatching. Large, lizardlike, distinguished by its size and broad, flattened snout. Ground color brown with 8 rows of enlarged shields running down back. Limbs stout, well-developed webs between toes. Tail massive and dangerous. Young have yellow and black crossbands.

Habitat

Usually near water: ponds, swamps, rivers.

Habits

Often seen submerged in water amidst aquatic vegetation. Hibernates in the North in burrows during cold weather.

Voice

Adults bellow frequently and young make high-pitched noises when in danger.

Food

Young—insects, small vertebrates such as frogs or fish; adults—fish, turtles, snakes, and small mammals, wild and domestic. Large adult can take deer or cattle.

Reproduction

Lays 2–88 eggs; average 42.

Range

Both coastal and inland waters; in Coastal Plain from se. N.C. s. through Fla., and w. to coastal Tex.; n. to s. Ark. Introduced in the Rio Grande of Tex.

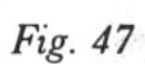

Fig. 47 American Alligator

Turtles

Order Testudines

Turtles are extraordinary animals, distinguished for their ancient ancestry, conservatism, and tenacity. Today man is their number one enemy, and due to his predation and destruction of habitats, many species in recent years have suffered a reduction in numbers. There are in the world today only about 255 species, divided into eleven families. Of these, 42 species are covered here from the eastern United States. Turtles are hardly as diverse as lizards or snakes, but they are widespread in warm and temperate climates and immediately recognizable.

Features of Turtles

A turtle is simply a reptile with a shell. The shell may be rigid and made of a bony layer overlaid by large horny scales called scutes; or it may be pliable and covered with a leathery skin. Vertebral column and ribs between neck and sacrum, in all turtles except the Leatherback, are fused with the bone outside of the hip and shoulder girdles, a situation unique among vertebrates. The clavicle is incorporated into the plastron. Partly because of this bizarre arrangement, it has been said that no scientist could make an accurate restoration from fossil remains alone. It is known from embryological studies that the turtle starts growing normally, but during development the girdles migrate inward, and muscles that normally occur above the vertebral column degenerate. The bones of the shell consist of a middorsal row and variable rows to the side and around the edge above, and both paired and unpaired series below, with a bridge often connecting them at the sides. The bones of this dorsal carapace and ventral plastron fuse into a solid shell in most turtles. Sometimes a hinge is present from side to side on the plastron to effect complete closure of the extremities into the protective shell. The seams between scutes are not arranged in line with the bone sutures below, thereby adding strength to the shell construction. In sea turtles the bone has been reduced to lighten the mass.

A turtle has four well-developed limbs and the shape of the feet reflects the environment in which it lives. Terrestrial tortoises have stout elephantine limbs with claws and no webbing. Aquatic freshwater species often have webbed feet and long claws, and pelagic sea turtles have limbs modified into true flippers with elongated and bound digits. The forelimbs are always larger than hindlimbs, and act as paddles, while the hindlimbs are used in steering. Another characteristic of all living turtles, which proves quite effective in procuring food and repelling predators, is the presence of a horny beak covering the jaw bones. The solidly built head has some features unusual in reptiles. The palate is fused to the braincase, and the skull is cut away from behind to accommodate the large muscles necessary to move the jaws.

Turtles, of course, are air breathers and have lungs. The mechanics of inhalation and exhalation have been solved by the retention of abdominal muscles within the immovable shell. Respiration is aided in aquatic forms by the mouth and cloaca, which possess respiratory surfaces similar to that of a gill, capable of absorbing oxygen from water passed over them. This permits a turtle to stay underwater a long time without surfacing for air, although it also uses anaerobic metabolism.

Turtles do not hear well, but they do smell and see well, and they are sensitive to touch on both shell and body. They are the most intelligent of the reptiles. Most eastern American turtles are silent, although the Leatherback, if so inclined, can utter a bellow audible for a quarter mile; some other species emit hisses, barks, or grunts.

Identification

Terms used in describing the turtle shell are illustrated below. A turtle that has retreated into its shell might emerge again when placed in a container of water. Note coloration of shell and skin, retractability of the head, number and configuration of the scutes, keels and knobs on the carapace, and size. Sizes given in the text approach the maximum, although average lengths may be smaller. When an individual far exceeds the maximum length of most specimens, both that figure and the record length are given. For measurement of length, take the straight line distance along the midline of the carapace.

Habitat

Turtles like the edges of bodies of water, such as lakes, ponds, rivers, and swamps. They also like to bask on offshore logs or rocks. Terrestrial turtles are commonly found in open woods, at the edges of fields, and in abandoned pastures and gardens. Turtles are most active in spring, when they move about looking for mates or nesting sites after emerging from their winter hibernation. They are also likely to be seen after rains. Summer heat will drive them into retirement, or in some cases into aestivation. In early fall, turtles are again on the move and more readily seen than in midsummer.

Successful strategies for finding turtles include using field glasses to observe turtles as they bask in ponds, seining lakes, attaching bucket traps to basking sites, and diving under water to catch them by hand. They may be revealed by using a bright flashlight while wading through shallow water at night. Baited traps can also be used. An experienced companion can often point out well-known haunts and habits in your area, and the curator of your local zoo or natural history museum can give you an idea where and when to find them.

Measurements

The sizes given are for maximum lengths of *carapace*.

Fig. 48

Parts of a Typical Turtle

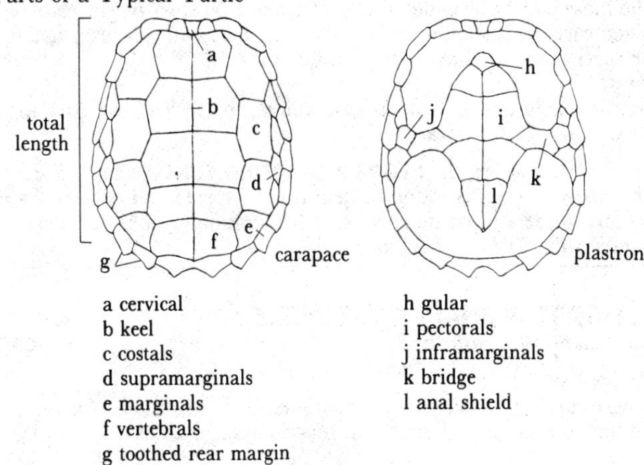

SNAPPING TURTLES

Family Chelydridae

Members of this family have a small, cross-shaped plastron, a greatly enlarged, nonretractile head, powerful jaws with a hooked beak, paired barbels, and a very long tail with erect bony scales. The family is confined to the Americas from Canada to Ecuador, and is represented in the East by two species. Authorities disagree about whether a third species exists in Florida.

SNAPPING TURTLE

Chelydra serpentina **64:1**

Description
Length, to 12 in. (30.5 cm); record size 18½ in. long and 86 lbs. (47.0 cm and 39.0 kg). Shell brown to olive above, lighter below. Large head covered with skin, not plates; eyes can be seen from directly above (not so in Alligator Snapping); neck and legs warty; tail has upperside saw-toothed. Carapace has 3 ridges of knobby plates, usually flatter in older specimens, rear margin toothed, and often algae-covered; plastron small and cross-shaped. Young have sharper keels on carapace than adults.

Similarities
Alligator Snapping Turtle is larger, keels on carapace are higher; has extra row of scutes on sides of carapace.

Habitat
A common species, found in any aquatic situation, preferably with mud and abundant vegetation.

Habits
Vicious predator, actively seeks prey at night, but during day may lie on bottom at a depth where head can reach air for breathing. Basking uncommon. When encountered on land will lunge at humans.

Food
Plants, earthworms, crayfish, clams, fish, frogs, salamanders, snakes, turtles, birds, mammals, and carrion.

Reproduction
Lays 11–83 round, hard eggs, average 25, May–September (usually in June) in hole about 6 in. (15.2 cm) deep in well-drained soil some distance from shore; incubation 8–18 weeks; young may overwinter in nest.

Remarks
The most widely distributed and consumed North American turtle. When carried, it should be held by hindlimbs with plastron facing the carrier and well away from human legs.

Range
From Canada and e. U.S. to Tex. and w. to the Rockies. Also in Mexico and S. America.

Note: A subspecies, the **FLORIDA SNAPPING TURTLE,** *Chelydra serpentina osceola,* is found in peninsular Florida and is considered by some to be a separate species. It has granular scales and long, pointed tubercules on back of head.

ALLIGATOR SNAPPING TURTLE

Macroclemys temmincki **64:2**

Description
Length, to 26 in. (66.0 cm); record weight 219 lbs. (99.3 kg). Similar to Snapping Turtle, but much larger and heavier, and eyes

cannot be seen from directly above. Check for extra row of 3–5 supramarginal scutes at each side above marginals. Light gray plastron small; brown carapace has 3 keels, persistent in old individuals; rear margin deeply and evenly toothed. Head covered with plates; nose pointed; beak with pronounced hook; fleshy pink extension, or "lure," on tongue. Tail very long, about length of plastron. Young rougher than adults.

Similarities
Snapping Turtle has keels lower, head covered with skin.

Habitat
Deep water of rivers, oxbows, sloughs, canals, swamps, sluggish streams.

Habits
Forages for food at night but by day "fishes" by lying motionless on bottom and wiggling lure with mouth open.

Food
Omnivorous, including fish, shellfish, carrion.

Reproduction
Lays 15–50 eggs, April–July in hole 4–7 in. (10.2–17.8 cm) deep, closer to water than Snapping Turtle.

Remarks
Largest freshwater turtle in the world; consumed in the South. Can give dangerous bite.

Range
Se. U.S. from sw. Ga. and n. Fla. w. to e. Tex., n. through se. Kans., and up Miss. R. to sw. Ind. and se. Iowa.

MUD AND MUSK TURTLES

Family Kinosternidae

Distributed widely in the New World, members of this family have odoriferous glands at either side of bridge region, have ten or eleven scutes on the plastron, and the pectoral scutes are not in contact with the marginals. Their jaws have smooth edges and they have claws on their limbs. Two genera are present in the United States, and they can be separated as follows: mud turtles, Genus *Kinosternon,* have a triangular pectoral scute and two transverse hinges across the large plastron; musk turtles, Genus *Sternotherus,* have squarish pectoral scutes and a single, indistinct hinge across the reduced plastron.

EASTERN MUD TURTLE

Kinosternon subrubrum **65:4**

Description
Length, to 4 in. (10.2 cm). Two distinct transverse hinges on plastron, and marginal scutes all approximately same height. Carapace unstriped, yellow to black, below yellow to orange or reddish. Head skin variably marked with 2 light stripes behind eye, sometimes broken into spots. Carapace smooth, unkeeled or with slight median keel, depressed; plastron wider, hinged at both ends. Male has nail at tip of a thicker tail and horny scales on inner side of thigh and below knee. Young have ridges on plates above and a large, yellow to orange spot on each marginal plate.

Habitat
Meadow brooks, ponds, ditches; also brackish and salt water.

Habits
Can exude offensive musk (but less obnoxious than that of Stinkpot); often forages on land.

Food
Omnivorous; aquatic insects, algae, crustaceans, carrion.
Reproduction
Lays 2–5 eggs under logs or in hole in ground, June–July.
Range
Sw. Conn., s. throughout se. U.S., excluding Appalachian region; n. into s. Ill. and w. Ind. and w. to s.-cen. Okla. and e. Tex., with a locality in w.-cen. Mo.

Note: The **STRIPED MUD TURTLE,** *Kinosternon bauri,* is distinguished by light stripes on the head and carapace. It occurs in extreme southern Georgia, south through Florida.

YELLOW MUD TURTLE
Kinosternon flavescens **65:2**

Description
Length, to 6 in. (15.2 cm). The 9th and 10th marginal scales are much higher than the 8th. Head, neck, and throat show considerable bright yellow. Carapace brown to olive with no markings; plastron yellowish-brown. Carapace smooth, unkeeled, low, broad. Male has horn-tipped tail and 2 patches of horny scales on inner surface of hindleg. In young, 9th and 10th marginals lower than 7th and 8th, but 9th has pointed top.
Habitat
Pools, ponds, marshes, canals, cattle tanks; prefers a mud bottom.
Habits
Basks; may forage on land, particularly in rainy season; migrates overland if pools dry up; hibernates in burrows during winter, aestivates under soil during drought. Can emit a strong odor.
Reproduction
Lays 2–4 eggs, July–August.
Range
S. Nebr. and e. Colo., s. to n. Mexico, and w. to Ariz. Disjunct populations in extr. se. Kans. and adj. Mo., and in w. Ill. and adj. states; also in cen. Ill.

STINKPOT
Sternotherus odoratus **65:3**

Description
Length, to 4½ in. (11.4 cm). Lacks distinct hinge below and has 2 conspicuous light stripes on each side of dark head and neck, 1 above and 1 below eye. Chin and throat have little projections of skin (barbels). Above dark brown or black, below lighter, yellow to brown; light stripes on sides of head may be indistinct. Carapace high, unkeeled in adults, often algae-covered; scutes do not overlap. Male has blunt nail on end of tail and horny scales on inner surface of hindleg. Young have a higher, keeled carapace.
Habitat
Any permanent water.
Habits
Pugnacious, thoroughly aquatic but may climb up trees above water; stalks prey on bottom at night.
Food
Aquatic plants, carrion, mollusks, insects, and crayfish.
Reproduction
Lays 1–9 hard, pebbly eggs, May–August under logs, brush piles, at base of a tree, or in shallow hole dug in loose soil.

Remarks
Stinkpot is the common name for this species, although Common Musk Turtle is also used. Name derives from 2 pairs of glands at bases of limbs just below carapace, which emit a powerful, disagreeable, musky odor.
Range
S. Ont. and most of e. U.S. from s. Maine w. into s. Wis., e. Kans., and e. Tex.

Note: In the **RAZOR-BACKED MUSK TURTLE,** *Sternotherus carinatus,* a keel is always present, and the upper part of the carapace forms a very pointed top. It occurs in the southern part of the Mississippi Valley, to southeastern Oklahoma and eastern Texas.

LOGGERHEAD MUSK TURTLE
Sternotherus minor

Description
Length, to 4½ in. (11.4 cm). Above brown, usually with irregular darker markings; below yellow and unmarked. Skin gray-brown with dark brown or black markings, head with dark stripes or spots. Carapace smooth, high, with some indication of median keel throughout life and sometimes 2 lateral keels; scutes are overlapping (imbricate). First vertebral scute does not touch second marginal; 10th and 11th marginal scutes raised; 1 scute on throat. Barbels on chin only. Male has spine on tail and horny scales on legs.
Similarities
Stinkpot has barbels on neck.
Habitat
Lakes, cypress swamps, and streams; found near submerged logs and debris.
Habits
Aquatic; bites when disturbed and should be held by rear of shell.
Range
Extr. w. Va., cen. Tenn. s. through n. Fla. and w. Miss.

Note: A subspecies, *Sternotherus minor minor,* lacks stripes, but has dark spots over a light ground on the neck and head. Males have a greatly enlarged head; young have 3 distinct keels, which are lost with age. It occurs in central and southern Georgia through northern Florida and southeastern Alabama.

Note: The **STRIPE-NECKED MUSK TURTLE,** *Sternotherus minor peltifer,* has a neck with dark stripes over a light ground color. There is a single middorsal keel in adults. It occurs in extreme southwestern Virginia, perhaps in adjacent Kentucky, through eastern Tennessee, northwestern Georgia, and south and west through Alabama to extreme eastern Louisiana.

SOFTSHELL TURTLES
Family Trionychidae

Members of this peculiar family, so different from all other freshwater turtles, split off from the main evolutionary line sometime in the Cretaceous, perhaps 100 million years ago. They are all aquatic. They have a flat, leathery shell with cartilaginous, and therefore pliable, edges. Softshells can breathe (in part at least) through the leathery skin of this shell and through highly vascularized folds in the pharynx (neck region) by absorbing oxygen from the surrounding water as it is flushed in and out of the nostrils or mouth. The nostrils are at the end of a long, tubelike extension of the snout. Males are smaller than females. Specimens should be handled with care by the rear of the carapace and away from the body, because the long neck allows a great range for powerful jaws, and the clawed, webbed feet are quite dangerous. Softshells are distributed in Africa, North America, Asia, and through many localities in the Pacific to New Guinea.

SPINY SOFTSHELL

Trionyx spiniferus **64:3**

Description
Length, to 18 in. (45.7 cm), but males to only 9 in. (22.9 cm). Has spines or warts on front margin of leathery shell and ridges projecting from the septum separating nostrils. Carapace gray, low, flat, almost round, and with spotted dark marginal line, but no ridge. Prominent markings in juveniles of dark rings or dots (or white dots, depending on area of origin) fade into lichenlike markings in adult females. Two light lines along each side of head through eye and back from jaw. Feet strongly patterned. Carapace has gritty or sandy surface.

Habitat
A freshwater species, in large rivers, ponds, lakes, small streams.

Habits
Lies in ambush for prey under mud or sand at bottom of quiet, shallow water; extends tubular nostrils to surface periodically to breathe, but can submerge a long time; a great basker, sometimes in groups, on shore or floating at surface.

Food
Plants, insects, crayfish, mollusks, earthworms, frogs, some carrion.

Reproduction
Lays 4-32 round eggs, June-July in flask-shaped hole up to 10 in. (25.4 cm) deep near water; hatchlings August-October.

Remarks
Vicious to handle; can give nasty bite and scratches with its sharp claws.

Range
Most of cen. U.S. n. to cen. Minn. and cen. Wyo., and in e. U.S. to w. N.Y. and to Atlantic Coast in Ga.; s. to Fla. panhandle and w. along Gulf Coast. Isolated populations in the East in ne. N.Y. and adj. N.H. and Canada, and in N.J. Sporadic distribution in the West and n. Mexico.

SMOOTH SOFTSHELL

Trionyx muticus **64:4**

Description
Length, to 14 in. (35.6 cm), but males average half that size. Has leathery shell with smooth front margin above. Carapace nearly circular, olive to brown; plastron clear and light; light line behind eye. Old females develop a lichenlike mottling, while males and juveniles have spots and streaks. Nostrils circular, no ridge projects from septum. Feet without markings.

Habitat
A freshwater species, in permanent water: large, slow rivers; lakes; clear, sandy creeks.

Habits
Similar to those of Spiny Softshell, but more aquatic; less of a basker, slightly less pugnacious.

Food
Crayfish, insect larvae and other invertebrates, fish, frogs, tadpoles, and some vegetation.
Reproduction
Lays 4–33 round eggs, spring or late summer, in holes excavated in sandbars or banks; incubation about 70 days.
Range
W. Pa. and e. Nebr. s. to Gulf Coast in w. Fla. and e. Tex. (farthest westward record in e. N.Mex.); and following large rivers n. into Minn. and s. N.Dak.

Note: The large **FLORIDA SOFTSHELL,** *Trionyx ferox,* is the heaviest and bulkiest of all North American turtles, but has the most limited range. It is dark brown with large, vague, dark spots and a bumpy carapace. It occurs in the Coastal Plain and all of Florida except the Keys.

Freshwater and Box Turtles

Subfamily Emydinae

This subfamily comprises the largest number of turtle species. The toes are webbed, or partly so, and the species are generally aquatic or semiaquatic, although the box turtles have been secondarily adapted to a life on land. The plastron contains twelve scutes, with the pectoral scute connected to the marginals. They are widely distributed in the Northern Hemisphere and absent from Australia and southern parts of Africa.

WOOD TURTLE

Clemmys insculpta *Fig. 49*

Description
Length, to 9 in. (22.9 cm). Has strongly keeled carapace, and orange on underside of legs and throat. Above dark, sometimes with black and yellow markings. Carapace low, broad, keeled; scutes on back sculptured with concentric grooves into pyramidal blocks, rear edges flaring. Plastron rigid, scutes yellow, each with black blotch at outer rear section. Legs and neck dark, sometimes with red dots. Beak notched. Male has thicker tail, longer claws, and larger scutes on forelimbs than female.
Habitat
Likes to swim but may be terrestrial and may feed some distance from water; found in bogs and marshes.
Habits
Sometimes hibernates in Muskrat den.
Food
Algae, fish, tadpoles, insect larvae, mollusks, earthworms; in captivity, berries, meat.
Reproduction
Lays 4–12 eggs in June.

Fig. 49

Wood Turtle

Remarks
Unusually intelligent; makes a good pet and will eat from the hand.
Range
Ne. U.S. and extr. se. Canada, s. through most of N.Y. to n. Va. and W.Va. and w. around Great Lakes in north of range to e. Iowa and Minn.

SPOTTED TURTLE

Clemmys guttata **65:1**

Description
Length, to 5 in. (12.7 cm). Small low carapace has black ground color with various numbers of conspicuous yellow spots; plastron light with black blotches usually on outer part of plates. Head, neck may show yellow markings. Male has brown eye and chin, concave plastron. Female has orange eye, yellow chin, and flat or convex plastron.
Habitat
Bogs, brooks, pastures.
Habits
Emerges from hibernation early in spring, stays out March-June; during day basks and forages actively, nesting females may be out of water at night. Often wanders across roads and is killed by careless (or malicious) motorists.
Reproduction
Lays 1-5 eggs in June.
Range
S. Maine, w. through most of lower Mich. and into e. Ill., but s. of Pa. Restricted to Atlantic Coastal Plain through cen. Fla.

BOG TURTLE

Clemmys muhlenbergi *Fig. 50*

Description
Length, to 4½ in. (11.4 cm). Has big, bright orange ear patch and low keel. Ground color above brown to black, plastron lighter, both usually with some red or yellow markings, but not dotted. Skin often mottled with orange or reddish. Carapace long, domed, straight-sided, wider at rear; scutes somewhat sculptured; plastron long and broad. Male has narrower rear notch on plastron than female.
Habitat
Sphagnum bogs, wet meadows, small streams; often seen on land. Man's destruction of bog habitats has made this species rare.
Habits
Aestivates as well as hibernates; likes to burrow.

Bog Turtle

Fig. 50

Blanding's Turtle, p. 331

Food
Berries, insects, but apparently omnivorous; feeds in captivity on chopped meat, earthworms, greens, fruit.
Reproduction
Lays 3–5 eggs, June–July.
Range
Very restricted and patchy in e. U.S., including cen. N.Y., R.I., w. Mass., N.J., parts of Pa., and the meeting of Va., Tenn., and N.C.

EASTERN BOX TURTLE
Terrapene carolina *Fig. 51*

Description
Length, to 6 in. (15.2 cm); record 8½ in. (21.6 cm). Slightly keeled, oblong, boxlike carapace, highest in the middle, plastron dark or mottled, but never striped. Shell brown to olive and yellowish (see below for description of subspecies variations). Head, neck, limbs irregularly marked with yellow. No bridge; wide plastron with single hinge; can completely enclose head, limbs, and tail; tail short. Male has bright red eyes, female yellow-brown. Young without hinged plastron.
Habitat
Open woodlands, wood edges, fields, stream banks, moist areas.
Habits
In hot, dry weather, burrows in mud or seeks shade; active in morning and after rain.
Food
Young mostly carnivorous (earthworms, snails, insect larvae); adults mostly herbivorous (berries, fungi, fruit); in captivity fresh meat, canned dog food, fruits, greens.
Reproduction
Usually mates soon after emergence from hibernation, but nests throughout summer; lays 2–8 eggs, May–July, in flask-shaped hole in loose soil 3 in. deep; incubation 69–136 days. Young may emerge and then hibernate, or overwinter in nest.
Remarks
Gentle, makes a good pet. Many are crushed by cars while crossing highways. This is the familiar turtle of woodlands, prairies, and plains. The plastron has a single hinge that allows complete closure of front and rear parts of the shell.
Range
S. New England, w. to lower Mich., s. Ill., and cen. Kans., s. along Atlantic and Gulf coasts to cen. Tex.

Note: Where the following 4 subspecies meet, intergradation occurs and precise identification may be impractical.

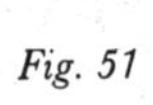

Fig. 51

Eastern Box Turtle

Western Box Turtle, p. 324

Terrapene carolina carolina, is found in northern part of range west to the Mississippi River, and south to northern Florida. The shell is highest at the rear and is decorated with yellow spots. The **GULF COAST BOX TURTLE,** *Terrapene carolina major,* is restricted to the Coast from Louisiana to northern Florida. The carapace is dark on this, the largest, subspecies. The **FLORIDA BOX TURTLE,** *Terrapene carolina bauri,* occurs in peninsular Florida and the Keys. There are radiating yellow lines on the blue-black carapace; plastron is unmarked yellow. The hindfoot usually has 3 toes. The **THREE-TOED BOX TURTLE,** *Terrapene carolina triunguis,* occurs from Missouri and Kansas south, east to extreme southwestern Georgia, and west to Texas. The hindfoot usually has 3 toes (but sometimes 4). Head is unicolor brown or olive with bright spots; male sometimes has a red head.

WESTERN BOX TURTLE

Terrapene ornata *Fig. 51*

Description

Length, to 5⅞ in. (14.9 cm). Carapace flat-topped, unkeeled. Plastron hinged, with striking pattern of light yellow lines on darker background; carapace has radiating yellow lines over dark to reddish-brown; 4 toes on hindfoot. Male has red iris and enlarged first toe on hindfoot. Young almost round; flatter and darker than adults.

Habitat

Short-grass plains, prairies, savannas; sandy, semiarid regions; woodlands, swamps.

Habits

Burrows, emerges in morning, at dusk, and during rains; many killed crossing highways.

Food

Insects, earthworms, some carrion; berries, melons, and other plants. Captives also take mealworms, fresh meat.

Reproduction

Lays 2–8 eggs, usually in June.

Remarks

This species has an interesting courtship pattern. The male chases the female (if a turtle may be said to "chase"); upon reaching her, he raises himself on his hind legs and hurls the front of his plastron at the rear of her carapace, emitting from each nostril a stream of fluid, which he sprays on her back. After half an hour of such wooing, the female yields.

Range

Cen. U.S. from w. Ind., s. Wis., and extr. se. Wyo. s. and w. to Gulf Coast in La. and w. to se. Ariz., into adj. n. Mexico.

DIAMONDBACK TERRAPIN

Malaclemys terrapin *Fig. 52*

Description

Length, to 9 in. (22.9 cm). Handsome polygonal scutes with concentric ridges on carapace and spotted skin distinguish this species. Carapace low, broad, slightly wider to rear; plastron oblong. Head and forelimbs white with dark spots, lips white. Female 2–3 in. (5.1–7.6 cm) longer than male, head heavier and blunter, shell deeper.

Habitat

Brackish water along coast, tidal shores, salt marshes.

Food
Crustaceans, mollusks, insects, carrion.
Reproduction
Lays 4–18 eggs, averaging 10 or less, with average clutch numbers decreasing from north to south; laid in holes in sand to depth of 6 in. (15.2 cm), April–July; often laid on a barrier beach.
Remarks
This species has been divided into several subspecies that differ in external characters, but which form a continuum of races of smaller size with smoother vertebral scutes in the North, to those of larger size with rougher, keeled vertebrals on the Gulf Coast. Once heavily consumed in fine restaurants, these turtles have only recently recovered from overcollecting.
Range
Coastal and adj. waters from Cape Cod through most of Gulf Coast.

Fig. 52

Diamondback Terrapin

MAP TURTLE

65:10
Graptemys geographica
Fig. 53

Description
Length, to 10¾ in. (27.3 cm). Carapace low, flat, low keel lacks distinct knobs; rear margin slightly toothed; netlike patterns of yellowish lines and circles over dark green color resemble streams on a map (best seen when moist); marginals below yellow with round olive marks. Plastron smooth, usually plain solid yellow or cream color, but males and juveniles may have dark line along seams of scutes. Head wide, large, jaws strong; green or yellow spot of variable shape behind eye. Young have wider, stronger keels on carapace.
Habitat
Rivers and large bodies of water, also less abundantly in ponds and marshes.
Habits
Timid, hard to catch; basks on rocks, logs, and (which is unusual) on sand or grass beaches. Active into October.
Food
Mollusks (eaten especially by females), crayfish, insect larvae, and carrion.

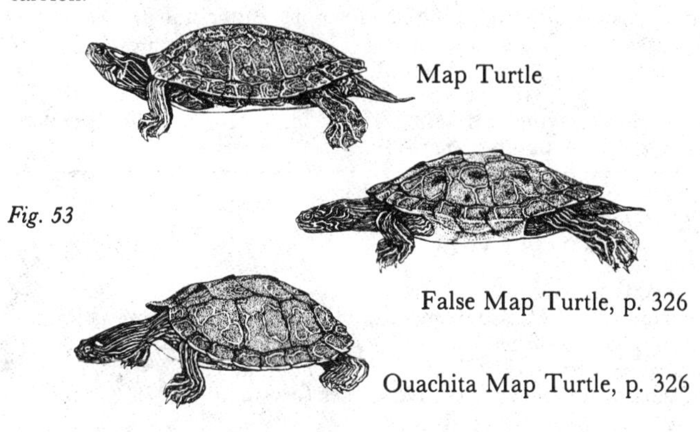

Map Turtle

Fig. 53

False Map Turtle, p. 326

Ouachita Map Turtle, p. 326

Reproduction
Lays 10–20 eggs, June–July in flask-shaped hole; hatchlings emerge August–September, or May–June from overwintering clutches.
Remarks
Does not thrive in captivity.
Range
Nw. Vt. and n. N.Y., w. through s. Mich., Wis., and e. Minn.; s. through Miss. Basin to ne. Okla., La., and cen. Ala.

Note: A number of species of map turtles occur in the southeastern United States, not all of which can be covered here. Those with narrow, sharp-edged jaws include the **RINGED MAP TURTLE,** *Graptemys oculifera;* the **YELLOW-BLOTCHED MAP TURTLE,** *Graptemys flavimaculata;* and **BLACK-KNOBBED MAP TURTLE,** *Graptemys nigrinoda,* all of which are found in much the same area, from Louisiana to western Florida, as are the **ALABAMA MAP TURTLE,** *Graptemys pulchra,* and the **BARBOUR'S MAP TURTLE,** *Graptemys barbouri.* The last two species have large jaw surfaces suitable for coping with crustaceans and shellfish, much as in the Map Turtle, above. The **FALSE** and **OUACHITA MAP TURTLES,** described below, have narrow jaws that are adapted for a more vegetarian diet and for insects. Two additional species are found only in central Texas—**CAGLE'S MAP TURTLE,** *Graptemys caglei,* and **TEXAS MAP TURTLE,** *Graptemys versa.*

FALSE MAP TURTLE

65:9

Graptemys pseudogeographica

Fig. 53

Description
Length, to 11 in. (27.9 cm). Recognized by variable orange-yellow crescent behind eye and alternating green and yellow lines on throat that reach eye. Carapace with low keel without prominent knobs; rear margin saw-toothed. Color olive with 1 or more dark blotches encircled with yellow or orange toward rear of each plate, often faded in older specimens. Marginals from below marked with concentric blotches, and plastron with concentric yellow and dark green circles, which may fade to mottling in adults.
Habitat
Very aquatic; likes rivers and lakes, usually in slow-moving water with dense vegetation.
Habits
Wary; often basks in large groups with other map turtle species; strong swimmers.
Food
Young are carnivorous; adults omnivorous, eating aquatic plants, mollusks, insects, and carrion.
Range
From Wis. and Minn. s. to La. and e. Tex. in large river drainages emptying into Miss. R., including Mo. R. w. to N.Dak., and Ohio R. e. to Ohio and W.Va.

OUACHITA MAP TURTLE

Graptemys ouachitensis

Fig. 53

Description
Length, to 10¼ in. (26.0 cm). Light blotch behind eye meets lines running length of head above; sometimes broken with 1–9 stripes that come in contact with eye. Throat has four yellow spots, or, infrequently, yellow transverse stripes (in Sabine River). Carapace

and plastron similar in color to False Map Turtle, with dark spots on rear of each scute encircled by latticework of yellow and orange. Color variable, may be faded in older specimens, or even dark and almost patternless. As in False Map Turtle, carapace has high keel with prominent black knobs at center of vertebral scutes.

Habitat

Rivers only, in sections with swift currents and aquatic vegetation.

Habits

Basks; females feed on surface while males feed below. Spends night on bottom.

Food

More herbivorous than other map turtles, but also takes insects, carrion, and a few mollusks.

Reproduction

Lays 8–15 eggs, May–July.

Range

Wis. and Minn., s. through La. in Miss. Basin, e. to Ind., and w. to Okla. A distinctive population in Sabine R. of La. and Tex.

PAINTED TURTLE

Chrysemys picta **65:6**

Description

Length, to 7 in. (17.8 cm); record 9⅞ in. (25.1 cm). Has smooth-edged oval shell; a notch, bordered by cusps, in upper jaw; and red and yellow marks over limbs, neck, and carapace. Carapace smooth, low, gently arched; margin rounded and often with light yellow or red markings. Marginal scutes always contain red. Plastron has 6 pairs of yellow or buff plates; plastron sometimes marked with central dark blotch. Neck and limbs have red streaks. Male's shell lower and foreclaws longer than female's.

Habitat

A common species, found on quiet lake shores, in ponds, slow streams, marshes, ditches, or even brackish tidal water; to 6000 ft. (1828.8 m).

Habits

Timid, but easily tamed; often basks in groups or floats with head sticking up through water amid lilies or duckweed; sometimes migrates short distances across land.

Food

Adults omnivorous—plants, insects, mollusks, crustaceans, mosquito larvae, carrion; juveniles carnivorous; feed under water or by skimming surface with mouth open.

Reproduction

In courtship male swims backward in front of female tickling her face with his claws. Lays 2–20 elliptical, soft eggs, May–mid-July, 2–5 in. deep within 10 yards (9.1 m) of water; incubation 8–12 weeks or until following spring; female often lays 2 clutches.

Range

Most of e. and n. U.S. and in extr. s. Canada, s. in the East to La.; but absent from s. Coastal Plain and from most of Southwest including Tex.

Note: Four eastern subspecies can be distinguished in the field. **EASTERN PAINTED TURTLE,** *Chrysemys picta picta,* with plastron unmarked, has seams across carapace continuous in a straight line and with wide, light edges. It occurs in Nova Scotia, southern Maine, and southeastern Quebec, south through northeastern Alabama (but not in Coastal Plain). Intergrades with Midland subspecies (see below).

SOUTHERN PAINTED TURTLE, *Chrysemys picta dorsalis,* with plastron unmarked, has carapace seams staggered, and with wide, light edges; a wide red or yellow stripe down middle of back. It occurs in Mississippi Valley from southern Illinois south, east to central Alabama, and west to extreme southeastern Oklahoma.

MIDLAND PAINTED TURTLE, *Chrysemys picta marginata,* with plastron narrow, has dusky figure down center, carapace seams staggered and with dark edges; sometimes with narrow, pale stripe down back. It occurs from southern Maine through southern Ontario, west to eastern Wisconsin, and south through Tennessee. Intergrades with Western and Eastern subspecies over broad areas.

WESTERN PAINTED TURTLE, *Chrysemys picta belli,* has plastron with large branching pattern; carapace seams are staggered with dark edges, but light lines decorate centers of scutes. It occurs in sw. Ont., upper Mich. and Ill., and w. through much of n. Central Plains and Pacific Northwest. Intergrades with Midland subspecies (see above).

COOTER

Pseudemys floridana **65:7**

Description
Length, to 15⅞ in. (40.3 cm). Carapace dark with network of light markings, but no "C" on 2nd costal. Plastron unspotted, yellow to orange or greenish; marginals below may have eye-shaped markings. Head and neck have parallel yellow lines that are variously broken and interconnected. Front of broad-domed, oval carapace is higher, rear margin toothed. Furrows present over shell. Lower jaw flattened.

Similarities
River Cooter has figure "C" on 2nd costal scute.

Habitat
Permanent bodies of water; large lakes, ponds, swamps, springs, or rivers.

Habits
Basks, often in groups, on logs, rocks; slides off quickly when approached.

Food
Young carnivorous, adults vegetarian.

Reproduction
Lays 12–29 eggs in flask-shaped hole in loose, open soil.

Range
Coastal Plain from e. Va. s. and w. to cen. Okla., n. through s. Miss. Valley to s. Ill.

SLIDER

Pseudemys scripta **65:8**

Description
Length, to 8 in. (20.3 cm). Carapace dark with yellowish and dusky stripes running crosswise on costal and marginal scutes, lengthwise on first two vertebrals. Marginals yellow below with dark spots between sutures. Plastron yellow with variable dark markings. Carapace smooth, weakly keeled; often algae-covered; margin slightly toothed. Beak notched, but not cusped; edges of both jaws smooth, lower jaw rounded. Male smaller, foreclaws larger. Old males often blackish, also lose red behind ear when present. Young have carapace nearly round and with vivid markings.

Habitat
Quiet water in a variety of situations.
Habits
Aquatic and wary, wanders through aquatic vegetation; basks 2- or 3-deep on logs, or floats with head just above water; disappears into bottom mud if frightened.
Food
Young are carnivorous, adults omnivorous.
Reproduction
Lays 2–22 soft eggs, April–July in hole 4–8 in. (10.2–20.3 cm) deep near water; incubation 8–14 weeks or more; some young overwinter in the nest.
Range
Most of se. U.S., se. Va., s. through n. Fla., w. through most of Ill. and into N.Mex., with isolated localities in Mich., Ohio, W.Va., s. Fla., and into Mexico.

Note: The **RED-EARED SLIDER,** *Pseudemys scripta elegans,* the commonest of the Slider's 3 subspecies in the East, is the only turtle with a wide red band behind eye. The **YELLOW-BELLIED SLIDER,** *Pseudemys scripta scripta,* and **CUMBERLAND SLIDER,** *Pseudemys scripta troosti,* subspecies, with a yellow patch and line respectively, occur in the southeastern United States.

RIVER COOTER

Pseudemys concinna

Description
Length, to 12 in. (30.5 cm); record 16⅜ in. (41.6 cm). Carapace variously marked but always with light "C" whose arms touch rear margin of 2nd costal. Plastron spotted variously, always has more spotting on bridge and marginals than Cooter. Head lines not broken and interconnected. Carapace broad and somewhat flattened; rugose, furrowed. Lower jaw flattened.
Similarities
Cooter's shell high-domed, has less spotting on bridge and marginals.
Habitat
Rivers, streams, springs, less frequently ditches; in fairly quiet water with vegetation.
Habits
Shy; often seen basking.
Reproduction
Lays 18 soft eggs, May–July.
Range
S. U.S., from se. Va. through w. Fla., w. to se. N.Mex. and n. Mexico, and n. in Miss. Valley to s. Ill. and Ind.

RED-BELLIED TURTLE

Pseudemys rubriventris **65:5**

Description
Length, to 15¾ in. (40.0 cm). Has much red on carapace and plastron. Reddish band through costals and marginals above and below, but some individuals dark overall; plastron with pinkish or reddish. Carapace highest in middle, widest at rear, often furrowed lengthwise, rear margin virtually smooth. Similar to Cooters and Slider, but shell more oblong, and beak markedly notched. Jaw edges saw-toothed, lower jaw flattened. Foreclaws of male straight, longer than those of female. Young are rounder and brighter-colored.

Habitat
Lakes, ponds, streams (clear, muddy, or brackish).
Habits
Throughout most of its range, the only basking turtle. Wary, basks in places difficult to reach and dives into deep water when disturbed.
Food
Omnivorous.
Reproduction
Lays 12–35 eggs in flask-shaped nest 4 in. (10.2 cm) deep, often in cornfields.
Remarks
Young make good pets but should be released, as with all other animals, if not thriving. (Be sure to release your pet exactly where you found it.)
Range
Two populations—in the North limited to Plymouth Co. and Naushon Is. in e. Mass.; in the South, continuous from s. N.J. s. to N.C. and w. through Md. to extr. e. W.Va.

Note: Two species, the **FLORIDA RED-BELLIED TURTLE,** *Pseudemys nelsoni,* of the Florida peninsula, and the **ALABAMA RED-BELLIED TURTLE,** *Pseudemys alabamensis,* of southwestern Alabama, are similar to the Red-bellied Turtle, but have much more limited distributions.

CHICKEN TURTLE

Deirochelys reticularia *Fig. 54*

Description
Length, to 6 in. (15.2 cm); record 10 in. (25.4 cm). Has very long, striped neck (almost as long as carapace), and vertically striped rump. Carapace ground color dark with yellow lines and often a yellow border on margin. Plastron hingeless, yellow to orange, sometimes with pattern along seams. Dark skin has yellowish or greenish lines.
Habitat
Lowland cypress swamps, lakes, ponds, or ditches and other temporary water.
Habits
Semiterrestrial, often found wandering overland.
Food
Plants, tadpoles, crayfish; adults eat more vegetation.
Reproduction
Lays 5–8 oblong, leathery eggs.
Range
Coastal Plain from se. Va. to e. Tex. and up Miss. Valley to se. Mo., but absent from Miss. R. itself.

Fig. 54

BLANDING'S TURTLE

Emydoidea blandingi *Fig. 50*

Description
Length, to 9 in. (22.9 cm). Only turtle with combination of a dotted carapace, hinged plastron, and bright yellow patch on lower jaw and under neck. Dots may blend together. Plastron yellow, with large scutes, each with black blotches at outer rear corners; closure incomplete, hind end notched. Carapace long, narrow, elevated. Beak notched; feet webbed. In young, carapace is dark and unmarked, plastron black with yellow margin.

Habitat
Ponds, backwaters, small streams, moist land.

Habits
Timid; if surprised basking, will plunge to bottom and remain there for hours; a good swimmer.

Food
Omnivorous; includes crustaceans, insects.

Reproduction
Lays 6–11 eggs, June–July; hatchlings appear September.

Remarks
Makes a good pet.

Range
Spotty and discontinuous distribution, in scattered localities along Atlantic Coast from N.S. to s. N.Y., and more or less continuously from extr. sw. Que. and s. Ont. around cen. Great Lakes area w. to cen. Nebr. and possibly s. to s. Ind.

Tortoises

Subfamily Testudininae

This group is closely related to the Emydinae. Found on most continents except Asia and Australia, they have elephantine hindlimbs and usually high, domed shells. They are strictly terrestrial and are commonly sold as pets.

GOPHER TORTOISE

Gopherus polyphemus *Fig. 54*

Description
Length, to 10 in. (25.4 cm); record 14½ in. (36.8 cm). Recognize this tortoise by the broad, domed (but somewhat flattened) carapace, stoutly elephantine hindlimbs, and hingeless plastron. Carapace brown to black, scutes with concentric growth rings; plastron yellow. All feet unwebbed. Forelimbs shovellike, used for burrowing, covered with hard scales that remain exposed when tortoise retreats into shell. Unable to close up tight because hinge in plastron lacking; it joins elbows in front of head and seals front opening with forearms.

Habitat
Sandy areas often far from water; strictly terrestrial.

Habits
An accomplished burrower, making tunnels up to 40 ft. (12.2 m) long. Male defends burrows against other males, but females may pass into them freely.

Food
Vegetarian; grazes on grasses; fruits and berries; may accept meat in captivity.

Reproduction
Lays 4–7 eggs in sandy hole dug far from burrow, infrequently at burrow entrance.
Remarks
This "gopher" usually shares its burrow with all manner of uninvited but tolerated guests: rattlesnakes, indigo snakes, frogs, opossums, burrowing owls, and insects. *Do not reach into a burrow with your hand* under any circumstances.
Range
Coastal Plain from s. S.C. through suitable habitat in Fla. w. to e. La.

SEA TURTLES

Family Cheloniidae

In these turtles the shell has evolved to be lighter in weight, the short, heavy neck cannot be completely drawn back into the shell, and the forelimbs are paddlelike. They are completely marine, the females returning to land to lay their eggs in holes they excavate on sandy beaches above the high-water mark. They are found in all tropical and warm temperate oceans.

GREEN TURTLE

64:6
Chelonia mydas
Fig. 55

Description
Length, to 48 in. (121.9 cm); record 55 in. (139.7 cm), and weight 650 lbs. (294.8 kg). The only sea turtle with 4 large plates (costals) on each side of back, *and* 1 pair of plates (prefrontals) behind nostrils (between eyes). Ground color mottled brown or gray. Carapace broad, low, and unkeeled; plates do not overlap. In males forelimb usually has 1 claw only, shell usually longer and tapering, and tail very long (8 in. or 20.3 cm). Young have white edges on carapace and limbs.
Similarities
Hawksbill's plates overlap.
Habitat
Pelagic, migrating long distances in open ocean; feeds on vegetation in shallows near mouths of rivers.
Habits
Sometimes basks on surface; normal speed in water to 1.4 mph. (2.3 km/hr.), but clumsy on the beaches where females lay eggs.
Food
Adult mainly herbivorous; young more carnivorous.
Reproduction
Lays 10–200 (average 107) eggs in hole about 18 in. (45.7 cm) deep.
Range
Warm tropical and subtropical seas, occasionally n. to Mass.

Fig. 55

HAWKSBILL

Eretmochelys imbricata **64:5**

Description
Length, average of 24 in. (61 cm), and weight 50 lbs. (22.7 kg); to length 36 in. (91.4 cm), and weight 160 lbs. (72.6 kg). The only sea turtle with overlapping plates (except when old), 2 pairs of prefrontals on snout, and 4 pairs of big costals. Plates variegated in color. Carapace shield-shaped, keeled, with toothed margin. Forelimb has 2 claws. Male has longer claws than female.
Habitat
Reefs and shallow coastal waters.
Food
Omnivorous; captives thrive on fish and meat.
Reproduction
Lays 53–206 eggs, normally in June–July.
Remarks
Plates formerly used to make tortoiseshell ornaments.
Range
Tropical warm seas, found only occasionally off ne. U.S. coast.

ATLANTIC RIDLEY

Lepidochelys kempi **64:8**

Description
Length, to 29½ in. (74.9 cm), and weight to 110 lbs. (49.9 kg). Carapace heart-shaped or round, gray. Inframarginal plates 4 or more on each side and costals 5. Large head has 2 pairs of prefrontals.
Similarities
Loggerhead has 3 inframarginals, Green Turtle and Hawksbill have 4.
Habitat
Shallow waters of tropical shoreline; nests in small numbers on southern Texas coast.
Habits
Irascible, violent when caught, cannot be kept on its back as can other sea turtles.
Food
Crabs and other animal food, details not well known.
Reproduction
Only one major nesting area exists, on northeast coast of Mexico, concentrated on a small stretch of beach. At peak of breeding season up to 40,000 Ridleys may gather off this beach.
Remarks
A separate species inhabits Pacific Ocean.
Range
Gulf of Mexico and up into coastline of n. U.S. states.

LOGGERHEAD **64:7**

Caretta caretta *Fig. 55*

Description
Length, to 45 in. (114.3 cm); record length 48 in. (121.9 cm), and weight 500 lbs. (226.8 kg). Color reddish-brown. Has 3 inframarginal plates and 5 or more costals. Head very large with 2 pairs of prefrontals; forelimb has 2 claws. Male has tapering shell and long, thick tail. Young have 3 keels above, 2 below.
Similarities
Ridley has 4 or more inframarginals; Green Turtle and Hawksbill have 4 costals.

Habitat
Coastal bays; brackish streams; creeks and, of course, the high seas. Most widespread and common of the sea turtles.
Habits
Sometimes vicious if molested; an oceanic wanderer; basks at surface of water and infrequently on beaches.
Food
More carnivorous than other sea turtles, consuming crabs, conchs, fish, clams, oysters, sponges, jellyfish, turtlegrass.
Reproduction
Lays 60–300 round, soft, leathery eggs, April–August on our southern coasts, and more recently reported on a New Jersey beach (formerly commonly nested to Virginia). Incubation 1–3 months; hatchlings emerge from nest en masse, usually at night, and head straight for the water.
Remarks
This hardy and tenacious species is the only sea turtle that still nests regularly on the South Atlantic coast. It is almost as pelagic as the Leatherback, and there have been reports of size up to 7 ft. (2.1 m) and 1000 lbs. (453.5 kg).
Range
Atlantic Ocean n. to N.S.

LEATHER-BACKED TURTLES

Family Dermochelyidae

The Leatherback is the sole member of this family.

LEATHERBACK

Dermochelys coriacea **64:9**

Description
Length, to 72 in. (182.9 cm), and weight to 1600 lbs. (725.6 kg). The largest living turtle and only sea turtle with a leathery shell lacking large plates. Smooth black carapace has 7 long ridges; plastron mainly white, has 5 long ridges; limbs lack claws.
Habitat
Pelagic, coming on land only to breed.
Habits
Strong, fights with both jaws and flippers. Able to live in summer in northern waters and warm its body to 80°F (26.7°C), even in 45°F (7.2°C) water, by a countercurrent system of blood vessels to and from extremities. (This idea is disputed by some herpetologists, who maintain that Leatherbacks in northern waters are strays that will eventually die.)
Voice
Utters loud sounds when agitated.
Food
Jellyfish and other marine life.
Reproduction
Lays 50–170 round, soft eggs, June–July in hole 1–3 ft. (0.3–0.9 m) deep excavated at night by female in sand above high-water mark; incubation 8–10 weeks.
Remarks
The leathery shell of this species makes its classification uncertain. Therefore, it may deserve a family of its own.
Range
Warm seas, straying to Newfoundland.

Lizards and Snakes

Order Squamata

Lizards

Suborder Sauria

Features of Lizards

A lizard is a scaly reptile that usually has visible eardrums, movable eyelids, a nonexpandable jaw, and four legs, each with five clawed toes. Most of them also have several rows of scales on the underside. But in certain lizards some of these characteristics may be missing. For example, glass lizards do not have legs and superficially they resemble a snake—but snakes do not have ears, eyelids, or legs, and they have an enormously expandable jaw and usually just one row of broad belly scales.

Do not confuse lizards with salamanders. Salamanders are amphibians and have a moist, scaleless skin and unclawed toes.

The tail is often fragile and breaks off when a lizard is caught; the discarded fragment wriggles for several minutes. When a tail is lost in this way, a new, shorter one grows in its place, but the vertebral column is replaced by cartilage, and the scale and color patterns may be noticeably unlike the originals.

Many lizards have excellent eyesight, especially adapted for detecting moving objects, and a good sense of smell. Despite the general presence of external ears, many seem oblivious to sound, and lizards have little, if any, voice. Teeth that are generally all alike are present on jaw edges and sometimes on the palate. They are commonly used to grab and hold their prey rather than for chewing. No poisonous lizards occur in the eastern United States, but in fact only two poisonous lizards are known, one being the Gila Monster of the American Southwest. Lizards mainly eat insects, spiders, snails, worms, and other lizards, but are also known to be herbivorous. Most eastern lizards are terrestrial and diurnal.

Fig. 56

Parts of a Typical Lizard

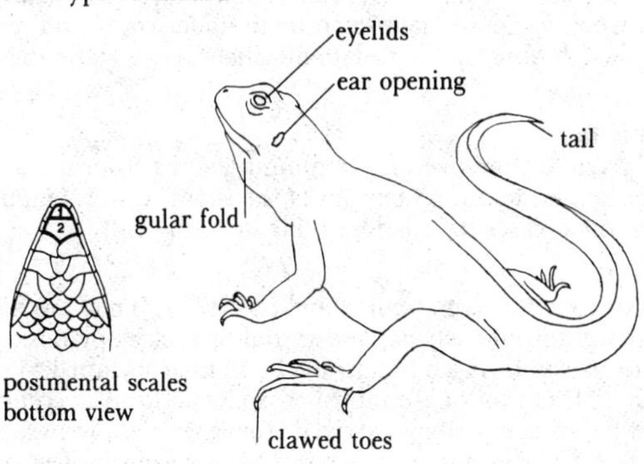

LIZARDS

Identification
Lizards in the eastern United States are a small group. With the exception of some of the skinks, the individual species are relatively easy to distinguish. First, determine which species occur in your state. Where several species are found, they may fall into certain groupings. If a species has spiny scales, it is either one of the fence or horned lizards (the only lizard with "horns"). If a species has granular scales, it is an anole. If a species has smooth, polished scales, it is probably a skink. The only skink with diagonal scale rows on its sides is the Great Plains Skink. Only three lizards with legs have the tail more than twice the length of the body, namely the Collared Lizard, the Green Anole, and the Six-lined Racerunner (which is slender with a pointed snout). The Collared Lizard is our only species that habitually runs on its hind legs.

Nonspiny lizards with legs can also be divided by preferred habitat:

Moist, Humid, Wooded Places	*Dry, Sandy, Rocky, Grassy Places*
Ground Skink	Collared Lizard
Five-lined Skink	Lesser Earless Lizard
Broad-headed Skink	Six-lined Racerunner
Southeastern Five-lined Skink	Great Plains Skink
Coal Skink	Many-lined Skink
Green Anole	Prairie Skink

Habitat
Almost everyone has seen some lizards. The anoles; the collared, horned, and fence lizards; and the racerunners are relatively conspicuous. Others, for instance the skinks and glass lizards, are wary or secretive and are seldom seen unless searched for.

Look for lizards basking or running along stone walls and fences, on stumps or logs or in brush piles, amid dead leaves, around old piles of sawdust, in deserted buildings and clearings, on limestone hills and rocky outcrops, in canyons, and on flat sandy or rocky soil with sparse vegetation. Look for some on tree trunks, for others in moist situations near water. Study lizards through binoculars, adjusted for close focus. You can approach many species to within a few feet; from there, good binoculars can bring them into beautiful scale-counting view.

Most lizards hibernate in the colder months. They are apt to be particularly active soon after emergence in the spring. Lizards can stand a good deal of heat. They prefer sunny days, particularly mornings. Cool or cloudy days keep them under cover and many retreat to their hiding places in late afternoon even while the sun is still up.

Measurements
The sizes given are the overall maximum lengths from *snout to end of tail.* The second measurement gives the snout to vent length because in many cases lizards lose their tails.

Range
Lizards are often seen amid ruins and in deserts. They are also found in moist and dry woods, and in flat or rocky uplands. Lizards are particularly abundant in the southern United States. In Florida, in addition to long-established endemic forms, recently introduced lizard species have arrived from the West Indies, particularly colonizing the area around Miami; and in Texas the desert faunas from the southwestern United States and Mexico swell the number of species. Included in this section are the most common and abundant species.

IGUANAS

Family Iguanidae

This is the dominant family of New World lizards. Its members are varied in form, habits, and habitat, but in all the teeth lie in grooves on the inner surface of the jaws. All have a scaly body and four well-developed limbs. Some reach a length of five feet (1.5 m) and are ornamented with crests and dewlaps. Some forms have taken to water; however, those of the eastern United States are somewhat less exotic. Iguanids have a characteristic habit of bobbing up and down, which signals territoriality or willingness to mate. The family ranges from Canada to Argentina, with many species in the West Indies as well as in Madagascar and the Fiji Islands of the South Pacific.

GREEN ANOLE

Anolis carolinensis *Fig. 57*

Description
Length, to 8 in. (20.3 cm); snout-vent length to 3 in. (7.6 cm). Recognize these "chameleons" by the male's pink or red throat fan, or dewlap, and the expanded pads near tips of digits. Color green to brown, and true to its misnomer (chameleons occur in the Old World), can effect rapid color change. Throat fan is longitudinal flap of skin extended by cartilaginous rod during mating or territorial displays. Small granular scales over body. Toe pads, or expanded lamellae, allow anole to gain purchase on arboreal perch.

Habitat
Various situations, including wooded areas and cypress swamps, but also close to human dwellings; a climber, found in trees and large bushes or on fence posts, but juveniles found on ground.

Habits
Acrobatic and visible, have complex social displays; sleep on vegetation, often high in trees.

Food
Insects and other invertebrates.

Reproduction
Lays a single egg, buries it in soft earth, every 10–14 days, May–January, but egg rarely discovered.

Remarks
The only native representative of a very interesting and diverse genus found mainly in the West Indies and Central and South America, where about 300 species are known. Through recent human activities, other anole species have been introduced into very limited localities in the United States; they are variously distributed in peninsular Florida, mainly near Miami.

Range
N.C. throughout Fla. and w. to se. Okla. and e. Tex. Disjunct locality in extr. s. Tex.

Fig 57

Green Anole

COLLARED LIZARD

Crotaphytus collaris **66:2**

Description
Length, to 14 in. (35.6 cm); snout-vent length 4½ in. (11.4 cm). Stout-bodied lizard with a long, thin, tapering tail; large head; and thin neck with 2 black collars. Much variation in color; males often green or yellow with yellowish to red dots, throat often orange, inside throat black. Blue spots on back, black collars broken in front. Females often gray, with red spots. Both sexes brightest for short time only during breeding season. Dark bands on tail and big hindlegs. Young duller, crossbands of cream and gray more conspicuous.
Habitat
Arid and semiarid regions, limestone-topped hills and bluffs, prairie rocks and canyons, unshaded hillsides.
Habits
Ground-dwelling, wary, agile, a good jumper and bipedal runner, but pugnacious when cornered.
Food
Grasshoppers, spiders, moths, beetles; other lizards and small snakes; flowers.
Reproduction
Lays 2–21 (usually fewer than 12) eggs, May–July in sand or tunnels to depth of 4 in. (10.2 cm).
Remarks
The sight of one of these highly colored lizards running on its hind legs, using its long tail for a balancer, suggests a tiny dinosaur.
Range
S. Mo. and nw. Ark., w. to Utah and Ariz., and s. through n.-cen. Mexico.

EASTERN FENCE LIZARD

Sceloporus undulatus **66:9**

Description
Length, to 7¼ in. (18.4 cm); snout-vent length to 3¼ in. (8.3 cm). Above dark gray or brown with curved, barlike pattern across back and tail, or longitudinal dark stripes. Stout body, scales keeled. Male shows blue under sides and on throat; female more boldly marked above.
Similarities
Collared Lizard larger and has collars; Lesser Earless Lizard stouter, shorter-tailed, and paler.
Habitat
The most common iguanid lizard, found in dry, piny, and deciduous woods; clearings, brushlands, grasslands, sandy areas; on fences, old houses, brush heaps, fallen tree trunks or rocks.
Habits
Climbs; a fast runner and artful dodger about tree trunks; most active on sunny days; often seen basking on fence posts; defends territories and has complex social system.
Food
Insects, spiders.
Reproduction
Lays 3–13 eggs, May–July to depth of 4 in. (10.2 cm) in sandy soil or rotten logs or under bark; incubation about 8 weeks.
Range
Most of cen. and s. U.S. from s. N.Y., n. Mo., and s. S.Dak., s. and w. to Utah and Ariz., and into n.-cen. Mexico. Absent from s. half of peninsular Fla., s. La., and extr. s. Tex.

Note: The **FLORIDA SCRUB LIZARD,** *Sceloporus woodi,* somewhat resembles the Eastern Fence Lizard, but its lateral stripe is brown rather than black, and the markings, though similar, are more clearly defined. This species occurs in 4 distinct and local areas in the Florida peninsula: on the east-central and southwestern coasts, and in 2 interior areas, 1 in the north and the other in the center.

LESSER EARLESS LIZARD

Holbrookia maculata **66:10**

Description
Length, to 5 in. (12.7 cm); snout-vent length to 2$\frac{7}{16}$ in. (6.2 cm). Tail is about equal in length to body; no ear opening. Above pale, back has gray stripe bordered with dark spots; 2 short, black bars low on each side; below plain whitish. Body short and plump but somewhat flattened, scales tiny, dorsal scales smaller than ventrals. Fleshy fold between chest and throat (gular fold).
Habitat
Plains, dry rocky or sandy areas with sparse vegetation; chalk beds.
Habits
Particularly active on hot days; makes short dashes; inquisitive; often seen in the shade of vegetation.
Food
Grasshoppers, true bugs, spiders, beetles, grubs.
Reproduction
Lays 2–8 eggs, July–early August; hatchlings August–September.
Range
Central Plains and sw. U.S. from s. S.Dak. and extr. se. Wyo. s. through n. Tex. and w. through much of Ariz. In cen. and w. mainland Mexico. Questionable record from Mo.

TEXAS HORNED LIZARD

Phrynosoma cornutum *Fig. 58*

Description
Length, to 4 in. (10.2 cm); record 7⅛ in. (18.2 cm); snout-vent length to 5⅛ in. (13.0 cm). Only lizard with horns projecting from back of head. Above tan to brown with light-bordered, black blotches to either side and light streak down middle of back; below pale with black spots. Spiny, short-tailed, squat.
Habitat
Dry, flat land with sparse vegetation.
Habits
Most active during heat of day, though in hottest summer months active morning and evening; found alongside ant trails in the open; reluctant to move when approached by a predator, but will eventually waddle away in a comical gait.
Food
Ants; also other insects, and spiders.
Reproduction
Lays 14–37 (average 26) eggs, May–July.

Short-horned Lizard, p.340

Fig. 58

Texas Horned Lizard

Remarks
These horned lizards have no problems of predation during periods of exposure because of their concealing coloration and spiny, widened bodies, which make them hard to spot and to swallow. However, they are not free from predation by some birds, such as shrikes and hawks, and by large snakes.
Range
S.-cen. U.S. from se. Colo., Kans., and extr. sw. Mo. s. through Tex. and nw. La., and w. in e. and s. N.Mex. to se. Ariz. through n. Mexico. Introduced in Fla. and elsewhere in the South.

SHORT-HORNED LIZARD
Phrynosoma douglassi *Fig. 58*

Description
Length, to 4½ in. (11.4 cm); snout-vent length to 2½ in. (6.4 cm). Has ridge on back of head that lacks large spines, and short tail, banded above. Spiny, squat.
Similarities
Texas Horned Lizard has prominent horns and more distinct blotches and line on back.
Habitat
Semiarid, short-grass plains; hardpan, light soil, sandy and rocky terrains.
Food
Ants; also grasshoppers, beetles, insect larvae.
Reproduction
Bears 5–31 (average 16) living young in August.
Range
Only marginally in eastern U.S.; s. Alta., Mont., and se. N.Dak., s. to Ariz. and N.Mex. and into n.-cen. Mexico; disjunct populations in w. Tex., in Pacific Northwest.

GLASS LIZARDS
Family Anguidae

Members of this family in North America have a groove of granular scales along each side, presumably to allow for expansion of the armored body after big meals or during egg development. In the eastern United States only the legless glass lizards occur, whereas in the West the alligator lizards, with four well-developed limbs, are found. This family comprises residents of the New World and Eurasia, but not of the tropics of Asia or Africa.

EASTERN GLASS LIZARD
Ophisaurus ventralis *Fig. 59*

Description
Length, to 42⅝ in. (103.8 cm); snout-vent length to 12 in. The largest legless lizard, superficially resembling a snake, but with movable eyelids and ventral scales same size as dorsal scales. Broad, dark band on side and white at edges of scales. Lateral fold to either side below. Above brown to olive with no middorsal stripe in adults and with white bars on neck. Below pale, no markings below lateral fold.
Similarities
Slender Glass Lizard has narrow dark lines on side, white in middle of scales.

Habitat
A burrower; usually in moist, grassy areas in pine flatwoods and in hummocks; in places with loose soil, cavities under roots or stones, or in burrows.

Habits
Basks in mild sun, comes out after rain. Moves by undulating from side to side stiffly.

Food
Grasshoppers; other insects, spiders, snails, earthworms.

Reproduction
Lays 4–17 white eggs with flexible shell, June–July; female broods eggs under cover. (See details of incubation habits under Slender Glass Lizard, below.)

Remarks
If this or following species is attacked or captured, its tail, which is ⅔ of total length, comes off, wriggles, and breaks into pieces, thus diverting attention and enabling owner to escape. Body regrows new, shorter tail.

Range
Coastal Plain from N.C. through Fla. and w. to e. La. Disjunct populations in e.-cen. Mo. and extr. se. Okla.

Fig. 59

SLENDER GLASS LIZARD

Ophisaurus attenuatus *Fig. 59*

Description
Length, to 42 in. (106.7 cm); snout-vent length to 11⅜ in. (38.9 cm). Has narrow dark lines along each lower side, white marks in centers of scales (may form a series of stripes). Usually with stripe along middle of back, and stripes below lateral fold on lower sides of body.

Similarities
Eastern Glass Lizard has broad band on side, white at edges of scales.

Habitat
Found in drier situations than Eastern Glass Lizard—oak and pine woodland, fields and grassy areas, brush piles.

Reproduction
Lays 6–17 eggs. Brooding female of this species and Eastern Glass Lizard curls body around eggs and guards them against small predators during incubation. From time to time she basks in sun and returns to warm eggs with her newly warmed body; pays no attention to hatchlings.

Range
Se. U.S. from se. Va. through most of Fla. and w. to cen. Tex.; a broad northward extension up the Miss. Valley to w. Ind. in east of range, s. Wis. in north, and se. Nebr. to w.

Note: The **ISLAND GLASS LIZARD,** *Ophisaurus compressus,* is distinguished from the Eastern and Slender Glass Lizards by the lack of dark stripes *below* the lateral groove. It does, however, have one solid dark stripe on each side of its body. It occurs in coastal South Carolina and Georgia, and in all of peninsular Florida except for the extreme southern tip.

SKINKS

Family Scincidae

The skinks make up a very large, cosmopolitan family. Eastern species are ground lizards with smooth, flat, polished-looking scales and short, but fast, legs. They can easily slip through a hand when picked up. They feed diurnally on insects and spiders, hibernate in winter, and lay eggs that the female often broods.

GROUND SKINK

Scincella lateralis **66:8**

Description
Length, to 5⅛ in. (14.0 cm); snout-vent length to 1$^{15}/_{16}$ in. (4.9 cm). The smallest eastern North American lizard, with tiny limbs; look for transparent "window" in eyelid. Back golden-brown, sides of body from snout to base of tail dark brown, below light, often with yellow. Scales smooth. Tail breaks off readily if seized.
Habitat
Has wide ecological tolerance; found in dry woods, pine or deciduous; in clearings, under leaves, bark, fallen tree trunks; also moist places near streams.
Habits
Very secretive, crawls quietly amid leaves or runs with snakelike sideways movements; may jump into shallow water.
Food
Small insects, earthworms and other invertebrates.
Reproduction
Lays 1–7 hard eggs, June–August in rotten logs, stumps, humus.
Range
Se. U.S. from s. N.J. w. through s. Ohio, Ill., and Mo. into e. Kans., and s. through most of Fla. to s. Tex., with a record in n. Mexico. Absent from most of Appalachian Region.

FIVE-LINED SKINK

Eumeces fasciatus **66:3**
Figs. 60, 61

Description
Length, to 8$^{1}/_{16}$ in. (20.5 cm); snout-vent length to 3⅜ in. (8.6 cm). Any lizard seen in New England or Michigan will be this species. This and following two species, however, are difficult (sometimes impossible) to tell apart in the field. In the hand, look for 7 upper labial scales with 4th below the eye, and 2 enlarged postlabials. Middle row of scales underneath original tail widened; 5 light lines down black back (but old male loses these lines, body turns olive-brown, head and throat turn reddish); 26–30 rows of scales around middle of body, rows at side parallel to rows on back. Head of breeding male wider, redder than that of female. Young have yellow head stripes and blue tails that turn brown with age.
Habitat
Dry to damp woods, old sawdust and brush piles, dead leaves, rotten logs, stumps; on shores of Great Lakes in decayed driftwood.
Habits
Very agile; basks on logs and stumps, changing position frequently; seldom climbs trees (two following species do); bites if captured.
Food
Insects, earthworms, spiders, snails.
Reproduction
Distinguishes its mate by behavior rather than appearance. Courting male rushes around with open mouth toward a

conspecific. If the lizard attacks back, it is a male; if it runs away or stays still, male grasps female's back in his mouth and mates with her. Lays 4–15 eggs, May–July in rotten logs, stumps, or loose soil; brooded by female for 4–7 weeks.

Range

Isolated and old records in Mass.; from se. N.Y. and adj. cen. Vt. s. to n. Fla., w. through s. and cen. Pa., s. Ont., around Lake Mich. to cen. Wis.; s. Ill. to extr. se. Nebr., e. Kans. to e. Tex. Isolated populations reach to N. Dak., Minn., and Iowa.

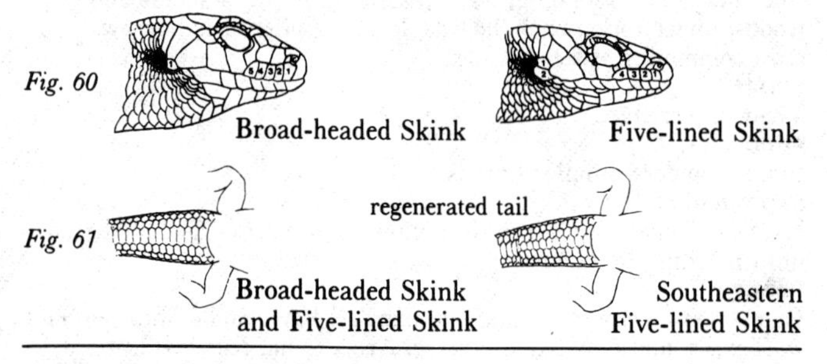

Fig. 60

Fig. 61

BROAD-HEADED SKINK

66:4

Eumeces laticeps

Figs. 60, 61

Description

Length, to 12¾ in. (32.4 cm); snout-vent length to 5⅝ in. (14.3 cm). Large, with 8 upper labials, the 5th below the eye, widened ventral scales underneath original tail. Similar to Five-lined and old Southeastern Five-lined Skinks, but body much thicker; males have broader yellowish or reddish heads than males of other two species. One or 2 small postlabials or none at all. Thirty to 32 scale rows around middle of body. Young sometimes have 7 instead of 5 stripes.

Habitat

Arboreal, but also found in habitat similar to that of Five-lined Skink.

Reproduction

Lays 13–16 eggs, June–July.

Range

Extr. se. Pa. through s. W.Va. and s. Ohio, w. to se. Kans., and s. to Atlantic and Gulf coasts, but only to cen. Fla.

SOUTHEASTERN FIVE-LINED SKINK

Eumeces inexpectatus

Fig. 61

Description

Length, to 8½ in. (21.6 cm); snout-vent length to 3⅓ in. (8.9 cm). Striped, scales underneath original tail all same size (middle row not widened as in Five-lined and Broad-headed species). Hard to distinguish in field from two preceding species if of this size, but light stripes on head are orange-red; sublateral light line may be present. In juveniles, blue color of tail extends beyond hip.

Habitat

Much as for two preceding species, but often in drier habitats.

Habits

More arboreal than Five-lined, less so than Broad-headed.

Reproduction

Lays 11 eggs.

Range

Coastal Plain from s. Md. and cen. Va. to Miss. R. in cen. Miss., n. through cen. Tenn. into s. Ohio.

GREAT PLAINS SKINK

Eumeces obsoletus **66:1**

Description
Length, to 13¾ in. (34.9 cm); snout-vent length to 5⅝ in. (14.3 cm). Adults light gray to brown above with black-bordered scales; below unmarked cream. Young black with blue tail and rows of gold dots on top of head, white spots on labial scales.
Habitat
Flat limestone rocks on grassy hillsides, also in grasslands and woods; under rocks, bark, boards, logs, but often digs burrows; most common in eastern Kansas.
Habits
Vicious; secretive.
Food
Insects, spiders, smaller lizards.
Reproduction
Lays 7–17 eggs, June–July in hollows under rocks. Female broods until hatching in August.
Range
Extr. e. Mo. through s. Nebr. to extr. se. Wyo. and s. into cen. and w. Tex., s. and e. N.Mex., and se. Ariz., to ne. Mexico. Isolated records in ne. Mo.

MANY-LINED SKINK

Eumeces multivirgatus **66:5**

Description
Length, to 7⅝ in. (19.4 cm); snout-vent length to 2⅞ in. (7.3 cm). Slender, has many light lines running length of body, but check for light line on 3rd scale row from midline. Wide, light middorsal stripe from head to tail; chin area, tail light, underparts gray, 24–26 rows of scales around middle of body. Tail 1½ times body length. Young have blue or lavender tail and only 5 light stripes, but range does not overlap that of other five-lined skinks.
Habitat
Short-grass prairies; under boards in urban vacant lots, under cow chips in prairie-dog towns.
Habits
Lays 3–5 eggs, otherwise little is known.
Range
Sw. S. Dak. and se. Wyo., s. through w. Nebr., and across Colo. to se. Utah, ne. Ariz., and much of N.Mex. Also found in disjunct populations in w. Tex.

COAL SKINK

Eumeces anthracinus **66:7**

Description
Length, to 7 in. (17.8 cm); snout-vent length to 2¾ in. (7.0 cm). Broad stripe on back bordered by white stripe and below it a wide dark brown lateral stripe 2½–4 scales wide, bordered below by another white stripe. Four light stripes above. Olive-brown band on back covers 4 or more scale rows; bordering white stripe extends onto tail but not onto head. Below bluish or gray. Scale rows 24 around middle of body. One postmental scale on chin. Young often all black with reddish spots on head; tail blue-violet.
Habitat
Under stones, brush piles, leaves, logs, or on wooded hillsides near water; also near swamps.

Food
Small insects.
Reproduction
Lays 8–9 eggs, usually late June; brooded by female 4–5 weeks.
Range
Spotty, unevenly distributed from w. N.Y. s. to w. Fla. and w. to cen. Kans. and ne. Tex., but absent entirely in Ill., Ind., and Tenn. Discontinuity of range indicates that this species once occupied a larger area.

PRAIRIE SKINK

Eumeces septentrionalis **66:6**

Description
Length, to 8⅛ in. (20.6 cm); snout-vent length to 3¼ in. (8.3 cm). Somewhat similar to Coal Skink but lateral dark stripe covers at most 2 scale rows, and 2 postmental scales present. Rows of scales 28 around middle of body, 2 white stripes on each side, 4 dark stripes on back, the 2 middle ones lighter than 2 outer ones. Light lines on 4th or 5th scale row from midline.
Similarities
In the hand, Many-lined Skink shows light lines on 3rd scale row from midline. Coal Skink has lateral dark line covering 2½–4 scale rows.
Habitat
Flat rocks on grassy hillsides and in oak woodlands; sandy or gravelly areas.
Habits
Wary, fast; usually found under some cover.
Food
Small insects, snails, spiders.
Reproduction
Lays 5–13 eggs, May–June under boards, logs, bark, stone, or in loose, moist soil; incubation about 45 days.
Range
Extr. se. N.Dak., Minn., and w. Wis., s. through e. Tex. Scattered localities in n. Ill., e. Wis., and s. Man.

WHIPTAILS

Family Teiidae

This largely South American family contains varied forms, ranging from tiny, elongate, limbless creatures to large lizards called tegus (Genus *Tupinambis*), three feet (0.9 m) in length, which prey on small birds and mammals. North American species are active and small. Parthenogenesis, or asexual reproduction in an all-female species, is common in some of the western forms and has made their classification complex.

SIX-LINED RACERUNNER

Cnemidophorus sexlineatus **66:11**

Description
Length, to 10½ in. (26.7 cm); snout-vent length to 3⅜ in. (8.6 cm). The only North American 4-limbed lizard with a very long tail, small scales above, and large rectangular plates below. Above dark with 3 light lines along each side of back; below white (with green or blue in adult male). Graceful; head, neck, and body almost same width; pointed snout. Tail twice or more length of body and with rings of large scales around it.

Habitat
Various terrains up to 2000 ft. (609.6 m) but prefers dry regions, flat or hilly; sandy or rocky soil; short grass, thin woods, dusty roadsides.
Habits
Reaches amazing speeds; stops and starts suddenly; digs its own burrows with forelegs; home range about ¼ acre (0.1 hectare); lives in colonies.
Food
Insects, spiders, snails.
Reproduction
Lays 1–6 eggs, June–August under rocks or logs or in holes in sandy soil a few inches deep. Incubation about 8 weeks.
Range
S. Md., e. Va., and e. Ind., s. throughout most of se. U.S.; follows Miss. R. into Wis. and Minn., and w. to s. S.Dak., se. Wyo., and e. N.Mex.

Snakes

Suborder Serpentes

Features of Snakes
A snake, or serpent, is a limbless reptile with expandable jaws; slender, curved teeth; no ear openings or movable eyelids; and a single row of large belly scales. The flickering of its forked tongue picks up dust particles that give the snake sensations of both taste and smell. Its sense of smell is excellent. So, in general, is its sight. A snake "hears" by means of vibrations that its body picks up, not through the air, but through the ground. Snakes cast their skins, usually several times a year, by crawling out of them head first. These shed skins, or sheds, are often seen in suitable snake habitats.

Snakes in the eastern United States fit well the general description given in the introduction; however, certain groups, such as the boas in the West, have flaps at each side of the vent, a condition that is in contrast to the limblessness of most of the others. However, in no case do snakes have vestiges of forelimbs or a pectoral girdle. In snakes, one lung is always reduced, and the remaining lung extends far down the length of the body. Their long, flexible form is well adapted for the undulating waves that move along the body

Fig. 62
Parts of a Typical Snake

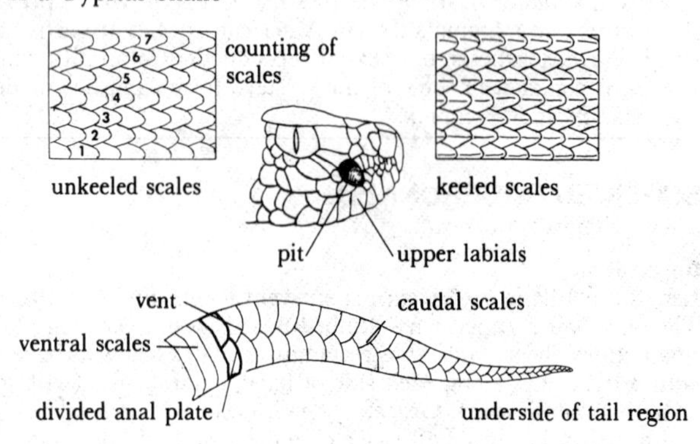

and propel the snake forward whenever resistance is met in the substrate. Without rocks, stones, furrows on logs, or even soil, the snake would not be able to move.

The distensible lower jaws spread apart to accommodate oversized prey, and the skin between the scales is very elastic. The teeth are not used to chew food, but make very effective grappling hooks to hold prey, usually still struggling, in the mouth. Although certain teeth may be specialized for the injection of venom, many species (in fact the majority) kill their prey by constricting it with folds of the body, or by engulfing it while it is still alive.

Identification
Individual snakes of the same species often vary widely in color in different parts of their range, or even in the same locality. Thus pattern—lines, blotches, or spots; shape of body—long, short, slender, or stout; and shape of head are important for identification, often more so than color. Indeed, colors may vary from light and patterned in young, to dark in old individuals. Unlike most other groups of animals, subspecies of snakes are often easily recognizable and may differ quite widely in appearance within the same species. Whenever this is the case in this chapter, the subspecies are discussed.

Habitat
Favorite snake habitats are the edges of watercourses and bodies of water, on stone walls, rocky ledges, and causeways (especially the marsh side), in tangles of downed timber, on stumps and logs, and at the edges of bogs. Many serpents are nocturnal, and they are often attracted by highways, which retain warmth after the sun goes down. However, serpents like warmth only to a point; they cannot long tolerate hot sun. Therefore, the hours of sunshine early in the season, and early in the day later on, are the best times to find snakes on the surface. They tend to disappear in daytime in the hottest months, when their activity is restricted to the hours of darkness. In the fall, just prior to hibernation, they become more common again by day.

Habits
Snakes have no voice, though some are able to make a hissing sound, and certain pit vipers can buzz with the rattles on the ends of their tails. Snakes live exclusively on animal food: frogs, toads, salamanders, rabbits, rats, mice, shrews, birds, and worms (there are no vegetarian serpents). They digest slowly, and after feeding they will often be sluggish for several days; meals may exceed 50 percent of their own body weight. Snakes hibernate in the colder months. In spring, they come out of hibernation and become active, starting once more to feed after the winter's fast, and to mate.

Measurements
Length refers to the *straight line measurement* from the tip of the snout to the vent. Tail length is not included.

BLIND SNAKES

Family Leptotyphlopidae

The pinkish, semitransparent members of this family look much like earthworms. Their tiny degenerate eyes, only a pair of black dots at one end of the body, are not used, and their scales are the same size all around the body, not enlarged on the belly as in other North American snakes.

TEXAS BLIND SNAKE

Leptotyphlops dulcis **67:1**

Description
Length, to 10¾ in. (27.3 cm). Very slender, pale, resembles an earthworm more than a vertebrate, has reduced eyes. Scales on underside not enlarged; body elongate, tail short. Brown above and white or pink below.
Habitat
In sandy areas and prairies.
Habits
A burrower, most often found under stones or logs, but sometimes abroad in early evening.
Food
Termites, ants, insect larvae.
Reproduction
Lays 2–7 eggs.
Range
S.-cen. U.S. from extr. sw. Kans. and cen. Okla. s. through Tex. and w. into se. Ariz. and Mexico.

COMMON HARMLESS SNAKES

Family Colubridae

This one family embraces all the North American nonpoisonous snakes except for the Texas Blind Snake (see above). All members have eyes and broad belly scales; most have a spine on the end of the tail. None has pits behind the nostrils, as do the pit vipers, nor the permanently erect front upper fangs of the coral snakes. Some species have grooved rear fangs that conduct venom to prey seized in the mouth, but in the East these species are harmless to humans.

WATER AND CRAYFISH SNAKES

Genera *Nerodia, Regina,* and *Clonophis*

These snakes have stout bodies and keeled scales. They may have wide heads, often quite large, very distinct from the neck. All swim well, have a cold-blooded diet, and usually do not kill their prey by constriction. They are often vicious when captured. All North American species bear living young, but closely related European and Asiatic members of this group lay eggs. The name *Natrix* has now been restricted to the Eurasian forms and American species are classified in different genera.

GREEN WATER SNAKE

Nerodia cyclopion **67:8**

Description
Length, to 60 in. (153.4 cm). Has ring of small scales around eye. Appearance often plain, but above dark green or olive-brown with many narrow black bands across back; alternating black blotches on sides, most pronounced to rear; below yellowish to brownish, especially anteriorly, and clear or spotted. Old individuals are uniform dull olive above. Head long, upper lip swollen; eyes placed high and forward.
Similarities
Resembles the Diamondback Water Snake, but lacks diamond pattern.
Habitat
Small lakes, marshes, swamps.

Habits
Vicious; climbs branches; will lie with head flattened on ground, giving head triangular appearance.
Food
Fish, amphibians.
Reproduction
Bears 7–19 living young, July–September.
Range
Coastally from s. S.C. to se. Tex., continuously throughout Fla. and up Miss. Valley to extr. w. Ky., s. Ill., and se. Mo.

PLAIN-BELLIED WATER SNAKE

Nerodia erythrogaster **67:6**

Description
Length, to 48 in. (121.4 cm); record 62 in. (157.5 cm). Unmarked yellow or reddish belly, above is usually patternless. Head slightly wider than neck. Scale rows 21–25, lower labials 10. Most adults gray, brown, or black above; top of head dark. Young have brownish blotches.
Similarities
Northern Water Snake has smaller eyes.
Habitat
Near water, usually ponds, swamps, or lakes.
Habits
Hard to catch, shy; much more wary than Northern Water Snake; active at night.
Food
Fish, amphibians, crayfish.
Reproduction
Bears 5–27 living young, August–October.
Range
Mostly se. U.S. from s. Delmarva Peninsula along Atlantic and Gulf Coastal Plains (excluding peninsular Fla.) to Tex. N. in cen. U.S. to Kans. and s. Mich.

Note: Four subspecies in the eastern U.S. are the following:

RED-BELLIED WATER SNAKE, *Nerodia erythrogaster erythrogaster,* is uniform rusty brown above, occasionally with faint banding; below unmarked vermilion. It occurs in the Coastal Plain in southern half of Delmarva Peninsula and continuously from southern Delaware to southeastern Alabama and northern Florida.

YELLOW-BELLIED WATER SNAKE, *Nerodia erythrogaster flavigaster,* is plain gray above, plain yellow below. It is found in north-central Georgia and extreme western Florida to eastern Texas and extreme southeastern Oklahoma, and north in Mississippi Valley to western Illinois and southeastern Iowa.

COPPER-BELLIED WATER SNAKE, *Nerodia erythrogaster neglecta,* is darker than Red-bellied Water Snake, red to orange below. There are scattered populations from southern Michigan and central Ohio to southeastern Illinois, western Kentucky, and northwestern Tennessee, and to southeastern Iowa.

BLOTCHED WATER SNAKE, *Nerodia erythrogaster transversa,* has dark dorsal saddles and dark spots on sides; adults essentially retain a juvenile pattern; plain yellow below. It ranges west of Mississippi Valley from western Missouri and northwestern Arkansas south and west through central Texas and into northern Mexico and eastern New Mexico.

NORTHERN WATER SNAKE

Nerodia sipedon **68:8**

Description
Length, to 53 in. (134.6 cm). Variable ground color gray to brown with dark crossbands anteriorly (usually about 30) that are wider than interspaces; alternating dorsal and side blotches toward rear of body. Below white or yellow with red to gray half-moons and dark dots that are more widely spaced at front of body and continue onto tail. Pattern varies in subspecies and becomes obscure in old individuals. Heavy-bodied; short-tailed; head distinct from neck. Lacks stripe from eye to corner of mouth. Very young are brightly marked, patterned jet-black on gray.

Habitat
Near water, especially streams, marshes, rivers, and lakes.

Habits
Diurnal to nocturnal depending on temperature, but active at temperatures lower than those tolerated by other snakes.

Food
Fish and anuran amphibians.

Reproduction
Bears 8–48 living young, with record litters of up to 99, June–September.

Range
Most of e. U.S. and se. Canada, but absent from n. New England and Atlantic Coastal Plain; in Gulf Coastal Plain from extr. w. Fla. to Miss. R.; nw. only to se. Minn. and extr. s. S.Dak. and to Colo. in Platte Valley.

Note: Four subspecies occur in the East:

A subspecies, *Nerodia sipedon sipedon,* as described above. It is found in North Carolina, Tennessee, Missouri, and northeastern Oklahoma northwards except in Mississippi River and Ohio River valleys in southern Illinois and Indiana; throughout most of northeastern U.S. and to southern Quebec.

LAKE ERIE WATER SNAKE, *Nerodia sipedon insularum (Fig. 63),* has no pattern, is plain grayish above, plain whitish below, or with faint evidence of pattern. It occurs in Pt. Pelée and Put-in-Bay archipelago in Lake Erie (between southern Ontario and northern Ohio).

CAROLINA SALT MARSH SNAKE, *Nerodia sipedon williamengelsi,* has dark overall coloration with black half-moons on belly. Found in salt-marsh areas, on outer banks of North Carolina and in nearby Pamlico and Core sounds.

MIDLAND WATER SNAKE, *Nerodia sipedon pleuralis (Fig. 63),* has brown to reddish ground above with narrower dark bands, overall light appearance. Below cream with reddish-brown belly spots in 2 rows. It occurs in southern Illinois and Indiana, west to eastern Oklahoma and northern Arkansas, and east of Mississippi River in lowland areas to northern Georgia and northern South Carolina.

Fig. 63

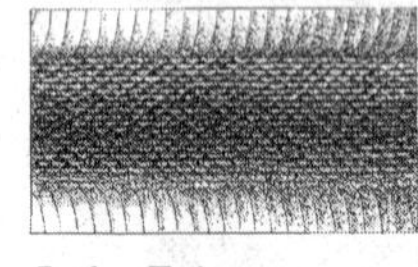

Lake Erie Water Snake

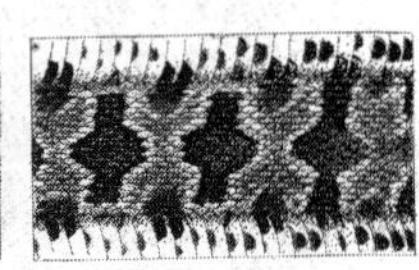

Midland Water Snake

SOUTHERN WATER SNAKE

Nerodia fasciata

Description
Length, to 42 in. (106.7 cm); record 62½ in. (158.8 cm). Broadly banded throughout length of body and has dark stripe from eye to corner of mouth. Belly spots squarish in most subspecies, but check below.

Similarities
Cottonmouth has no red spots below; Salt Marsh Snakes (see below) often lack bands and have longitudinal stripes.

Habitat
Permanently aquatic situations, ponds, marshes, swamps, lakes, also in temporary water, more often near shallow, still water.

Habits
Nocturnal, activity often correlated with large choruses of frogs. Like the Cottonmouth, defends itself when cornered, but bite is not venomous.

Food
Frogs, small fish, salamanders, tadpoles. (Salt Marsh Snakes feed on fish and crabs.)

Reproduction
Bears 9–57 living young, born June–August.

Range
Coastal Plain from N.C. to e. Tex. and up Miss. R. valley to extr. s. Ill.

Note: Four subspecies are found in the East:

BANDED WATER SNAKE, *Nerodia fasciata fasciata,* has dark crossbands over a tan to reddish-brown ground color, with crossbands rarely broken. Below yellow with reddish-brown blotches alternating irregularly from side to side, each blotch limited to 1 ventral. It is found in the Coastal Plain, except peninsular Florida, from North Carolina to Mississippi.

FLORIDA WATER SNAKE, *Nerodia fasciata pictiventris,* is similar to above, with gray, brown, or red crossbands, wormlike belly markings, and dull blotches between crossbands at sides. It occurs in all of eastern and peninsular Florida and into adjacent southeastern Georgia.

BROAD-BANDED WATER SNAKE, *Nerodia fasciata confluens,* has dark crossbands very wide above, almost running together, outlined with orange (or red) and white. Below yellowish-white with large, squarish, dark spots covering 3 or more scales. It occurs in the Mississippi Valley from extreme southern Illinois south; eastward only narrowly, and westward to southeastern Oklahoma and eastern Texas.

GULF COAST SALT MARSH SNAKE, *Nerodia fasciata clarki,* is dark gray to brown with 2 prominent longitudinal yellow stripes on each side and a duller middorsal stripe. Below blackish with a midventral light stripe. Attains a length of only 36¾ in. (93.3 cm), occurs in salt marsh habitats, and may intergrade with the freshwater species in areas of intermediate salinity. It is found on the Gulf Coast from central Florida to southern Texas.

DIAMONDBACK WATER SNAKE

Nerodia rhombifera **67:7**

Description
Length, to 48 in. (121.9 cm); record 63 in. (160.0 cm). No other water snake has diamond pattern above. Above yellow to olive with pattern of interconnecting diamonds of dark brown lines enclosing areas of ground color; dark bands at sides extend from bottom of diamonds to edge of belly. Below yellowish, ventral scales edged with crescent-shaped dark spots, especially near tail. Body large, heavy in adults.

Habitat
Near water; in open and under cover on lake shores and marshes; on branches overhanging water; by cattle tanks.

Habits
Aquatic, nocturnal; a good climber and vicious fighter. Emerges early from hibernation.

Food
Fish, frogs, turtles.

Reproduction
Bears 14–62 living young, August–November.

Range
Sw. Ind., s. Ill., extr. se. Iowa, Mo., and s. and e. Kans., s. to Gulf Coast, and into e. Mexico. Absent from Ozark Plateau and Ouachita Mts.

BROWN WATER SNAKE

Nerodia taxispilota **67:9**

Description
Length, to 69 in. (175.3 cm). Largest water snake, distinguished by rectangular black-bordered blotches on rusty or brownish-green upperparts. Dark blotches on sides alternate with rectangles above, but pattern may be irregular posteriorly. Below pied, white or buff with numerous dark markings. Young have paler ground color with black bands. Body very stout; head long, narrow; tail tapers rather abruptly.

Similarities
Cottonmouth either plain dark or brown or with wide, dark crossbands on paler ground.

Habitat
Near water; rivers, creeks, swamps; on branches overhanging water; sometimes frequents brackish water.

Habits
Arboreal as well as aquatic.

Food
Frogs, fish.

Reproduction
Bears 14–58 living young, June–November, but principally in August.

Range
Coastal Plain from se. Va. through Fla to se. Ala.

GRAHAM'S CRAYFISH SNAKE

Regina grahami **67:3**

Description
Length, to 28 in. (71.1 cm); record 47 in. (119.4 cm). Distinguished by broad yellow band on first 3 scale rows on each side, bordered below with thin, black, zigzag stripe. Above brown with paler dorsal stripe; belly yellowish, sometimes with median dark dots or darkish stripe. Head small, narrower than body, but wider than neck.

Habitat
Edges of creeks in prairie country; sloughs, ponds, ditches; often under logs, boards, or rocks, or in crayfish holes; occasionally seen on branches overhanging water.
Habits
Generally nocturnal; emerges early from hibernation, dens up late.
Food
Primarily crayfish; also fish, amphibians.
Reproduction
Bears 9–39 living young, July–September.
Remarks
Most snakes sleep in shelter of some protective nook, but this species and other water snakes often sleep on a branch over water.
Range
Ill. to se. Nebr., southward w. of Miss. R. to La. and e. Tex., and e. into n. Miss. Absent from Ozark Plateau and Ouachita Mts.

GLOSSY CRAYFISH SNAKE
Regina rigida **67:5**

Description
Length, to 31⅜ in. (79.7 cm). Dark, shiny, with traces of black lines down back and 2 rows of prominent dark spots on yellow belly. Top of head dark, upper lip yellow; underside of tail unmarked or may have irregular mid-stripe. No light stripes on sides. Scales have polished look. Body small, stout, head small.
Habitat
Edges of lakes, ponds, streams, swamps, marshes; often in mud.
Habits
Very aquatic; timid; a mud burrower.
Food
Hard-shelled crayfish, which it immobilizes by body constriction.
Range
Coastal Plain from N.C. through se. Okla. and e. Tex., but only in n. half of Fla. Disjunct population in New Kent Co., Va.

QUEEN SNAKE
Regina septemvittata **67:4**

Description
Length, to 24 in. (61 cm); record 36¼ in. (92.1 cm). Relatively slender, with 4 dark stripes on yellow belly and very small head. Above dark olive-brown with yellow band on side; upper lips and nose yellow; 3 faint narrow black stripes on mid-back. Belly pattern often distinct only anteriorly. Head barely distinct from neck; tail slender, tapering to a point.
Habitat
Swift, rocky, and shallow streams; edges of lakes, ponds, ditches, canals, sloughs; old quarries; often under rocks, branches, or debris, overhanging water.
Habits
Spends much time in water, tends to congregate before hibernating.
Food
Crayfish that have just molted.
Reproduction
Bears 5–23 living young, July–September.
Range
S. Great Lakes region from w. N.Y. to se. Wis., s. to Gulf Coast from panhandle of Fla. to Miss., e. to Del. and se. Pa. in mid-Atlantic region. Disjunct part of range in w. Mo. and Ark.

KIRTLAND'S WATER SNAKE

Clonophis kirtlandi *Fig. 64*

Description
Length, to 24 in. (61 cm). Above with 4 rows of blotches down body, mid-belly red with 2 distinct rows of black spots. Top of head black; lips, chin, and throat yellowish. Upright dark spots on sides alternate with blotches on back; background fairly dark, thus giving generally dark appearance. Head small, not distinct from neck; tail slender.

Habitat
Prairie country near streams and ponds; under cover in marshes, damp woods, ravines, pastures.

Habits
Probably nocturnal; when alarmed flattens itself and becomes immobile.

Food
Minnows, amphibians, earthworms.

Reproduction
Bears 4–13 living young, August–September.

Range
S. Great Lakes area from se. Wis. and s. Mich. s. to w. Ky., e. to w. Pa., and w. through cen. Ill. into Mo.

Fig. 64

Fox Snake, p. 376

Butler's Garter Snake, p. 358

Kirtland's Water Snake

Note: The **STRIPED CRAYFISH SNAKE,** *Regina alleni,* is similar in coloring to the garter and ribbon snakes, but can be distinguished by the fact that it has unkeeled scales and a divided anal plate. It occurs in extreme southern Georgia and the entire Florida peninsula.

SWAMP SNAKES

Genus *Seminatrix*

BLACK SWAMP SNAKE

Seminatrix pygaea *Fig. 65*

Description
Length, to 18½ in. (47 cm). Shiny black above and reddish below with regular prominent black bars to sides of ventrals only. Scale rows 17; anal divided; scales smooth. Above smooth, with narrow light line through middle of lower scale rows.

Habitat
Swamps, ponds, lakes with emergent vegetation, especially where water hyacinths are abundant.

Food
Invertebrates, tadpoles, small frogs, salamanders, small fish.

Reproduction
Bears 3–11 living young, August–October.

Range
Coastal Plain from N.C. through all Fla. except for w. tip of panhandle.

Fig. 65 Black Swamp Snake

PINE WOODS AND EARTH SNAKES

Genera *Rhadinaea* and *Virginia*

PINE WOODS SNAKE

Rhadinaea flavilata *Fig. 66*

Description
Length, to 15⅞ in. (40.3 cm). Upper lip yellow and dark line from snout through eye to mouth. Scales smooth; scale rows 17 at midbody; upper labials 7; anal divided. Above tan to yellowish or red, sometimes with obscure and irregular longitudinal stripes. Below yellow.

Habitat
Damp pine flatwoods, under rotting logs or stones.

Habits
Secretive.

Food
Small amphibians and reptiles.

Reproduction
Lays 2–4 eggs, May–August; female may produce 2 clutches per egg-laying period.

Remarks
The Pine Woods Snake is rear-fanged and mildly venomous, but poses no danger to humans. It is the sole member of the Genus *Rhadinaea* in the U.S. Over 20 related species occur southward through central South America.

Range
Coastal Plain marginally in states from N.C. to Miss. R. and throughout Fla. except s. tip of peninsula.

Fig. 66 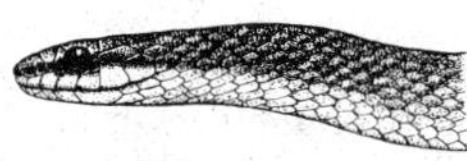 Pine Woods Snake

ROUGH EARTH SNAKE

Virginia striatula **68:4**

Description
Length, to 12¾ in. (32.4 cm). Head small, thick, sharply pointed, distinct from neck. Brown or reddish, with strongly keeled scales and yellow or pink belly. In younger individuals a light band across back of head. Upper labials 5; anal divided.

Similarities
Worm Snake has smooth scales, pink belly, but is darker above, and head is not distinct from neck. The Red-bellied Snake has dark head.

Habitat
Rocks, hillsides, woods, rubbish piles; near cities; wooded bottomlands, under bark, logs, or trash.

Habits
A burrower, secretive; stays under cover, but sometimes found on roads.

Food
Earthworms, snails, insects.

Reproduction
Bears 2–8 living young in August.

Range
Se. U.S., along Coastal Plain from s. Va. to cen. Tex., but absent in peninsular Fla. and s. La.; n. to s. Tenn., cen. Mo., se. Okla., and se. Kans.

SMOOTH EARTH SNAKE
Virginia valeriae **68:5**

Description
Length, to 13¼ in. (33.7 cm). Small, brown or gray, with cream-colored belly and smooth or weakly keeled scales. Back may be spotted with lines of tiny black dots, a faint line through each scale. Head wider than neck, but small, flat, narrow; snout pointed. Upper labials, 6; 2 small scales behind eye; anal divided.
Habitat
Fields, pastures, dry or moist woodlands; vacant lots, abandoned buildings, under tarpaper, rubbish, leaves, boards.
Habits
A burrower; secretive; stays under cover.
Food
Earthworms, snails, insects.
Reproduction
Bears 4–12 living young in August.
Range
Much of e. U.S. from N.J. and Pa. s. through s. Fla., and w. in lower elevations to s. Iowa, cen. Okla., and cen. Tex. Absent from Gulf Coast w. of Miss. R.

BROWN AND RED-BELLIED SNAKES
Genus *Storeria*

BROWN SNAKE
Storeria dekayi **68:2**

Description
Length, to 13 in. (33.0 cm); record 20¾ in. (52.7 cm). Two lines of small, dark spots down brown back, and dark brown blotches on each side of head behind eye. Scales keeled, 17 scale rows on body; anal divided. Above gray to brown with faint, light middorsal stripe. Young are darker and have light rings around neck; distinguish from young Ringneck Snake by keeled scales.
Habitat
Moist and dry situations, under stones, boards, leaves, trash heaps, roofing paper; rurally and in urban vacant lots.
Habits
Nocturnal and secretive.
Food
Earthworms, snails, slugs, insects, spiders, treefrogs.
Reproduction
Bears 3–27 living young, July–September.
Remarks
Formerly known as Dekay's Snake.
Range
Extr. s. Canada and e. U.S. from s. Maine through Fla., w. to cen. Minn. and extr. ne. S.Dak., se. Nebr., and cen. Tex.; also into e. Mexico.

RED-BELLIED SNAKE
Storeria occipitomaculata **68:3**

Description
Length, to 16 in. (40.6 cm). Plain reddish belly, and 3 light spots on nape. Fifteen scale rows; scales keeled; anal divided. Ground color gray, black, brown, to red-orange; head dark; faint middorsal

stripe. Belly edged with slate-gray. Sometimes small dark spots form lines down back.

Habitat

Moist woodlands, especially those with rocks; under stones, logs, boards, leaves, bark; in abandoned dwellings.

Food

Insects, earthworms, slugs.

Reproduction

Bears 1–13 living young, June–September.

Range

E. U.S. and s. Canada from N.S. to extr. se. Sask., e. N.Dak., and s. through extr. e. Tex., with patchy distribution in Ind. and Mo. Absent from extr. n. Maine and s. Fla. A disjunct population in S. Dak. and Wyo.

LINED SNAKES

Genus *Tropidoclonion*

LINED SNAKE

Tropidoclonion lineatum **67:2**

Description

Length, to 21 in. (53.3 cm). Greenish-white underparts with 2 rows of dark triangular dots running length of body. Scales keeled. Above brownish with 3 yellowish stripes; 2 rows of dots flank middorsal stripe. Light lateral stripe on 2nd and lower part of 3rd scale row. Head small, with 3 yellowish stripes; 2 rows of dots flank middorsal stripe. Head small, pointed, not distinct from neck; tail short.

Habitat

Near water; in old fields, rocky places, under boards and litter; urban vacant lots.

Habits

Secretive.

Food

Earthworms, insects.

Reproduction

Bears 2–12 living young in August.

Range

Continuously from se. S.Dak. and w. Iowa s. through e. Nebr. to cen. Tex. Spotty distribution in east of range to Ill. and Mo. and in west to Colo. and N.Mex.

GARTER SNAKES

Genus *Thamnophis*

Garter snakes are characterized by three narrow, light stripes on a darker ground color above, but coloration is variable. Typically, their heads are distinct from their necks and the scales are keeled. Scale rows are fewer than twenty-seven, and a single anal plate is found in most.

SHORT-HEADED GARTER SNAKE

Thamnophis brachystoma *Fig. 67*

Description

Length, to 22 in. (55.9 cm). Only garter snake with 17 rows of dorsal scales throughout length of body; yellow or orange lateral stripes on 2nd and 3rd scale rows toward front. Scales keeled; anal single. Body stout, tapers forward to long, thin neck not distinct from head; head very small. Above brown or black, light stripe on back may or may not be conspicuous. Black blotches between lateral stripes reduced to thin bars or absent.

Habitat

Meadows and pastures near water; among stones near marshes; not in undisturbed areas of forest.

Food

Earthworms, plus frogs, fish, toads, mice, insects.

Reproduction

Numbers of living young and season of birth not yet known.

Range

In nw. Pa. from Meadville area n. to Lake Erie and eastward, barely into s. N.Y.; along Allegheny R. drainage.

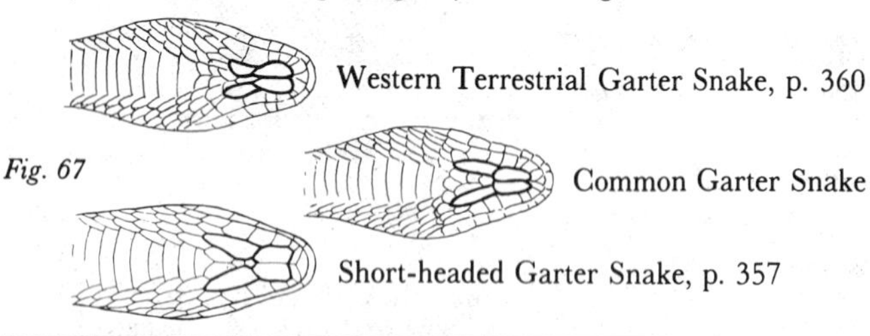

Fig. 67

BUTLER'S GARTER SNAKE

Thamnophis butleri *Fig. 64*

Description

Length, to 27¼ in. (69.2 cm). Small head, not distinct from neck. Above, 3 yellow stripes on black or brown ground, lateral stripes on upper 2nd, all of 3rd, and lower 4th scale rows; dark blotches sometimes present between stripes; top of head black. Below greenish-white with some black spots on edges. Nineteen rows of dorsal scales at neck, 17 just before anus. Tail short, stout, tapers to point.

Habitat

Near water, swamp edges, stream banks; in dry grass.

Habits

Mild-tempered.

Food

Earthworms, leeches, amphibians, insects.

Reproduction

Bears 4–16 living young, June–September.

Range

Great Lakes area in extr. s. Ont., e. Mich., ne. Ind., and w. and cen. Ohio. A disjunct population in se. Wis.

CHECKERED GARTER SNAKE

Thamnophis marcianus **67:12**

Description

Length, 24 in. (61 cm); record 42½ in. (108 cm). Yellow crescent on each side of head. Pale; above straw-colored with indistinct yellowish back and side stripes and dark, square blotches between stripes. Lateral stripe on 3rd scale row anteriorly, on 2nd (or 2nd and 3rd) to rear. Black vertical stripes on labial scales. Below whitish, often with small dark spots on ends of ventrals. Scales keeled, anal single.

Habitat
Near water, stream edges; river bottoms in desert regions.
Food
Mice, frogs, toads, fish, lizards, earthworms.
Reproduction
Bears 6–18 living young, July–August.
Range
Sw. U.S. from sw. Kans., s. through cen. and w. Tex., and w. to extr. se. Calif. Also through most of cen. Mexico and to the south.

PLAINS GARTER SNAKE

Thamnophis radix **67:11**

Description
Length, to 28 in. (71.1 cm); record 40 in. (101.6 cm). Three distinct bright stripes with 2 rows of squarish black spots between them; lateral stripes on 3rd and 4th scale rows in front, on 2nd and 3rd rows to rear. Above light olive to dark brown or black, side stripes may be lighter than yellow-orange stripe on back; upper lip yellow, barred with black. Below white to greenish with black spots on ends of ventrals. Scales keeled; anal single, body stout; head broad, distinct; tail ¼, but no more, of total length.
Similarities
Checkered Garter Snake has lateral stripe on 3rd scale row only anteriorly.
Habitat
Wet prairies, roadside ditches, shores of bodies of water and watercourses.
Habits
Aggressive; like most garter snakes, attempts to bite upon capture.
Food
Frogs, fish, toads, mice, earthworms, insects.
Reproduction
Bears 5–92 (generally not more than 40) living young, late July–September.
Range
N. Plains states w. of Great Lakes from Wis., Ind., and Ill., w. through Kans. to ne. N.Mex.; in north of range through w. Minn. and s. Man. to se. Alta.; also in cen. Ohio and extr. nw. Ark.

COMMON GARTER SNAKE

Thamnophis sirtalis **67:10, 68:7**
Fig. 67

Description
Length, to 26 in. (66 cm); record to 48¾ in. (123.8 cm). Variably colored, with light lateral stripes not involving 4th scale row, but on 2nd and 3rd scale rows anteriorly. Above green to orange-brown or black with 3 yellowish or greenish stripes, sometimes broken into alternating square black spots. Below greenish-white or yellow, infrequently with black spots on tips of ventrals.
Similarities
Ribbon snake more slender, has longer tail, lacks alternating spots between stripes.
Habitat
Widely distributed in various habitats, including fields, meadows, marshes, roadsides, gardens; often near water.
Habits
First snake to emerge in spring, last to disappear in fall.
Food
Earthworms, amphibians, mice, young birds, insects, spiders, fish.

Reproduction
Bears 3–96 living young, June–September.
Remarks
The northernmost-ranging reptile in North America, this snake reaches about 60° N latitude in the Northwest Territories. It has an interesting aggregative behavior during spring in northern parts. In Manitoba, up to 10,000 Red-sided Garter Snakes may gather to overwinter in limestone sinks. Immediately after emergence in the spring they may form balls of up to 200 intertwined individuals to conserve heat in response to low temperatures. Makes a good pet.
Range
Throughout e. U.S. and s. Canada, although distribution patchy in w. Tex., Okla., and Kans.; also widely distributed in the Northwest and in Calif.

Note: The major subspecies in the East are:

EASTERN GARTER SNAKE, *Thamnophis sirtalis sirtalis,* has no red on sides and has 2 rows of dark spots distinct on upperside. It occurs in Mississippi Valley eastward.

RED-SIDED GARTER SNAKE, *Thamnophis sirtalis parietalis,* generally has red skin in between scales on sides (best seen when stretched), and 2 rows of dark spots at each side coalesce. It ranges west of Mississippi Valley.

WESTERN TERRESTRIAL GARTER SNAKE 67:14
Thamnophis elegans *Fig. 67*

Description
Length, to 35½ in. (90.2 cm). A discontinuous yellow back stripe and dull yellow stripe on each side on 2nd and 3rd scale rows. Above brownish; may have rows of dark squarish spots; yellow spot in front of eye, black band across neck.
Habitat
Near or in water, flowing or still, but also in terrestrial places.
Food
Small vertebrates, including frogs and fish; earthworms, insects.
Reproduction
Bears 8–19 living young, June–September.
Remarks
Eastern subspecies is called Wandering Garter Snake, *T. e. vagrans.*
Range
Widely distributed through the West, including w. S.Dak. and Nebr.; also in extr. w. Okla.

EASTERN RIBBON SNAKE
Thamnophis sauritus 67:13

Description
Length, to 25 in. (64.3 cm); record 40 in. (101.6 cm). Lateral stripe on 3rd and 4th scale rows only, and dark lower lateral stripe that usually runs onto sides of ventral scales. Ground color brown to black with 3 prominent stripes along body, and occasionally vague dark spots between stripes. Spots to rear on top of head sometimes present, but dull and never fused. Relatively large, long-tailed, the tail more than ¼ total length; scale rows 19 over most of body.
Habitat
Marshes, lake shores, stream edges, meadows.

Habits
Semiaquatic, quick, able to climb to low vegetation.
Food
Amphibians, fish.
Reproduction
Bears 3–20 living young, July–August.
Remarks
Will bite; secretes fluid from anal glands when caught.
Range
E. of Miss. R. from s. Maine, s. Ont., and lower Mich., s. throughout Fla. to Miss. R., from Ky. to La. Disjunct populations in s. N.S. and ne. Wis.

WESTERN RIBBON SNAKE

Thamnophis proximus *Fig. 68*

Description
Length, to 30 in. (76.2 cm); record 48½ in. (123.2 cm). Similar to Eastern Ribbon Snake, but with 2 bright spots, usually fused, at rear of top of head, and only a narrow lower lateral dark stripe. Ground color of brown to black or olive with 3 stripes running along body. Middorsal stripe yellow, orange, or red. Lateral stripe on 3rd and 4th scale rows. Ventral scales lack dark pigment. Large, long tail; scale rows 19; scales keeled; anal single.
Habitat
Near most permanent water—streams, swamps, lakes, and rivers.
Habits
Diurnal; may try to bite on capture.
Food
Frogs, toads, salamanders, fish.
Reproduction
Bears 4–27 living young, July–September.
Range
S. Wis. and Ind., s. through Miss. Valley and w. to e. Nebr., se. Colo., and e. N.Mex.; also into e. Mexico.

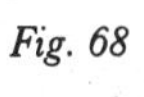
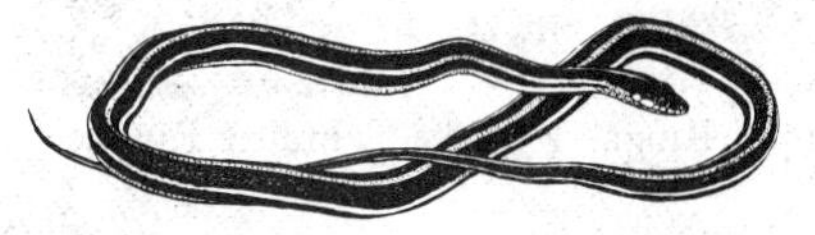

Fig. 68 Western Ribbon Snake

SMALL WOODLAND SNAKES

Genera *Diadophis* and *Carphophis*

RINGNECK SNAKE

Diadophis punctatus **68:1**

Description
Length, usually to 18 in. (45.7 cm). Unicolor above with yellow or orange ring around neck (though ring sometimes broken or even absent). Above dark gray to black; below yellow to orange or reddish, sometimes with one or more rows of black dots. Scales smooth, plain; anal divided; body slender, head not distinct from neck, top of head flat. Hatchlings blackish with distinct yellow around neck.
Habitat
Under bark or logs in woods, near water, rock-covered hillsides, forest paths, field edges.

Habits
Nocturnal; secretes foul-smelling defensive substance; in some populations, coiling of tail and exposing of red underside when disturbed is characteristic.

Food
Subspecies in the East feed on earthworms, insects, salamanders, small frogs and toads; larger western subspecies feed on lizards and snakes, and may be mildly venomous.

Reproduction
Lays 1–10 eggs, June–August; hatchlings August–September.

Range
Most of e. U.S. and in extr. se. Canada w. to se. S. Dak., se. Colo., through much of Southwest. Also n. on Pacific Coast and s. into Mexico.

Note: There are 5 subspecies in the East:

SOUTHERN RINGNECK SNAKE, *Diadophis punctatus punctatus (Fig. 69),* has yellow neck ring broken, series of half-moon-shaped black dots down center of yellow or orange belly. It ranges in Coastal Plain from New Jersey through Florida and Upper Keys to Alabama.

KEY RINGNECK SNAKE, *Diadophis punctatus acricus,* has no neck ring. It occurs on Florida Keys only.

PRAIRIE RINGNECK SNAKE, *Diadophis punctatus arnyi (Fig. 69),* has neck ring normally unbroken, black dots scattered or in pairs over yellow belly. It occurs in northern Mississippi Valley from extreme southeastern Minnesota to Missouri, west through eastern Nebraska, eastern New Mexico, and northern Texas.

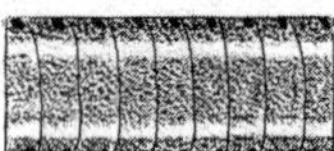
Northern Ringneck Snake

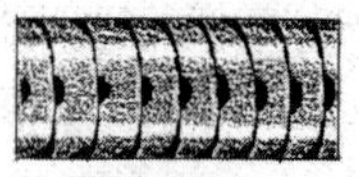
Southern Ringneck Snake

Fig. 69

Mississippi Ringneck Snake

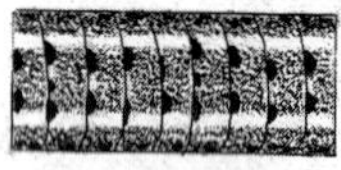
Prairie Ringneck Snake

NORTHERN RINGNECK SNAKE, *Diadophis punctatus edwardsi (Fig. 69),* has neck ring 1–3 scales wide, only occasionally broken; belly plain yellow or orange, sometimes with faint black dots down middle. It occurs in Appalachians and Piedmont from northeastern Alabama and northern Georgia, north through southern Canada, and west to southern Illinois through Wisconsin around most of Great Lakes, to northeastern Minnesota.

MISSISSIPPI RINGNECK SNAKE, *Diadophis punctatus stictogenys (Fig. 69),* like *D. p. edwardsi,* above, but belly strongly spotted. Occurs in lower Mississippi Valley from western Alabama to eastern Texas and north to extreme southern Illinois.

WORM SNAKE

Carphophis amoenus **68:6**

Description
Length, to 14¾ in. (37.5 cm). Small, slender, dark above and pink below without any markings or pattern. Scales smooth, opalescent. Above brown to black; pale belly color extends to 3rd scale row at

each side. Thirteen scale rows; 5 upper labials; head not distinct from neck, snout pointed, eye very small; tail short, pointed. Young darker than adults and have red belly.

Habitat

Under logs, stones, slabs, leaves, in wooded areas, abandoned fields, pastures.

Habits

Semiburrowing; has limited home range.

Food

Earthworms primarily, but also slugs, insects, snails.

Reproduction

Lays 1–12 eggs, late June–July; hatchlings August–September.

Remarks

Its common name refers not only to dietary specialization but also appropriately describes appearance.

Range

E.-cen. U.S. from w. Mass. s. coastally to Ga. and w. below Great Lakes area to se. Nebr. in the North and extr. ne. Tex. and n. La. Absent in most of w. Miss. Valley, but present on Gulf Coast from Miss. R. to Ala.

MUD AND RAINBOW SNAKES

Genus *Farancia*

MUD SNAKE

Farancia abacura **68:16**

Description

Length, to 81 in. (205.7 cm). Large, stout, shiny black above, black and coral-red below, the red forming bars or triangles on lower sides. Scales smooth, glossy. Body rounder than in rat snakes; head long, barely distinct from neck; spinelike scale at end of tail. Young may show signs of red crossbands on back.

Similarities

Rainbow Snake has stripes.

Habitat

Marshes, swamps, ditches.

Habits

A mud burrower; nocturnal; can jab with the spine of its tail, but spine is not poisonous as many who fear snakes would believe.

Food

Mainly amphiumas and sirens, but also frogs, fish.

Reproduction

Lays 4–104 eggs, July–September; female remains coiled around eggs during incubation; hatchlings, September–October.

Remarks

Like many other brilliantly colored animals, this snake has a characteristic warning posture that displays the red belly. In actuality a harmless snake, it is sometimes confused with poisonous snakes or with a fictitious "hoop snake," which rolls like a wheel while holding its tail with its mouth.

Range

Coastal Plain from s. Va. to e. Tex. and up Miss. Valley to s. Ill.

RAINBOW SNAKE

Farancia erytrogramma **68:15**

Description

Length, to 66 in. (167.6 cm). Three red stripes above and 3 rows of black spots on red belly. Ground color above black; scales smooth, glossy; body heavy, head not distinct from neck, spine at end of short tail.

Habitat

Mud or sand; sandy fields near marshes, springs.

Habits

A burrower; also aquatic; rarely surfaces; able to jab with spine on its tail.

Food

Eels; also frogs, fish, amphiumas, sirens, worms.

Reproduction

Lays 20–52 eggs, July–August in hole 4–6 in. (10.2–15.4 cm) deep in sandy field; hatchlings in September.

Range

Coastal Plain of North America from s. Md. to Miss. R. In s. Fla. present only near Lake Okeechobee.

HOGNOSE SNAKES

Genus *Heterodon*

Hognose snakes have an upturned, keeled snout. When threatened, a hognose snake puffs itself out, flattens the head, hisses, and strikes but never bites. If these menacing tactics fail, it feigns death by rolling over on its back and letting its tongue loll out. Thus fooled, potential predators relax their attention, and the snake escapes. However, the hognose snake betrays itself by its behavior when turned upright; it promptly repeats its performance and rolls belly-up.

WESTERN HOGNOSE SNAKE

Heterodon nasicus *Fig. 70*

Description

Length, to 35¼ in. (89.5 cm). Blotched with dark olive or brown above and underside of tail as black as belly. Yellowish ground above, blotches alternating with those on sides. Body heavy, scales keeled; head broad, barely distinct from neck; tail very short.

Similarities

Eastern Hognose darker, uniformly colored, has larger eye, snout less sharply upturned.

Habitat

Dry, sandy areas; prairies.

Habits

Diurnal. See also genus description, above.

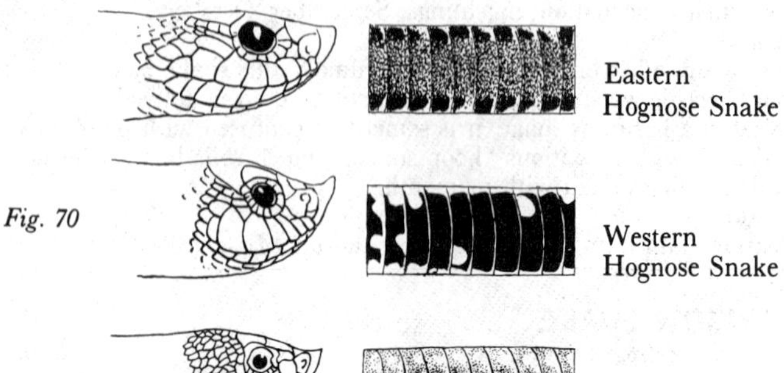

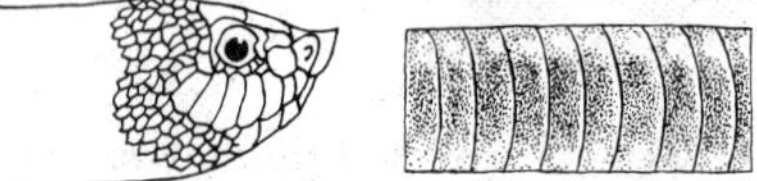

Fig. 70

Food
Toads, frogs, lizards, snakes, rodents, shrews, small birds.
Reproduction
Lays 5–24 eggs in August.
Range
Central Plains from s.-cen. Canada to Mexico, with easternmost localities in Ill., and a patchy distribution in Minn., Iowa, Mo., and ne. Tex.; w. to cen. Mont., cen. Colo., and se. Ariz.

EASTERN HOGNOSE SNAKE

68:9

Heterodon platyrhinos — *Fig. 70*

Description
Length, to 45½ in. (115.6 cm). Underside of tail lighter than belly. Above color very variable, sometimes uniform gray, brown, black, or green; if blotched, patterns may be gray, black, brown, or even reddish on a lighter ground. Dark line across top of head, dark line from eye to corner of mouth. Below whitish or yellowish with black markings. Body heavy, scales keeled, head flat on top, eyes large.
Similarities
Western Hognose has snout more upturned.
Habitat
Dry, sandy areas, open woods, uplands, hillsides, fields.
Habits
See genus description, above.
Food
Toads, frogs, insects, lizards.
Reproduction
Lays 4–61 eggs, June–July in earth, rotten logs, etc.; female often coils about eggs while they incubate, 6–9 weeks.
Range
S. Ont. and e. U.S. from se. N.H. through most of Fla. and w. through s. N.Y., lower Mich., and most of Wis. to cen. Minn., e. and s. Iowa, and ne. Nebr.; also se. S.Dak. and Kans. s. to Gulf Coast and e. Tex.

SOUTHERN HOGNOSE SNAKE

Heterodon simus — *Fig. 70*

Description
Length, to 24 in. (61 cm). Generally light in color with uniformly gray belly, the same color continuing under tail. Belly either uniformly colored or speckled with gray-brown. Dorsal ground gray, brown, yellowish with some red; dark blotches with smaller blotches laterally.
Similarities
Eastern Hognose is darker, larger, has snout scale less sharply upturned.
Habitat
Sandy lowland areas, usually in the open.
Habits
See genus description, above.
Food
Toads.
Reproduction
Lays 6–10 eggs, about which little is known.
Range
Coastal Plain from cen. N.C. to s. Miss., but absent s. of Lake Okeechobee in Fla.

SCARLET SNAKE

Genus *Cemophora*

SCARLET SNAKE

Cemophora coccinea **69:8**

Description
Length, to 20 in. (50.8 cm); record 32¼ in. (81.9 cm). Pale, unmarked belly and wide blotches of red on back and sides, with narrow black edges separated by narrow yellow or white in the formula black-yellow-black-red-black. Snout and top of head red with black band across eyes. Body slender and round, scales smooth; head not distinct from neck; snout pointed; eye small. In young, yellow is replaced by white.
Similarities
Pattern of banded Scarlet Kingsnake. See under Coral Snake. Blotches of Scarlet Snake do not go onto ventral scales.
Habitat
Underground in earth, sandy soil, or muck; on surface under slabs, stones, logs.
Habits
Secretive; a burrower.
Food
Young mice, lizards, small snakes, snake eggs, insect larvae.
Reproduction
Lays 3–8 eggs in June.
Range
Se. U.S.; cen. N.J. and Delmarva Pen.; s. Md., s. along Coastal Plain to extr. e. Tex.; northern limits in Va., Ky., extr. s. Ind., Ill., extr. s. Mo., and Okla. Disjunct populations at periphery of range, especially s. Tex., cen. Mo., Ky., and Va.

KINGSNAKES AND MILK SNAKES

Genus *Lampropeltis*

This large group of snakes varies widely in pattern, color, and size. Their heads are typically small and barely distinct from neck, the bodies are cylindrical, the scales smooth, and they have a single anal scale. The various species are often strikingly colored in patterns of rings or bands. All are powerful constrictors, and some species are notable in their immunity to the venom of rattlesnakes. All lay eggs.

COMMON KINGSNAKE

Lampropeltis getulus **69:4**

Description
Length, varies according to subspecies. A highly variable species: ground color from black or dark blue to yellow, with speckled, banded, or striped pattern. Top of head usually dark, lips white with dark edges. Body stout.
Habitat
Woods, hayfields, pastures, meadows, roadsides; near water; around farm buildings; rocky places.
Habits
Nocturnal and diurnal, above and below ground; uses holes made by other animals.

Food
Snakes, reptiles' and birds' eggs, young birds, rodents, lizards, amphibians.
Reproduction
Lays 5–17 eggs, June–July; hatchlings August–September.
Remarks
Immune to venom of native poisonous snakes.
Range
From s. N.J. in the East and Nebr. in the West, s. throughout e. U.S.; into the Southwest and Mexico. Subspecies also in West.

Note: There are 5 subspecies in the East. Color pattern and ranges are as follows:

EASTERN KINGSNAKE, *Lampropeltis getulus getulus,* length, to 82 in. (208.3 cm), has white chain pattern above on a black background; below blotched with black and white. It occurs in eastern coastal and Appalachian U.S. from southern New Jersey through North Carolina south to central Florida and coastal Alabama. Intergrades with Florida Kingsnake (see below).

FLORIDA KINGSNAKE, *Lampropeltis getulus floridana,* length, to 66 in. (167.6 cm), is generally pale with vague bands. It is found in peninsular Florida, intergrading with Eastern Kingsnake in central Florida, and separate populations on the eastern Georgia and Florida borders and in south-central area of Panhandle.

BLACK KINGSNAKE, *Lampropeltis getulus niger (Fig. 71),* length, to 58 in. (147.3 cm), is black with traces of yellow or white spots, or bands, above; throat white. It occurs in southern Ohio, western West Virginia, and southeastern Illinois, south to northeastern Mississippi, northern Alabama, and northwestern Georgia.

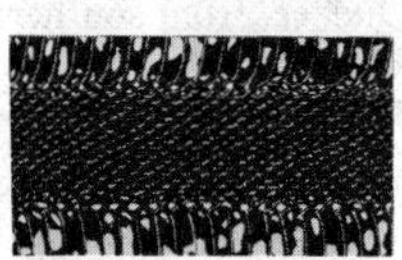

Fig. 71

Black Kingsnake

SPECKLED KINGSNAKE, *Lampropeltis getulus holbrooki* **(69:7),** length, to 72 in. (182.9 cm), has yellow center in each dark scale, giving a salt-and-pepper appearance. It occurs in the Mississippi Valley from central Illinois and western Alabama, west to Nebraska and eastern Texas. Intergrades with Desert Kingsnake (see below).

DESERT KINGSNAKE, *Lampropeltis getulus splendida,* length, to 60 in. (152.4 cm), has dark saddles above, and light-centered scales with occasional dark spots on each side. Black below. It intergrades with Speckled Kingsnake over large area from southern Nebraska to central Texas, but is found farther west than Speckled Kingsnake into southeastern Arizona and northern Mexico.

MILK SNAKE

69:3

Lampropeltis triangulum

Fig. 72

Description
Length, varies according to subspecies. A highly variable species; usually has dark-bordered red, brown, or gray blotches, sometimes forming rings around body over a white, gray, or yellowish ground. Body slender.

Habitat
Fields, open woods, river bluffs, rocky hillsides, prairies, pine-barren bogs, sandy soil; under cover or in open.
Habits
Secretive, but may be encountered during day in early spring.
Food
Rodents, snakes, lizards, frogs, eggs and young birds, fish, earthworms.
Reproduction
Lays 5–16 parchmentlike, adhesive eggs, June–July; hatchlings August–October.
Range
Most of e. U.S. from s. Maine w. into Mont. and s. to Atlantic and Gulf Coasts, into Mexico. Subspecies in w. U.S. and w. coast of Mexico.

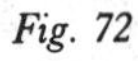
Fig. 72

Milk Snake, p. 367

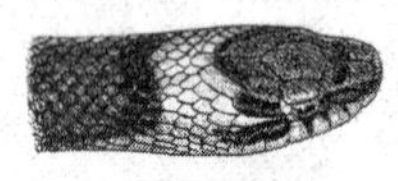
Red Milk Snake

Note: There are 7 subspecies in the area covered by this chapter, 3 east of the Mississippi, along with Eastern and Red Milk Snakes, and 4 others to the west.

EASTERN MILK SNAKE, *Lampropeltis triangulum triangulum,* length, to 52 in. (132.1 cm), has reddish-brown saddles with black borders on a milky to gray background, black and white checkered belly. There is a light triangle or Y on back of head and neck. The young are more brilliantly colored. It occurs in northeastern U.S., southern Ontario, and Quebec, from southern New England south through Appalachians and west to southeastern Minnesota and Northern Illinois. Intergrades with Scarlet Kingsnake (see below).

SCARLET KINGSNAKE, *Lampropeltis triangulum elapsoides* **(69:6),** length, to 27 in. (68.6 cm), broad red bands bordered with black, on a yellow or white ground; red and black encircle body. Found in southeastern U.S., intergrading with Eastern Milk Snake from southern New Jersey to northern North Carolina, and in central Kentucky and Tennessee; along Coastal Plain from North Carolina to Mississippi River, north to Kentucky. (See Coral Snake.)

RED MILK SNAKE, *Lampropeltis triangulum syspila (Figs. 72, 73),* length, to 42 in. (106.7 cm), resembles Eastern Milk Snake, but saddles are red; black spots on edge of belly; no triangle. It is found in central Mississippi Valley from southern Indiana west to extreme southeastern South Dakota and south to northwestern Mississippi and northeastern Oklahoma.

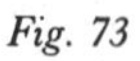
Fig. 73

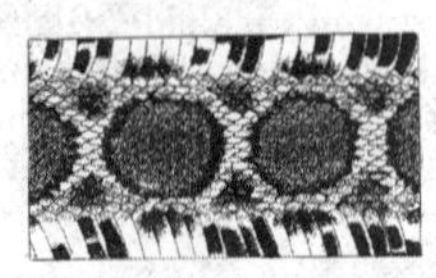
Red Milk Snake

CENTRAL PLAINS MILK SNAKE, *Lampropeltis triangulum gentilis,* length, to 36 in. (91.4 cm), has red bands bordered by black which encircle body, over a white background. Black head with lighter snout. It occurs in southern Nebraska, western and central Kansas, western and central Oklahoma, and northern Texas into north Colorado.

LOUISIANA MILK SNAKE, *Lampropeltis triangulum amaura,* length, to 31 in. (78.7 cm), has broad red crossbands on body; black head, but snout normally light; black over much of belly. It ranges west of Mississippi in southwestern Arkansas, southeastern Oklahoma, Louisiana, and eastern Texas.

PALE MILK SNAKE, *Lampropeltis triangulum multistrata,* is similar to above, but with orange snout; body rings and center of belly white. Found in western South Dakota and western and central Nebraska into extreme northeastern Colorado and north-central Montana.

MEXICAN MILK SNAKE, *Lampropeltis triangulum annulata,* length, to 39 in. (99.1 cm), has wide red rings, a black snout and belly. Black rings usually wide. It occurs in central Texas southward into Mexico, with many subspecies extending into Central and South America.

The **SHORT-TAILED SNAKE,** *Stilosoma extenuatum,* may be confused with the Milk Snakes because of its similar habits and habitat. However, its coloration is distinct, with alternating black and orange blotches, and it is particularly remarkable for being extremely slender. It occurs only in northern and central peninsular Florida.

PRAIRIE KINGSNAKE

Lampropeltis calligaster **69:5**

Description

Length, to 52⅛ in. (132.4 cm). Smooth scales, very narrow transverse markings on back and dark blotches on yellow belly. Above dark blotches, black-edged and somewhat dumbbell-shaped; two rows of alternating round blotches on flank; pattern becomes obscure in older individuals; may appear as 4 indistinct stripes or be all brownish. Dark band across front of head and from eye to angle of mouth, arrowhead mark on back of head. Also an all-black phase. Head blunt, anal plate single.

Similarities

Glossy Snake is unmarked below.

Habitat

Prairie country, in pastures, fields, roadsides, hay or grain shocks, open woods.

Food

Rodents, frogs, toads, lizards, small snakes.

Reproduction

Lays 6–13 adhesive eggs, June–August in holes in fields; hatchlings from August on.

Range

Se. U.S. from s.-cen. Md. s. to n. Fla. and w. to se. Nebr., Kans. and e. Tex. *Lampropeltis calligaster calligaster* is found in the Mississippi Valley west from central Kentucky and western Indiana through central Illinois, southern Iowa, and Kansas, south to western Louisiana and eastern Texas.

Note: The **MOLE KINGSNAKE,** *Lampropeltis calligaster rhombomaculata,* a burrower, is found east of the Mississippi Valley from southern Maryland through eastern and central Virginia and into central Tennessee, south to northern Florida and central Mississippi and Louisiana, to the Mississippi River.

GREEN SNAKES

Genus *Opheodrys*

ROUGH GREEN SNAKE

Opheodrys aestivus **68:10**

Description

Length, to 45⅝ in. (115.9 cm). Unicolor bright green dorsum, keeled scales; underparts yellow to green. Body slender, head distinct, but narrow, tail long.

Similarities

Smooth Green Snake has stouter body, shorter tail.

Habitat

Bushes, small trees in streamside and roadside locations where it is well camouflaged.

Habits

Arboreal, diurnal, even-tempered.

Food

Insects, spiders, snails, salamanders.

Reproduction

Lays 3–12 long, smooth, hard, adhesive eggs, July–August; hatchlings August–September.

Range

S. N.J., w. to e. Kans., s. throughout e. U.S. except Appalachians of W.Va. to cen. Tex. and into e. Mexico.

SMOOTH GREEN SNAKE

Opheodrys vernalis *Fig. 74*

Description

Length, to 26 in. (66 cm). Green above with lighter green or yellowish belly, scales smooth.

Similarities

Rough Green Snake has slender body, longer tail, keeled scales.

Habitat

Grassy fields, meadows, low bushes, bogs.

Habits

Less arboreal than Rough Green Snake; secretive; gentle.

Food

Much as for Rough Green Snake.

Reproduction

Lays 2–8 thin, cylindrical, blunt eggs, June–September (mainly July–August); hatchlings August–early October.

Range

Primarily ne. U.S. and se. Canada from N.S. w. to extr. se. Sask., s. to s. N.C., the Appalachians in Tenn., n. Mo., Kans., and Nebr. Also in disjunct populations in n. and se. Tex., western states, and Mexico.

Fig. 74 Smooth Green Snake

RACERS AND COACHWHIP

Genera *Coluber* and *Masticophis*

RACER

Coluber constrictor **68:11**

Description
Length, to 73 in. (185.4 cm). Large snake, uniformly bluish or black above and below (except in Yellow-bellied and Buttermilk subspecies) with smooth, satiny, lustrous scales. Throat and chin white. Body long, slender; tail very long; scales smooth, anal plate divided, scale rows 15 just in front of anus. Young pale gray above with dorsal saddles—large brownish blotches on back and black spots on sides.
Habitat
Varied, often near water, but includes brush, stone walls, rocky outcrops, where it may hibernate; trash piles, roadsides, swamps, cultivated fields.
Habits
Agile; will fight when cornered and can give a vicious, but not dangerous, bite; may be found in bushes and trees.
Food
Reptiles, birds, mammals, amphibians, insects.
Reproduction
Lays 1–28 eggs, June–July in soil or rotten wood; hatchlings late July–September.
Range
Most of e. U.S. except in extr. North, w. to Tex. and into Pacific Northwest. Also along e. coast of Mexico.

Note: Four subspecies of the East are:

BLACK RACERS, *Coluber constrictor constrictor* and *C. c. priapus,* are gun-metal black above and below. They occur in southern New England, south through most of Florida, and west through Ohio Valley and south-central Mississippi Valley, southeastern Missouri, extreme southeastern Oklahoma, and extreme northeastern Texas; in Coastal Plain west only to Mississippi River.

BLUE RACER, *Coluber constrictor foxi,* is bluish above, lighter below. Found in southern Ontario and west and central Ohio, west through southern Wisconsin and Illinois to southeastern Minnesota and eastern Iowa. A disjunct population occurs in southern upper Michigan.

EASTERN YELLOW-BELLIED RACER, *Coluber constrictor flaviventris* **(68:12),** is bluish to brownish above and yellowish below. It occurs in Iowa, Missouri, and northwestern Arkansas, west through the western part of the Dakotas to Montana, south to northern and eastern Texas, and into the Central Plains. Also coastally in western Louisiana. Subspecies in the West and in Mexico.

BUTTERMILK RACER, *Coluber constrictor anthicus,* is a racer with contrasting light spotting variably sprinkled over body. It ranges west of Mississippi River in extreme southern Arkansas, Louisiana, and eastern Texas.

There are seven "black" snakes in the East:

BLACK RACERS are all black above and below except for white chin, have smooth scales with a gunmetal gloss, a slender body with long, narrow head.

BLACK RAT SNAKE is all black above and whitish below, has highly polished, keeled scales, and a stouter body, with a broader, flatter, more angular head.

PINE SNAKE has scales strongly keeled, is all black, but often with brown markings on lips and trace of a pattern. Its range is limited: from southern Alabama to extreme eastern Louisiana.

BLACK KINGSNAKE is all black but shows some yellow or white above.

INDIGO SNAKE has large glossy scales and patterned belly. Black phase of **EASTERN HOGNOSE** is all black above; it has an upturned nose. Black phase of **PRAIRIE KINGSNAKE** is all black above and below; it has no white chin or light marks on back.

COACHWHIP

Masticophis flagellum **68:17**

Description
Length, to 60 in. (152.4 cm); record 102½ in. (260.4 cm). A large, slender whipsnake, 2-toned, dark toward the head, light toward the rear, both above and below. Ground color reddish to gray or black. Tail markings resemble braided coachwhip; tail very long and thin. Scales smooth; 17 scale rows in front to about 12 near vent; anal scale divided; 8 labials above. Head narrow with large eyes. Populations from central Kansas south and west lack 2-toned appearance and are light brown or pinkish over entire body, somewhat paler below or with crossbands. Young have dark crossbands on anterior part of body.

Habitat
Warm, dry uplands; semiarid regions; pine flatwoods, pastures, fields, roadsides.

Habits
A fast mover; terrestrial, diurnal.

Food
Rodents, snakes, lizards, young birds, and turtles.

Reproduction
Lays 4–24 pebbly surfaced eggs, June–July in holes in fields.

Remarks
Can give a nasty bite when handled.

Range
Se. U.S. from s. N.C. through most of Fla. and w. to Tenn. and Miss. Hiatus in distribution along Miss. R. Western range from s. Mo. and w. La. to sw. Nebr., Wyo., and e. N.Mex. continuing into sw. U.S. and most of Mexico.

INDIGO SNAKES

Genus *Drymarchon*

INDIGO SNAKE

Drymarchon corais *Fig. 75*

Description
Length, to 103½ in. (262.9 cm). Large-bodied with smooth, shiny scales blue to black dorsally; variable ventral coloration of alternating dark markings to either side of rear of scales, or with dark markings at edge of each. Sometimes reddish on chin, throat, and sides of head.

Habitat
Moist localities in pine woods, dry glades, tropical hummocks, flatwoods; also sandy hills, pine barrens, and oak ridges.
Habits
Slow-moving; may overwinter in tortoise burrows. Mild-tempered, but when cornered can give vicious appearance.
Food
Snakes, frogs, small mammals, birds, young turtles.
Reproduction
Lays 5–11 eggs, April–May.
Range
Disjunct population in s. Coastal Plain from se. Ga. through most of peninsular Fla. and Keys, and on Ala.–Fla. border; main population in s. Tex., s. through Mexico; Central and South America.

Fig. 75 Indigo Snake

PINE, BULL, AND GOPHER SNAKES
Genus *Pituophis*

BULLSNAKE
Pituophis melanoleucus

69:1
Fig. 76

Description
Length, to 72 in. (182.9 cm); record 100 in. (254.0 cm). Variably colored ground, white to yellow or gray, with 30–60 dark blotches. Body stout, head small, snout pointed, tail ends in a spine. Scales keeled on back, smooth on sides, a single anal scale, and approximately 29 scale rows.
Habitat
Prairies, fields, roadsides; farmlands, farm buildings; dry, sandy pine woods; occasionally swamps, rhododendron slopes.
Habits
Diurnal; when cornered hisses and vibrates its tail.
Food
Small rodents, birds, and birds' eggs.
Reproduction
Lays 3–24 eggs, May–August; hatchlings August–September.
Remarks
Takes well to captivity; is highly beneficial to the farmer in pest control. The Gopher Snake of the West is a subspecies.
Range
As described for subspecies below and throughout w. U.S. and Mexico.

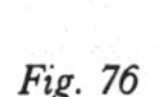

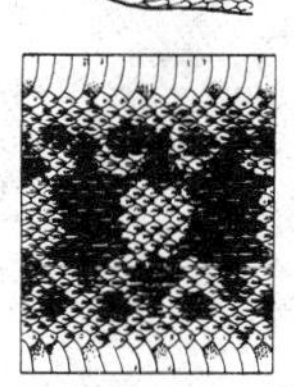

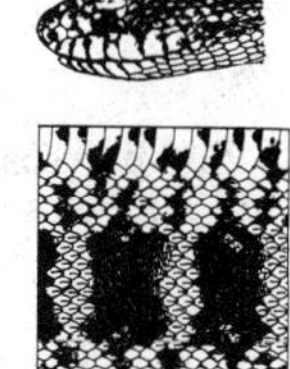

Fig. 76 Bullsnake

Pine Snake, p. 374

Note: *Pituophis melanoleucus sayi* has dark yellow head with dark band across top, dark bar below eye; alternating lateral black, brown, or dark reddish blotches on a yellow ground. Venter is yellow with 2 rows of dark blotches. Central Plains continuously from s. Wis., e. Ill., and n. and w. Mo., w. through w. N.Dak. to s. Alta. and through cen. and w. Tex. into Mexico. Disjunct populations in nw. Minn., e. Ill., and Ind.

PINE SNAKE, *P. m. melanoleucus* (*Fig. 76*), and other subspecies—have fewer blotches on back and alternating small blotches on sides; venter is marble-white, but in some areas blotches above are obscure, or individuals totally dark. They occur in western Virginia and adjacent southern West Virginia to southern Kentucky, south and east to the Coastal Plain in extreme southern South Carolina through eastern Louisiana. Through all Florida except southern tip. Disjunct populations in New Jersey Pine Barrens and west-central Louisiana into adjacent eastern Texas.

RAT SNAKES

Genus *Elaphe*

Rat snakes are distinguished by their flattened abdomens, an indication of climbing habits. They have stout bodies, broad and distinct heads, and boxlike snouts. All are powerful constrictors, vibrate the tail, and usually try to bite if caught. They show a diet preference for frogs and lizards when young, and rodents, birds, and eggs when older. All members of the genus lay eggs.

CORN SNAKE

Elaphe guttata **68:14**

Description

Length, to 48 in. (121.9 cm); record 72 in. (182.9 cm). Forty rich, reddish, saddle-shaped blotches, sometimes faded, above a reddish-brown or yellowish ground. Belly checkered dark and light with a pair of stripes under tail. Red stripe across forehead, red bar edged with black from eye through mouth; neck bands meet in middle of head, forming a spear pointing to snout; 2 rows of small blotches flank saddles. Body stout, head small, scales weakly keeled on at least 5 rows.

Similarities

Milk Snake has smaller head that is less distinct from body and weaker colors.

Habitat

Woods, wood edges, cornfields, outbuildings, roadsides, prairies, plains, mixed terrain, and even towns (in the South).

Habits

A ground snake, but can climb; nocturnal. Look for it on highways by night, under cover by day.

Reproduction

Lays 3–21 adhesive eggs, June–July.

Remarks

Makes a good pet.

Range

S. U.S. in N.J. Pine Barrens, s. Delmarva Pen., and s. Md. and Va., s. through Fla. Keys, and w. to se. Nebr. and into Colo., s. through Tex. into Mexico. Absent from most of Miss. R. and Ohio R. valleys n. of La.

Note: A subspecies, the **GREAT PLAINS RAT SNAKE,** *Elaphe guttata emoryi* (*Fig. 77*), record length to 60¼ in. (153 cm), is similar to the Corn Snake, but has no red; its blotches are dark on a gray background. (See Prairie Kingsnake.)

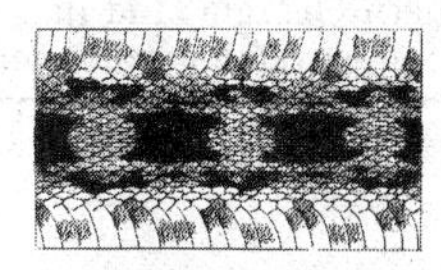

Gray Rat Snake

Fig. 77

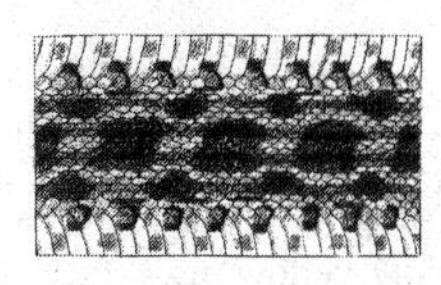

Great Plains Rat Snake

RAT SNAKE

Elaphe obsoleta

Description

Length, to 72 in. (182.9 cm); record 101 in. (256.5 cm). Highly variable, with upper surface blotched, striped, or all black. No spear on head, no regular stripes under tail. Body stout and flat ventrally in cross section; tail short, ending in a spine. Dark band from in front of eyes to mouth. Young often with H-shaped saddles on back over gray to brown ground.

Habitat

Trees, shrubs, ledges, hollow logs, fence rows, or stumps in a variety of situations, dry and moist.

Habits

A climber; emerges early from hibernation.

Food

As common to genus, plus small mammals in general.

Reproduction

Lays 5–44 eggs, June–July, in earth or logs; hatchlings late August–October.

Range

From extr. s. New England and into s. Ont., through the East and Midwest to sw. Wis., se. Nebr., e. Kans., and s. Tex.

Note: There are four major subspecies in the East:

GRAY RAT SNAKE, *Elaphe obsoleta spiloides* (*Fig. 77*), is gray above with brown saddles, which are often H-shaped, particularly on neck; below is white, peppered with gray forward, all gray to rear. Essentially it retains juvenile coloration. It occurs in Mississippi Valley to southern Illinois, and on Gulf Coast in Mississippi through Florida panhandle.

YELLOW RAT SNAKE, *Elaphe obsoleta quadrivittata,* has 4 dark stripes running length of the body, which has a yellow to light brown ground color. It occurs in southern North Carolina through most of peninsular Florida.

TEXAS RAT SNAKE, *Elaphe obsoleta lindheimeri,* is much like Gray Rat Snake above, but dark blotches, over the gray to yellow ground, are not as well defined. It occurs in all but the northern part of Louisiana to eastern Texas.

BLACK RAT SNAKE, *Elaphe obsoleta obsoleta* **(67:13),** black above with much white below, and has highly polished, keeled scales.

Body is stouter, head broader, and body less uniformly black than Racer; has reddish skin between scales, white spots sometimes showing in edges of back scales. A northern subspecies, occurring in Mississippi Valley south only to southern Illinois, but south through Appalachians to northern Georgia, and in the west of range, to northern Louisiana and extreme northeastern Texas.

FOX SNAKE

Elaphe vulpina *Fig. 64*

Description
Length, to 54 in. (137.2 cm); record 70½ in. (179.1 cm). Above light brown or straw with 28–51 large brown blotches on mid-back, flanked by 2 rows of smaller blotches. Head conspicuously chestnut, red, or brown, with neck bands not joining into a spear on head, but appearing to. Below yellow and black in keyboard design. Stout body, scales weakly keeled; tail stout, ending in a sharp spine. Young have more red.
Habitat
Prairies, wood edges, fields.
Habits
A ground snake, but can climb.
Food
Small mammals.
Reproduction
Lays 7–29 adhesive eggs, June–August in hollow stumps, logs, sawdust piles, manure heaps; hatchlings August–October.
Remarks
Makes a good pet.
Range
S. Ont. and adj. Mich. and n. Ohio; separate range from nw. Ind. around Lake Mich. into upper Mich. and w. through n. Mo. and s. Minn. to e. Nebr. and se. S.Dak.

GLOSSY SNAKES

Genus *Arizona*

GLOSSY SNAKE

Arizona elegans **69:2**

Description
Length, to 36 in. (91.4 cm); record 54⅝ in. (138.7 cm). Series of dark brown blotches on light brown ground above, and a white, unmarked belly. Ground color ranges from brown to yellow; 2 rows of small, dark lateral blotches, dark streak from eye to angle of jaw. Scales polished, smooth; anal single; snout scale projects backward. Head narrow, pointed.
Habitat
Sandy soil in woods, fields, plains.
Habits
Nocturnal.
Food
Rodents, moles, and probably lizards.
Reproduction
Lays 3–23 eggs.
Range
Extr. sw. Nebr., e. Colo., s. through w. Kans., w. Okla., and most of cen. and w. Tex., and in sw. U.S. and n. Mexico.

GROUND SNAKES

Genus *Sonora*

GROUND SNAKE

Sonora episcopa *Fig. 78*

Description

Length, to 15⅛ in. (38.4 cm). Smooth, shiny scales, 15 scale rows around body and anal scale divided. Head broad, flattened above; barely distinct from body. Color and markings extremely variable—may be gray, dark brown, or red above, sometimes uniformly brownish. Dark saddles or bands variably present, sometimes black rings encircle body, sometimes just a black collar on neck. Below cream-white or greenish, sometimes uniformly light.

Habitat

Prairies, dry places, buried in soil, or under stones or boards; roadsides, timbered hillsides.

Habits

Nocturnal.

Food

Ants, spiders, centipedes, scorpions.

Reproduction

Lays 4–6 eggs in June; hatchlings in August.

Range

Sw. Mo. and nw. Ark., w. to se. Colo., and s. through Okla. and e. N.Mex. to cen. and w. Tex. and into n. Mexico.

Fig. 78

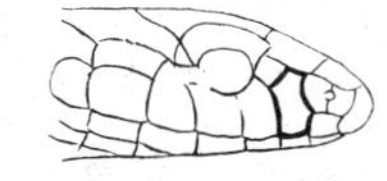

Flat-headed Snake, p. 378

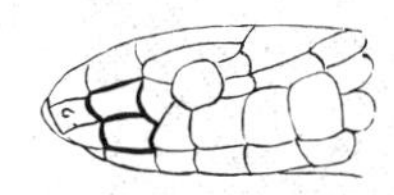

Ground Snake

LONG-NOSED SNAKE

Genus *Rhinocheilus*

LONG-NOSED SNAKE

Rhinocheilus lecontei *Fig. 79*

Description

Length, to 41 in. (104.1 cm). Red, white, and black pattern of blotches along body. Pattern of red, white, or yellow with 26–46 dark blotches, giving speckled or mottled effect. Sometimes black above with red squares—or orange half-rings; reddish pattern may be dull. Below yellow or white with some black at sides. Scales smooth, snout moderately long, conical; plates under tail form single row.

Fig. 79

Long-nosed Snake

Habitat
Sandy deserts, rocky slopes, arid brush.
Habits
A burrower and constrictor; nocturnal.
Food
Small rodents, lizards, young snakes.
Reproduction
Lays 4–9 eggs.
Range
Sw. and cen. Kans., w. Okla., s. into Mexico and into the West.

BLACK-HEADED SNAKES

Genus *Tantilla*

These are small snakes with uniform body color, fifteen rows of smooth scales, and a small, dark head. They are secretive burrowers that range from the southern United States down into South America.

SOUTHEASTERN CROWNED SNAKE

Tantilla coronata *Fig. 80*

Description
Length, to 13 in. (33 cm). Black head and collar, separated by a light band. Above brownish, below pearl- to purplish-gray. Eye small.
Habitat
Rocky or wooded hillsides under flat stones, logs, boards, bark.
Food
Insects, earthworms.
Reproduction
Lays 2–3 oval eggs in rotten wood of stump or log.
Range
E. U.S. e. of Miss. R. from s. Va., s. to cen. Ga., and w. in lowland areas to w. Ky. and extr. s. Ind. and along Gulf Coast from w. half of Fla. panhandle.

Southeastern Crowned Snake

Plains Black-headed Snake

Flat-headed Snake

Fig. 80

FLAT-HEADED SNAKE

Tantilla gracilis *Figs. 78, 80*

Description
Length, to 9⅝ in. (24.4 cm). Above shiny and brownish, gradually becoming only slightly darker on head; no collar. Below pinkish-orange, white toward head. Smooth scales; anal divided. Head flattened, barely wider than neck.
Habitat
High, flat grassy plains; rocky, sandy hillsides; under limestone rocks; by stream banks.
Habits
Probably nocturnal.

Food
Insects, spiders, centipedes, slugs.
Reproduction
Lays 1–4 eggs, mid-June–mid-July.
Remarks
Inoffensive when captured.
Range
Sw. Ill., s. Mo., and se. Kans., s. to n. La. and cen. and s. Tex. Other records in ne. Mo., n. Tex., and Mexico.

PLAINS BLACK-HEADED SNAKE
Tantilla nigriceps *Fig. 80*

Description
Length, to 14¾ in. (37.5 cm). Convex or V-shaped black cap, with apex pointing to rear. Above light brown or yellow; below white to pinkish. Head flattened, barely distinct from neck; black area not bordered by white band.
Similarities
Head color of Flat-headed Snake blends with, and is only slightly darker than, body color.
Habitat
Under flat rocks on dry hillsides.
Habits
Nocturnal; secretive.
Food
Insects, spiders, earthworms.
Range
Sw. Nebr., cen. Kans., and w. Colo., s. through cen. Tex., w. to se. Ariz., and into n. Mexico.

NIGHT SNAKE
Genus *Hypsiglena*

NIGHT SNAKE
Hypsiglena torquata *Fig. 81*

Description
Length, to 20 in. (50.8 cm). Above with numerous dark spots, lateral alternating with dorsal; dark blotch on either side of neck. Dorsal ground color gray, pale brown, or yellowish; below light, unmarked, scales smooth; head flat, pointed, distinct from neck; eye pupil vertical.
Habitat
Extremely varied in arid and semiarid regions.
Food
Lizards, small snakes, frogs, earthworms.
Reproduction
Lays 3–12 eggs, April–July; hatchlings from June on.
Remarks
Venomous, fangs at rear of mouth, but not fatal to humans.
Range
E. of 100th meridian occurs only in se. Kans., w. Okla., and w. and cen. Tex.; also w. to Colo. and into s. Mexico.

Fig. 81  Night Snake

CORAL SNAKES

Family Elapidae

These snakes are venomous. In the New World, this family is represented by only two genera, with most species restricted to Central and South America. They all have a conspicuous pattern of red, black, and white or yellow rings around the body. The upper front fangs are permanently erect and there are no pits behind the nostril.

EASTERN CORAL SNAKE

Micrurus fulvius **69:10**

Description
Length, to 30 in. (76.2 cm); record 47½ in. (120.7 cm). Body rings in the formula black-yellow-red-yellow-black. Face black, wide band of yellow across middle of head. Yellow rings around body narrow, tail has no red rings. Scales smooth; head flat, blunt, not distinct from body. Young have similar pattern but paler colors.
Habitat
Underground, often in fields; in or near water.
Habits
Primarily diurnal.
Food
Small snakes, lizards, frogs.
Reproduction
Lays 2–4 soft eggs, May–July in loamy soil, humus, or logs; incubation 3 months.
Range
Se. N.C., w. to cen. Tex. in Coastal Plain, and on into e. Mexico; n. into cen. Ark.; absent in most of e. La. in Miss. Valley.

Note: Three snakes in the East have alternating body bands of red, yellow, and black. They may be distinguished as follows:

EASTERN CORAL SNAKE (above): Black-yellow-red-yellow-black (yellow narrow); face black; bands encircle body. The poisonous Coral Snake is the only one of these that wears a black mask on its face and has yellow rings touching red.

SCARLET SNAKE: Black-yellow-black-red-black (black and yellow narrow); bands do not encircle body; face red.

SCARLET KINGSNAKE: Black-yellow-black-red-black (black and yellow narrow); bands encircle body; face red.

Pit Vipers

Subfamily Crotalinae

This subfamily belongs to the family Viperidae, whose other members are found exclusively in the Old World; the Crotalinae are found extensively in the New World and Asia. These venomous snakes have a heat-sensitive pit on either side of the head, behind, and much larger than, the nostrils. Each species has vertical pupils, a distinct, triangular head, and keeled scales. They all bear living young. Bites from these snakes may prove fatal to man; sometimes even severed heads can bite.

COPPERHEADS AND COTTONMOUTHS

Genus *Agkistrodon*

COPPERHEAD

Agkistrodon contortrix

69:9
Fig. 83

Description
Length, to 36 in. (91.4 cm); record 53 in. (134.6 cm). Distinguish this snake by its hourglass-shaped dark crossbands enclosing light lateral areas and coppery brown, red-brown, or pink head. Ground color above light brown or pink; below whitish with dark blotches; tongue red at base and with lighter tips. Young have bright yellow tail that is waved to attract food.

Habitat
Dry upland woods, rocky hillsides, ledges; mountains; moves to lowlands in summer.

Habits
Bites only if startled, though more aggressive in hot weather.

Food
Rodents, small birds, frogs, toads, lizards.

Reproduction
Bears 2–17 living young, August–October; 2–10 clutches per season.

Range
E. U.S. from sw. Mass. s. to extr. n. Fla., w. to extr. s. Iowa, e. Kans., and into cen. Tex., to Rio Grande.

Note: Four subspecies are:

NORTHERN COPPERHEAD (*Fig. 82*), *Agkistrodon contortrix mokasen,* has dark bands narrow on the back, wide on the sides, often with dark spots on the interspaces of ground color. It occurs in Massachusetts and Central Illinois, south to higher elevations in northeastern Mississippi and northern Georgia, and west to the Mississippi River.

OSAGE COPPERHEAD, *Agkistrodon contortrix phaeogaster,* is similar to northern subspecies, but the dark crossbands and immaculate lighter ground color contrast sharply; the belly is often dusky. It occurs west of the Mississippi River in extreme southeastern Iowa, northern and central Missouri into extreme southeastern Nebraska, eastern Kansas, and extreme northeastern Oklahoma.

SOUTHERN COPPERHEAD, *Agkistrodon contortrix contortrix* (*Fig. 82*), is similar, but has dark bands wider laterally, narrower dorsally; ground color is paler. It occurs in the lower Mississippi Valley to southern Illinois, and in the southern Coastal Plain from extreme southeastern Virginia to Texas, but is absent from most of Florida.

BROAD-BANDED COPPERHEAD (*Fig. 82*), *Agkistrodon contortrix laticinctus,* is brighter, has chestnut-brown crossbands very wide, broadest on sides. It occurs in the Great Plains from southern Kansas to central Texas.

Northern Copperhead

Southern Copperhead

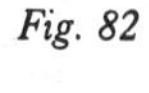
Fig. 82

Broad-banded Copperhead

COTTONMOUTH

Agkistrodon piscivorus **69:11**

Description
Length, to 48 in. (121.9 cm); record 74 in. (188 cm). Largest pit viper without a rattle. Has flattened head, eyes not visible from above, and dark band from eye to angle of mouth bordered above by pale line. Above brown with indistinct black bands that are wider laterally. Below yellow, with or without dark markings. Old adults may be entirely black on body. Inside of mouth white, but this is not unique. Head chunky, tail short. Young brilliantly marked and banded with dark on reddish-brown, very much resembling close relatives, the copperheads.
Similarities
Deep, broad, spade-shaped head, eyes invisible from above, and light lip marking distinguish Cottonmouth from water snakes.
Habitat
Near lakes, sloughs, streams, swamps, marshes, and other aquatic situations.
Habits
Aquatic, usually nocturnal; gaping widely at foe is characteristic.
Food
Snakes, turtles, amphibians, fish, and mammals.
Reproduction
Bears 1–15 living young, August–October.
Range
Se. U.S. along Coastal Plain from se. Va. to cen. Tex., and n. through w. Tenn., w. Ky., and e. Okla. to Mo. and extr. s. Ill.

RATTLESNAKES

Genera *Sistrurus* and *Crotalus*

These snakes are all venomous, and all have a series of interlocking, horny links on the tail that form a rattle. A new link is added at each shedding of the skin, but due to wear and tear rarely are more than ten segments found on an animal in the field. They cannot be used to determine the animal's age. When disturbed, a rattlesnake vibrates the tail, thus making the characteristic buzzing noise for which they are famed. Members of the Genus *Sistrurus* are small and have enlarged plates on the head. Members of the Genus *Crotalus* have small scales on top of the head between the eyes and have good-sized rattles.

MASSASAUGA

Sistrurus catenatus **69:14**

Description
Length, to 30 in. (76.2 cm); record 39½ in. (100.3 cm). Small, brownish, with squarish dark blotches on back. (However, some individuals are almost black with markings almost or entirely blended into background.) Two or 3 series of small, dark blotches on sides. Below dark and heavily blotched. Tail short, stout, and ringed with dark brown and ground color. Rattle well developed, medium-size.
Habitat
Damp lowlands; swamps, edges of streams, ponds, marshes, meadows; fence rows and fields in summer.
Habits
Not usually aggressive.
Remarks
Venom deadly.

Food
Rodents, frogs, toads, birds, insects.
Reproduction
Bears 2–19 living young, late July–September.
Range
S. Ont. and w. N.Y., w. through lower Mich., n. Ind., and s. Wis. to extr. se. Nebr. and e. and s. Kans., s. through much of Tex., and w. to se. Colo., N.Mex.; also in extr. n. Mexico.

PIGMY RATTLESNAKE

Sistrurus miliarius *Fig. 83*

Description
Length, to 22 in. (55.9 cm); record 31 in. (78.7 cm). Small, with 3 rows of vivid black blotches with reddish interspaces middorsally. Ground color variable, gray to tan, even bright red-orange in some North Carolina populations, and almost black in Florida. Black stripe runs back from eye, middorsal blotches wider than long. Below dark, blotched. Body stout, tail long, slender; rattle very small.
Habitat
Wet or moist locations; grass or woodlands; abandoned buildings, debris.
Habits
Short-tempered, aggressive.
Food
Mice, lizards, frogs, small birds.
Reproduction
Bears 2–32 living young, August–September.
Range
In Coastal Plain from N.C., w. to e. Tex. and e. Okla., and n. to s. Mo. and extr. sw. Ky.

Fig. 83

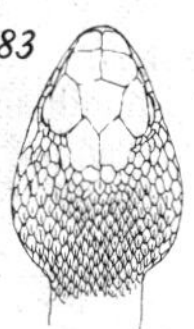

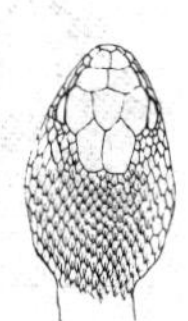

Pit Vipers

Pigmy Rattlesnake Timber Rattlesnake Copperhead, p.381

TIMBER RATTLESNAKE

69:13

Crotalus horridus *Fig. 83*

Description
Length, to 60 in. (152.4 cm); record 74 in. (188.0 cm). Has phases varying from yellow or gray with irregular brown or black wavy crossbands, to almost black in Northern populations, with bands not discernible. (Phases are independent of sex, age, or season.) Coastal Plain and Mississippi Valley populations with yellowish or reddish dorsal band lengthwise on neck and body. Below yellowish, with many minute black spots. Young are never black.
Habitat
Dry uplands, rocky ledges, hillsides, usually with southern exposure; brambles, second growth, wastelands, swamp edges.
Habits
Shy, not aggressive. If sufficiently disturbed, may sound rattle as warning.
Food
Small mammals, birds.

Reproduction
Bears 3–17 living young, August–October.
Range
S. New England, s. through Appalachians throughout most of e. U.S. In s. N.J., but absent from cen. N.J. to Va., and absent around Great Lakes except in N.Y. and scattered localities in Ohio and Ont. Northward extension in west of range into se. Minn. and Wis., and westward extension to extr. se. Nebr. through e. Tex.

EASTERN DIAMONDBACK RATTLESNAKE

Crotalus adamanteus *Fig. 84*

Description
Length, to 96 in. (243.8 cm). Yellow-edged, dark diamonds with light centers along back, and vertical yellow stripe in front of each nostril. Ground color olive to brown above; below yellow or white with some brown pigment at sides. Dark stripe from eye back to jaw. Large, heavy-bodied.
Habitat
Pine flatwoods and wooded hillsides, often resident in Gopher Tortoise burrows. Dry areas generally.
Habits
May emerge on sunny days in winter to bask or search for food.
Food
Small mammals.
Remarks
Largest rattlesnake in the United States; extremely dangerous when aroused to anger.
Range
Coastal Plain e. of Miss. R. from s. N.C. through Fla. Keys w. to extr. e. La.

Fig. 84

Eastern Diamondback Rattlesnake

WESTERN RATTLESNAKE

Crotalus viridis **69:12**

Description
Length, to 57 in. (144.8 cm). Greenish-gray rattlesnake (but sometimes yellow or red) with dark lozenge-shaped saddles down middle of back. Saddles edged with white and flanked by 2 rows of alternating smaller dark blotches; below yellow, mottled. Sides of head have 2 white diagonal lines; tail has light and dark rings.
Habitat
Prairies, plains, badlands, open rocky ground.
Habits
If disturbed will usually sound rattle as warning.
Food
Rodents, small birds, frogs.
Reproduction
Bears 3–21 living young.
Range
Primarily Great Plains and the West, but occurs e. of 100th meridian in sw. N.Dak. s. through w. Tex., with the farthest eastward extension into Iowa near its border with S.Dak. and Nebr.; also into Mexico.

Amphibians

Consulting Editor
Patricia Gardner Haneline
Andrew Mellon Fellow
Department of Biological Sciences, The University of Pittsburgh

Illustrations
Plates 70–72, 73, 75 John Cameron Yrizarry
Plate 74 Jennifer Emry-Perrott
Text Illustrations John Cameron Yrizarry and
Jennifer Emry-Perrott

Amphibians

Class Amphibia

Amphibians are cold-blooded vertebrates that have limbs and moist skin containing mucous glands. This is in contrast to the fins of fish and the dry, scaly skin characteristic of reptiles. The major difference between amphibians and reptiles, however, is in their reproductive habits. Amphibians generally lay small eggs surrounded by a jellylike capsule that swells when in contact with water; reptiles, on the other hand, possess an amniotic egg containing a large yolk, three extraembryonic membranes, and a leathery or hard shell. Amphibians are tied to moist places because of their skin and eggs, and furthermore usually go through a free-swimming aquatic larval stage after hatching. All but three of the eastern amphibian species possess four clawless limbs as adults. A distinguishing feature between orders of adult amphibians is the presence of a tail in salamanders and its absence in frogs and toads. However, because of their short bodies and their long legs adapted for jumping, or saltatory, locomotion, frogs will never be mistaken for anything else. Aquatic salamanders might be mistaken for eels at first glance, but you can tell the difference easily.

Amphibians were the first four-limbed land dwellers, or tetrapods. They arose in the Devonian period about 340 million years ago and were the dominant land animals for over 100 million years, reaching their peak of diversity in the Carboniferous. Amphibians evolved from the rhipidistian fishes of the order Crossopterygii, or lobe-finned fishes, whose sole surviving member is a rare coelacanth found only in deep water off the east coast of Africa. The first fossil amphibian known is *Ichthyostega* from the Upper Devonian of Greenland. Unlike the fins of their fish ancestors, amphibian limbs, fully articulated with pectoral and pelvic girdles, are used in supporting the body and in propulsive movements in walking or running. Early amphibians attained a large size, up to fifteen feet (4.6 meters) in the giants, and some had bony external armor covering parts of the body. The three surviving orders are in comparison small and secretive and are considered to be very closely related to one another, which is reflected in their grouping together in the Subclass Lissamphibia. Besides the two familiar orders of the frogs and toads, and the salamanders, there are the limbless, wormlike caecilians, some of which possess scales embedded in the skin. Caecilians occur only in the tropics.

Distribution and Diversity

There are approximately 2500 species of amphibians worldwide. Amphibians occupy almost all land masses except the Arctic and Antarctic, but overall species diversity generally increases as one approaches the tropics. The factors ultimately limiting amphibian distribution are moisture and temperature.

Habits

Amphibians breathe through skin, gills, or lungs, or a combination of them. Needed moisture is also absorbed through the skin. Like fishes and reptiles they are cold-blooded, or ectothermic; that is, their body temperature is maintained at high levels by external sources and is not internally controlled, as in birds and mammals.

Reproduction

The typical pattern of reproduction is egg laying in or near water, although some species utilize moist areas on land to deposit eggs that hatch into miniature adults. The phenomenon of reproduction around water often provides a herpetologist with the best opportunity to observe and collect amphibians, which are encountered either migrating to the breeding site or about the ponds, streams, and rivers themselves. Larval salamanders have external gills that disappear during metamorphosis and mouthparts similar to those of the adult. Larval frogs, called tadpoles, lack visible external gills and have a horny beak. Both types of larvae have a tail which is used in swimming, but in frogs this structure is later resorbed. Identification of larval forms, particularly in the early stages, is difficult and much too complex for this volume.

Range and Scope

Discussed in this chapter are 107 amphibians, including fifty-nine species of salamanders and forty-eight species of frogs and toads. Coverage includes the North American land areas between the Canadian–United States border and southern Florida, and from the 100th meridian east to the Atlantic coastline.

Nomenclature

The common and scientific names in this chapter follow *Standard Common and Current Scientific Names for North American Amphibians and Reptiles.*

USEFUL REFERENCES

Amphibians and reptiles are generally considered together in books, and the reader is referred to the reptile references found on pages 309-311 in order to avoid duplication. Additional references are found below.

Altig, R. 1970. A Key to the Tadpoles of the Continental United States and Canada. *Herpetologica* 26: 180–270.

Bishop, S. C. 1943. *Handbook of Salamanders.* Ithaca, N.Y.: Comstock.

Blair, A. P. 1968. Amphibians. In Blair, Blair, Brodkorb, Cagle, and Moore, *Vertebrates of the United States.* New York: McGraw-Hill.

Cochran, D. M. 1961. *Living Amphibians of the World.* Garden City, N.Y.: Doubleday.

Logier, E. B. S. 1952. *Frogs, Toads and Salamanders of Eastern Canada.* Toronto: Clarke, Irwin.

Orton, G. L. 1952. Key to the Genera of Tadpoles in the United States and Canada. *American Midland Naturalist,* 47: 382–95.

SSAR Committee on Common and Scientific Names. 1978. *Standard Common and Current Scientific Names for North American Amphibians and Reptiles.* Available from D. H. Taylor, Dept. of Zoology, Miami University, Oxford, Ohio.

Wright, A. A., and Wright, A. H. 1949. *Handbook of Frogs and Toads of the United States and Canada.* 3rd ed. Ithaca, N.Y.: Comstock.

GLOSSARY

Adpressed Front and back limbs extended toward each other, flat against sides; for measuring lengths.

Anterior Toward the front (of the body).

Cloaca Area into which internal wastes discharge.

Costal groove Vertical furrow on the flanks of salamanders, set off by the skin between successive grooves known as intercostal fold.

Cranial crests Raised area framing inner border of upper eyelids in toads.

Dewlap Skin fold hanging from neck region.

Diurnal Daytime.

Dorsolateral Pertaining to the uppersides of the animal.

Frontals Bony membranes that form the forehead.

Gular fold Fold of skin across posterior section of throat in salamanders.

Intercostal fold Skin fold bounded on each side by a costal groove.

Larva (*pl.* larvae) The earliest form that certain species take; unlike the parent (e.g., tadpole).

Lateral Of or pertaining to the side of the body.

Maxillary bone Bone on each side of head forming side border of upper jaw and bearing most of the upper teeth.

Nasolabial groove Groove from nostril to edge of upper lip.

Neotenic Retaining gills or other larval features throughout entire life.

Nuptial pad A patch of darkly pigmented, roughened skin that appears on certain digits during breeding season.

Occipitals Bony membranes that form the posterior part of the skull.

Parotoid One of two large wartlike glands found on the rear of the head in toads.

Parietal bones A pair of membrane bones in the roof of the skull between the frontals and occipitals.

Posterior Toward the rear (of the body).

Postocular Behind the eye.

Postorbital Behind the eye.

Reticulate Resembling the form or appearance of a net.

Riparian Relating to the bank of a river, lake, or pond.

Rugose Rough, wrinkled.

Tubercle Any of various small knoblike prominences.

Tympanum Thin round or oval external ear covering; prominent in frogs and many lizards.

Vent Opening on the surface of the cloaca.

Venter Belly.

Vertebral stripe Stripe down the midline of the back.

Vertical pupil Pupil that is elliptical; long axis is vertical.

Vocal sac Loose skin on the throat that can distend, forming a chamber which echoes the animal's vocalization.

Vomerine teeth Teeth anchored to the vomerine bones that form part of the forepart of the roof of the mouth.

Salamanders

Order Caudata

Salamanders are seldom seen because of their generally small size, secretive habits, and slow movements. They are encountered only with active searching, and even then, with their concealing coloration, they may elude the naturalist.

Salamanders have a tail throughout life (in contrast to adult frogs), and have moist, relatively smooth skin, no external ear opening, and never more than four toes on their unclawed front feet. Lizards usually have five toes on their front feet, all their toes are clawed, and they have a dry, scaly skin. Many salamanders can regrow a lost limb or tail, and the latter is often sacrificed when grabbed by a predator.

Identification

Features generally useful in identification are illustrated below. To start with, check for presence or absence of gills, nasolabial grooves from nostril to jaw (used in chemical sensing of the environment), costal grooves and their number (the grooves around the side of the body), and the gular, or throat, fold. Other important features include numbers of limbs and digits, and coloration, particularly if there are bands, stripes, or spots on the body; however, color and pattern are often variable. The species descriptions give common variations, but other listed diagnostic features should all be checked. The color often becomes darker and the pattern more obscure with age.

Fig. 85

Parts of a Typical Salamander

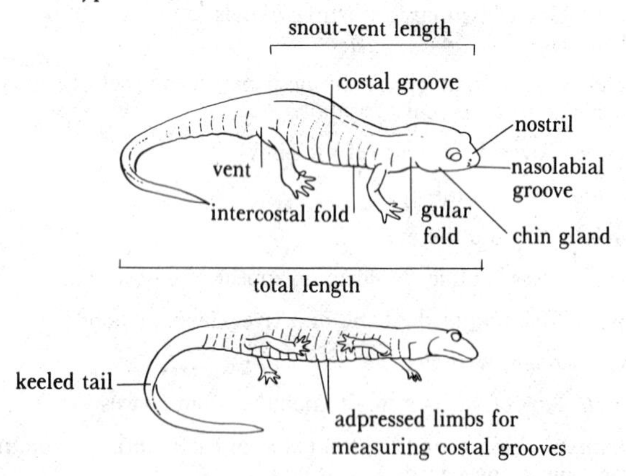

Measure the animal if possible, because size or body proportions may be important. The lengths given in the descriptions are the straight line *distances from the tip of the snout to the posterior edge of the vent.* Because the tail is often broken, the total length is not a reliable measurement. Amphibians grow continuously throughout life, although rapid growth occurs only in the first few years. Sizes, therefore, are the *maximum* size of any individual known, and in cases where the record size far exceeds that of most adults, both record and average are given.

The identification of salamanders provokes dissatisfaction even among herpetologists. One problem is the dearth of important distinguishing characteristics to measure and count, such as the useful scale patterns in reptiles. Many obvious differences in color or pattern are due to variations within one species, well illustrated in humans by variation in hair color. Other problems are the small size and secretiveness of salamanders, making it hard to observe all features without an examination in the hand.

Perhaps the worst problem in identification is one experienced by the salamanders themselves. At times they cannot tell who belongs to their own species when choosing a mate and they produce hybrid offspring derived from parents of two different, but usually closely related, species. Hybrids are often less likely to survive and reproduce themselves, but in certain areas they are successful.

Habitat and Habits
Salamanders prefer a moist habitat and avoid direct sunlight. Some spend their entire lives in water; some spend most of their lives on land. Some are arboreal, others are cliff dwellers; some, usually blind, dwell in caves or in deep wells. Salamanders live from near sea level to an altitude of over 13,000 feet (3962.4 m). Where the winters are cold they hibernate, but they may be active underground. Unlike snakes and lizards, however, they prefer cool weather to warm, and some are active in icy water close to freezing.

Look for terrestrial salamanders in the spring and fall, particularly at or near the pools where they breed. After breeding they usually return to their hiding places under rocks, logs, bark, leaf litter, and debris left lying by human activities. Specific habitats of each species are listed in the descriptions, but the successful salamander hunter will overturn all cover, and then, of course, carefully replace it in its original position. Most species are nocturnal.

Food
Salamanders are entirely carnivorous, and among other items will feed on aquatic or terrestrial insects, or other small invertebrates and their eggs and larvae, and on small vertebrates such as salamander larvae. In captivity they will eat earthworms, beef, or other meat, and insects when these are available.

Reproduction
When male and female encounter each other during the mating season they usually go through an elaborate and species-characteristic courtship ritual accompanied by tactile and chemical sensory cues. You might observe this behavior in captive animals, given the proper sexes, the mating season, and no distractions from the observer. Fertilization is internal, except in the Hellbender, whose male fertilizes the eggs as they are released from the body of the female. In a few species the sperm is transferred to the female through the touching of vent to vent, but in most species after the courtship dance the male will deposit small spermatophores that contain sperm in a little sac on top of a jellylike pyramid. The female follows the male and picks up these sperm sacs with the sides of her vent and introduces them into her spermatheca, a specialized storage receptacle where sperm may be stored for several years. One clutch of eggs laid by the female can potentially be fertilized by sperm from two different males. It should be noted that mating and egg laying may not occur in the same season.

Eggs are deposited singly, in strings, or in masses either in the water or on land, where they are suspended from the roof of a moist retreat. The eggs may be brooded by the female. In due time, depending on the temperature, the eggs in aquatic situations hatch into larval salamanders with gills. Lower ambient temperatures will decrease the growth rate in these stages. The length of the larval period varies from a few days to over a year or two. In terrestrial species the larval period is passed within the egg, and upon hatching the salamander has only rudimentary gills, if any, which are shed shortly after birth. Another adaptation, neoteny, allows salamanders to spend their entire lives with gills and other larval features while coming to reproductive maturity. This last strategy is often found in an inhospitable or underground environment where the perils of transformation are too great.

SIRENS

Family Sirenidae

Sirens are large aquatic salamanders with three pairs of external gills and forelimbs of moderate size. Hind limbs are entirely lacking.

LESSER SIREN

Siren intermedia

Description
Length, to 27 in. (68.6 cm). Has no hind limbs and 3 pairs of external bluish gills. Body long, slender. Above dark brown, black, olive, or gray flecked with small black spots; below lighter and unspotted. Costal grooves 31–34 in area of overlap with Greater Siren, but 38 farther west.
Similarities
Greater Siren is generally larger.
Habitat
Shallow, warm, quiet ponds; ditches; cypress swamps and sometimes rivers and streams; under stones, logs, in matted vegetation.
Habits
Aestivates in cocoon when water dries up.
Remarks
May emit a distress cry when seized.
Range
Coastal Plain from se. N.C. to cen. Fla., w. to e. Tex. and into Mexico, and n. in Miss. Valley through cen. Ill. and sw. Mich.

GREATER SIREN

Siren lacertina **72:2**

Description
Length, to 30 in. (76.2 cm); record 38½ in. (97.8 cm). Distinguished by its large size, absence of hind legs, and 3 pairs of external greenish gills. Dorsum olive to light gray, lighter sides blotched with greenish-yellow. Venter paler with small flecks or mottling of greenish-yellow. Snout yellow or light brown. Costal grooves 36–39. Young darker.
Habitat
Shallow roadside ditches, under rocks in streams, weedy ponds and pools, muddy swamps.
Habits
Nocturnal; may be found during day under objects in water, or burrowed in mud, or in dense vegetation.
Range
Coastal Plain from e. Va. and adj. Md. through Fla. to s. Ala.

Note: The **DWARF SIREN,** *Pseudobranchus striatus*, has brown and yellow stripes on its body. It occurs from extreme southern South Carolina southward through Florida.

AMPHIUMAS

Family Amphiumidae

Large eellike aquatic salamanders with small eyes, and four tiny limbs, without external gills and eyelids. Graceful in water but clumsy on land. Females lay eggs in strings under logs or rocks near water's edge or in water, then remain coiled around them through incubation period. CAUTION: Amphiumas will bite viciously when handled.

THREE-TOED AMPHIUMA

Amphiuma tridactylum *Fig. 86*

Description
Length, to 30 in. (76.2 cm); record 41¾ in. (106.1 cm). Three toes present on most of the tiny limbs. Has distinctly bicolored body, somewhat lighter below. Other distinguishing features are dark throat patch and oval outline of tail base in cross section; boundary between lighter ventral and dark dorsal coloration sharply defined.
Habitat
Lowland streams, pools, swamps, and cypress sloughs; also in water in nearby hilly regions.
Reproduction
Mates December to June; eggs found April to September in southern part of range, and August through midwinter in Tennessee.
Remarks
Mud Snake and sometimes Cottonmouth are major predators.
Range
Miss. Valley from extr. w. Ky. and se. Mo., s. through w. Ala., and w. to extr. se. Okla., and e. Tex.

Fig. 86

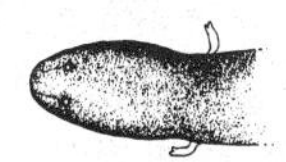

Two-toed Amphiuma

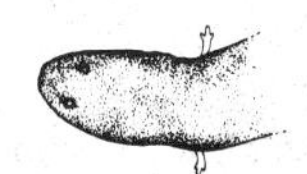

Three-toed Amphiuma

TWO-TOED AMPHIUMA

72:1

Amphiuma means *Fig. 86*

Description
Length, to 30 in. (76.2 cm); record 45¾ in. (116.2 cm). Has dark overall coloration and only 2 toes on the four tiny limbs. Body dark brown to slate with belly only slightly lighter than dorsum, color change gradual from side to below; no obvious throat patch. Tail keeled above and rounder in cross section at base.
Similarities
Three-toed Amphiuma is distinctly bicolored, with somewhat lighter venter, and has 3 toes on most of its tiny limbs.
Habitat
Lowland swamps, bayous, along sloughs, in drainage ditches, muddy pools, sluggish streams, and wet meadows.
Habits
Lives and feeds in crayfish burrows, often deep in the soil.
Remarks
Major predators are mud and rainbow snakes.
Range
Coastal Plain from se. Va., s. through Fla., and w. to extr. e. La.

HELLBENDER AND MUDPUPPIES

Families Cryptobranchidae and Necturidae

All are aquatic forms with four short, stout legs and small eyes. The family Cryptobranchidae is represented in the United States by the Hellbender, which has a wrinkled skin and prominent lateral fold and lacks external gills. Its closest relatives live in Japan and China, where the Japanese Giant Salamander reaches over five feet (1.5 m) in length and is considered a great delicacy.

Members of the family Necturidae, the mudpuppies and waterdogs, have three pairs of dark red external gills. They are restricted to North America.

HELLBENDER

Cryptobranchus alleganiensis **72:4**

Description
Length, to 20 in. (50.8 cm); record 29⅛ in. (74.0 cm). Large, aquatic, with 1 pair of circular gill slits in the adult; has loose, wrinkled skin with prominent lateral fold. Legs short, stout; toes 4 on front- and 5 on hind-limbs. Head broad, flat; snout rounded. Color brownish, sometimes with darker blotches, especially in Ozark region.

Habitat
Rocky, fast-flowing waters of large streams and rivers under 2500 ft. (762.0 m) elevation.

Habits
Forages on bottom of water at night.

Reproduction
Up to 450 eggs laid in strings in saucer-shaped excavation below rock under water. Male guards nest, but cannibalism of eggs and larvae documented.

Range
Susquehanna R. drainage and w. N.Y. and Pa., s. to n. Ga., w. to s. Ill. along Ohio R. and Tenn. R. drainages, and to extr. ne. Miss. Disjunct populations in tributaries of Mo. R. and Black R. in Mo. and Ark.

DWARF WATERDOG

Necturus punctatus

Description
Length, to 7 7/16 in. (18.9 cm). Small, slender. Color deep brown to gray, unspotted, or nearly so; throat white; belly bluish-white. Inconspicuous dark line passes through eye. Has external feathery red gills. Head and snout depressed; 4 toes on all 4 feet.

Habitat
Slower parts of streams near sandy or muddy banks and in ditches; hidden in bottom debris.

Range
Atlantic Coastal Plain from se. Va. to cen. Ga.

Note: The **ALABAMA WATERDOG,** *Necturus alabamensis,* occurs in west-central Georgia and the Florida panhandle to eastern and southwestern Mississippi. It is a variable reddish-brown and is distinguished by an unmarked central belly area.

MUDPUPPY

Necturus maculosus **72:3**

Description
Length, to 17 in. (43.2 cm). Color dark olive or grayish-brown covered with blue-black spots. Has 4 short, stout limbs, each with 4 toes, and 3 pairs of external feathery red gills retained throughout life. Body and head flattened; tail compressed from side to side, keeled. Males in breeding season have swollen ridge around cloaca. Young have dark back bordered by narrow yellow line.
Habitat
Strictly aquatic; in clear or muddy streams, bayous, polluted canals, ditches, and reservoirs.
Habits
Active all year.
Food
Fish and fish eggs, also crayfish, insects, insect larvae, snails, and other amphibians.
Reproduction
Mates in fall; eggs laid singly, stuck to underside of an object in water.
Range
Most of ne. U.S. and se. Canada from s. Que. and nw. Vt., s. to n. Ga., and w. around most of Great Lakes region to se. Man., s. to e. Okla. and n. La. Introduced in New England.

MOLE SALAMANDERS

Family Ambystomatidae

Large, stout-bodied salamanders distinguished by a lack of nasolabial grooves, presence of costal grooves, and lack of external gills in adults. Four sturdy limbs are capable of digging. Larvae have bushy external gills and a keel on the back and tail. Adults terrestrial except during the breeding season, and most frequently encountered migrating across roads on warm, rainy nights. Two species are terrestrial fall breeders and brooders; the remainder breed in temporary or semipermanent ponds that lack fish early in the spring. Males have swollen vents during the breeding season. Food includes worms and arthropods.

SPOTTED SALAMANDER

Ambystoma maculatum **72:9**

Description
Length, to 9¾ in. (24.8 cm). Two rows of large, round, yellowish spots on black, bluish-black, slate, or brown back. Spots somewhat irregular and extending from behind eyes to tip of tail; rarely cream; first pair may be orangish. Ventral surface slate-gray, although sometimes lighter. Costal grooves usually 12 (11–13); tail at least ½ length of body.
Habitat
Hardwood and mixed deciduous forests.
Habits
Breeds in early spring; commonly found over its range throughout summer.
Range
Most of e. U.S. and se. Canada except absent from s. N.J., s. Delmarva Pen., most of Fla. and s. La. Western boundaries in Wis., forested parts of Ill., and se. Kans. s. to ne. Tex.

MARBLED SALAMANDER

Ambystoma opacum **72:10**

Description
Length, to 5 in. (12.7 cm). Shiny black, with irregular white bands, giving it a marbled appearance. Body thick, short; legs short, stout; tail about 40% of total length, tapered and somewhat compressed. Light dorsal markings variable but usually bands meeting laterally and enclosing black spots along back. In some individuals crossbands are lacking or incomplete, or lateral markings are interrupted. Light bands brighter in male than in female. Below black.
Habitat
Under bark, logs, stones, or rubbish, along margins of ponds, streams, and swamps during breeding season; at other times in dry and shady areas away from water.
Reproduction
Breeds in fall; terrestrial breeding and brooding of eggs. Adults do not become aquatic.
Range
Most of e. U.S. from se. N.H. to n. Fla., w. to s. Ill., se. Okla., and e. Tex. Disjunct populations on Lake Michigan border in se. Wis. and Ind.-Mich. area.

RINGED SALAMANDER

Ambystoma annulatum **72:12**

Description
Length, to 9¼ in. (23.5 cm). Slender and elongate. Color blackish-brown, with 4–7 crossbands or rings of yellow. Rings on body may be incomplete above, resembling spots; light line below at side; rings never extend onto belly. Below slate-gray, with or without white dots. Costal grooves 15; tips of adpressed limbs separated by 4 costal grooves.
Habitat
Under logs in mud; in tunnels beneath brush and leaves, in fields near ponds, swamps, marshes.
Reproduction
Breeds following first heavy rain in the fall; lays eggs in loose masses under water.
Range
On Ozark Plateau and in Ouachita Mts. in se. and cen. Mo., nw. Ark., and e. Okla.

TIGER SALAMANDER

Ambystoma tigrinum **72:7**

Description
Length, to 8½ in. (21.6 cm), record 13 in. (33 cm). Large, dark, with light, irregular dorsal blotches, bars, or reticulations above, and yellowish with black marbling below. Body stout, head broad, legs stout and long, eyes small. Tail laterally compressed. Geographic variation in body pattern and color of spots; yellow blotches sometimes tinged with olive or orange, may form bands on lower sides and tail. Costal grooves 14 or fewer.
Habitat
A spring breeder often encountered at night or during rains traveling to breeding sites of permanent shallow lakes, ponds, ditches, or pools of rivers.

Reproduction
Eggs laid singly or in masses of up to 110.
Remarks
Neoteny, or the retention of some larval features into adult life, is common in many western populations where Tiger Salamanders retain their gills and aquatic life style.
Range
E. U.S., exclusive of New England and the Appalachian region, from s. N.Y. to cen. Fla., w. through Great Plains of Canada and U.S., and s. into e. Mexico. Rare in La. and adj. areas. (Also a disjunct population in Calif.)

JEFFERSON SALAMANDER

Ambystoma jeffersonianum **72:11**

Description
Length, to 8¼ in. (20.9 cm). Long-limbed. Color dorsally dark brown, brownish-gray, or lead-color with lighter belly; sometimes scattered blue flecks on limbs and lower sides; area around vent usually gray. Tips of adpressed limbs meet or overlap by 1–5 intercostal folds.
Similarities
Blue-spotted Salamander has shorter limbs and toes and more flecks. Female Jefferson Salamanders are similar to Silvery Salamander, and positive identification can be made only in the laboratory.
Habitat
Eastern deciduous forest.
Remarks
This is a bisexual diploid species (that is, both males and females and a normal set of paired chromosomes are present—in this case 2x14, or a total of 28).
Range
Se. Vt., s. to n. N.J., and through n. Va.; w. from s. of Lake Erie to s. Ind. and cen. Ky.

SILVERY SALAMANDER

Ambystoma platineum

Description
Length, to record 7⅞ in. (20 cm). Coloration much like Jefferson Salamander's, but bluish flecks are scattered on dorsum, and may have larger spots on venter and dorsolateral surface. Area around vent usually gray.
Habitat
Eastern deciduous forest; in ponds or under logs, stones, or debris in moist areas.
Remarks
Although similar in appearance to Jefferson Salamander females, this is a triploid species (one extra set of chromosomes is present in the cells, making a total of 3x14, or 42), composed almost entirely of females.
Range
Within range of Jefferson Salamander in s. Mich. through s. Ohio, e. Ind. to extr. n. Ky.; also in n. N.J. and w. Mass.

BLUE-SPOTTED SALAMANDER

Ambystoma laterale

Description
Length, to 5⅛ in. (13 cm). Black or grayish-black dorsum with numerous large bluish-white flecks covering body, especially along lower sides. Venter slightly lighter than dorsum. Area around vent is black. Limbs shorter than in Jefferson Salamander, which may be confused with it, but the adpressed limbs barely meet, if at all, or at most overlap by 1 intercostal fold.

Similarities
Shares same relationship with Tremblay's Salamander as does Jefferson Salamander with Silvery Salamander. Positive identification can be made only in the laboratory.

Habitat
Eastern deciduous forest.

Remarks
This is a bisexual, diploid species (see under Jefferson Salamander).

Range
Atlantic Coast from s. Que., s. to N.Y., w. through Canada to se. Man., and in U.S. in extr. nw. Ohio, n. Ind., and ne. Ill., nw. to ne. Minn. Disjunct localities in Labrador; Long Island, N.Y.; ne. N.J.; and e. Iowa.

TREMBLAY'S SALAMANDER

Ambystoma tremblayi

Description
Length, to record of 6⅜ in. (16.2 cm). Dorsum dark gray to gray-black with lighter venter. Bluish-white markings on body most numerous along lower sides. Area around vent is black.

Similarities
Resembles female Blue-spotted Salamander. Light markings more diffuse than in Silvery Salamander.

Habitat
Eastern deciduous forest.

Remarks
Members of this triploid species (see under Silvery Salamander) are probably all female.

Range
Within range of Blue-spotted Salamander, known from scattered localities in N.S., Maine, Mass., s. Que., and from nw. Ohio and ne. Ind. n. through Mich. and ne. Wis.

SMALL-MOUTHED SALAMANDER

Ambystoma texanum **72:6**

Description
Length, to 7 in. (17.8 cm). Medium-sized, recognized by slender head and small mouth. Dorsum brownish-black or gray with lighter buff or grayish lichenlike blotches. Below paler, sometimes with light markings. Dark, unspeckled individuals may occur in northeastern part of range. Costal grooves 14 or 15; adpressed limbs separated by 2½ or 3 intercostal folds.

Habitat
Very diverse, from tall-grass prairie and farmed areas to woodland and dense hardwood forest; beneath logs or rocks, or buried in moist soil.

Habits
Breeds in early spring in temporary ponds or spring-fed streams and pools, but not in ponds with other mole salamanders.
Range
E. Ohio and se. Mich. to extr. se. Nebr., s. to Gulf from w. Ala. to e. Tex., but not in the Ozark Plateau or the Ouachita Mts.

MOLE SALAMANDER

Ambystoma talpoideum **72:8**

Description
Length, to $4\frac{13}{16}$ in. (12.2 cm). Small, stout; very large, broad, flattened head. Dorsum of adults from dark gray to blue-black or brown, flecked with light gray or blue-gray markings. Below blue-gray with tiny white flecks.
Habitat
Woodlands at low elevation; adults live in burrows or under logs, debris, and leaf litter, often around a breeding pond.
Range
Atlantic Coastal Plain from cen. S.C. through n. Fla., w. to e. Tex., and n. in Miss. Valley to s. Ill. Disjunct populations in w. N.C., e. Tenn., and se. Okla.

Note: The **MABEE'S SALAMANDER,** *Ambystoma mabeei*, is distinguished by light flecks on the sides only. It occurs in the coastal Carolinas.

FLATWOODS SALAMANDER

Ambystoma cingulatum *Fig. 87*

Description
Length, to $5\frac{1}{16}$ in. (12.9 cm). Recognized by slender body, small head, and a black to brown dorsum with gray reticulations. Some individuals may lack reticulations. Ventral coloration dark with pattern of numerous light flecks or larger spots.
Habitat
Pine flatwoods. Adults often found beneath logs near cypress swamps.
Habits
One of two land-breeding mole salamanders.
Reproduction
Breeding takes place before mid-February, and when rains fill up depression where eggs are laid, larvae get head start on development.
Range
Atlantic Coastal Plain from S.C. to Okefenokee Swamp, Fla., to se. Miss.

Fig. 87

Flatwoods Salamander

NEWTS

Family Salamandridae

In North America two groups of this widespread family occur, one over most of the East, the other confined to the Pacific Coast. Other members of the family live in eastern Asia and Japan, Europe, and North America. These brightly colored salamanders make attractive pets and in captivity feed on live insects, earthworms, and pieces of liver or other meat. Newts lack nasolabial grooves and costal grooves, and generally have thick, warty skins. Poisonous substances may be secreted from glands in the skin when these salamanders are handled.

NEWT

Notophthalmus viridescens **72:5**

Description
Length, to 5½ in. (14.0 cm). This salamander may be encountered in three stages. *Aquatic adult:* Skin smooth to finely granular, yellowish-green above with red spots, below yellow with many scattered black spots; tail strongly keeled. Vent of male enlarges during breeding season. *Immature terrestrial eft:* Length, to 3¼ in. (8.3 cm); skin warty and rough, orange-red above with red spots, yellow to orange below. *Larva*: With gills.

Habitat
Adult found in ponds, small lakes, marshes, swamps, quiet pools of streams, especially where vegetation is abundant. Eft frequents wooded areas in moist places.

Habits
Larva with gills is aquatic; after transformation it emerges onto land, where it remains as an eft for 2–3 years, then re-enters water and becomes permanently aquatic. Some newts omit the red eft stage.

Reproduction
Eggs laid singly on aquatic vegetation.

Range
Se. Canada and e. U.S., w. to n.-cen. Minn. and Lake Superior region in the North and to se. Kan. and e. Tex. in the South.

Note: The **STRIPED NEWT,** *Notophthalmus perstriatus*, has a continuous red dorsolateral stripe. It occurs in southern Georgia and northern Florida.

LUNGLESS SALAMANDERS

Family Plethodontidae

These salamanders have costal grooves and a groove from nose to lip. They lack lungs, relying on skin and mouth for respiration. Those east of the 100th meridian usually have five toes on the hind feet (with two species having four), and four toes on the front feet. This is the most diverse family of salamanders, and except for two species confined to limestone areas in southern Europe, the family is restricted to the Americas. The eastern United States has the greatest generic diversity; however, the greatest number of species occurs in Central and South America. Reproduction in most species is terrestrial, and eggs are laid in secluded places on land. They develop directly into miniature adults without going through an aquatic larval stage in most cases, although the dusky salamanders and other groups have larvae. A salamander's larval stage is often subject to the heaviest predation, and in salamanders with terrestrial egg laying the average number of eggs laid per clutch or year is much smaller than in those with aquatic egg laying because fewer are lost.

DUSKY AND SHOVEL-NOSED SALAMANDERS

Genera *Desmognathus* and *Leurognathus*

This group of salamanders is confined to the eastern United States. They all have a light line from eye to angle of the jaw, but the line in the Shovel-nosed Salamander is not well defined. The head points downward because the muscles above and behind it are quite enlarged. Unlike most other vertebrates, these salamanders have little mobility of the lower jaw, and the skull is raised to open the mouth. (These species may be rendered incapable of eating if their mouths are opened incorrectly or too roughly, so they should be treated with care when undergoing field identification.) The tongue is a pad attached in front. Most species are aquatic, but the smaller species have become increasingly terrestrial both in habitat and breeding sites. Food consists mainly of aquatic arthropods, but larval and small salamanders are also eaten.

PIGMY SALAMANDER

Desmognathus wrightì **70:2**

Description

Length, to 2 in. (5.1 cm). The smallest species of *Desmognathus*, with dark herringbone pattern on light dorsal band, which is light tan to reddish and with dark borders; belly flesh-colored. Tail round, slender, tapering.

Habitat

Spruce-fir forest at elevations of 2750 ft. (838.2 m) and up; under stones, bark, moss, and leaf litter.

Habits

Purely terrestrial; may overwinter in seepages and springs.

Reproduction

Eggs develop into miniature adults.

Range

High altitudes from Whitetop Mt., Va., s. along Tenn.-N.C. border.

Note: The **SEEPAGE SALAMANDER,** *Desmognathus aeneus*, occurs in the extreme south of the range of the Pigmy Salamander. It too is tiny, and can be distinguished by its kidney-shaped gland under the chin, smaller than that of the Pigmy Salamander.

MOUNTAIN DUSKY SALAMANDER

Desmognathus ochrophaeus **71:2**

Description

Length, to 4⅜ in. (11.1 cm). Small, with a straight-edged (or wavy in the South), light dorsal band from head onto tail and a round tail in cross section. Body slender, tail slim, tapering. *Light phase:* Color very variable—tan, gray, red, brown; within dorsal band there may be a row of dark spots. Infrequently, reddish patches on legs and/or cheeks. Sides blackish with tiny white spots (sometimes missing); below bluish-gray (darker than in Dusky Salamander). *Dark phase:* Uniformly black above, dark below.

Habitat
Moist woodland often near streams, under logs, rocks, and leaf litter, even under bark of trees. This species is not tied to streams as are most members of the genus.
Habits
Nocturnal.
Range
Mostly in higher elevations from cen. and s. N.Y., s. through n. and w. Pa. to Ga., w. to e. Ohio, e. Ky., and extr. e. Tenn. Disjunct population in ne. Ala.

DUSKY SALAMANDER

Desmognathus fuscus **71:3**

Description
Length, to 5⁹⁄₁₆ in. (14.1 cm). A tail keeled dorsally immediately behind vent and oval in cross section, and dorsum plain or blotched, along with bright and speckled lower sides, help distinguish this confusing species. Costal grooves 14; 2–5, usually fewer than 4, intercostal folds between adpressed limbs. Back tan to dark brown or green; plain, or commonly with 6–7 pairs of light blotches, often forming a stripe down back with somewhat wavy edges. Color of stripe variable. Below cream with brown blotches or spots. Edges of jaw in males not as sinuous as in Mountain Dusky Salamander.
Habitat
Shallow water under rocks in debris at stream edge, and in seepage areas, swamps.
Habits
Spends much time in water; active year-round but burrowed below frozen ground in gravel or limestone; mostly nocturnal; a good jumper when pursued.
Food
Terrestrial and aquatic insects, insect larvae, and worms.
Remarks
See Southern Dusky Salamander.
Range
New England and adj. Canada, s. through Appalachian region to Gulf Coast from w. Fla. to Miss. R. Absent from s. N.J. and Coastal Plain of se. Va. to cen. Fla. Populations w. of Miss. R. in ne. Ark., n.-cen. La., and adj. Ark.

SEAL SALAMANDER

Desmognathus monticola **71:5**

Description
Length, to 5⅞ in. (14.9 cm). Color and markings variable, light buff to grayish-brown above with many strong, dark-bordered pale blotches; belly pale to yellow and plain; lower sides mottled, often with a row of white spots from armpit to groin. Eyes large; limbs strong; tail short and compressed, keeled above.
Habitat
Under bark, logs, stones in or near mountain brooks or shallow muddy streams. Cool hemlock-shaded ravines on damp Appalachian mountainsides are favorite haunts.
Range
Sw. Pa., s. through the Appalachians to Ga. and Ala. Populations in s. Ga. and adj. w. Fla.

BLACK MOUNTAIN SALAMANDER

Desmognathus welteri **70:12**

Description
Length, to $6\frac{11}{16}$ in. (17 cm). Large, stout, similar to Seal Salamander, but with mottled body, small, dark dorsal blotches, and black on tips of toes. Black or brown blotches over back of lighter brown. Change of coloration from back to belly gradual.
Habitat
Occurs only at higher elevations; found under stones in or near brooks, springs, and roadside ditches in wooded areas.
Range
Black Mountain in Harlan Co., Ky., and nearby areas in w. Va., and e. Ky. Also in Warren Co. in cen. Ky.

SOUTHERN DUSKY SALAMANDER

Desmognathus auriculatus *Fig. 88*

Description
Length, to 6⅜ in. (16.2 cm). Generally darker than Dusky Salamander, with white or reddish spots prominent along sides in 1 or 2 rows, and below dark with white flecks; dorsally black with traces of blotches, or patternless. Usually more than 4 intercostal folds between adpressed limbs. Tail keeled above and compressed from side to side.
Habitat
Around springs, swamps, cypress heads, muddy streams, and pools in floodplains below Fall Line.
Habits
Often found in burrows.
Remarks
This species may hybridize with the Dusky Salamander where the two occur together in the South, and animals with intermediate characteristics may be found.
Range
Coastal Plain from extr. se. Va., s. to cen. Fla., and w. to se. Tex.

Fig. 88 Southern Dusky Salamander

BLACK-BELLIED SALAMANDER

Desmognathus quadramaculatus **70:1**

Description
Length, to 8¼ in. (21.0 cm). Large, with a dark or black belly. Tail stout, less than ½ total length and sharply keeled dorsally. Upperparts blackish with greenish-yellow blotches, sometimes with rusty dorsal band; small light spots in 2 rows on sides from armpit to groin.
Habitat
Along mountain streams above 2500 ft. (762.0 m), beneath stones and logs in or by water. May be introduced at lower elevations by fishermen releasing bait.
Habits
Highly aquatic, but in south of range may live on rock faces; nocturnal.
Range
S. Appalachian Mts. from s.-cen. W. Va., s. to nw. S.C. and n. Ga.

SHOVEL-NOSED SALAMANDER

Leurognathus marmoratus **70:6**

Description
Length, to 5¾ in. (14.6 cm). Resembles the Dusky Salamanders, but can be distinguished in the hand by opening its mouth. It lacks *round* nostrils, visible near front of the roof of mouth on each side. White stripe from eye to angle of jaw is obscure. Color and pattern variable; back usually dark brown to blackish with two rows of light-colored blotches that are dull and blend into habitat. Snout often darker than body. Two rows of spots along side of body; below grayish with a light central area. Tail long and keeled.
Habitat
Found in mountain brooks and streams with sandy or gravelly bottoms and rocks to hide under. Often occurs in same places where the Black-bellied Salamander is found.
Habits
Strictly aquatic.
Range
Elevations over 1000 ft. (304.8 m) in s.-cen. Va., s. along the N.C. and Tenn. borders into extr. nw. S.C. and ne. Ga.

WOODLAND SALAMANDERS

Genus *Plethodon*

A high number of costal grooves, long and slender bodies, rounded tails, and pronounced gular folds characterize the Woodland Salamanders. These salamanders are very abundant in mostly terrestrial places—in leaf mold and under rotting logs and rocks—in the forests of the eastern United States, and other members of the genus occur in the Pacific Northwest and in New Mexico. The space needed for a salamander to crawl under an object is quite small, and there are always other burrowing animals which help make pathways. The eggs are usually suspended from above in logs or from the stone, which arrangement seems to result from gyrations of the female salamander as she lays them. Presumably, getting them off the ground helps to protect them from the elements and infections. Females may guard the clutch of eggs. Embryos and newly hatched individuals are hard to tell apart.

SOUTHERN RED-BACKED SALAMANDER

Plethodon serratus

Description
Length, to 4 in. (10.2 cm). Costal grooves 18–20. Edges of dorsal stripe often serrated. Red often present on anterior belly and also on back in the infrequent unstriped variant. Chin light. Check Red-backed Salamander for description of the two color phases. Rarely has individuals of the lead-backed color phase. Males have large gland under chin.
Similarities
Zigzag Salamander has dorsal stripe with wavy edges, but fewer costal grooves (normally 18).
Habitat
Under logs and stones in forests and also in fields; much as for Red-backed Salamander, but often at higher elevations.
Range
Five disjunct localities—the s. Appalachians in extr. se. Tenn. and sw. N.C.; the Piedmont of w. Ga. and adj. Ala.; se. Mo.; uplands of se. Okla. and w.-cen. Ark.; and cen. La.

Plates

Note on the Bird Plates

On each of the bird plates following (Plates 1 through 48) each species is designated by a different number. These guidelines will be of help in using the numbering system:

1. When a simple number (**1, 2, 3,** etc.) appears with an illustration, the bird depicted is either an adult male (designated ♂) or an adult female (designated ♀); in many instances, there are no immediately visible distinctions between males and females.

2. When the number is followed by a letter (**1a, 1b,** etc.) the bird depicted is a variant form. The major variants are immature or juvenal plumages, geographical races or subspecies, or color phases.

PLATE 1
LOONS AND GREBES

1 Common Loon, **1a** summer, **1b** winter, p. 10. **2** Arctic Loon, **2a** summer, **2b** winter, p. 10. **3** Red-throated Loon, **3a** summer, **3b** winter, p. 11. **4** Red-necked Grebe, **4a** summer, **4b** winter, p. 12. **5** Western Grebe, p. 13. **6** Horned Grebe, **6a** summer, **6b** winter, p. 12. **7** Eared Grebe, **7a** winter, **7b** summer, p. 12. **8** Pied-billed Grebe, **8a** winter, **8b** summer, **8c** immature, p. 13.

PLATE 2
PELAGIC BIRDS

1 Parasitic Jaeger, **1a** light phase immature, **1b** light phase, **1c** dark phase, p. 91. **2** Pomarine Jaeger, **2a** light phase, p. 91. **3** Long-tailed Jaeger, **3a** light phase, p. 92. **4** Sooty Shearwater, p. 15. **5** Greater Shearwater, p. 15. **6** Leach's Storm-Petrel, p. 16. **7** Northern Fulmar, **7a** dark phase, **7b** light phase, p. 14. **8** Cory's Shearwater, p. 14. **9** Wilson's Storm-Petrel, p. 16. **10** Manx Shearwater, p. 15.

PLATE 3
PELICANS, CORMORANTS, AND ALLIES

1 Magnificent Frigatebird, **1a** immature, p. 20. **2** Anhinga, **2a** breeding, p. 20. **3** Gannet, **3a** subadult, **3b** immature, p. 18. **4** Brown Pelican, **4a** immature, p. 17. **5** Double-crested Cormorant, **5a** immature, **5b** breeding, p. 19. **6** Great Cormorant, **6a** immature, **6b** breeding, p. 18. **7** Neotropical Cormorant, **7a** breeding, p. 19.

PLATE 4
HERONS

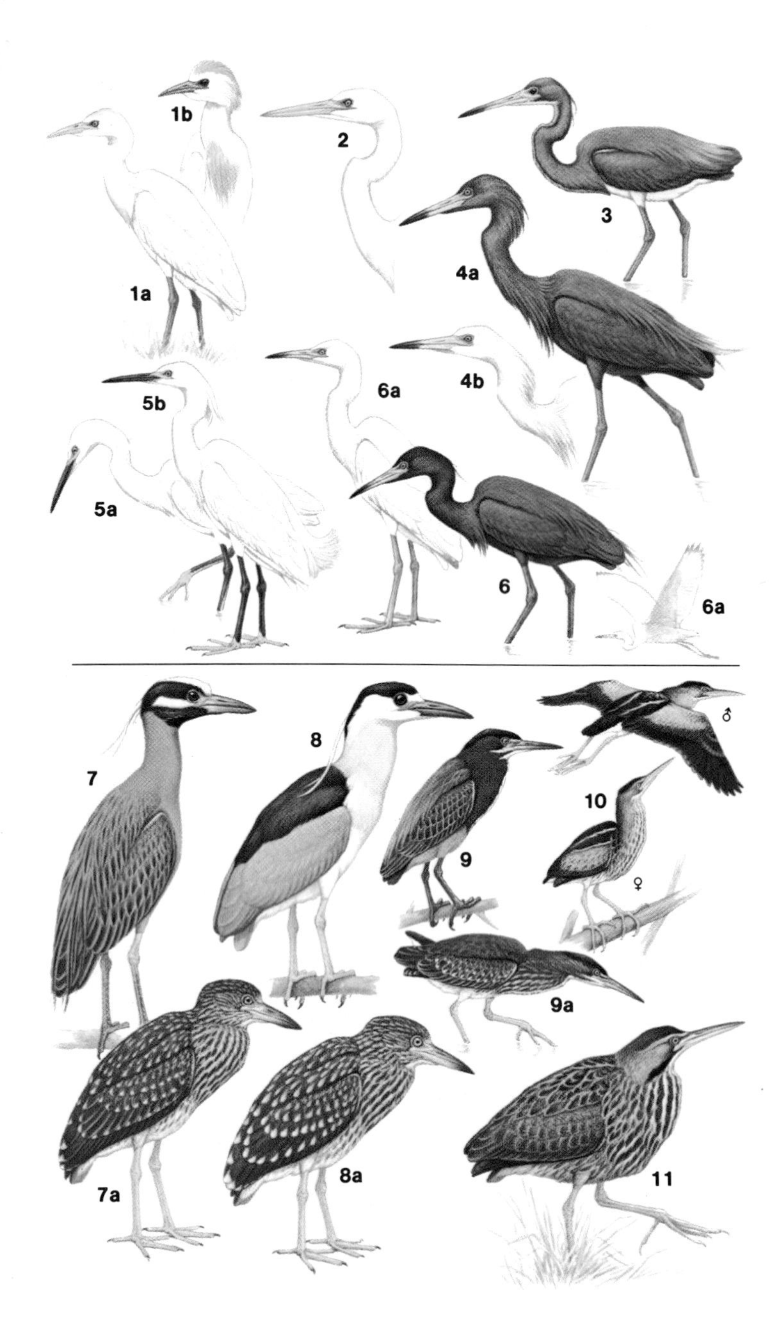

1 Cattle Egret, **1a** immature, **1b** breeding, p. 24. **2** Great Egret, p. 21. **3** Louisiana Heron, p. 23. **4** Reddish Egret, **4a** dark phase, **4b** white phase, p. 23. **5** Snowy Egret, **5a** immature, **5b** breeding, p. 22. **6** Little Blue Heron, **6a** immature, p. 22. **7** Yellow-crowned Night Heron, **7a** immature, p. 24. **8** Black-crowned Night Heron, **8a** immature, p. 24. **9** Green Heron, **9a** immature, p. 21. **10** Least Bittern, p. 25. **11** American Bittern, p. 25.

PLATE 5
HERONS, IBISES, STORK, CRANES

1 Great Blue Heron, **1a** breeding, **1b** "Great White" form, p. 21.
2 Great Egret, **2a** breeding, p. 21. **3** Sandhill Crane, p. 65.
4 Whooping Crane, p. 65. **5** White Ibis, **5a** immature, p. 27.
6 Wood Stork, p. 26. **7** Limpkin, p. 66. **8** Glossy Ibis, **8a** breeding,
8b immature, p. 26. **9** Roseate Spoonbill, **9a** immature, p. 28.

PLATE 6
GEESE AND SWANS

1 Snow Goose, **1a** "White" form, **1b** "White" form immature, **1c** "Blue" form, **1d** "Blue" form immature, p. 31. **2** White-fronted Goose, **2a** immature, p. 31. **3** Brant, p. 30. **4** Whistling Swan, **4a** immature, p. 29. **5** Canada Goose, **5a** large race, **5b** small race, p. 30.

PLATE 7
POND DUCKS

1 Wood Duck, p. 36. **2** Fulvous Whistling-Duck, p. 31.
3 Eurasian Wigeon, p. 35. **4** American Wigeon, p. 35. **5** Common Teal, p. 34.
6 Northern Shoveler, p. 36. **7** Blue-winged Teal, p. 34. **8** Gadwall, p. 33.
9 Black Duck, p. 32. **10** Mallard, **10a** "Mottled" race, p. 32.
11 Northern Pintail, p. 33.

PLATE 8
POND DUCKS IN FLIGHT–FEMALES

1 Black Duck, p. 32. **2** Mallard, p. 32. **3** Gadwall, p. 33.
4 Northern Pintail, p. 33. **5** Common Teal, p. 34.
6 Blue-winged Teal, p. 34. **7** Northern Shoveler, p. 36.
8 American Wigeon, p. 35. **9** Wood Duck, p. 36.

PLATE 9
BAY AND SEA DUCKS IN FLIGHT—FEMALES

1 Redhead, p. 37. **2** Canvasback, p. 39. **3** Greater Scaup, p. 38.
4 Lesser Scaup, p. 38. **5** Common Goldeneye, p. 39. **6** Bufflehead, p. 40.
7 Hooded Merganser, p. 44. **8** Red-breasted Merganser, p. 45.
9 Oldsquaw, **9a** winter, p. 41. **10** King Eider, p. 42.
11 White-winged Scoter, p. 43. **12** Surf Scoter, p. 43.

PLATE 10
BAY AND SEA DUCKS

1 Barrow's Goldeneye, p. 40. **2** Bufflehead, p. 40. **3** Common Goldeneye, p. 39. **4** Harlequin Duck, p. 41. **5** Oldsquaw, **5a** winter, **5b** summer, p. 41. **6** Black Scoter, p. 44. **7** Surf Scoter, **7a** subadult, p. 43. **8** White-winged Scoter, **8a** subadult, p. 43. **9** King Eider, **9a** immature, p. 42.

PLATE 11
POCHARDS, MERGANSERS, RUDDY DUCK

1 Redhead, p. 37. **2** Canvasback, p. 39. **3** Ring-necked Duck, p. 37.
4 Lesser Scaup, p. 38. **5** Greater Scaup, p. 38. **6** Hooded Merganser, p. 44.
7 Red-breasted Merganser, p. 45. **8** Common Merganser, p. 45.
9 Ruddy Duck, **9a** summer, **9b** winter, p. 46.

PLATE 12
VULTURES, EAGLES, CARACARA, OSPREY

1 Turkey Vulture, p. 47. **2** Black Vulture, p. 47.
3 Bald Eagle, **3a** immature, p. 55. **4** Crested Caracara, p. 57.
5 Osprey, p. 56. **6** Golden Eagle, **6a** immature, p. 55.

PLATE 13
ACCIPITERS, BUTEOS, HARRIER

1 Northern Goshawk, **1a** immature, p. 49. **2** Cooper's Hawk, **2a** immature, p. 50. **3** Sharp-shinned Hawk, **3a** immature, p. 50. **4** Broad-winged Hawk, **4a** immature, p. 52. **5** Red-shouldered Hawk, **5a** immature, **5b** southern Florida forms, p. 52. **6** Northern Harrier, **6a** immature, p. 56. **7** Red-tailed Hawk, **7a** immature, p. 51. **8** Swainson's Hawk, **8a** immature, **8b** light phase, p. 53. **9** Rough-legged Hawk, **9a** light phase, p. 54.

PLATE 14
HAWKS IN FLIGHT

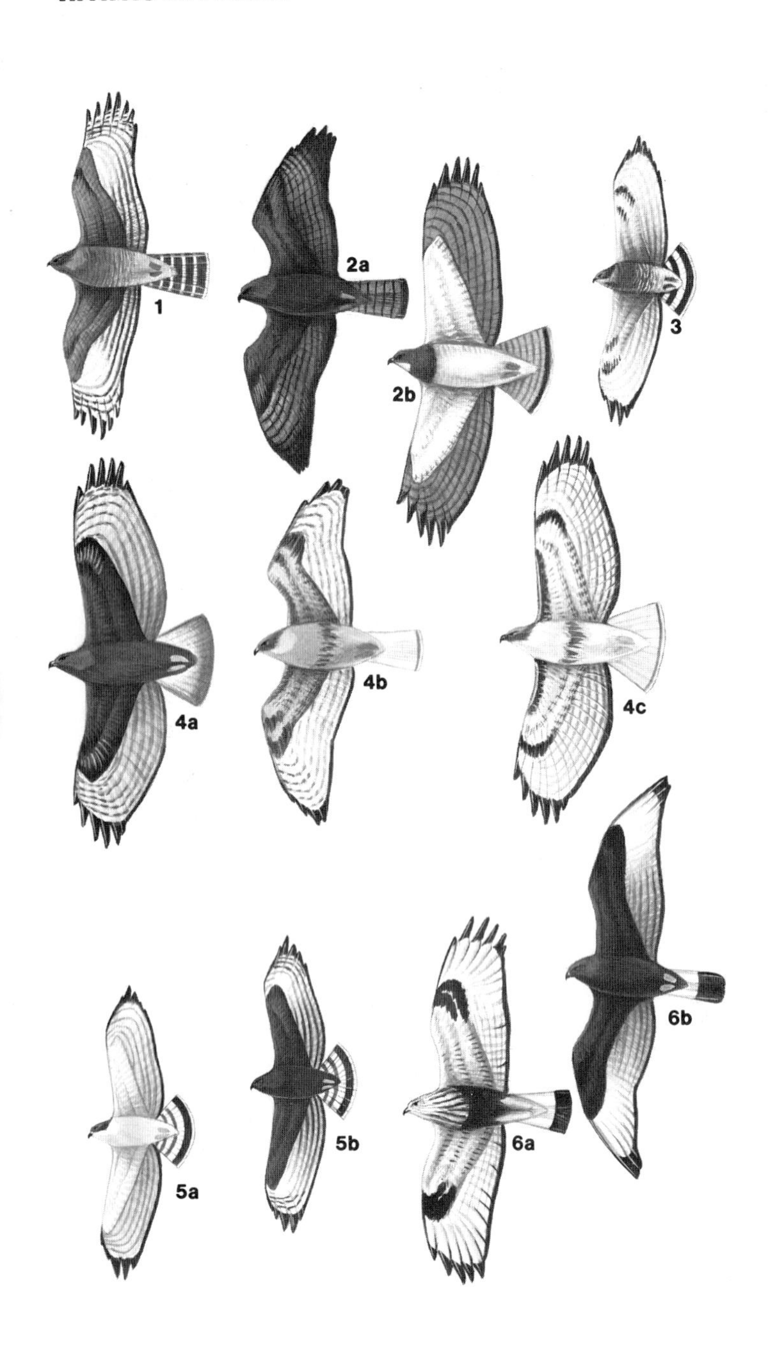

1 Red-shouldered Hawk, p. 52. **2** Swainson's Hawk, **2a** dark phase, **2b** light phase, p. 53. **3** Broad-winged Hawk, p. 52. **4** Red-tailed Hawk, **4a** "Harlan's" race, dark phase, **4b** typical western race, **4c** eastern race, p. 51. **5** Short-tailed Hawk, **5a** light phase, **5b** dark phase, p. 53. **6** Rough-legged Hawk, **6a** light phase, **6b** dark phase, p. 54.

1 Crested Caracara, p. 57. **2** Osprey, p. 56. **3** Snail Kite, p. 48.
4 Swallow-tailed Kite, p. 49. **5** Northern Harrier, **5a** immature, p. 56.

PLATE 16
FALCONS, KITES, GROUSE, QUAIL

1 American Kestrel, p. 59. **2** Merlin, p. 58. **3** Mississippi Kite, **3a** immature, p. 49. **4** White-tailed Kite, **4a** immature, p. 48. **5** Gyrfalcon, **5a** gray phase immature, **5b** gray phase, p. 57. **6** Peregrine Falcon, **6a** immature, p. 58. **7** Snail Kite, **7a** immature. p. 48. **8** Spruce Grouse, p. 60. **9** Ruffed Grouse, **9a** gray phase, **9b** tail of rufous phase, p. 60. **10** Bobwhite, p. 63. **11** Gray Partridge, p. 63.

PLATE 17
OPEN COUNTRY GROUSE AND PHEASANT

1 Rock Ptarmigan, **1a** winter, **1b** summer, p. 61. **2** Willow Ptarmigan, **2a** summer, **2b** winter, **2c** spring, p. 61. **3** Greater Prairie Chicken, **3a** displaying, p. 62. **4** Sharp-tailed Grouse, **4a** displaying, p. 62. **5** Ring-necked Pheasant, p. 63.

PLATE 18
RAILS

1 Purple Gallinule, **1a** immature, p. 69. **2** American Coot, **2a** immature, **2b** chick, p. 70. **3** Common Gallinule, **3a** immature, **3b** chick, p. 69. **4** Clapper Rail, **4a** chick, p. 67. **5** King Rail, p. 66. **6** Virginia Rail, **6a** immature, p. 67. **7** Sora, **7a** immature, p. 68. **8** Black Rail, p. 68. **9** Yellow Rail, p. 68.

PLATE 19
SHOREBIRDS—FULL BIRDS IN SPRING PLUMAGES

1 Piping Plover, **1a** fall, p. 72. **2** Semipalmated Plover, **2a** fall, p. 72.
3 Killdeer, p. 73. **4** Snowy Plover, p. 73. **5** Wilson's Plover, p. 73.
6 Black-bellied Plover, p. 74. **7** American Golden Plover, p. 74.
8 Ruddy Turnstone, p. 76. **9** Upland Sandpiper, p. 77.
10 Buff-breasted Sandpiper, p. 85. **11** Marbled Godwit, p. 87.
12 Willet, p. 80. **13** Whimbrel, p. 76. **14** Long-billed Curlew, p. 77.
15 Black-necked Stilt, p. 90. **16** Common Oystercatcher, p. 71.
17 American Avocet, **17a** fall, p. 90.

PLATE 20
SHOREBIRDS—ALL BIRDS IN SPRING PLUMAGES

1 Western Sandpiper, p. 84. **2** Least Sandpiper, p. 82.
3 Semipalmated Sandpiper, p. 84. **4** White-rumped Sandpiper, p. 81.
5 Baird's Sandpiper, p. 82. **6** Sanderling, p. 84. **7** Dunlin p. 83.
8 Pectoral Sandpiper, p. 81. **9** Curlew Sandpiper, p. 83.
10 Purple Sandpiper, p. 80. **11** Ruff, p. 79. **12** Red Knot, p. 80.
13 Hudsonian Godwit, p. 87. **14** Common Snipe, p. 75.
15 Short-billed Dowitcher, p. 86. **16** Wilson's Phalarope, p. 88.
17 American Woodcock, p. 75. **18** Red Phalarope, p. 88.
19 Northern Phalarope, p. 89.

PLATE 21
SHOREBIRDS—MOST BIRDS IN FALL PLUMAGES

1 Semipalmated Sandpiper, p. 84. **2** Least Sandpiper, p. 82.
3 Western Sandpiper, p. 84. **4** Sanderling, p. 84. **5** Ruddy Turnstone, p. 76.
6 White-rumped Sandpiper, p. 81. **7** Purple Sandpiper, p. 80.
8 Curlew Sandpiper, p. 83. **9** Dunlin, p. 83. **10** Red Knot, p. 80.
11 American Golden Plover, p. 74. **12** Black-bellied Plover, p. 74.
13 Stilt Sandpiper, **13a** spring, p. 85. **14** Short-billed Dowitcher, p. 86.
15 Hudsonian Godwit, p. 87. **16** Solitary Sandpiper, p. 78. **17** Willet, p. 80.
18 Spotted Sandpiper, **18a** spring, p. 77. **19** Wilson's Phalarope, p. 88.
20 Lesser Yellowlegs, p. 79. **21** Greater Yellowlegs, p. 78.
22 Northern Phalarope, p. 89. **23** Red Phalarope, p. 88.

PLATE 22
FALL SHOREBIRDS IN FLIGHT

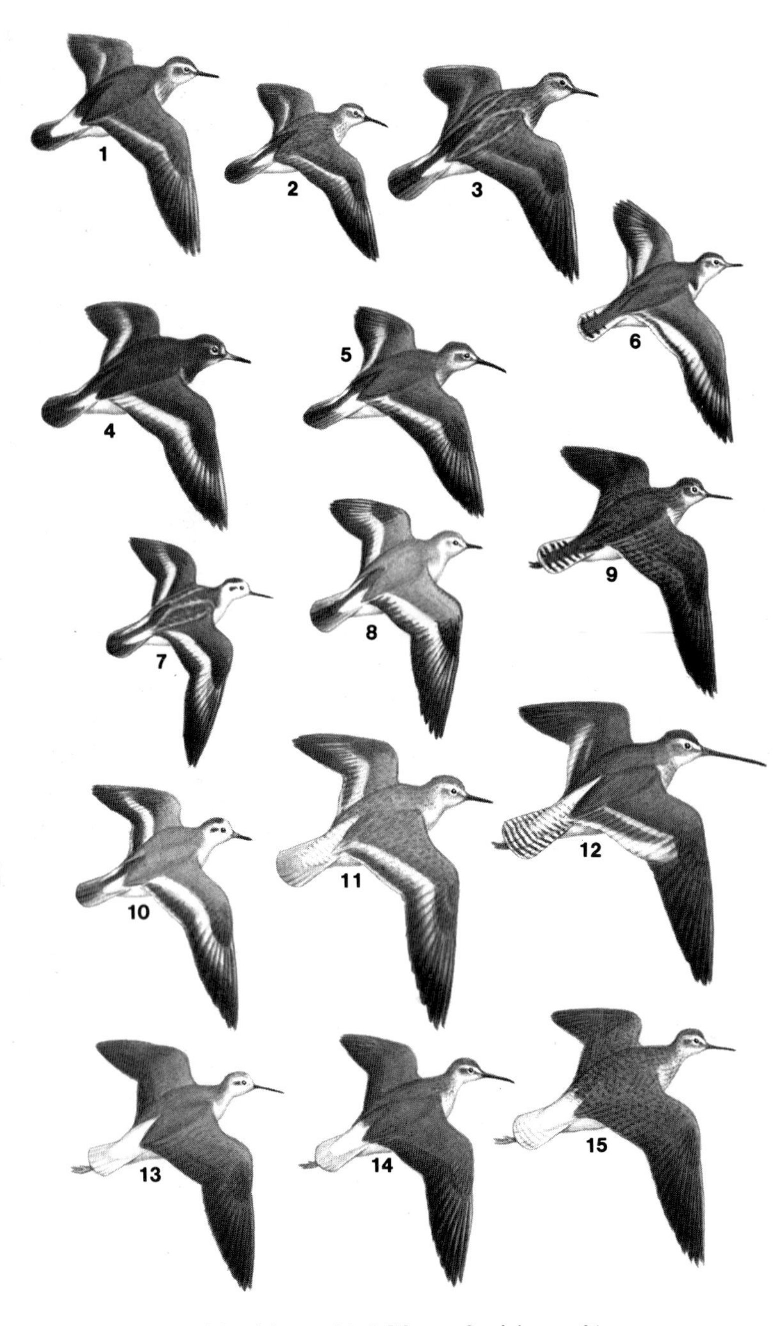

1 White-rumped Sandpiper, p. 81. **2** Western Sandpiper, p. 84. **3** Pectoral Sandpiper, p. 81. **4** Purple Sandpiper, p. 80. **5** Dunlin, p. 83. **6** Spotted Sandpiper, p. 77. **7** Northern Phalarope, p. 89. **8** Sanderling, p. 84. **9** Solitary Sandpiper, p. 78. **10** Red Phalarope, p. 88. **11** Red Knot, p. 80. **12** Short-billed Dowitcher, p. 86. **13** Wilson's Phalarope, p. 88. **14** Stilt Sandpiper, p. 85. **15** Lesser Yellowlegs, p. 79.

PLATE 23
SHOREBIRDS IN FLIGHT

1 Semipalmated Plover, **1a** spring, p. 72. **2** Piping Plover, **2a** spring, p. 72. **3** Killdeer, p. 73. **4** Buff-breasted Sandpiper, p. 85. **5** Upland Sandpiper, p. 77. **6** American Golden Plover, **6a** fall, **6b** spring, p. 74. **7** Common Snipe, p. 75. **8** Black-bellied Plover, **8a** fall, **8b** spring, p. 74. **9** American Woodcock, p. 75. **10** Ruddy Turnstone, **10a** spring, p. 76.

PLATE 24
LARGER GULLS

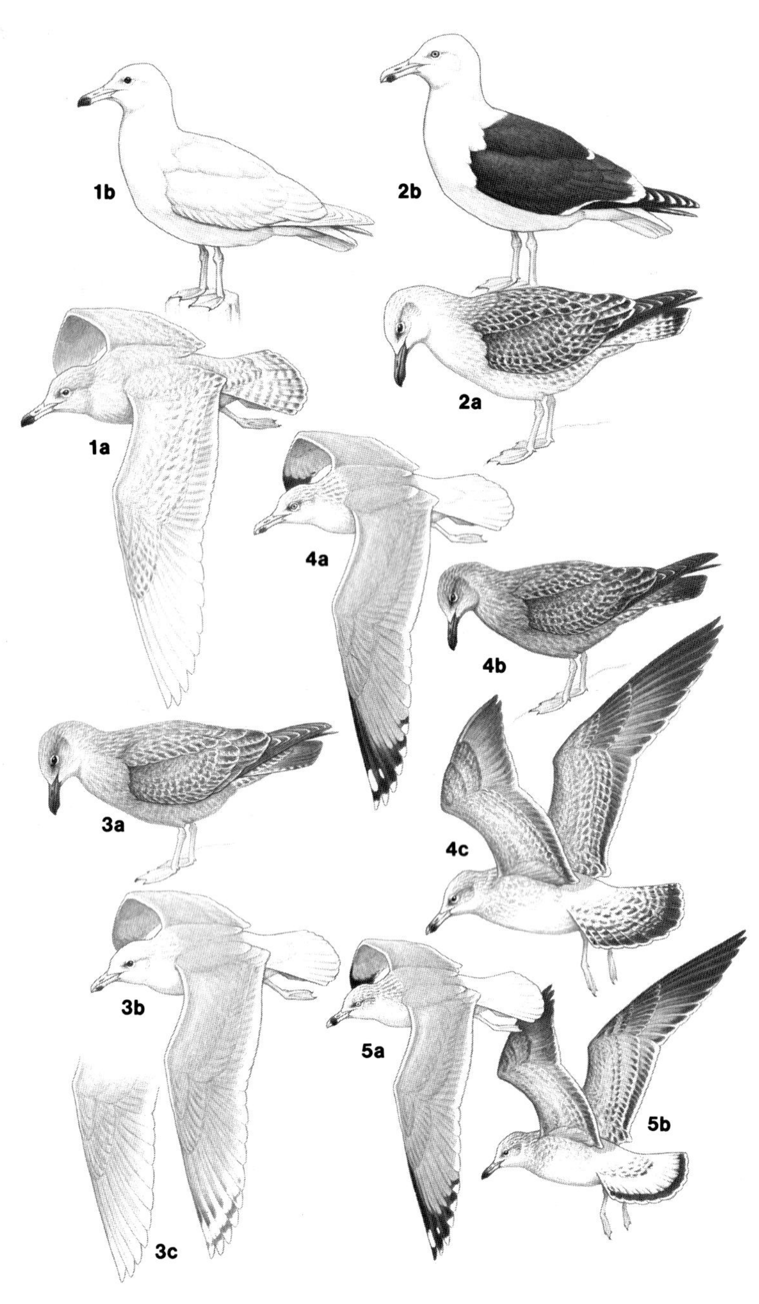

1 Glaucous Gull, **1a** first winter immature, **1b** second winter immature, p. 93.
2 Great Black-backed Gull, **2a** first winter immature, **2b** winter, p. 96.
3 Iceland Gull, **3a** first winter immature, **3b** winter Kumlien's race,
3c nominate race, p. 94. **4** Herring Gull, **4a** winter,
4b first winter immature, **4c** second winter immature, p. 95.
5 Ring-billed Gull, **5a** winter, **5b** first winter immature, p. 95.

PLATE 25
SMALLER GULLS

1 Laughing Gull, **1a** winter, **1b** summer, **1c** first winter immature, p. 97.
2 Franklin's Gull, **2a** winter, **2b** summer, **2c** first winter immature, p. 97.
3 Bonaparte's Gull, **3a** summer, **3b** winter, **3c** first winter immature, p. 98.
4 Sabine's Gull, **4a** first winter immature, **4b** summer, p. 100.
5 Black-headed Gull, **5a** winter, p. 96. **6** Little Gull, **6a** winter,
6b first winter immature, p. 98. **7** Black-legged Kittiwake, **7a** winter,
7b first winter immature, p. 99.

PLATE 26
TERNS AND GULLS

1 Least Tern, **1a** fall, **1b** spring, p. 103. **2** Black Tern, **2a** fall, p. 105. **3** Common Tern, **3a** fall, **3b** spring, p. 102. **4** Forster's Tern, **4a** fall, **4b** spring, p. 101. **5** Gull-billed Tern, **5a** fall, **5b** spring, p. 101. **6** Arctic Tern, **6a** spring, p. 102. **7** Roseate Tern, **7a** spring, p. 103. **8** Sandwich Tern, **8a** fall, p. 104. **9** Black-legged Kittiwake, **9a** spring, p. 99. **10** Ring-billed Gull, **10a** spring, p. 95. **11** Herring Gull, **11a** spring, p. 95. **12** Royal Tern, **12a** fall, p. 104. **13** Caspian Tern, **13a** fall, p. 105.

PLATE 27
TERNS—ADULTS IN SPRING PLUMAGES

1 Royal Tern, p. 104. **2** Caspian Tern, p. 105. **3** Black Tern,
3a immature, p. 105. **4** Sandwich Tern, p. 104. **5** Gull-billed Tern, p. 101.
6 Roseate Tern, p. 103. **7** Forster's Tern, **7a** immature, p. 101.
8 Arctic Tern, p. 102. **9** Common Tern, **9a** immature, p. 102.
10 Least Tern, **10a** immature, p. 103.

PLATE 28
ALCIDS

1 Common Murre, **1a** summer, **1b** winter, p. 107. **2** Black Guillemot, **2a** summer, **2b** immature, **2c** winter, p. 108. **3** Common Puffin, **3a** summer, **3b** winter, p. 109. **4** Dovekie, **4a** winter, **4b** summer, p. 108. **5** Razorbill, **5a** summer, **5b** immature, **5c** winter, p. 107. **6** Thick-billed Murre, **6a** winter, p. 107.

PLATE 29
OWLS

1 Burrowing Owl, p. 116. **2** Boreal Owl, p. 119. **3** Screech Owl, **3a** rufous phase, **3b** gray phase, p. 115. **4** Saw-whet Owl, **4a** immature, p. 119. **5** Hawk Owl, p. 116. **6** Short-eared Owl, p. 118. **7** Long-eared Owl, p. 118. **8** Barn Owl, p. 114. **9** Barred Owl, p. 117. **10** Snowy Owl, p. 115. **11** Great Horned Owl, p. 115. **12** Great Gray Owl, p. 117.

PLATE 30
PIGEONS, SWIFTS, NIGHTJARS

1 Rock Pigeon, p. 110. **2** White-crowned Pigeon, p. 110. **3** Mourning Dove, p. 110. **4** Common Ground Dove, p. 111. **5** Chimney Swift, p. 122. **6** Common Nighthawk, p. 122. **7** Chuck-will's-widow, **7a** tail, p. 120. **8** Poor-will, **8a** tail, p. 121. **9** Whip-poor-will, **9a** tails, p. 121.

PLATE 31
WOODPECKERS

1 Pileated Woodpecker, p. 125. **2** Common Flicker, **2a** eastern "Yellow-shafted" form, **2b** western "Red-shafted" form, p. 125. **3** Downy Woodpecker, p. 128. **4** Hairy Woodpecker, p. 127. **5** Red-bellied Woodpecker, **5a** immature, p. 126. **6** Red-cockaded Woodpecker, p. 128. **7** Black-backed Woodpecker, p. 129. **8** Yellow-bellied Sapsucker, **8a** immature, p. 127. **9** Three-toed Woodpecker, p. 129. **10** Red-headed Woodpecker, **10a** immature, p. 126.

PLATE 32
SWALLOWS AND CUCKOOS

1 Purple Martin, p. 139. **2** Barn Swallow, p. 139. **3** Tree Swallow, **3a** immature, p. 137. **4** Bank Swallow, p. 138. **5** Rough-winged Swallow, p. 138. **6** Mangrove Cuckoo, p. 113. **7** Cliff Swallow, p. 139. **8** Yellow-billed Cuckoo, p. 112. **9** Black-billed Cuckoo, **9a** immature, p. 113.

PLATE 33
FLYCATCHERS

1 Great Crested Flycatcher, p. 131. **2** Say's Phoebe, p. 132.
3 Eastern Phoebe, p. 132. **4** Scissor-tailed Flycatcher, p. 131.
5 Cassin's Kingbird, p. 131. **6** Western Kingbird, p. 130.
7 Eastern Kingbird, p. 130. **8** Gray Kingbird, p. 131.
9 Eastern Wood Pewee, p. 135. **10** Western Wood Pewee, p. 135.
11 Olive-sided Flycatcher, p. 136. **12** Alder/Willow Flycatcher, p. 133.
13 Least Flycatcher, p. 134. **14** Acadian Flycatcher, p. 133.
15 Yellow-bellied Flycatcher, p. 133.

PLATE 34
NUTHATCHES, TITMICE, WRENS AND OTHERS

1 Brown Creeper, p. 147. **2** White-breasted Nuthatch, p. 146. **3** Brown-headed Nuthatch, p. 147. **4** Tufted Titmouse, p. 145. **5** Red-breasted Nuthatch, p. 146. **6** Boreal Chickadee, p. 145. **7** Carolina Chickadee, p. 144. **8** Short-billed Marsh Wren, p. 150. **9** Long-billed Marsh Wren, p. 150. **10** Black-capped Chickadee, p. 144. **11** Winter Wren, p. 148. **12** Blue-gray Gnatcatcher, p. 157. **13** Bewick's Wren, p. 149. **14** House Wren, p. 148. **15** Ruby-throated Hummingbird, p. 123. **16** Rock Wren, p. 151. **17** Carolina Wren, p. 149.

PLATE 35
MIMIDS, CORVIDS, SHRIKES, KINGFISHERS

1 Brown Thrasher, p. 152. **2** Northern Mockingbird, **2a** immature, p. 151.
3 Gray Catbird, p. 152. **4** Sage Thrasher, p. 153.
5 Loggerhead Shrike, p. 162. **6** Blue Jay, p. 141.
7 Northern Shrike, **7a** winter, **7b** immature, p. 161. **8** Scrub Jay, p. 141.
9 Gray Jay, **9a** juvenal, p. 140. **10** Clark's Nutcracker, p. 143.
11 Black-billed Magpie, p. 141. **12** Belted Kingfisher, p. 124.

PLATE 36
THRUSHES AND WAXWINGS

1 Varied Thrush, p. 154. **2** American Robin, **2a** juvenal, p. 153.
3 Bohemian Waxwing, p. 161. **4** Cedar Waxwing, **4a** juvenal, p. 160.
5 Hermit Thrush, p. 154. **6** Eastern Bluebird, **6a** juvenal, p. 156.
7 Mountain Bluebird, **7a** juvenal, p. 156. **8** Veery, p. 156. **9** Swainson's Thrush, p. 155. **10** Gray-cheeked Thrush, p. 155. **11** Wood Thrush, p. 154.

PLATE 37
WARBLERS–SPRING PLUMAGES

1 American Redstart, p. 187. **2** Bay-breasted Warbler, p. 178. **3** Chestnut-sided Warbler, p. 178. **4** Black-throated Blue Warbler, p. 174. **5** Northern Parula, p. 172. **6** Cerulean Warbler, p. 176. **7** Orange-crowned Warbler, p. 171. **8** Tennessee Warbler, p. 172. **9** Worm-eating Warbler, p. 168. **10** Swainson's Warbler, p. 168. **11** Ovenbird, p. 181. **12** Northern Waterthrush, p. 182. **13** Louisiana Waterthrush, p. 182.

PLATE 38
WARBLERS—SPRING PLUMAGES

1 Yellow-rumped Warbler, **1a** eastern "Myrtle" form, **1b** western "Audubon's" form, p. 175. **2** Magnolia Warbler, p. 173. **3** Cape May Warbler, p. 174. **4** Canada Warbler, p. 186. **5** Blackburnian Warbler, p. 177. **6** Black-throated Green Warbler, p. 175. **7** Yellow-throated Warbler, p. 177. **8** Kirtland's Warbler, p. 180. **9** Black-throated Gray Warbler, p. 176. **10** Blackpoll Warbler, p. 179. **11** Black-and-White Warbler, p. 167. **12** Pine Warbler, p. 179. **13** Palm Warbler, **13a** eastern race, **13b** western race, p. 181. **14** Prairie Warbler, p. 180.

PLATE 39
WARBLERS—SPRING PLUMAGES

1 Kentucky Warbler, p. 183. **2** Common Yellowthroat, p. 184.
3 Yellow-breasted Chat, p. 185. **4** Wilson's Warbler, p. 186.
5 Hooded Warbler, p. 185. **6** Yellow Warbler, p. 173.
7 Golden-winged Warbler, p. 169. **8** "Lawrence's" Hybrid, p. 169.
9 Prothonotary Warbler, p. 167. **10** "Brewster's" Hybrid, p. 169.
11 Nashville Warbler, p. 171. **12** Blue-winged Warbler, p. 170.
13 Mourning Warbler, p. 184. **14** Connecticut Warbler, p. 183.
15 Bachman's Warbler, p. 170.

PLATE 40
FALL WARBLERS AND VIREOS

1 Yellow Warbler, **1a** immature, p. 173. **2** Wilson's Warbler, **2a** immature, p. 186. **3** Hooded Warbler, **3a** immature, p. 185. **4** Connecticut Warbler, **4a** immature, p. 183. **5** Nashville Warbler, **5a** immature, p. 171. **6** Mourning Warbler, **6a** immature p. 184. **7** Black-throated Blue Warbler, **7a** immature, p. 175. **8** Common Yellowthroat, **8a** immature, p. 184. **9** Orange-crowned Warbler, **9a** immature, p. 171. **10** Warbling Vireo, p. 166. **11** Tennessee Warbler, **11a** immature, p. 172. **12** Philadelphia Vireo, p. 166. **13** Solitary Vireo, p. 165. **14** Red-eyed Vireo, p. 165. **15** Bell's Vireo, p. 164. **16** Yellow-throated Vireo, p. 164. **17** Black-whiskered Vireo, p. 166. **18** White-eyed Vireo, **18a** immature, p. 163.

PLATE 41
FALL WARBLERS AND KINGLETS

1 Bay-breasted Warbler, **1a** immature, p. 178. **2** Blackpoll Warbler, **2a** immature, p. 179. **3** Pine Warbler, **3a** immature, p. 179. **4** Black-throated Green Warbler, **4a** immature, p. 175. **5** Kirtland's Warbler, p. 180. **6** Blackburnian Warbler, **6a** immature, p. 177. **7** Cape May Warbler, **7a** immature, p. 174. **8** Yellow-rumped Warbler, **8a** immature eastern "Myrtle" form, **8b** immature western "Audubon's" form, p. 175. **9** Palm Warbler, **9a** immature, p. 181. **10** Magnolia Warbler, **10a** immature, p. 173. **11** Prairie Warbler, **11a** immature, p. 180. **12** Northern Parula, p. 172. **13** Ruby-crowned Kinglet, p. 158. **14** Chestnut-sided Warbler, **14a** immature, p. 178. **15** Golden-crowned Kinglet, p. 158.

PLATE 42
BLACKBIRDS AND STARLING

1 Yellow-headed Blackbird, p. 190. **2** Red-winged Blackbird, **2a** immature, p. 191. **3** Starling, **3a** fall, **3b** spring, **3c** immature, p. 162. **4** Brown-headed Cowbird, **4a** juvenal, p. 194. **5** Rusty Blackbird, **5a** spring, **5b** fall immature, p. 192. **6** Brewer's Blackbird, **6a** spring, **6b** fall immature, p. 192. **7** Common Grackle, **7a** "Purple" form, **7b** "Bronzed" form, p. 194. **8** Boat-tailed Grackle, p. 193.

PLATE 43 ORIOLES, TANAGERS, FINCHES—SPRING PLUMAGES

1 Northern Oriole, **1a** western "Bullock's" race, **1b** "Bullock's" immature, **1c** eastern "Baltimore" race, **1d** "Baltimore" immature, p. 191.
2 Spot-breasted Oriole, **2a** immature, p. 191. **3** Orchard Oriole, **3a** immature, p. 190. **4** Scarlet Tanager, p. 195. **5** Western Tanager, p. 195.
6 Summer Tanager, **6a** immature, p. 196. **7** Blue Grosbeak, p. 198.
8 Painted Bunting, p. 199. **9** Indigo Bunting, p. 199.
10 Lazuli Bunting, p. 200. **11** Black-headed Grosbeak, p. 197.
12 Rose-breasted Grosbeak, **12a** immature, p. 197.

PLATE 44
FINCHES

1 Cardinal, p. 196. **2** Pine Grosbeak, **2a** immature, p. 202.
3 Evening Grosbeak, p. 200. **4** Red Crossbill, **4a** juvenal, p. 204.
5 Purple Finch, p. 201. **6** White-winged Crossbill, p. 205.
7 House Finch, p. 200. **8** Common Redpoll, **8a** immature, p. 202.
9 Pine Siskin, p. 204. **10** American Goldfinch, **10a** winter,
10b summer, p. 203. **11** Lesser Goldfinch, p. 203.

PLATE 45
SPARROWS, JUNCOS, TOWHEES

1 Black-throated Sparrow, p. 211. **2** House Sparrow, **2a** summer, **2b** immature, p. 187. **3** Lark Sparrow, p. 212. **4** Dark-eyed Junco, **4a** eastern "Slate-colored" form, **4b** "Slate-colored" juvenal, **4c** Black Hills "White-winged" form, **4d** no. Rocky Mt. "Pink-sided" form, **4e** western "Oregon" form, p. 211. **5** Green-tailed Towhee, p. 206. **6** Rufous-sided Towhee, **6a** western "Spotted" form, **6b** juvenal, p. 205.

PLATE 46
SPARROWS

1 Fox Sparrow, p. 216. **2** Vesper Sparrow, p. 210.
3 Lincoln's Sparrow, p. 216. **4** Song Sparrow, p. 217. **5** Savannah Sparrow,
5a "Ipswich" race, p. 208. **6** Henslow's Sparrow, p. 207.
7 Baird's Sparrow, p. 207. **8** Sharp-tailed Sparrow, **8a** Interior race,
8b n.e. "Acadian" race, **8c** Coastal race, p. 209. **9** LeConte's Sparrow, p. 209.
10 Seaside Sparrow, p. 208. **11** Grasshopper Sparrow, p. 206.
12 Clay-colored Sparrow, **12a** immature, p. 213.
13 Chipping Sparrow, **13a** spring, **13b** immature, p. 213.

PLATE 47
SPARROWS

1 Harris' Sparrow, **1a** spring, **1b** immature, p. 214.
2 White-throated Sparrow, **2a** immature, p. 215. **3** White-crowned Sparrow, **3a** immature, p. 215. **4** Bachman's Sparrow, p. 211. **5** Field Sparrow, p. 214. **6** Tree Sparrow, p. 212. **7** Swamp Sparrow, **7a** spring, **7b** immature, p. 217. **8** Rufous-crowned Sparrow, p. 210.

PLATE 48 OPEN COUNTRY SONGBIRDS— MALES IN SPRING PLUMAGES

1 Bobolink, p. 188. **2** Eastern Meadowlark, p. 189. **3** Lark Bunting, p. 206.
4 Sprague's Pipit, p. 160. **5** Water Pipit, **5a** fall, p. 159. **6** Dickcissel, p. 198.
7 Horned Lark, **7a** "Yellow-faced" form, **7b** "White-faced" form,
7c juvenal, p. 136. **8** Snow Bunting, **8a** winter, **8b** immature, p. 220.
9 Lapland Longspur, **9a** fall, p. 218. **10** Chestnut-collared Longspur,
10a fall, p. 219. **11** Smith's Longspur, **11a** fall, p. 219.
12 McCown's Longspur, **12a** fall, p. 218.
13 Gray-crowned Rosy Finch, **13a** immature, p. 202.

PLATE 49
BATS

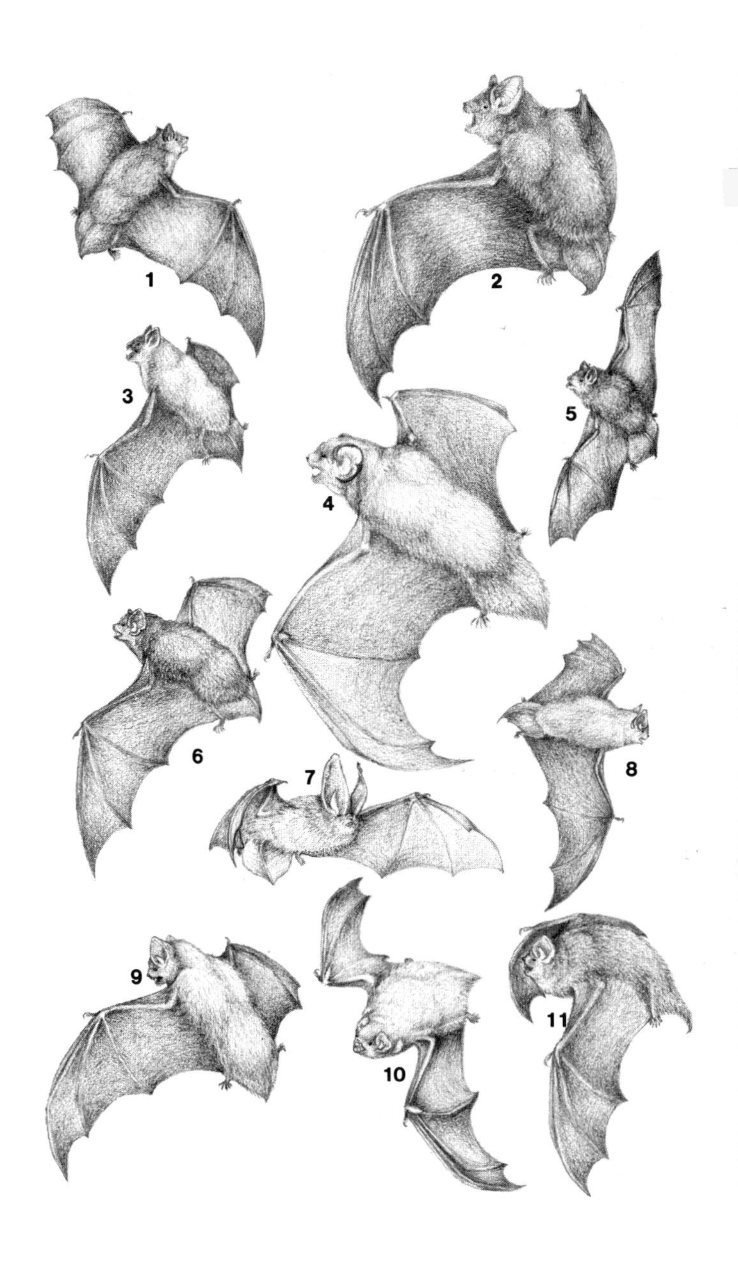

1 Evening Bat, p. 239. **2** Big Brown Bat, p. 237. **3** Small-footed Myotis, p. 236. **4** Hoary Bat, p. 238. **5** Little Brown Myotis, p. 235. **6** Silver-haired Bat, p. 237. **7** Townsend's Big-eared Bat, p. 240. **8** Eastern Pipistrelle, p. 237. **9** Northern Yellow Bat, p. 239. **10** Red Bat, p. 238. **11** Seminole Bat, p. 238.

PLATE 50
RABBITS AND SQUIRRELS

1 Red Squirrel, p. 252. **2** Gray Squirrel, **2a** melanistic, p. 251.
3 Southern Flying Squirrel, p. 252. **4** Fox Squirrel, **4a** eastern,
4b melanistic, **4c** mid-Atlantic, p. 251. **5** Rock Squirrel, p. 250.
6 Woodchuck, p. 248. **7** Black-tailed Prairie Dog, p. 250.
8 Thirteen-lined Ground Squirrel, p. 249. **9** Spotted Ground Squirrel, p. 249.
10 Least Chipmunk, p. 247. **11** Eastern Chipmunk, p. 247.
12 Eastern Cottontail, p. 243. **13** Black-tailed Jackrabbit, p. 246.
14 New England Cottontail, p. 243. **15** Swamp Rabbit, p. 243.
16 Snowshoe Hare, **16a** summer, **16b** winter, p. 244.

TRACKS—SMALLER MAMMALS

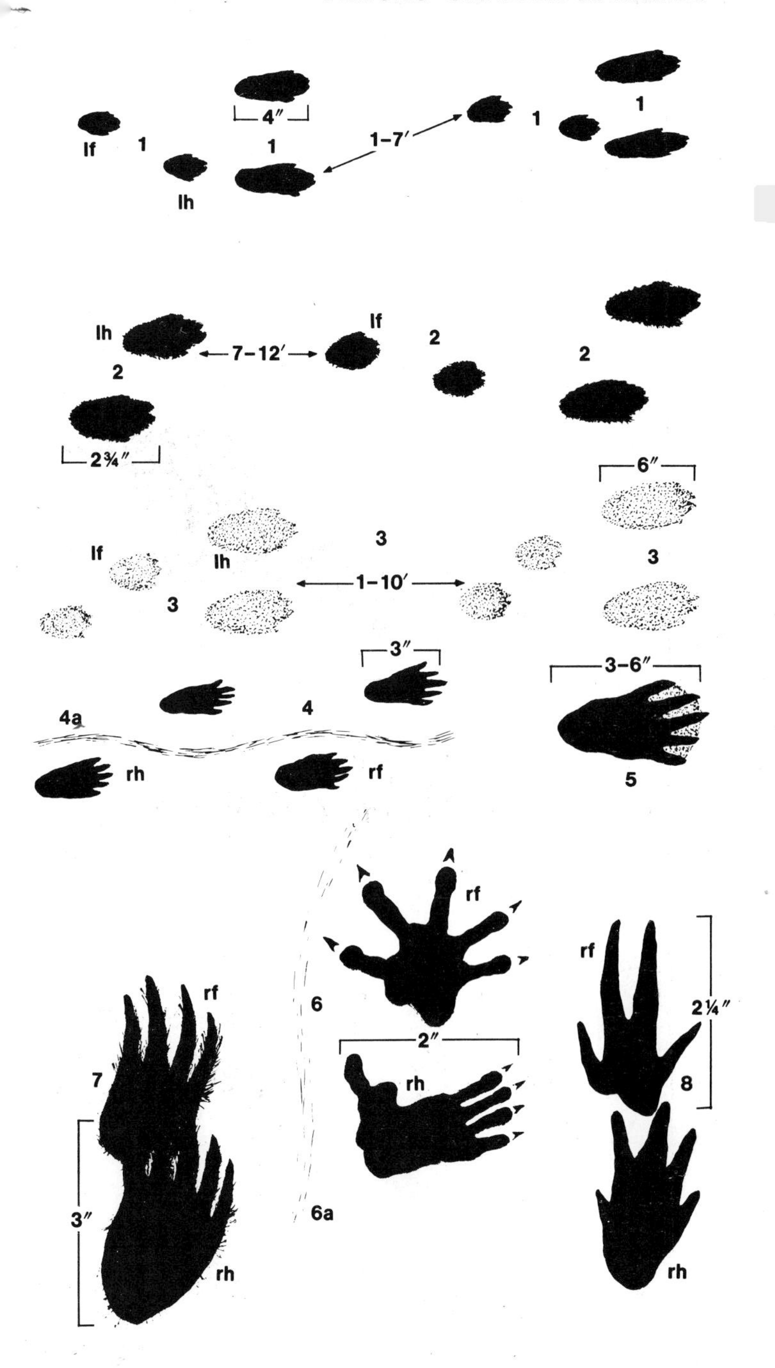

1 Cottontail, p. 243. **2** Jackrabbit, p. 245.
3 Snowshoe Hare, p. 244. **4** Muskrat, **4a** tail mark, p. 265.
5 Beaver, hind foot covers 4 in. (10.2 cm) front to next track, p. 256.
6 Virginia Opossum, **6a** tail mark, p. 227. **7** Porcupine, p. 269.
8 Armadillo, p. 241.

PLATE 52
MICE, SHREWS, MOLES, ALLIES

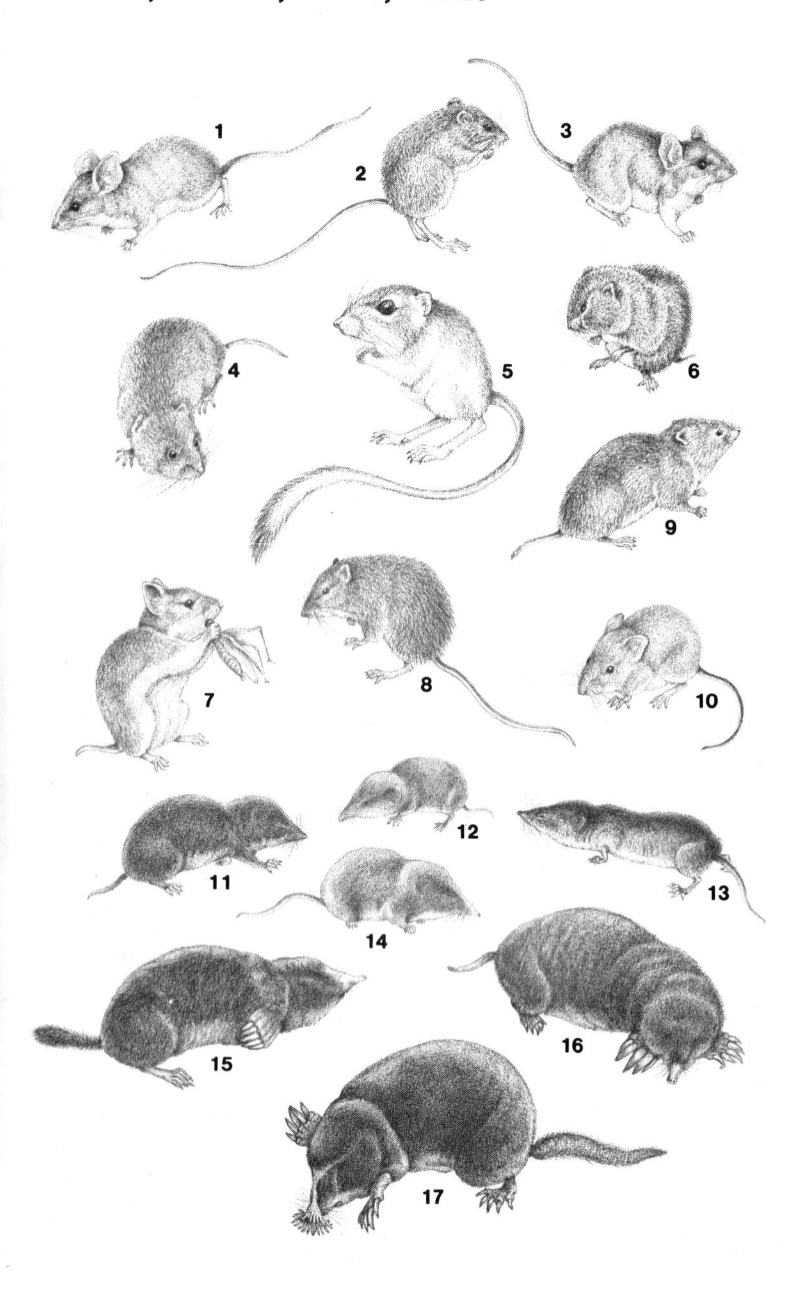

1 White-footed Mouse, p. 260. **2** Meadow Jumping Mouse, p. 269. **3** Deer Mouse, p. 258. **4** Meadow Vole, p. 263. **5** Ord's Kangaroo Rat, p. 255. **6** Northern Bog Lemming, p. 266. **7** Northern Grasshopper Mouse, p. 260. **8** Hispid Pocket Mouse, p. 255. **9** Southern Red-backed Vole, p. 262. **10** Plains Harvest Mouse, p. 257. **11** Short-tailed Shrew, p. 231. **12** Least Shrew, p. 232. **13** Arctic Shrew, p. 230. **14** Masked Shrew, p. 228. **15** Hairy-tailed Mole, p. 233. **16** Eastern Mole, p. 233. **17** Star-nosed Mole, p. 233.

PLATE 53
MICE AND RATS

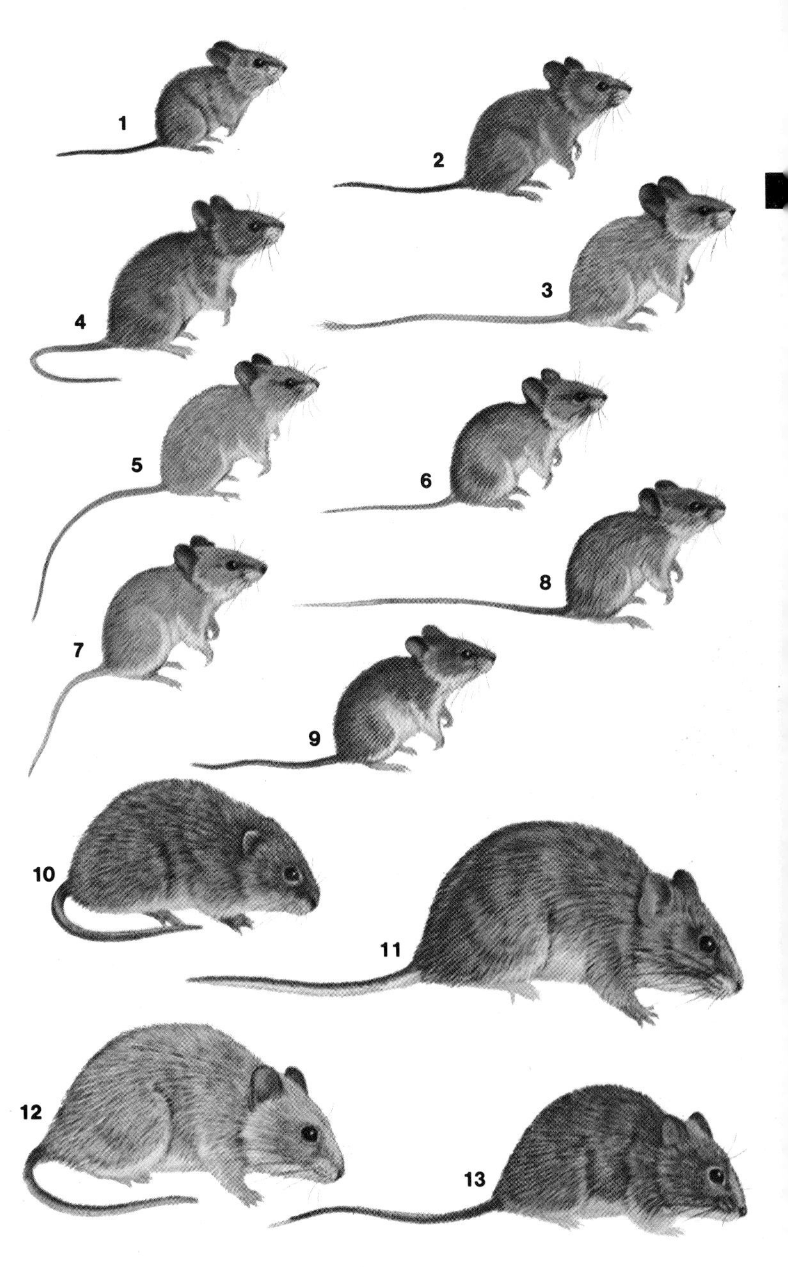

1 Northern Pygmy Mouse, p. 260. **2** House Mouse, p. 268. **3** Florida Mouse, p. 259. **4** Texas Mouse, p. 259. **5** Fulvous Harvest Mouse, p. 258. **6** Eastern Harvest Mouse, p. 257. **7** Oldfield Mouse, p. 259. **8** Woodland Jumping Mouse, p. 269. **9** Golden Mouse, p. 260. **10** Hispid Cotton Rat, p. 261. **11** Eastern Woodrat, p. 261. **12** Southern Plains Woodrat, p. 262. **13** Marsh Rice Rat, p. 257.

PLATE 54
TRACKS–SMALL AND HOOFED ANIMALS

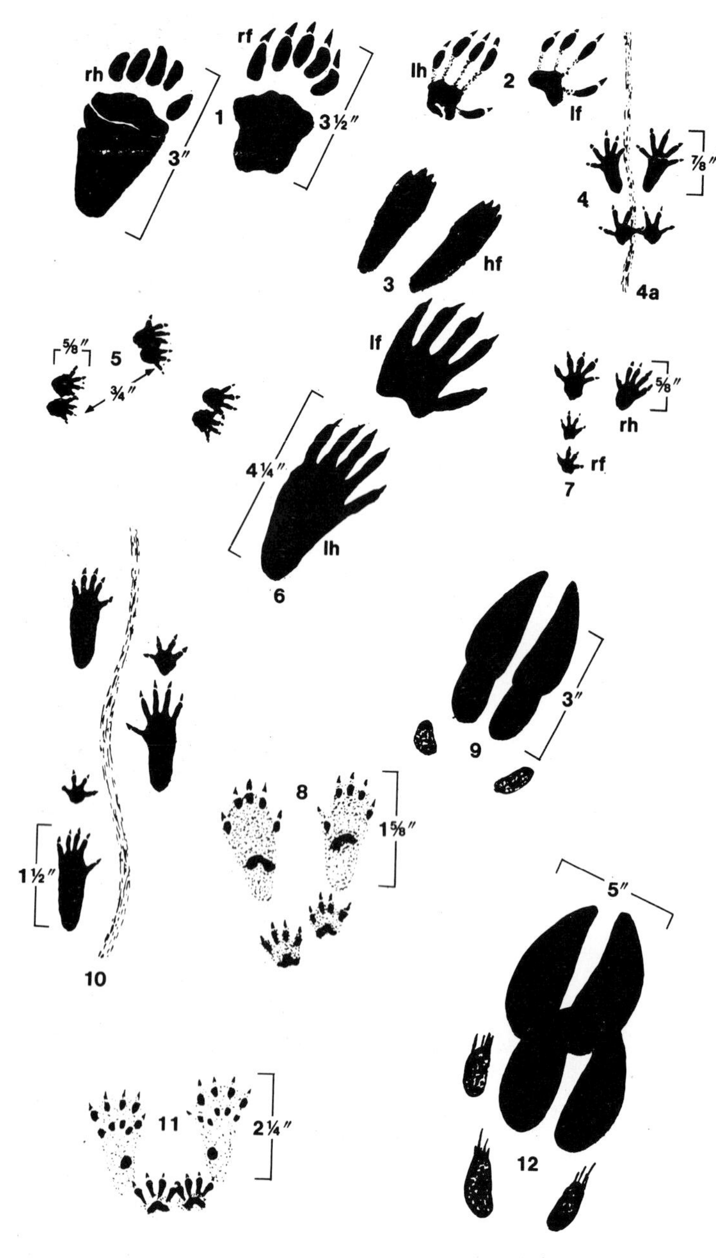

1 Badger, p. 294. **2** Woodchuck, p. 248. **3** Kangaroo Rat, p. 255. **4** Deer Mouse, **4a** tail mark, p. 258. **5** Grasshopper Mouse, running, p. 260. **6** Raccoon, p. 289. **7** Meadow Vole, bounding, p. 263. **8** Chipmunk, p. 247. **9** Deer, walking, p. 302. **10** Norway Rat, walking, p. 268. **11** Gray Squirrel, p. 251. **12** Bison, p. 305.

PLATE 55
VOLES AND LEMMINGS

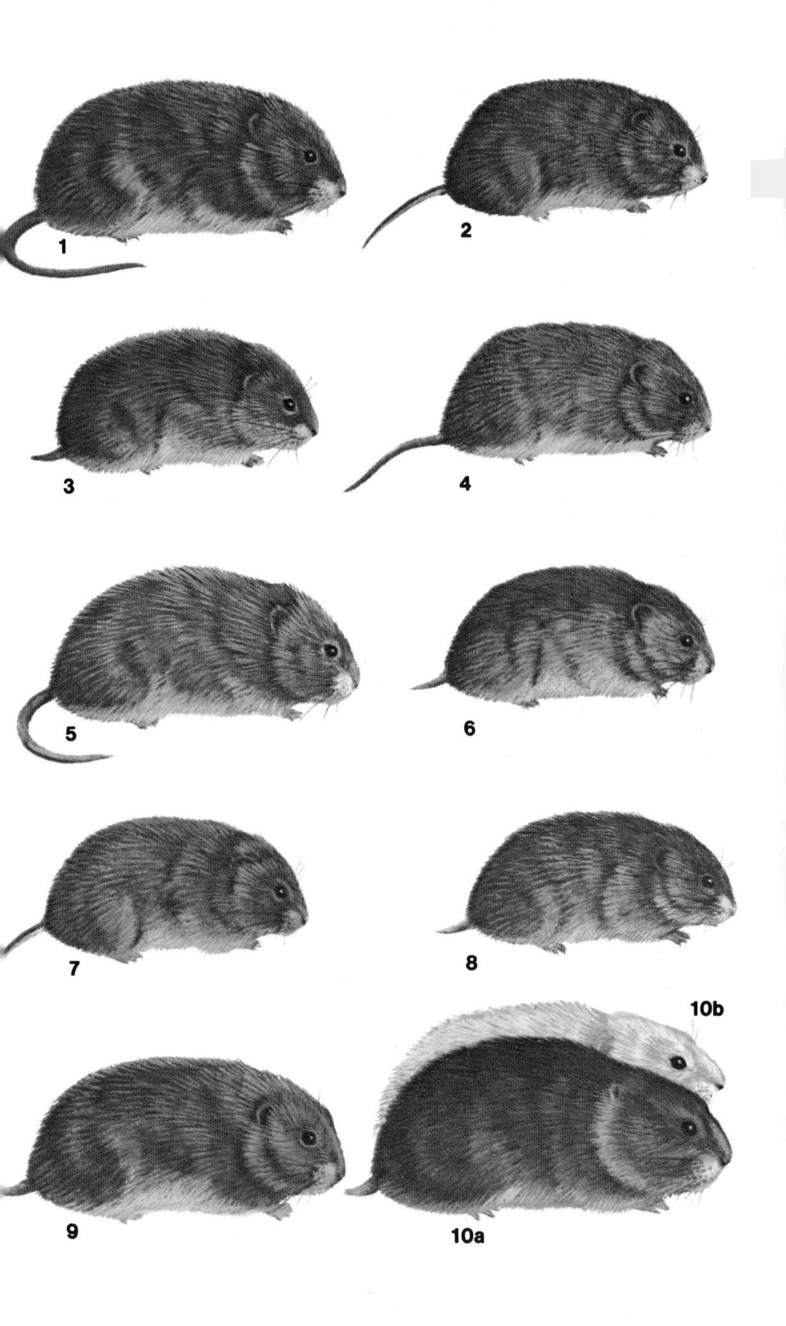

1 Beach Vole, p. 263. **2** Heather Vole, p. 262. **3** Woodland Vole, p. 264. **4** Meadow Vole, p. 263. **5** Rock Vole, p. 263. **6** Southern Red-backed Vole, p. 262. **7** Prairie Vole, p. 263. **8** Southern Bog Lemming, p. 266. **9** Brown Lemming, p. 265. **10** Collared Lemming, **10a** summer, **10b** winter, p. 266.

PLATE 56
TOOTHED WHALES

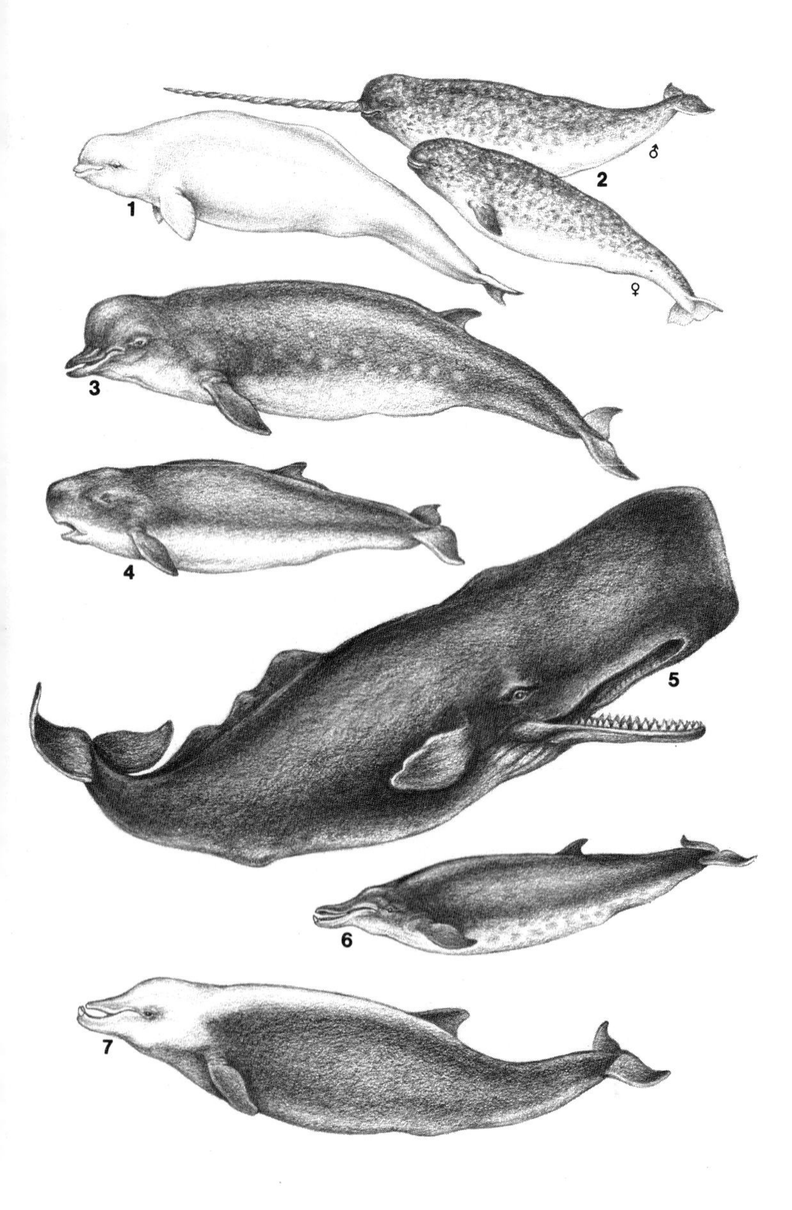

1 White Whale, p. 274. **2** Narwhal, p. 275. **3** North Atlantic Bottle-nosed Whale, p. 273. **4** Pygmy Sperm Whale, p. 274. **5** Sperm Whale, p. 273. **6** True's Beaked Whale, p. 272. **7** Goose-beaked Whale, p. 273.

PLATE 57
DOLPHINS AND OTHER WHALES

1 White-beaked Dolphin, p. 277. **2** Bottle-nosed Dolphin, p. 276. **3** Atlantic White-sided Dolphin, p. 277. **4** Harbor Porpoise, p. 278. **5** Killer Whale, p. 278. **6** Common Dolphin, p. 276. **7** Grampus, p. 277. **8** Common Pilot Whale, p. 278.

PLATE 58
TRACKS—WEASELS AND ALLIES

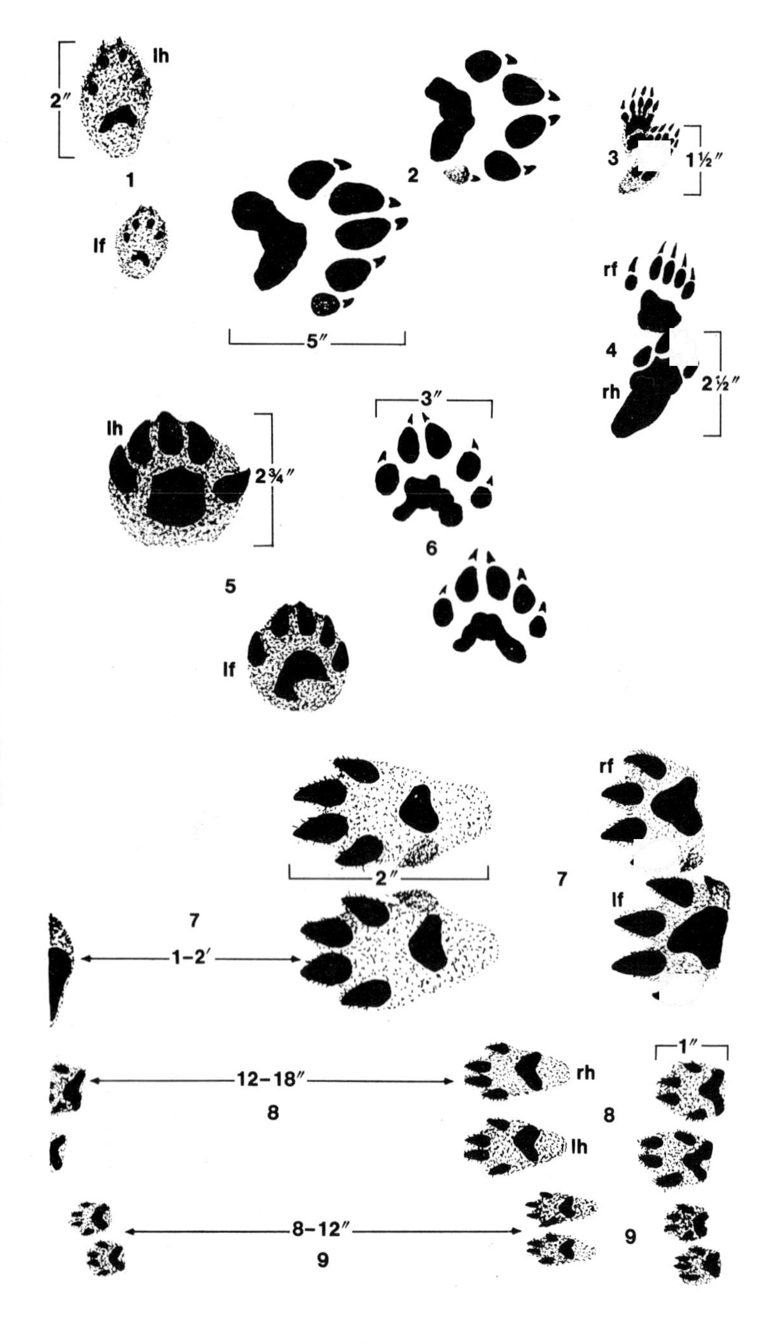

1 Marten, p. 290. **2** Wolverine, p. 293. **3** Spotted Skunk, p. 294. **4** Striped Skunk, p. 295. **5** River Otter, p. 295. **6** Fisher, p. 291. **7** Mink, p. 293. **8** Long-tailed Weasel, p. 292. **9** Least Weasel, p. 292.

PLATE 59
FOXES, WEASELS, ALLIES

1 Coyote, p. 283. **2** Red Fox, **2a** "cross" phase, p. 286. **3** Swift Fox, p. 285.
4 Least Weasel, **4a** summer, **4b** winter, p. 292. **5** Gray Fox, p. 286.
6 Ermine, **6a** summer, **6b** winter, p. 291. **7** Marten, p. 290.
8 Black-footed Ferret, p. 293. **9** Fisher, p. 291. **10** Wolverine, p. 293.

PLATE 60
TRACKS—SMALL AND HOOFED ANIMALS

1 Badger, p. 375. **2** Woodchuck, p. 307. **3** Kangaroo Rat, p. 324.
4 Deer Mouse, **4a** tail mark p. 331. **5** Grasshopper Mouse, running, p. 334.
6 Raccoon, p. 370. **7** Meadow Vole, bounding, p. 341. **8** Chipmunk, p. 303.
9 Deer, walking, p. 384. **10** Norway Rat, walking, p. 348.
11 Western Gray Squirrel, p. 315. **12** Bison, p. 387.

PLATE 61
CARNIVORES

1 Striped Skunk, p. 295. **2** Eastern Spotted Skunk, p. 294.
3 Black Bear, **3a** black phase, **3b** cinnamon phase, p. 287.
4 Long-tailed Weasel, **4a** winter, **4b** intermediate, **4c** summer, p. 292.
5 Mink, p. 293. **6** River Otter, p. 295. **7** Raccoon, p. 289. **8** Badger, p. 294.

PLATE 62
SEALS

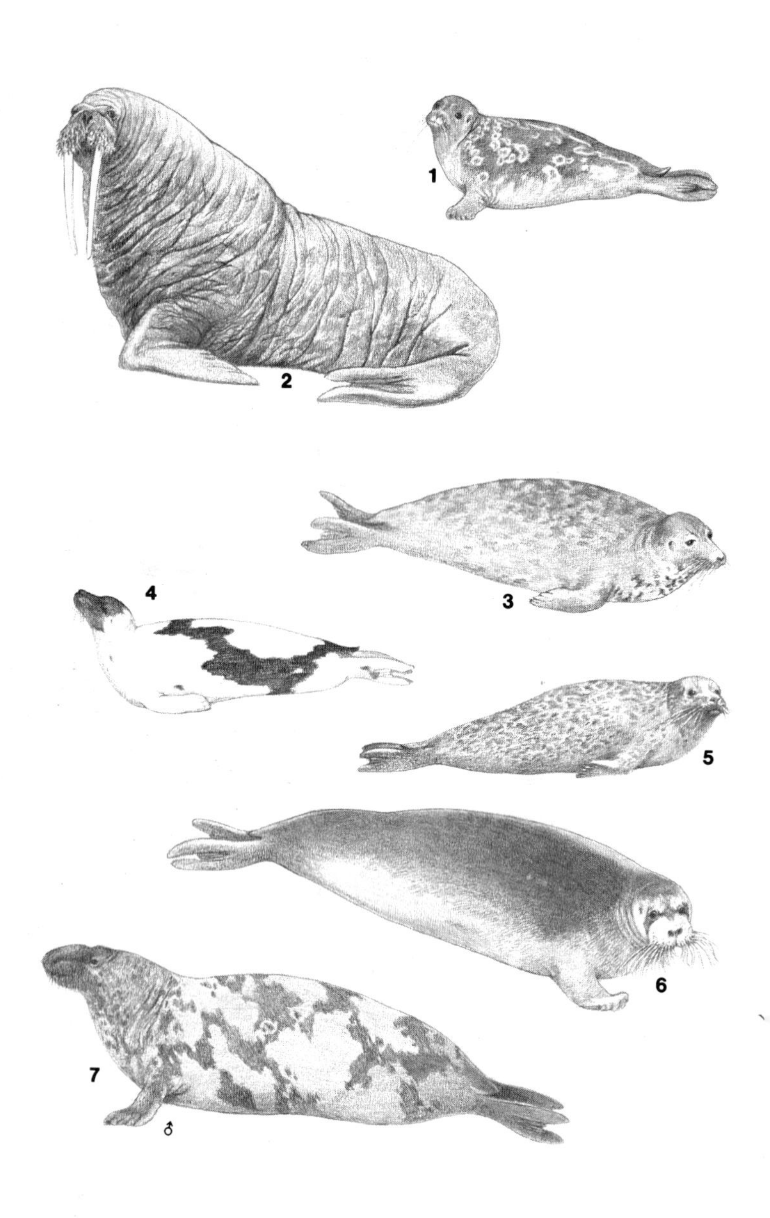

1 Ringed Seal, p. 296. **2** Walrus, p. 288. **3** Gray Seal, p. 297. **4** Harp Seal, p. 296. **5** Harbor Seal, p. 297. **6** Bearded Seal, p. 297. **7** Hooded Seal, p. 298.

PLATE 63
DEER AND PRONGHORN

1 Moose, p. 303. **2** Elk, p. 302. **3** Pronghorn, p. 304. **4** Caribou, p. 304.
5 Mule Deer, p. 302. **6** White-tailed Deer,
6a summer, **6b** winter, p. 303.

PLATE 64
SNAPPING TURTLES, SOFTSHELLS, SEA TURTLES

1 Snapping Turtle, p. 316. **2** Alligator Snapping Turtle, p. 316.
3 Spiny Softshell, p. 320. **4** Smooth Softshell, p. 320.
5 Hawksbill, p. 333. **6** Green Turtle, p. 332. **7** Loggerhead, p. 333.
8 Atlantic Ridley, p. 333. **9** Leatherback, p. 334.

PLATE 65
OTHER TURTLES

1 Spotted Turtle, p. 322. **2** Yellow Mud Turtle, p. 318. **3** Stinkpot, p. 318. **4** Eastern Mud Turtle, p. 317. **5** Red-bellied Turtle, p. 329. **6** Painted Turtle, p. 327. **7** Cooter, p. 328. **8** Slider, p. 328. **9** False Map Turtle, p. 326. **10** Map Turtle, p. 325.

PLATE 66
LIZARDS, SKINKS, WHIPTAILS

1 Great Plains Skink, **1a** immature, p. 344. **2** Collared Lizard, p. 338. **3** Five-lined Skink, **3a** immature, p. 342. **4** Broad-headed Skink, p. 343. **5** Many-lined Skink, p. 344. **6** Prairie Skink, p. 345. **7** Coal Skink, p. 344. **8** Ground Skink, p. 342. **9** Eastern Fence Lizard, p. 338. **10** Lesser Earless Lizard, p. 339. **11** Six-lined Racerunner, p. 345.

WATER, GARTER, OTHER SNAKES

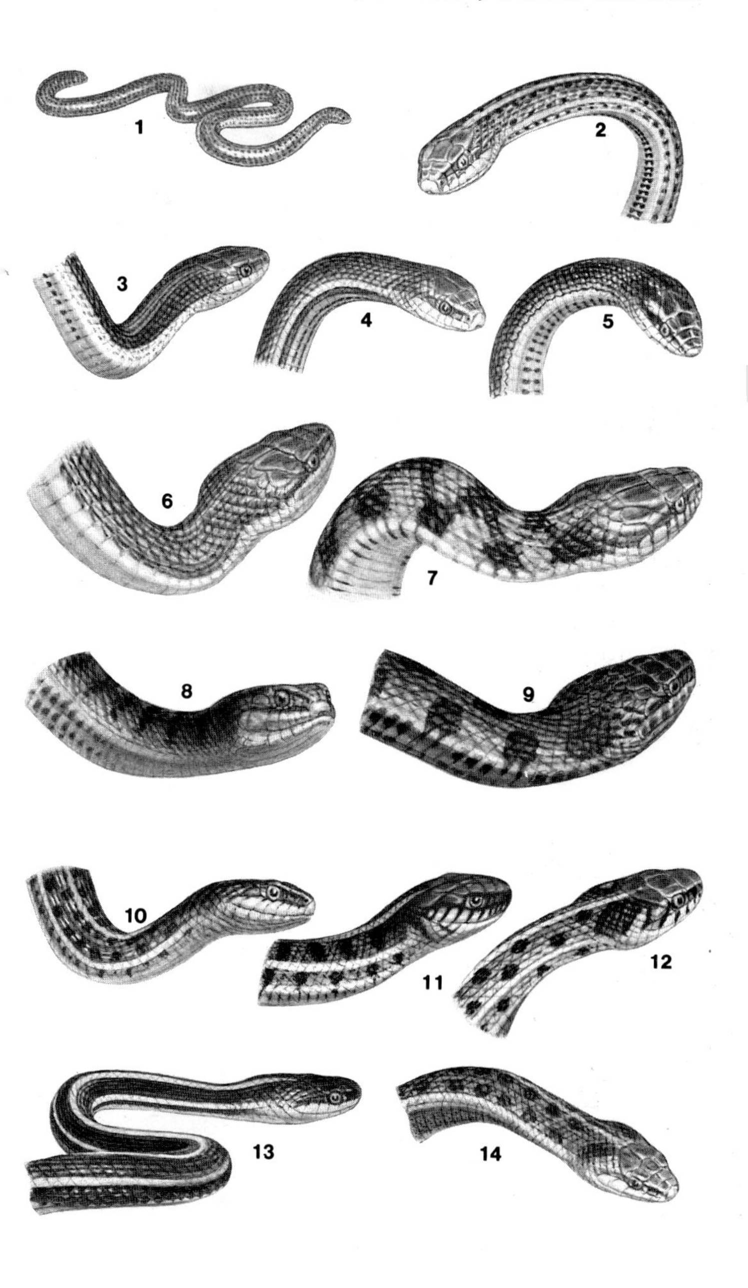

1 Texas Blind Snake, p. 348. **2** Lined Snake, p. 357. **3** Graham's Crayfish Snake, p. 352. **4** Queen Snake, p. 353. **5** Glossy Crayfish Snake, p. 353. **6** Plain-bellied Water Snake, p. 349. **7** Diamondback Water Snake, p. 352. **8** Green Water Snake, p. 348. **9** Brown Water Snake, p. 352. **10** Common Garter Snake, p. 359. **11** Plains Garter Snake, p. 359. **12** Checkered Garter Snake, p. 358. **13** Eastern Ribbon Snake, p. 360. **14** Western Terrestrial Garter Snake, p. 360.

PLATE 68
SNAKES

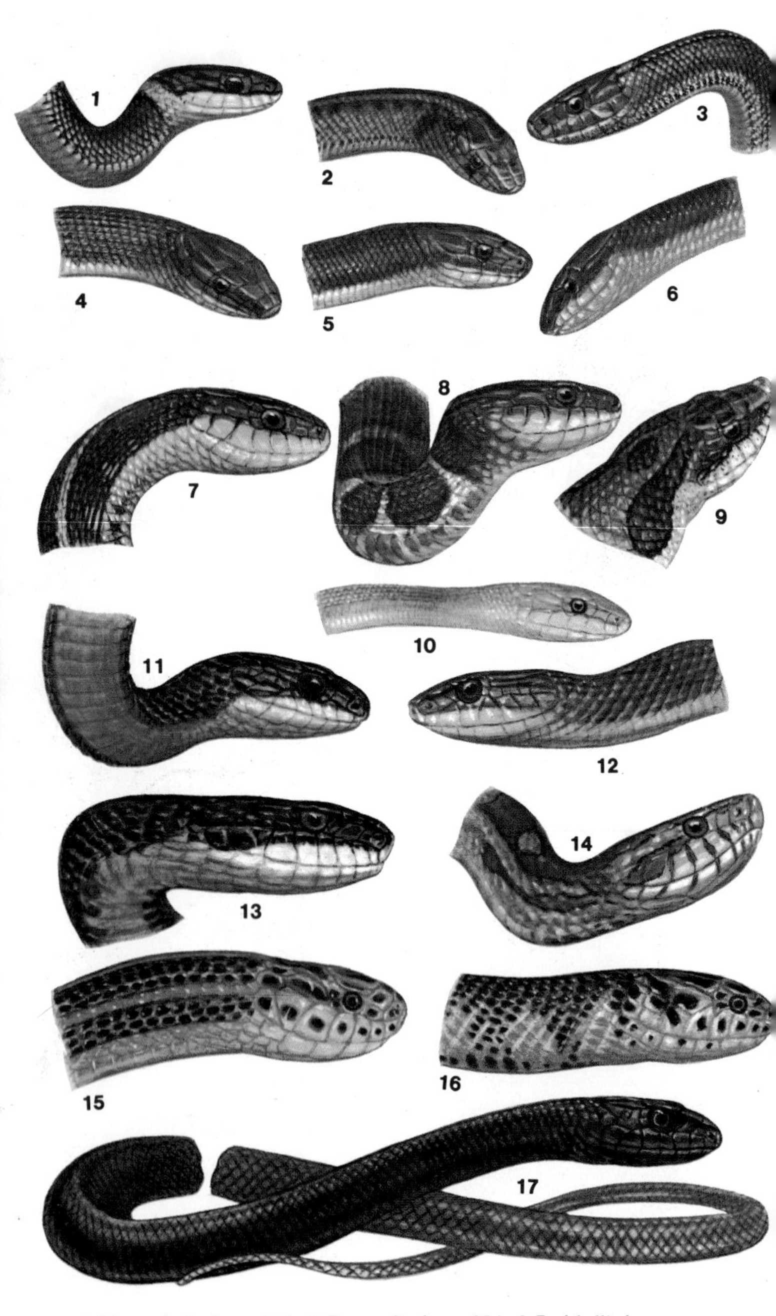

1 Ringneck Snake, p. 361. **2** Brown Snake, p. 356. **3** Red-bellied Snake, p. 356. **4** Rough Earth Snake, p. 355. **5** Smooth Earth Snake, p. 356. **6** Worm Snake, p. 362. **7** Common Garter Snake, p. 359. **8** Northern Water Snake, p. 350. **9** Eastern Hognose Snake, p. 365. **10** Rough Green Snake, p. 370. **11** Racer, p. 371. **12** Eastern Yellow-bellied Racer, p. 371. **13** Black Rat Snake, p. 375. **14** Corn Snake, p. 374. **15** Rainbow Snake, p. 363. **16** Mud Snake, p. 363. **17** Coachwhip, p. 372.

PLATE 69
SNAKES

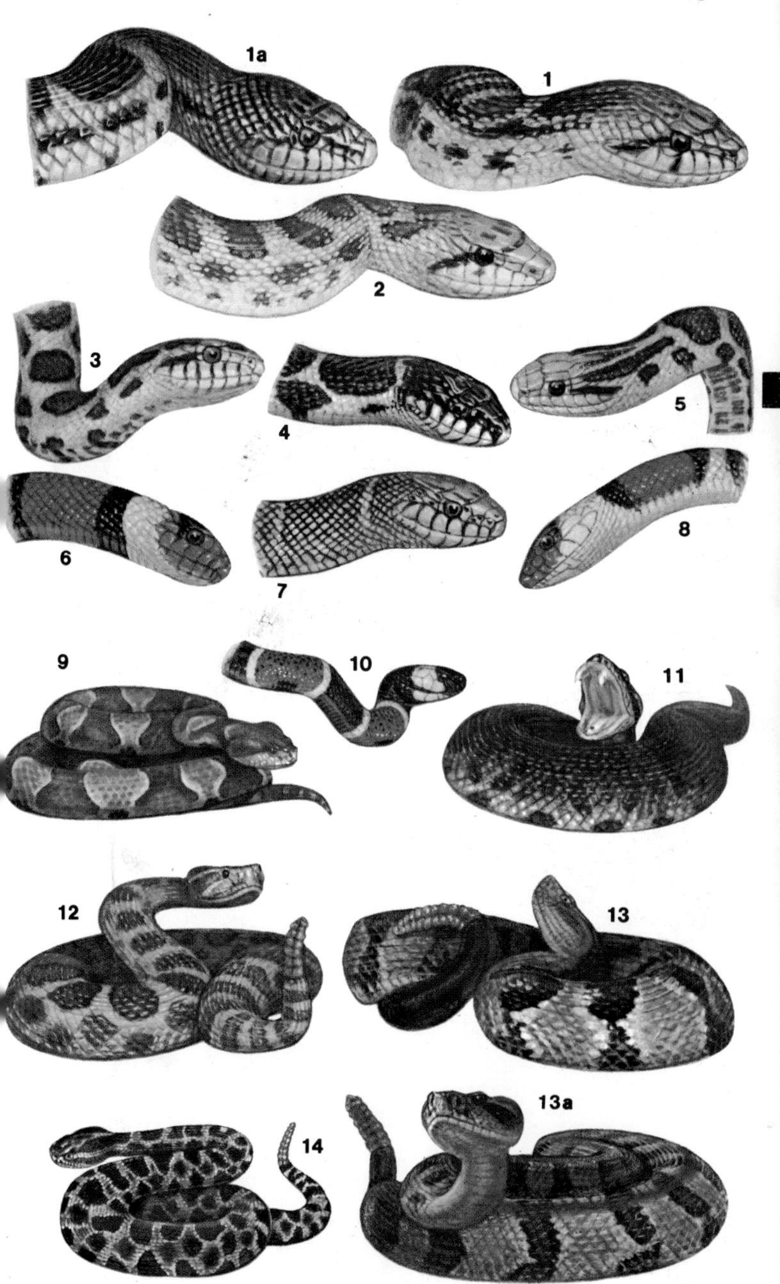

1 Bullsnake, **1a** Pine race, p. 373. **2** Glossy Snake, p. 376. **3** Milk Snake, p. 367. **4** Common Kingsnake, p. 366. **5** Prairie Kingsnake, p. 369. **6** Scarlet Kingsnake, p. 368. **7** Speckled Kingsnake, p. 367. **8** Scarlet Snake, p. 366. **9** Copperhead, p. 381. **10** Eastern Coral Snake, p. 380. **11** Cottonmouth, p. 382. **12** Western Rattlesnake, p. 384. **13** Timber Rattlesnake, **13a** variant, p. 383. **14** Massasauga, p. 382.

PLATE 70
SALAMANDERS

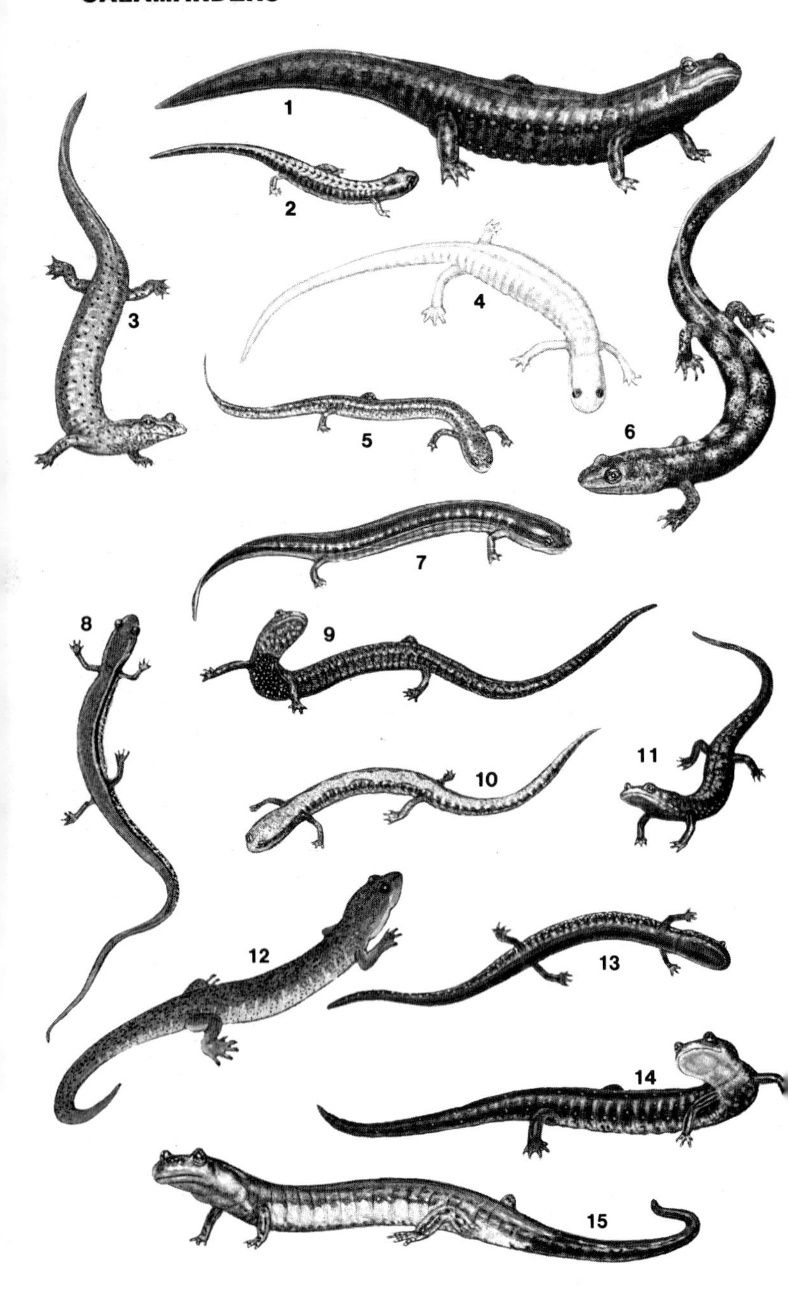

1 Black-bellied Salamander, p. 403. **2** Pigmy Salamander, p. 401.
3 Spring Salamander, p. 412. **4** Grotto Salamander, p. 409.
5 Many-ribbed Salamander, p. 412. **6** Shovel-nosed Salamander, p. 404.
7 Many-lined Salamander, p. 409. **8** Dwarf Salamander, p. 411.
9 Ravine Salamander, p. 405. **10** Zigzag Salamander, p. 406.
11 Weller's Salamander, p. 405. **12** Black Mountain Salamander, p. 403.
13 Cheat Mountain Salamander, p. 407. **14** Wehrle's Salamander, p. 408.
15 Yonahlossee Salamander, p. 408.

PLATE 71
SALAMANDERS

1 Red-backed Salamander, **1a** "Lead-backed" phase, p. 405. **2** Mountain Dusky Salamander, p. 401. **3** Dusky Salamander, **3a** variant, p. 402 **4** Two-lined Salamander, p. 411. **5** Seal Salamander, p. 402. **6** Spring Salamander, p. 412. **7** Slimy Salamander, p. 408. **8** Jordan's Woodland Salamander, p. 407. **9** Green Salamander, p. 410. **10** Red Salamander, **10a** old, p. 413. **11** Four-toed Salamander, p. 409. **12** Mud Salamander, p. 414. **13** Cave Salamander, p. 410. **14** Long-tailed Salamander, p. 411.

PLATE 72
SALAMANDERS

1 Two-toed Amphiuma, p. 393. **2** Greater Siren, p. 392. **3** Mudpuppy, p. 395. **4** Hellbender, p. 394. **5** Newt, aquatic adult, **5a** Eft, terrestrial stage, p. 400. **6** Small-mouthed Salamander, p. 398. **7** Tiger Salamander, p. 396. **8** Mole Salamander, p. 399. **9** Spotted Salamander, p. 395. **10** Marbled Salamander, p. 396. **11** Jefferson Salamander, p. 397. **12** Ringed Salamander, p. 396.

PLATE 73
FROGS AND TOADS

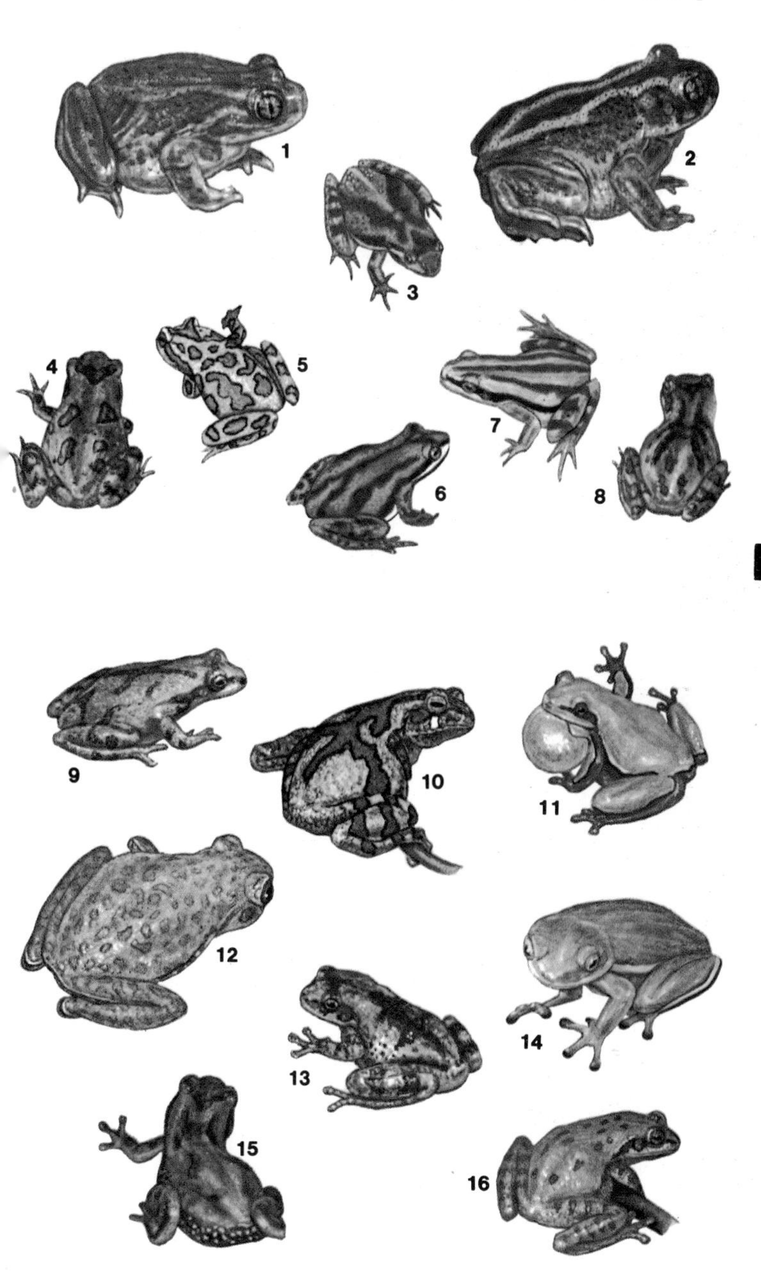

1 Plains Spadefoot, p. 418. **2** Eastern Spadefoot, p. 417. **3** Southern Cricket Frog, p. 427. **4** Strecker's Chorus Frog, p. 430. **5** Spotted Chorus Frog, p. 430. **6** Striped Chorus Frog, p. 428. **7** Brimley's Chorus Frog, p. 429. **8** Mountain Chorus Frog, p. 429. **9** Spring Peeper, p. 423. **10** Gray Treefrog, p. 425. **11** Pine Barrens Treefrog, p. 424. **12** Barking Treefrog, p. 423. **13** Bird-voiced Treefrog, p. 425. **14** Green Treefrog, p. 423. **15** Pine Woods Treefrog, p. 424. **16** Squirrel Treefrog, p. 425.

PLATE 74
FROGS AND TOADS

1 Western Spadefoot, p. 418. **2** Canadian Toad, p. 421. **3** Great Plains Narrow-mouthed Toad, p. 431. **4** Oak Toad, p. 420. **5** Southern Toad, p. 419. **6** Ornate Chorus Frog, p. 429. **7** Northern Leopard Frog, p. 435. **8** Southern Chorus Frog, p. 428. **9** Northern Cricket Frog, p. 426. **10** Plains Leopard Frog, p. 434. **11** Pig Frog, p. 433. **12** River Frog, p. 433.

FROGS AND TOADS

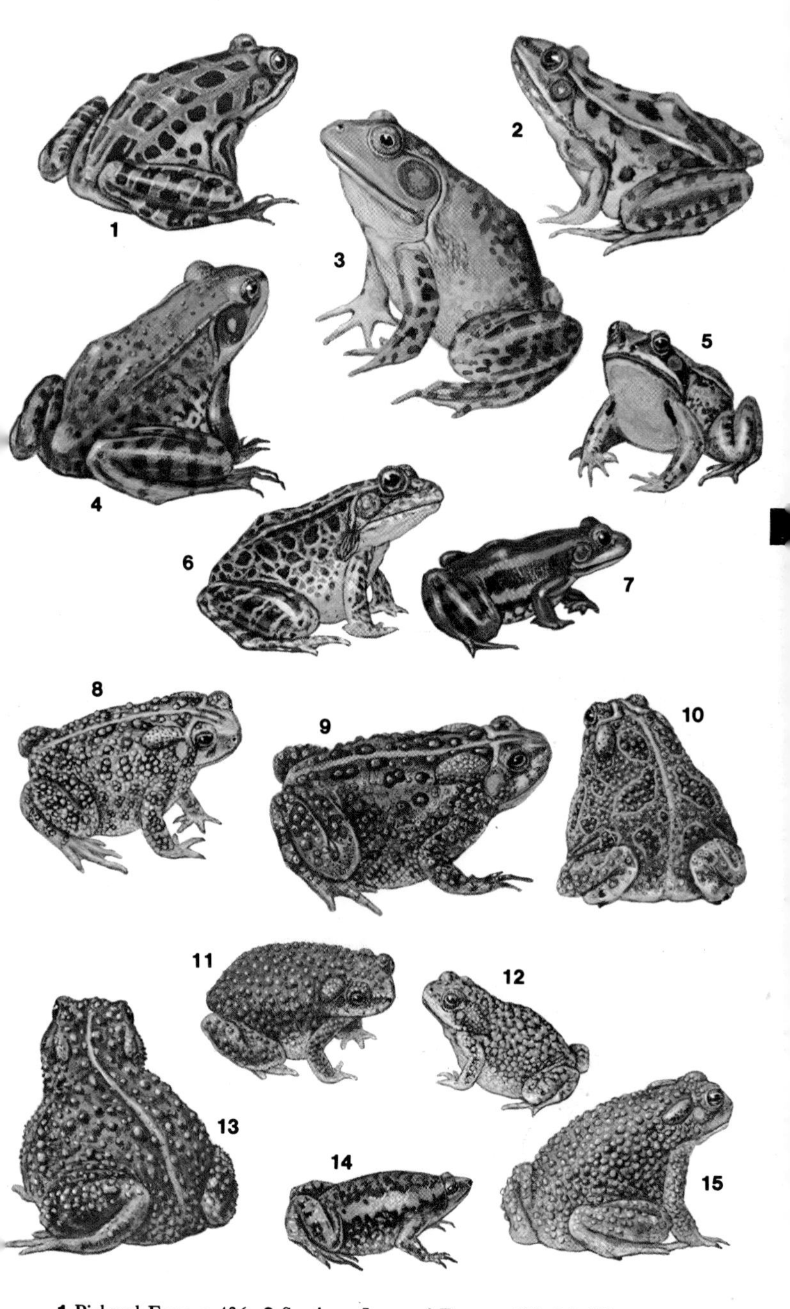

1 Pickerel Frog, p. 436. **2** Southern Leopard Frog, p. 435. **3** Bullfrog, p. 432.
4 Green Frog, p. 432. **5** Wood Frog, p. 437. **6** Crayfish Frog, p. 434.
7 Carpenter Frog, p. 434. **8** Woodhouse's Toad, p. 419. **9** American Toad, p. 419.
10 Great Plains Toad, p. 420. **11** Red-spotted Toad, p. 422.
12 Green Toad, p. 422. **13** Western Toad, p. 420.
14 Eastern Narrow-mouthed Toad, p. 431. **15** Texas Toad, p. 421.

PLATE 76
SALTWATER FISHES

1 Northern Kingfish, p. 485. **2** Atlantic Silverside, p. 474. **3** Atlantic Mackerel, p. 494. **4** Spot, p. 485. **5** Redfish, p. 499. **6** Scup, p. 487. **7** Northern Puffer, p. 509. **8** Tautog, p. 489. **9** Haddock, p. 468. **10** Silver Hake, p. 470. **11** Pollack, p. 468. **12** Atlantic Cod, p. 467. **13** Menhaden, p. 463. **14** Tilefish, p. 477. **15** Atlantic Halibut, p. 505.

PLATE 77
OTHER SALTWATER FISHES

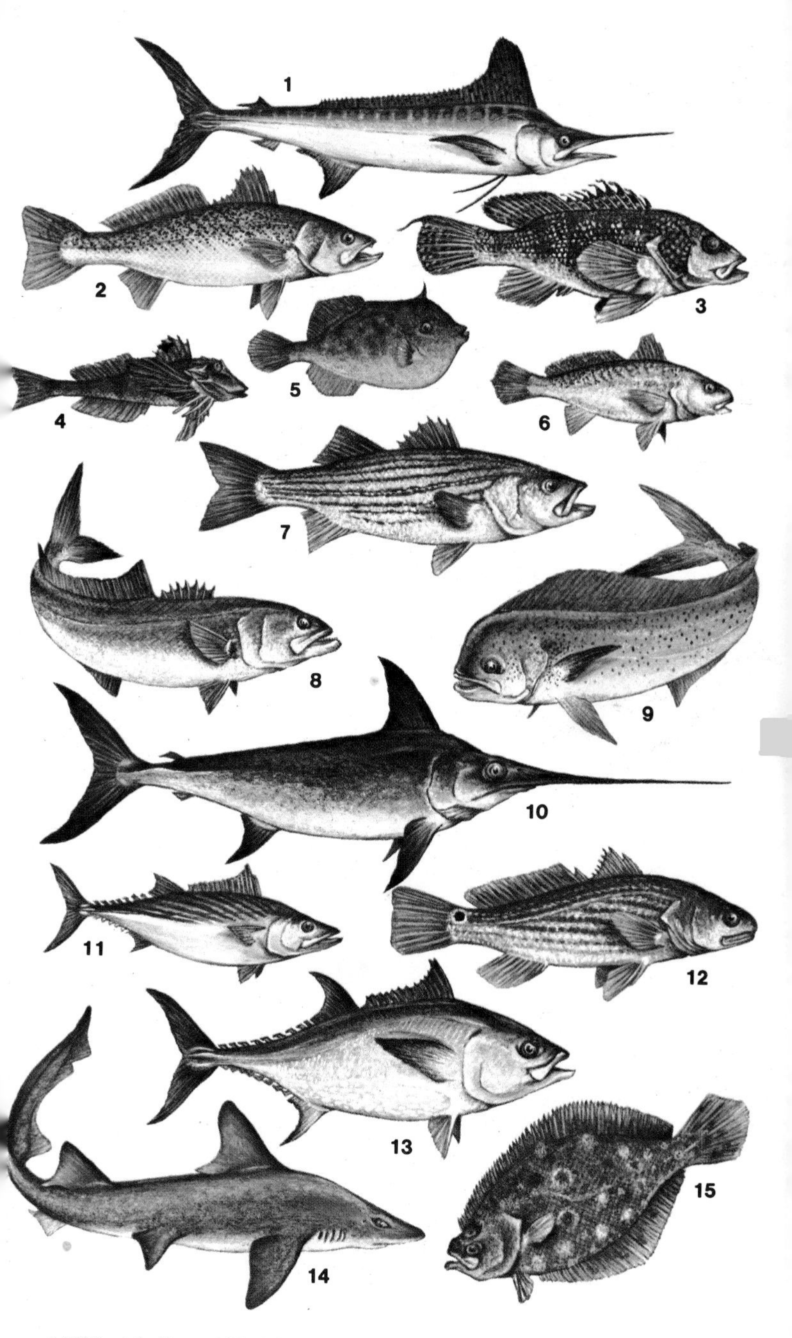

1 White Marlin, p. 497. **2** Weakfish, p. 486. **3** Black Sea Bass, p. 476. **4** Northern Sea Robin, p. 500. **5** Orange Filefish, p. 508. **6** Atlantic Croaker, p. 484. **7** Striped Bass, p. 475. **8** Bluefish, p. 477. **9** Dolphin, p. 482. **10** Swordfish, p. 498. **11** Atlantic Bonito, p. 496. **12** Red Drum, p. 484. **13** Bluefin Tuna, p. 495. **14** Smooth Dogfish, p. 455. **15** Summer Flounder, p. 503.

PLATE 78
SHARKS

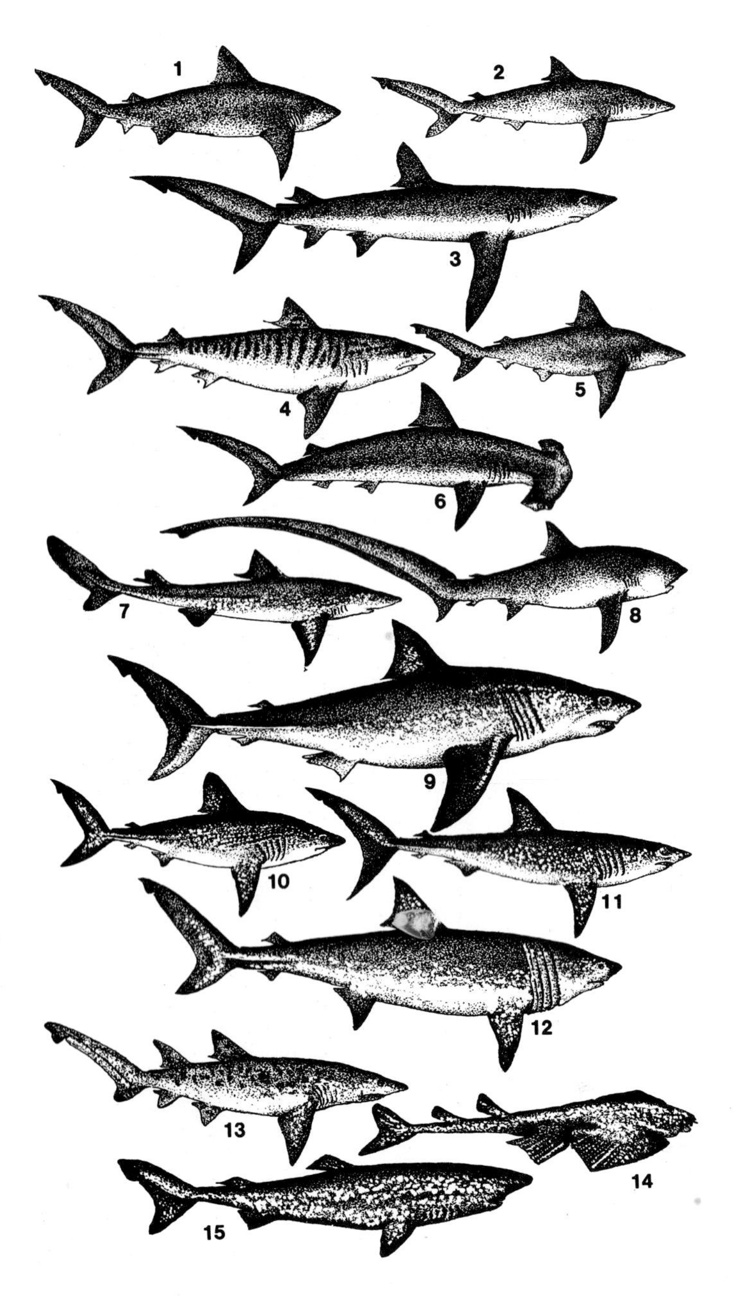

1 Bull Shark, p. 454. **2** Dusky Shark, p. 455. **3** Great Blue Shark, p. 454. **4** Tiger Shark, p. 455. **5** Sandbar Shark, p. 455. **6** Smooth Hammerhead, p. 456. **7** Spiny Dogfish, p. 456. **8** Thresher Shark, p. 454. **9** White Shark, p. 453. **10** Porbeagle Shark, p. 452. **11** Shortfin Mako Shark, p. 453. **12** Basking Shark, p. 453. **13** Sand Tiger, p. 451. **14** Angel Shark, p. 457. **15** Greenland Shark, p. 457.

PLATE 79
SKATES AND RAYS

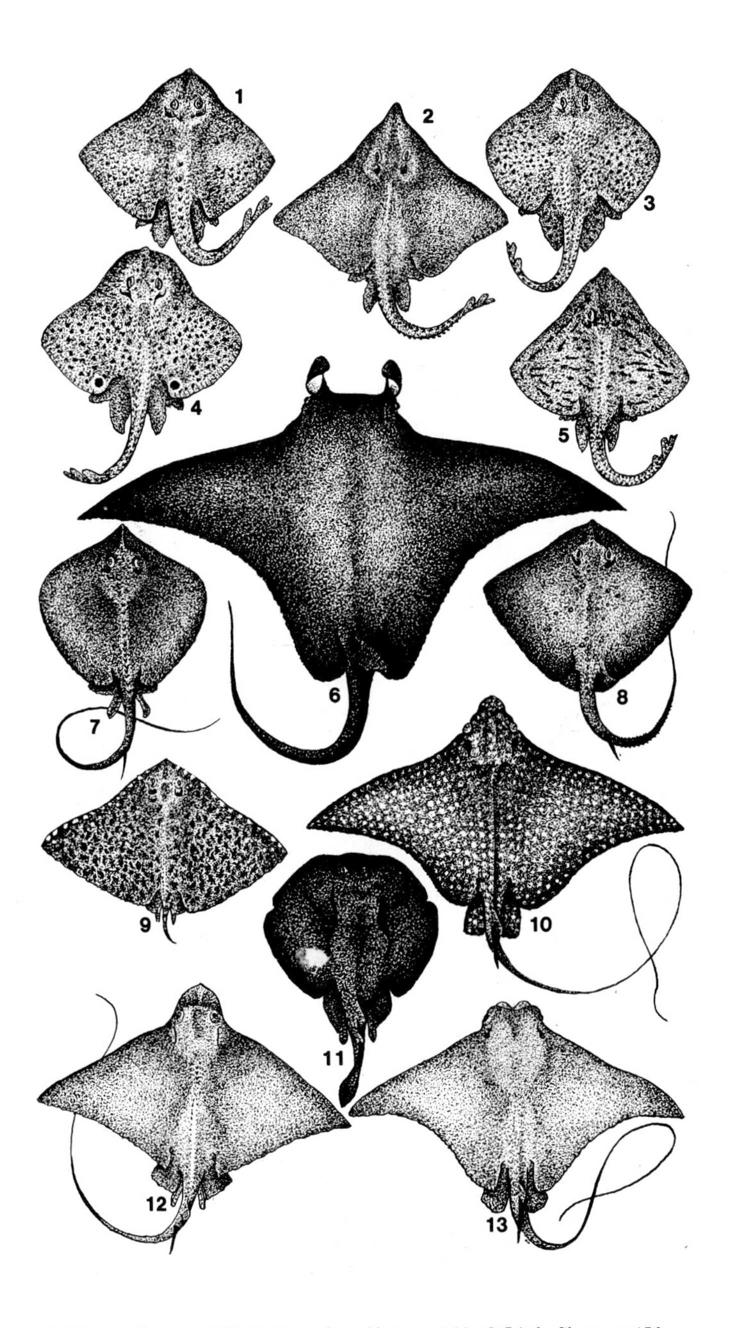

1 Thorny Skate, p. 459. **2** Barn-door Skate, p. 458. **3** Little Skate, p. 458. **4** Winter Skate, p. 459. **5** Clearnose Skate, p. 458. **6** Atlantic Stingray, p. 460. **7** Atlantic Manta, p. 461. **8** Roughtail Stingray, p. 460. **9** Smooth Butterfly Ray, p. 459. **10** Spotted Eagle Ray, p. 460. **11** Atlantic Torpedo, p. 457. **12** Bullnose Ray, p. 460. **13** Cownose Ray, p. 461.

PLATE 80 KILLIFISHES, PERCHLIKE FISHES, EELS, OTHERS

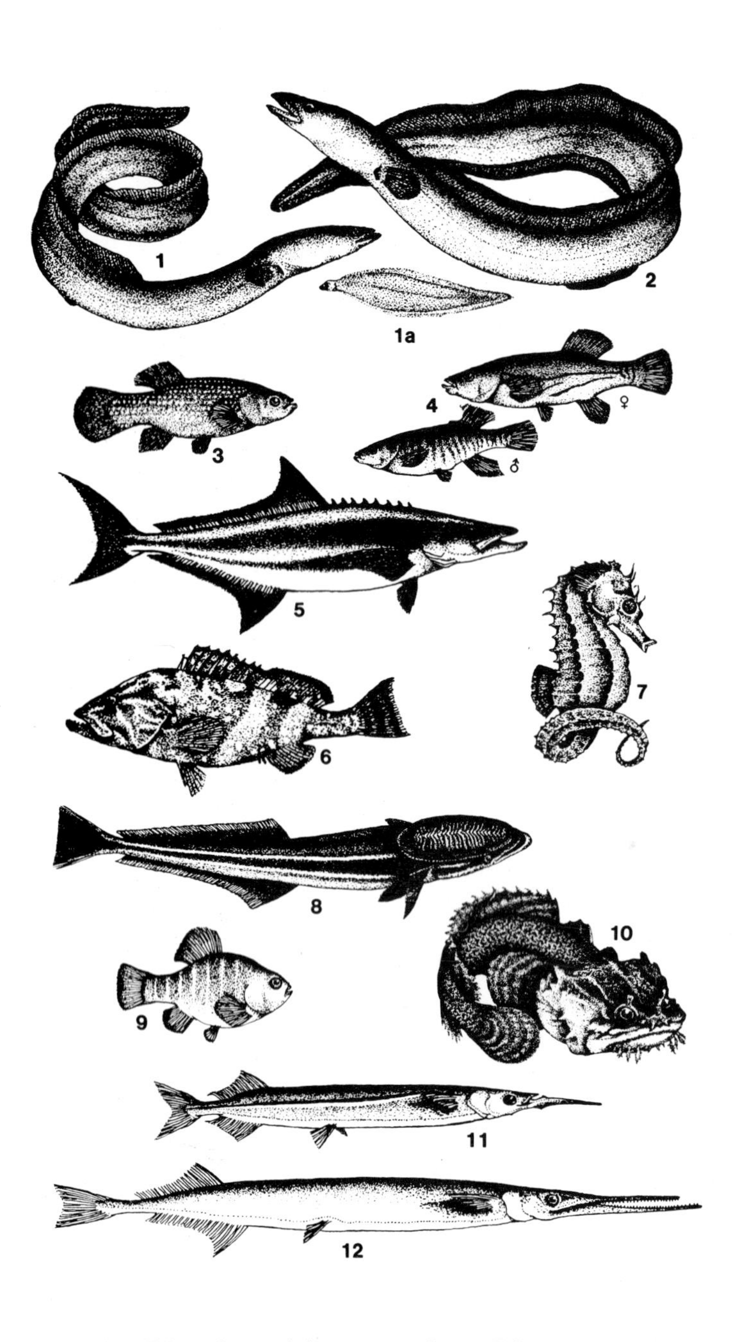

1 American Eel, **1a** leptocephalus, p. 465. **2** Conger Eel, p. 466. **3** Mummichog, p. 473. **4** Striped Killifish, p. 473. **5** Cobia, p. 477. **6** Red Grouper, p. 476. **7** Lined Sea Horse, p. 475. **8** Sharksucker, p. 478. **9** Sheepshead Minnow, p. 473. **10** Oyster Toadfish, p. 470. **11** Halfbeak, p. 472. **12** Atlantic Needlefish, p. 472.

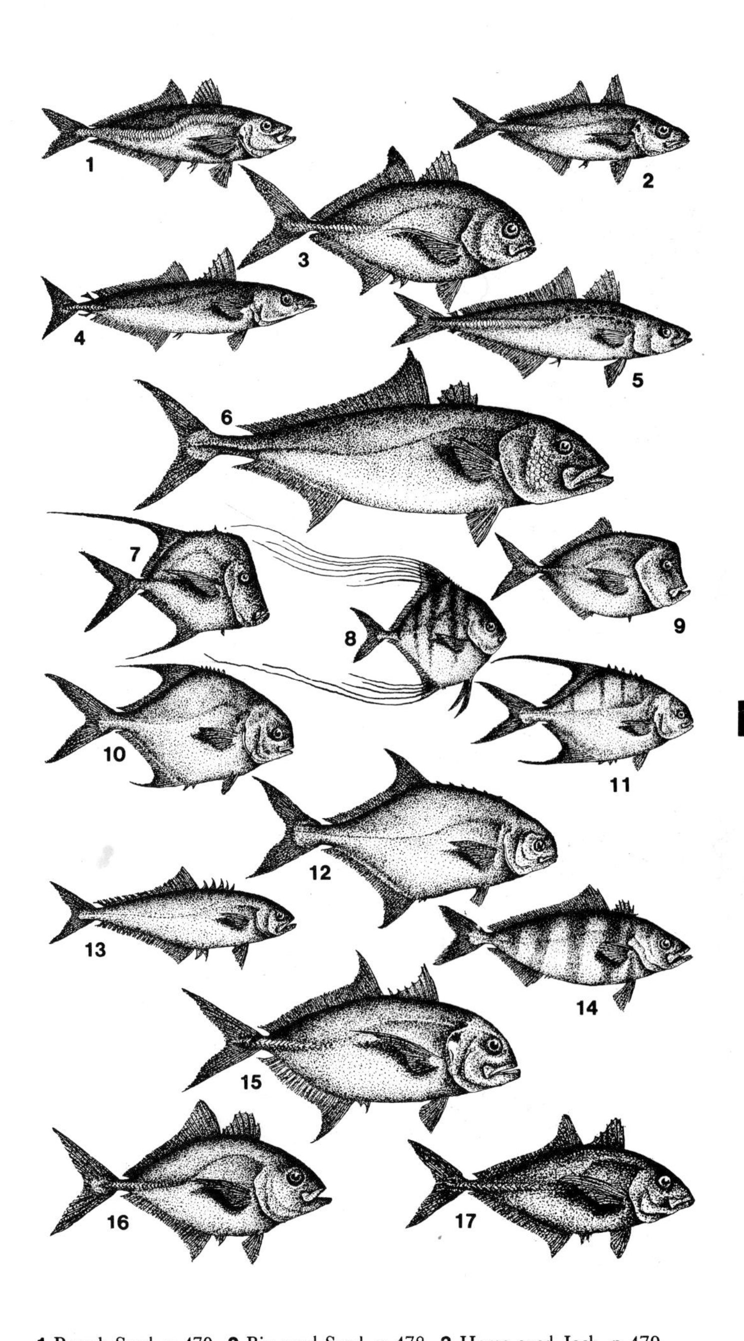

1 Rough Scad, p. 479. **2** Big-eyed Scad, p. 478. **3** Horse-eyed Jack, p. 479. **4** Mackerel Scad, p. 479. **5** Round Scad, p. 479. **6** Great Amberjack, p. 482. **7** Lookdown, p. 480. **8** African Pompano, p. 480. **9** Atlantic Moonfish, p. 480. **10** Permit, p. 481. **11** Palometa, p. 481. **12** Florida Pompano, p. 481. **13** Leatherjacket, p. 481. **14** Banded Rudderfish, p. 482. **15** Crevelle Jack, p. 480. **16** Yellow Jack, p. 479. **17** Blue Runner, p. 480.

PLATE 82
MACKERELS AND CROAKERS

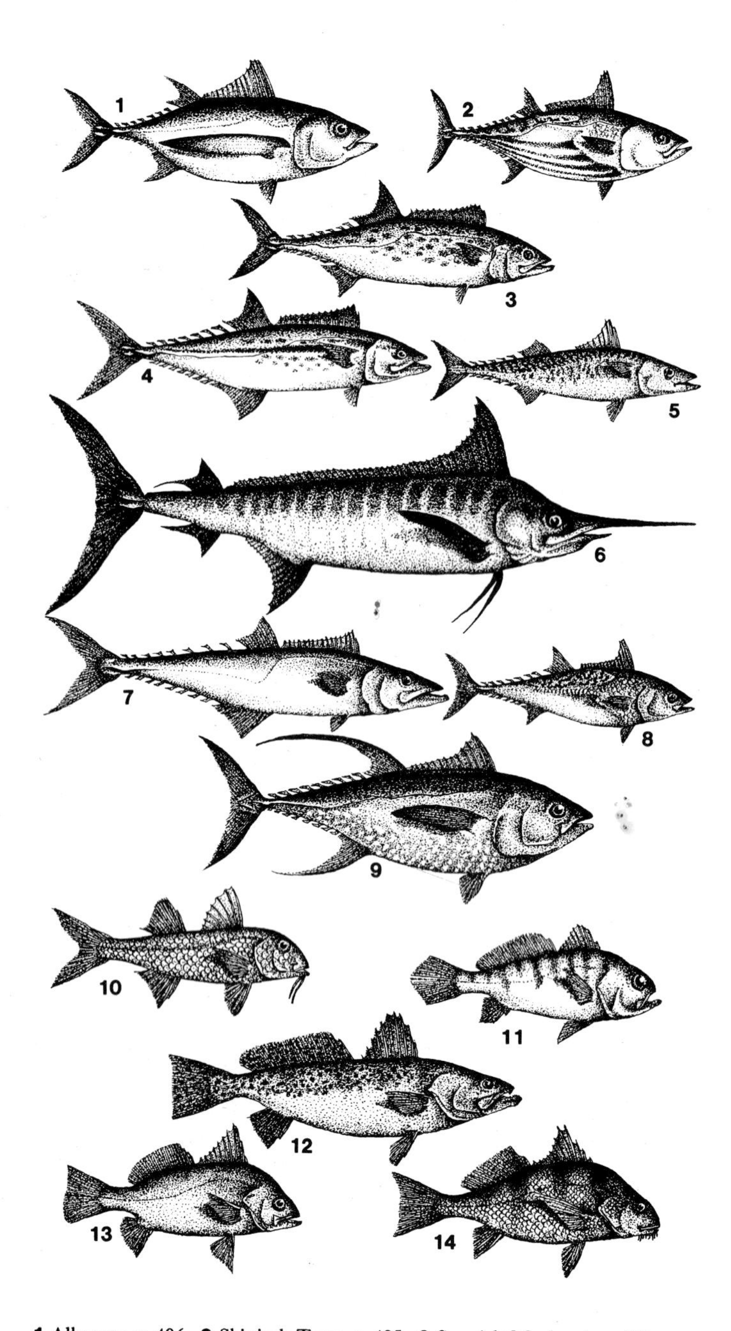

1 Albacore, p. 496. **2** Skipjack Tuna, p. 495. **3** Spanish Mackerel, p. 493. **4** Cero, p. 493. **5** Chub Mackerel, p. 494. **6** Blue Marlin, p. 497. **7** King Mackerel, p. 494. **8** Little Tuna, p. 495. **9** Yellowfin Tuna, p. 496. **10** Red Goatfish, p. 488. **11** Banded Drum, p. 484. **12** Spotted Sea Trout, p. 486. **13** Silver Perch, p. 484. **14** Black Drum, p. 486.

PLATE 83
PERCHLIKE FISHES

1 Wrymouth, p. 492. **2** Atlantic Wolffish, p. 491. **3** Great Barracuda, p. 489. **4** Striped Mullet, p. 490. **5** Northern Sennet, p. 489. **6** White Mullet, p. 490. **7** Sheepshead, p. 487. **8** Pinfish, p. 487. **9** Bermuda Chub, p. 488. **10** Cunner, p. 488. **11** Atlantic Cutlassfish, p. 492. **12** American Sand Lance, p. 492.

PLATE 84
BOTTOM FISHES

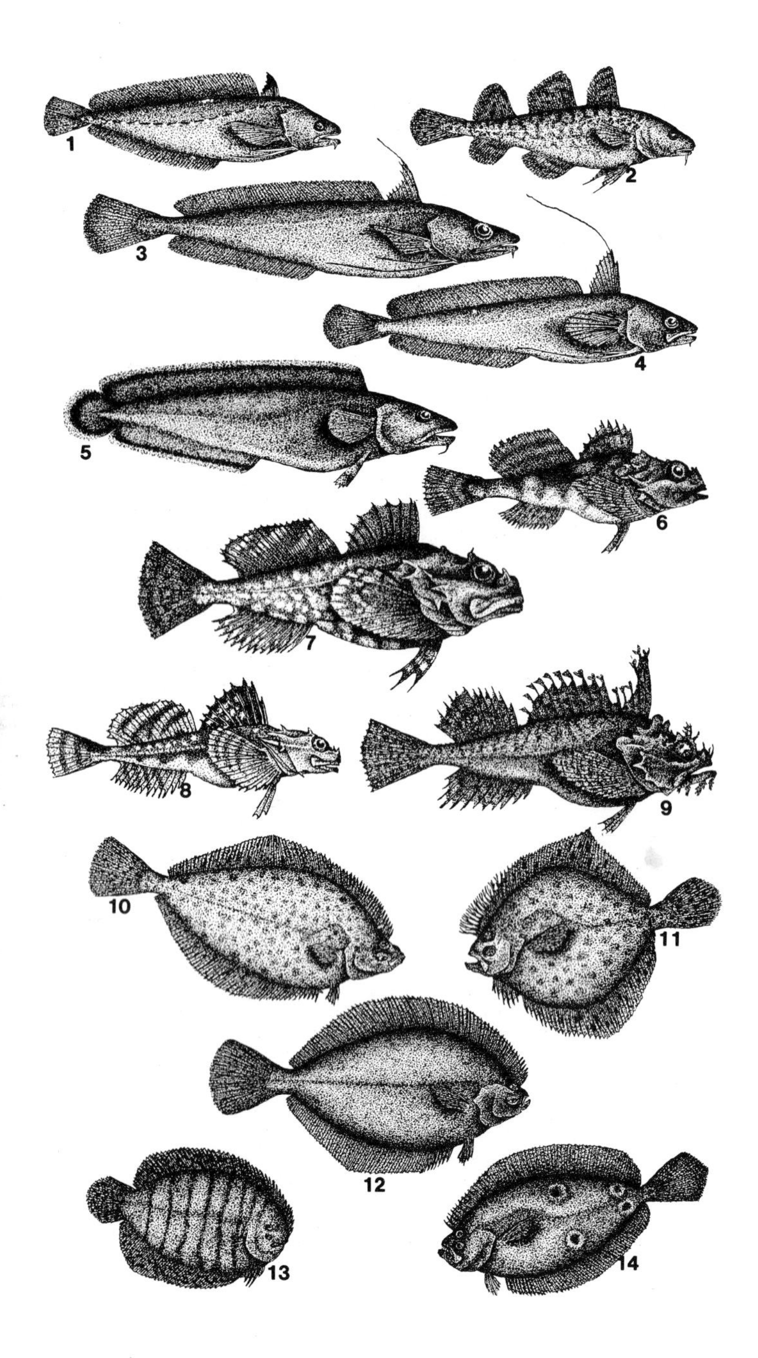

1 Spotted Hake, p. 469. **2** Atlantic Tomcod, p. 467. **3** White Hake, p. 469. **4** Red Hake, p. 469. **5** Cusk, p. 470. **6** Grubby, p. 501. **7** Shorthorn Sculpin, p. 501. **8** Longhorn Sculpin, p. 501. **9** Sea Raven, p. 501. **10** Yellowtail Flounder, p. 506. **11** Windowpane, p. 503. **12** Winter Flounder, p. 506. **13** Hogchoker, p. 507. **14** Fourspot Flounder, p. 503.

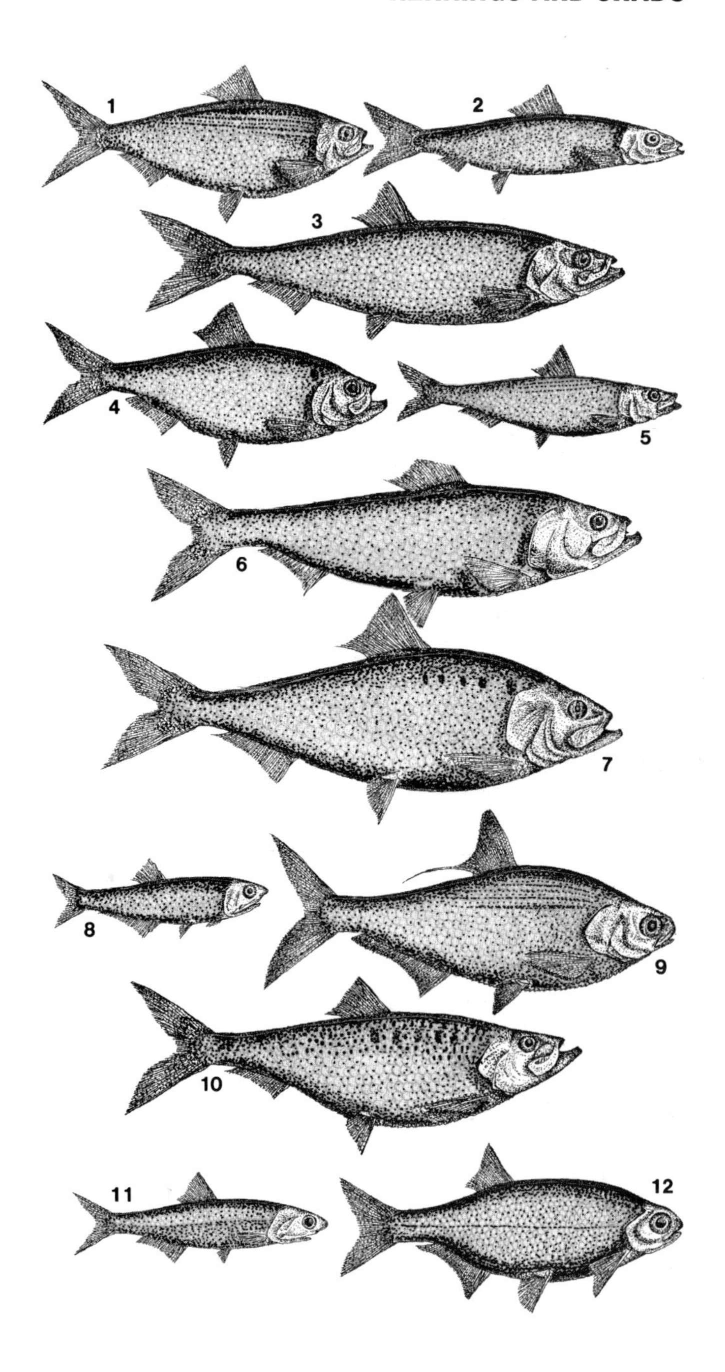

1 Blueback Herring, p. 517. **2** Round Herring, p. 464.
3 Atlantic Herring, p. 464. **4** Alewife, p. 518. **5** Spanish Sardine, p. 464.
6 Skipjack Herring, p. 517. **7** American Shad, p. 518.
8 Bay Anchovy, p. 465. **9** Gizzard Shad, p. 516. **10** Hickory Shad, p. 517.
11 Striped Anchovy, p. 465. **12** Mooneye, p. 516.

PLATE 86
FRESHWATER FISHES

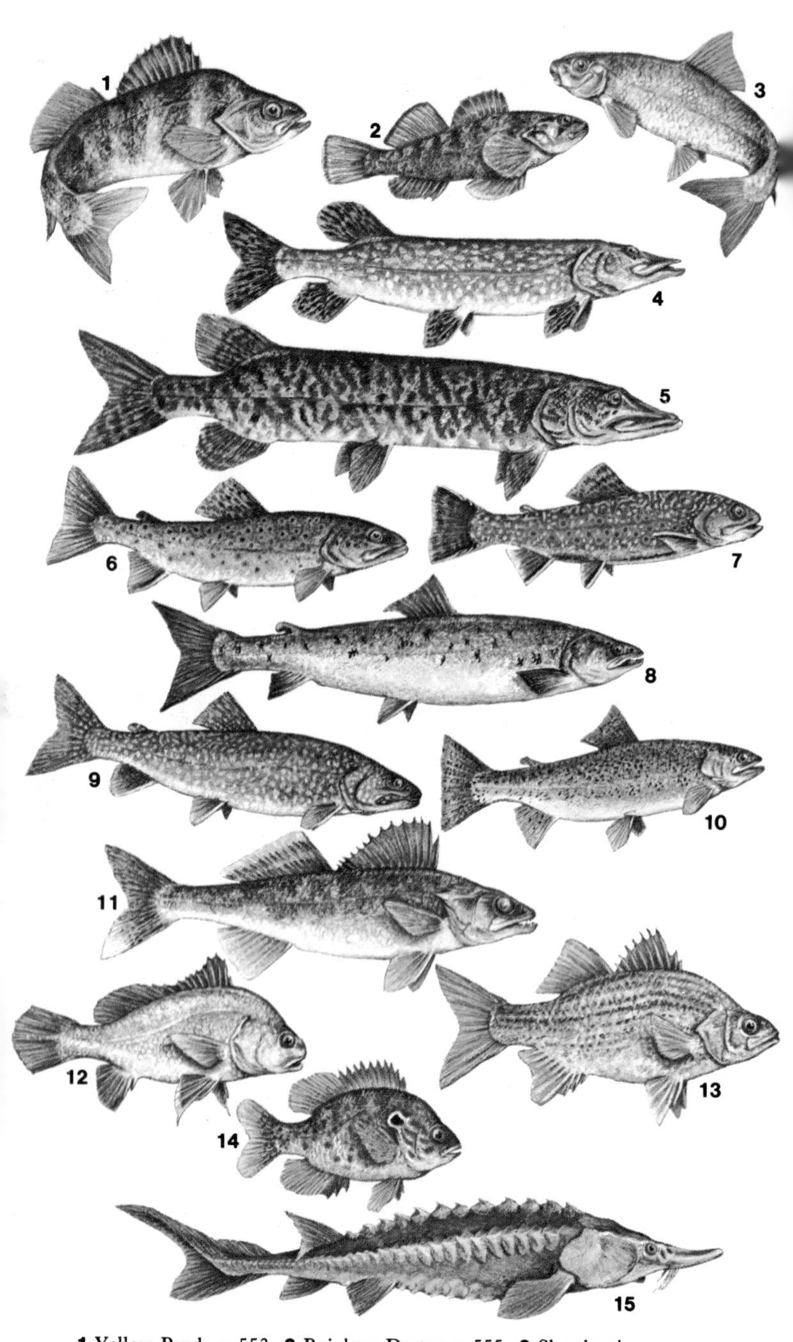

1 Yellow Perch, p. 553. **2** Rainbow Darter, p. 555. **3** Shorthead Redhorse, p. 531. **4** Northern Pike, p. 525. **5** Muskellunge, p. 526. **6** Brown Trout, p. 519. **7** Brook Trout, p. 520. **8** Atlantic Salmon, p. 519. **9** Lake Trout, p. 520. **10** Rainbow Trout, p. 520. **11** Walleye, p. 554. **12** Freshwater Drum, p. 556. **13** White Bass, p. 547. **14** Orange Spotted Sunfish, p. 552. **15** Lake Sturgeon, p. 513.

PLATE 87
FRESHWATER FISHES

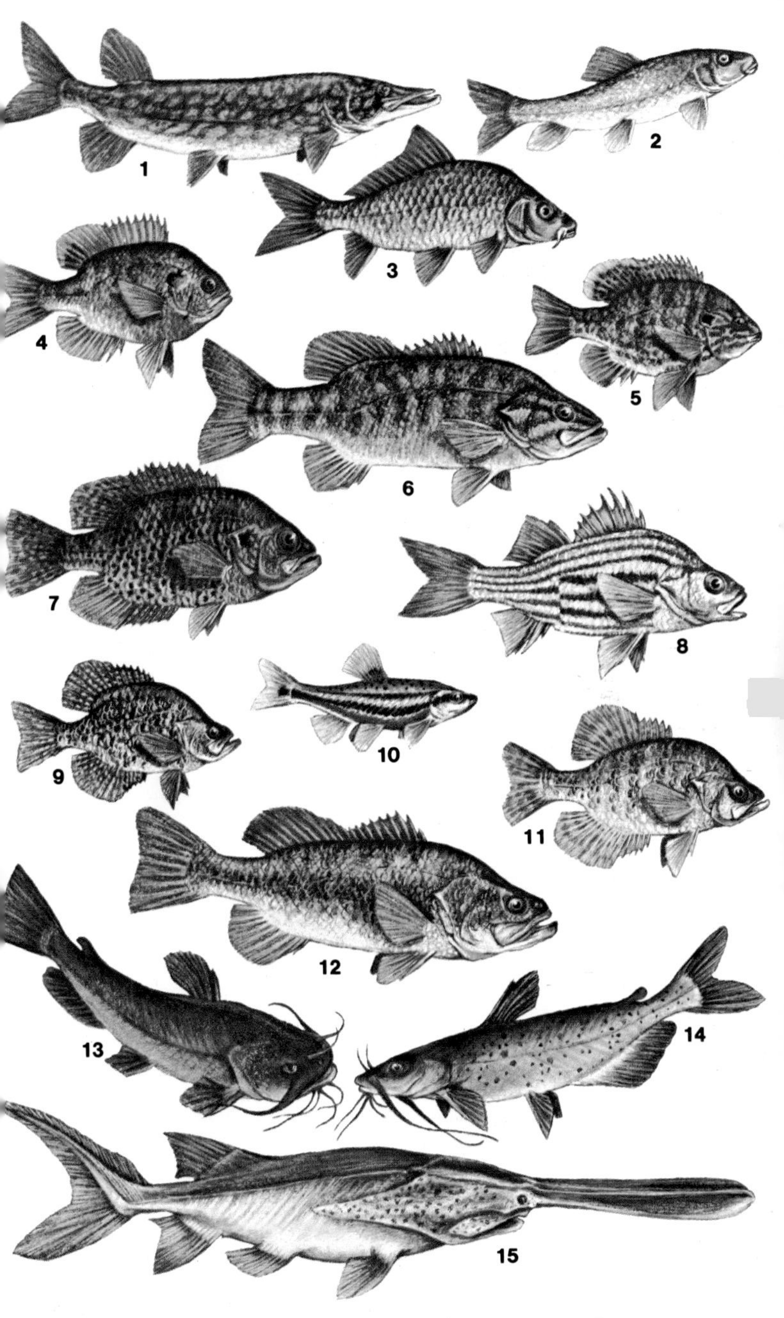

1 Chain Pickerel, p. 525. **2** White Sucker, p. 527. **3** Carp, p. 532.
4 Bluegill, p. 551. **5** Pumpkinseed, p. 552. **6** Smallmouth Bass, p. 549.
7 Rockbass, p. 548. **8** Yellow Bass, p. 547. **9** Black Crappie, p. 552.
10 Southern Redbelly Dace, p. 534. **11** White Crappie, p. 553.
12 Largemouth Bass, p. 549. **13** Black Bullhead, p. 540.
14 Channel Catfish, p. 538. **15** Paddlefish, p. 513.

PLATE 88
STURGEONS AND GARS

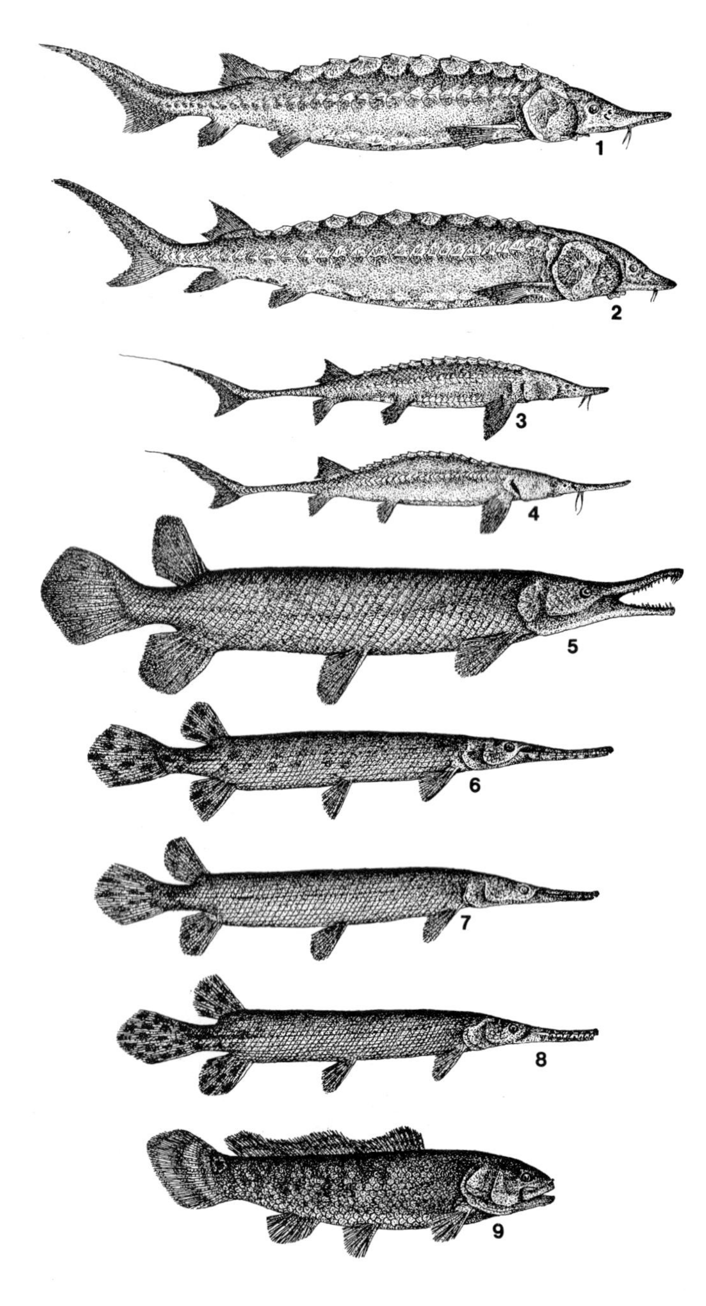

1 Atlantic Sturgeon, p. 512. **2** Shortnose Sturgeon, p. 513.
3 Shovelnose Sturgeon, p. 512. **4** Pallid Sturgeon, p. 512.
5 Alligator Gar, p. 515. **6** Longnose Gar, p. 514, **7** Shortnose Gar, p. 515.
8 Spotted Gar, p. 514. **9** Bowfin, p. 514.

PLATE 89
HERRINGLIKE, CODLIKE, PERCHLIKE FISHES

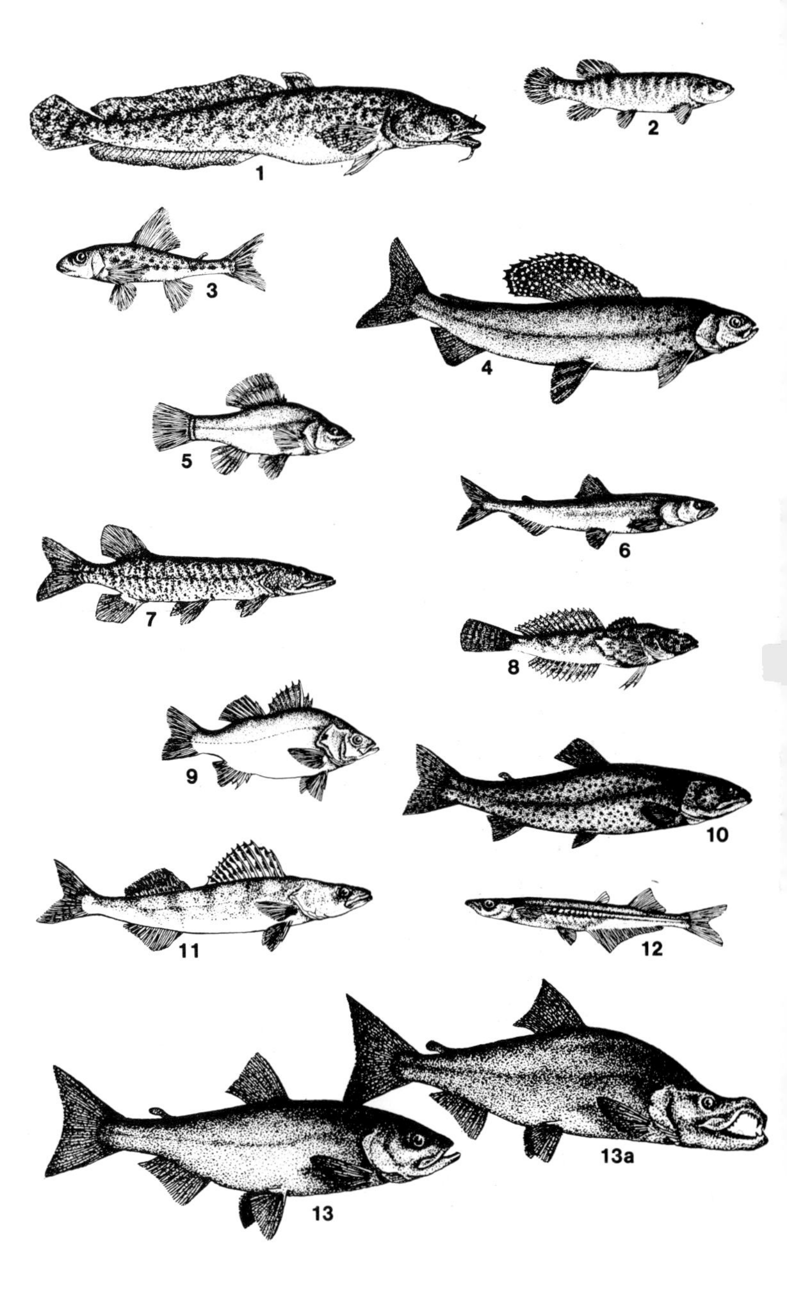

1 Burbot, p. 543. **2** Central Mudminnow, p. 524. **3** Troutperch, p. 542. **4** Arctic Grayling, p. 523. **5** Pirate Perch, p. 543. **6** Rainbow Smelt, p. 524. **7** Redfin Pickerel, p. 525. **8** Mottled Sculpin, p. 556. **9** White Perch, p. 547. **10** Cutthroat Trout, p. 520. **11** Sauger, p. 553. **12** Brook Silverside, p. 545. **13** Sockeye Salmon, **13a** breeding, p. 519.

PLATE 90
CHUBS AND SUCKERS

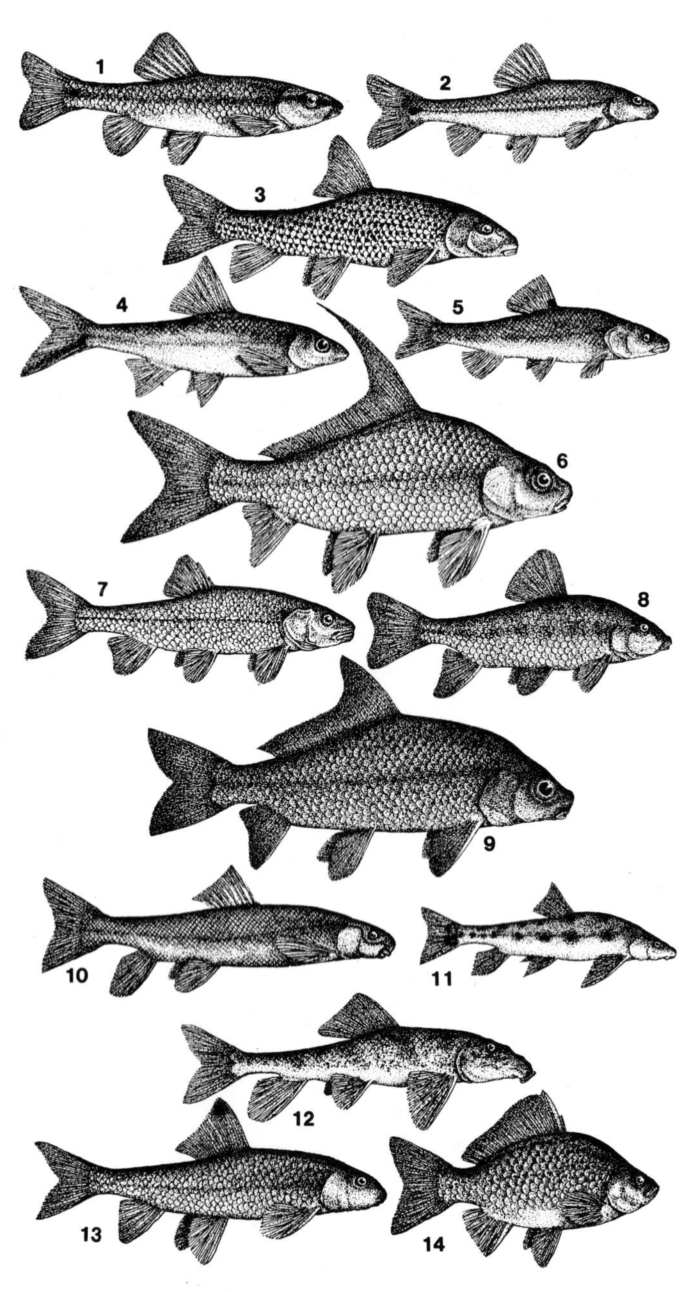

1 River Chub, p. 534. **2** Suckermouth Minnow, p. 536.
3 Spotted Sucker, p. 526. **4** Silver Chub, p. 535. **5** Creek Chub, p. 534.
6 Quillback, p. 529. **7** Fallfish, p. 533. **8** Lake Chubsucker, p. 527.
9 Black Buffalo, p. 529. **10** Mountain Sucker, p. 528.
11 Torrent Sucker, p. 531. **12** Northern Hog Sucker, p. 527.
13 Golden Redhorse, p. 532. **14** Goldfish, p. 533.

PLATE 91
DACE, SHINERS, MINNOWS

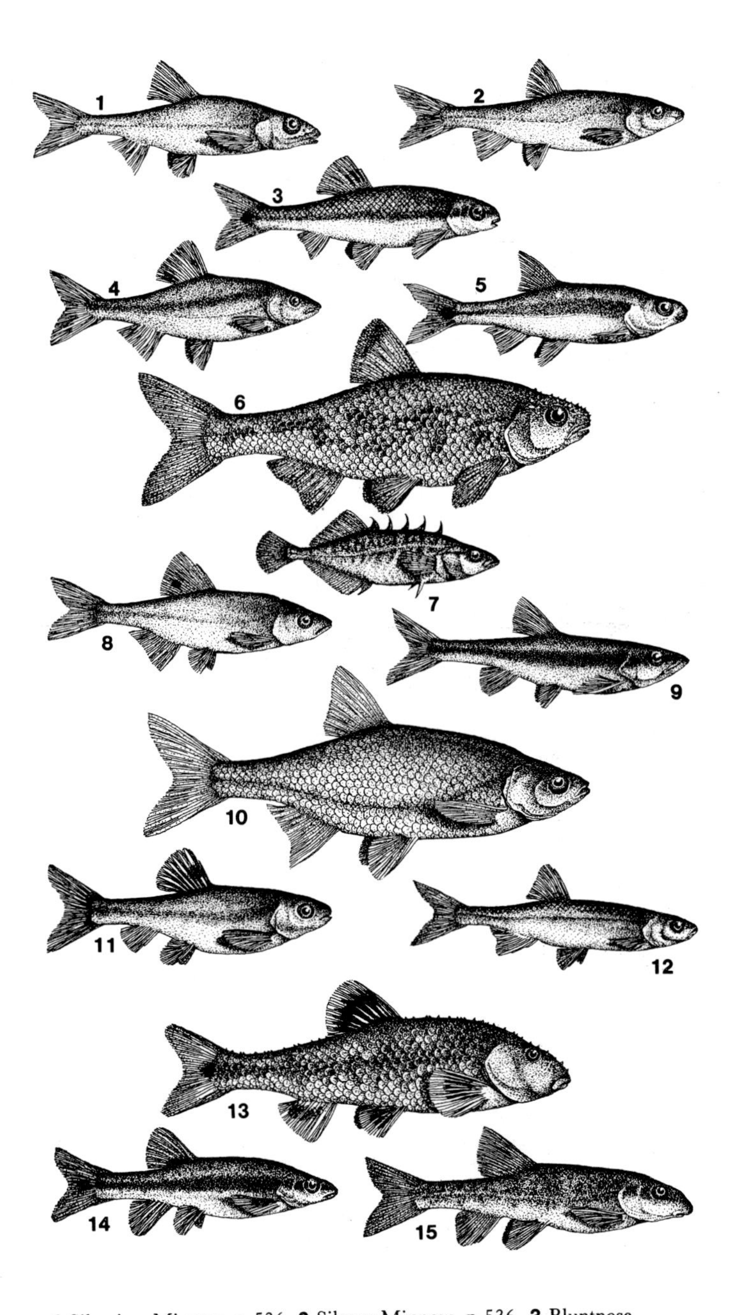

1 Silverjaw Minnow, p. 536. **2** Silvery Minnow, p. 536. **3** Bluntnose Minnow, p. 537. **4** Satinfin Shiner, p. 538. **5** Spottail Shiner, p. 538. **6** Common Shiner, p. 538. **7** Brook Stickleback, p. 546. **8** Spotfin Shiner, p. 538. **9** Redside Dace, p. 533. **10** Golden Shiner, p. 533. **11** Fathead Minnow, p. 536. **12** Emerald Shiner, p. 537. **13** Stoneroller, p. 537. **14** Blacknose Dace, p. 535. **15** Longnose Dace, p. 535.

PLATE 92
CATFISHES AND SUNFISHES

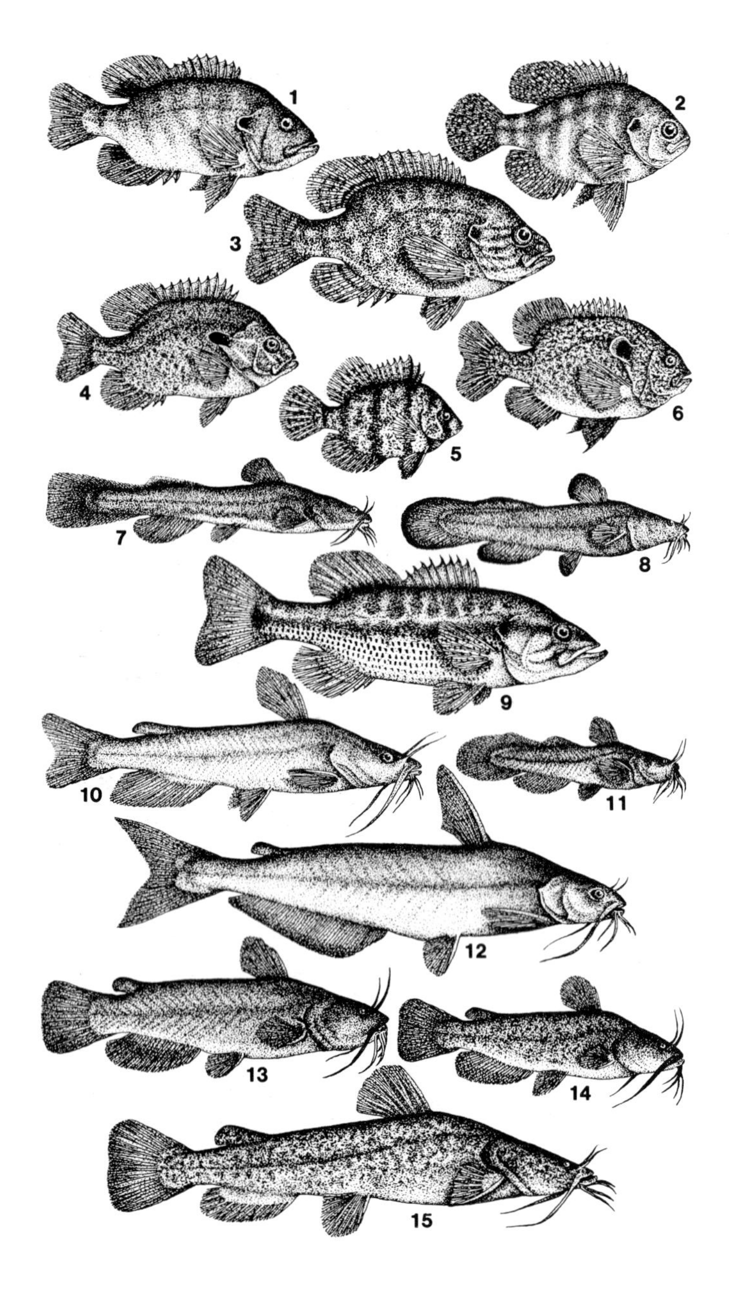

1 Green Sunfish, p. 551. **2** Banded Sunfish, p. 548. **3** Warmouth, p. 551. **4** Redbreast Sunfish, p. 550. **5** Blackbanded Sunfish, p. 548. **6** Longear Sunfish, p. 550. **7** Stonecat, p. 541. **8** Margined Madtom, p. 541. **9** Spotted Bass, p. 549. **10** White Catfish, p. 539. **11** Tadpole Madtom, p. 541. **12** Blue Catfish, p. 539. **13** Yellow Bullhead, p. 539. **14** Brown Bullhead, p. 540. **15** Flathead Catfish, p. 540.

PLATE 93
BLINDFISHES, KILLIFISHES, DARTERS

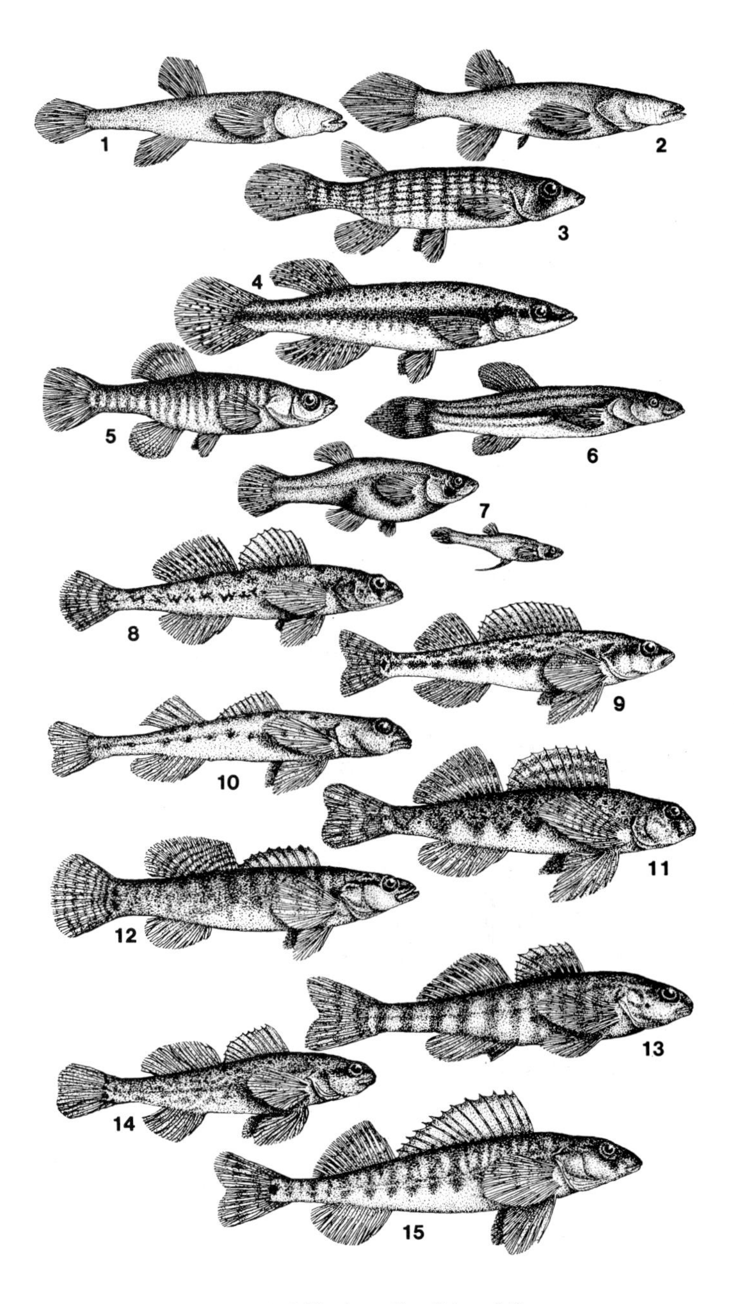

1 Southern Cavefish, p. 542. **2** Northern Cavefish, p. 542.
3 Starhead Topminnow, p. 545. **4** Blackstripe Topminnow, p. 545.
5 Banded Killifish, p. 544. **6** Swampfish, p. 542. **7** Mosquitofish, p. 545.
8 Johnny Darter, p. 555. **9** Blackside Darter, p. 554.
10 Eastern Sand Darter, p. 554. **11** Greenside Darter, p. 555.
12 Fantail Darter, p. 555. **13** Banded Darter, p. 555.
14 Northern Swamp Darter, p. 555. **15** Logperch, p. 554.

PLATE 94 LORICATES, KEYHOLE LIMPETS, LIMPETS, TOP SHELLS

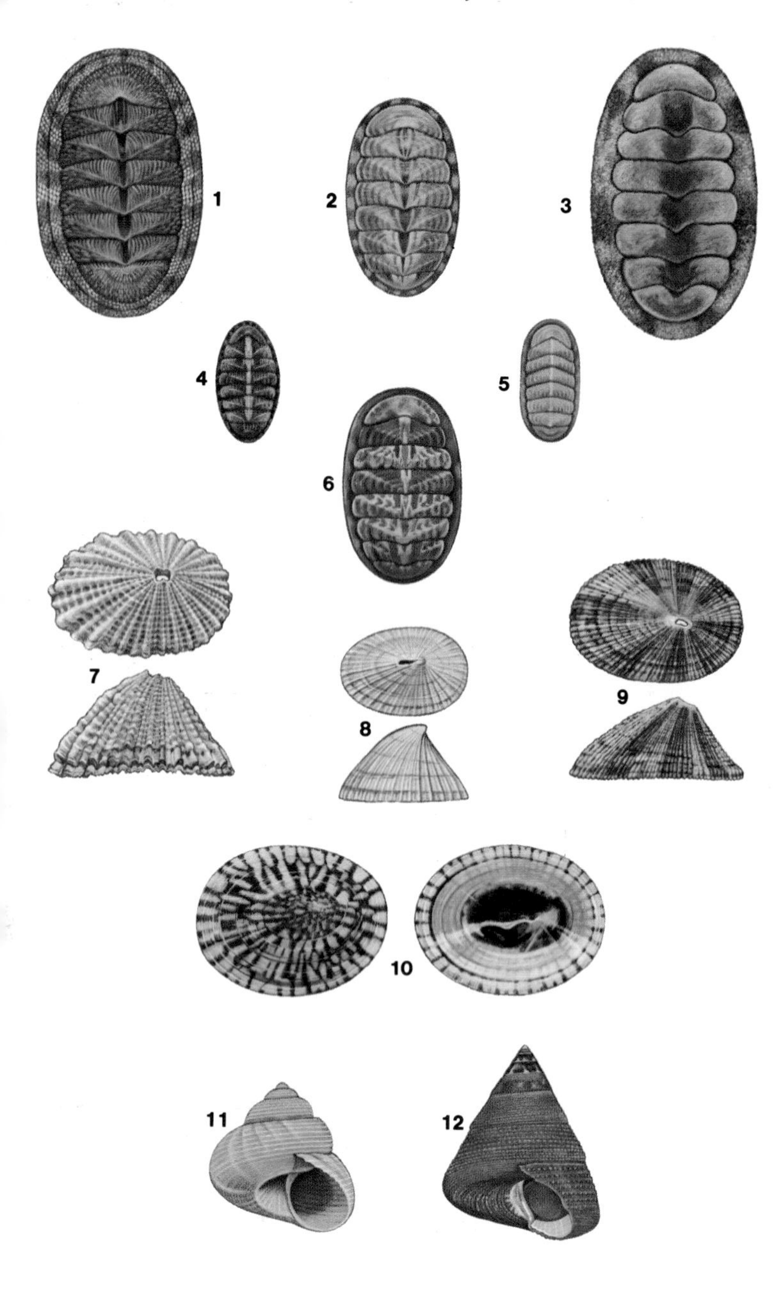

1 Tuberculate Chiton × ½, p. 565. **2** Red Chiton × 1, p. 566. **3** Fuzzy Chiton × ½, p. 565. **4** Eastern Chiton × 1, p. 565. **5** White Chiton × 1, p. 566. **6** Mottled Chiton × 1, p. 566. **7** Lister's Keyhole Limpet × 1, p. 568. **8** Linnaeus' Keyhole Limpet × 1, p. 567. **9** Little Keyhole Limpet × 1, p. 568. **10** Atlantic Plate Limpet × 1, p. 568. **11** Greenland Top Shell × 2, p. 569. **12** Mottled Top Shell × 1, p. 569.

PLATE 95
TURBANS, NERITES, PERIWINKLES, CHINK SHELLS

1 American Star Shell × 1, p. 570. **2** Long-spined Star Shell × 1, p. 569. **3** Knobby Turban × 1, p. 569. **4** Bleeding Tooth × 1, p. 570. **5** Zebra Nerite × 2, p. 571. **6** Tessellated Nerite × 1, p. 570. **7** Variegated Nerite × 1, p. 570. **8** Rough Periwinkle × 2, p. 572. **9** Smooth Periwinkle × 2, p. 571. **10** Salt Marsh Periwinkle × 1, p. 572. **11** Knobby Periwinkle, p. 572. **12** Little Chink Shell × 2, p. 572. **13** Common Periwinkle × 1, p. 571.

PLATE 96 CERITHS, WENTLETRAPS, ALLIED FAMILIES

1 Common Sundial × 1, p. 573. **2** Adams' Miniature Cerith × 2, p. 574. **3** Florida Horn Shell × 1, p. 573. **4** Variable Bittium × 5, p. 574. **5** Emerson's Cerith × 2, p. 574. **6** Dwarf Horn Shell × 2, p. 574. **7** Common Worm Shell × 1, p. 573. **8** Angled Wentletrap × 1, p. 575. **9** Brown-banded Wentletrap × 1, p. 575. **10** Atlantic Modulus × 1, p. 575. **11** Common Purple Sea Snail × 1, p. 576.

PLATE 97
BOAT SHELLS, MOON SHELLS, ALLIED FAMILIES

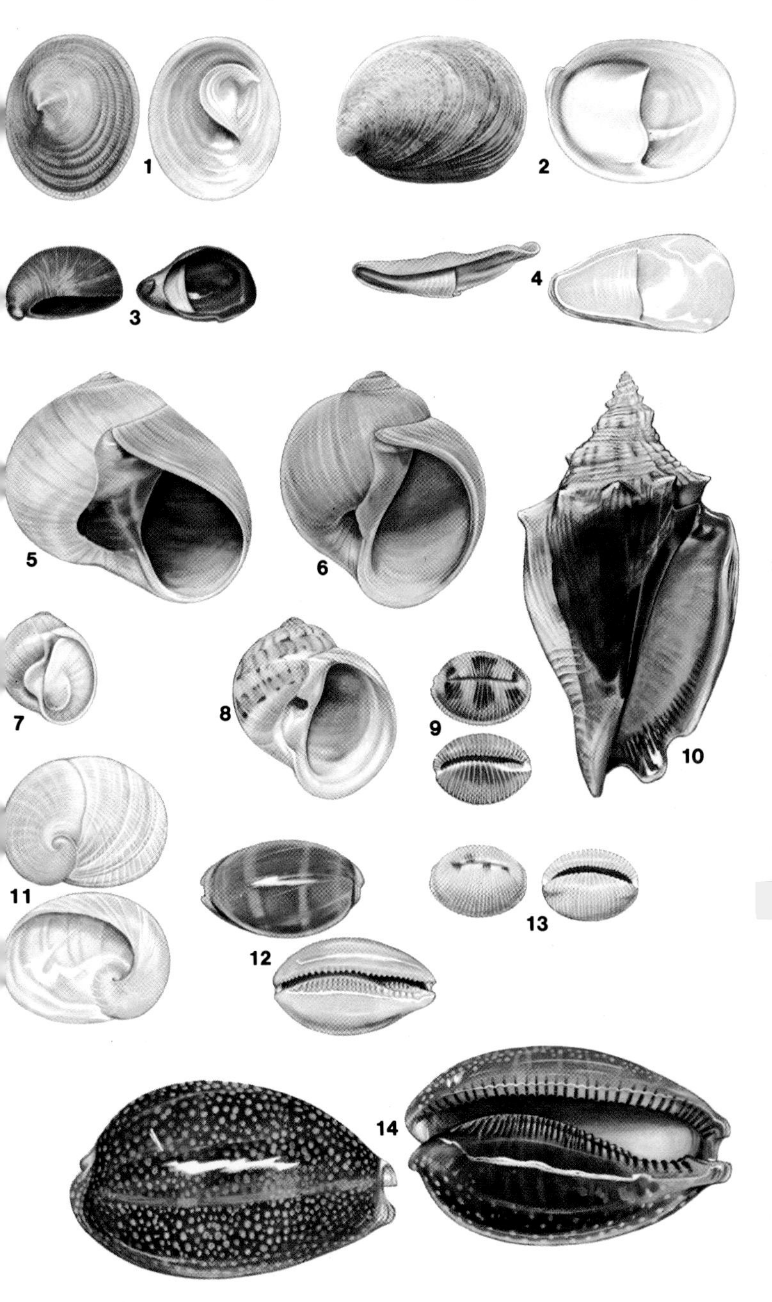

1 Striate Cup-and-saucer × 1, p. 576. **2** Common Boat Shell × 1, p. 576.
3 Little Boat Shell × 1, p. 577. **4** Flat Slipper Shell × 1, p. 576.
5 Shark Eye × 1, p. 578. **6** Common Moon Shell × 1, p. 578.
7 Miniature Natica × 2, p. 579. **8** Colorful Moon Shell × 1, p. 579.
9 Coffee Bean Trivia × 1, p. 577. **10** Florida Fighting Conch × 1, p. 578.
11 Common Baby's Ear × 1, p. 579. **12** Gray Cowry × 1, p. 580.
13 Four-spotted Trivia × 2, p. 577. **14** Deer Cowry × 1, p. 579.

PLATE 98
TRITON, EGG, HELMET, FROG AND ROCK SHELLS

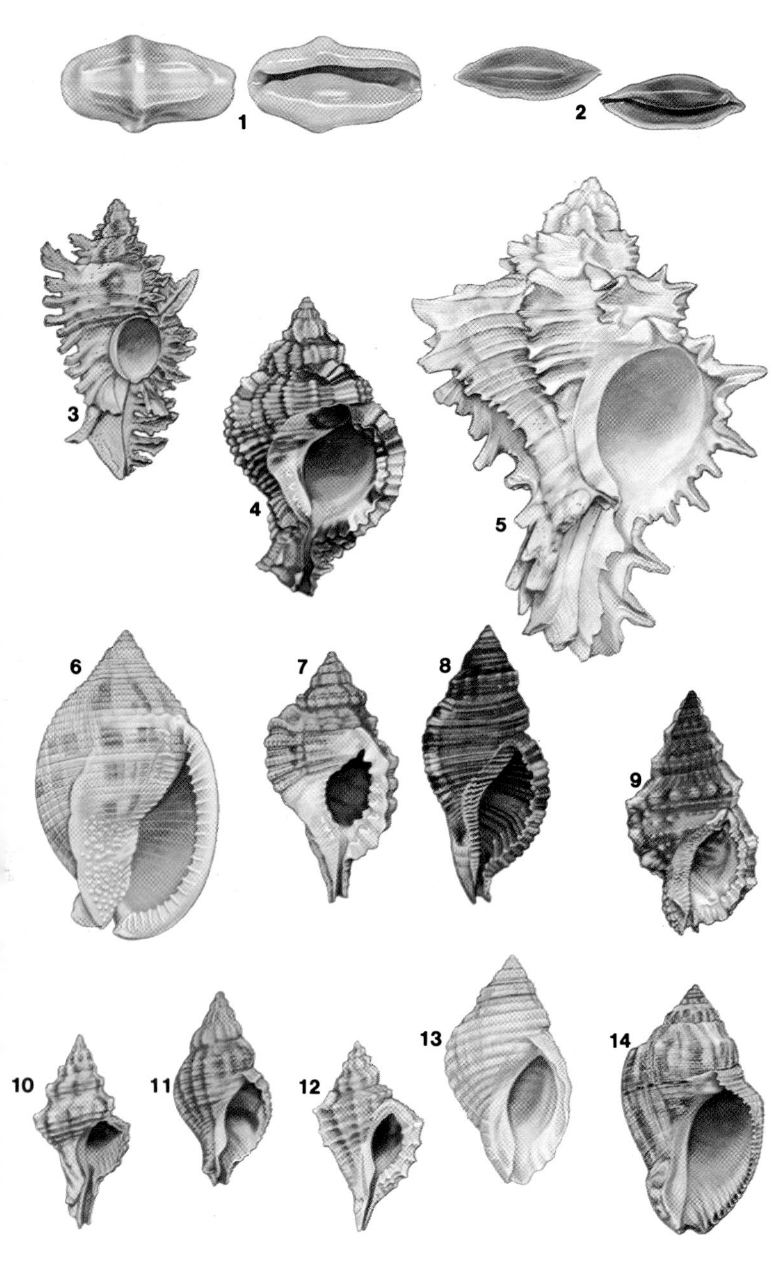

1 Flamingo Tongue × 1, p. 580. **2** Single-toothed Simnia × 1, p. 580. **3** Lace Murex × 1, p. 582. **4** Apple Murex × 1, p. 582. **5** Giant Eastern Murex × ½, p. 582. **6** Scotch Bonnet × 1, p 580. **7** White-mouthed Triton × 1, p. 581. **8** Hairy Triton × 1, p. 581. **9** Granular Frog Shell × 1, p. 581. **10** Florida Drill × 1, p. 583. **11** Oyster Drill × 1, p. 583. **12** Thick-lipped Drill × 1, p. 582. **13** Rock Purple × 1, p. 583. **14** Florida Rock Shell × 1, p. 583.

PLATE 99
WHELKS, DOG WHELKS, DOVE AND TULIP SHELLS

1 Lightning Whelk × ⅓, p. 586. **2** Channeled Whelk × ⅓, p. 585.
3 Knobbed Whelk × ⅓, p. 586. **4** Waved Whelk × ⅓, p. 584.
5 Mud Snail × 1, p. 587. **6** Mottled Dove Shell × 1½, p. 584.
7 New England Dog Whelk × 1, p. 587. **8** Greedy Dove Shell × 1½, p. 584.
9 Mottled Dog Whelk × 1½, p. 586. **10** Brown-corded Neptune × ⅓, p. 585.
11 Crown Conch × ⅓, p. 586. **12** Florida Banded Tulip × ⅓, p. 588.
13 Tulip Shell × ⅓, p. 587. **14** Stimpson's Whelk × ⅓, p. 585.
15 Florida Horse Conch × ¼, p. 588.

PLATE 100
OLIVES AND ALLIED FAMILIES; BUBBLES, PYRAMS

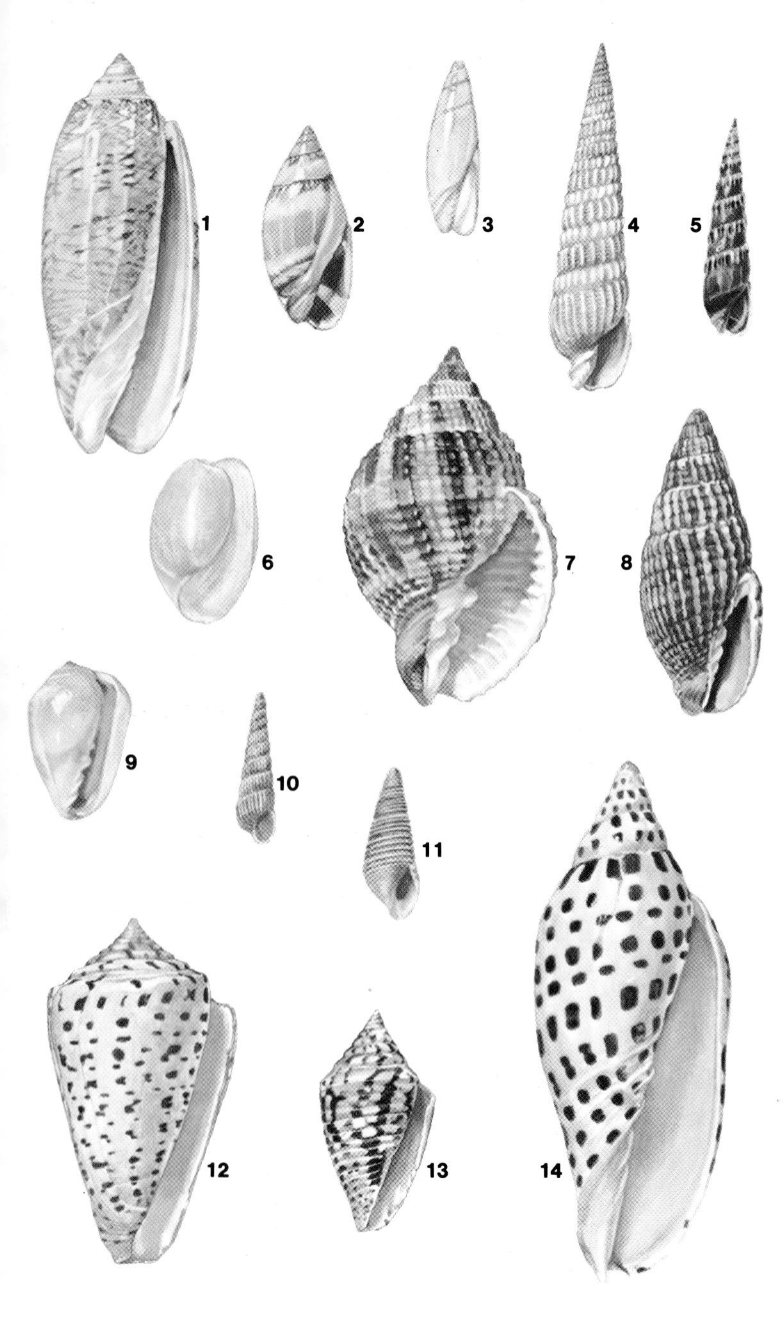

1 Lettered Olive × ¾, p. 588. **2** Minute Dwarf Olive × 1½, p. 588. **3** Rice Dwarf Olive × 1½, p. 589. **4** Atlantic Auger × 1, p. 591. **5** Sallé's Auger × 1, p. 591. **6** Eastern Paper Bubble × 1½, p. 591. **7** Common Nutmeg × 1, p. 589. **8** Beaded Miter × 1, p. 589. **9** Common Atlantic Marginella × 1½, p. 590. **10** Interrupted Turbonille × 3, p. 592. **11** Impressed Odostome × 3, p. 592. **12** Alphabet Cone × ¾, p. 590. **13** Jasper Cone × 1, p. 590. **14** Junonia × 1, p. 589.

PLATE 101
TUSK SHELLS, AWNING, NUT AND ARK CLAMS

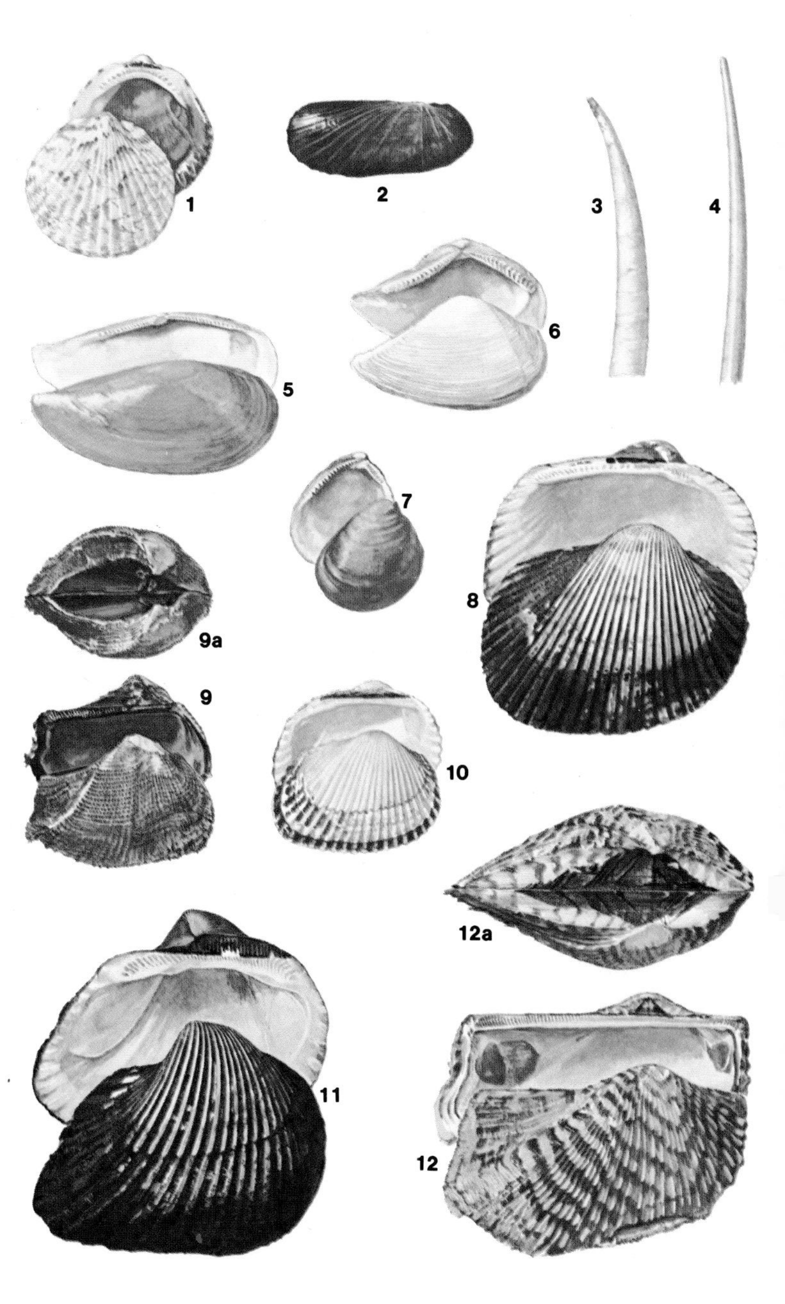

1 Comb Bittersweet × ¾, p. 597. **2** Atlantic Awning Clam × 1¼, p. 594.
3 Stimpson's Tusk × 1, p. 593. **4** Ivory Tusk × 1, p. 593.
5 File Yoldia × ¾, p. 595. **6** Pointed Nut Clam × 2, p. 595.
7 Atlantic Nut Clam × 2, p. 594. **8** Blood Ark × ½, p. 596.
9 Mossy Ark × ¾, **9a** hinge, p. 595. **10** Transverse Ark × ¾, p. 596.
11 Heavy Ark × ¾, p. 596. **12** Turkey Wing × ¾, **12a** hinge, p. 595.

PLATE 102 MUSSELS, PEN SHELLS, TREE AND SPINY OYSTERS

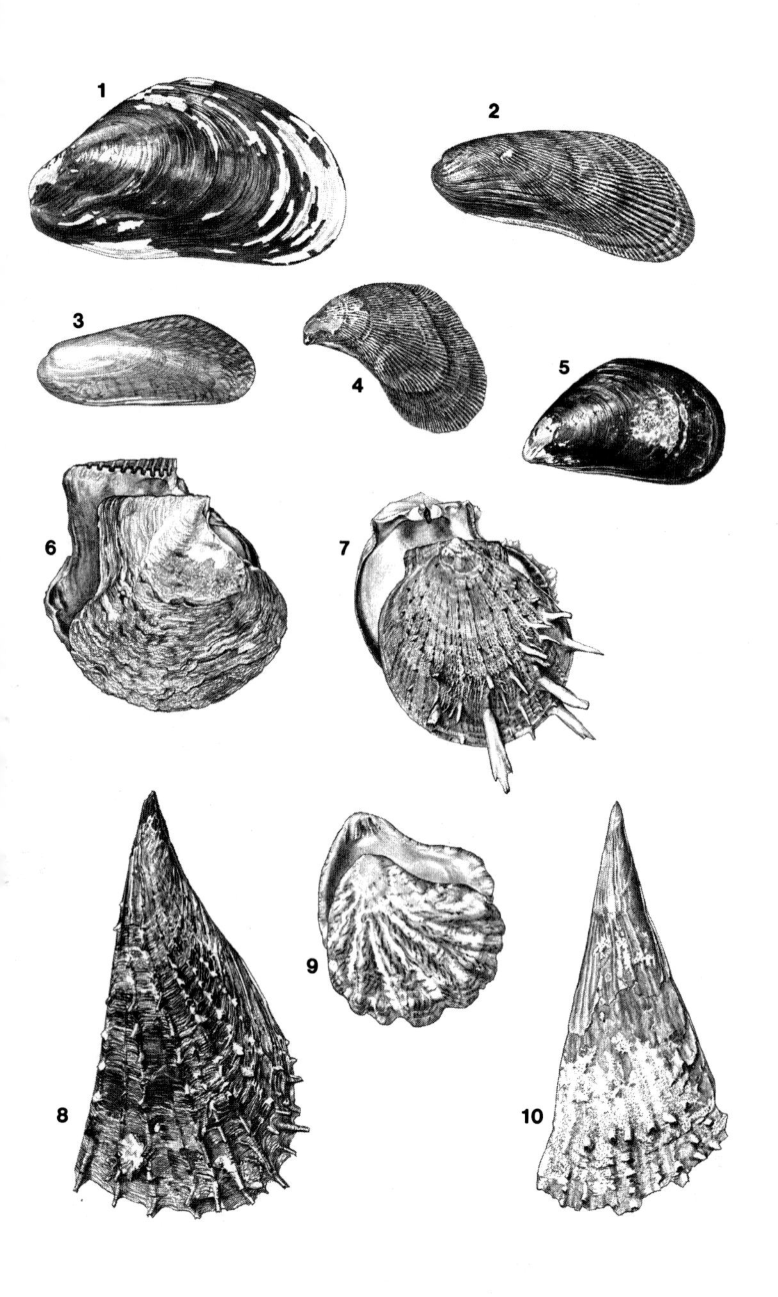

1 Horse Mussel × ½, p. 598. **2** Ribbed Mussel × ½, p. 598.
3 Paper Mussel × 1½, p. 597. **4** Hooked Mussel × ½, p. 597.
5 Blue Mussel × ½, p. 597. **6** Flat Tree Oyster × ½, p. 599.
7 Spiny Oyster × ½, p. 600. **8** Stiff Pen Shell × ⅓, p. 599.
9 Kitten's Paw × 1, p. 601. **10** Flesh Pen Shell × 1, p. 598.

SCALLOPS, JINGLE SHELL, OYSTERS, LUCINES

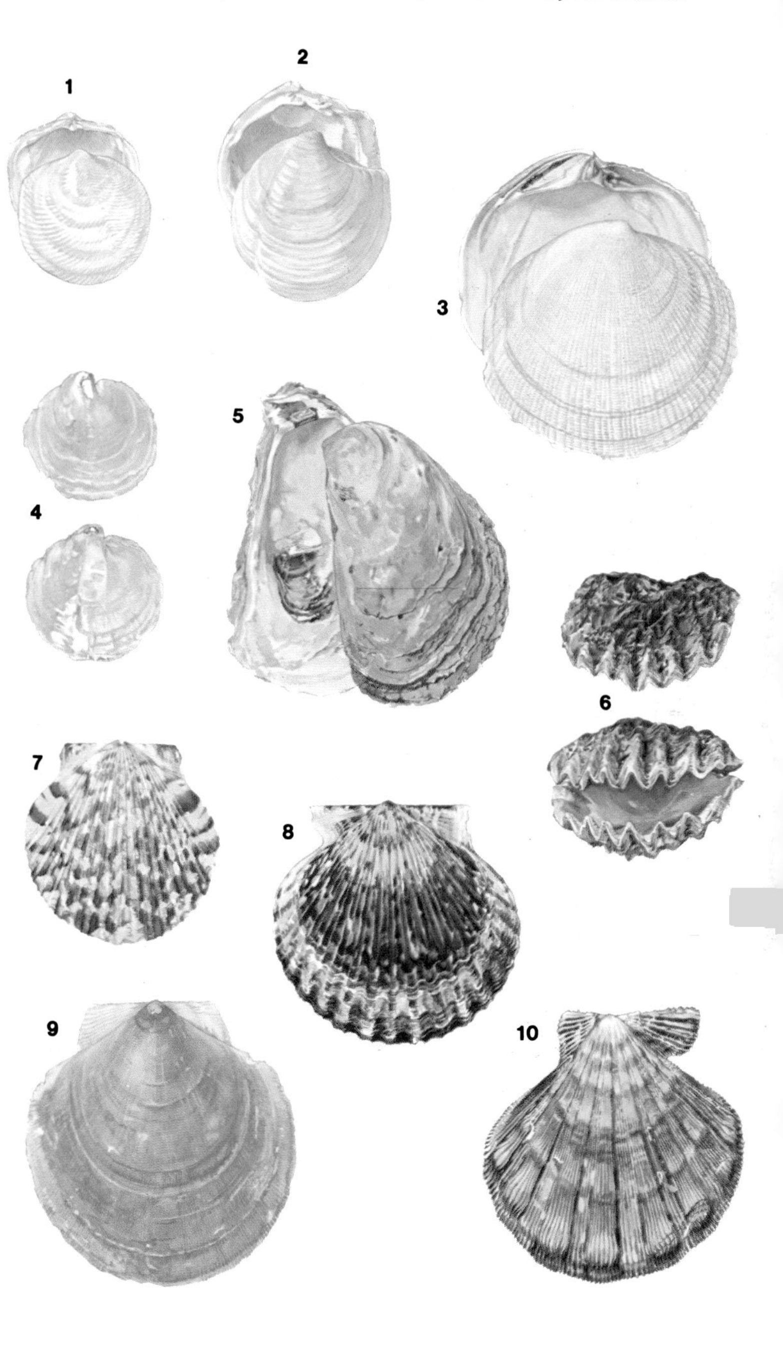

1 Crosshatched Lucine × 1, p. 603. **2** Pennsylvania Lucine × ½, p. 603. **3** Great White Lucine × ½, p. 603. **4** Common Jingle Shell × ½, p. 601. **5** Virginia Oyster × ½, p. 602. **6** Coon Oyster × ½, p. 602. **7** Calico Scallop × ½, p. 600. **8** Bay Scallop × ½, p. 600. **9** Deep Sea Scallop × ⅓, p. 599. **10** Lion's Paw × ⅓, p. 600

PLATE 104
COCKLES AND OTHER CLAMS

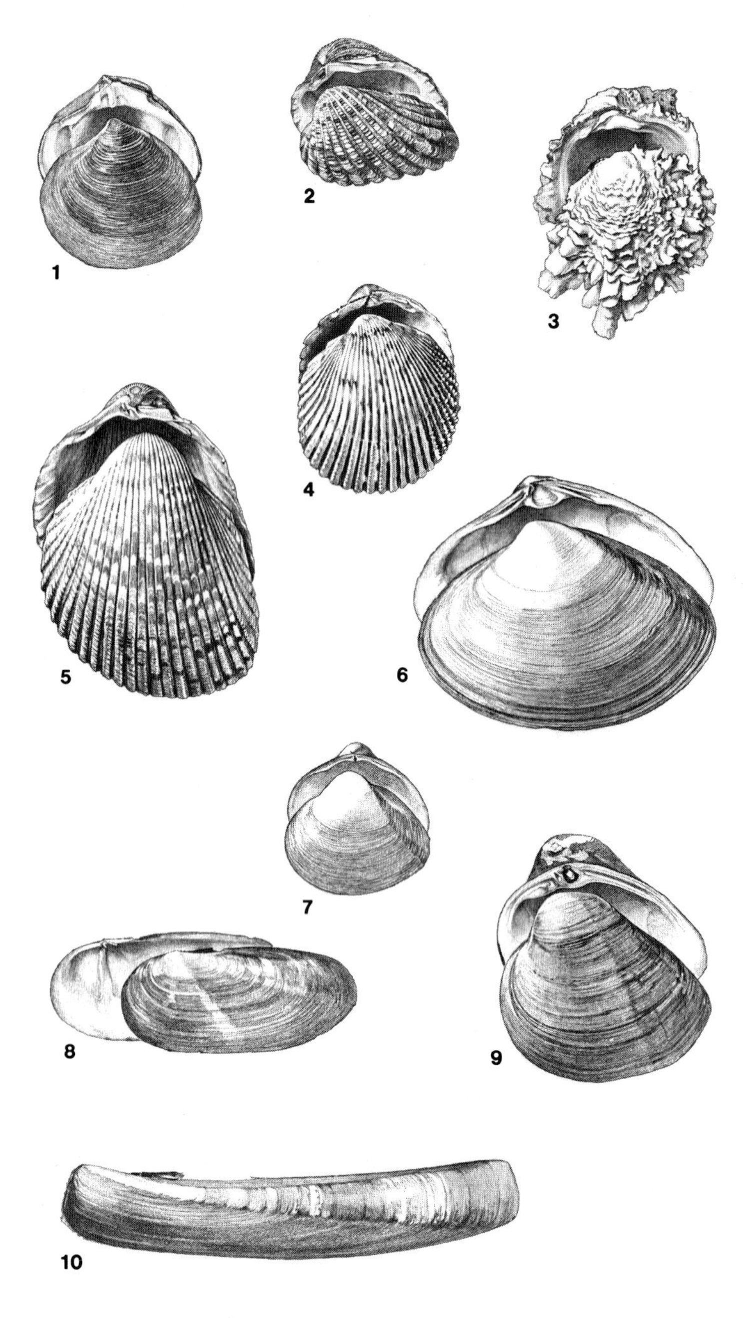

1 Boreal Astarte × ½, p. 604. **2** Broad-ribbed Cardita × ½, p. 602.
3 Leafy Jewel Box × ½, p. 604. **4** Common Cockle × ½, p. 605.
5 Great Heart Cockle × ½, p. 604. **6** Atlantic Surf Clam × ½, p. 611.
7 Dwarf Surf Clam × 1, p. 611. **8** Ribbed Pod × ½, p. 607.
9 Wedge Rangia × ½, p. 611. **10** Common Razor × ½, p. 607.

PLATE 105
TELLINS AND OTHER CLAMS

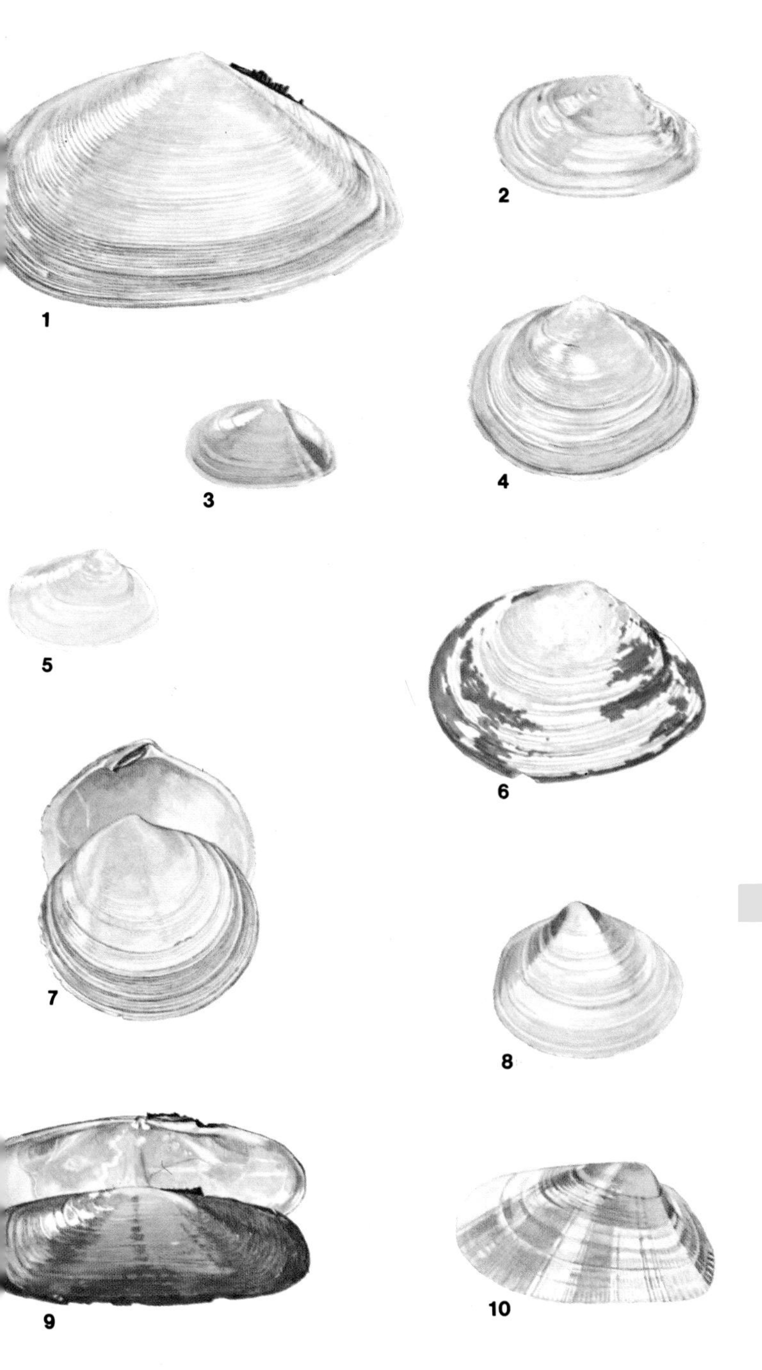

1 Lined Tellin × ¾, p. 607. **2** Tenta Macoma × 1½, p. 609.
3 Northern Tellin × 1½, p. 608. **4** Baltic Macoma × 1, p. 608.
5 DeKay's Tellin × 1½, p. 608. **6** Chalky Tellin × 1, p. 608.
7 White Semele × 1, p. 610. **8** False Coquina × 1½, p. 609.
9 Purplish Razor × 1, p. 610. **10** Southern Coquina × 1½, p. 609.

PLATE 106
VENUS CLAMS AND OTHER CLAMS

1 False Angel Wing × ¾, p. 606. **2** Arctic Rock Borer × ¾, p. 613. **3** Cross-barred Chione × ¾, p. 606. **4** Disk Dosinia × ½, p. 605. **5** Black Clam × ½, p. 610. **6** Common Basket Clam × 1½, p. 612. **7** Common Soft-shelled Clam × ½, p. 612. **8** Northern Quahog × ½, p. 605. **9** Sunray Shell × ½, p. 606. **10** Checkerboard Clam × ½, p. 606.

PLATE 107
PIDDOCKS AND OTHER CLAMS

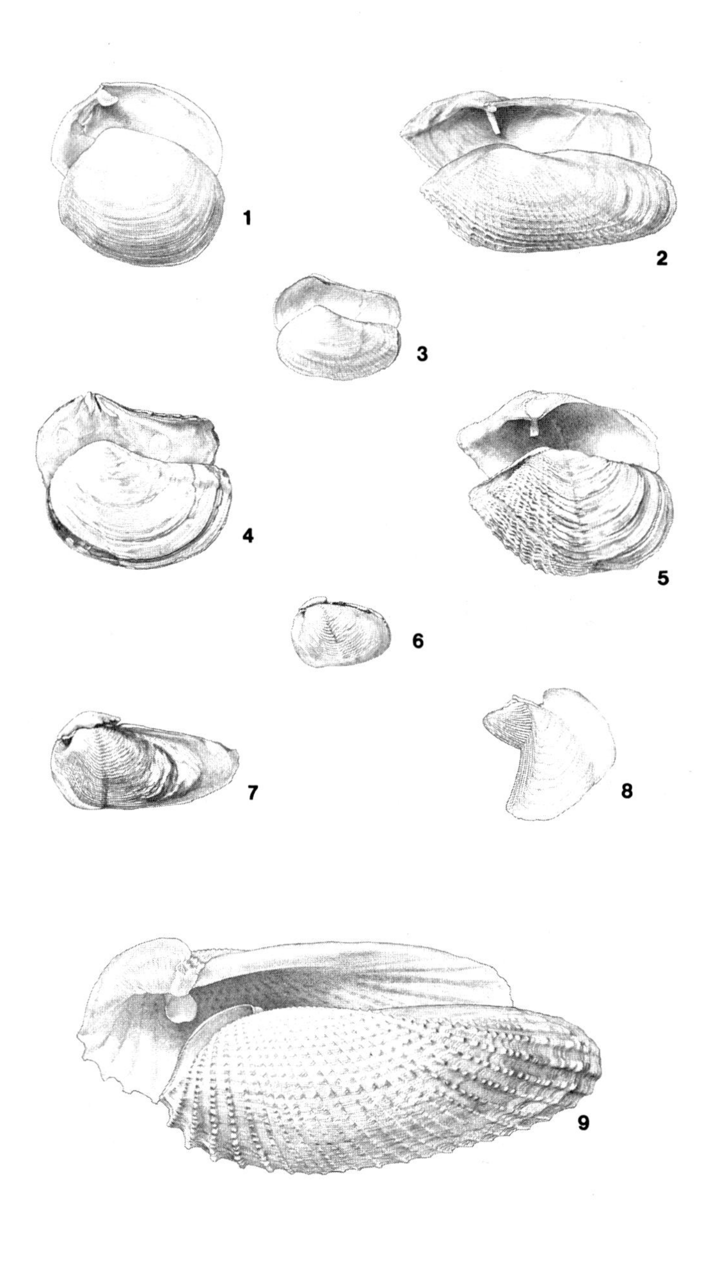

1 Unequal Spoon Shell × ¾, p. 616. **2** Fallen Angel Wing × ¾, p. 613. **3** Glassy Lyonsia × 1, p. 615. **4** Gould's Pandora × ¾, p. 616. **5** Great Piddock × ¾, p. 614. **6** Wedge-shaped Piddock × 1, p. 614. **7** Striate Wood Piddock × 1, p. 614. **8** Naval Shipworm × ½, p. 615. **9** Common Angel Wing × ½, p. 613.

PLATE 108
CEPHALOPODS

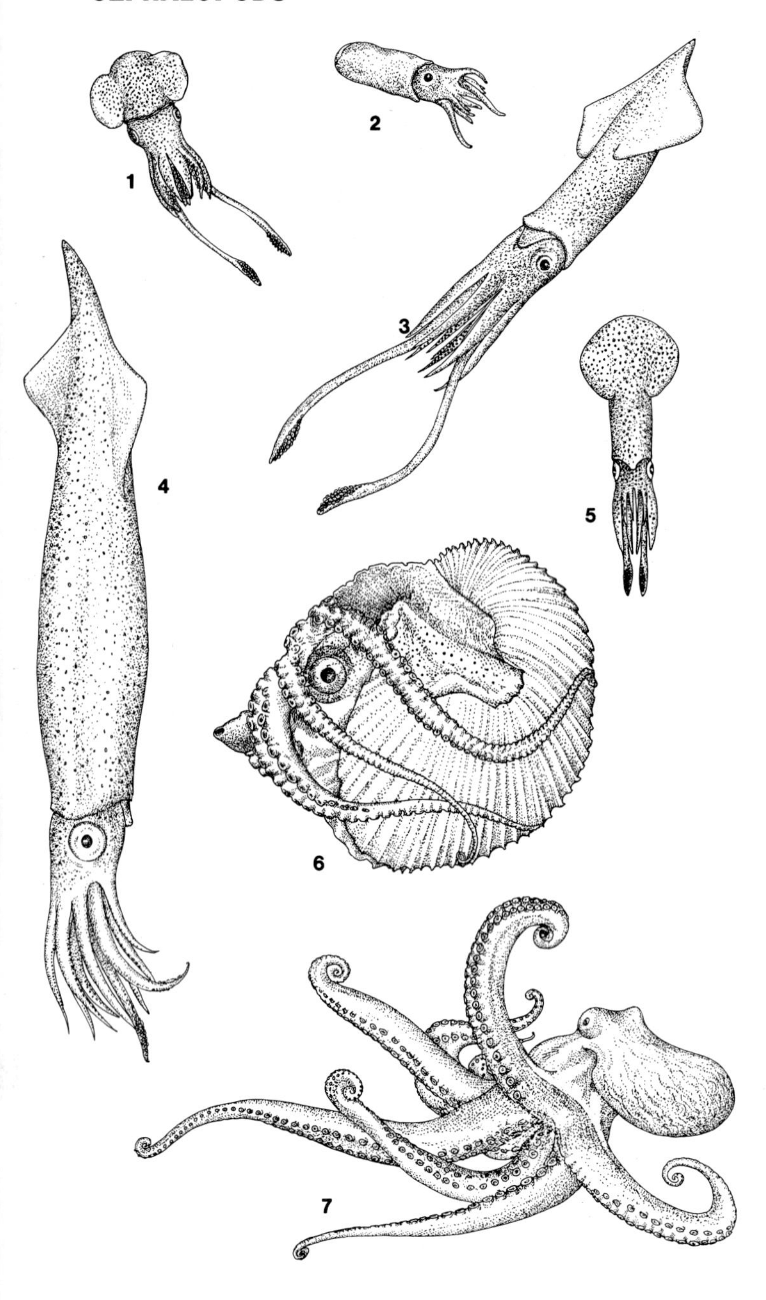

1 Atlantic Bob-tailed Squid × ⅜, p. 617. **2** Ram's Horn × ⅓, p. 617.
3 Common Squid × 1, p. 618. **4** Sea Arrow × ¼, p. 618.
5 Dwarf Squid × ¼, p. 618. **6** Paper Nautilus × ¼, p. 619.
7 Atlantic Octopus × ¼, p. 619.

PLATE 109
SEASHORE LIFE

1 *Dactylometra quinquecirrha,* Sea Nettle, p. 630.
2 *Cyanea capillata* (Jellyfish), p. 631.
3 *Physalia physalis,* Portuguese Man-of-war, p. 630.
4 *Aurelia aurita,* Moon Jelly, p. 631. **5** *Tubularia crocea* (Hydroid), p. 628.
6 *Microciona prolifera,* Red Sponge, p. 627.
7 *Mellita quinquiesperforata,* Keyhole Urchin, p. 643. **8** *Arbacia punctulata,* Purple Sea Urchin, p. 643. **9** *Henricia sanguinolenta* (Starfish), p. 640.
10 *Asterias forbesii* (Starfish), p. 641. **11** *Ophiothrix angulata* (Brittle Star), p. 642. **12** *Ophioderma brevispinum,* Green Brittle Star, p. 642.
13 *Leptogorgia virgulata,* Whip Coral, **13a** pod detail, p. 632.
14 *Astrangia danae* (Coral), p. 634. **15** *Orchestia platensis,* Beach Flea, p. 657.
16 *Pentamera pulcherrima* (Sea Cucumber), p. 645.

PLATE 110
SEASHORE LIFE

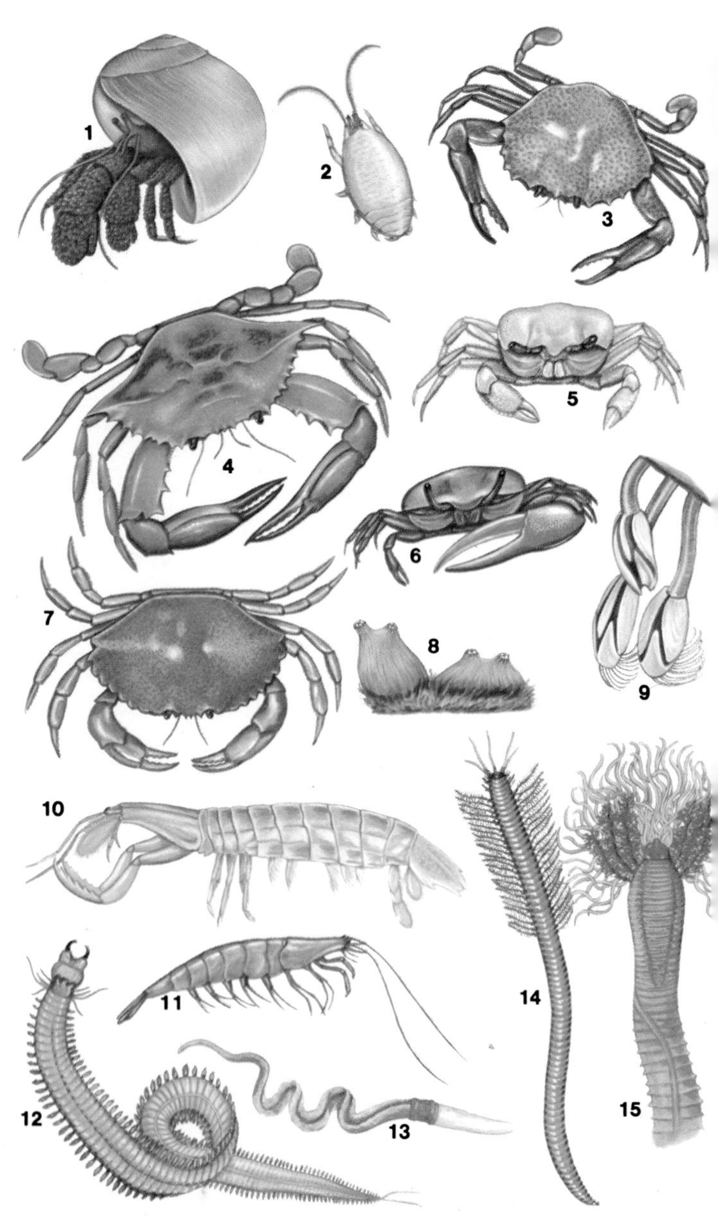

1 *Pagurus pollicaris* (Hermit Crab), p. 661. **2** *Emerita talpoida,* Mole Crab, p. 660. **3** *Ovalipes ocellatus,* Lady Crab, p. 663. **4** *Callinectes sapidus,* Blue Crab, p. 663. **5** *Ocypode quadrata,* Ghost Crab, p. 664. **6** *Uca pugilator,* Sand Fiddler Crab, p. 665. **7** *Cancer irroratus,* Rock Crab, p. 663. **8** *Styela partita* (Sea Squirt), p. 667. **9** *Lepas fascicularis,* Goose Barnacle, p. 653. **10** *Squilla empusa,* Mantis Shrimp, p. 666. **11** *Crangon septemspinosa,* Sand Shrimp, p. 658. **12** *Nereis virens,* Green Clam Worm, p. 647. **13** *Saccoglossus kowalevskii* (Acorn Worm), p. 638. **14** *Diopatra cuprea* (Segmented Worm), p. 648. **15** *Amphitrite ornata* (Segmented Worm), p. 649.

RED-BACKED SALAMANDER

Plethodon cinereus **71:1**

Description
Length, to 3⅝ in. (9.2 cm); record 5 in. (12.7 cm). Identified by straight-edged gray or red middorsal band and mottled belly with equal amounts of black and yellow or white pigment; no red below except occasionally between front legs. Body relatively stout; costal grooves 18–20. Legs short; intercostal folds between adpressed limbs 7–9. Two distinct color phases occur with variable frequency throughout range: *Red-backed* has dorsal band of some shade of red; sides brown, gray, or black. *Lead-backed* is uniformly dark gray or black above without a middorsal stripe, but with brassy red or white flecks. Males have a large gland under chin.
Habitat
A common species, found in and under rotted logs, in stumps, and among forest floor litter in relatively cool and mesic (moderately moist) coniferous and hardwood forests.
Habits
Active during above-freezing weather in Middle Atlantic States, but below ground at low elevations during midsummer heat.
Range
Ne. U.S. and se. Canada from N.S. s. through e. and n. N.C. and adj. border of Tenn., w. through Great Lakes to w.-cen. Minn., e. Ill., and extr. n. Ky.

WELLER'S SALAMANDER

Plethodon welleri **70:11**

Description
Length, to 3⅛ in. (7.9 cm). Small; has irregular dorsal band of silvery or golden blotches over black dorsal ground color. Sides sometimes with light spots. Throat grayish-white; belly dark slate with scattered small white spots. Costal grooves 16; 3–4 intercostal folds between adpressed limbs. Small gland under chin in males.
Habitat
Under logs, stones, or talus in forests, usually between 2300–5700 ft. (701.0–1737.4 m).
Range
Extr. nw. N.C. and extr. ne. Tenn., n. into s. Va.

RAVINE SALAMANDER

Plethodon richmondi **70:9**

Description
Length, to 5⅝ in. (14.3 cm). Distinguished by wide dorsal band and large number of costal grooves (20–22). Legs small, thin. Dorsally flecked with white and bronze to form a brown band, bordered by bluish-black; lower sides and cheeks spotted with white. Belly darker with fine, light mottling, as in Lead-backed Salamander. Often red pigment on cheeks, front legs, and anterior sides.
Habitat
Slopes of valleys and ravines.
Range
W. Pa., s. to nw. N.C. and ne. Tenn., and w. to se. Ind. In the southeast, range separated from Valley and Ridge Salamander by New R. in Va. and W.Va.

ZIGZAG SALAMANDER

Plethodon dorsalis **70:10**

Description
Length, to 4⅜ in. (11.1 cm). Similar to Red-backed Salamander, but red dorsal band has wavy black edges. Occasionally edges are straight, but then band on body is narrower than in Red-backed. Stripe sometimes absent, but flecks of red pigment in its place. Red often present on sides. Dorsally, brassy to silvery flecks to small white spots; ventrally mottled with equal amounts of white, red, and black pigment. Band down tail with straight edges. Slender body, head and legs. Normally 18 costal grooves; 6–7 intercostal folds between adpressed limbs. In males gland under chin small, round or oval.

Habitat
Often close to a moist retreat in a rock formation, escarpments, talus slopes, under logs and stones in woods. In and near mouths of caves, mostly under 1000 ft. (304.8 m) elevation.

Habits
Rarely found in summer at low elevations.

Range
Cen. Ind. and s. Ill., s. to nw. Ga. and n. Ala. Disjunct localities in s. Ohio, N.C., S.C., Ala., Miss., and the Ozark Plateau in ne. Ark. and adj. states.

VALLEY AND RIDGE SALAMANDER

Plethodon hoffmani *Fig. 89*

Description
Length, to 5⅜ in. (13.7 cm). Has unusually high number of costal grooves (20–21, usually 21); 11 intercostal folds between adpressed limbs. Gray mottled venter with red pigment confined to narrow area; light chin. Rarely has striped phase. Usually has brassy flecks dorsally. Feet slightly webbed.

Habitat
Wooded slopes.

Range
Valley and Ridge physiographic province of cen. Pa., s. through Md. and along Va.-W. Va. border to New R.

Valley and Ridge Salamander

Fig. 89

White-spotted Salamander

JORDAN'S WOODLAND SALAMANDER

Plethodon jordani **71:8**

Description
Length, to 5 in. (12.7 cm); record 7¼ in. (18.4 cm). Large, black, usually without dorsal spots. Dark brown to black ground color with lighter belly and chin. Some populations have red on cheeks and legs, lateral white or yellow spots, or brassy dorsal flecks; also red dorsal spots in young. Costal grooves 16, sometimes 15–17; 1–3 intercostal folds between adpressed limbs.
Similarities
Slimy Salamander has dorsal white spots, but variants of each species may look confusingly similar even to experts.
Habitat
Under rotten logs, bark, and leaf litter; sometimes under stones in heavily wooded areas.
Remarks
In southern part of range this species may hybridize with Slimy Salamander in areas of overlap.
Range
Normally above 3000 ft. (914.4 m) elevation in s. Appalachian Mts. from sw. Va. through w. N.C. and e. Tenn. to nw. S.C. and extr. ne. Ga.

CHEAT MOUNTAIN SALAMANDER

Plethodon nettingi **70:13**

Description
Length, to $4\frac{13}{16}$ in. (12.2 cm). Dark, flecked above with light, brassy red or yellow in a bandlike pattern. Below slate-gray to purplish-black with slight light spotting. Lighter under throat. Costal grooves 18–19, 7 intercostal folds between adpressed limbs.
Similarities
Weller's Salamander has 2–3 fewer costal grooves, shorter tail. Ravine Salamander has longer tail, 2–4 more costal grooves.
Habitat
Under rocks and logs in forests and in north-facing talus slopes, above about 3000 ft. (914.4 m).
Range
Three disjunct populations, each given subspecific status, at high elevations: e.-cen. W.Va.; the Peaks of Otter region of cen. Va.; and in Shenandoah National Park in nw. Va.

WHITE-SPOTTED SALAMANDER

Plethodon punctatus *Fig. 89*

Description
Length, to $6\frac{13}{16}$ in. (17.3 cm). Many large spots on the back; usually with 18, sometimes 17, costal grooves. Large dorsal spots like lateral yellow-white or yellow spots in coloration. Throat pale.
Similarities
Slimy Salamander has no yellow dorsal spots. Wehrle's Salamander has dorsal white spots smaller and less numerous and has red or brassy pigment on dorsum.
Habitat
High elevations, usually over 2800 ft. (853.4 m).
Range
Shenandoah and North Mts. along nw. Va.-W.Va. border.

WEHRLE'S SALAMANDER

Plethodon wehrlei **70:14**

Description
Length, to 6⅝ in. (16.0 cm). Somewhat resembles Slimy Salamander, but has light throat and light spots on lower sides only. Above dark brown or black, sometimes with whitish specks or paired red disklike spots. Lower sides bluish-brown with white markings. Belly light gray or gray mottled. Throat light. Costal grooves usually 17; 2–3 intercostal folds between adpressed limbs.
Habitat
Under rocks or stones in talus.
Range
Sw. N.Y., s. to sw. Va. and into N.C., w. to se. Ohio.

SLIMY SALAMANDER

Plethodon glutinosus **71:7**

Description
Length, to 8⅛ in. (20.6 cm). Shiny black above, dark slate below, and usually with varying amounts of white, yellow, or brassy spots on sides and back. Throat dark or light slate, somewhat mottled. Spotting above may be on sides, head, back, tail, and limbs; or individuals may lack spotting altogether. Costal grooves 16; 1–3 intercostal folds between adpressed limbs. Body cylindrical, legs long.
Similarities
Blue-spotted Salamander lacks nasolabial groove, has bluish-white flecks, and rounded, rather than compressed, tail.
Habitat
Under large logs or stones in wooded areas, in crevices of rocks, and in caves, under moist humus.
Habits
Nocturnal; burrows deeply in dry season; active throughout year in the South during wet weather, but rarely seen in winter in the North. Secretes a sticky substance from skin glands.
Remarks
See under Jordan's Woodland Salamander.
Range
S. N.Y. and extr. w. Conn., s. through N.J. and n. Md. to cen. Fla., w. to Miss. R. in s. Ill. to La., and across Miss. R. in s. Mo. to nw. Ark. and e. Okla. Disjunct localities in s. N.H. and adj. Mass., n. Ind., cen. La., and cen. Tex.

YONAHLOSSEE SALAMANDER

Plethodon yonahlossee **70:15**

Description
Length, to 7½ in. (19.1 cm). Dark gray, with bright chestnut, sometimes in a band, dorsally from head to base of tail. Head large; sides of head, body, and tail blotched with white. Venter black with some white spots, especially on sides. Dark gray hind legs, tail, and sometimes above lateral white spots. Costal grooves 16; 0–1 intercostal fold between toe-tips of adpressed limbs.
Habitat
Under logs or stones on wooded slopes between and in nearby grassland.
Range
Blue Ridge Mts., in w. N.C., extr. ne. Tenn., and into s. Va. at elevations from 2700–5700 ft. (823.0–1737.4 m) in the n., as low as 1400 ft. (426.7 m) in se., part of range.

FOUR-TOED, MANY-LINED, GROTTO, AND GREEN SALAMANDERS

Genera *Hemidactylium, Stereochilus, Typhlotriton,* and *Aneides*

FOUR-TOED SALAMANDER

Hemidactylium scutatum **71:11**

Description
Length, to 4 in. (10.2 cm). Four toes on hind feet and a strong constriction at base of keeled tail. Head, body, and tail mottled reddish-brown, lower sides grayish; belly china-white with dark dots, especially near sides. Body short, costal grooves 13–14.
Habitat
Under logs, stones, leaf litter, or moss in wooded or open areas near sphagnum ponds, swamps, or bogs; larval life spent in pools or bogs.
Habits
Nocturnal, terrestrial.
Remarks
Constriction on tail marks a zone for tail breakage. In nature a predator may be fooled into eating the broken tail, which continues to wriggle, and in the process ignore the fleeing salamander.
Range
Continuous distribution from s. Maine through n. Ga. and Ala., w. through s. Ont. and Mich. to Wis. and w. Ky. to extr. ne. Miss. Spotty occurrence in N.S., n.-cen. Fla., La., and e.-cen. Mo.

GROTTO SALAMANDER

Typhlotriton spelaeus **70:4**
Fig. 90

Description
Length, to 5 5/16 in. (13.5 cm). Adults, pale pinkish and blind. No gills; head flattened and wide; degenerate eyes appear as black spots. Body color grayish-lavender above, with dashed dorsal and lateral markings; below white. Costal grooves 16–19. Larvae darker, with functional eyes and external gills.
Habitat
In and near caves and underground limestone passages; larvae in cold open streams and springs.
Range
Sw. Mo. and nw. Ark. w. to extr. se. Kans. and ne. Okla.

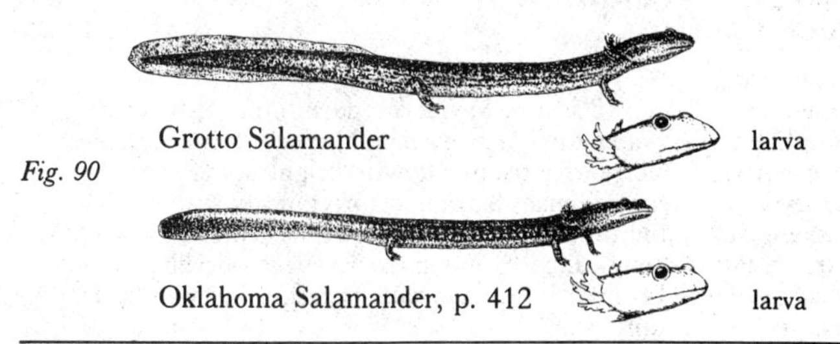

Fig. 90

MANY-LINED SALAMANDER

Stereochilus marginatus **70:7**

Description
Length, to 4½ in. (11.4 cm). Recognized by narrow lines of yellow and brown along sides. Head small and pointed, with small sensory pores, remnants of lateral line system. Neck indistinct. Tail

flattened sideways with prominent dorsal keel. Dull yellow above, middorsal line dark; belly yellow with a few dark spots. On each side a dark line from the nostril runs onto tail with three lines below. Lines on tail have a meshlike appearance. Intercostal folds between adpressed limbs 8–9.

Habitat

Slow streams and pools in pine forests and in cypress and gum swamps; under dead leaves, rubbish, and logs in damp places and in sphagnum moss in the water.

Range

Se. Va. at least to s. Ga. in Coastal Plain.

GREEN SALAMANDER

Aneides aeneus **71:9**

Description

Length, to 5½ in. (14.0 cm). This is the only truly green salamander. Expanded toe tips and the yellowish-green lichenlike blotches on the black back, tail, and limbs will also distinguish this beautiful salamander. Underparts pale. Depressed body; costal grooves 14–15. Limbs quite long and overlap by 1-3 intercostal folds. Tail long and round.

Habitat

Damp, but not wet, crevices in shaded rock outcrops and beneath rocks, bark, or logs in forests, up to 4400 ft. (1341.1 m) in N. Carolina. Rarely encountered.

Reproduction

Female broods eggs and young.

Range

Allegheny and Cumberland Mts. from sw. Pa. and extr. s. Ohio to n. Ala. and ne. Miss. Also in Blue Ridge Mts. in N.C., nw. S.C., and extr. ne. Ga.

BROOK SALAMANDERS

Genus *Eurycea*

These salamanders are generally aquatic, but may be found near water in moist woodlands. They have slender bodies and moderate to long tails that are flattened sideways. Ground color generally yellow or orange. All have aquatic larvae.

CAVE SALAMANDER

Eurycea lucifuga **71:13**

Description

Length, to 7⅛ in. (18.1 cm). Slender, reddish-orange, with long, whiplike tail (greater than ½ body length). Body cylindrical; legs long. Ground color varies from yellow to bright orange-red. Unspotted venter yellowish; black spots irregularly scattered over dorsum, sides, and tail, but forming no distinct pattern such as side bars on tail of Long-tailed Salamander. Eyes well developed. Costal grooves 14–15. Males have raised margins around cloacal opening at breeding time.

Habitat

In lighter regions of caves, limestone crevices around springs, under logs, stones, in damp places outside caves.

Range

Limestone areas from n. Va. to n. Ga., w. through s. Ind. and extr. ne. Miss.; in s. Ill. and w. to Ozark Plateau in Mo. and ne. Okla.

TWO-LINED SALAMANDER

Eurycea bislineata **71:4**

Description
Length, to 4¾ in. (12.1 cm). Light dorsal band bordered by two black lines that continue onto tail as a row of spots; varies from yellow to yellow-orange and has middle line of dark spots, sometimes continuous. Above yellow to brownish or greenish; belly always brilliant yellow. Well-developed eyes. Tail long, compressed.

Habitat
In or near brooks and seepages, in moist soil, under stones, logs.

Habits
Aquatic, nocturnal; forages on land in rain.

Food
Insects and snails found at stream's edge.

Range
Se. Canada, excluding N.S., s. to n. Fla., and w. below Lake Erie to e. Ill., e. of Miss. R., and s. to La.

LONG-TAILED SALAMANDER

Eurycea longicauda **71:14**

Description
Length, to 7¾ in. (19.7 cm). Yellowish ground color with small dark spots over yellowish to orange dorsum and continuing onto upper surfaces of limbs. Belly pale yellow to white, unspotted. Tail very long, almost ½ again length of body, with dark vertical bars on both sides. Eyes well developed. Costal grooves 14.

Habitat
In and under rotting logs, under stones near edges of streams, in underground caves and crevices of rocks.

Habits
During breeding season males develop swollen snouts with long projections called cirri.

Range
S. N.Y., s. through most of Ga. and Fla. panhandle, w. through s. Ind. and La.; also n. in Ill. along Miss. R. and w. of river through Ozark Plateau to s. Mo. and e. Okla.

Note: A subspecies, the well-named **DARK-SIDED SALAMANDER,** *Eurycea longicauda melanopleura*, occurs in the Ozarks and north in Illinois along the Mississippi River.

DWARF SALAMANDER

Eurycea quadridigitata **70:8**

Description
Length, to 3$\frac{9}{16}$ in. (9.1 cm). Yellow dorsal color with dark side stripe from snout to tip of tail, and 4 toes on hind foot distinguish this small, slender species. Well-developed eyes; tail squarish in cross section. Costal grooves 14–17.

Habitat
Under logs and debris in low, damp places, especially low pine flatwoods.

Range
Atlantic Coastal Plain from se. quarter of N.C., s. to Lake Okeechobee in Fla., and w. to e. Tex. Also in nw. S.C. and nw. Ark. and in Mo.

MANY-RIBBED SALAMANDER

Eurycea multiplicata **70:5**

Description
Length, to 3 9/16 (9.1 cm). Some individuals are neotenic and may become larger; they are often pale gray stippled with darker pigment. Costal grooves 19–20. Body elongate. Broad brown median stripe often bordered by dark line on each side; occasionally a narrow dark vertebral line also present; lower sides have silvery flecks. Below pale gray to lemon-yellow.
Habitat
Under logs, gravel, stones, debris in streams, in springs, both in the open and in caves.
Habits
Mainly aquatic; neoteny associated with caves and cave streams.
Range
Extr. se. Kans. and in Ozark and Ouachita Mts. in sw. Mo., most of nw. Ark., and e. Okla.

OKLAHOMA SALAMANDER

Eurycea tynerensis *Fig. 90*

Description
Length, to 3⅛ in. (7.9 cm). Dorsum grayish with reticulate pattern of cream and black; pale, weakly pigmented body. Costal grooves 19–21.
Habitat
Loose gravel in small, cold, spring-fed streams.
Habits
Aquatic.
Reproduction
Eggs laid in fall or late spring.
Remarks
A neotenic species that retains gills throughout life.
Range
In drainage of Grand R. and Ill. R. in sw. Mo., nw. Ark., and adj. e. Okla.

SPRING, RED, AND MUD SALAMANDERS

Genera *Gyrinophilus* and *Pseudotriton*

Members of this group of large salamanders are found mainly near or in water. The Red and Mud Salamanders are reddish with black spots. The Spring and West Virginia Cave Salamanders are duller and darker and have a light line from eye to nostril.

SPRING SALAMANDER

Gyrinophilus porphyriticus **70:3, 71:6**

Description
Length, to 8⅝ in. (21.9 cm). Large, yellowish-brown to purplish-brown, with flesh-colored belly and a light line from eye to nostril bordered below by a darker one. Dorsal colors lack luster; markings variable, from dark mottlings to spots that may be in a row on the back and sides. Body stout and rounded. Head broad, flattened, snout rounded. Tail keeled above.
Habitat
Hill or mountain country in cool springs or streams, under logs, large stones, in moist areas away from streams; caves.

Habits
Nocturnal, will leave streams on rainy nights to forage nearby.
Food
Both terrestrial and aquatic insects; larvae, which reach up to 4 in. (10.2 cm) in length, will eat other salamander larvae, and immatures will eat Two-lined Salamander.
Range
Sw. Maine and s. Que., s. through Appalachians to Fall Line in S.C., Ga., and Ala., w. through N.Y. and Pa. to s. Ohio and cen. Tenn.

Note: The **TENNESSEE CAVE SALAMANDER,** *Gyrinophilus palleucus,* a neotenic cave dweller, is found only in caves of central and southeastern Tennessee, northern Alabama, and northwestern Georgia.

WEST VIRGINIA CAVE SALAMANDER

Gyrinophilus subterraneus

Description
Length, to 4½ in. (11.4 cm). (Larvae attain size of adults.) Similar to Spring Salamander but restricted in range, with smaller eyes and 2–3 rows of pale spots along sides. Brownish-pink, darker dorsally, and with reticulated pattern over back and sides; light pink below with red gills. Head wider behind eyes; heavy-bodied. Transformed adults may lose pattern and have thinner head and body. Costal grooves 17, 7 between toe tips of adpressed limbs.
Habitat
Mud banks and stream in the one cave where they occur.
Food
Cave invertebrates.
Range
Known only from General Davis Cave, Greenbrier Co., W. Va.

RED SALAMANDER

Pseudotriton ruber **71:10**

Description
Length, to 7⅛ in. (18.1 cm). Red, with many irregular black spots; color darkens with age. Tail thick at base, keeled above. Newly transformed, it is bright coral red with black spots and plain pink belly. With age, it becomes purplish-brown above with dark spots fusing to give mottled appearance; belly becomes dull orange and develops black spots.
Similarities
Distinguished from somewhat larger Mud Salamander by generally redder color; brassy, gold, or silver-flecked iris; and irregular, more fused dark spots on back.
Habitat
In spring-fed brooks and in crevices and burrows in loose, moist soil nearby. Also, in more terrestrial situations, under logs, boards, and stones.
Habits
Less aquatic than Mud Salamander. As a self-protective device may mimic in coloration red eft, which is distasteful to predators.
Range
S. N.Y., s. to Ga. and Fla. panhandle, w. to Miss. R., and ne. to s. Ohio and Ind.

MUD SALAMANDER

Pseudotriton montanus **71:12**

Description
Length, to 6 in. (15.2 cm), record 8⅛ in. (20.6 cm). Similar to Red Salamander, but brownish above. Iris dark; snout short, rounded. Body stout; legs short, stout. Tail keeled above at tip. Above some shade of brown or reddish with darker spots; sides and belly salmon to flesh color, immaculate or flecked with round dark markings. Old animals become darker with spots less prominent.

Habitat
In or near muddy springs, in wet lowlands under logs, bark. Larvae aquatic. Adults usually beneath objects in water or along stream banks.

Habits
These salamanders mimic distasteful red efts.

Range
Coastal Plain e. of Miss. R. from s. N.J. s. to n. Fla., and w. to La.; in s. Ohio, w. W.Va. through e. and cen. Ky. and e. Tenn. Also in disjunct localities in Pa., n. Ala., and cen. Miss.

Frogs and Toads

Order Anura

Adult frogs and toads are distinguished from other amphibians by having short, squat bodies; four limbs, the hind limbs longer and more powerful than the forelimbs; and no tail. The use of the words "toad" and "frog" is often confusing. A true toad is a member of the Family Bufonidae, Genus *Bufo,* while a frog is any anuran; hence, a toad is also a frog. However, common names for several other groups incorporate the word "toad." Toads are recognizable by their warty skin, whereas most other frogs have relatively smooth skin. Toads also often have large parotoid glands at the shoulder.

The greatest diversity of amphibians occurs in the tropics. In the eastern United States there are seven families, fifteen genera, and sixty species, forty-seven of which are covered here. The total for all frogs is well over 2500 species divided into sixteen families, although authors will differ in their arrangements and numbers.

Identification

Identification can be made with certainty for some forms only with the specimen in hand. Frogs and toads of the same species often vary widely in color depending on the environment, temperature, and humidity (the cooler and moister the environment, the darker the color). Note size or shape, the presence or absence of webbing or enlarged pads on the feet, exposed and hidden markings, or the presence of glands. Females, as a rule, are larger than males and may have a greatly distended belly full of eggs in the breeding season. Males often have dark throats and special clasping pads on their thumbs, fingers, forelegs, or chest. Sizes given reflect the *record adult sizes* and the length is measured from the *tip of the snout to the vent.* When record sizes are appreciably larger than the average maximum, such lengths are given separately.

Fig. 91

Parts of a Typical Frog and Toad

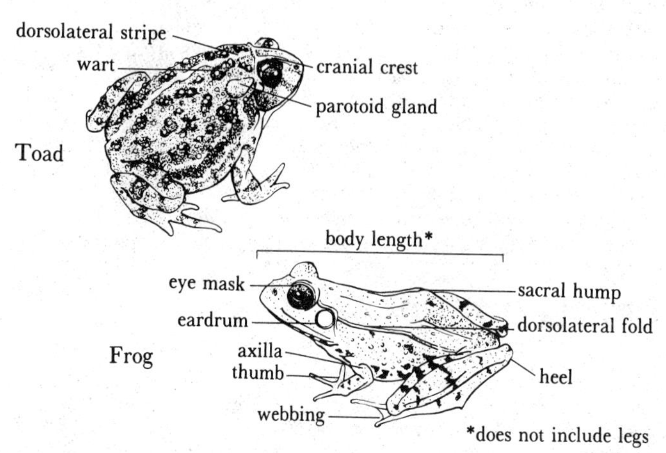

Habitat

Look for frogs in wet areas, under cover or in water. They are easily caught by hand but a net may assist in water. Some anurans, especially toads, are found during the day in mountainous, forested, or prairie situations some distance from water.

Vocalization

Vocalizations in frogs are usually limited to males, but in some species the female may utter a nonmating call, or both sexes may be silent. In addition to other visual, olfactory, and tactile cues, the mating call is a major means of mate and species recognition. Vocalizations may be divided into the following categories: mating calls, in males, often in a chorus heard over miles; release calls, given by a male or unreceptive female when clasped by a breeding male; a territorial call, which may not necessarily be associated with breeding activity; and various calls to indicate distress or give a warning. The cessation of calling may also be a signal.

Food

Adults eat insects and other invertebrates, and bullfrogs are known to eat anything that moves and is of a suitable size, even the young bullfrogs. Tadpoles are usually vegetarian. Major changes in metamorphosis from tadpole to adult are in the shortening of the digestive tract and the disappearance of larval mouthparts.

Reproduction

Eastern species generally reproduce by laying eggs in aquatic sites, although within the order Anura egg laying on land and live-bearing are known. The Greenhouse Frog, introduced in Florida, lays eggs that hatch into miniature adults under moist vegetation.

Frogs are somewhat less formal than salamanders when they encounter a potential mate, and the major device that prevents a mixing of species is the mating call. At the breeding grounds the males mount and clasp the females, holding this position of amplexus from a few minutes to over a day, at the end of which the male fertilizes the eggs as they are extruded from the female. Eggs are laid under water or on the surface, depending on the species. They may be laid singly, in long strings or cables, in globular masses, or in flat films. They may float or sink or be attached to vegetation. Individual clusters range from a few to hundreds; the total number laid by one female may go as high as 20,000 in the Bullfrog.

Eggs hatch into tadpoles in a short period, developmental rates often being dependent on the temperature. The tadpoles have horny beaks and often toothlike structures in rows around the mouth, and they breathe through gills covered by a flap of skin, the operculum. In general, most toad tadpoles are small and black and transform at very small sizes; at the other extreme, the true frogs have tadpoles that are larger and greenish, and which may live more than a year before transforming. At transformation they develop lungs, four limbs appear, the tail is resorbed, and structures related to feeding change. They spend less time in deeper water and more time on the edges of ponds and streams.

SPADEFOOT TOADS

Family Pelobatidae

Spadefoots have generally smoother skin than other toads, vertical eye pupils, teeth in the upper jaw, and parotid glands small or absent. The hind feet have horny black "spades." They breed in temporary pools after rains, and their offspring may develop through egg and larval stages in as few as twelve days.

EASTERN SPADEFOOT

Scaphiopus holbrooki

73:2
Figs. 92, 93

Description
Length, to 2⅞ in. (7.3 cm). The only spadefoot throughout the East. Has small, round parotoid glands. Above brownish, blackish, greenish, or yellowish, usually with 2 light stripes on the back forming hourglass pattern; other lines often continuing to the sides; sometimes with red spots. Below whitish.

Habitat
Deciduous or coastal pine forest, in sandy or light soils.

Habits
Digs burrows and is seen at surface only at night when breeding or foraging.

Voice
An explosive, harsh *ker-r-aw,* short in duration and repeated at intervals.

Reproduction
Breeds in temporary pools after heavy rains.

Range
Mass. and se. N.Y., throughout most of Fla., and w. to cen. Okla. and Tex., with large gap separating eastern and western populations in Miss. Flood Plain; also absent in higher areas in s. Appalachians.

Fig. 92

American Toad, p. 419

Southern Toad, p. 419

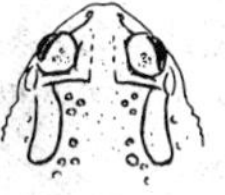

Woodhouse's Toad, p. 419

Oak Toad, p. 420

Great Plains Toad, p. 420

Canadian Toad, p. 421

Green Toad, p. 422

Red-spotted Toad, p. 422

Texas Toad, p. 421

Plains Spadefoot, p. 418

Eastern Spadefoot

Fig. 93

Western Spadefoot, p. 418

Eastern Spadefoot

Plains Spadefoot, p. 418

PLAINS SPADEFOOT
Scaphiopus bombifrons

73:1
Figs. 92, 93

Description
Length, to 2½ in. (6.4 cm). Has protruding bump between eyes and a wedge-shaped black spade on each hindfoot. Lacks parotoid glands. Above light- to olive-gray with irregular dark spots; below whitish. Area between eyelids narrow. Male has dusky throat.
Habitat
Sandy soil of arid regions and prairies.
Habits
Burrows in sand and forages at night; seen mainly in breeding season after heavy rains.
Voice
Harsh squawking repeated about once or twice per sec.
Reproduction
Breeds in temporary pools.
Range
Great Plains as far e. as cen. Mo., from sw. Man. s. to cen. Okla. and e. Ariz. in the s. U.S. and into adj. Mexico. A disjunct population along Gulf of Mexico from extr. s. Tex. s.

Note: The **COUCH'S SPADEFOOT,** *Scaphiopus couchii,* which occurs in eastern Texas and southwestern Oklahoma, is distinguished by its sickle- rather than wedge-shaped spade.

WESTERN SPADEFOOT
Scaphiopus hammondi

74:1
Fig. 93

Description
Length, to 2½ in. (6.4 cm). Distinguished by short, wedge-shaped black spade on hind foot and lack of boss between eyes. Dorsal color variable, tan or brown, sometimes with olive or green, and with small, light spots with dark edges. Usually several broad, light-colored bands running irregularly down back. Venter light with dark markings. Width of eyelid usually greater than distance between eyes.
Habitat
Lowland and foothill streams and temporary ponds in areas of loose soil.
Voice
Likened to purr of a cat, a *waa* lasting just under 1 sec., repeated about 80 times per min.
Range
Sw. U.S. from w. Okla. and cen. Tex., into sw. U.S., cen. mainland Mexico, and Baja Calif.

TRUE TOADS
Family Bufonidae

This is an almost cosmopolitan family of thirteen genera, with only the largest, Genus *Bufo,* occurring in North America. They have warty skin, large parotoid glands, horizontal pupils, and often prominent and species-characteristic cranial crests. Teeth are lacking in the upper jaw. Eggs are laid in strings in the water. Identification of young toads is often complicated because cranial crests develop quite some time after metamorphosis, and because the newly transformed young are so small. Breeding males have a dark or discolored throat.

AMERICAN TOAD

Bufo americanus

75:9
Fig. 92

Description
Length, to 4⅜ in. (11.1 cm). Has bony crests behind eye not touching parotoid gland or connected to it only by small spur; dark-spotted belly, and 1–2 warts in each large, dark dorsal spot. Color variable from gray to brown or reddish with dark spots on cream-colored belly. Large dark spots enclose up to 2 warts on the back; a light line frequently runs down back.

Similarities
Woodhouse's Toad also has light dorsal line, but spots on back enclose 3 or more warts.

Habitat
A common species, found in gardens, cultivated fields, and in wooded and open areas.

Habits
Active at dusk and into the night.

Voice
A high musical trill lasting 10–30 sec., about an octave lower than that of Southern Toad.

Reproduction
Breeds in shallow streams or ponds.

Range
E. Canada to e. U.S. from s. Labrador s. to n. Ga., w. to e. Man., and s. through n. La., e. Okla., and adj. w. Tex.

SOUTHERN TOAD

Bufo terrestris

74:5
Fig. 92

Description
Length, to 3 in. (7.6 cm); record 4 7/16 in. (11.3 cm). Cranial crests form prominent knobs behind eyes to middle of head and do not touch parotoid gland. Belly unmarked or with a few spots near the arm. Coloration dorsally variable: gray, brown, or red, for example.

Habitat
Most common toad in the South, abundant in loose soil.

Voice
High-pitched trill, about an octave higher than and with a pulse rate twice as large as that of American Toad; call duration 8 sec. or less.

Reproduction
Breeds in late spring in small permanent ponds, woodland ponds, or flooded depressions.

Range
Coastal Plain from se. Va. through Fla. to La.

WOODHOUSE'S TOAD

Bufo woodhousei

75:8
Fig. 92

Description
Length, 2–4 in. (5.1–10.2 cm). Distinguished by prominent bony crests that are connected with elongate parotoid glands behind eyes; by light belly with no or only one spot; and by conspicuous white stripe down back and dark dorsal spots, usually with 3 or more warts (may have fewer than 3 warts in populations from the Dakotas west). Above yellow-brown, greenish, grayish, or blackish. Below yellowish, occasionally with a breast spot. Parotoid glands smooth, narrow, 1½ times length of eyelid.

Habitat
Gardens, fields, woods, deserts; valleys in sandy, lowland areas.
Habits
Active day and night.
Voice
Suggests remotely the low *baa* of a sheep; 1–2½ sec. in length, repeated at 5–13-sec. intervals.
Reproduction
Breeds later in the year than American Toad, in shallow, quiet water.
Range
Much of e. and cen. U.S. from s. N.H. and se. N.Y. to n. S.C. and n. and w. Ga., w. to lower Mich., Ill., and the Great Plains; along Gulf Coastal Plain from extr. w. Fla. to s. Tex.; into much of w. U.S. and n.-cen. Mexico.

Note: The **WESTERN TOAD,** *Bufo boreas* (**75:13**), may occasionally wander east within the range of Woodhouse's Toad.

OAK TOAD

Bufo quercicus

74:4
Fig. 92

Description
Length, to 1⁵⁄₁₆ in. (3.3 cm). Small, with prominent light line from tip of snout down middle of back. Brownish ground color with light vertebral line dissecting two blotches; white patch between legs ventrally that ranges from cream to orange in color. Dorsum evenly warty; cranial crests weak in back; parotoid glands round to oval and not parallel.
Habitat
Roadside ditches, ponds, pools in areas of sandy soil.
Habits
Generally adapted to digging and rarely encountered.
Voice
High-pitched *peeps,* usually heard in chorus; calls night and day in warm, rainy weather. Vocal sac sausage-shaped when inflated.
Range
Atlantic Coastal Plain from se. Va. through Fla., and w. to Miss. R. in La.

GREAT PLAINS TOAD

Bufo cognatus

75:10
Fig. 92

Description
Length, to 4½ in. (11.4 cm). Has symmetrically placed rows of *large* dark spots on back, distinctly outlined with cream or yellow. Small, oval parotoid glands, cranial crests united to form knob between eyes, and paired, dark, distinctly outlined spots distinguish this species. Body large, broad; above brownish-yellow, greenish, or grayish, sometimes with narrow light stripe down back. Numerous small warts in each blotch. Below light, unmarked. Inner tubercle on foot long, blackened, and with free cutting edge; toes webbed.
Habitat
Desert and long- and short-grass prairies; farmlands; irrigation ditches.
Habits
Nocturnal but sometimes active by day.
Voice
Sustained, low-pitched trill in chorus, likened to sound of a riveting machine. Vocal sac sausage-shaped when inflated.

Reproduction
Up to 20,000 eggs laid in long strings, in early spring.
Range
Great Plains, extending e. in cen. Mo., from w. Minn. s. to cen. Okla. and nw. Tex., n. from cen. Mont. into extr. s. Canada, and through much of sw. U.S.; also into w. and cen. Mexico.

CANADIAN TOAD

74:2
Bufo hemiophrys
Fig. 92

Description
Length, to 3¼ in. (8.3 cm). Large boss between eyes from rear of eyelid to snout. Dorsally has brownish or greenish ovoid spots each containing 2–5 warts on either side of a cream to yellow midline stripe. Underparts buffy with small dark spots, a prominent one on lower throat. Snout short.
Habitat
Streams and pond margins, usually near mud flats or sandy beaches.
Habits
Breeds in May and calls into the summer.
Voice
A low trill similar to that of American Toad, but lower in pitch and shorter, given about twice a min.
Other name
Dakota Toad.
Range
Nw. Minn., ne. S.Dak., and e. and n. N.Dak., w. to Mont., and n. to the District of Mackenzie.

TEXAS TOAD

75:15
Bufo speciosus
Fig. 92

Description
Length, to 3½ in. (8.9 cm). Two dark, sharp-edged "spades" on foot, the inner one sickle-shaped. Dorsum greenish-gray, reddish, or yellowish, with or without olive spots. Lacks light vertebral stripe. Below yellowish or white and unspotted. Head ridges weak or missing. Faint cranial crests. Parotoid glands small, oval, and separated by 1½–2 times their own width. Male has buffy throat with olive center.
Habitat
Arid, semiarid, or cultivated areas, mesquite woodlands, short-grass plains.
Voice
Like call of Great Plains Toad, with a loud, shrill ½–1-sec. trill, but higher pitched and with a faster trill rate. Vocal sac sausage-shaped when inflated.
Reproduction
Breeds after rains in temporary or permanent pools.
Range
W. Okla. and adj. extr. s. Kan., s. through most of Tex., se. N.Mex., and into n. Mexico.

Note: The **GULF COAST TOAD,** *Bufo valliceps,* is distinguished by its broad, dark lateral stripe. This toad occurs along the Gulf Coast east to Mississippi.

GREEN TOAD

Bufo debilis

75:12
Fig. 92

Description
Length, to 2⅛ in. (5.4 cm). Small; has short legs and a greatly enlarged, kidney-shaped or elongate gland behind eye with indistinct cranial crests near eye. Above greenish or yellowish with many black spots and yellow warts. Below pale with pinkish or bluish cast, and may be spotted. Head narrow, body flat. Neck glands oblique on shoulders and almost as long as head.
Habitat
Short-grass prairie at higher elevations in dry areas.
Habits
Nocturnal, seldom seen except after rains.
Voice
Cricketlike; steady low trill, 3–7 sec. with interval of 5–9 sec.
Reproduction
Breeds in rain pools and ditches in spring and summer; eggs laid singly or in short strings.
Range
Sw. Kans. to se. Ariz., s. through cen. and w. Tex. to Gulf Coast and extr. ne. Mexico.

RED-SPOTTED TOAD

Bufo punctatus

75:11
Fig. 92

Description
Length, to 3 in. (7.6 cm). Only eastern toad with *round* parotoid glands, and they are smaller than the eye. Has poorly defined cranial crests. Above gray, brown, or reddish with rusty warts, and without vertebral stripe. Underparts light, plain or with some small spots. Head and body flat; eyes widely spaced.
Habitat
Prairies and deserts, in rocky areas near water.
Habits
Nocturnal, active after heavy rainfall.
Voice
Clear, high-pitched, pleasing birdlike trill lasting 4–10 sec. with almost equal intervals between trills.
Reproduction
Eggs laid singly and adhere to objects in water.
Range
Sw. U.S. from w. Kans., w. Okla., and cen. Tex., w. to Calif., and s. through much of Mexico.

TREEFROGS AND GRASS FROGS

Family Hylidae

North American members of this family are small, thin, and narrow-waisted. They have no neck glands or cranial crests. The tips of the digits are frequently enlarged into adhesive pads that enable these frogs to cling to smooth surfaces, even upside down. They are often found in bushes and trees. The species described all breed in the spring or summer, except the Spring Peeper, which breeds at the end of winter. In the South some species are active year-round.

SPRING PEEPER

Hyla crucifer **73:9**

Description
Length, to 1⅜ in. (3.5 cm). The only small treefrog in the North. Can be identified by the distinct, but often incomplete, dark X on its back. Above variably greenish-gray to reddish-brown; dark band between eyes and an irregular stripe from snout through eye and tympanum; underparts light. Skin smooth, snout pointed. Toe disks small but distinct. Male has brown throat.
Habitat
Marshes, swamps, pools, on low bushes and grasses in wooded areas.
Habits
Nocturnal, but often calls during day; hibernates under leaves.
Voice
A musical series of ½-sec. *peeps* like a high, shrill whistle, rising slightly at the end; often heard in a metallic chorus in early spring in the North. Calls from ground at edge of, and from elevated position in, tussocks or bushes.
Reproduction
Early breeder, peaks February–March.
Range
E. U.S. and se. Canada, but absent from s. half of peninsular Fla.; w. to se. Man. in n. of range and e. Tex. in s. Introduced in Cuba.

GREEN TREEFROG

Hyla cinerea **73:14**

Description
Length, to 2½ in. (6.4 cm). The only uniformly green treefrog, with or without a light lateral stripe, often dark-bordered. Individuals of the small Squirrel Treefrog may be confusingly similar, however. Above bright green, often with tiny yellow dots; below white or yellow. Skin smooth, head pointed.
Habitat
Males call from emergent vegetation in open, permanent bodies of water; pools, edges of waterways.
Voice
Likened to a cowbell, a call of *quonk* or *quank,* repeated 30–60 times per min.
Reproduction
500–1000 eggs laid, adhering to floating vegetation.
Range
Atlantic Coastal Plain of se. U.S. from Delmarva Pen. s. through Fla., and w. to cen. Tex.; also n. to s. Ill. and into se. Okla.; disjunct populations in cen. Ky. and Tenn. Introduced in Puerto Rico.

BARKING TREEFROG

Hyla gratiosa **73:12**

Description
Length, to 2¾ in. (7.0 cm). The largest native treefrog. Has evenly granular skin, a white stripe on upper jaw, and large disks on all digits. Above dark olive, gray, or green, usually with conspicuous round dark spots. Below creamy; groin and armpit yellow. Head short, broad. Prominent fold on breast; hands large, toes webbed.
Habitat
Trees of hummocks and in pine barrens, in small permanent and temporary bodies of water.

Habits
Arboreal, but also burrows.
Voice
Calls from water in a floating position, deep, low *boonk* or *moonk* repeated 30–60 times per min.; also a "bark" call from position in trees.
Remarks
Skin has strong-smelling secretion.
Range
Atlantic Coastal Plain from se. Va., s. through n. and cen. Fla., and w. to Miss. R. in La. Disjunct populations in s. N.J. (introduced), Ky., Tenn., n. Ala., and nw. Ga.

PINE BARRENS TREEFROG

Hyla andersoni **73:11**

Description
Length, to 2 in. (5.1 cm). Has uniform green back with a light-bordered plum-colored stripe on side and concealed bright orange markings on legs. Below pale with yellow spots. Body stout, skin smooth; tympanum about ⅓ size of eye. Toes partly webbed; toe pads small. Males have dark throats.
Similarities
Green Treefrog is larger and more slender.
Habitat
Specialized habitat of acid hillside seepage bogs, shrub bogs, and white cedar swamps; apparently needs acidic water conditions for development of larvae.
Habits
Hardly to be found except during breeding season, May–July in the North and April–September in Florida.
Voice
Calls from trees and bushes, a loud *quaak, quaak, quaak* . . . lasting about 20 sec.
Range
Separate populations below Fall Line in s. N.J. Pine Barrens; in se. N.C. and adj. S.C.; in e. Ga.; and in nw. Fla.

PINE WOODS TREEFROG

Hyla femoralis **73:15**

Description
Length, to 1¾ in. (4.4 cm). Conspicuous yellow spots on back of thigh. Above reddish-brown, but sometimes greenish-gray, with dark star on back; dark triangle between eyes. Below whitish, throat sometimes dark. Back granular, belly rough. Head short, snout rounded. Body slender, toe disks well developed.
Habitat
Pine woods.
Habits
Lives in tops of pine trees, but calls from bushes or small trees over water from few inches above surface to head height.
Voice
Succession of harsh, high notes likened to "an amateur playing with a telegraph key" (Conant).
Reproduction
Breeds in temporary pools or cypress ponds following heavy rains.
Range
Atlantic Coastal Plain from se. Va., s. through n. and cen. Fla., and w. to Miss. R. in La. Records from cen. Ala.

BIRD-VOICED TREEFROG

Hyla avivoca **73:13**

Description
Length, to 2⅙ in. (5.5 cm). Resembles a small, slender Gray Treefrog, but is more slender and has pale yellow or white coloration in groin, and smoother skin. Above brownish, grayish, or green with variable dark markings on the back; limbs barred. Black-bordered light spot below eye, dark bar across head. Below whitish, groin yellowish-green. Skin fairly smooth above, but granular below.

Habitat
Flooded wooded swamps, river valleys, and lake shores.

Voice
Calls from vines or bushes at about waist to head height, in repeated, piping, birdlike whistles.

Range
E. Miss. Valley from extr. s. Ill. and w. Ky., s. to Fla. panhandle and La.; crosses Miss. R. in cen. La. Disjunct populations cen. Ga. to adj. S.C. Also a record in Okla.

GRAY TREEFROG

Hyla versicolor and *Hyla chrysoscelis* **73:10**

Description
Length, to 2⅜ in. (6 cm). Moderate-sized, with rough skin and a distinct white spot below eye; usually a large dark blotch on back. Above light brown to gray or green (can change color quickly), the green phase usually lacking the large dark blotch on back; legs usually banded; below light, rear of thigh and under hind parts bright orange-yellow. Skin rough, warty; head short.

Habitat
Ubiquitous in permanent and temporary water; found on trees, bushes, fences, vines in and at edges of woodlands.

Habits
Nocturnal, but may call on humid days; arboreal and acrobatic.

Voice
Calls in early evening from trees or bushes, but closer to water's edge later in evening; a melodious, uniform trill, long in *versicolor* and short in *chrysoscelis*, repeated.

Reproduction
Breeds for a short time in early spring.

Range
Most of e. U.S. and extr. se. Canada, from s. Maine s. to n. Fla., w. to s. Man. and e. Tex.

SQUIRREL TREEFROG

Hyla squirella **73:16**

Description
Length, to 1⅝ in. (4.1 cm). Colors vary and can change rapidly. A difficult species to identify. Above green, brownish, or pearly gray with or without spots. Dark band between eyes, white line on upper lip, light line from below eye to shoulder; underparts light with no color on concealed surfaces of legs. Skin smooth above. Head short; snout rounded.

Habitat
Breeds in temporary and permanent water in a variety of situations: around buildings, gardens, fields, open woods.

Habits
Seen mainly during breeding season; calls by day during rains; occasionally transported outside its natural range while hidden on cultivated plants.
Voice
Calls from edge of shallow water or from emergent twigs or stems; harsh, but muffled, *aak, aak, aak* repeated once or twice per sec.
Range
Coastal Plain from se. Va., s. through Fla., w. along Gulf Coast to se. Tex. Disjunct populations in extr. se. Okla., e.-cen. Miss., and cen. Ala. Introduced in the Bahamas.

LITTLE GRASS FROG

Limnaoedus ocularis *Fig. 94*

Description
Length, to 11/16 in. (1.8 cm). The smallest North American frog. Has dark stripe through eye and down side; triangle between eyes continuing as a line down back; also variable lines on each side. Color varies: brown, green, reddish.
Habitat
Moist areas around ponds.
Habits
Climbs only to low vegetation, breeds year-round in the South, and winter through summer in the North.
Voice
High-pitched tinkling, repeated.
Range
Coastal Plain from se. Va. through most of Fla. with western boundary extr. se. Ala. and the cen. panhandle.

Fig. 94

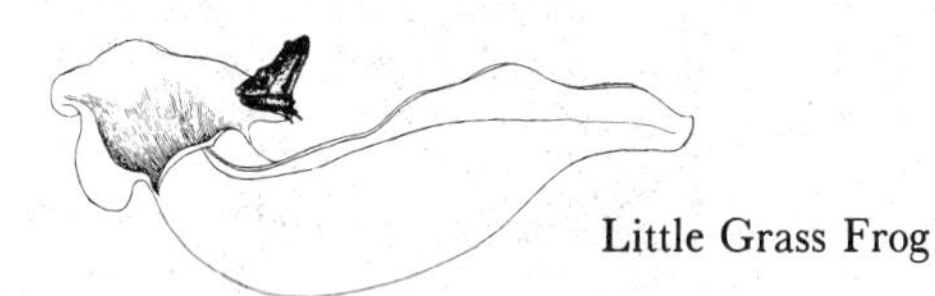

Little Grass Frog

CRICKET AND CHORUS FROGS

Genera *Acris* and *Pseudacris*

Cricket frogs, Genus *Acris,* are not arboreal and lack expanded digit tips, but have long toes and considerable webbing for their semiaquatic habits at the water's edge. They have smooth skin below. Active both night and day as weather permits, they breed fairly late in the season.

Chorus frogs, Genus *Pseudacris,* are small with pointed snouts, reduced disks, and little webbing. They usually have a dark stripe from nostril through eye and a pale stripe above the upper lip. The skin is smooth above and granular below. Chorus frogs breed in winter or early spring.

NORTHERN CRICKET FROG

74:9

Acris crepitans *Fig. 95*

Description
Length, ⅝–1⅜ in. (1.6–3.5 cm). Dark triangle between eyes. Uppermost black lengthwise stripe on rear of thigh irregular,

bordered above and below by white stripe. Body stout with rounded snout. Uniformly gray or brown, or with green, brown, yellow, or reddish markings bordering a stripe extending from triangle down middle of back; belly white. Alternating light and dark bars on upper lip and light stripe from eye to angle of jaw. Dorsum warty. Chin of males may be spotted and yellowish during breeding season.

Similarities
Southern Cricket Frog has dorsum less warty, toes less webbed, thigh stripes with smooth edges.

Habitat
Edges of small, shallow streams and ponds.

Habits
Forages mostly on land; some individuals active well into fall.

Voice
Long series of rapid, sharp notes *kick, kick, kick, kick,* like striking two stones together; calls from rim of pond in shallow water or on floating vegetation.

Food
Terrestrial arthropods such as spiders and insects.

Reproduction
Breeds April to July around lakes and ponds, marshes, small temporary bodies of water.

Remarks
Enters water when frightened.

Range
E. and cen. U.S. except New England and Appalachian region, and generally absent from Coastal Plain from N.C. s. through Fla. Northern boundaries in se. N.Y., lower Mich., s. Wis., and se. S.Dak., s. through extr. w. Fla. to w. Tex. and into extr. n. Mexico. Occurs primarily at low elevations.

Fig. 95

Southern Cricket Frog

Northern Cricket Frog

SOUTHERN CRICKET FROG

73:3

Acris gryllus

Fig. 95

Description
Length, to 1¼ in. (3.2 cm). Has clean-edged dark stripe (2 stripes in Florida) on rear of thigh. Above with variable markings; dark brown or black triangle between eyes. Body slender with pointed snout; 1st toe not fully webbed, disks on digits very small and indistinct.

Similarities
Northern Cricket Frog has rough-edged thigh stripe, toes more webbed, dorsum wartier.

Habitat
Grassy edges of small streams, ponds, and swamps.

Voice
Short, explosive, rasping notes in a series, similar to the Northern Cricket Frog.

Range
Coastal Plain from se. Va., s. through Fla., and w. to Miss. R. from extr. sw. Tenn. to La.

STRIPED CHORUS FROG

Pseudacris triseriata **73:6**

Description
Length, to 1½ in. (3.8 cm). Five dark stripes, sometimes not continuous, down speckled or unicolor back and sides; frequently a dark triangle between eyes with apex meeting the median dorsal stripe. Dorsal ground color usually light grayish-brown to sooty-gray (but sometimes olive to dark green) with darker stripes. A dark gray or black stripe through eyes. Belly white, sometimes lightly spotted. Dark band present on, at most, ½ dorsal surface of lower leg. Body slender, snout round. Male has yellow vocal sac.
Similarities
Southern Chorus Frog also slender, but less robust.
Habitat
Temporary pools, flooded meadows, and roadside ditches in wooded areas.
Voice
Calls only in spring, from grassy pond margins to more open areas, may call day and night on cloudy days; a rising series of short trills, *prreeep-preep . . . preep,* resembling the sound produced by running thumbnail over teeth of a comb, somewhat resembles Spring Peeper, but notes more trilled.
Remarks
Populations of this species may hybridize freely with Southern Chorus Frog west of Alabama, and individuals with intermediate characters will occur.
Range
Widely distributed in e. and cen. U.S., from n. N.J. s. to e. Tex., but absent from most of Coastal Plain. In n. and w. N.Y. and extr. s. Que. and s. Ont. through Ohio, Ind., and w. to Northwest Territories, Idaho, Utah, and Ariz.

SOUTHERN CHORUS FROG

Pseudacris nigrita **74:8**

Description
Length, to 1¼ in. (3.2 cm). White stripe on lip (but in peninsular Florida black and white spots). Dorsal pattern of 5 longitudinal black stripes, sometimes broken into spots; dark bands on lower leg are broad with narrow bands of light in between. Head narrow, snout pointed.
Similarities
Striped Chorus Frog has well-defined triangle between eyes and stripes not as dark.
Habitat
Water in pine flatwoods and more open lowland areas; pools and swamps; also lives in sandy soils.
Voice
Calls from grassy pond margins, sometimes concealed; call similar to Chorus Frog but with a slower repetition rate.
Remarks
See under Chorus Frog.
Range
Coastal Plain from se. N.C., s. through Fla. and w. to se. Miss.

ORNATE CHORUS FROG

Pseudacris ornata **74:6**

Description
Length, to 1⁷⁄₁₆ in. (3.7 cm). Broad dark stripe from tip of snout through eye, interrupted at shoulder. Dark spot anterior to and above groin, which is washed with yellow or orange. Usually with 2 wide, irregular stripes down back and a sharply defined stripe, or often spots, on each side. Dark triangle between eyes. Ground color varies from green to the usual reddish-brown. Heavy-bodied with short legs.
Habitat
Ponds, roadside ditches, and flooded fields in well-drained areas.
Voice
Shrill, metallic *peep-peep* repeated up to 80 times per min., like Spring Peeper, but without rise at the end. Male sits in clumps of grass or on floating debris.
Reproduction
Very early breeder, December–March.
Range
Coastal Plain from se. N.C., s. to cen. Fla., and w. to extr. e. La.

BRIMLEY'S CHORUS FROG

Pseudacris brimleyi **73:7**

Description
Length, to 1¼ in. (3.2 cm). Distinguished by a dark line on side from snout through eye to groin, by 3 broad, clean-edged stripes down back and sides, and by dark spots prominent on chest. Above yellowish- or reddish-brown with small warts. Middorsal stripe brown; light line from upper jaw to shoulder; dark stripes lengthwise on legs. Below yellow with some dark spots. Snout long, round; no triangle between eyes. Disks on digits slightly dilated.
Habitat
Wooded pools, swamps, along rivers.
Voice
A short, rasping trill.
Range
Coastal Plain from se. Va to e. Ga.

MOUNTAIN CHORUS FROG

Pseudacris brachyphona **73:8**

Description
Length, to 1½ in. (3.8 cm). Dark triangle between eyes is not white-edged behind. Usually has a pair of dorsolateral half-moons on its back with ends curving outward to sides; sometimes absent or replaced by dark flecks. Above gray or brown, with no middorsal stripe or spots; dark bands down leg. Skin somewhat rough.
Habitat
In hilly wooded, or partially wooded, areas; swampy areas along small streams with abundant vegetation; and in seepage or roadside ditches.
Voice
Calls day and night in rasping, fast notes; high-pitched, continuous *wrrink* in chorus audible for ¼ mile (0.4 km).
Range
S. Appalachians from sw. Pa., s. to nw. Ga. and n. Ala. Disjunct populations in e. W.Va. and extr. ne. Miss.

STRECKER'S CHORUS FROG

Pseudacris streckeri **73:4**

Description
Length, to 1⅞ in. (4.8 cm). Body stout, squat; coloration variable, but usually has a dark spot below eye and V-shaped mark between eyes. Above brownish to gray or green; dark stripe from snout to shoulder and then irregularly along body. Legs banded; below whitish to buffy. Head short, stout; tympanum small. Toes only slightly webbed and with small disks. Male has dark throat.
Habitat
Prairies, pastures, swampy areas, and flatwoods.
Habits
Usually nocturnal.
Voice
Shrill, clear, and with carrying power; sounds like a bell, sometimes likened to an ungreased wooden wheel. Sound in chorus resembles that of distant flock of geese.
Reproduction
Breeds in winter in the South and in early spring in southwestern Illinois and adjacent areas; in ditches, sloughs, and small ponds.
Range
E. and cen. Okla., s. to Gulf Coast, rarely into ne. Ark. and ne. La.; disjunct populations in cen. Ill.; and in vicinity of se. Mo. and adj. Ill. and Ark.

SPOTTED CHORUS FROG

Pseudacris clarki **73:5**

Description
Length, to 1¼ in. (3.2 cm). Dark-edged green spots irregularly placed, sometimes forming 3 broken stripes over lighter dorsal ground color of grayish or olive, and a light line on upper lip. Below white. Sometimes dark triangle, not bordered by white, between eyes. Dark band from nostril through eye and ear, widening down sides; legs barred above. Skin smooth dorsally, body slender. Snout projects beyond lower jaws. Males have dark throats.
Habitat
Open prairie grasslands and marshes at edge of woodlands.
Habits
Nocturnal, secretive, shy.
Voice
Males call from clumps of grass in shallow temporary water; repeated raucous and grating low-pitched notes at the rate of 2 notes per sec. with a rest between notes equal to note length.
Range
S.-cen. Kans., s. through cen. Tex. and extr. ne. Mexico. Record from Mont.

NARROW-MOUTHED TOADS

Family Microhylidae

This family is widespread in the Southern Hemisphere, but in North America is limited to two genera and three species, and is primarily found in the South. Ant eaters, they have a small mouth, unwebbed toes, and a narrow, pointed snout; a fold is generally seen across the head behind the eyes.

EASTERN NARROW-MOUTHED TOAD

Gastrophryne carolinensis **75:14**

Description
Length, to 1½ in. (3.8 cm). Has narrow, pointed head, white belly covered with dark mottling, and dorsum some shade of brown to reddish-brown, sometimes with irregular dark markings. Broad light stripes may be present dorsolaterally. Fold of skin across head behind eyes; toes unwebbed. Males generally smaller, with vocal pouch, blackish throat, and a gland on chest.
Habits
Active from early April to September or early October.
Voice
½-2½ sec. in duration, a nasal buzz, like bleating of a lamb, loud in chorus; typically no *peep*.
Food
Mostly ants, also other insects.
Reproduction
Breeds in clear temporary bodies of water and around the shallows of lakes, flood plains; lays up to 850 eggs in a mass on water's surface. May hybridize with Great Plains Narrow-Mouthed Toad.
Range
Se. U.S. from s. Md. s. to Key West, Fla. w. through s. Ky., s. Ill., and s. Mo. and s. to e. Tex. Records in Iowa and s. Tex. Introduced in the Bahamas.

GREAT PLAINS NARROW-MOUTHED TOAD

Gastrophryne olivacea **74:3**

Description
Length, to 1⅝ in. (4.1 cm). Has olive, gray, or tan dorsum, generally plain, but may be marked with black dots or small streaks. Belly unspotted, white. Throat, chest, and lower sides may be marked. Fold of skin across head behind eyes; toes unwebbed.
Similarities
Eastern Narrow-mouthed Toad has brown to reddish-brown dorsum with dark markings and belly with dark mottling.
Habitat
Dry rocky upland areas in open woods or woodland edges, also river flood plains and cultivated fields.
Habits
Nocturnally active during humid or rainy weather; may be active by day during rains. Hides during day under surface rocks or in rodent burrows.
Voice
A *peep*, then a nasal buzz, like the bleating of a goat, 1–4 sec. in duration.
Reproduction
Breeds May–August.
Remarks
May hybridize with Eastern Narrow-mouthed Toad.
Range
Extr. w. Mo. and se. Nebr. through e. and cen. Tex. to s. Ariz. and in Mexico s. along both Atlantic and Pacific Coasts of mainland.

TRUE FROGS

Family Ranidae

Members of this family are generally widespread and successful, absent only from Antarctica. Typically they occupy semiaquatic situations. Although the family is diverse in the Old World, only the Genus *Rana* is found in North America. These frogs have long legs with broadly webbed hind feet and a large distinct eardrum or tympanum. They lack cranial crests, parotoid glands, and disks on the tips of their digits, and their skin is usually quite smooth. Usually a prominent dorsolateral fold is present down each side of the back. In some species the folds may be discontinuous, and the last half inch or so may be displaced toward the midline. Females lay a large number of eggs.

The Leopard Frogs have confounded scientists for decades. Because of their wide distribution and overall similarity, differences found in physiological tolerances, external appearance, and mating call among animals from various parts of the country were ascribed to variations within one species. The true story, and the story is by no means yet complete, is that several closely related species exist, and they may be found in the same breeding sites where their ranges overlap, with little or no hybridization. The species of the eastern United States, at least, are probably well-enough studied that new forms will not be found.

GREEN FROG

Rana clamitans **75:4**

Description
Length, to 4 in. (10.2 cm). Has dorsolateral folds not reaching groin. Above green to brownish, sometimes spotted with black; legs faintly barred; below white, occasionally with spots. Male has yellow throat and larger eardrums; female has white throat with spots.

Similarities
Resembles Bullfrog, but is smaller.

Habitat
Swamps, brooks, streams, ponds, and edges of lakes.

Habits
Solitary; hibernates in mud.

Voice
Low-pitched, like sound of banjo or double bass; throat swollen when calling.

Range
Se. Canada and e. U.S. except s. Fla. and cen. Ill., w. to extr. se. Man., cen. Minn., and e. Tex.

BULLFROG

Rana catesbeiana **75:3**

Description
Length, to 8 in. (20.3 cm). The largest North American frog. Lacks dorsolateral folds and has white belly, often mottled. Above green (especially forward), brown, or black with irregular dark spots; below whitish, often mottled; hind legs with dark crossbars. Skin fairly smooth; ridge running above, but stopping just behind, tympanum. Tympanum as large as eye in female, much larger in male.

Habitat
Permanent bodies of water: marshes, ponds, lakes, rivers.
Habits
Lives near deep water, hibernates in soft mud on bottom of ponds and lakes; larvae live at least a year before metamorphosis.
Voice
Deep, hoarse, *jug o'rum, more rum* resounds throughout breeding ponds.
Food
Invertebrates; also any vertebrates—fish, other frogs, snakes, baby turtles, mice, ducklings—that it can catch and swallow, including its own young.
Range
Extr. se. Canada throughout e. U.S. except s. Fla. and n. New England, and w. to Wis., se. Wyo., and w. Tex. Also in numerous localities throughout w. U.S. and s. along coastal Mexico, where it has been introduced. Introduced in Cuba, Puerto Rico, Hispaniola, and Jamaica.

PIG FROG

Rana grylio **74:11**

Description
Length, to 6⅜ in. (16.2 cm). Large frog with webbing almost to tip of longest toe, no dorsolateral folds, and pointed snout. Above brown to blackish-brown with black spotting; below heavily colored with dark reticulations. Feet fully webbed; tympanum as large as, or larger than, eye.
Similarities
Bullfrog has blunter snout and longest toe extends well beyond webbing.
Habitat
Permanent open bodies of water with emergent vegetation.
Voice
Calls from water or from among masses of aquatic plants; loud, resonant sound resembling grunting of a pig.
Range
S. Coastal Plain from s. S.C. through Fla. and w. to extr. se. Tex. Introduced in the Bahamas.

RIVER FROG

Rana heckscheri **74:12**

Description
Length, to 5 5/16 in. (13.5 cm). Large greenish-black frog; venter gray with light spots. Throat gray, white spots along lip. Skin rough. No dorsolateral folds, or else poorly defined. Toes not fully webbed; tympanum large.
Similarities
Pig Frog and Bullfrog have smoother skin and *light* venters with dark markings.
Habitat
Swampy areas; shallows of streams and rivers.
Voice
Deep, variable, but likened to a snore and grunt.
Range
Atlantic Coastal Plain from se. S.C., s. through n. Fla., and w. to se. Miss.

CRAYFISH FROG

Rana areolata **75:6**

Description
Length, to 4½ in. (11.4 cm). Recognize this secretive frog by 3 or 4 rows of round, dark spots with light gray edges overlying a meshlike brownish or gray dorsal pattern, rough skin texture, and continuous dorsolateral folds. Venter whitish and unspotted, legs barred with brown and gray. In males paired vocal sacs over the shoulder are conspicuous, bluish, like folds of skin when collapsed.
Habitat
Often in crayfish burrows, or "gopher" holes (made by tortoises), by pools in lowland meadows and flood plains.
Habits
Nocturnal, subterranean; rarely seen except at breeding pools in early spring (March and April).
Voice
Loud, deep snoring, deafening in chorus.
Range
Atlantic Coastal Plain from se. N.C. to cen. Fla., w. to extr. e. La. Also from cen. Miss., n. to s. Ill. and sw. Ind., and w. through n. Mo.; absent in Ozark region and e. La., but s. from se. Kans. to Gulf Coast in w. La. and Tex.

CARPENTER FROG

Rana virgatipes **75:7**

Description
Length, to 2⅝ in. (6.7 cm). Lacks dorsolateral folds, but has 4 golden-brown stripes down the back contrasting with olive to black ground color. Below yellowish or white with dark spots; back of thighs with broad black stripes; throat yellow. Skin smooth; toes not fully webbed. Male has 1 round vocal pouch on each side. Young lighter overall than adults.
Habitat
Sphagnum bogs, cedar ponds; swamps and quiet streams; under mats of vegetation.
Habits
Aquatic, seen at water's edge or on or under logs; also among vegetation.
Voice
Similar to hammering of a nail (hence the name) repeated 3–6 times.
Range
Atlantic Coastal Plain; in Pine Barrens of cen. N.J. and s. Delmarva Pen.; continuously s. from e. Va. to extr. ne. Fla.

PLAINS LEOPARD FROG

Rana blairi **74:10**

Description
Length, to 4⅜ in. (11.1 cm). Recognized by discontinuous light dorsolateral folds—toward hind legs they break and are displaced to midline. Has light spot in center of tympanum and only a few dark spots on sides. Irregular dark spots dorsally on gray, tan, or green ground color; hind limbs barred; belly white.Vocal sacs small and paired above shoulder.
Habitat
Plains and prairies.

Voice
Call lasts less than 1 sec. at 60°F (15.6°C) and consists of 2–6 notes delivered at a rate of about 3 per sec.
Reproduction
Breeds in slow or standing water, even in temporary ponds, in early spring and into summer.
Range
W. Ind. to e. and s. Nebr., e. Colo., and s. through ne. N.Mex. and Okla. to cen. Tex.

NORTHERN LEOPARD FROG

Rana pipiens **74:7**

Description
Length, to 4⅜ in. (11.1 cm). Above brown or green with rounded dark spots between dorsolateral folds; folds continuous, with other smaller folds between them. White stripe on lip. Males lack external vocal sacs, though internal pouches sometimes seen.
Similarities
Southern and Plains Leopard Frogs have fewer spots on lower sides.
Habitat
Permanent water; swamps, marshes, streams; also in meadows in summer.
Voice
Highly variable, but long, low-pitched trills lasting from 1–3 or more sec.; pulse rate about 20 per sec. at 60°F (15.6°C).
Reproduction
Breeds with warm temperatures of spring, about March or April in eastern U.S.
Range
S. Canada from Labrador to District of Mackenzie; and n. U.S. s. to cen. Conn., n. Pa., cen. Ky., s. Iowa, and s. Nebr. into N.Mex. and Ariz. Disjunct populations in Md., cen. Ill., and e.-cen. Okla. Occurs in most western states.

SOUTHERN LEOPARD FROG

Rana sphenocephala **75:2**

Description
Length, to 3½ in. (8.9 cm); record 5 in. (12.7 cm). Has continuous dorsolateral folds, a pointed snout, and few spots on sides of body; also often light spots on tympanum. Above green or brown with dark brown spots; light line from shoulder to snout. Males have large external vocal sacs at each side above shoulder.
Habitat
In and around shallows of rivers, lakes, streams, marshes, or ponds.
Voice
At 60°F (15.6°C) the short trill lasts less than 1 sec., and consists of 4 or more pulses with a pulse rate of 13 per sec. or less.
Range
Se. N.Y., s. through Ga. and Fla., but absent from Appalachian region; n. through s. Ind. to s. and e. Mo. and se. Kans., w. in Plains and Gulf Coast to cen. Tex. Introduced in the Bahamas.

PICKEREL FROG

Rana palustris **75:1**

Description
Length, to 3 7/16 in. (8.7 cm). Resembles a small brown Leopard Frog, but has 2 regular rows of closely spaced, square to rectangular, dark brown or black spots between the dorsolateral folds, and underside of thighs is conspicuously bright yellow or orange. Gray or tan above, spots outlined with light color. Dark mark from eye to nostril and light line on lip. Hind limbs banded. Vocal sac above shoulders on each side.
Habitat
Grassy meadows, streams, ponds in cool, clear water.
Habits
Active by day and night and much of the year; much time spent out of, but near, water.
Voice
Long, low-pitched trills; at end of call frequency changes.
Remarks
Poisonous secretions of the skin affect other animals, even other species of frogs kept in the same container.
Range
Se. Canada to S.C., w. to Wis., extr. se. Minn. in the North, and e. Tex. in the South; absent from most of Coastal Plain and prairies of Ill. and adj. states.

MINK FROG

Rana septentrionalis *Fig. 96*

Description
Length, to 3 in. (7.6 cm). In some ways a small edition of the Green Frog, but with buff-bordered black markings. Above buff or brown and green with mottled or spotted pattern; vermiculated pattern on legs; below white or yellowish. Skin smooth, snout rounded, body stout, eyes close together. Dorsolateral folds short or absent. Feet fully webbed.
Habitat
Rivers, lakes, or peaty ponds; near water lilies. A common frog of the northern forests.
Habits
Aquatic and solitary.
Voice
High, fast quacking with a metallic ring, *cuck-cuck-cuck-cuck.*
Remarks
Gives off strong minklike odor when annoyed.
Range
An exclusively northern distribution. E. Canada and extr. n. U.S. from Maritime Provinces and New England s. to n. N.Y., w. above Great Lakes to upper Mich., Wis., and ne. Minn., and to se. Man.

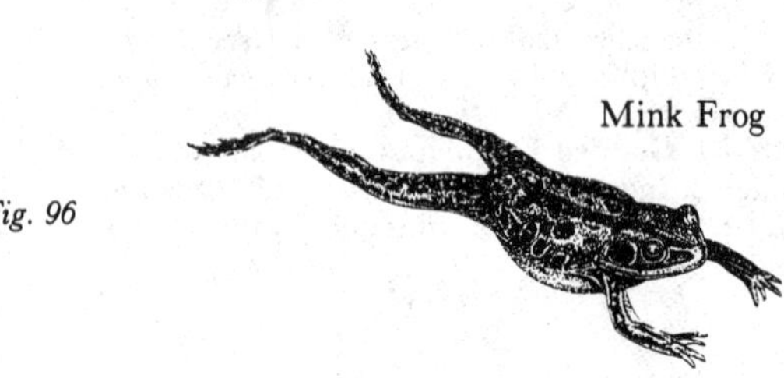
Mink Frog

Fig. 96

WOOD FROG

Rana sylvatica **75:5**

Description

Length, to 3¼ in. (8.3 cm). The only species with a wide black mask extending from snout to tympanum, and a dark line from eye to tip of snout. Above reddish-brown (either dark or light), occasionally gray; middorsal white line may be present. Dark markings on sides but rarely on back; below whitish. Dark mask has narrow light line below it. Head pointed; prominent light dorsolateral folds. Males with paired vocal sacs.

Habitat

Damp woods.

Habits

Terrestrial, hibernating in logs, stumps, under stones, rather than under water.

Voice

A short, sharp, rasping clacking almost like clucking (not quacking) sound of domestic ducks. Shorter than that of Leopard Frogs, but not as loud or deep; calls only during breeding season.

Reproduction

Breeds in ponds very early in spring, even before ice has completely melted; then returns to land.

Range

Most of Alaska and Canada, even n. of Arctic Circle and most of ne. U.S. s. to Md.; in Appalachians of n. Ga. and ne. Tenn. Isolated populations in Wyo. and Colo., in highlands of Ark. and Mo., and in n. Idaho.

Fishes

Consulting Editor
Franklin C. Daiber
Professor
College of Marine Studies
University of Delaware

Illustrations
Plates 76–93 Nina L. Williams
Text Illustrations by Jennifer Emry-Perrott and Nina L. Williams

Fishes

Superclass Pisces*

There are in the world today more than 25,000 different species of fishes, more than there are of all other vertebrates put together. Included in this chapter are 164 species of saltwater and 141 freshwater fishes, or 305 species in all. These represent the food and game fishes that occur regularly in appreciable numbers and that can be readily identified in the field.

A fish is an aquatic, cold-blooded vertebrate with jaws and paired fins that breathes chiefly by means of gills. Most fishes also have scales. Fishes are grouped into two classes. The first, the Chondrichthyes, consists of sharks, skates, and rays that have a skeleton of cartilage. The second, the Osteichthyes, includes all the modern fishes that have a skeleton of bone.

Also included in this chapter are representatives of two more primitive classes of vertebrates, the lancelets (hagfishes) and the lampreys. These are also cold-blooded and aquatic, but they lack jaws, paired fins, scales, and plates; in the lancelets, lateral gill openings are also absent and the brain is rudimentary. Both are ancient groups of which only a few species, often parasitic among the lampreys, have survived to the present.

In degrees varying by species, fishes as a whole possess all five senses of sight, smell, taste, touch, and hearing, although they hear without external ears. Their mental powers are low. An air bladder serves in many species as a depth regulator and in some, notably the croakers, as a means of making a noise. In some species the males are the only sound makers, and some naturalists think the males use sound in the breeding season to attract the females, just as some birds use song. The subject of sound in fishes has been studied in recent years, yet more is to be learned.

Most fishes swim by flexing the body through lateral motions of the caudal peduncle and tail. Fishes with forked tails are the fastest swimmers. The dorsal and anal fins serve as a keel; the pectorals deflect the fish up or down. In a few species, however, the pectorals provide the driving power; and certain eellike forms progress by moving the whole body sideways in a series of undulations.

Fishes vary greatly in size, from the 2-inch (5.1-cm) length of the Naked Goby to the 45 feet (13.7 m) of the Basking Shark. Many continue to grow through much of their lives. Thus in any one species in any particular area, the larger fish are usually the older fish. In the data in this chapter, the maximum recorded size is given, whether caught by hook, gaff, or net. In some cases the all-tackle record is given as well. The average, of course, is substantially smaller.

Many fishes are strikingly beautiful in color, particularly the fishes of the coral reefs. There, seen through a glass-bottomed boat, finny gems of orange and blue, crimson, green, and violet flash by in kaleidoscopic profusion. In more northern waters, however, except in some darters and mummichogs in the breeding season, colors in general are not bright. Some fish go through spectacular color changes when dying; but colors usually fade quickly upon death. In certain species color may vary with the environment, but in most

* Also Classes Leptocardii and Agnatha.

Fig. 97
Parts of Typical Fishes

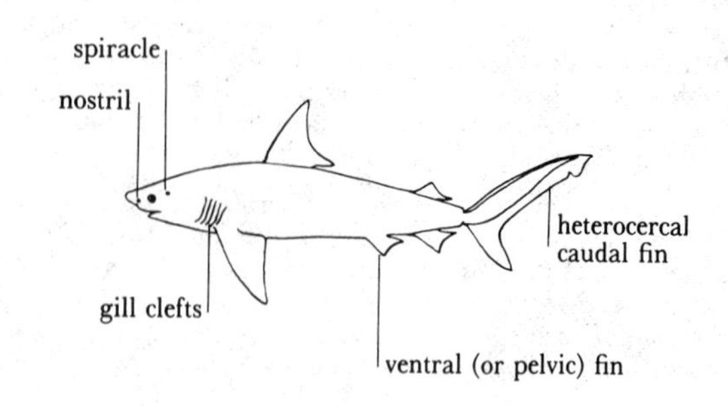

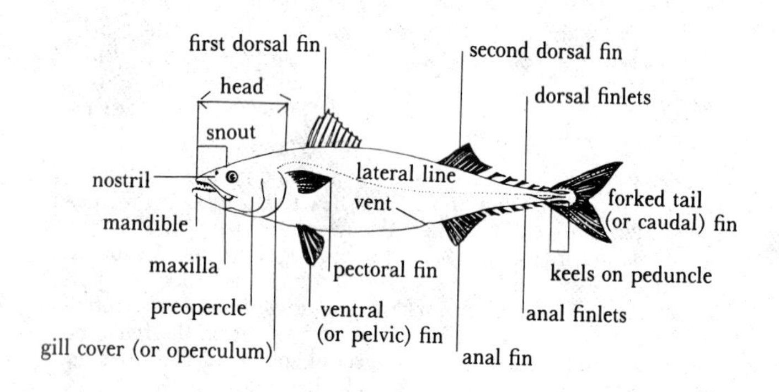

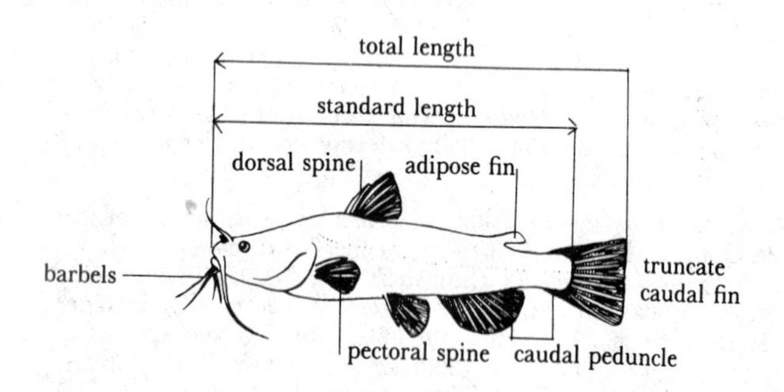

cases fish of the same kind tend to look much alike. There is some individual variation, but generally not as much as, for example, there is in snakes. However, many species go through a larval stage quite different in appearance from the adult.

Habitat
Almost all groups of animals, as they evolve, tend to spread out by adaptive radiation into every available habitat. Fishes, being the oldest vertebrates, have had time to spread out widely. Some habitats, because they are richer in food or provide shelter, or for other reasons, naturally support more individuals and species of fish than do others. In the northern oceans, for example, on the fringe of the ice pack, waters rich in minerals support in abundance the food chains upon which many saltwater fishes depend.

The more varied the habitat, of course, the more species of fishes can live in it. Thus the coral reef boasts more varieties than the open sea. The tropics, in general, have many species but relatively few individuals of each; the boreal, or northern, seas count fewer species but many of these occur in stupendous numbers. In all waters, except the very deepest, whether the bottom be of sand, mud, shell, or rock, whether it be bare or covered with weed, there are always fishes.

In fresh waters as well, virtually all aquatic habitats are inhabited by some species of fish. Muddy river mouths appeal to catfish, clear streams to trout. Sunfish prefer calm, weedy pools, while salmon leap upstream through swift-flowing rapids and over waterfalls.

Habits
The natural history of fishes is beyond the scope of this chapter; however, a few cases of special interest are mentioned. In a group as old as the fishes almost every imaginable way of life has been tried. At this time, one can see the culmination of all the experiments of the past that have succeeded.

Different habitats require different habits and different bodily adaptations. In the flatfishes that lie on their sides in the sand, one eye has migrated from the lower side to join the other on top of the head. The burrfish can puff itself up with air or water into a porcupinelike ball, highly unattractive to predators. The habits of other fishes are equally fascinating and almost infinitely varied.

Many species are migratory, moving north or inshore in spring and south or offshore in fall. Most have a favored range of temperature, pressure, and saltiness or muddiness of the water.

Food
Many fishes are carnivorous. Most carnivorous fishes feed on other kinds of fishes or on marine invertebrates such as jellyfish, copepods, squid and other mollusks, and on plankton, the minute animal and plant life that floats in the top layers of the sea. Some fishes, however, are herbivorous and feed on seaweeds or on the microscopic one-celled diatoms. Other fishes are scavengers and help keep the sea and rivers clean by feeding on carrion and waste.

Reproduction
In most fishes, as in all the land vertebrates, the sexes are separate. Female fish produce eggs, the males milt. In most species fertilization is external, the males discharging milt over the eggs as they are extruded by the female. However, in the sharks, skates, and rays, and in some killifishes and gambusia, fertilization is

internal. Sharks and rays bring forth their young alive. Skates lay eggs. In these three groups the pelvic fins of the male are modified into claspers that serve as organs of intromission. In males of the viviparous killifishes and gambusia, the anal fin is so modified.

In the fishes that practice external fertilization, the eggs vary in size from 1/50 of an inch (0.05 cm) in diameter in the Naked Goby to 7/8 of an inch (2.2 cm) in the Gafftopsail Catfish. Eggs of different species may variously float or sink, be free in the water, or become attached to weeds, snags, or rocks. They may be separate or they may adhere to each other in globular or stringy masses. Noteworthy is the floating egg mass of the Goosefish, which may be 2 feet (61 cm) wide and 40 feet (12.2 m) long and weigh up to 30 pounds (13.6 kg).

The number of eggs of any one species has little or nothing to do with the number of individuals of that species in the sea. Population sizes are controlled by the environment, in particular by the availability of food. Given ample food supplies and an increase in the amount of suitable territory, any species, no matter how few eggs it may lay, will soon have occupied the additional territory and stocked it to the limit.

Only a few saltwater fishes pay any attention to their eggs after they are laid. Among those that do, the most famous are the sea catfishes, whose males for as long as 65 days incubate the eggs and shelter the young in their mouths. Many freshwater species, on the other hand, such as the Largemouth and Smallmouth Bass, build nests of sticks or stones or scooped-out sand. The males often assist incubation by stirring the water around the eggs. They zealously guard the eggs during incubation and sometimes the young after they are hatched. The male seahorse and pipefish brood the eggs in an abdominal pouch, where they are deposited by the female.

Evolution

The notochord, regarded as the prototype of the spinal column and the distinguishing feature of the first vertebrates, is believed by some authorities to have arisen in animals that lived in flowing water. The first fishlike creatures presumably had only notochords; later ones developed spinal columns of cartilage or bone.

The earliest such creatures of which there is any record were the armored, jawless ostracoderms. They originated in the late Silurian period, flourished during the Devonian, and then died out, leaving as the sole relics of their class the hagfishes and lampreys. Then jaws were developed—one of the great steps in the history of evolution—by the placoderms, which are all now extinct but which gave rise to the sharks and the true fishes. The well-jawed, well-toothed sharks appeared in the Devonian period and for the rest of the Paleozoic era dominated the sea. They have continued, in an aggressive though subordinate capacity, to roam the saltwater regions ever since.

The bony fishes also appeared in the Ordovician or late Silurian period, but the more advanced forms did not become conspicuous until the Mesozoic era. From then on, their varieties and their numbers multiplied and they successfully withstood all competition, including that of the marine reptiles, the mosasaurs and ichthyosaurs, and of the marine mammals, the whales, all of which were members of more advanced groups that invaded the sea. Today, if success were to be reckoned by variety and numbers, the bony fishes would be judged the most successful backboned animals on the globe. They display a greater variety of species, and an astronomically greater number of individuals, than all other vertebrates combined.

Conservation

Fishes are of great importance to man. We eat probably several billion of them a year; we use many for oil, fertilizer, and other purposes; and we catch them for sport. Angling for sport, in its various branches, has increased considerably in recent years. Some 25,000,000 North Americans buy licenses each year to fish the fresh waters.

Fishes need conservation just as land animals and birds do. Even though commercial fisheries often take only a relatively small harvest of the great bounty of the sea, there are many problems to consider, such as the overexploitation of fishery resources by the efficient use of large commercial trawls and factory ships, and the pollution caused by industrial wastes. These and many other conservation questions engage the attention of federal and state fishery authorities.

In fresh waters the situation is quite different. Entire river systems in the past have been polluted and the fish destroyed by the dumping of toxic factory wastes. Much of this still goes on. The East Coast run of shad is a pitiful fraction of what it was 100 years ago. The Atlantic Salmon has been extirpated from southern New England. Similar situations occur elsewhere. Anglers, too, can soon overfish a stream and do so in heavily fished areas. Most states have had to institute elaborate hatchery and restocking programs in order to offset the effects of pollution and of heavy angling pressure. Such programs, if wisely conducted, are beneficial; but much must still be done to stop pollution.

The question of conservation leads inevitably to the question of the sea as a new food frontier for man, because fish are notably rich in vitamins, minerals, and proteins and make an excellent human food. The resources of the oceans and even the topography of their floors are still most inadequately understood. Scientists recognize that there is a finite limit to fishery resources and that the situation is presently at its upper limit or even has exceeded the capacity of fish populations to renew themselves.

Where to Look for Fishes

To begin knowing what kinds of fishes are available to find, exploration of local markets is a good start. Here one can become familiar with the species in a particular locale. During the warmer months, scuba and snorkeling can then prove productive in any body of water, whether fresh or salt water.

Range and Scope

Included in this chapter are all species of food and game fishes that occur regularly in appreciable numbers, and many other readily identifiable in the field (a) that occur annually within a depth of 25 fathoms (45.7 m) off the coasts of North America, principally from southern Florida north and west to the delta of the Mackenzie River in Canada, and (b) that occur annually in the fresh waters of continental North America east of the 100th meridian between the Canadian border and southern Florida. Ranges apply to the species as a whole only within the area of geographic coverage of the chapter. Sizes given are the maximum average sizes. Occasionally the record size is given.

The Atlantic Coast from Cape Hatteras to Cape Cod is temperate and has, in general, temperate-water species of fishes. North of

Cape Cod, the fish fauna changes, in large part, to circumpolar Arctic forms. Few temperate species or individuals are found north of the elbow of Cape Cod; few boreal fishes are south of it.

Classification and Nomenclature

Within recent decades, numerous changes have been proposed for the scientific names of many fishes. There is as yet no firm agreement among scientists as to the proper generic placement for many fishes or even wholehearted agreement as to the relative order in which some fish families should appear.

In this chapter, the classification of fishes follows Nelson, *Fishes of the World,* 1976, with slight modifications. The scientific nomenclature has been brought up to date and the English names follow, with some modifications, the 1970 American Fisheries Society List of Common Names.

USEFUL REFERENCES

Bigelow, H., and Schroeder, W. C. 1953. *Fishes of the Gulf of Maine.* Washington: U.S. Government Printing Office.

Breder, C. M., Jr. 1948. *Field Book of Marine Fishes of the Atlantic Coast.* New York: Putnam.

Clay, W. M. 1975. *The Fishes of Kentucky.* Frankfort: Kentucky Department of Fish and Wildlife Resources.

Dahne, R. A. 1950. *Saltwater Fishing.* New York: Holt.

Eddy, S. 1957. *Freshwater Fishes.* Dubuque: W. C. Brown.

Eddy, S., and Underhill, J. C. 1974. *Northern Fishes.* Minneapolis: University of Minnesota Press.

Harlan, J. R., and Speaker, E. B. 1956. *Iowa Fish and Fishing.* Des Moines: State Conservation Commission.

Hildebrand, S. F., and Schroeder, W. C. 1927. *Fishes of Chesapeake Bay.* Bulletin U.S. Bureau of Fisheries, Vol. 43.

Hubbs, C. L., and Lagler, K. F. 1964. *Fishes of the Great Lakes Region.* Ann Arbor: Univ. of Michigan Press.

LaMonte, F. 1950. *North American Game Fishes.* New York: Doubleday.

Leim, A. H., and Scott, W. B. 1966. *Fishes of the Atlantic Coast of Canada.* Ottawa: Fisheries Research Board of Canada, Bulletin 155.

Moore, G. A., 1968. Fishes. In Blair, Blair, Brodkorb, Cagle, and Moore, *Vertebrates of the United States.* New York: McGraw-Hill.

Nelson, J. S. 1976. *Fishes of the World.* New York: John Wiley & Sons.

Pflieger, W. L. 1975. *The Fishes of Missouri.* St Louis: Missouri Dept. of Conservation.

Scott, W. B., and Crossman, E. J. 1973. *Freshwater Fishes of Canada.* Ottawa: Fisheries Research Board of Canada, Bulletin 184.

Trautman, M. D. 1957. *The Fishes of Ohio.* Columbus: Ohio State Univ. Press.

GLOSSARY

Abdominal ridge Belly ridge from region below gills to anus.

Adipose fin A fleshy fin, without rays, between dorsal and caudal fin.

Anadromous Ascending rivers from salt water to spawn.

Anal fin Fin situated between anus and caudal fin.

Anterior Toward the front of the body (usually toward the head or cranial end).

Barbel Fleshy appendage projecting from upper jaw, lower jaw, or chin region.

Body depth Greatest vertical distance through body, not including fins.

Branchiostegal rays Bony rays that support the membranes under the head below the opercular bones.

Caudal fin Tail fin.

Caudal peduncle Posterior end of body from last ray of anal fin to base of caudal fin.

Cephalic fins Detached part of the pectorals on the heads of certain rays.

Circuli A series of concentric ridges on the scales.

Cirrus (*pl.* cirri) Skin flap projecting outward from head or body.

Claspers Extension from the paired pelvic fins of male sharks, rays, and chimaeras that serve as copulatory organs.

Compressed Flattened from side to side (laterally compressed) or from top to bottom (dorsoventrally compressed).

Ctenoid scales Bony fish scales with small toothlike projections from the posterior edge.

Cycloid scales Bony fish scales with a smooth posterior edge.

Dermal denticles Spinelike "scales" of a cartilaginous fish; also termed placoid scales.

Disk The flat, circular or diamond-shaped forepart of skates and rays, formed by fusion of pectoral fins to head.

Dorsal Toward the upper back region of body.

Dorsal fin Fin situated on top of the back, not including the adipose fin.

Epipelagic The oceanic zone into which enough light penetrates for photosynthesis.

Furcate Forked.

Gill arch Bony or cartilaginous structure to which the gill filaments and rakers are attached.

Gill cover Bony or cartilaginous gill cover; synonomous with *operculum.*

Gill filaments Fleshy red protuberances on the outer sides of the gill arch that serve in respiration.

Gill rakers Cartilaginous (sometimes bony) protuberances on the inner sides of the gill arch that direct food into the gullet.

Gill slits Openings between gill arches for passage of water, visible externally in cartilaginous fishes, covered by an operculum in bony fishes.

Heterocercal Unequally lobed; said of caudal fin when upper lobe is larger than lower.

Keel A longitudinal fin extending along the center of the bottom of the body.

Lateral line Sensory organ composed of a canal which connects a series of openings along sides of body; or a posterior extension of the sensory canals on the head.

Lobule Either a small lobe or a subdivision of a lobe.

Lunate Between crescent and halfmoon in shape.

Mandible Lower jaw, composed of from four to seven bones in fishes.

Maxilla The bone lying on each side of the two halves of the upper jaw of fishes.

Medial Toward the center of the body.

Mesopelagic Relating to ocean depths from about 600 to 3000 feet.

Nostrils Paired, cuplike structures on the snout of fishes that do not connect with the mouth cavity for respiration but serve in the reception of chemical stimuli.

Notochord A flexible, rodlike structure that supports the body of vertebrate embryos and the adults of jawless fishes. (In most higher vertebrates, the notochord is later replaced by the vertebral column.)

Operculum Bony or cartilaginous gill cover.

Palatines Paired bones of the palate, lateral and posterior to the upper jaw bones.

Parr marks Dark vertical markings on sides of young fish, especially trout.

Pectoral fins Paired fins attached to the shoulder or pectoral girdle, located just posterior to gill openings. They may be fused with pelvic fins into a sucking disk (in the clingfish) or completely absent (as in eels).

Pelvic fins Paired fins attached to pelvic girdle, located on belly between throat and anus. Both fins may be united into a sucking disk (in gobies), be completely absent (in eels and eellike fishes), or form barbels (as in cusk-eels and brotulas).

Peritoneum The membrane lining the visceral or abdominal cavity.

Pharyngeal teeth Teeth attached to bones of the paired fifth gill (pharyngeal) arch, located immediately anterior to esophagus or gullet. (Pharyngeal tooth counts are listed in order from left to right.)

Placoid scales Spinelike scales of a cartilaginous fish; also termed dermal denticles.

Posterior Towards the rear of the body; usually toward caudal fin.

Precaudal pit Depression that is found immediately anterior to the base of the caudal fin in some sharks.

Preopercle Paired bones of the posterior cheek region, anterior to bones of gill cover (operculum).

Premacilla Bone at front of upper jaw, or forming entire upper jaw.

Preorbital Region between eye and tip of snout.

Prickles Small, fine, sometimes curved spines in place of scales.

Ray Flexible (cartilaginous) supports of the soft fins.

Roe Fish eggs.

Scute An external scale.

Shagreen Rough, hard-scaled skin of some sharks.

Spinous rays The fins of advanced bony fishes, which are sharp, bony, and usually inflexible, and contain spines.

Spiny Composed of sharp, inflexible spines.

Spiracle A vestigial gill slit present behind eyes of sharks, rays, and a few primitive bony fishes. In rays it is large and serves for intake of water during respiration.

Striae Small ridges or lines.

Tubercle Small rounded lump; a modified scale, hard or soft.

Vent Anus.

Viviparous Gives birth to living young.

Vomer Bone at the midline of the palate, immediately behind the upper jawbone.

Saltwater Fishes

LANCELETS
Class Leptocardii

Lancelets
Order Amphioxiformes

LANCELETS
Family Branchiostomidae

VIRGINIA LANCELET
Branchiostoma caribaeum *Fig. 98*

Description
Size, to 2¼ in. (5.7 cm). No jaws; mouth a longitudinal slit below. Body long, tapering, compressed, colorless, naked; many small gill slits, eyes rudimentary, fins only slightly developed.
Habitat
Shallow, sandy bottoms.
Other name
Amphioxus.
Range
Chesapeake Bay and s.

Virginia Lancelet

Fig. 98

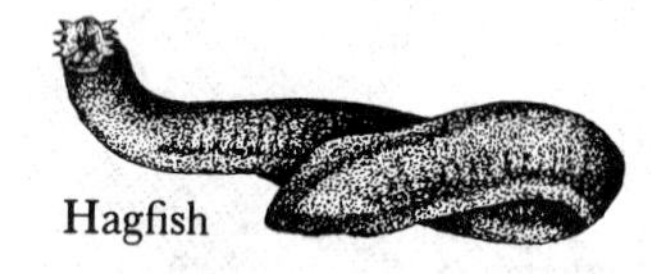
Hagfish

LAMPREYS
Class Agnatha

Hagfishes
Order Myxiniformes

HAGFISHES
Family Myxinidae

Members of this family are eel-shaped, jawless fishes lacking paired fins and scales. They generally inhabit cool marine waters. They are scavengers, feeding on dead or dying fish. When disturbed or handled, hagfishes secrete a great quantity of slime. Each egg is contained in a horny capsule.

HAGFISH
Myxine glutinosa *Fig. 98*

Description
Size, to 31 in. (78.7 cm). Eellike body; no jaws or eyes. Skeleton boneless, skin without scales; single finlike fold running around length of body. Mouth lipless and with tentacles, star-shaped when closed; tongue studded with horny teeth, mucous sacs along sides on abdomen.
Habitat
Soft mud, where it lies embedded with tip of snout projecting.
Food
Carrion; also parasitic.
Range
Arctic seas to N.C.

SHARKS, SKATES, AND RAYS
Class Chondrichthyes

Members of this class are distinguished from the bony fishes by having the entire skeleton cartilaginous. Fertilization is internal, effected by a pair of rodlike copulating organs developed from the pelvic fins in the male. Most species give birth to live young.

Sharks are distinguished from the skates and rays by the free eyelids, at least partly lateral gill openings, and pectoral fins not attached to the sides of the head. Most sharks have distinctive teeth.

Sharks
Order Lamniformes

Sharks of Order Lamniformes have two dorsal fins (with the exception of one genus) without spines, an anal fin, and five gill slits. Gill rakers are generally absent; spiracles are generally present.

SAND SHARKS
Family Odontaspididae

SAND TIGER
Odontaspis taurus **78:13**

Description
Size, to 9 ft. (2.7 m) and 250 lb. (113.4 kg). Last gill opening in front of pectorals, base of first dorsal in front of pelvics; no keel on peduncle, upper tail lobe notched and twice the size of lower.
Habitat
Relatively shallow inshore waters.
Food
Fishes, squid, crustaceans.
Range
Maine and s.

MACKEREL SHARKS

Family Lamnidae

Members of this family are distinguished by the following characteristics: The body is spindle-shaped and the teeth are large and sharp. The last gill opening is set in front of pectorals, which are large and half as long as they are high. The first dorsal fin is high, and its base is wholly in front of the small pelvics; the second dorsal and anal are very small. The caudal peduncle is slender and keeled and the tail crescent-shaped, less than one-third the total length of the fish. This family goes back to the Cretaceous.

Fig. 99
Shark Teeth

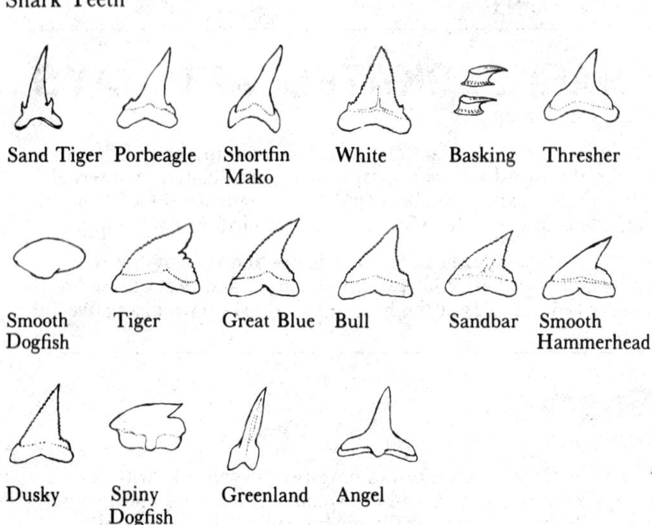

PORBEAGLE SHARK

Lamna nasus **78:10**

Description
Size, to 12 ft. (3.7 m); all-tackle record, 271 lb. (122.9 kg). Body stout and heavy-shouldered; first dorsal fin large, as high as long and originating slightly behind armpits or pectorals; furrow on upper and lower surface of root of tail. Tail fin has secondary longitudinal keel toward the front; second dorsal directly above anal.

Habitat
Surface waters offshore.

Habits
Surfaces on calm days.

Food
Fishes.

Other name
Mackerel Shark.

Range
Newfoundland to S.C.

SHORTFIN MAKO SHARK

Isurus oxyrhynchus **78:11**

Description
Size, to 11 ft. 5 in. (348.0 cm); all-tackle record, 1000 lb. (453.5 kg). Body moderately slender. First dorsal fin large and triangular, about as high as long; no secondary lateral keels on caudal. First dorsal originates above or behind rear base of the pectorals; and second dorsal originates in front of the anal.
Habitat
Surface of warm seas.
Food
Fishes.
Range
Newfoundland and s.

WHITE SHARK

Carcharodon carcharias **78:9**

Description
Size, to 21 ft. (6.4 m); all-tackle record, 2536 lb. (1150.1 kg). Body stout, lead-white in large individuals, darker in smaller ones. First dorsal originates over armpits of sickle-shaped pectorals; second dorsal slightly in front of anal.
Habitat
Temperate and tropical deep water; also in shallow waters inshore.
Food
Large fishes, porpoises.
Other name
Maneater Shark.
Range
Newfoundland to Brazil.

BASKING SHARK

Cetorhinus maximus **78:12**

Description
Size, to 40 ft. (12.2 m). Gill openings so large they almost meet under the throat. Body thick, snout short, teeth very small. Base of first dorsal fin wholly in front of small pelvics; second dorsal and anal very small, former slightly in front of latter. Caudal peduncle keeled; tail crescent-shaped. Color dark metallic gray fading to light belly.
Habitat
Surface waters inshore in summer, presumably deeper offshore in winter.
Habits
Often basks in sun at surface.
Food
Plankton.
Range
Newfoundland to N.C.

THRESHER SHARKS

Family Alopiidae

THRESHER SHARK

Alopias vulpinus **78:8**

Description
Size, to 20 ft. (6.1 m); all-tackle record, 922 lb. (418.1 kg); tail as long as its head and body. Body stout, snout blunt; first dorsal fin about as high as long, set midway between sickle-shaped pectorals and small pelvics; second dorsal and anal fins extremely small, caudal peduncle thick. Color blue-gray to purplish back fading to white belly; fins dark gray to purplish.
Habitat
Surface of temperate seas.
Food
Small schooling fishes.
Range
N.S. and s.

REQUIEM SHARKS

Family Carcharhinidae

The members of this family have sharp, saw-edged or flat and hard teeth, and eyes with nictitating, or winking, membranes. The last gill openings are above or behind the base of the pectorals; the first dorsal fin lies above the similar-sized anal. The tail, notched in many species, is less than one-third total length, and the upper lobe is longer.

GREAT BLUE SHARK

Prionace glauca **78:3**

Description
Size, to 12 ft. 7 in. (3.8 m). First dorsal nearer the pelvics than the pectorals. Body color deep blue, slender, thickest at mid-length; snout very long. Pectorals narrow, curved, very long.
Habitat
Open sea.
Food
Small fishes.
Range
Warm seas, n. to Newfoundland.

BULL SHARK

Carcharhinus leucas **78:1**

Description
Size, to 10 ft. (3 m). Short, broadly rounded snout; smooth back with no ridge between the dorsals. Dorsal profile more convex than the ventral, head very wide, eyes small. First dorsal originates over armpit of long pectorals; anal is more concave than second dorsal; upper lobe of tail more than twice length of lower.
Habitat
Relatively shallow inshore waters.
Range
Brazil to N.C., rarely to N.Y.

SMOOTH DOGFISH

Mustelus canis

77:14

Description
Size, to 5 ft. (1.5 m). Body slender, flattened below; snout blunt, spiracles present. Both dorsals spineless, first dorsal originating over rear angle of pectorals; pectorals large, higher than long; anal fin smaller than, and to rear of, second dorsal; no pit at root of tail.
Habitat
Inshore.
Food
Larger crustaceans.
Remarks
Edible.
Range
Bay of Fundy and s.

TIGER SHARK

Galeocerdo cuvieri

78:4

Description
Size, to 18 ft. (5.5 m); all-tackle record, 1382 lb. (626.7 kg). First dorsal above the armpit of the pectorals; caudal peduncle keeled, notched above and below. Head large, snout blunt, spiracles present; mouth broad, with long grooves along upper jaw. First dorsal high, nearly as large as pectorals; second dorsal ⅓ to ¼ size of first; upper lobe of tail 3 times length of lower.
Habitat
Surface of warm seas.
Food
Sea turtles, other sharks, fishes, carrion.
Range
Tropical and temperate seas occasionally n. to N.S.

SANDBAR SHARK

Carcharhinus milberti

78:5

Description
Size, to 8 ft. (2.4 m). Triangular snout; dorsals connected by a low ridge on the back. First dorsal fin originates over armpit of pectoral; length of front edge less than ½ distance from base of fin to snout. Body stout, heavy forward; broad first dorsal and pectorals.
Habitat
Inshore waters.
Food
Bottom fishes and crustaceans.
Range
Cape Cod to Brazil; relatively common along the coasts of N.Y. and N.J.

DUSKY SHARK

Carcharhinus obscurus

78:2

Description
Size, to 14 ft. (4.3 m). Body slender, tapering; snout rounded; dorsal fins connected by low ridge on back. First dorsal large and originating over rear corner of pectoral; length of front edge at least ½ distance from base of fin to snout; anal fin slightly longer than second dorsal.

Habitat
Inshore.
Food
Fishes, squid.
Range
Cape Cod to Brazil.

HAMMERHEAD SHARKS
Family Sphyrnidae

SMOOTH HAMMERHEAD
Sphyrna zygaena **78:6**

Description
Size, to 11 ft. (3.4 m). Hammer-shaped head; widely separated eyes at the far edges.
Habitat
Open seas near surface.
Food
Fishes.
Range
Worldwide in tropical and temperate waters n. to Cape Cod, rarely to N.S.

Dogfish and Angel Sharks
Order Squaliformes

These sharks have two dorsal fins, with or without spines, have no anal fins, and have five or six gill slits.

DOGFISH SHARKS
Family Squalidae

SPINY DOGFISH
Squalus acanthias **78:7**

Description
Size, to 4 ft. (1.2 m). Single spine in front of each of 2 dorsal fins; no anal fin. Head flattened and tapering to blunt tip; first dorsal nearer to pectorals than to pelvics, second dorsal originating well behind pelvics; no notch on tail. Color gray to brownish back fading to lighter belly.
Habitat
Cold water, inshore in winter, offshore in summer.
Habits
Travels in large schools.
Food
Fishes, squid, worms.
Range
N.S. to Cuba.

GREENLAND SHARK
Somniosus microcephalus **78:15**

Description
Size, to 21 ft. (6.4 m). Dorsal fins small, almost equal in size; upper tail lobe not much larger than lower. Body tapering to rear but cylindrical in front, snout rounded. Pectorals small, hardly larger than pelvics; second dorsal originating to rear of origin of pelvics; no spines on dorsals, no anal.
Habitat
Near the bottom of cold seas.
Range
Polar seas s. to Cape Cod.

ANGEL SHARKS
Family Squatinidae

ANGEL SHARK
Squatina dumerili **78:14**

Description
Size, to 5 ft. (1.5 m). Head broadly rounded, eyes on top of head. Both dorsal fins brush-shaped and similar; lower tail lobe larger than upper.
Similarities
Distinguished from all other sharks by its flattened body. Distinguished from the skates by the following: pectoral fins not attached to sides of head; gill openings extending from lower surface up onto sides of neck.
Habitat
Warm waters.
Range
Cape Cod and s.

SKATES AND RAYS
Order Rajiformes

The members of this order have flattened bodies and spiracles. They lack anal fins. If dorsal fins are present, they are located on the tail. Skates and rays are distinguished from sharks in the following respects: they do not have free eyelids, gill openings are below only, and the edges of the pectoral fins are attached to the sides of the head in front of the gill openings.

ELECTRIC RAYS
Family Torpedinidae

ATLANTIC TORPEDO
Torpedo nobiliana **79:11**

Description
Size, to 5 ft. (1.5 m). Body almost completely rounded. Skin smooth, eyes very small and set far forward. Pelvic fins somewhat in front of first dorsal; 2 dorsal fins situated at forward end of thick, spineless, triangular tail; electric organ, consisting of hexagonal tubes between pectoral fins and head, capable of giving a strong shock.

Habitat
Ocean bottom.
Habits
Bottom feeder.
Food
Fishes.
Other name
Electric Ray.
Range
N.S. to Fla. Keys.

SKATES
Family Rajidae

Skates have the skin more or less covered with spines and a muscular tail usually supporting two dorsal fins and sometimes a caudal fin. Distinguish skates from rays by their concave, instead of convex, pelvic fins and by the absence of a tail spine. The family goes back to the Cretaceous.

CLEARNOSE SKATE
Raja eglanteria **79:5**

Description
Size, to 3 ft. (0.9 m). Brown above, white below; sides of snout form a right angle, outer corners of pectoral fins angular. Row of small, stout, star-shaped spines along back and tail; 1-4 rows of spines on each side of tail; groups of large spines opposite and behind eyes; 1-5 spines on each shoulder.
Habitat
Sandy bottoms.
Other name
Brier Skate.
Range
Fla. to Mass.

LITTLE SKATE
Raja erinacea **79:3**

Description
Size, to 21 in. (53.3 cm). No row of spines along midline of back (except in young); sides of snout meet at obtuse angle and bulge opposite eyes; about 50 series of teeth in each jaw.
Habitat
Smooth shallow bottoms.
Other name
Common Skate.
Range
N.S. to Va.

BARN-DOOR SKATE
Raja laevis **79:2**

Description
Size, to 5 ft. (1.5 m). Large, brown with some light spots, paler below. Skin nearly smooth, spines few and small; sides of snout meet at acute angle and bulge opposite eyes; row of spines along tail.

Habitat
Smooth bottoms, often in deep water. Tolerates or enjoys an unusually wide range of temperature; is found in waters from 32° to over 68°F (0° to over 20° C).
Range
N.S. to N.C.

WINTER SKATE

Raja ocellata **79:4**

Description
Size, to 34 in. (86.4 cm). Large white spot with a black center near the rear corner of each pectoral fin. No row of spines along midline of back; sides of snout meet at obtuse angle and bulge opposite eyes; about 90 series of teeth in each jaw.
Habitat
Smooth bottoms.
Range
Newfoundland to N.C.

THORNY SKATE

Raja radiata **79:1**

Description
Size, to 40 in. (1.0 m). One large spine in front of each eye and 1 behind, 3 on each shoulder; small spines scattered elsewhere above; sides of snout meet at obtuse angle, 30–50 series of teeth in each jaw.
Habitat
Smooth bottoms in cool waters.
Range
Hudson Bay to N.Y.

STINGRAYS

Family Dasyatidae

Stingrays lack dorsal fins. They are distinguished from skates by their convex, instead of concave, pelvic fins and their long tails, often bearing venomous spines. They bear living young.

SMOOTH BUTTERFLY RAY

Gymnura micrura **79:9**

Description
Size, to 2 ft. (61.0 m) in width. This ray and its relative the Spiny Butterfly are the only rays with tail shorter than body.
Habitat
Ocean bottom.
Habits
Bottom feeder.
Food
Crustaceans.
Reproduction
Usually 2 living young.
Range
Cape Cod to Brazil.

Note: The **SPINY BUTTERFLY,** *Gymnura altavela,* gets up to 6 ft. (1.8 m) in width.

ROUGHTAIL STINGRAY

Dasyatis centroura **79:8**

Description
Size, to 10 ft. (3.1 m). Snout forms obtuse angle, pectorals angled; tail about 2½ times length of body and keeled below only.
Habitat
Ocean bottom.
Habits
Bottom feeder.
Range
Cape Cod and s.

ATLANTIC STINGRAY

Dasyatis sabina **79:6**

Description
Size, to 3½ ft. (1.1 m). Sides of snout meet at right angle, pectorals rounded; tail less than twice as long as body and keeled both above and below.
Habitat
Ocean bottom.
Habits
Bottom feeder.
Range
Chesapeake Bay and s.

EAGLE RAYS

Family Myliobatidae

The members of this family are distinguished by pectorals that stop on the sides of the head; cephalic, or head, fins that are not below the level of the body and are not long and earlike; a high-domed crown; eyes and spiracles on the sides of the head; and a long, whiplike tail, with one dorsal fin near its root. Eagle rays are swimmers as well as bottom feeders. They bear living young. The family goes back to the Cretaceous.

BULLNOSE RAY

Myliobatis freminvillei **79:12**

Description
Size, to 5 ft. (1.5 m). Body not covered with small spots; teeth in several flat series.
Habitat
Ocean bottom.
Habits
Bottom feeder.
Food
Crabs, lobsters, larger mollusks.
Range
Cape Cod to Brazil.

SPOTTED EAGLE RAY

Aëtobatus narinari **79:10**

Description
Size, to 12 ft. (3.7 m) and 450 lb. (204.1 kg). Body covered with small spots; teeth in single series.

Habitat
Ocean bottom.
Habits
Bottom feeder.
Food
Shelled mollusks.
Reproduction
4 living young.
Range
Va. coast (common) to Brazil (rare).

COWNOSE RAY
Rhinoptera bonasus **79:13**

Description
Size, to 7 ft. (2.1 m). Pectoral fins stop on sides of head, 1–2 rounded head fins in front, crown high-domed with eyes and spiracles on sides of head; teeth large flat grinding plates in mosaic pattern. Tail armed with poisonous spines.
Habitat
Bottom and higher levels.
Habits
Bottom feeder.
Food
Mollusks, especially soft clams.
Range
Nantucket to Brazil.

MANTAS
Family Mobulidae

ATLANTIC MANTA
Manta birostris **79:7**

Description
Size, to 22 ft. (6.7 m) wide and 3500 lb. (1587.3 kg). Head fins curve out in front. Crown high-domed, eyes on sides instead of on top; pectoral fins stop on sides of head.
Habitat
Not confined to the bottom.
Habits
A swimmer.
Food
Small pelagic animals.
Range
Brazil to Va., rarely to Cape Cod.

BONY FISHES
Class Osteichthyes

These are the modern, bony fishes, distinguished from the sharks, skates, and rays by the possession of true bones. Generally they have erectile fins with stiff or soft rods between which stretch membranes, often transparent. Normally, their bodies are covered with thin overlapping scales, but sometimes the body is naked. The gill opening is usually covered with a bony gill cover, or operculum. An air bladder is generally present. We call them modern, but actually they appeared before the sharks and go back in time to the Lower Devonian, 340,000,000 years ago.

Bonefishes

Order Elopiformes

These species lack spines in the vertical fins, have the ventral fins in an abdominal position, have a single dorsal fin, and usually have a well-forked tail. Gill openings are wide. The body is elongate and more or less compressed.

TARPONS

Family Elopidae

LADYFISH

Elops saurus *Fig. 100*

Description
Size, to 3 ft. (0.9 m). Bluish above, silvery on sides, whitish beneath; plate between branches of lower jaw. Dorsal lacks prolonged ray of the tarpon and is depressible into a scaly sheath; belly not compressed or covered with specialized scales.
Habitat
Warm waters.
Food
Crustaceans and smaller fishes.
Other names
Ten Pounder, Big-eyed Herring.
Range
Fla., n. infrequently to Cape Cod in fall.

Fig. 100

TARPON

Megalops atlantica *Fig. 101*

Description
Size, to 8 ft. (2.4 m) and 315 lb. (142.9 kg); all-tackle record, 283 lb. (128.3 kg); scales to 3 in. (7.6 cm). Large scales, long dorsal ray, curved anal. Plate between branches of lower jaw.
Habitat
Estuaries, coast.
Food
Fishes, crabs.
Range
Gulf of Maine to Brazil.

Fig. 101

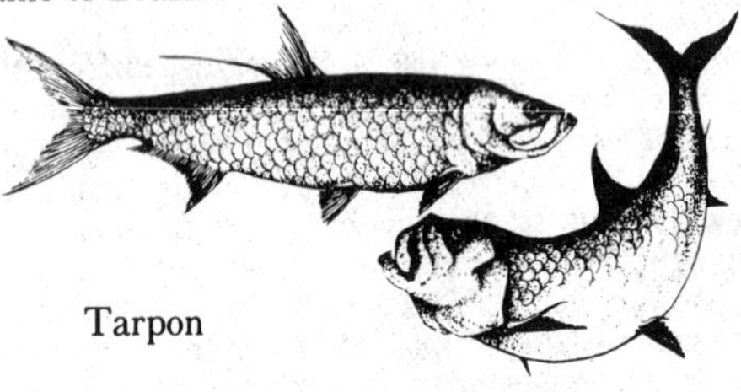

Tarpon

BONEFISHES
Family Albulidae

BONEFISH
Albula vulpes *Fig. 100*

Description
Size, to 31 in. (78.7 cm); record, 18 lb. (8.2 kg). Face unique; mouth small, teeth hard, flat in back of mouth; snout piglike; scales small.
Habitat
Relatively shallow water.
Habits
A solitary bottom feeder.
Food
Small mollusks.
Other name
Ladyfish.
Range
Warm seas; n. sparingly to Cape Cod.

Herringlike Fishes
Order Clupeiformes

These species are silvery and have the ventral fins in an abdominal position. They have branchiostegal rays, usually fewer than fifteen, and the abdomen often has keeled scutes along the ventral midline. Most are plankton feeders with numerous long gill rakers.

HERRINGS
Family Clupeidae

These are compressed, silvery, fork-tailed fishes without a lateral line, with very small teeth or none, and with very long gill rakers. The family goes back to the Cretaceous.

MENHADEN
Brevoortia tyrannus **76:13**

Description
Size, to 18 in. (45.7 cm) and 1 lb. 13 oz. (0.8 kg). Large scaleless head, nearly ⅓ total length; no teeth. Scales comb-toothed, mouth large, lower jaw projects; dorsal above or slightly behind very small ventrals; tail deeply forked.
Habitat
Near surface.
Habits
Occurs in vast compact schools.
Food
Plankton.
Age
To 10 years.
Other names
Pogy, Mossbunker, Fatback.
Range
N.S. to Fla., Mar. to Nov.

ATLANTIC HERRING

Clupea harengus **85:3**

Description
Size, to 17 in. (43.2 cm) and 1½ lb. (0.7 kg). Body deeper than thick; scales large, loose, rounded; mouth large, lower jaw projects, oval patch of small teeth in roof of mouth. Dorsal midway along body and directly above smaller ventrals, belly sharp.
Habitat
Open sea.
Habits
Travels in vast schools.
Food
Plankton.
Reproduction
Spawned in spring, summer, and fall; spring and fall the chief seasons; hatch in 22 days at 45°F (7°C).
Age
To 20 years.
Remarks
An important food fish, and one of the most abundant fishes in the world. Young are marketed as "sardines."
Other name
Sea Herring.
Range
N. Atlantic; most abundant in winter.

SPANISH SARDINE

Sardinella anchovia **85:5**

Description
Size, to 6 in. (15.2 cm). Body long, sides of nape striped; dorsal originates distinctly forward of ventral.
Habitat
Warm coastal waters.
Food
Copepods and other plankton.
Range
Cape Cod to Rio de Janeiro; erratic.

ROUND HERRING

Etrumeus teres **85:2**

Description
Size, to 15 in. (38.1 cm). Dorsal wholly in front of the ventrals. Slender; depth only ⅙ length; belly rounded, not sharp.
Habitat
Offshore marine waters.
Range
Bay of Fundy to Gulf of Mexico, but rare n. of Cape Cod.

ANCHOVIES

Family Engraulididae

These fishes have a small, compressed body, an undershot, cleft mouth, and usually a silvery lateral band. They are found in large schools often close to shore. The family goes back to the Tertiary.

STRIPED ANCHOVY
Anchoa hepsetus

85:11

Description
Size, to 6 in. (15.2 cm). Lateral band well defined, as broad as eye.
Habitat
Shallow water.
Habits
Schools.
Food
Copepods and small mollusks.
Range
N.S. to Brazil; first of May to late fall.

BAY ANCHOVY
Anchoa mitchilli

85:8

Description
Size, to 4 in. (10.2 cm). Lateral band diffuse or indistinct, scarcely wider than pupil of eye.
Habitat
Sandy shores, estuaries; usually abundant.
Range
Maine to Tex., but rare n. of Cape Cod.

Eels
Order Anguilliformes

These are fishes with jaws and snakelike bodies. They are naked or have only tiny scales. Their small gill openings are on the sides. They have no ventral fins, sometimes no pectorals, no spines in the fins, and no distinct tail fin.

The amazing life story of the eel is a classic of natural history. The species spawns in deep water southwest of the Bermudas. The transparent young, called leptocephali and long thought to be a different species, slowly make their way to the mouth of the freshwater rivers whence their parents came. Still translucent, they become elvers and finally grow into dark-colored eels. The females then ascend these rivers. After several years they are ready to spawn. In the fall, they descend the rivers and are joined by the males in the estuaries (the males do not go up the rivers). The adult males and females seek their traditional breeding grounds, where they spawn once and die.

TRUE EELS
Family Anguillidae

AMERICAN EEL
Anguilla rostrata

80:1

Description
Size, to 5 ft. (1.5 m). Distinguished from the Conger by the fact that it is often found in fresh water and by its small, imbedded scales.
Habitat
Shallow water, fresh or salt.
Habits
Hibernates in mud.

Food
Shrimps, crabs, mollusks, worms, small fishes.
Range
Labrador to the Guianas; common.

CONGER EELS
Family Congridae

CONGER EEL
Conger oceanicus **80:2**

Description
Size, to 12 ft. (3.7 m) and 8 lb. (3.6 kg). Distinguished from the American Eel by an entire absence of scales.
Habitat
Salt water offshore, crevices and nooks in rocks.
Food
Fishes, worms.
Range
Gulf of Maine to Brazil.

Catfishes
Order Siluriformes

SEA CATFISHES
Family Ariidae

The sea catfishes, which are the saltwater family of this largely freshwater order, have a transverse mouth surrounded by barbels. Each dorsal and pectoral fin bears a stout spine. The family goes back to the Eocene.

GAFFTOPSAIL CATFISH
Bagre marinus *Fig. 102*

Description
Size, to 2 ft. (0.6 m). Lower jaw with 2 barbels. Long streamers on dorsal and pectoral fins and on barbel on the jaw.
Habitat
Coastal estuaries.
Reproduction
Eggs are incubated within mouth of the male.
Range
Cape Cod to Panama, but rare n. of Del.

Fig. 102

SEA CATFISH
Arius felis *Fig. 102*

Description
Size, to 1 ft. (0.3 m). Lower jaw with 4 barbels, no streamers on pectoral or dorsal spines.
Habitat
Harbors of southern states.
Habits
Active mostly at night.
Reproduction
Males incubate eggs, numbering up to 55, in their mouths. The young may reach 4 in. (10.2 cm) in length before leaving paternal shelter.
Range
Cape Cod (rarely) to West Indies.

Codfishes
Order Gadiformes

These fishes have an elongate body, usually with long dorsal and anal fins. The pelvics, if present, are below or in front of the pectorals. There are cycloid scales in most species.

CODFISHES
Family Gadidae

These fishes have soft fins only. Their ventrals are under or in front of their pectorals, not behind them. The only fishes with three dorsal fins are found in this family. Most are bottom feeders. This family includes some of the most important food fishes.

ATLANTIC TOMCOD
Microgadus tomcod **84:2**

Description
Size, to 15 in. (38.1 cm) and 1¼ lb. (0.6 kg). Resembles a young Atlantic Cod except for its ventral fins, which have a filament as long as the rest of the fin.
Habitat
Less than 3 fathoms (5.5 m) deep; moves inshore in fall.
Food
Invertebrates, fishes.
Other name
Frostfish.
Range
Labrador to Va.

ATLANTIC COD
Gadus morhua **76:12**

Description
Size, to 6 ft. (1.8 m) and 211 lb. (95.7 kg); all-tackle record, 57 lb. (25.9 kg). Speckled gray-green or reddish above, white below; head large, mouth wide, teeth small and numerous.
Habitat
Continental shelf and submarine banks, often lying a fathom (1.8 m) or so from the bottom.

Habits
Rapacious.
Food
Mollusks, crustaceans, squid, fishes.
Age
At least to 9 years.
Other names
Rock Cod, Scrod.
Range
N. Atlantic, particularly the Grand Banks, s. to N.Y. rarely to Va.

POLLACK

Pollachius virens **76:11**

Description
Size, to 3½ ft. (1.1 m) and 35 lb. (15.9 kg); all-tackle record, 36 lb. (16.3 kg). Dorsals, 3; projecting lower jaw. Greenish-blue; lateral line light; 2 anals, tail slightly forked.
Habitat
Cool water, any moderate level.
Habits
More active than Cod or Haddock.
Reproduction
Spawns Oct. to Dec., at depth of 15–50 fathoms (27.4–91.4 m); 200,000–400,000 eggs hatch in 9 days at 43° F (6.1° C).
Age
To 19 years.
Remarks
A very important commercial fish.
Other name
Boston Bluefish.
Range
N. Atlantic, rarely s. to Chesapeake Bay.

HADDOCK

Melanogrammus aeglefinus **76:9**

Description
Size, to 44 in. (1.1 m) and 37 lb. (16.8 kg), but usually 3–4 lbs. (1.4–1.8 kg). Distinguished by 3 dorsal fins, a black lateral line, and a dusky blotch over each pectoral fin. Purplish-gray above; mouth smaller than in other codfishes and with shorter gape; 2 anals.
Habitat
Bottoms in cold water, 10–75 fathoms (18.3–137.2 m).
Food
Mollusks, worms, squid, fishes.
Age
To 14 years.
Remarks
A great food fish. The Georges Bank–South Channel area has often yielded 37,000,000 fish a year, but numbers are now greatly reduced.
Other name
Scrod.
Range
N. Atlantic s. to N.J.

HAKES

Genus *Urophycis*

These hakes have two dorsal fins, the second much longer than the first, one anal fin, and long ventrals.

SPOTTED HAKE

Urophycis regius **84:1**

Description
Size, to 16 in. (40.6 cm) and 1½ lb. (0.7 kg). No streamer on first dorsal fin.
Habitat
Shallow water.
Habits
Bottom feeder.
Food
Fishes, squid, crustaceans.
Range
Fla. to Cape Cod, rarely to Maine.

WHITE HAKE

Urophycis tenuis **84:3**

Description
Size, to 4 ft. (1.2 m) and 40 lb. (18.1 kg), average 8 lb. (3.6 kg). Short streamer on first dorsal. Distinguish from Red Hake by ventrals that do not reach beyond vent, smaller scales, and mouth gaping to back of eye.
Habitat
Soft bottoms, to 300 fathoms (0.6 km).
Habits
Most active at night. Locates much of its food by swimming slowly over the bottom and trailing the thin tips of its ventrals, which act as feelers.
Food
Shrimps, squid, fishes.
Other names
Boston Hake, Ling, Mud Hake.
Range
Newfoundland to Cape Hatteras.

RED HAKE

Urophycis chuss **84:4**

Description
Size, to 30 in. (76.2 cm) and 7 lb. (3.2 kg). Long streamer on first dorsal. Distinguish from White Hake by ventrals that reach beyond vent, larger scales, and mouth gaping only to pupil of eye.
Habitat
Soft bottoms, to 300 fathoms (0.6 km).
Food
Shrimps, squid, fishes.
Other names
Ling, Squirrel Hake.
Range
Newfoundland to Chesapeake Bay.

CUSK

Brosme brosme **84:5**

Description
Size, to 3½ ft. (1.1 m) and 27 lb. (12.2 kg). Only 1 dorsal fin, extending entire length of back; dorsal and anal fins continuous with tail fin, the boundaries marked merely by notches.
Habitat
Cool waters over rocks or gravel from 10 fathoms (18.3 m) down.
Habits
Solitary, a bottom feeder.
Food
Crustaceans, mollusks.
Remarks
Although Cusk is excellent eating, landings of this species have decreased in recent years with the decrease in long-line fishing and increase in otter trawling. This is because the cusk frequents rough bottoms that are not reached by the trawl.
Range
N. Atlantic, s. to Cape Cod, rarely to N.J. in deeper water.

SILVER HAKE

Merluccius bilinearis **76:10**

Description
Size, to 2½ ft. (0.8 m) and 5 lb. (2.3 kg). Distinguished from other hakes by the absence of chin barbels and by its ordinary and not elongated ventrals.
Habitat
Surface to 300 fathoms (0.6 km), sandy or pebbly bottoms.
Food
Fishes.
Remarks
A fine food fish, but soft.
Other name
Whiting.
Range
Grand Banks to N.Y.

Toadfishes

Order Batrachoidiformes

TOADFISHES

Family Batrachoididae

OYSTER TOADFISH

Opsanus tau **80:10**

Description
Size, to 15 in. (38.1 cm). Body tapering, belly plump; head, mouth, and eye large; head covered with fleshy flaps. Pectorals large and fanlike; 2 dorsals separated by a notch, soft dorsal 5 times as long as spiny; anal shorter than second dorsal, with outer ends of rays free; tail rounded. Somewhat resembles the sculpins, but ventrals are set under the throat in front of pectorals, not below or behind.
Habitat
Shallow inshore waters with hiding places.

Habits
Vocally active during the summer spawning season.
Remarks
This fish can severely pinch the fingers while one is trying to extract the hook.
Range
Maine to Fla.

Frogfishes
Order Lophiiformes

ANGLERS
Family Lophiidae

GOOSEFISH
Lophius americanus *Fig. 103*

Description
Size, to 4 ft. (1.2 m) and 50 lb. (22.7 kg). Colossal mouth. Body soft, flattened, and tadpolelike, tapering back from enormous head to small, broom-shaped tail; teeth many, curved; lower jaw projects, gill openings very small. Above 3 thin, stiff spines; dorsals separated, second longer than first; pectorals with wristlike joint at base.
Habitat
Any bottom, tide line to 365 fathoms (0.7 km).
Habits
Uses end tab on first dorsal spine as a fishing lure while it lies in wait on the bottom.
Food
Almost anything edible and marine, including sea birds.
Other names
Angler, Allmouth.
Range
Newfoundland to Brazil.

Fig. 103

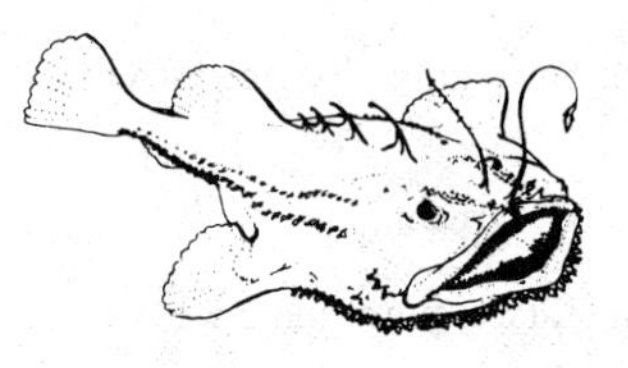

Goosefish

Needlefishes, Halfbeaks, Killifishes, and Silversides

Order Atheriniformes

Fishes of this order are varied. Gill cover and preopercular margins are without spines or serrations. Generally surface feeding fishes, many species are found in fresh water.

NEEDLEFISHES

Family Belonidae

ATLANTIC NEEDLEFISH

Strongylura marina **80:12**

Description
Size, to 4 ft. (1.2 m). Greenish above, silvery on sides, white below; both jaws extended into a long, toothed beak.
Habitat
Coastal, often ascends rivers part way.
Habits
Thrusts itself up through ocean surface like a spear; can be dangerous to man.
Food
Fishes.
Other name
Billfish.
Range
Maine to Fla., but rare n. of Cape Cod.

HALFBEAKS

Family Exocoetidae

HALFBEAK

Hyporhamphus unifasciatus **80:11**

Description
Size, to 1 ft. (0.3 m). Unique for its long lower jaw and short upper jaw.
Habitat
Tropic waters and Gulf Stream.
Food
Small crustaceans, mollusks, plant material.
Range
Maine to Argentina, more common in tropics.

KILLIFISHES

Family Cyprinodontidae

These are small fishes with a very small mouth, ventrals on the abdomen, one dorsal fin set to the rear, and a thick peduncle. The brilliant colors of the spawning males rival the beauty of many tropical reef fishes.

MUMMICHOG

Fundulus heteroclitus **80:3**

Description
Size, to 6 in. (15.2 cm). Body stout and covered with large rounded scales; snout blunt, teeth sharp, back and belly rounded. Anal fin of male larger than in female and used as an intromittent organ when spawning. Male assumes brilliant orange and blue colors in breeding season.
Habitat
Quiet water, less than 100 yards (91.4 m) from shore; usually associated with mud bottoms of tidal marsh creeks.
Food
Omnivorous to carnivorous.
Remarks
Provides good bait.
Other names
Killifish, Saltwater Minnow, Chub, Mummy.
Range
Gulf of St. Lawrence and s.

STRIPED KILLIFISH

Fundulus majalis **80:4**

Description
Size, to 7 in. (17.8 cm). Male takes on bright colors in breeding season. Distinguished from Mummichog by black bars, transverse on male, lengthwise on adult female; and by more pointed snout and more slender body.
Habitat
Quiet water close to shore; usually associated with sandy bottoms.
Habits
Can flop several feet back into the water if stranded by outgoing tide.
Food
Mollusks, crustaceans, fishes, insects.
Other name
Striped Mummichog.
Range
N.H. to Fla.

SHEEPSHEAD MINNOW

Cyprinodon variegatus **80:9**

Description
Size, to 3 in. (7.6 cm). Back high, arched; sides flat. Males assume brilliant blue above and orange below in breeding season. Distinguished from Mummichog by greater depth of body, large wedge-shaped teeth, and square tail.
Habitat
Shallow waters of inlets.
Other names
Broad Killifish, Pupfish.
Range
Cape Cod to Mexico.

SILVERSIDES

Family Atherinidae

ATLANTIC SILVERSIDE

Menidia menidia **76:2**

Description
Size, to 6 in. (15.2 cm). Two dorsals, long anal, and no adipose fin. Head and body slender and eye large.
Habitat
Shallow water near shore.
Food
Small crustaceans, mollusks, plankton, bottom organisms.
Range
Va. to N.S. Abundant.

Pipefishes and Cornetfishes

Order Syngnathiformes

This order includes species with long snouts ending in a small mouth.

CORNETFISHES

Family Fistulariidae

BLUESPOTTED CORNETFISH

Fistularia tabacaria *Fig. 104*

Description
Size, without lash, to 6 ft. (1.8 m). Long whiplike lash growing from the center of forked tail. Body long, silvery; scales small, inconspicuous.
Habitat
Drifts into eastern range in the Gulf Stream.
Food
Small fishes.
Range
Cape Cod to Brazil, but rare out of the tropics.

Fig. 104

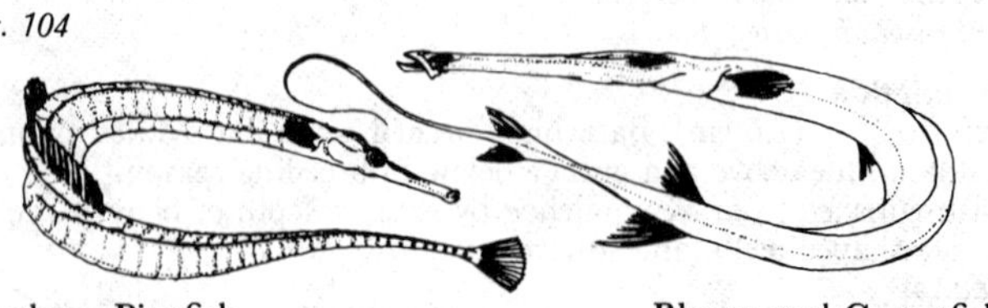

Northern Pipefish   Bluespotted Cornetfish

PIPEFISHES

Family Syngnathidae

These fishes have segmented body plates; the gill opening is a small pore and the gills are small, rounded tufts.

The mode of reproduction among these fishes suggests, remotely, that of marsupial mammals, but here it is the male that has the pouch on his abdomen. The female deposits her eggs in this pouch. There they are incubated and hatched and the young are protected until they can shift for themselves. The family goes back to the Eocene.

NORTHERN PIPEFISH

Syngnathus fuscus *Fig. 104*

Description
Size, usually 6 in. (15.2 cm), but up to 12 in. (30.5 cm). Axis of head in line with axis of body, tail not prehensile.
Habitat
Shallow water among seaweed.
Habits
Not very active; may sometimes be caught by hand.
Food
Small crustaceans.
Range
Halifax to N.C., most common near N.Y.

LINED SEA HORSE

Hippocampus erectus **80:7**

Description
Size, to 7½ in. (19.1 cm). Axis of head at angle to that of body, tail prehensile.
Habitat
Shallow water among seaweed.
Habits
Curls tail about stalk, swims vertically by undulating dorsal fin.
Reproduction
Spawns in summer; up to 150 young.
Food
Tiny organisms.
Range
Mass. to S.C., rarely to N.S.

Perchlike Fishes

Order Perciformes

This large and varied order is the most diversified fish order today. Its members have spines on the first dorsal fin and the anal fin. The pelvic fins, if any, are usually located on the throat, not the abdomen. Scales are ctenoid or absent.

TEMPERATE BASSES

Family Percichthyidae

STRIPED BASS

Marone saxatilis **77:7**

Description
Size, average 20–48 in. (50.8–121.9 cm), 3–40 lb. (1.4–18.1 kg); maximum 6 ft. (1.8 m), 125 lb. (56.7 kg); all-tackle record, 73 lb. (33.1 kg). Longitudinal stripes; 3 spines and 12 rays on anal fin. Body stout, mouth oblique and gapes back to eye; lower jaw projects. Dorsals about equal in length, anal similar to second dorsal and originates below it; tail wide, slightly forked.

Habitat
Coastal waters.
Food
Fishes and larger crustaceans.
Age
To 23 years.
Remarks
This species has occasioned much controversy. The anglers want it taken only on hook and line; commercial fishermen say it is hard to prevent a few from getting caught occasionally in nets at river mouths designed for other species.
Other names
Rockfish, Striper.
Range
Gulf of St. Lawrence to Fla.; most common from Cape Cod to N.C.

SEA BASSES

Family Serranidae

These carnivorous fishes are highly valued as food. Both spiny and soft-rayed portions of the dorsal are well developed, either as separate fins or divided by a deep notch. The ventral fins are under the pectorals; the anal is about as long as the soft part of the dorsal. There is a large caudal peduncle and a broad tail. The family goes back to the Eocene.

RED GROUPER

Epinephelus morio **80:6**

Description
Size, to 3 ft. (0.9 m). Brownish, bases of jaws reddish, sides with pale blotches, dark dots around eye. Dorsals adjoin, second dorsal spine highest.
Habitat
Close to bottom of inlets and reefs.
Remarks
Large and commercially valuable as food fish; a well known frequenter of reefs.
Range
Brazil to Va., straggling to Cape Cod.

BLACK SEA BASS

Centropristes striata **77:3**

Description
Size, to 2 ft. (0.6 m) and 7½ lb. (3.4 kg); all-tackle record, 8 lb. (3.6 kg). Dorsal fins continuous; in large fish, upper ray of the tail fin lengthened. Scales large but head scaleless; flat spine near rear angle of gill cover. Front dorsal saw-toothed in outline, pectorals broad and rounded, anal with 3 short spines at front, tail rounded.
Habitat
Shallow waters offshore over rocky bottoms.
Habits
Bottom feeder over rocky ground.
Food
Crustaceans, mollusks, fishes, squid.
Range
N. Fla. to Cape Ann, rarely in Maine; May to November.

TILEFISHES

Family Branchiostegidae

TILEFISH

Lopholatilus chamaeleonticeps **76:14**

Description
Size, to 3½ ft. (1.1 m) and 50 lb. (22.7 kg). Fleshy, finlike flap on nape in front of dorsal fin. Mouth wide, both jaws with 2 rows of teeth, eye set high in head. Spiny and soft portions of dorsal fin continuous and of even height.
Habitat
Continental slope.
Food
Crabs, squid, shrimps, shelled mollusks, worms, fishes.
Range
Chesapeake Bay to Maine.

BLUEFISHES

Family Pomatomidae

BLUEFISH

Pomatomus saltatrix **77:8**

Description
Size, to 3½ ft. (1.1 m) and 27 lb. (12.2 kg); all-tackle record, 24 lb. (9 kg). Blue above, white below; mouth large, lower jaw projects, teeth prominent. First dorsal low, spiny, and depressible in a groove; second dorsal concave and twice as long as first. Anal fin, originating to rear of second dorsal, shaped like it and preceded by 2 short spines; tail broad and forked.
Habitat
Warm water inshore or on offshore banks.
Habits
One of the most ferocious fishes of its size.
Food
Fishes and squid.
Other names
Snapper (for the young), Tailor.
Range
Gulf of Mexico to Cape Cod, straggling to Maine; arrives April to May off N.Y.

COBIAS

Family Rachycentridae

COBIA

Rachycentron canadum **80:5**

Description
Size, to 5 ft. (1.5 m); all-tackle record, 102 lb. (46.3 kg). Body spindle-shaped, broadly striped; mouth large, lower jaw projects. First dorsal made up of low, separate spines; second dorsal long and concave; anal originates behind second dorsal, but is similar in outline. Young Cobias up to 4 in. (10.2 cm) have convex tails; later tails become concave.

Habitat
Offshore or near wharves in bays and inlets.
Food
Crabs and fishes.
Other names
Sergeantfish, Cabio, Crabeater.
Range
Cosmopolitan in warm waters n. to Chesapeake Bay, rarely to Cape Cod.

REMORAS
Family Echeneidae

SHARKSUCKER
Echeneis naucrates **80:8**

Description
Size, to 38 in. (96.5 cm). Unique in having on top of its head a flat oval sucking disk with 20–28 cross-ridges; 3 other species with 14–18 ridges occur far offshore. Body very slender, nearly round in cross section, tapering to very slender caudal peduncle. Belly as dark as back. Pectorals pointed and set high, ventrals pointed and directly under pectorals, dorsal and anal fins long and tapering; tail convex in young, concave in old.
Habitat
Open sea.
Food
Attaches itself by its sucking disk to the underside of sharks and tarpons, and feeds on scraps from its host's meals.
Other name
Remora.
Range
Warm seas, rarely n. to Cape Cod.

JACKS
Family Carangidae

In this family the first, spiny, dorsal is much shorter than the soft-rayed second, and it may be reduced to a few short spines or be lacking. The anal fin is preceded by three short spines, which may become a permanent finlet or be lost in old age. The caudal peduncle is very slender and the tail deeply forked. Most species have a prominent hard keel on the rear portion of the lateral line, which serves to strengthen the peduncle. These are largely oceanic fishes. The family goes back to the Eocene.

BIG-EYED SCAD
Selar crumenophthalmus **81:2**

Description
Size, to 2 ft. (0.6 m). Eyes large; lateral line slightly arched but without spots, heavily scaled to rear only; 2 spines in front of anal.
Habitat
Warm seas.
Other name
Goggle-eyed Scad, Cigarfish.
Range
Atlantic Coast n. to Cape Cod, straggling to N.S.

ROUND SCAD

Decapterus punctatus **81:5**

Description
Size, to 1 ft. (0.3 m). First dorsal originates over middle of pectorals; single finlet behind dorsal and anal. Distinguished from Mackerel Scad by toothed jaws, longer second dorsal, and spots along arched lateral line, rear half of which is covered with large keeled scales.
Habitat
Open ocean.
Range
Warm Atlantic, n. to Cape Cod, rarely to N.S.

MACKEREL SCAD

Decapterus macarellus **81:4**

Description
Size, to 1 ft. (0.3 m). Distinguished from Round Scad by absence of spots along lateral line, untoothed jaws, and less developed scales on peduncle.
Habitat
Open ocean.
Range
Warm waters n. to Cape Cod, rarely to N.S.

ROUGH SCAD

Trachurus lathami **81:1**

Description
Size, to 1 ft. (0.3 m). Strongly arched lateral line heavily scaled along its entire length; 2 small detached spines in front of anal fin. No finlets behind dorsal or anal; tail deeply forked.
Habitat
Open ocean.
Other name
Saurel.
Range
Atlantic Coast n. to Cape Cod, rarely to Maine.

YELLOW JACK

Caranx bartholomaei **81:16**

Description
Size, to 15 in. (38.1 cm). Body deep; dorsal and anal fins completely covered with small scales, last rays not notably enlarged.
Range
Brazil to Nantucket, rarely to N.S.

HORSE-EYED JACK

Caranx latus **81:3**

Description
Size, to 2 ft. (0.6 m). Body deep, eye large, lateral line strongly arched. Dorsal and anal fins scaly only on elevated portions.
Food
Small fishes.
Range
West Indies n. to Va.

BLUE RUNNER

Caranx crysos **81:17**

Description
Size, to 22 in. (55.9 cm) and 4 lb. (1.8 kg). Breast scaly; lateral line strongly arched and with bony plates increasing in size to rear. Dorsal and anal fins scaly only on elevated portions; anal preceded by finlet of 2 short spines.
Food
Shrimps, fishes.
Other name
Hard-tailed Jack.
Range
Brazil to Cape Cod, rarely to N.S.

CREVELLE JACK

Caranx hippos **81:15**

Description
Size, to 2½ ft. (0.8 m). Well-rounded forehead, black spot on gill cover, scimitar-shaped pectoral fins. Body oblong; lateral line arched, with strongly keeled scales to rear; 2 pairs of small canine teeth in lower jaw. Breast naked, with only a small triangular patch of scales in front of ventrals.
Food
Fishes.
Range
Warm seas n. to Cape Cod, rarely to Maine.

AFRICAN POMPANO

Alectis crinitus **81:8**

Description
Size, to 7 in. (17.8 cm). Unique for long fin streamers. Body short, very deep, head convex.
Other names
Threadfin, Threadfish.
Range
Tropical waters, straying to Cape Cod.

LOOKDOWN

Selene vomer **81:7**

Description
Size, to 1 ft. (0.3 m). Oblique head; front rays of dorsal and anal fins lengthened. Body deep, thin; mouth low; first dorsal reduced to 7 or 8 short spines, pectorals sickle-shaped.
Habitat
Warm waters.
Range
Argentina to N.S.

ATLANTIC MOONFISH

Vomer setapinnis **81:9**

Description
Size, to 1 ft. (0.3 m). Deep, thin body; concave head; low dorsal and anal fins. Pectorals short; first dorsal reduced to 8 short spines; second dorsal and anal low and even in height.

Habitat
Warm seas.
Other names
Horsefish, Bluntnose, Shiner, Dollarfish.
Range
Uruguay to Cape Cod, straggling to N.S. The young appear off N.Y. in late summer, brought up by the Gulf Stream.

PALOMETA

Trachinotus goodei **81:11**

Description
Size, to 1 ft. (0.3 m). Body deep, thin; black crossbars on sides. First dorsal and anal prolonged past middle of tail fin.
Habitat
Clear waters with sandy bottoms; rare.
Range
Argentina to Mass.

PERMIT

Trachinotus falcatus **81:10**

Description
Size, to 18 in. (45.7 cm). Profile arched; dorsal fin spear longer than head. Lacks black crossbars of Palometa on sides, and dorsal- and anal-fin spines rarely reach to tail.
Habitat
Shallow water over sandy bottoms in bays and near islands.
Range
Brazil to Cape Cod.

FLORIDA POMPANO

Trachinotus carolinus **81:12**

Description
Size, to 18 in. (45.7 cm) and 2 lb. (0.9 kg). Profile not strongly arched; spines of dorsal and anal fins very short.
Habitat
Shallow water close to shore.
Range
Gulf and Atlantic coasts n. to Cape Cod; only young common above Va.

LEATHERJACKET

Oligoplites saurus **81:13**

Description
Size, to 12 in. (30.5 cm). Rear parts of soft dorsal and anal fins divided into 12 low brushlike finlets. Scales embedded in corrugated skin giving fish a leathery look; lateral line nearly straight. First dorsal reduced to 5 short separate spines, and preceded by 2 stout spines; caudal peduncle slightly keeled.
Habitat
Clear, swift water, often in bays or inlets.
Habits
Often leaps from the water as it pursues its prey.
Range
Tropical waters n. to N.Y., sometimes to Cape Cod.

BANDED RUDDERFISH
Seriola zonata **81:14**

Description
Size, to 3 ft. (0.9 m). Body thin, nose pointed. Anal half as long as second dorsal and, in young, preceded by 1–2 short spines. Young have sides crossbarred with 5–6 broad dark bands which fade or disappear with age.
Habits
Travels in schools following drift or large fish.
Other names
Shark Pilot, Slender Amberjack.
Range
Gulf of Mexico to N.S.

GREAT AMBERJACK
Seriola dumerili **81:6**

Description
Size, to 6 ft. (1.8 m); all-tackle record, 120 lb. (54.5 kg). Resembles smaller Banded Rudderfish but has longer second dorsal (36–38 rays instead of 30–34).
Habitat
Warm waters over offshore flats and reefs, around wrecks.
Remarks
Both fish and roe make good eating.
Range
West Indies to N.J., rarely to Cape Cod.

DOLPHINS
Family Coryphaenidae

DOLPHIN
Coryphaena hippurus **77:9**

Description
Size, to 6 ft. (1.8 m); all-tackle record, 76 lb. (34.5 kg). Distinguished by high arched head, long dorsal fin, brilliant color, swiftness, and spectacular leaps. Sides gold with blue polka dots, tail golden; changes color rapidly when dying. Body flattened sideways; head massive, blunt (forehead higher in male); small comblike teeth in jaws and roof of mouth. Ventrals below pectorals, and ½ height and ½ length of dorsal; tail deeply forked.
Habitat
Near drift in warm open seas, notably Gulf Stream.
Habits
Solitary or in schools; often feed under kelp beds.
Food
Fishes, particularly flying fishes.
Remarks
Both fish and roe make good eating.
Range
Warm seas n. to Cape Cod, occasionally to N.S.

TRIPLETAILS

Family Lobitidae

TRIPLETAIL

Lobotes surinamensis *Fig. 105*

Description
Size, to 3 ft. (0.9 m). Rear dorsal and anal fins prominently extended to the rear, suggesting a triple tail. Young light, often mottled; adults almost black. Body large, deep, compressed; scales small, rough; lower jaw projects, head concave above eye. First dorsal of stout spines joins second dorsal; tail rounded.
Habitat
Drifting debris in channels; near wrecks, pilings.
Food
Mainly fishes.
Range
Warm waters, Argentina to Cape Cod.

Fig. 105

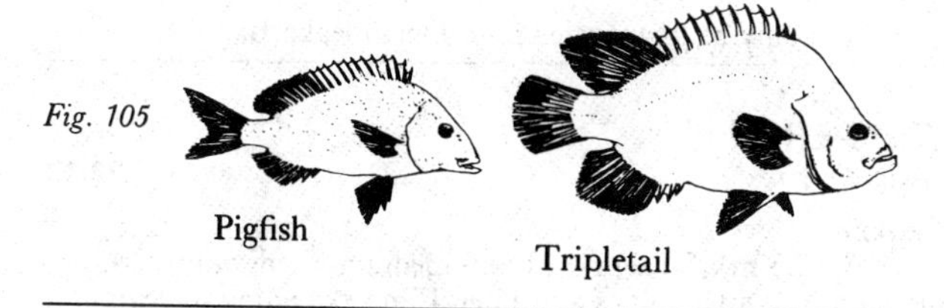

GRUNTS

Family Pomadasyidae

PIGFISH

Orthopristis chrysoptera *Fig. 105*

Description
Size, to 15 in. (38.1 cm). Noted for piglike mouth, continuous dorsals, forked tail, and pointed pectorals. Body compressed, mouth small with lengthwise groove on chin and projecting lower jaw; teeth small, pointed. Fins more or less scaly, scaly flap at base of ventrals.
Habitat
Sandy shores (most commonly); grassy flats.
Food
Annelids, crustaceans, fishes.
Range
Mass. to Fla.

CROAKERS

Family Sciaenidae

In croakers the lateral line extends to the back edge of the caudal fin and the anal fin has only one or two spines. The spiny and soft-rayed dorsals are often continuous. Most members of the family have air bladders which can vibrate, producing the sounds from which the name is derived. Many species are valued for sport and food.

BANDED DRUM

Larimus fasciatus **82:11**

Description
Size, to 10 in. (25.4 cm). Body somewhat compressed, back arched; mouth large, oblique; lower jaw projects, teeth in both jaws; tail wedge-shaped.
Habitat
Offshore.
Range
Mass. to Fla., only as a straggler n. of Chesapeake Bay.

SILVER PERCH

Bairdiella chrysura **82:13**

Description
Size, to 1 ft. (0.3 m). Straight tail and medium-sized, only slightly oblique mouth; 2 anal spines and lateral line extending to rear edge of tail.
Habitat
Sandy shores.
Range
N.Y. to Tex.; season, N.Y., June to Dec.

RED DRUM

Sciaenops ocellata **77:12**

Description
Size, to 5 ft. (1.5 m); all-tackle record, 83 lb. (37.6 kg). Black spot at base of tail; lateral line extending to the back edge of tail. Reddish stripes, upper jaw projects.
Habitat
Sandy shores.
Food
Mostly crustaceans.
Other names
Channel Bass, Redfish.
Range
Mass. to Fla.

ATLANTIC CROAKER

Micropogon undulatus **77:6**

Description
Size, to 12 in. (30.5 cm). Distinguished from its relatives by row of small barbels on each side of lower jaw.
Habitat
Sandy bottoms.
Food
Mostly worms, crustaceans, fishes taken largely from the bottom.

Reproduction
Spawn in cold half of the year.
Other name
Hardhead.
Range
Mass. to Argentina, not common n. of N.J.; great fluctuations in population size.

SPOT
Leiostomus xanthurus **76:4**

Description
Size, to 14 in. (35.6 cm). Black spot behind gill cover; lateral line extending to back edge of tail. Snout blunt, mouth small. Spiny dorsal triangular, with top rounded; second dorsal only ½ as high; tail concave.
Habitat
Inshore waters.
Range
Gulf of Maine to Tex. Quite variable in numbers; have some years of great runs.

SOUTHERN KINGFISH
Menticirrhus americanus *Fig. 106*

Description
Size, to 15 in. (38.1 cm). Closely resembles Northern Kingfish, but normally light in color, and longest spine of first dorsal does not reach front of second dorsal.
Habitat
Shallow, sandy waters; surf; inlets.
Food
Invertebrates and small fishes.
Other name
Southern King Whiting.
Range
N.Y. to Argentina.

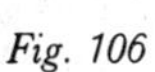
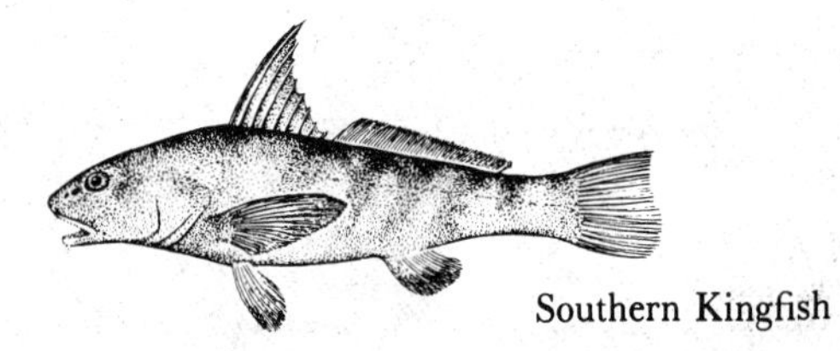
Fig. 106 Southern Kingfish

NORTHERN KINGFISH
Menticirrhus saxatilis **76:1**

Description
Size, to 17 in. (43.2 cm) and 3 lb. (1.4 kg). Has chin barbel. Third spine of the first dorsal greatly extended and, in adults, threadlike; reaches past second dorsal. Dark in color. Lower half of tail fin rounded, upper half concave.
Habitat
Sandy shores.

...ts
...tom feeder.
...ood
Crabs, squid, shrimps, young fishes.
Other names
King Whiting or Northern Whiting, Minkfish.
Range
Maine to Fla.

BLACK DRUM
Pogonias cromis **82:14**

Description
Size, to 4 ft. (1.2 m) and 146 lb. (66.2 kg); all-tackle record, 94 lb. (42.6 kg). Deep body, high arched back, many barbels. Scales large, first dorsal triangular, second dorsal oblong, pectorals long and pointed, anal fin with 2 spines, tail almost square. Young characterized by 4–6 broad black bands on the sides.
Habitat
Inlets and surf of sandy shores.
Food
Shellfish.
Range
Cape Cod to Argentina.

WEAKFISH
Cynoscion regalis **77:2**

Description
Size, to 32 in. (81.3 cm); all-tackle record, 17½ lb. (7.9 kg). Snout pointed, mouth large, lower jaw projects, 2 canine teeth in upper jaw. First dorsal triangular, second longer and rectangular; soft rays of dorsal and anal scaled; tail broad and slightly concave. Lacks the spots on the fins and tail of the spotted sea trout.
Habitat
Shallow water.
Habits
Travels in schools.
Food
Crabs, shrimps, fishes.
Reproduction
Many thousands of eggs, hatching 48 hours at 65°F (18°C), in June and July.
Age
To 12 years.
Other names
Gray Weakfish, Sea Trout.
Remarks
This is a very popular and important species both commercially and for sport. Its jaws are weakly set in its head and are sometimes pulled out by the fisherman's hook, whence the name is derived.
Range
N.S. to Fla.

SPOTTED SEA TROUT
Cynoscion nebulosus **82:12**

Description
Size, to 7 lb. (3.2 kg); all-tackle record, 15 lb. (6.8 kg). Similar to weakfish in appearance and habits, but has black spots on the second dorsal and tail, and soft rays of dorsal anal fins are scaleless.

Habitat
Coastal bays and estuaries.
Remarks
One of the important food fishes of Chesapeake Bay.
Other name
Spotted Squeteague.
Range
N.Y. to Tex., but scarce n. of Del.; season, Chesapeake Bay, Mar. to Dec.

PORGIES
Family Sparidae

SCUP
Stenotomus chrysops **76:6**

Description
Size, to 18 in. (45.7 cm) and 4 lb. (1.8 kg). Body thin, depth ½ length; scales large, thick; head short, forehead concave, eye set high. Pectorals long, dorsal long and preceded by a spine; anal and dorsal fins can be laid back in grooves; tail deeply concave.
Habitat
Sandy bottoms inshore and offshore.
Habits
Ground feeder.
Food
Crustaceans, worms, fishes, squid.
Age
To at least 5 years.
Other name
Northern Porgy.
Range
N.S. to e. Fla.

PINFISH
Lagodon rhomboides **83:8**

Description
Size, to 10½ in. (26.7 cm). Similar to Scup but with molars in each jaw, shorter pectorals, notched incisors, and a widely forked tail.
Habitat
Shallow water near pilings.
Range
Cape Cod to Cuba, but common only Del. and s.

SHEEPSHEAD
Archosargus probatocephalus **83:7**

Description
Size, to 30 in. (16.2 cm) and 20 lb. (9.1 kg). Body has 7 broad vertical bands. Lacks concave forehead of Scup.
Habitat
Near pilings and jetties close to the bottom; tidal streams and bays.
Food
Mollusks and crustaceans crushed by teeth.
Reproduction
Eggs hatch in 40 hours at 77°F (25°C).
Range
N.S. to Fla., rare n. of Va.

GOATFISHES
Family Mullidae

RED GOATFISH
Mullus auratus **82:10**

Description
Size, to 8 in. (20.3 cm). Red with a long double barbel. Dorsal fins high and well separated, ventrals beneath pectorals, anal beneath second dorsal; tail well forked.
Habitat
Sandy bottoms.
Range
N.S. to West Indies.

SEA CHUBS
Family Kyphosidae

BERMUDA CHUB
Kyphosus sectatrix **83:9**

Description
Size, to 18 in. (45.7 cm). Dorsals continuous, low, almost even height; 2 spines in front of low anal fin. Body deep, compressed, dull in color; both jaws with small incisor teeth; tail forked.
Habitat
Shoals and bars.
Food
Omnivorous, including offal.
Other name
Rudderfish.
Range
Mass. to Brazil.

WRASSES
Family Labridae

These are chiefly tropical fishes with large scales. Their mouths are small and thick-lipped, with strong canine teeth. Their dorsal fins are continuous, usually with a long spiny portion. The family goes back to the Paleocene.

CUNNER
Tautogolabrus adspersus **83:10**

Description
Size, to 15 in. (38.1 cm) and 2½ lb. (1.1 kg). Distinguished from Tautog by extended snout, saw-toothed preopercle, and scaly gill cover.
Habitat
Rocky bottoms to 35 fathoms (64 m), and wharves.
Food
Omnivorous.
Reproduction
Spawns June to early July.
Age
To 7 years.

Other name
Bergall.
Range
Newfoundland to Chesapeake Bay.

TAUTOG
Tautoga onitis **76:8**

Description
Size, to 36½ in. (92.7 cm) and 22½ lb. (10.2 kg); all-tackle record, 21 lb. (9.5 kg). Distinguished from Cunner by blunt snout, untoothed preopercle, and naked gill cover.
Habitat
Shallow water over rocky bottoms near wrecks, pilings.
Habits
When not feeding, have the odd habit of often lying crowded together on their sides in a hole amid rocks.
Food
Mollusks, crustaceans.
Age
Probably to 10 years.
Other name
Blackfish.
Range
N.B. to S.C.

BARRACUDAS
Family Sphyraenidae

These fishes have strong jaws and teeth and a small first dorsal of a few spines, set well forward of the second. The family goes back to the Eocene.

GREAT BARRACUDA
Sphyraena barracuda **83:3**

Description
Size, to 10 ft. (3 m); all-tackle record, 103 lb. (46.7 kg). Scales and teeth large, lower jaw projects and has fleshy tip. First dorsal arises behind root of ventrals.
Habitat
Bays, inlets, and surface water over reefs.
Food
Fishes.
Range
Brazil to Fla., straggling to Cape Cod.

NORTHERN SENNET
Sphyraena borealis **83:5**

Description
Size, to 12 in. (30.5 cm). Tiny version of Great Barracuda, lacking fleshy tip on jaw.
Habitat
Coastal inshore waters.
Food
Fishes.
Range
Cape Cod and s.; N.Y., June to Nov.; Woods Hole, July to Dec.

MULLETS

Family Mugilidae

These are medium-sized bay fishes with silvery, cylindrical bodies, small toothless mouths, a small first dorsal of four slender spines, and a longer, soft-rayed second dorsal. The family goes back to the Oligocene.

STRIPED MULLET

Mugil cephalus **83:4**

Description
Size, to 2½ ft. (0.8 m). Sides striped; rear dorsal and anal fins almost scaleless.
Habitat
Warm and temperate seas.
Food
Young feed on plankton, adults on detritus and associated microscopic benthic animals.
Remarks
As diet changes from plankton to detritus, intestine gets longer and more convoluted in order to cope with new digestive problems.
Range
N. to N.Y., occasionally to N.S.

WHITE MULLET

Mugil curema **83:6**

Description
Size, to 3 ft. (0.9 m). Sides plain, unstriped; rear dorsal and anal fins scaled.
Habitat
Warm and temperate seas.
Food
Same as Striped Mullet.
Range
Cape Cod to Brazil, more abundant to s.

STARGAZERS

Family Uranoscopidae

NORTHERN STARGAZER

Astroscopus guttatus *Fig. 107*

Description
Size, to 12 in. (30.5 cm). Eyes on top of head; large vertical mouth. Electric organ behind eye.
Habitat
Sand bottoms.

Fig. 107

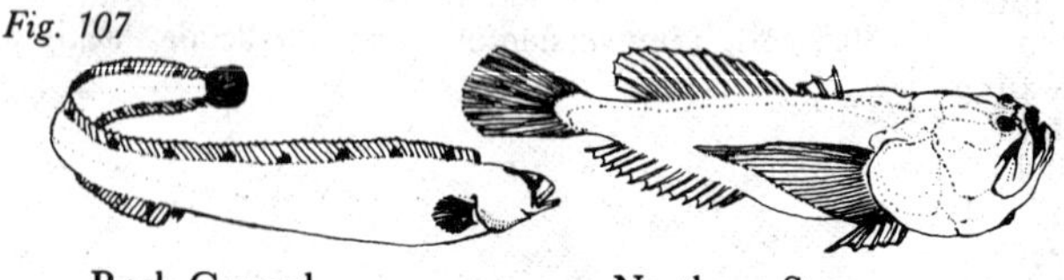

Rock Gunnel   Northern Stargazer

Habits
Remains buried for long intervals with only eyes and nostrils protruding.
Remarks
Specimens from 6 in. (15.2 cm) up of this unusual-looking species can give an electric shock either to stun their prey or for self-defense.
Range
N.Y. to Va.

WOLFFISHES

Family Anarhichadidae

ATLANTIC WOLFFISH

Anarchicas lupus **83:2**

Description
Size, to 5 ft. (1.5 m) and 40 lb. (18.1 kg). Body tapers from savage-looking head to small tail; tusks large, projecting; teeth formidable. Dorsal, of flexible spines of uniform height, extends from nape to base of tail. Pectorals large and rounded; no ventrals; anal ½ height and length of dorsal, and with angular rear corner.
Habitat
Hard bottoms and seaweed.
Habits
Solitary bottom feeder.
Food
Hard-shelled echinoderms, mollusks, and crustaceans.
Remarks
This large fish is dangerous to handle when caught and is reported to have attacked waders in rock pools.
Range
S. Labrador to Nantucket.

GUNNELS

Family Pholidae

ROCK GUNNEL

Pholis gunnellus *Fig. 107*

Description
Size, to 12 in. (30.5 cm). Unique for low, long spiny dorsal of uniform height, with similar half-length anal, both separated from small, rounded tail by only a notch. Body eellike, scales very small; head short, naked; mouth small and oblique, several rows of teeth in upper jaw, 1 in lower. Pectorals small, rounded; ventrals inconspicuous and reduced to 1 short spine and 1 ray apiece.
Habitat
Tide pools and seaweed down to 40 fathoms (73.2 m) with gravelly ground or shell beds.
Habits
Swims like an eel, curls about eggs.
Food
Worms, crustaceans, mollusks.
Other name
Rock Eel.
Range
Hudson Strait to Delaware Bay.

WRYMOUTHS
Family Cryptacanthodidae

WRYMOUTH
Cryptacanthodes maculatus **83:1**

Description
Size, to 3 ft. (0.9 m). Oblique mouth, continuous vertical fins, oval tail. Body naked, eellike, much flattened sidewise; head flat-topped, eyes high, mouth oblique; dorsal, caudal, and anal fins continuous; dorsal spiny, pectorals low; no ventrals.
Habitat
Bottom fish.
Habits
Burrows in mud.
Food
Fishes, mollusks, crustaceans.
Other name
Ghostfish.
Range
Labrador to N.J.

SAND LANCES
Family Ammodytidae

AMERICAN SAND LANCE
Ammodytes americanus **83:12**

Description
Size, to 8½ in. (21.6 cm). Distinguished from other eel-shaped fishes by forked tail. Head long, nose pointed; mouth large, toothless, with lower jaw projecting. Dorsal long, low, soft-rayed, and more than twice length of low anal; pectorals low, pointed; no ventrals.
Habitat
Sandy shores.
Habits
Occurs in dense schools; swims like an eel; burrows with great speed into sandy beach, where it remains at low tide.
Food
Small crustaceans.
Other name
Sand Eel.
Range
Labrador to Cape Hatteras.

CUTLASSFISHES
Family Trichiuridae

ATLANTIC CUTLASSFISH
Trichiurus lepturus **83:11**

Description
Size, to 5 ft. (1.5 m). Pointed whiplike tail; no caudal fin. Body naked, bandlike, tapering. Eye large; snout pointed; mouth with long barbed fangs, 4 in upper jaw, 2 in lower jaw, which projects.

Dorsal soft-rayed, extending from nape to near tip of tail; no ventrals; long anal of very low spines.

Habitat

Surface waters of warm seas.

Food

Fishes.

Other names

Hairtail, Scabbardfish, Ribbandfish.

Range

Mass. to Argentina.

MACKERELS AND TUNAS

Family Scombridae

These are swift, powerful, spindle-shaped fishes, frequently of large size. Distinctive characteristics are small finlets behind both dorsal and anal fins; and from one to three keels on the thin caudal peduncle. Mackerels have velvety skins, small scales, no bony scales on the rear lateral line, both soft and spiny dorsal fins, and deeply forked or crescent-shaped tails. The family goes back to the Eocene. All eastern mackerels are highly regarded as food for man.

SPANISH MACKEREL

Scomberomorus maculatus **82:3**

Description

Size, to 36 in. (91.4 cm) and 10 lb. (4.5 kg). Yellow spots on upper sides; blue above, silvery below. Mouth oblique and with 32 large conical jaw teeth; lateral line bends in center. Second dorsal concave, pectorals scaleless, ventrals originating behind origin of first dorsal, anal symmetrical with second dorsal; tail deeply concave.

Habitat

Warm surface waters offshore; bays, inlets.

Food

Fishes.

Range

Maine to Brazil.

CERO

Scomberomorus regalis **82:4**

Description

Size, average 5–10 lb. (2.3–4.5 kg). Gradually sloping lateral line; narrow brown stripe from behind pectoral fin to tail. Mouth oblique and with 40 large conical jaw teeth. Spots on sides in rows below lateral line. Upper half of first dorsal deep blue; pectorals scaled; ventrals below (instead of behind) origin of first dorsal, symmetrical with second dorsal.

Habitat

Warm waters offshore.

Other name

Painted Mackerel.

Range

Brazil to N.C., rarely to Cape Cod.

KING MACKEREL

Scomberomorus cavalla **82:7**

Description
Size, to 5 ft. (1.5 m) and 100 lb. (45.4 kg); all-tackle record, 77 lb. (34.9 kg). Lateral line with abrupt bend downward in the middle; mouth oblique and with 40 large conical jaw teeth; first dorsal concave and uniform in color.
Habitat
Warm waters inshore or offshore.
Remarks
This is one of the fastest and most game of all fish, a furious striker and a savage fighter. It can make long twisting underwater runs and leap as high as 10 feet (3 m) in the air. It will often follow a hooked fish to the boat and be caught in the process.
Other names
Cero, Kingfish.
Range
Brazil to N.C., rarely to Gulf of Maine.

ATLANTIC MACKEREL

Scomber scombrus **76:3**

Description
Size, to 22 in. (55.9 cm) and 4 lb. (1.8 kg). Widely separated dorsal fins; sides uniformly silver below lateral line. Scales very small; teeth small, sharp; 2 small longitudinal keels (but no median lateral keel) on the caudal peduncle. First dorsal with 11 or more spines, which can be laid down in a groove on the back; 4–6 finlets behind second dorsal, 5 behind anal.
Habitat
Open sea.
Habits
Found in vast schools of many thousands. A swift swimmer; uses so much oxygen that in warm water it must keep swimming all the time to maintain the necessary flow of water to its gills.
Food
Plankton and small fishes.
Age
To at least 8 years.
Remarks
One of the greatest food fishes.
Range
N. Atlantic from Labrador to Cape Hatteras; season, N.Y., May to July; Woods Hole, May to Dec.

CHUB MACKEREL

Scomber japonicus **82:5**

Description
Size, to 25 in. (63.5 cm). Fine black network of lines on back, lower part of sides mottled with dusky blotches, eye larger, gap between dorsals smaller, 9–10 spines in first dorsal.
Habitat
Temperate waters, inshore or offshore.
Range
N.S. to Brazil; season, N.Y.: June to Sept., but irregular. Varies greatly in numbers.

SKIPJACK TUNA

Euthynnus pelamis **82:2**

Description
Size, to 34½ in. (87.6 cm); all-tackle record, 31 lb. (14.1 kg). Has 4 long stripes on sides. Body robust; lateral line curves and is above side stripes. First dorsal high, slopes sharply down and back; second dorsal symmetrical with anal; tail slightly concave.
Habitat
High seas.
Food
Flying fishes.
Other name
Oceanic Skipjack.
Range
Tropical waters n. to Gulf of Maine.

LITTLE TUNA

Euthynnus alleteratus **82:8**

Description
Size, to 3 ft. (0.9 m) and 20 lb. (9.1 kg). Dorsals triangular and concave; anal originates behind second dorsal; tail concave. Black stripe above the gently sloping lateral line.
Habitat
Open sea.
Food
Fishes.
Other name
False Albacore.
Range
Gulf of Maine to Brazil.

BLUEFIN TUNA

Thunnus thynnus **77:13**

Description
Size, to 14 ft. (4.8 m) and 1600 lb. (725.6 kg); all-tackle record, 977 lb. (443.1 kg). Very large; plain colors; continuous dorsals, 9 or more dorsal finlets. Lower gill rakers, first arch, 24–28. Steel-blue above, silver below, no markings; body robust, scaled; jaw teeth small. First dorsal triangular; can be laid down. Second dorsal higher than long, concave and with top pointed. Short second dorsal and anal do not reach peduncle, which bears strong median longitudinal keel; anal finlets 8–9.
Habitat
Open sea.
Habits
Frequently jumps or splashes about at surface; younger ones of same size school together; very large ones often solitary.
Food
Menhaden, mackerel, other fishes and squid.
Remarks
One of the great food resources of the deep.
Other names
Horse Mackerel; small individuals are called School Tuna.
Range
N. to Newfoundland.

YELLOWFIN TUNA
Thunnus albacares **82:9**

Description
Size, to 6 ft. (1.8 m); all-tackle record, 265 lb. (120.2 kg). Lower gill rakers, first arch, 18–22. Very long second dorsal and anal fins which reach to the caudal peduncle. Pectorals long and slim; tail concave.
Habitat
Offshore.
Range
N.S. to Md. and s.

BLACKFIN TUNA
Thunnus atlanticus *Fig. 108*

Description
Size, average to 25 lb. (9.3 kg); all-tackle record, 44 lb. (20 kg). Pectoral more than 4/5 but less than 1 1/5 length of head, usually does not reach anal. Lower gill rakers, first arch, 15–18; upper, 3–7; total, 20–24. Dorsal finlets dusky brown, anal finlets dusky steel; rear edge of tail not white.
Habitat
Offshore.
Range
Mass. to Brazil.

Fig. 108 Blackfin Tuna

ALBACORE
Thunnus alalunga **82:1**

Description
Size, to 5 ft. (1.5 m), average 15–30 lb. (6.8–13.6 kg); all-tackle record, 55 lb. (24.9 kg). Lower gill rakers, first arch, 15–22. Very long pectorals reaching to the third dorsal finlet. Rear edge of tail white. Body stout and with small scales; first dorsal long and sloping; second dorsal and anal concave, pointed.
Habitat
Near surface offshore.
Habits
Migratory, erratic; schools.
Food
Small fishes, plankton, squid.
Range
N.J. to Lesser Antilles.

ATLANTIC BONITO
Sarda sarda **77:11**

Description
Size, to 3 ft. (0.9 m) and 12 lb. (5.4 kg). Faint, iridescent stripes that slant upward to the rear. Body stout, scaly; mouth and jaw

teeth large; median longitudinal keel on peduncle. Dorsals join; first dorsal triangular, long, tapering. Second dorsal, anal, and caudal concave; dorsal finlets, 7–8; anal finlets, 7.

Habitat
Open sea, notably the Gulf Stream.

Range
Abundant n. to Cape Cod, rare to N.S.

MARLINS

Family Istiophoridae

Marlins differ from the swordfishes in having a shorter, rounded (not flattened) sword; narrow, often embedded, scales; and a very long first dorsal, capable of being depressed into a groove. Their edibility varies from poor to good, depending upon locality and whether fresh or smoked. The family goes back to the Eocene.

BLUE MARLIN

Makaira nigricans **82:6**

Description
Size, to 15 ft. (4.6 m) and 1000 lb. (453.5 kg); all-tackle record, 756 lb. (342.9 kg). Scales thornlike; no lateral line; caudal peduncle with 2 small longitudinal keels. Dorsal and anal fins, 2 each; first dorsal very long, second dorsal and second anal low, short. Pectorals long, pointed; ventrals reduced to 2 long spines; tail concave, much broader than long. Differs from White Marlin in its thick body; darker coloration, blue above, silver below; and pointed dorsal fin that is brilliant cobalt.

Habitat
Gulf Stream and environs.

Habits
Generally solitary; often leaps from water.

Food
Mackerel, mullet, and other fishes.

Remarks
One of the great game fishes.

Range
N. to Mass.

WHITE MARLIN

Tetrapturus algidus **77:1**

Description
Size, to 8½ ft. (2.6 m); all-tackle record, 161 lb. (73.0 kg). Similar to larger Blue Marlin but body slender, belly white, lateral line conspicuous, and first dorsal rounded.

Habitat
1–25 miles (1.6–40.2 km) offshore.

Habits
Solitary or in schools.

Food
Other fishes.

Range
Warm Atlantic n. to N.S.

SWORDFISHES
Family Xiphiidae

SWORDFISH
Xiphias gladius **77:10**

Description
Size, to 16 ft. (4.9 m); all-tackle record, 1182 lb. (536.0 kg). Lacks scales and ventral fins. Sword flattened toward the tip; first dorsal short, high, and sharklike. Sword nearly ⅓ total length of fish; mouth large, toothless, and with pointed lower jaw; peduncle with longitudinal keel. Dorsal and anal fins, 2 each; first anal similar to first dorsal but far to rear, second dorsal and second anal very small and near caudal peduncle; tail concave.
Habitat
Open sea.
Habits
Usually solitary, sometimes suns at surface.
Food
Squid, other fishes.
Other name
Broadbill.
Range
Warm Atlantic n. to Newfoundland; season, N.Y., June to Sept.; N.S., July to Sept.

BUTTERFISHES
Family Stromateidae

BUTTERFISH
Peprilus triacanthus *Fig. 109*

Description
Size, to 12 in. (30.5 cm) and 1½ lb. (0.7 kg). No pelvics; long, low anal; a forked tail. Body deep and very thin, scales small, head short, snout blunt, mouth small, teeth weak. Dorsal and anal fins somewhat symmetrical.
Habitat
Near surface over sandy bottoms.
Food
Plankton, small fishes, squid, crustaceans.
Remarks
A good pan fish.
Range
N.S. to Cape Hatteras.

Fig. 109 Butterfish

GOBIES

Family Gobiidae

Most gobies occur abundantly on mud flats and in tidepools and some in freshwater streams. Several are important to fishermen as bait. Gobies are distinguished by their pelvic fins that are completely joined. These fishes are usually small-sized carnivorous bottom feeders.

NAKED GOBY

Gobiosoma bosci

Description
Size, to 2 in. (5.1 cm). Pelvic fins form a sucking disk. Body naked, cylindrical, and without caudal peduncle. Dorsal fins short and high, first spiny, second soft and longer than first; anal short and high; tail rounded.
Habitat
Grassy flats.
Food
Small crustaceans.
Reproduction
Eggs become elliptical in shape; often attached to oyster shell.
Range
Cape Cod and s.; particularly abundant in Chesapeake Bay.

SCORPIONFISHES

Family Scorpaenidae

REDFISH

Sebastes marinus **76:5**

Description
Size, to 2 ft. (0.6 m) and 13½ lb. (6.1 kg). All red; projecting lower jaw. Head concave; mouth large and oblique; bony knob at tip of lower jaw; gill covers large, pointed near top. Dorsal fin continuous, spiny part longer, soft part higher; pectorals large, lozenge-shaped; anal fin has 3 spines and 7 or 8 rays; tail small, concave.
Habitat
Cool water to 260 fathoms (0.5 km).
Food
Crustaceans, small fishes.
Age
To 20 years.
Remarks
Valuable commercially.
Other name
Ocean Perch.
Range
N. Atlantic, s. to Cape Cod, rarely to N.J.

SEA ROBINS

Family Triglidae

The members of this family resemble sculpins, with their broad heads, slender bodies, and large fanlike pectoral fins. They differ, however, in that their entire heads are armored with rough bony and spiny plates and each pectoral fin is preceded by three detached fingerlike rays. These rays bend downward at the tips; as the fishes rest on the bottom they drag themselves forward by hooking the rays into the sand. The family goes back to the Eocene.

NORTHERN SEA ROBIN

Prionotus carolinus **77:4**

Description
Size, to 16 in. (40.6 cm). Snout flat and depressed; 2 short spines over each eye; 1 spine on each cheek, side of neck, and shoulder. Spiny and soft dorsal fins barely separated, first higher, second longer; pectorals long, rounded, and extending past origin of second dorsal; ventrals long; anal symmetrical with second dorsal. Distinguished from Striped Sea Robin by absence of longitudinal stripe on body and by concave tail.
Habitat
Bottoms.
Food
Crustaceans, squid, small fishes.
Reproduction
Spawns in summer at N.Y.
Other name
Carolina Sea Robin.
Range
Bay of Fundy to S.C., common Cape Cod to Hatteras.

STRIPED SEA ROBIN

Prionotus evolans *Fig. 110*

Description
Size, to 18 in. (45.7 cm). Pectorals very long, overlapping most of second dorsal. Distinguished from Northern Sea Robin by a dusky stripe on each side below the lateral line, and by a straight instead of a concave tail.
Range
Mass. to Fla.

Fig. 110

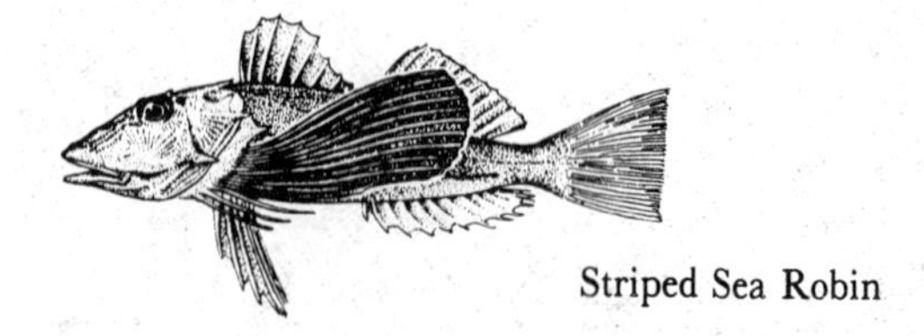

Striped Sea Robin

SCULPINS

Family Cottidae

Sculpins resemble sea robins, having broad heads, slender bodies, and large fanlike pectoral fins. They differ, however, in that the tops of their heads are naked, instead of armored, and in that they have no detached rays preceding the pectoral fins. They are chiefly Arctic fishes, and voraciously bite on any bait. The family goes back to the Oligocene. They are sometimes used as bait for lobster pots.

LONGHORN SCULPIN

Myoxocephalus octodecemspinosus **84:8**

Description
Size, to 18 in. (45.7 cm). Body elongate, slender; head blunt, flattened; 3 spines on the preopercle, upper one at least 4 times as long as the one below. First dorsal sooty with irregular markings, second dorsal with 3 or 4 dark crossbars.
Habitat
Open ocean.
Habits
Bottom feeder.
Food
Crustaceans, mollusks, squid; scavenger.
Range
Coastal waters, Newfoundland to Va.

Note: The **SHORTHORN SCULPIN,** *Myoxocephalus scorpius* **(84:7),** has a more rounded body and a short spine on the preopercle.

GRUBBY

Myoxocephalus aenaeus **84:6**

Description
Size, to 6½ in. (16.5 cm). Differs from other sculpins in the following combination of characteristics: longitudinal ridge with 2 spines along top of head over each eye; 2 spines between nostrils; 6 short spines on each side of face. Spiny dorsal fin originates slightly in front of upper corner of gill opening and is separated from soft dorsal by deep notch. Soft dorsal lower than spiny and about equal in length, ventral fins of 3 rays each.
Habitat
Open oceans.
Habits
Bottom feeder.
Range
Coastal waters, Gulf of St. Lawrence to N.J.

SEA RAVEN

Hemitripterus americanus **84:9**

Description
Size, to 25 in. (63.5 cm) and 7 lb. (3.2 kg). Fleshy tabs on head. Brown or purplish, sometimes bright yellow; skin prickly, both jaws with large teeth. Spiny dorsal highest at front, jagged in outline, higher and longer than soft dorsal; ventrals have 3 rays; anal longer than second dorsal; tail brush-shaped.
Habitat
Rock, pebble, or sand bottom, 2–50 fathoms (3.7–91.4 m).
Food
Invertebrates, small fishes.
Range
Newfoundland to Cape Cod, some to Chesapeake Bay.

FLYING GURNARDS
Family Dactylopteridae

FLYING GURNARD
Dactylopterus volitans *Fig. 111*

Description
Size, to 1 ft. (0.3 m). Can skim through the air on its pectorals or walk in the water with its ventrals as if on stilts. Eye large and set high; armor of head extends back in a spine to beyond origin of spiny dorsal. First 2 spines of dorsal separate; pectorals enormous and reach caudal peduncle when laid back; first 5–6 rays of each pectoral, with membrane, form a separate fin; tail concave.
Habitat
Warm waters.
Range
N. to N.Y., rarely to Cape Cod.

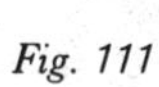
Fig. 111

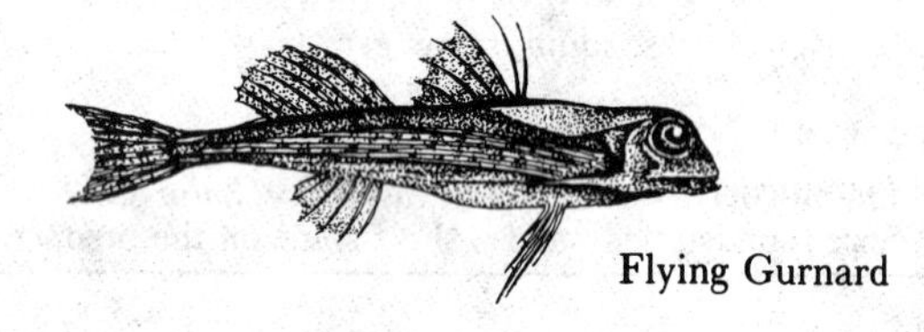
Flying Gurnard

Clingfishes
Order Gobiesociformes

CLINGFISHES
Family Gobiesocidae

SKILLETFISH
Gobiesox strumosus *Fig. 112*

Description
Size, 4 in. (10.2 cm). Dark, tadpole-shaped body; naked skin; large ventral sucking disk.
Habitat
Clings by sucking disk to submerged objects, such as piles, rocks, and shells.
Food
Small crustaceans.
Range
N.J. to Brazil.

Fig. 112

Skilletfish

Flatfishes

Order Pleuronectiformes

Members of this order are bottom fishes and have the body flattened so that the fish rests on one side on the ocean floor. The underside is white, and adults have both eyes on the upper side of the head. The young have one eye on each side and swim like other fishes; but in the course of a few weeks one eye travels over the top of the head to join its mate and the fish rests on one side. The order goes back to the Paleocene. All are edible, and several species are taken in great numbers each year by commercial fishermen.

LEFTEYE FLOUNDERS

Family Bothidae

Members of this family are sinistral, that is, they have the eyes and coloring on the left side of the body in adults. Pelvic fins are asymmetrically located on abdominal ridge.

SUMMER FLOUNDER

Paralichthys dentatus **77:15**

Description
Size, to 46 in. (1.2 m) and 26 lb. (11.8 kg); all-tackle record, 20 lb. (9.1 kg). Eyes on left; mouth large; single dorsal fin extends from above eyes to peduncle. Pectoral and ventrals small; anal extending from beneath origin of pectoral to peduncle; tail somewhat wedge-shaped. Differs from Fourspot Flounder in having many small dark spots.
Habitat
Any kind of bottom; moves offshore in winter.
Food
Crustaceans, mollusks, fishes.
Remarks
One of the tastiest flatfishes.
Other names
Fluke, Summer Fluke.
Range
Cape Cod to Fla.

FOURSPOT FLOUNDER

Paralichthys oblongus **84:14**

Description
Size, to 15 in. (38.1 cm). Similar to Summer Flounder but with 4 large, oblong pink-edged black spots.
Habitat
7–150 fathoms (12.8–274.3 m).
Food
Mollusks, small crustaceans, fishes.
Range
Cape Cod to S.C.

WINDOWPANE

Scophthalmus aquosus **84:11**

Description
Size, to 18 in. (45.7 cm) and 2 lb. (0.9 kg). First 10 or 12 rays of dorsal fins partially free and branched at the tips. Body almost

round, translucent; eyes on left. Ventral fins as wide at base as at tip; left (upper) ventral a continuation of anal; right (lower) ventral smaller and nearer throat; tail rounded.

Habitat

Sandy shores to 40 fathoms (73.2 m).

Food

Mollusks, small crustaceans, fishes.

Age

To at least 7 years.

Range

Cape Cod to S.C.

EYED FLOUNDER

Bothus ocellatus *Fig. 113*

Description

Size, to 8 in. (20.3 cm). The most rounded flatfish in the range; eyespots on back. Eyes on left, lateral line arched in front. Dorsal and anal fins of virtually uniform height throughout; tail small and rounded.

Habitat

Sandy shores.

Range

Va. to Brazil, occasionally n. to N.Y.

Fig. 113

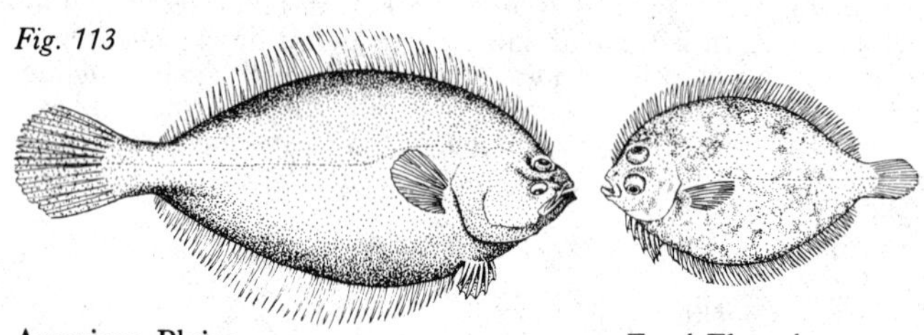

American Plaice Eyed Flounder

SMALLMOUTH FLOUNDER

Etropus microstomus

Description

Size, to 6 in. (15.2 cm). Small, elongate; small mouth; very close-set eyes. Eyes on left; lateral line without arch in front; 1 row of teeth in each jaw, tail rounded.

Habitat

Shallow coastal waters.

Remarks

Anglers often characterize small species such as this one, and the young of larger species, as "postage stamps." Large individuals and species they call "bathmats."

Range

N.Y. to Brazil.

RIGHTEYED FLOUNDERS

Family Pleuronectidae

Members of this family usually have the eyes and coloration on the right side. A number of exceptions do occur, however. Pelvic fins are symmetrically located, one on each side of the abdominal ridge.

GREENLAND HALIBUT

Reinhardtius hippoglossoides

Description
Size, to 40 in. (1 m) and 25 lb. (11.3 kg). Eyes on right; body long, lateral line almost straight; mouth large, with jaws and teeth nearly equally developed on both sides. Dorsal extends from nape to peduncle, highest at center; anal originates behind pectoral but otherwise symmetrical with dorsal; tail slightly concave.
Habitat
To 530 fathoms (1 km).
Range
Arctic and n. N. Atlantic; occurs regularly only n. of Maine.

ATLANTIC HALIBUT

Hippoglossus hippoglossus **76:15**

Description
Size, to 9 ft. (2.7 m) and 700 lb. (317.5 kg); rare over 450 lb. (167.9 kg). Largest and thickest flatfish in range. Eyes to right, well separated; body thick, long; mouth large, gaping back to eyes; teeth sharp, curved; lateral line arched above pectoral fin. Dorsal extends from above eye to caudal peduncle, rising to low angle at center; upper pectoral pointed, lower rounded; ventrals small, symmetrical, and in front of pectorals; anal originates just behind pectorals, similar to dorsal in shape; tail concave.
Habitat
Sand, clay, or gravel bottom down to 500 fathoms (0.9 km).
Habits
Occurs in schools; voracious.
Age
Probably to 50 years.
Remarks
A valuable food fish.
Range
Newfoundland to Cape Cod.

AMERICAN PLAICE

Hippoglossoides platessoides *Fig. 113*

Description
Size, to 32½ in. (82.6 cm) and 14 lb. (6.4 kg). Dorsal and anal fins smoothly rounded; tail rounded. Body long, lateral line nearly straight, scales saw-toothed on eyed (upper) side; mouth large, gaping back to eyes; teeth sharp, conical. Pectoral on eyed side longer and more rounded than that on blind side; ventrals symmetrical and close in front of anal; anal originates in front of pectorals, shaped like dorsal, and preceded by short sharp spine.
Habitat
Sandy or muddy bottoms.
Food
Invertebrates.
Reproduction
To 60,000, spawned March to June, hatch in 12 days at 39°F (3.9° C).
Age
To 30 years.
Other names
Rough Dab, Rough Sole.
Range
Labrador to Cape Cod.

YELLOWTAIL FLOUNDER

Limanda ferruginea **84:10**

Description

Size, to 22 in. (55.9 cm). Eyes on right, barely separated; body oval, top of head concave, snout pointed, mouth and teeth small; lips thick, fleshy, no mucous pits on underside. Pectoral fin on eyed side longer than that on blind side, ventrals symmetrical, anal smoothly rounded, tail slightly convex. Resembles the American Plaice, but dorsal fin comes to high point near the middle, and lateral line is arched in front.

Habitat

Rocky bottoms.

Other name

Yellowtail.

Range

Labrador to N.J.

WINTER FLOUNDER

Pseudopleuronectes americanus **84:12**

Description

Size, to 22 in. (55.9 cm). Body thick, oval; scales rough above, smooth below; eyes well separated and with rough scales between; mouth small; lips thick, fleshy; lateral line nearly straight, no mucous pits below. Dorsal fin almost uniform in height; tail broad, rounded. Resembles the American Plaice, but anal fin comes to apex just to rear of the middle.

Habitat

Sand and mud bottoms to 20 fathoms (36.6 m); ascends rivers almost to fresh water.

Food

Shrimps, mollusks.

Remarks

This is the most common and meatiest shallow-water flounder in the range.

Other names

Blackback, Sole, Lemon Sole.

Range

Labrador to Ga.

SMOOTH FLOUNDER

Liopsetta putnami

Description

Size, to 1 ft. (0.3 m) and 1½ lb. (0.7 kg). Both dorsal and anal fins come to an apex near the middle. Body thick, oval; no scales between eyes; mouth small; lips thick, fleshy; lateral line straight; no mucous pits on blind side of head.

Habitat

Soft mud bottom, 2–5 fathoms (3.7–9.1 m) deep.

Reproduction

Spawns in winter.

Range

Arctic Atlantic to R.I. The most plentiful flatfish along the Strait of Belle Isle.

WITCH FLOUNDER
Glyptocephalus cynoglossus

Description
Size, to 25 in. (63.5 cm). Right-eyed, small-mouthed; dorsal and anal fins of even height, with many rays. Above brownish-gray. Body thin, long; scales smooth; head small, with about 12 large open mucous pits on blind side. Pectorals and ventrals symmetrical, pectoral fin membrane dark on eyed side.
Habitat
Muddy sand or clay, 10–150 fathoms (18.3–274.3 m).
Food
Small invertebrates.
Other name
Gray Sole.
Range
Gulf of St. Lawrence to Cape Cod, offshore to Va.

SOLES
Family Soleidae

HOGCHOKER
Trinectes maculatus **84:13**

Description
Size, to 8 in. (20.3 cm). Unique among eastern flatfishes in having no pectorals. Eyes on right, small and set flat; body and head rounded; scales very rough on both sides, slimy with mucous; no snout or caudal peduncle; mouth small, gape shorter and more crooked on blind side than on eyed side. Dorsal and anal highest to rear and extending to base of tail; right ventral continuous with anal; tail rounded.
Habitat
Brackish bays and estuaries.
Food
Worms, small crustaceans.
Range
Mass. to Panama.

Triggerfishes, Filefishes, and Others
Order Tetrodontiformes

TRIGGERFISHES AND FILEFISHES
Family Balistidae

PLANEHEAD FILEFISH
Monocanthus hispidus

Description
Size, to 10 in. (25.4 cm). Distinguished by its 1-spined first dorsal and the long thread of the first ray of the second dorsal. Body deep, thin; scales small, eye set high, gill openings nearly vertical. Dorsal spine with rear edge barbed; pectorals short, rounded, and set below gill openings; external ventral spine conspicuous; soft dorsal, anal, and tail fins rounded.

Habitat
Warm waters, particularly inshore and about pilings.
Remarks
Filefish skin is so hard and rough that the early settlers are said to have dried it and used it for sandpaper, whence the name.
Range
N.S. to Brazil.

ORANGE FILEFISH

Aluterus schoepfi **77:5**

Description
Size, to 2 ft. (0.6 m). Distinguished by its 1-spined first dorsal and rounded second dorsal, without extended first ray. Body thin, oval; scales small, eye not set high, gill openings oblique. Pectorals short and rounded, no external ventral spine, tail narrower than in Planehead.
Habitat
Around wharves.
Habits
Often swims head down.
Food
Marine vegetation.
Range
N.S. to Brazil.

BOXFISHES

Family Ostraciidae

These fishes are encased in hard, rigid shells with openings for fins, gills, eyes, mouth, and the thin, longish caudal peduncle. There are no spiny dorsal or ventral fins, but there are spines on the ventral ridges to the rear. Young are almost spherical in cross section, adults are triangular. The family goes back to the Eocene. Its members make excellent food.

TRUNKFISH

Lactophrys trigonus *Fig. 114*

Description
Size, to 9 in. (22.9 cm). Body deep, profile concave behind eye. Anal and soft dorsal small, rounded, and pointing rearward; tail slightly rounded. Distinguished from the Scrawled Cowfish by the lack of spines over its eyes.
Habitat
Shallow marine waters.
Other name
Boxfish.
Range
New England to Gulf of Mexico.

Fig. 114

Scrawled Cowfish

Trunkfish

SCRAWLED COWFISH

Lactophrys quadricornis *Fig. 114*

Description
Size, to 1 ft. (0.3 m). Forward spine over each eye. Body deep, profile convex behind eye. Soft dorsal, anal, and caudal with nearly straight rear margins.
Habitat
Warm waters.
Other name
Horned Trunkfish.
Range
Brazil to the Carolinas, rarely in the Gulf Stream to Mass.

SWELLFISHES

Family Tetraodontidae

Like porcupine fishes, swellfishes can inflate themselves with air or water until they become almost spherical, and their teeth are fused into a strong parrotlike beak. They lack spiny dorsal or ventral fins. The family goes back to the Miocene.

SMOOTH PUFFER

Lagocephalus laevigatus *Fig. 115*

Description
Size, to 2 ft. (0.6 m). Body somewhat elongated, rounded in front, tapering to rear. Pectorals broad, dorsal and anal fins short and somewhat pointed, ventral surface with a few scattered prickles. Tail concave, with skin folds below. Distinguished from Northern Puffer by absence of dark spots along side below.
Habitat
Shallow sandy shores.
Range
N. to Cape Cod.

Fig. 115 Smooth Puffer Striped Burrfish, p. 510

NORTHERN PUFFER

Sphoeroides maculatus **76:7**

Description
Size, to 10 in. (25.4 cm). Dark spots along side below; body somewhat elongated, tapering from above eye fairly steeply to front, gradually to rear. Pectorals broad, dorsal and anal fins small and angular; no skin folds below tail; tail rounded.
Habitat
Sandy shores, tide line to a few fathoms.
Habits
Partially buries itself in sand; a group will attack, apparently in concert, and dispose of a blue crab.
Food
Crustaceans, mollusks.
Range
Fla. to N.S.

PORCUPINE FISHES

Family Diodontidae

STRIPED BURRFISH

Chilomycterus schoepfi *Fig. 115*

Description
Size, to 10 in. (25.4 cm). Inflatable, spiny body. Body oval, covered with short, stout spines; teeth fused into a strong parrotlike beak. No spiny dorsal or ventrals; pectorals broad; soft dorsal and anal broad and rounded; tail slender and rounded.
Habitat
Shallow inshore waters.
Habits
This fish, like the swellfishes, can inflate a saclike part of its gullet with air or water, swelling the abdomen to such a degree that the fish becomes almost spherical.
Other name
Spiny Boxfish.
Range
N. to Cape Cod.

OCEAN SUNFISHES

Family Molidae

OCEAN SUNFISH

Mola mola *Fig. 116*

Description
Size, to 10 ft. 11 in. (3.3 m) and 2000 lb. (907 kg). Body oblong, flattened sidewise; ½ as deep as long, ¼ as thick as deep; skin 1½ in. (3.8 cm) thick; mouth very small, teeth fused into bony beak; eye small and in line with mouth; bones soft. No spiny dorsal or ventrals; pectorals small; soft dorsal and anal fins high, pointed, symmetrical; no caudal peduncle; tail very short, broad, scalloped.
Habitat
Open sea; all temperate and tropical seas.
Food
Jellyfish, small crustaceans.
Other name
Headfish.
Range
Atlantic Coast, drifting n. to Newfoundland; Pacific.

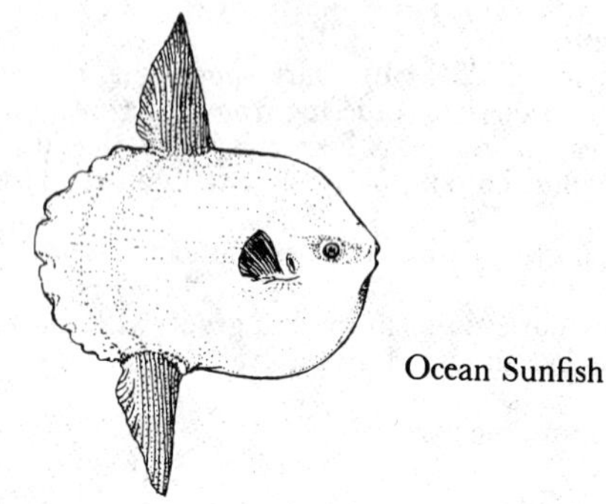

Fig. 116 Ocean Sunfish

Freshwater Fishes

LAMPREYS

Class Agnatha

Members of this class are eellike in shape and lack scales, jaws, gill covers, and paired fins. The mouth is circular and adapted for sucking. Behind the head are seven paired gill clefts. Eggs are deposited in spring in fresh water in or at the head of riffles, where they hatch into blind larvae which live buried in the bottom for four to six years. The larvae feed on microorganisms; the adults of some species are often parasitic on fish, to which they attach themselves by their suctorial mouth. The class goes back to the Silurian. There are seven freshwater lampreys in the range; five have an undivided dorsal fin; four are not parasitic.

Lampreys

Order Petromyzoniformes

LAMPREYS

Family Petromyzonidae

SILVER LAMPREY

Ichthyomyzon unicuspis *Fig. 116a*

Description
Size, to 1 ft. (0.3 m). Silvery above, lighter below; teeth usually single-pointed; parasitic.
Habitat
Large rivers and lakes.
Habits
Lives only one year as an adult, then spawns and dies.
Range
Minn. to N.Y., Miss. R. Valley s. to Mo.

Fig. 116a

SEA LAMPREY

Petromyzon marinus *Fig. 116a*

Description
Size, to 30 in. (76.2 cm). Dorsal fin divided in 2. Slate-blue to brown, speckled with blue or black; inner teeth have 2 points.
Habitat
Ocean, rivers, and large lakes. Ascends into fresh water to spawn and die; the larvae on maturing normally enter salt water.
Range
Atlantic Ocean to Great Lakes and Finger Lakes and s. to n. Fla.

BONY FISHES

Class Osteichthyes

The fishes in this class have jaws, paired fins, and a single gill slit and gill cover. The class goes back to the Devonian.

Sturgeons and Paddlefishes

Order Acipenseriformes

Members of this order have a long snout, inferior mouth, and two to four barbels, threadlike sense organs, on the undersurface of the snout. The tail lobes are unequal in size, the upper being the larger. The skeleton is largely of cartilage. The order goes back to the Cretaceous.

STURGEONS

Family Acipenseridae

The bodies of these fishes are covered with five rows of bony plates. They reach 12 ft. (3.7 m) and 300 lb. (136.1 kg). Adults have four barbels but are toothless. They are bottom dwellers. When they achieve sexual maturity at about twenty years of age, they spawn in spring or early summer, ascending streams or moving to shallow water.

Members of the genus *Acipenser* have a cone-shaped snout, a small opening, or spiracle, between the eye and the upper corner of the gill cover, and a caudal peduncle without plates.

SHOVELNOSE STURGEON

Scaphirhynchus platorhynchus **88:3**

Description
Size, to 3 ft. (0.9 m), average 4 lb. (1.8 kg); rarely to 8 lb. (3.6 kg). No spiracle; belly and slender caudal peduncle covered with small bony scalelike plates.
Habitat
Streams and lakes with clear bottoms.
Range
Miss. R. drainage.

Note: The similar **PALLID STURGEON,** *Scaphirhynchus albus* **(88:4)**, of the upper Mississippi drainage has the belly largely naked.

ATLANTIC STURGEON

Acipenser oxyrhynchus **88:1**

Description
Size, to 12 ft. (3.7 m). Sharp snout; 7–14 dorsal plates. Olive-gray or brown, paler below. Snout almost as long as head; anal fin entirely below middle of dorsal fin. Barbels short, midway between snout tip and mouth; lateral plates, 26–30.
Habitat
Mud bottoms off coast.
Food
Mud-dwelling marine invertebrates and some fishes.
Range
Coastal waters of the Atlantic; ascends rivers.

SHORTNOSE STURGEON

Acipenser brevirostris **88:2**

Description
Size, to 30 in. (76.2 cm). Blunt snout; 11 dorsal plates. Color brownish, paler below. Snout about ¼ length of head; anal fin below front of dorsal; lateral plates, 32.
Habitat
Salt water; ascends streams to spawn.
Range
Cape Cod to Fla., rarer in n.

LAKE STURGEON

Acipenser fulvescens **86:15**

Description
Size, to 8 ft. (2.4 m). Very blunt snout in adult; 13 dorsal plates. Above dark olive; sides paler or reddish, especially in young, often spotted with black. Anal fin below middle of dorsal fin; lateral plates, 29–42.
Habitat
Lakes and large rivers.
Food
Mollusks, aquatic insect larvae, crayfish, some fishes, aquatic vegetation.
Range
Great Lakes, upper Miss. R., Sask. R., and Hudson Bay drainages, s. to Ala. and Mo.

PADDLEFISHES

Family Polyodontidae

Paddlefishes, like sturgeons, are an ancient group represented by only two living species—one in North America and the other in the valley of the great Yangtze River in China. The fins, like those of sturgeons, are archaic and sharklike, and the scales have degenerated to a patch on the upper lobe of the caudal fin. The long, paddle-shaped snout, whose precise function is still unknown, sets this fish apart from all others in North America. Although of large size, the paddlefish feeds throughout life on microscopic plants and animals.

PADDLEFISH

Polyodon spathula **87:15**

Description
Size, to 6 ft. (1.8 m) and 180 lb. (87.6 kg). Large paddle-shaped snout. No scales except for small patch on tail; gill covers have long backward projecting tips. Very young lack the paddle.
Habitat
Large silty rivers; oxbow and flood-plain lakes.
Food
Feeds by straining plankton from water by its efficient gill rakers.
Range
Great Lakes and Miss. R. systems.

Bowfins
Order Amiiformes

BOWFINS
Family Amiidae

BOWFIN
Amia calva **88:9**

Description
Size, to 31 in. (78.7 cm) and 8 lb. (3.6 kg). Long dorsal fin. Cycloid (rounded) scales; bony gular plate between angle of lower jaws.
Habitat
Lakes and sluggish waters.
Habits
Sometimes gulps air at the surface.
Food
Fishes, crayfish.
Reproduction
In spring, male builds circular nest among weeds, where he guards eggs and young.
Range
Minn. to Que. and, w. of Appalachians, s. to Tex. and Fla.

Gars
Order Semionotiformes

GARS
Family Lepisosteidae

Gars are long cigar-shaped fishes, olive above, gray below, and with thick ganoid (diamond-shaped) scales. They are predators and have beaklike jaws with sharp, pointed teeth. They frequent large streams and rivers and shallow weedy lakes, where they spawn in spring. They can use atmospheric oxygen and may bask on the surface. The family goes back to the Cretaceous.

LONGNOSE GAR
Lepisosteus osseus **88:6**

Description
Size, to 5 ft. (1.5 m); rod and reel record, 50 lb., 5 oz. (22.8 kg). Thin snout twice the length of its head. In adult, black spots near tail and on vertical fins; 1 row of teeth in upper jaw.
Habitat
Large streams and rivers; shallow, weedy lakes.
Range
Mont. and Vt. and s. to Mexico and Fla.

SPOTTED GAR
Lepisosteus oculatus **88:8**

Description
Size, to 44 in. (111.8 cm). Large dark spots on top of head. Snout broad, longer than rest of head but not twice as long.

Habitat
Clear, weedy lakes and bayous.
Range
Lakes Mich. and Erie, and Miss. R. drainage to Gulf of Mexico.

SHORTNOSE GAR
Lepisosteus platostomus **88:7**

Description
Size, to 30 in. (76.2 cm). Snout short, longer than rest of head but not twice as long; large teeth in single series. Spots only on rear fins and rear part of body; the spots are distinct.
Range
Long, muddy rivers of Miss. R. drainage.

ALLIGATOR GAR
Lepisosteus spatula **88:5**

Description
Size, to 93 in. (236.2 cm); rod and reel record, 279 lb. (126.5 kg). In adult, 2 rows of large teeth in upper jaw. Snout relatively broad and short; body mottled toward tail.
Habitat
Large streams and rivers.
Range
Miss. R. drainage and Gulf of Mexico tributaries.

Mooneyes
Order Osteoglossiformes

MOONEYES
Family Hiodontidae

This is a small family of silvery, deep, thin-bodied fishes, averaging a foot in length, with a short round snout, a slanting, well-toothed mouth, and a lateral line. They lack head scales, gular plate, spines in fins, and an adipose fin. In females the lower edge of the anal fin is almost straight; in the male it is lobed.

GOLDEYE
Hiodon alosoides

Description
Size, to 20 in. (50.8 cm). Eye golden; dorsal fin rays 9–10; fleshy keel on belly extends forward of pelvic fins; anal fin originates before dorsal fin.
Habitat
Shallow, turbid lakes and rivers.
Food
Insects, mollusks, crayfish, fishes.
Reproduction
Moves upstream to quiet marshy shoals, where it spawns over gravel beds.
Range
N. from Man. and Sask., throughout the Mo. R. system, s. to Mont.

MOONEYE
Hiodon tergisus **85:12**

Description
Size, to 18 in. (45.7 cm). Eye silvery; dorsal-fin rays 11–12; belly usually not keeled; anal fin originates behind dorsal fin.
Habitat
Clearest, largest waters, preferably swift.
Food
Insects, mollusks, crayfish, small fishes, plankton.
Reproduction
In May to July ascends rivers and streams to spawn in shallow water.
Range
Lake Champlain system w. to the Great Lakes; Miss. R. drainage as far s. as Ark.

Herringlike Fishes
Order Clupeiformes

These fishes have soft fins, and pelvic fins are abdominal. The order goes back to the Triassic.

SHADS AND HERRINGS
Family Clupeidae

Fishes in this family are compressed and have a naked head, weak teeth, and a keeled belly. They have no adipose fin, gular plate, or lateral line. Many of them are marine; of these we treat here only those that go up freshwater streams and rivers to spawn. The family goes back to the Cretaceous.

Fishes of the genus *Alosa* are silvery, with jaws about equal or the lower jaw projecting. The upper jaw has a deep or shallow notch to receive the lower jaw; jaw teeth are absent or very weak. The dorsal fin is short and the last ray is never greatly lengthened. All, except the Glut Herring, have a silvery or pale peritoneum.

GIZZARD SHAD
Dorosoma cepedianum **85:9**

Description
Size, to 15 in. (38.1 cm). Last ray of dorsal fin greatly lengthened. Mouth small, toothless, upper jaw projecting, dark spot behind gill cover.
Habitat
Mud-bottomed, shallow water. Its numbers have increased in reservoirs in recent years, perhaps in part because it has a greater tolerance for turbid water than other species.
Food
Plankton.
Reproduction
Eggs laid in shallow water during spring.
Range
Canada to Fla. excluding New England.

SKIPJACK HERRING

Alosa chrysochloris **85:6**

Description
Size, to 15 in. (38.1 cm). Pale lower jaw; no spots behind gill cover. Sides of upper jaw meet in an obtuse angle and form a shallow notch to receive lower jaw; always has tongue teeth; sides plain.
Habitat
Swift, deep waters.
Habits
Large, swift schools of this species force schools of shiners, its prey fish, to jump into the air. It then lightly skips into the air in pursuit, hence its name.
Food
Plankton and small fishes.
Reproduction
Eggs laid in early spring.
Range
Miss. Valley, s. to Gulf of Tex.

HICKORY SHAD

Alosa mediocris **85:10**

Description
Size, to 15 in. (38.1 cm). Dark spots, 5–6, behind gill cover. Sides of upper jaw meet in an obtuse angle and form a shallow notch to receive lower jaw; tongue teeth sometimes present; silvery patch on cheek longer than wide; faint stripes lengthwise on sides.
Habitat
Marine; anadromous.
Range
N.S. to Fla.

BLUEBACK HERRING

Alosa aestivalis **85:1**

Description
Size, to 12 in. (30.5 cm). Dark or black peritoneum. Somewhat similar to Alewife, but darker, thinner, and with smaller eyes; tongue with single row of teeth and no black spot behind operculum.
Habitat
Anadromous.
Range
N.S. to Fla.

ALABAMA SHAD

Alosa alabamae

Description
Similar to American Shad, but much of lower jaw black; no dark spot behind gill cover; peritoneum pale.
Habitat
Freshwater rivers and streams.
Range
Ohio R. Valley and Miss. R. from Iowa to W.Va.; streams entering Gulf of Mexico from w. Fla. to Miss. R.; rare.

ALEWIFE

Alosa pseudoharengus **85:4**

Description
Size, to 15 in. (38.1 cm). Small patch of teeth on tongue, single black spot behind gill cover, and silvery patch on cheek longer than wide.
Habitat
Anadromous; also landlocked in deep water of large lakes.
Food
Plankton and small insects.
Reproduction
Eggs laid over shallow, sandy bottoms in late spring, early summer.
Range
Newfoundland to Fla.; is in the process of occupying the Great Lakes via the Welland Canal. The first Lake Erie record was 1931; the first from Lake Huron, 1934; Lake Superior, 1954.

AMERICAN SHAD

Alosa sapidissima **85:7**

Description
Size, to 3 ft. (0.9 m). Jaws about equal; sides of upper jaw join in an acute angle and form a deep notch. Lower jaw light; 1 or more dark spots behind gill cover; about 60 long, slender gill rakers on lower arm of first arch; silvery patch on cheek wider than long; peritoneum white.
Habitat
Anadromous.
Food
Plankton.
Reproduction
Eggs deposited in spring in upstream areas.
Range
Newfoundland to n. Fla.

Salmons, Trouts, and Pikes

Order Salmoniformes

SALMONS, TROUTS, AND WHITEFISHES

Family Salmonidae

These fishes have a long body with small cycloid scales and often an adipose fin. The salmon and trout have large mouths with large teeth. The lake herrings have a small mouth with weak or no teeth. These fishes feed on insects, plankton, and bottom organisms. They prefer water cooler than 70°F (21.1°C).

SALMON AND TROUT COMPARISON CHART

Species	*Body Spots*	*Tail Spots*	*Anal-Fin Rays*
Sockeye	none	none	13–17
Atlantic	red, with halos	none	9
Brown	red, with halos	none	8–12
Rainbow	dark	yes	10–13
Cutthroat	dark	yes	10–13
Brook	red or pink	none	10–13
Lake	pale	yes	10–13

SOCKEYE SALMON

Onchorhynchus nerka **89:13**

Description
Size, to 2 ft. (0.6 m) and 8 lb. (3.6 kg). Body silver or blue-black, unspotted; turns bright red with a greenish head when spawning in fall.
Habitat
Sea, rivers, and lakes; anadromous.
Reproduction
Eggs laid in streams or on shallow gravel shoals, after which adults die.
Range
Introduced from American Pacific.

ATLANTIC SALMON

Salmo salar **86:8**

Description
Size, to 5 ft. (1.5 m), rod and reel record, 79 lb. 2 oz. (35.9 kg); for landlocked variety, 28½ lb. (12.9 kg). Silvery to steel-blue; dark spots large and diffuse, hardly visible on tail; orange or reddish spots prominent and often with blue halos. Teeth on shaft of vomer weakly developed or absent, maxilla extends barely behind eye; fins pale to white; adipose fin small, olive in young; tail deeply forked.
Habitat
Anadromous.
Reproduction
Eggs laid in nest and covered during fall.
Other name
Ouananiche.
Range
Labrador to Maine; landlocked in cold lakes of e. Canada and New England; introduced elsewhere, as N.Y.

BROWN TROUT

Salmo trutta **86:6**

Description
Size, to 2 ft. (0.6 m) and over; rod and reel record, 39½ lb. (17.9 kg). Spotting similar to Atlantic Salmon, but adipose fin relatively large and orange in young, teeth on shaft of vomer well developed, and maxilla extends well behind eye.
Habitat
Clear water with cover.
Habits
Feeds chiefly at dusk or night.
Reproduction
200–5000 eggs laid in nest during fall.
Range
All states except those of the extreme se. and middle s.; widely introduced.

RAINBOW TROUT

Salmo gairdneri **86:10**

Description
Size, to 3 ft. (0.9 m); rod and reel record, 37 lb. (16.8 kg). Broad pink or red band on the side; spotted straight tail. Body has dark spots.
Habitat
Similar to Brown Trout, but prefers swifter current.
Reproduction
400–3000 eggs laid in nest on clear gravel in Apr. to May.
Other name
Steelhead (for anadromous specimens).
Range
100th meridian and e., as far s. as the mountainous areas of N.C. and Tenn.; widely introduced.

CUTTHROAT TROUT

Salmo clarki **89:10**

Description
Size, commonly to 15 in. (38.1 cm); rod and reel record, 41 lb. (18.6 kg). Red or pink streak on underside of lower jaw. Body spotted.
Habitat
Cold-water streams and lakes.
Reproduction
Eggs left in nest made on gravel in May or June.
Range
Upper Mo., Ark., Platte, Colo., and Rio Grande R. drainages.

BROOK TROUT

Salvelinus fontinalis **86:7**

Description
Size, to 32 in. (81.3 cm), rod and reel record, 14½ lb. (6.6 kg). Body dark with many blue-bordered red spots, dorsal fin with dark spots; tail not forked.
Habitat
Cold, small streams and ponds with cover.
Habits
Cautious, easily frightened away.
Reproduction
200–500 eggs deposited in nest in shallow water over a gravel bottom during late fall.
Range
Same as Brown Trout.

LAKE TROUT

Salvelinus namaycush **86:9**

Description
Size, to 5 ft. (1.5 m), rod and reel record, 63 lb. 2 oz. (28.6 kg). Sides with red or pale spots, dorsal fin with pale spots; tail deeply forked.
Habitat
Deep-water lakes; moves to shallow water in fall and winter.
Reproduction
Up to 20,000 eggs deposited over rocks at various depths in late fall.
Range
Canadian border states.

ROUND WHITEFISH

Prosopium cylindraceum *Fig. 117*

Description
Size, to 15 in. (38.1 cm). Single flap between the nostrils; 15–20 first-arch gill rakers. Body slender, head short, no jaw teeth.
Habitat
Shallow lake water.
Habits
A bottom feeder.
Food
Aquatic insects, mollusks, plankton.
Reproduction
Eggs laid in Nov. and Dec. in the mouths of streams and rivers.
Range
N. America from the Great Lakes and Maine n.

Note: The similar **MOUNTAIN WHITEFISH,** *Prosopium williamsoni,* of the upper Saskatchewan and Missouri rivers has 19 to 26 first-arch gill rakers.

Fig. 117

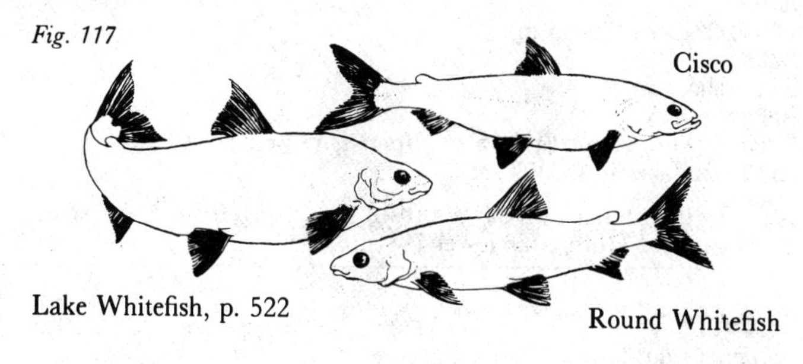

CISCO

Coregonus artedii *Fig. 117*

Description
Size, 8–12 in. (20.3–30.5 cm); rarely to 2 ft. (0.6 m). Body thin, deepest at middle; mouth at end of pointed snout; fins black-tipped; first-arch gill rakers, variable from lake to lake, 38–64.
Habitat
Moderately deep lake water.
Habits
May school.
Food
Plankton, insects, small fishes.
Reproduction
Eggs laid in late fall in shallow water over sand or gravel.
Range
Yukon to N.S. and s. to n. Ind. and Vt.

Note: The **NIPIGON CISCO,** *Coregonus nipigon,* dark green, with 54 to 64 first-arch gill rakers and living in the shallow waters of Lakes Nipigon and Winnipeg and some smaller Ontario lakes, has been combined with the Cisco as a single species.

SHORTNOSE CHUB
Coregonus reighardi

Description
Size, to 14 in. (35.6 cm). Body large, deepest at middle; head large, snout projects; mandible thick, blackish, and without knob at tip; first-arch gill rakers, 32–42; back, light green.
Habitat
13–50 fathoms (23.8–91.4 m).
Range
Lakes Ontario, Huron, Superior, Mich., and Nipigon.

SHORTJAW CISCO
Coregonus zenithicus

Description
Size, to 12 in. (30.5 cm). Body thin, green, deepest at middle; jaws equal, mandible thick, not blackish, and without knob at tip. First-arch gill rakers, 32–46.
Habitat
11–30 fathoms (20–55 m).
Habits
May school.
Range
Great Lakes (except Erie), also Nipigon, Winnipeg, Athabasca, and other lakes in N.W.T.

Note: In the similar **LONGJAW CHUB,** *Coregonus aplenae,* of Lakes Michigan and Huron, the lower jaw projects.

BLOATER
Coregonus hoyi

Description
Size, to 9 in. (22.9 cm). Body thin, deepest at middle; lower jaw projects, mandible thin and with knob at tip; fins black-tipped; first-arch gill rakers, usually 37–47.
Habitat
15–70 fathoms (27.4–128 m).
Range
Lakes Superior, Huron, Mich., Ontario, Nipigon, and others to n.

LAKE WHITEFISH
Coregonus clupeaformis *Fig. 117*

Description
Size, to 3 ft. (0.9 m), but average 15 in. (38.1 cm). Body thin, inferior mouth; snout rounded, mouth below and behind. Gill rakers 19–33, seldom less than 22.
Habitat
Shallow to moderate depths.
Habits
May school.
Food
Includes bottom organisms.
Reproduction
About 35,000 eggs laid in late fall over shallow shoals or streams.
Range
Great Lakes, Canada, and to n. New England.

DEEPWATER CISCO

Coregonus johannae

Description
Size, to 1 ft. (0.3 m). Body deepest to front of middle; first-arch gill rakers 25–36, usually fewer than 33.
Habitat
16–100 fathoms (29.3–183 m).
Range
Lakes Mich. and Huron; rare, not seen in Lake Mich. since 1951.

KIYI

Coregonus kiyi

Description
Size, to 1 ft. (0.3 m). Body small, thin, deepest to front of middle; mandible thin and with knob at tip; first-arch gill rakers, 34–47.
Habitat
30–100 fathoms (54.8–183 m).
Range
Lakes Superior, Mich., Ontario, Huron; numbers greatly reduced.

BLACKFIN CISCO

Coregonus nigripinnis

Description
Size, average 13 in. (33 cm). Body large, thick, deepest to front of middle; mandible thick and without knob at tip; fins often black; first-arch gill rakers, 41–43.
Habitat
15–100 fathoms (27.4–183 m).
Range
Lakes Superior, Huron, and others to n.

ARCTIC GRAYLING

Thymallus arcticus **89:4**

Description
Size, to 24 in. (61 cm); and 5 lb. (2.3 kg). Body long, compressed, purplish-gray with black spots; mouth small, teeth in jaw and roof of mouth; no spines in fins. Head naked; dorsal fin flaglike; adipose fin and lateral line present; tail forked. Smells like a cucumber.
Habitat
Clear cold streams with pools.
Habits
May school.
Food
Insects and their larvae.
Reproduction
About 6000 eggs laid, March to June, over sand and gravel in headwaters.
Range
Mo. R. tributaries above Great Falls, n. to Arctic coast w. of Hudson Bay, w. to Siberia; widely introduced.

SMELTS

Family Osmeridae

Members of this family are small, slender, silvery fishes inhabiting either marine, brackish or fresh water.

RAINBOW SMELT

Osmerus mordax **89:6**

Description
Size, average 7–8 in. (17.8–20.3 cm). Body slender; greenish with dark dots, silvery band on sides. Head naked, teeth in jaw; lateral line and adipose fin present, dorsal placed over pelvics. Smells like a cucumber.
Habitat
Coastal waters, ascending rivers and lakes in spring to breed; some landlocked.
Habits
Schools.
Food
Mainly plankton and insects, but some fishes.
Range
Atlantic Coast, Labrador to N.Y.; w. to Great Lakes. Introduced into Great Lakes drainage in 1912, this species is now well established in all the Great Lakes.

MUDMINNOWS

Family Umbridae

These are small red-brown fishes with an oblong body, blunt snout, scaled head, and rounded tail. They have no lateral line or adipose fin. They live in soft-bottomed sluggish or stagnant water, burrow into the mud when alarmed, and are extremely resistant to adverse conditions. They feed on insects, crustaceans, and some vegetation. Their eggs are laid in early spring.

CENTRAL MUDMINNOW

Umbra limi **89:2**

Description
Size, usually 2–4 in. (5.1–10.2 cm). Body with faint vertical bars, lower jaw light, dark bar at base of tail.
Habits
Extremely hardy and tolerant. This species, it is said, can even be frozen more or less solidly in ice, provided its own flesh is not frozen, and revived upon thawing.
Range
Man. to mouth of St. Lawrence R.; s. through upper Miss. R. Valley.

Note: The **EASTERN MUDMINNOW,** *Umbra pygmaea*, size, 2–4 in. (5.1–10.2 cm), has longitudinal streaks, a dark lower jaw, and lacks the vertical bars of the Central Mudminnow. It swims in coastal streams from New York to Florida

PIKES

Family Esocidae

These carnivorous fishes are distinguished by a duck-billed snout, pores on the head, sharp teeth, a long cylindrical body, a forked tail, and an incomplete lateral line. The dorsal and anal fins are set far back and are opposite each other; there is no adipose fin. Pikes live in vegetated waters and feed largely on fish and frogs. They spawn in the spring in shallow water and flooded marshes.

PIKE COMPARISON CHART

	Lower Half of			*Rod and Reel Record*	
Species	*Gill Cover*	*Cheek*	*Sides*	*lb.*	*oz.*
Redfin	scaled	scaled	dark bands	—	—
Chain	scaled	scaled	dark chains	9	3
Northern	unscaled	scaled	light spots	46	2
Muskellunge	unscaled	unscaled	dark markings	69	15

REDFIN PICKEREL

Esox americanus **89:7**

Description
Size, to 1 ft. (0.3 m). Gill cover and whole cheek (usually) fully scaled; dark oblique bands on the sides.
Habitat
Ponds, lakes, and streams.
Range
Esox americanus americanus, coastal plain from St. Lawrence R. s. to Fla., w. to Miss. *Esox americanus vermiculatus,* Great Lakes and Miss. R. drainages.

CHAIN PICKEREL

Esox niger **87:1**

Description
Size, usually 15–20 in. (38.1–50.8 cm); rod and reel record, 9 lb. 3 oz. (4.2 kg). Gill cover and whole cheek fully scaled; dark chainlike markings on sides.
Habitat
Ponds, lakes, and streams.
Range
St. Lawrence R. and Lake Ont. s. to Fla., e. of Appalachians; also lower Miss. Valley.

NORTHERN PIKE

Esox lucius **86:4**

Description
Size, usually 18–30 in. (45.7–76.2 cm); rod and reel record, 46 lb. 2 oz. (20.9 kg). Fully scaled cheeks, but no scales on lower half of gill cover. Pores on underside of lower jaw, 10–11. Body olive, sides with light barlike spots on a dark background, dark spots on fins.
Habitat
Ponds, lakes, and streams.
Habits
Summers in shallows, winters in deep water.
Range
N. North America, s. to Nebr. and Mo., w. of Appalachians.

MUSKELLUNGE
Esox masquinongy **86:5**

Description
Size, to 60 in. (152.4 cm); usually 28–48 in. (71.1–121.9 cm); rod and reel record, 69 lb. 15 oz. (31.7 kg). Lower half of both cheeks and gill covers scaleless. Greenish above, paler below; sides and fins with dark bars or spots on light background. Lower jaw with 6–9 pores on each side underneath.
Habitat
Ponds, lakes, and streams.
Habits
Summers in deep water, comes into shallow water in fall.
Reproduction
Eggs laid with much commotion, in flooded marshy areas.
Range
Great Lakes, St. Lawrence R., and n. Miss. R. systems.

Minnowlike Fishes
Order Cypriniformes

These fishes, in the East, have scaleless heads and abdominal pelvic fins. Their anterior vertebrae are modified so as to form a hearing aid (Weberian apparatus), connecting the air bladder to the ear. They go back to the Cretaceous.

SUCKERS
Family Catostomidae

In suckers the mouth is usually behind the point of the snout; the lips are thick and suckerlike. The teeth are located in the throat in a single comblike row. All fins lack spines. Suckers are bottom feeders and often move in large schools. They spawn in spring. The family goes back to the Eocene.

BLUE SUCKER
Cycleptus elongatus

Description
Size, to 3 ft. (0.9 m). Eyes in the rear half of head; head abruptly more slender than body. Body bluish, long and slender.
Habitat
Rivers, impoundments.
Range
Miss. R. drainage, s. from s. Minn. and Wis., into n. Mexico.

SPOTTED SUCKER
Minytrema melanops **90:3**

Description
Size, 9–15 in. (22.9–38.1 cm), rarely to 18 in. (45.7 cm), and 3 lb. (1.4 kg). Body silvery, lateral line incomplete; in adults, a black spot on each scale. Dorsal fin rays, usually 12; pelvic rays, 9.
Habitat
Clear lakes and ponds.
Food
Mollusks and insect larvae.
Range
Minn. to Pa., s. to Fla. and w. to Tex.

LAKE CHUBSUCKER

Erimyzon sucetta **90:8**

Description
Size, to 10 in. (25.4 cm) and 14 oz. (0.4 kg). Body deep, compressed, and with bands or blotches; mouth slanted; dorsal-fin rays, 11–12; lateral line missing.
Habitat
Sluggish streams, lakes and ponds.
Food
Insects and mollusks.
Reproduction
Eggs laid in spring.
Range
Minn. to Conn. and s. Apparently rare in midwest.

Note: The similar but smaller **CREEK CHUBSUCKER,** *Erimyzon oblongus,* has 9–10 dorsal-fin rays and inhabits clear streams.

NORTHERN HOG SUCKER

Hypentelium nigricans **90:12**

Description
Size, to 14 in. (35.6 cm). Head concave between eyes; lateral line complete; lower fins light orange or brown; dorsal-fin rays, 10–12; pectoral-fin rays, both sides, 34.
Habitat
Rocky riffles or clear, cool streams.
Food
Organic material, algae, insect larvae.
Reproduction
Eggs laid in small streams in April.
Range
Cen. U.S. and s. Canada, Great Lakes to Gulf of Mexico.

Note: The similar but smaller **ROANOKE HOG SUCKER,** *Hypentelium roanokense,* with 31 pectoral-fin rays (total), is confined to the upper Roanoke River system in Virginia.

WHITE SUCKER

Catostomus commersoni **87:2**

Description
Size, 12–20 in. (30.5–50.8 cm), rarely to 24 in. (61 cm), and 5 lb. (2.3 kg). Body long, slender; mouth large and behind end of snout; lateral line complete; dorsal-fin rays, 10–13.
Habitat
Bottoms of shallow lakes.
Food
Insects, mollusks, worms, plant material.
Reproduction
Some 50,000 eggs deposited on rock or gravel shoals in streams in spring.
Range
Throughout eastern range.

Note: The somewhat similar **LONGNOSE SUCKER,** *Catostomus catostomus,* has a long snout, a broad rosy band on its sides, prefers deeper and colder water, and ranges north from the Ohio River and Colorado.

MOUNTAIN SUCKER

Catostomus platyrhynchus **90:10**

Description
Size, usually 6–8 in. (15.2–20.3 cm). Head small; mouth rectangular and underslung, distinct notch at mouth corners; lateral line complete; scales smaller toward head; dorsal-fin rays, 9-11.
Habitat
Clear, cold streams.
Food
Slime, algae, some insects.
Reproduction
Eggs laid in late spring in shallow streams.
Range
Upper Mo. R. and Columbia R. drainages.

BUFFALOFISHES

Genus *Ictiobus*

Buffalofishes are deep-bodied, heavy-headed, and usually golden or reddish-brown. The subopercle, the area below the gill cover, is almost semicircular. They feed on mollusks, crustaceans, insect larvae, and plants. They are found in the Great Lakes and Mississippi River drainage. Buffalofishes are prolific and will often dominate a lake at the expense of other fishes. In Iowa, the State Conservation Commission removes almost 500,000 pounds (226,750 kg) a year to make room for more desirable species.

SMALLMOUTH BUFFALO

Ictiobus bubalus

Description
Size, to 30 in. (76.2 cm) and 15 lb. (6.8 kg). Body thin, back strongly arched; mouth small and horizontal, lips thick and streaked by parallel lines, upper lip far below eye.
Habitat
Deep, clear, swift waters of large rivers.
Reproduction
Eggs spawned at random over mud or vegetation in May.
Range
Same as for Bigmouth Buffalo.

BIGMOUTH BUFFALO

Ictiobus cyprinellus *Fig. 118*

Description
Size, to 40 in. (101.6 cm) and 50 lb. (22.7 kg). Large, slanted mouth at tip of snout; the upper jaw as long as snout. Tip of upper lip about level with lower margin of eye; lips thin, faintly streaked by parallel lines.

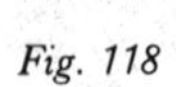

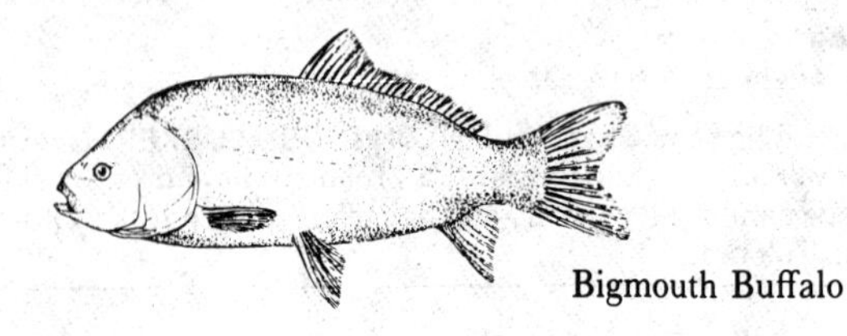

Fig. 118 Bigmouth Buffalo

Habitat
Large rivers, turbid lakes and sloughs.
Reproduction
A few hundred thousand eggs are spawned at random over muddy or vegetated bottoms in late April or early May.
Range
Miss. R. and Mo. R. drainages w. to Mont.; s. to La. and Tex.

BLACK BUFFALO
Ictiobus niger **90:9**

Description
Size, to 28 in. (71.1 cm), usually 14–20 in. (35.6 –50.8 cm), and 16 lb. (7.3 kg). Somewhat similar to Smallmouth Buffalo, but body more torpedo-shaped and back gently arched.
Habitat
Clear to moderately turbid waters, sluggish to swift.
Range
Throughout Miss. R. Valley, greater abundance in s.

CARPSUCKERS
Genus *Carpiodes*

Carpsuckers are somewhat similar to buffalofishes, but they have silvery bodies, the mouth is ventral, and the subopercle is almost triangular. They live in large rivers, feeding on bottom material. The meat of these fishes is well flavored but bony, so they are little sought by fishermen; but if these species were adequately harvested, they could provide much valuable high-protein food.

CARPSUCKER COMPARISON CHART

Species	*Long Front Dorsal Ray*	*Knob at Tip of Front Jaw*
Quillback	yes	no
River	no	yes
Highfin	yes	yes

QUILLBACK
Carpiodes cyprinus **90:6**

Description
Size, to 2 ft. (0.6 m), usually 10–15 in. (25.4–38.1 cm) and 9 lb. (4.1 kg). No knob at tip of mandible; long front dorsal-fin ray.
Habitat
Lakes, creeks, rivers, especially those with sandy bottoms.
Reproduction
Eggs spawned at random over mud in streams and bayous in late April or May.
Range
St. Lawrence R. system, across Ont. to Lake of the Woods; s. from Minn., Iowa, and Mo. to Tenn. R. Valley, e. to Chesapeake Bay drainages.

Note: The **PLAINS CARPSUCKER,** *Carpiodes forbesi,* has been combined with the Quillback.

RIVER CARPSUCKER

Carpiodes carpio *Fig. 119*

Description
Size, 12–18 in. (30.5–45.7 cm), rarely 24 in. (61 cm). Small knob at tip of mandible; lacking long front dorsal-fin ray.
Habitat
Large silty streams and rivers.
Reproduction
Eggs deposited at random, April to May.
Range
E. Mont. to w. Pa., s. to Mexico and Tenn.

Fig. 119

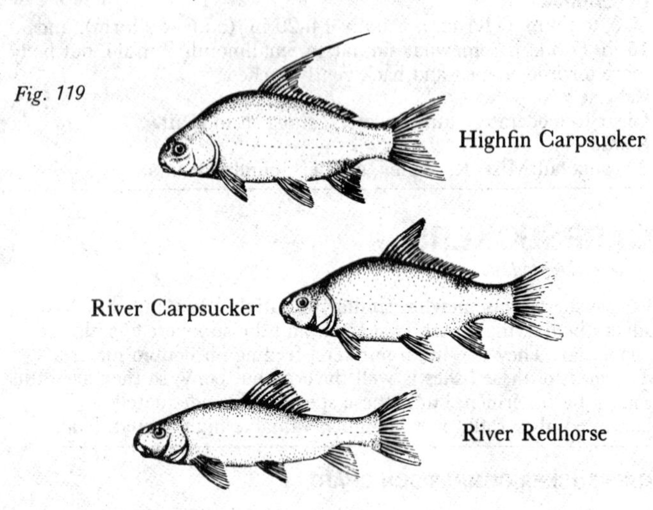

HIGHFIN CARPSUCKER

Carpiodes velifer *Fig. 119*

Description
Size, 10–12 in. (25.4–30.5 cm), rarely to 15 in. (38.1 cm), and 2 lb. (0.9 kg). Small knob at tip of mandible; long front dorsal-fin ray.
Habitat
Clear rivers.
Reproduction
Eggs laid in shallows in May.
Range
Miss. R. Valley (but not Great Lakes).

REDHORSES

Genus *Moxostoma*

This is a genus of silvery to reddish-brown fishes with a round body, a large head convex between the eyes, a short dorsal fin, and a mouth set behind the tip of the snout. These fishes feed principally on insect larvae and mollusks. Most spawn in spring in small streams. There are eighteen somewhat similar species in the East, of which we discuss four and illustrate two.

RIVER REDHORSE

Moxostoma carinatum *Fig. 119*

Description

Size, 12–22 in. (30.5–55.9 cm), rarely more than 1 lb. (0.45 kg). Large, long head; large mouth; dark spots at bases of scales. Snout blunt, tail pinkish.

Habitat

Deep waters of clear rivers.

Range

Upper St. Lawrence R., nw. Pa.; s. to Ala. and Ga., w. of Appalachians; and e., through n. Mich. and Ohio.

SILVER REDHORSE

Moxostoma anisurum

Description

Size, 10–16 in. (25.4–40.6 cm). Distinguished by convex dorsal fin with 15–16 rays and absence of dark spots at bases of scales. Body silvery, tail gray.

Habitat

Large, clear streams and lakes.

Range

Man. to St. Lawrence R. drainage; s. to n. Ala., w. of Appalachians; w. to 100th meridian.

SHORTHEAD REDHORSE

Moxostoma macrolepidotum **86:3**

Description

Size, to 18 in. (45.7 cm) and 2 lb. (0.9 kg). Short head; 8–9 pelvic-fin rays; front dorsal-fin rays do not reach end of last ray. Each upper-body scale has a black crescent at its base.

Habitat

Clear, shallow waters of lakes and rivers over clean bottoms.

Range

Upper St. Lawrence R. and s., e. of Appalachians, to S.C.; w. to 100th meridian.

Note: The **OHIO REDHORSE,** *Moxostoma breviceps,* has been combined with the Shorthead Redhorse.

TORRENT SUCKER

Moxostoma rhothoecum **90:11**

Description

Size, to 10 in. (25.4 cm). Head small with top convex, snout slightly pointed, lower lips partly covered with small fleshy projections.

Habitat

Mountain streams.

Range

James, Kanawha, and Potomac R. systems in Va. only.

Note: The similar **RUSTYSIDE SUCKER,** *Moxostoma hamiltoni,* occupies the upper Roanoke River system in Virginia.

GOLDEN REDHORSE

Moxostoma erythrurum **90:13**

Description
Size, 1–18 in. (2.5–45.7 cm) and 5 lb. (2.3 kg). Concave or straight dorsal fin with 11–14 rays; no dark spots at bases of scales. Body bronze to golden, fins red or orange, tail gray.
Habitat
Large, fairly clear streams.
Range
Drainages of Lakes Mich. and Huron, and Miss. R. system s. to Ark.

MINNOWS

Family Cyprinidae

Eastern minnows in general have thin lips, teeth in the throat, a forked tail, and a dorsal fin in the middle of the back with less than eleven rays. Only the Carp and Goldfish have any spines in the fins. Many species are hard to distinguish. The family goes back to the Paleocene.

CARP

Cyprinus carpio **87:3**

Description
Size, 1–2 ft. (30.5–61 cm), rarely to 3 ft. (0.9 m); rod and reel record, 55 lb. 5 oz. (25.1 kg). Strong saw-toothed spines at base of dorsal and anal fins; 2 barbels on each side of upper jaw. Scales large and with dark spot on base; back ruddy, belly lighter; mouth straight; lateral line present; Dorsal-fin rays, usually 18–20; anal-fin rays, 5–6.
Habitat
Warm, often very shallow waters, especially lakes.
Food
Insects, snails, crustaceans, vegetation.
Reproduction
Eggs laid June to July in shallow, weedy water.
Remarks
This is a native of Asia introduced into North America from Europe, where it is valued for food and raised in ponds for the market.
Range
S. Man. to N.S. and s.

BITTERLING

Rhodeus sericeus

Description
Size, to 2–4 in. (5.1–10.2 cm). Body with lateral band; short lateral line, first 5–6 scales; mouth small; dorsal-fin rays, 10; anal-fin rays, 8–9.
Habitat
Clear waters, usually with vegetation.
Reproduction
By means of a long ovipositor, the female Bitterling lays her eggs in the gills of a fresh-water mussel, where in time they hatch.
Range
Introduced from e. Europe and cen. Asia.

GOLDFISH

Carassius auratus **90:14**

Description
Size, typically 5–10 in. (12.7–25.4 cm). Strong saw-toothed spine at base of dorsal and anal fins; no barbels. Brown to dull gold; no spot on base of scales; mouth slanted.
Habitat
Dense aquatic vegetation.
Food
Insects, snails, crustaceans, plant material.
Reproduction
Eggs laid in spring in weedy shallow.
Range
Introduced; Mont. to Maine and s. Sporadic in Canada. Native to Asia.

GOLDEN SHINER

Notemigonus crysoleucas **91:10**

Description
Size, usually 3–5 in. (7.6–12.7 cm), up to 12 in. (30.5 cm). Body silvery or gold; scaleless belly keel; mouth very slanted; edge of anal fin concave.
Habitat
Clear, quiet water with much vegetation.
Food
Plankton, insects, plant material.
Reproduction
Eggs adhesive, scattered over submerged plants, June to August.
Range
N.S. w. to Dakotas; s. to Gulf of Mexico.

REDSIDE DACE

Clinostomus elongatus **91:9**

Description
Size, to 4 in. (10.2 cm). Sides red; body compressed, head pointed; mouth large, upper jaw separated from snout by a groove, lower lip fleshy; lateral line complete; anal-fin rays 8–9.
Habitat
Small, clear streams.
Remarks
A victim, too often, of wastes of factory, farm, and mine.
Range
E. Iowa and e. Mo. through Ill., s. Wis., Ind., Ohio, w. N.Y., and all but se. Pa.

Note: The coarser-scaled **ROSY DACE,** *Clinostomus vandoisulus,* with sides rosy, not red, ranges from New York to Georgia and west in tributaries of the Ohio and Tennessee rivers.

FALLFISH

Semotilus corporalis **90:7**

Description
Size, average 4 in. (10.2 cm), grows to over 12 in. (30.5 cm). Back bluish; scales large; head large, mouth at end of snout, upper jaw separated from snout by a groove with a small flaplike barbel. Differs from Creek Chub in having uncrowded scales and no black

spot on dorsal fin, which begins directly over pelvic fin; dorsal- and anal-fin rays, 8 each. The largest of native eastern minnows.
Habitat
Eddies of falls and rapids; also lakes.
Reproduction
Eggs laid in large pile of small stones.
Range
Drainages of James Bay, n. St. Lawrence R., e. Lake Ontario; Atlantic Coast to Va.

CREEK CHUB
Semotilus atromaculatus **90:5**

Description
Size, average 4 in. (10.2 cm) to 1 ft. (0.3 m). Somewhat resembles Fallfish, but scales crowded forward, black spot at base of dorsal fin, which begins behind forward edge of pelvics; dorsal-fin rays, 8.
Habitat
Clear, small streams with clean bottoms.
Food
Insects, crustaceans, mollusks, fishes, plant material.
Reproduction
Male builds a nest of stones in spring and later guards the eggs.
Remarks
Much used as a bait minnow for Bass, Walleye, and Channel Catfish.
Range
Mont. to Que. and s. to Gulf of Mexico.

Note: The related **PEARL DACE,** *Semotilus margarita,* of northern North America (south to Virginia) is duskier.

SOUTHERN REDBELLY DACE
Phoxinus erythrogaster **87:10**

Description
Size, to 3 in. (7.6 cm). Has 2 dark bands on each side with a red streak between. Scales tiny, lower lips fleshy, lateral line incomplete.
Habitat
Clear, cool gravelly creeks.
Habits
Easily frightened; may school.
Range
N. Ark. and n. through Iowa; e. to cen. Pa.; s. to Tenn.

RIVER CHUB
Nocomis micropogon **90:1**

Description
Size, 4–6 in. (10.2–15.2 cm), rarely to 10 in. (25.4 cm). Slender body; upper jaw completely separated from snout by a groove; slender round barbel at rear of maxilla. Color brownish.
Habitat
Clear streams.
Range
Tenn. and Va., n. to s. Mich. and Lake Ont. drainage.

SILVER CHUB

Hybopsis storeriana **90:4**

Description
Size, to 9 in. (22.9 cm). Resembles River Chub, but silvery; lower 3–4 rays of caudal fin unpigmented and immaculate white.
Habitat
Silty streams.
Range
Wyo., N.Dak., and from Lake Ont. s. to Red R. and Ala.

BLACKNOSE DACE

Rhinichthys atratulus **91:14**

Description
Size, to 4 in. (10.2 cm). Mouth at tip of snout; upper jaw not separated from snout by a groove; barbel at rear of maxilla. Dorsal fin begins behind origin of pelvics; dorsal-fin rays, 8; anal-fin rays, 7.
Habitat
Clear brooks with clean bottoms.
Range
N.S. and w. to s. Ont. and N.Dak.; s. to the lower Miss. R. drainages.

LONGNOSE DACE

Rhinichthys cataractae **91:15**

Description
Size, to 5 in. (12.7 cm). Somewhat similar to Blacknose Dace, but snout long and overhangs mouth.
Habitat
Fast water.
Range
Coast to coast; s. to Mexico and N.C.

CUTLIPS MINNOW

Exoglossum maxillingua *Fig. 120*

Description
Size, 3½–4½ in. (8.9–11.4 cm). Snout overhangs mouth; mandible 3-lobed, no groove between snout and mouth; dorsal fin origin over pelvic fin origin; lateral line complete; anal-fin rays, 7.
Habitat
Clear, gravelly creeks.
Range
Watersheds of St. Lawrence R. and e. Lake Ont., s. to Va., e. of Appalachians.

Note: The somewhat similar **TONGUETIED CHUB,** *Exoglossum laurae,* has a maxillary barbel. It occurs in isolated areas in the western parts of New York, Virginia, and West Virginia.

Fig. 120

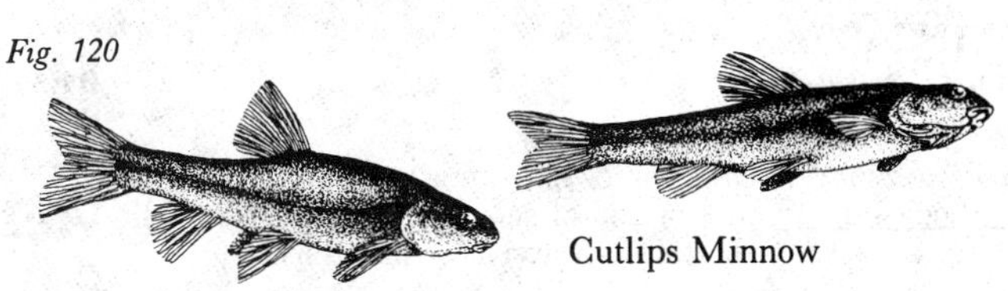

Cutlips Minnow

SUCKERMOUTH MINNOW

Phenacobius mirabilis **90:2**

Description
Size, to 4 in. (10.2 cm). Body slender, silvery; snout overhangs mouth; no maxillary barbels; lateral band black; dorsal fin starts in front of pelvics; peritoneum silvery.
Habitat
Riffles of large, turbid streams and rivers.
Range
S. Minn. and Wis., w. to include Okla., Kans., and Nebr.; e. to Ohio, w. Ky., and w. Tenn.; s. to the lower Miss. drainages.Originally a plains species, this fish has now become established quite far eastward with the spread of agriculture and muddy water. There are three other similar members of this genus in the range, in the Kanawha, Tennessee, and Alabama river systems.

SILVERJAW MINNOW

Ericymba buccata **91:1**

Description
Size, to 3 in. (7.6 cm). Olive back, silver sides; upper jaw separated from snout by a groove; no maxillary barbels, lower lip fleshy, underside of head with readily visible cavernous spaces; first ray of dorsal fin thin, small, and attached to first full ray; anal- and dorsal-fin rays, 8 each.
Habitat
Sandy brooks and small streams.
Range
Se. Mich. to w. Pa., s. to La. and w. Fla.; also Potomac R. system above Great Falls.

SILVERY MINNOW

Hybognathus nuchalis **91:2**

Description
Size, to 6 in. (15.2 cm). Dark back, silver sides; upper jaw separated from snout by a groove; no maxillary barbels, lower lip fleshy, underside of head without cavernous spaces; first ray of dorsal fin thin, small, and attached to first full ray; dorsal fin begins forward of pelvic. Peritoneum black, intestine long and coiled.
Habitat
Large, quiet waters.
Reproduction
Eggs laid in water with much vegetation.
Range
Sw. Que. and se. Ont., s. to Ga., La., Miss.; n. and w. in Miss. R. and Mo. R. systems, excluding the Great Lakes except e. Lake Ont., to Mont. and s. Ala.

FATHEAD MINNOW

Pimephales promelas **91:11**

Description
Size, to 3 in. (7.6 cm). Somewhat similar to Bluntnose Minnow, but body stout, mouth at end of snout and slanted, lateral line short and incomplete; tail spot absent or indistinct.

Habitat
Common; muddy brooks, small creeks, ponds, and small lakes.
Remarks
Highly adaptable; widely eaten by larger fish, and used by man for bait and for fish propagation.
Range
Appalachians and w. to Rocky Mts.

BLUNTNOSE MINNOW

Pimephales notatus **91:3**

Description
Size, to 3 in. (7.6 cm). Back dusky green; body slender; upper jaw separated from snout by a groove; no maxillary barbels, mouth below end of snout and almost straight, fleshy lower lip. First small ray of dorsal fin stout, blunt, and separated from first full ray; lateral line complete; dark tail spot conspicuous.
Habitat
Streams and lakes rich in plankton.
Reproduction
Eggs laid Apr. to Sept. on the underside of stones or boards; male guards and cleans eggs.
Range
S. Que. to Man. and Minn., s. to Iowa and e. of Miss. R. as far s. as Tenn., Ky., Va.

STONEROLLER

Campostoma anomalum **91:13**

Description
Size, to 7 in. (17.8 cm). Horny ridge extending to the edge of the lower lip, used to scrape food from the bottom. Upper jaw separated from snout by a groove, mouth below end of snout, no maxillary barbels; dorsal-fin rays, 8; anal-fin rays, 7; peritoneum black. In spring males become reddish or orange and grow large nuptial tubercles on the head and the body.
Habitat
Clear creeks to small rivers.
Range
Wyo. to w. N.Y., s. to Mexico and w. Fla.

SHINERS

Genus *Notropis*

This is the largest genus of American minnows. Its members in the East have the upper jaw separated from the snout by a groove. They have fleshy lower lips but no maxillary barbels. The first dorsal-fin half-ray is thin and tightly attached to the first of the eight full rays. Shiners feed mainly on plankton and have relatively short intestines. Members of this genus furnish bait for the angler and natural food for many of the larger game species, such as pike, bass, perch, and crappie. Of 101 species, 5 are illustrated, as follows, on Plate 91:

EMERALD SHINER, *Notropis atherinoides* **(91:12),** 2–3 in. (5.1–7.6 cm). Back emerald, sides silvery. Occurs in large clear open lakes and rivers, southern Canada and Great Lakes to Gulf Coast.

COMMON SHINER, *Notropis cornutus* **(91:6),** 2½–4 in. (6.4–10.2 cm). Silvery, dark blotches; no lateral band. Frequents clear streams, southern Alberta to Nova Scotia and south to Colorado and Virginia.

SATINFIN SHINER, *Notropis analostanus* **(91:4),** to 4 in. (10.2 cm). Found in coastal lowlands, St. Lawrence R. to North Carolina.

SPOTFIN SHINER, *Notropis spilopterus* **(91:8),** to 3 in. (7.6 cm). Spot on last rays of dorsal fin. Range extends from North Dakota to Montreal and south to northeast Oklahoma, Alabama, and Delaware.

SPOTTAIL SHINER, *Notropis hudsonius* **(91:5),** to 5 in. (12.7 cm). Silvery; lateral band; black spot on tail. Occurs in Canada and the Great Lakes, south to Iowa and Missouri (not in Kansas); also Georgia.

Catfishes

Order Siluriformes

FRESHWATER CATFISHES

Family Ictaluridae

Catfishes have scaleless skins, a lateral line, a single strong spine in both dorsal and pectoral fins, an adipose fin, and eight barbels—two on the snout, two on the jaw, and four on the chin. They have bristlelike teeth in bands in the upper jaw. Catfishes are virtually omnivorous and are principally active after dark. Caution should be exercised in handling them, as a poison gland at the base of the pectoral spines in some species can cause a painful wound. The poison, however, is no more dangerous to man than a wasp sting, and it does not affect the edibility of the fish. Spawning takes place in the spring and summer. The family goes back to the Miocene.

CATFISH COMPARISON CHART

Species	*Tail*	*Tail Lobes*	*Anal-Fin Rays*
Channel Catfish	forked	pointed	24–29
White Catfish	slightly forked	rounded	18–22
Blue Catfish	forked	pointed	30–36
Yellow Bullhead	rounded		25–26
Black Bullhead	straight*		17–24
Brown Bullhead	straight*		17–24
Flathead Catfish	long, straight		under 16
Stonecat	long, straight		16

* or slightly notched

CHANNEL CATFISH

Ictalurus punctatus **87:14**

Description

Size, usually 14–21 in. (35.6–53.3 cm), 2–4 lb. (0.9–1.8 kg); rod and reel record, 55 lb. (24.9 kg). Deeply forked tail with pointed lobes; 24–29 anal-fin rays. Body bluish or silvery, often with black spots; eyes large; anal fin slightly convex.

Habitat

Mainly large waters.

Reproduction

In summer, up to 20,000 eggs are deposited in a gelatinous mass in a nest which the male guards; the young travel in schools for several weeks.

Remarks

During the early 1950s, commercial fishermen took nearly 270,000 lb. (122,445 kg) annually from the Mississippi River.

Range

Widely distributed throughout the range.

WHITE CATFISH

Ictalurus catus **92:10**

Description

Size, to 2 ft. (0.6 m). Deeply forked tail with rounded lobes; 22–23 anal-fin rays. Bluish above, silvery below, unspotted (sometimes mottled). Lower jaw shorter than upper; bony ridge between head and dorsal fin incomplete; tips of anal-fin rays form an arc.

Habitat

Fresh to brackish streams, ponds, and bayous.

Remarks

Owners of private lakes often introduce this species. It grows to a good size and bites well.

Range

Coastal plains, s. N.Y. to Fla.; widely introduced elsewhere.

BLUE CATFISH

Ictalurus furcatus **92:12**

Description

Size, usually to 30 in. (76.2 cm), rarely to 40 in. (101.6 cm); rod and reel record, 94½ lb. (42.9 kg). Deeply forked tail with pointed tips; 30–36 anal-fin rays. Milky white to bluish, unspotted. Lower jaw shorter than upper, eyes small; back highest at front of dorsal fin; base of anal very long, tips of anal-fin rays almost in straight line.

Range

Large rivers of Miss. R. Valley and impoundments.

YELLOW BULLHEAD

Ictalurus natalis **92:13**

Description

Size, usually 8–12 in. (20.3–30.5 cm). Body brown, chunky; belly yellow; jaws about equal, chin barbels white. Rear edges of pectoral spines sharply barbed, anal fin usually with dark band parallel to base; anal-fin rays, 22–26.

Habitat

Shallow waters of large ponds, lakes, and streams.

Reproduction

In May and June, up to 3000 eggs laid in nests in shallow water; parents guard fry.

Range

Se. N.Dak. to Vt., s. to Mexico and Fla.

BLACK BULLHEAD

Ictalurus melas **87:13**

Description
Size, 5–10 in. (12.7–25.4 cm); rod and reel record, 8 lb. (3.6 kg). Slightly notched tail; unmottled body. Back dark, belly yellow to milky white, chin barbels black. Rear edges of pectoral spines smooth; bar at base of tail white; edge of anal fin dark; anal-fin rays, 15–19.
Habitat
Mud-bottomed lakes, ponds, and oxbows; also large rivers.
Reproduction
Nest saucer-shaped, in sand in shallow water; about 6000 eggs laid in May or June; the parents guard the eggs and fry.
Range
Mont. to w. N.Y., s. to Mexico and Fla.

BROWN BULLHEAD

Ictalurus nebulosus **92:14**

Description
Size, to 8–12 in. (20.3–30.5 cm). Square to slightly notched tail; mottled body. Jaws about equal; chin barbels black. Rear edges of pectoral spines sharply barbed; anal-fin rays, 17–24; no white bar at base of tail.
Habitat
Quiet, weedy, mud-bottomed lakes and ponds; also large rivers.
Reproduction
Nest saucerlike, protrudes out of mud in spring; about 10,000 eggs; the parents guard the eggs and young.
Range
Se. Man. to N.B., s. to Ark. and Fla.

FLATHEAD CATFISH

Pylodictis olivaris **92:15**

Description
Size, 15–35 in. (38.1–88.9 cm), rarely to 48 in. (121.9 cm); weight to 25 lb. (11.3 kg), uncommon to 50 lb. (22.7 kg). Long, greatly flattened head. Body brown; lower jaw longer than upper, mouth large and at end of snout; eyes very small. Dorsal fin begins forward of origin of pelvic fin; adipose fin as large as head and free at rear; tail rounded.
Habitat
Deep pools of large, slow-moving streams.
Reproduction
Eggs laid in June or July in a secluded nest; parents guard eggs and young.
Range
S.Dak. to w. Pa., s. in Miss. Valley to Gulf coastal plain.

MADTOMS

Genus *Noturus*

These small catfishes have a short lower jaw, horizontal mouth, poison glands at the base of their pectoral spines, a complete lateral line, and an adipose fin attached to the back and separated from the tail only by a notch. There are twelve to twenty-three anal rays.

STONECAT

Noturus flavus **92:7**

Description
Size, 6–8 in. (15.2–20.3 cm). Body yellow-olive, fins yellow-edged. Lower jaw shorter than upper; dorsal spine short, back edges separated from tail only by a notch; tail rectangular and with light border.
Habitat
Fast-water streams and riffles; also weedy lake-shore waters.
Reproduction
Eggs laid in June under stones; the parents guard the nest.
Range
S. Que. to n. New England, w. to Mont., s. to Miss. R. and Ohio R. valleys.

TADPOLE MADTOM

Noturus gyrinus **92:11**

Description
Size, to 3½ in. (8.9 cm). Body brown to yellowish, usually mottled; sides with a thin black streak. Jaws equal; eyes small; back humped at front of dorsal fin; tail rounded.
Habitat
Slow, clear streams, oxbows, and ponds with vegetation.
Range
S. Que. to n. New England, w. to Mont., s. to Fla. and Tex.

MARGINED MADTOM

Noturus insignis **92:8**

Description
Size, to 4 in. (10.2 cm). Body dark, sides with no black streak; lower jaw shorter than upper; fins usually black-edged; tail rounded.
Habitat
Under stones in riffles of creeks and rivers.
Range
Atlantic Coast from s. tributaries of Lake Ont. and N.H. to n. Ga.

Percopsiform Fishes

Order Percopsiformes

These fishes have ctenoid scales, spines in some of their fins, pectorals usually below the mid-body line, and a small mouth not separated from the snout by a groove. They are exclusively freshwater fishes and are found only in North America. The order goes back to the Eocene.

CAVEFISHES

Family Amblyopsidae

These unusual fishes are live-bearing and, in the adults, the anus is located under the throat. Many, such as the Southern Cavefish and Northern Cavefish, are blind; the power of sight is not needed in the darkness of an underground stream. Scales are small and embedded, the head is naked; no groove separates the snout from the mouth. The dorsal fin is positioned above the anal.

SWAMPFISH

Chologaster cornuta **93:6**

Description
Size, to 2 in. (5.1 cm). Brown body; no pelvic fins. Body with 3 dark longitudinal stripes and sprinkled with black specks. Eyes present but covered by transparent skin. Dorsal fin white, often black-edged; tail with black and white blotches; rays in tail branched, 9–11.
Habitat
Swamps and weedy streams.
Range
Atlantic coastal plain from s. Va. to n. Fla.

Note: The similar **SPRINGFISH,** *Chologaster agassizi,* of western Kentucky and southern Illinois, lacks the side stripes.

SOUTHERN CAVEFISH

Typhlichthys subterraneus **93:1**

Description
Size, to 2 in. (5.1 cm). No pelvic fins; body usually unpigmented. Blind.
Habitat
Streams of cave regions.
Range
Ky., Tenn., n. Ala., and the Ozarks.

NORTHERN CAVEFISH

Amblyopsis spelaea **93:2**

Description
Size, to 5 in. (12.7 cm). Small pelvic fins; body unpigmented. Blind.
Habitat
Subterranean cave streams.
Range
Mammoth Cave region of Ky. and s. Ind.

Note: The similar **ROSA'S BLINDFISH,** *Amblyopsis rosae,* without pelvic fins, is confined to caves in southwestern Missouri.

TROUTPERCHES

Family Percopsidae

TROUTPERCH

Percopsis omiscomaycus **89:3**

Description
Size, average 5 in. (12.7 cm). Noted for the peculiar translucence of its body, the adipose fin, and the fine saw-toothed edges to its scales. Body pale olive, sides with a silvery stripe. Head naked, mouth small and straight, teeth in jaw; 2 weak dorsal spines; adipose fin; pelvic fins and anus in rear half of body; 1 frail anal spine.
Habitat
Large lakes at medium depth.
Habits
Nocturnal.

Food
Insects and plankton.
Reproduction
Eggs laid in spring in tributary streams and lake shallows.
Range
Yukon to Labrador, s. to Kan. and the Potomac R.

PIRATE PERCHES
Family Aphredoderidae

PIRATE PERCH
Aphredoderus sayanus **89:5**

Description
Size, to 4 in. (10.2 cm). Dark vertical tail bar; no adipose fin. Body dark to pinkish; sides of head scaly, mouth large and slanted; pelvic fins in front half of body; anus (in adult) under throat. Spines short and weak; dorsal, 3; anal 2.
Habitat
Vegetated creeks with thick mud bottoms; rarely in large rivers and lakes.
Food
Small fishes, insects, and their larvae.
Reproduction
Nest built in spring; parents guard eggs and young.
Remarks
The odd position of the anus in the throat region distinguishes this species and the cavefishes from almost all other species.
Range
Se. Minn. to w. Ohio, s. to e. Tex. and the Gulf states; also Atlantic Coast from N.Y. to Fla.

Codlike Fishes
Order Gadiformes

CODFISHES
Family Gadidae

BURBOT
Lota lota **89:1**

Description
Size seldom exceeds 30 in. (76.2 cm) and 10 lb. (4.5 kg). Long second dorsal and anal; 1 chin barbel. Body olive or gray, marbled with black. Scales tiny, embedded; no spines in fins; pelvics under pectorals.
Habitat
Deep, cool lake water.
Habits
Retires to deeper water in summer.
Food
Other fishes, insects.
Reproduction
Eggs laid January to March in rivers or rocky lake bottoms.
Range
N. North America, s. to Kans., Mo., New England, and N.Y.

Cyprinodont and Atherine Fishes

Order Atheriniformes

Members of this order lack spines in their fins. The head as well as the body is scaled. The pelvic fins, when present, are small and positioned near the middle of the belly. There is no lateral line and the caudal fin is either straight or rounded (never forked).

KILLIFISHES

Family Cyprinodontidae

The fishes of this shallow-water family are frequently barred or striped and have large scales, extending onto their heads. The mouth is small, the snout often flattened and with a groove. Many are surface feeders. Of the thirteen species in the range, we treat four.

KILLIFISHES

Genus *Fundulus*

This is a genus of slender fishes with pelvic fins, three to six pores on the lower jaw, and no lateral line.

BANDED KILLIFISH

Fundulus diaphanus **93:5**

Description
Size, to 4 in. (10.2 cm). Dorsal fin starts in front of anal fin; dorsal-fin rays, 13–15; anal-fin rays, 10–12.
Habitat
Quiet, weedy waters of lakes, rivers, and estuaries.
Habits
Surface feeder; often travels in schools at surface.
Food
Insects, tiny crustaceans, vegetation.
Reproduction
Eggs laid in summer in weedy shallows.
Range
E. U.S. and Canada, w. to the Plains.

NORTHERN STUDFISH

Fundulus catenatus

Description
Size, to 7 in. (17.8 cm). Bluish body; orange spots. Dorsal fin starts above anal; dorsal-fin rays, 13–16; anal-fin rays, 15–18; female with short brown bar.
Habitat
Clear streams.
Range
Kans. and Ark., e. to Va. and Ala.

STARHEAD TOPMINNOW

Fundulus notti **93:3**

Description
Size, to 3 in. (7.6 cm). Dorsal fin starts well behind origin of anal; dorsal-fin rays, 6–8; anal-fin rays, 8–10.
Habitat
Clear, weedy backwaters.
Habits
Surface feeder.
Range
Lake Mich. drainage and Miss. R. Valley from Iowa to w. Tenn. and ne. Ark.

BLACKSTRIPE TOPMINNOW

Fundulus notatus **93:4**

Description
Size, to 3 in. (7.6 cm). Big black stripe on side. Dorsal fin starts well behind start of anal; dorsal-fin rays, 9; anal-fin rays, 11.
Habitat
Clear ponds, lakes, and streams; canals with some vegetation.
Range
E. Iowa to s. Mich. and Ohio, s. to e. Tex., Miss., and Tenn.

LIVEBEARERS

Family Poeciliidae

MOSQUITOFISH

Gambusia affinis **93:7**

Description
Size, to 2 in. (5.1 cm). Back dusky, sides silvery, body and fins with small black spots. Scales large; mouth separated from snout by a groove; dorsal fin starts behind origin of anal fin; dorsal-fin rays, 6–8.
Habitat
Clear vegetated water in ponds, pools, ditches, and marshes.
Food
Insects and crustaceans.
Reproduction
Bears live young. The modified anal fin of the male serves as an intromittent organ to introduce sperm into the body of the female, where the eggs are fertilized.
Range
E. from the Rio Grande to Atlantic; s. from Del. to Fla.

SILVERSIDES

Family Atherinidae

BROOK SILVERSIDE

Labidesthes sicculus **89:12**

Description
Size, to 4 in. (10.2 cm). Silver sides; 2 dorsal fins, concave anal fin. Scales cycloid and extending onto head; mouth forms a short beak; 2 dorsal fins separated; anal fin with 1 weak spine.

Habitat
Clear lakes and quiet parts of streams.
Habits
Surface swimmer, often seen in large schools; jumps and skips in air.
Food
Insects and plankton.
Reproduction
Eggs adhesive, laid in spring. Spawns after its first winter; few live through a second.
Range
Minn. to Appalachians, s. to e. Tex. and Fla.

Sticklebacks

Order Gasterosteiformes

STICKLEBACKS

Family Gasterosteidae

Sticklebacks are unique in the East for their several separate dorsal spines. The males build elaborate nests of plants and sticks and guard the eggs and young. Of the five eastern species, we discuss and illustrate one. The others, quite widely distributed in the range, have self-identifying names: Fourspine Stickleback, *Apeltes quadracus;* Ninespine Stickleback, *Pungitius pungitius;* Threespine Stickleback, *Gasterosteus aculeatus;* Blackspotted Stickleback, *Gasterosteus wheatlandi.*

BROOK STICKLEBACK

Eucalia inconstans **91:7**

Description
Size, to 2 in. (5.1 cm). Has 5 separate dorsal spines. Back dark, sides mottled, belly speckled.
Habitat
Clear, cold vegetated springs and brooks with mud bottoms.
Food
Small insects and crustaceans.
Range
Most of eastern range; s. Canada and s. to Iowa and N.Y.

Perchlike Fishes

Order Perciformes

Members of this large order have the head scaled, the pelvic fins forward, and spines in both dorsal and anal fins. The order goes back to the Cretaceous.

TEMPERATE BASSES

Family Percichthyidae

Eastern members of this family have pointed gill covers, two dorsal fins, three anal spines, and a lateral line which does not extend onto the slightly forked tail. Each pectoral has one spine and five rays.

WHITE BASS

Morone chrysops **86:13**

Description
Size, to 1 ft. (0.3 m) and 3 lb. (1.4 kg). Separate dorsal fins; lower jaw longer than upper. Silvery with 5–7 bold stripes on each side; dorsal spines graduated; anal rays, 11–12.
Habitat
Lakes and large rivers.
Habits
Usually travels in schools.
Food
Small fishes, insects, plankton, crayfish.
Reproduction
In spring, up to a million eggs spawned in gravelly shoals.
Remarks
The populations of this species and of the Yellow Bass fluctuate much more widely in numbers than do those of most other fishes.
Range
S. Minn. to w. N.Y., s. to Mexico and Gulf of Mexico; widely introduced.

WHITE PERCH

Morone americana **89:9**

Description
Size, usually 5–7 in. (12.7–17.8 cm); rod and reel record 4¾ lb. (2.2 kg). Dorsal fins joined; jaws almost equal; faint streaks on sides. Back olive, sides silvery and unmarked; mouth small, forehead long and sloping; anal spines not graduated; anal-fin rays, usually 8–10.
Habitat
Lakes and brackish water.
Food
Crustaceans, fishes, insects.
Reproduction
Eggs adhesive, laid in spring in freshwater shallows.
Remarks
A good pan fish.
Range
Atlantic Coast, N.S. to Ga.; widely introduced. May have started an invasion of the Great Lakes above Niagara; first Lake Erie specimen caught in 1953.

YELLOW BASS

Morone mississippiensis **87:8**

Description
Size, to 1 ft. (0.3 m). Similar to White Perch, but sides with 6–7 bold, discontinuous stripes.
Habitat
Lakes, oxbows, large rivers.
Food
Crustaceans, small fishes, insects, insect larvae.
Reproduction
In May, 1–2 million tiny eggs deposited over gravel or rocky reefs.
Range
Minn. to Ind., s. to e. Tex. and Ala.; widely introduced.

SUNFISHES

Family Centrarchidae

Sunfishes are carnivorous and have thin oblong or circular bodies, dorsal fins that are completely joined (except in *Micropterus*), a lateral line, and three or more anal spines. The males make a shallow depression for a nest and guard the eggs and fry.

BANDED SUNFISH

Enneacanthus obesus **92:2**

Description
Size, to 3 in. (7.6 cm). Olive with purple or golden spots; 5–8 broad vertical crossbars; gill cover notched and with a large dark spot. Dorsal spines, usually 9; tail rounded.
Range
Coastal plain from Mass. to Fla.

Note: The related **BLUESPOT SUNFISH,** *Enneacanthus gloriosus,* ranging from New Jersey to Florida, has bright blue spots on sides and fins.

BLACKBANDED SUNFISH

Enneacanthus chaetodon **92:5**

Description
Size, to 4 in. (10.2 cm). Has 4 vertical black bands and black forepart of first dorsal fin. Body yellow; mouth small, gill cover notched, tail rounded; dorsal-fin spines, 10.
Range
Coastal plain, N.J. to Fla.

ROCKBASS

Ambloplites rupestris **87:7**

Description
Size, 6–10 in. (15.2–25.4 cm) and 1 lb. (0.4 kg). Dark markings, red eye, 6 anal-fin spines. Body brassy, scales each with a dark spot; mouth large, lower jaw protrudes, gill cover notched. Lateral line complete; dorsal fin with 10–12 spines and much longer than anal.
Habitat
Rocky lake shallows, clear streams with rock bottoms.
Habits
Often schools; hibernates in winter.
Food
Insects, crayfish, fishes.
Reproduction
Nest built by male in gravelly shallows; about 5000 eggs laid in May or June; male guards nest and young.
Range
Great Lakes, St. Lawrence R. and Atlantic drainages, and Miss. R. Valley to Gulf of Mexico; widely introduced.

BLACK BASSES

Genus *Micropterus*

This is the genus of sunfishes in which the dorsals are separated by a deep notch.

SMALLMOUTH BASS

Micropterus dolomieui **87:6**

Description
Size, usually 8–15 in. (20.3–38.1 cm); rod and reel record, 11 lb. 14 oz. (5.4 kg). Shallow notch in dorsal fin; upper jaw rarely reaches rear of eye. Back olive, belly dusky silver, no stripe of dark spots on sides (but young have dark vertical bar on sides.). Shortest dorsal spine more than ½ length of longest; dorsal-fin rays, 12–15, pectoral-fin rays, 16–18; anal-fin rays, 11.
Habitat
Clear, rocky lakes and rivers.
Food
Fishes.
Reproduction
Male fans out nest in gravelly bottom, where up to five females may deposit total of 2000 eggs; male guards eggs and young; respawning within several days is frequent.
Range
Minn. to Que., s. to Ark. and n. Ala.; widely introduced elsewhere.

SPOTTED BASS

Micropterus punctulatus **92:9**

Description
Size, 8–17 in. (20.3–43.2 cm) and 3 lb. (1.4 kg). Shallow notch in dorsal fin; dark spot at base of tail; upper jaw rarely reaches rear of eye. Back yellow, sides white, prominent dark spot on gill cover, band of connected black blotches on side. Shortest dorsal spine more than ½ size of longest; dorsal-fin rays, 11–13; pectoral-fin rays, 15–16; anal-fin rays, 9–11.
Habitat
Slow streams and deep pools.
Food
Fishes.
Range
Most of s. U.S. from Fla. to Tex. and n. to Mo.

LARGEMOUTH BASS

Micropterus salmoides **87:12**

Description
Usually 10–18 in. (25.4–45.7 cm), rarely to 24 in. (61 cm), and 10 lb. (4.5 kg); rod and reel record, 22¼ lb. (10.1 kg). Dorsal fins almost divided by notch; upper jaw extending beyond eye. Back olive, sides and belly silver, side with a dark band (which becomes broken with age). Shortest dorsal spine less than ½ longest; dorsal-fin rays, 12–15; anal-fin rays, 11.
Habitat
Ponds, small lakes, oxbows; fairly tolerant of turbidity.
Food
Fishes, crayfish, frogs, insects.
Reproduction
Nest often on mud bottom with vegetation; eggs similar to Smallmouth Bass.
Range
S. Canada from Que. to Man., entire Great Lakes area and Miss. R. Valley to Gulf of Mexico.

SUNFISHES

Genus *Lepomis*

These sunfishes have deep, thin bodies, ten dorsal and three anal spines, a notched tail, and no teeth on the tongue. They are colorful and popular game fishes, and some species have been widely introduced.

REDBREAST SUNFISH

Lepomis auritus **92:4**

Description
Size, 5–7 in. (12.7–17.8 cm). Gill cover long, black, narrow, flexible. Back olive, belly orange-red, no black spot on hind dorsal rays. Teeth on roof of mouth; pectoral fins short, rounded.
Habitat
Rivers.
Food
Insects, mollusks, fishes.
Range
N.B. to Fla.; introduced w. to s. Okla.

LONGEAR SUNFISH

Lepomis megalotis **92:6**

Description
Size, to 9 in. (22.9 cm). Color variable, often brilliant; sides blue-spotted, belly yellow, iris red. Mouth small, but reaching to front edge of eye. Ear flap large, rear edge red or bluish-white and with thin, flexible projection. No teeth on roof of mouth; pectoral fins rounded; pectoral-fin rays, 10–13.
Food
Insects.
Range
Miss. R. Valley, from Minn. to Gulf states, n. to N.C.

REDEAR SUNFISH

Lepomis microlophus *Fig. 121*

Description
Size, usually 6–7 in. (15.2–17.8 cm), rarely to 10 in. (25.4 cm). Back olive, belly brassy, gill cover with red rear edge. Mouth barely reaching front edge of eye. Pectoral fin long and pointed; no distinct spots on rear of dorsal and anal fins.
Habitat
Clear, still waters with some vegetation.
Range
Miss. R. Valley, Iowa to Ind., and s. to Gulf of Mexico.

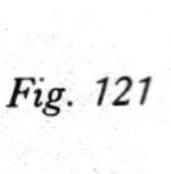
Fig. 121

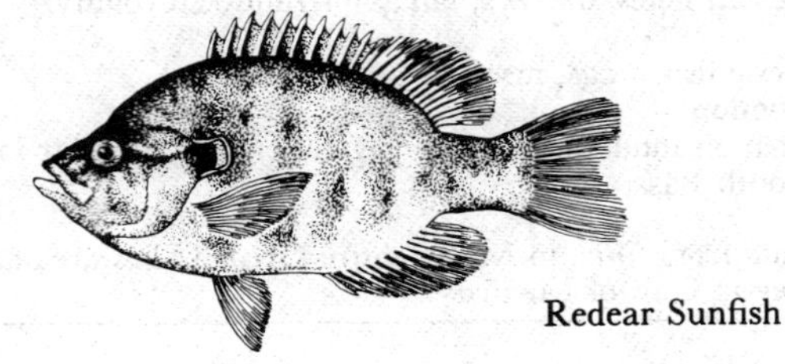
Redear Sunfish

WARMOUTH

Lepomis gulosus **92:3**

Description
Size, rarely to 10 in. (25.4 cm) and 1 lb. (0.4 kg). Anal spines, 3; notched tail. Mouth large, upper jaw extending to eye. Color, olive; 3–5 dark spokes around eye; tip of gill cover with white border. Pectoral fins short, rounded; dorsal-fin spines, usually 10; second dorsal fin with dark wavy lines.
Habitat
Vegetated ponds, oxbows, marshes, sluggish streams, preferably with muddy bottoms.
Food
Insects, snails, crustaceans, fishes.
Reproduction
Nest built on sand or gravel bottom.
Range
Kans., Mich., and N.Y., s. to Gulf of Mexico.

GREEN SUNFISH

Lepomis cyanellus **92:1**

Description
Size, to 8 in. (20.3 cm) and 15 oz. (0.4 kg). Body green; mouth large and extending past front edge of eye; gill cover stiff to smooth rear edge; lateral line complete. Pectoral fin short and rounded; dark spot on gill cover and on bases of dorsal and anal fins.
Habitat
Warm streams and ponds.
Remarks
A good pan fish.
Range
S.Dak. to N.Y. and s. to N.Mex. and Ala.; introduced elsewhere.

Note: This species frequently hybridizes with the Bluegill. The **BANTAM SUNFISH,** *Lepomis symmetricus,* of the Mississippi Valley from southern Illinois south, has the lateral line incomplete. The **SPOTTED SUNFISH,** *Lepomis punctatus,* has many bronze specks on an olive background, and occurs in Mississippi basin from southern Indiana south.

BLUEGILL

Lepomis macrochirus **87:4**

Description
Size, usually 4–9 in. (10.2–22.9 cm), rarely to 11 in. (27.9 cm); rod and reel record, 4¾ lb. (2.2 kg). Back olive with blue and orange on sides; mouth small and rarely reaching front edge of eye; 2 bluish bars extend back from mouth and chin. Rear edge of gill cover thin, lower edge usually bluish; dusky spot on last 3 dorsal rays. No teeth on roof of mouth; pectoral fins long and pointed; anal-fin rays, 10–12.
Habitat
Warm, weedy waters of bays, ponds, and lakes.
Food
Insects, fishes, crayfish.
Range
Drainages of Great Lakes, Miss. R., Gulf Coast, and Atlantic Coast, n. to N.J.

PUMPKINSEED

Lepomis gibbosus **87:5**

Description
Size, to 9 in. (22.9 cm) and 11 oz. (0.3 kg). Long, pointed pectoral fins; small red spot near edge of gill cover. Body very deep; mouth small, not reaching eye. Color orange to olive, sometimes with blue markings on cheek. No teeth on roof of mouth; gill cover stiff to smooth rear edge; rear of dorsal fin mottled.
Habitat
Clear, still water with many submerged plants.
Food
Insects, mollusks, fishes.
Range
N.Dak. to N.B., s. to Kans., Ky., and Ga.

ORANGE SPOTTED SUNFISH

Lepomis humilis **86:14**

Description
Size, to 4 in. (10.2 cm). Conspicuous orange spots on silvery sides. Back green; lobe of gill cover with broad red or orange margin; edges of gill cover white; rear edge thin with flexible projection. Mouth extending past front edge of eye; teeth on roof of mouth. Pectoral fins pointed; anal-fin rays, 9–10. The colors of the breeding male are exceptionally brilliant.
Habitat
Turbid lakes, ponds, and slow streams with silt-covered bottoms.
Food
Insects, small crustaceans, fishes.
Range
Dakotas to Ohio, s. to N.Mex. and Miss.

CRAPPIES

Genus *Pomoxis*

These members of the sunfish family have long heads with concave foreheads.

BLACK CRAPPIE

Pomoxis nigromaculatus **87:9**

Description
Size, usually 7–12 in. (17.8–30.5 cm); rod and reel record, 5 lb. (2.3 kg). Body greenish with black blotches; base of dorsal fin as long as distance from eye to dorsal fin; dorsal-fin spines, 6–10 (usually 7–8).
Habitat
Clear, quiet water with mud or sand bottoms and submerged vegetation.
Food
Small fishes, insects.
Reproduction
Nest in 3–6 ft. (0.9–1.8 m) of water on sandy or mud bottom among plants; eggs laid in late spring.
Range
Across s. Canada from upper St. Lawrence R. to Man.; s. through e. and cen. U.S. to Fla. and Tex. Widely stocked by state conservation departments; but it often suffers in competition with the White Crappie.

WHITE CRAPPIE

Pomoxis annularis **87:11**

Description
Size, to 1 ft. (0.3 m); rod and reel record, 5 lb. 3 oz. (2.4 kg). Body greenish or silvery, with indistinct vertical bars; mouth large, lower jaw heavy and protruding, gill cover notched. Base of dorsal fin shorter than distance from eye to dorsal fin; dorsal-fin spines, 5–7 (usually 6).
Habitat
Ponds, lakes, bayous, oxbows, and sluggish pools, preferably with vegetation; very tolerant of turbidity.
Habits
Often travels in schools.
Food
Insects, crustaceans, small fishes.
Reproduction
Eggs laid in late spring.
Remarks
An abundant, tolerant, and popular pan fish, much stocked by state conservation departments.
Range
E. S.Dak. to sw. N.Y., s. to Gulf of Mexico.

PERCHES

Family Percidae

These predaceous fishes have ctenoid scales, two completely separate dorsal fins, and one or two anal-fin spines.

YELLOW PERCH

Perca flavescens **86:1**

Description
Size, average 4–10 in. (10.2–25.4 cm), to 14 in. (35.6 cm); rod and reel record, 4 lb. 3½ oz. (1.9 kg). Prominent broad vertical bars on yellow sides. Body deep, compressed; no canine teeth; head slightly concave above eyes, making humped outline before first dorsal. Pelvics close together; first dorsal with 13–15 spines and well separated from second; anal-fin rays, 6–8.
Habitat
Vegetated lakes and streams.
Habits
Usually travels in schools.
Food
Plankton, insects, fishes.
Reproduction
Eggs laid in spring in long strands of gelatinous material.
Range
Alta. to N.S.; s. to ne. Kans., sw. Pa., and Atlantic Coast to S.C.

SAUGER

Stizostedion canadense **89:11**

Description
Size, commonly to 15 in. (38.1 cm); rod and reel record, 8 lb. 3 oz. (3.7 kg). Many round black spots (but no single large black blotch) on the dorsal fin. Body elongated, cheeks scaled, free edge of preoperculum toothed; dorsal-fin spines, 13–15; dorsal-fin rays, 17–20; anal-fin rays, 11–14.

Habitat
Shallow, turbid lakes and large rivers.
Food
Fishes.
Reproduction
Eggs laid in early spring in shallow water.
Range
Cen. Alta. to N.B.; s. to Colo. and Ala.

WALLEYE
Stizostedion vitreum **86:11**

Description
Size, average 13–20 in. (33–50.8 cm); rod and reel record, 22¼ lb. (10.1 kg). Distinct black blotch on the rear end of the first dorsal fin. Strong canine teeth; dorsal-fin rays, 19–22.
Habitat
Clear and moderately deep lakes; large rivers.
Food
Fishes, crayfish.
Reproduction
Eggs laid in early spring, in lake shoals or tributary systems.
Range
Athabaska Lake (Alta.–Sask. border), e. to Labrador and s. to N.C.; Great Lakes, Miss. R. Valley to Tenn. R.

DARTERS
Genus *Percina*

These are small elongate fishes with small mouths, complete lateral lines, and unforked tails. Their dorsal fins are well separated, with ten to seventeen spines in the first fin, and their pelvic fins are separated by a width at least equal to the base of a pelvic fin. There are usually two anal spines. The midline of the belly either is naked or has large spiny scales.

BLACKSIDE DARTER, *Percina maculata* (**93:9**), to 2–3 in. (5.1–7.6 cm). Mouth at end of snout, 7 oblong side blotches. Its range extends from southern Manitoba and Ontario south to northeast Texas and Alabama.

LOGPERCH, *Percina caprodes* (**93:15**), usual size, 3–4 in. (7.6–10.2 cm), to 7 in. (17.8 cm). Mouth below snout, dark vertical stripes, spot at base of tail. Occurs from central Alberta to southern Quebec south to east Texas, west of Appalachians.

SAND DARTER
Genus *Ammocrypta*

EASTERN SAND DARTER
Ammocrypta pellucida **93:10**

Description
Size, to 3 in. (7.6 cm). Body long, slender, pellucid, lower half almost scaleless; sides with 12–19 dusky blotches. Mouth small and separated from upper jaw by groove; dorsal fins well separated; dorsal-fin rays, 9–12; anal-fin rays, 8–10.

Habitat
Sandy bottoms.
Habits
Buries itself in sand up to its eyes, which scan the surrounding water for its prey; darts out after prey, then quickly reburies itself in the sand.
Food
Insects.
Reproduction
Eggs laid in spring.
Range
Se. Mich., Que., and Lake Champlain, s. to Ky. and W.Va.

OTHER DARTERS

Genus *Etheostoma*

Members of this genus are small; their bodies, including the midline of the belly, are well scaled. The lateral line may be absent, incomplete, or complete. The anal fin is smaller than the second, or soft, dorsal. Eighty species are recognized.

JOHNNY DARTER, *Etheostoma nigrum* (**93:8**), to 3 in. (7.6 cm). Occurs in streams of eastern North America west to the Plains.

BANDED DARTER, *Etheostoma zonale* (**93:13**), to 3 in. (7.6 cm). Its range includes clear streams east of Plains, Lake Michigan drainage to Pennsylvania, and south to Gulf of Mexico.

GREENSIDE DARTER, *Etheostoma blennioides* (**93:11**), to 4 in. (10.2 cm). This species is found in riffles of Mississippi River system and Ozark region, east to New York and Georgia.

FANTAIL DARTER, *Etheostoma flabellare* (**93:12**), to 3 in. (7.6 cm). Found in riffles from Minnesota and Quebec south to Oklahoma and Alabama.

NORTHERN SWAMP DARTER, *Etheostoma fusiforme* (**93:14**), to 2 in. (5.1 cm). Range extends from Maine to North Carolina.

RAINBOW DARTER, *Etheostoma caeruleum* (**86:2**), to 2½ in. (6.4 cm). Found from southern Minnesota and eastern Ontario to Arkansas and Alabama.

CROAKERS
Family Sciaenidae

FRESHWATER DRUM
Aplodinotus grunniens **86:12**

Description
Size, average 18–20 in. (45.7–50.8 cm). High back; long, continuous dorsal fins; short anal fin; lateral line extending onto the tail. Dusky above, sides silvery; body and head scaled, throat teeth coarse. Dorsal spines, 8–9; anal spines, 2, second long and heavy.
Habitat
Lake shallows, large rivers.
Habits
Produces a grunting sound.
Food
Mollusks, crayfish, insects.
Reproduction
Eggs laid in spring.
Remarks
An important commercial fish.
Range
Lakes and rivers as far n. as the St. Lawrence Basin to Man. and Mont.; s. to Tex.; peak numbers in Miss. R. and lower Mo. R. systems.

SCULPINS
Family Cottidae

MOTTLED SCULPIN
Cottus bairdi **89:8**

Description
Size, average 3 in. (7.6 cm). Body dark, mottled; head and mouth large, eyes high. Dorsal-fin spines, 7–9; dorsal rays, 16–18; pectoral-fin spines, 1; pectoral rays, 13–17; anal-fin rays, 10–14.
Habitat
Cool streams.
Food
Insects, crustaceans, small fishes.
Range
S. Canada to s. Appalachians; also Ozarks. Discontinuous into parts of Mo. R. and Columbia R. drainages.

Mollusks

Consulting Editor
Harald A. Rehder
Zoologist Emeritus
National Museum of Natural History
Smithsonian Institution

Illustrations
Plates 94–98 Klarie Phipps
Plates 99–107 Jennifer Emry-Perrott
Plate 108 Nancy Lou Gahan

Mollusks

Phylum Mollusca

The enormously successful phylum of mollusks, sometimes termed the "soft ones" from translation of the Latin name Mollusca, constitutes one of the most advanced groups of animals in the invertebrate world. To be sure, an oyster may not seem very advanced, but the squid that can remove the lid from a box with its tentacles to get the food within is certainly more intelligent than some of the lower vertebrates. Moreover, its eye is the most highly organized visual organ known among the invertebrates.

Nature of Mollusks

Mollusks, in general, are soft-bodied, unsegmented, shell-bearing invertebrates. Five classes are treated in this chapter: the gastropods, class Gastropoda; the bivalves, class Bivalvia; the scaphopods, class Scaphopoda; the chitons, class Polyplacophora; and the cephalopods, class Cephalopoda. The first four classes are the seashells, which make up two-thirds of all mollusk species. In these, the shell is external and conspicuous; in the cephalopods, it is internal and inconspicuous. In a few groups, such as the sea slugs, or nudibranchs, and a few chitons, the shell has been lost or never developed.

These five classes are, basically, as different from one another as are the five classes of the phylum Chordata, that were dealt with earlier in the book. In other words, in fundamental structure, a clam is as different from a conch shell as a mammal is from a bird, and a squid and chiton are as far removed from each other as man is from a mud turtle.

Classification

In general terms, mollusks are soft-bodied, unsegmented, and usually hard-shelled invertebrate animals. The living mollusks comprise seven classes; the five major ones discussed in this chapter are:

Class Polyplacophora

The chitons, also known as coat-of-mail shells, are primitive, ancestral-type mollusks of very sluggish habits. Elongated and flattened, the chitons bear a shelly armor of eight saddle-shaped plates (imbricating valves) arranged in an overlapping series along the back and held together by a surrounding girdle often bearing spiny hairs. On the underside is the broad, flat foot. Although most chitons inhabit shallow water close to shore, others live in depths as great as 2000 fathoms (3.7 km).

Class Gastropoda

The gastropods are the snails, most of which have shells. Most gastropods are univalves; that is, they have a single shell, or valve, which is usually spiral. A few exhibit two valves. Gastropods have a low order of sight, hearing, and possibly taste but well-developed senses of smell and touch. Most are carnivorous and predatory. They move over the ocean floor, and some are lifelong wanderers on the open sea.

Class Scaphopoda

The scaphopods comprise a small class of burrowing univalve mollusks having a tapering conical shell open at both ends and slightly arched, like a miniature elephant's tusk. The members of this class are commonly known as tooth or tusk shells, after their shape. From the larger end of the shell projects the foot and several slender tentacles. The animals live partially buried in sand in clear water at various depths, feeding on minute marine organisms.

Class Bivalvia
The bivalves, also known as pelecypods, are the clams, which have two valves joined along a hinge line by a tough ligament and held together by one or two strong muscles. Except in a short larval stage, most clams are sedentary throughout their lives.

Class Cephalopoda
The cephalopods are highly specialized, carnivorous mollusks with relatively keen sight and a well-developed nervous system. The shell, if present, is generally internal and lengthened. The head supports two large eyes and is armed with a powerful parrotlike beak; it is surrounded by long flexible tentacles, or arms, studded with powerful sucking disks. Most cephalopods can discharge a caustic, inky fluid in defense. They move rapidly through the water with a jet-propelled action, the tentacles streaming behind.

Fig. 122

Parts of a Typical Snail and Clam

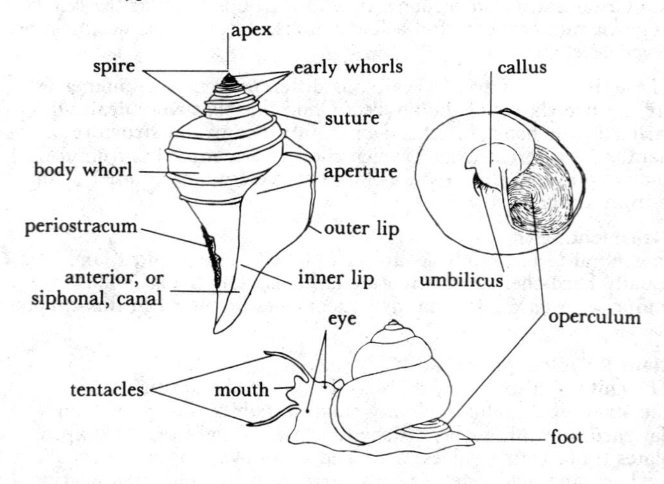

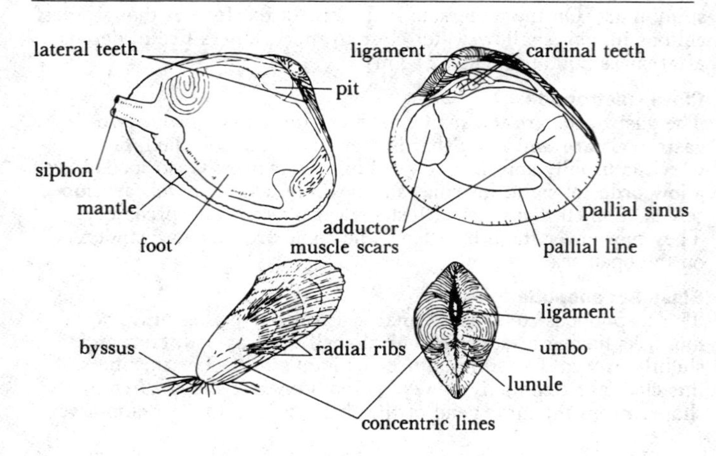

Habitat

There are some 80,000 to 100,000 species of mollusks in the world, a little over half of them marine. Marine species live from three miles below the surface of the ocean to the top of the springtide splash line on sea cliffs. In their different varieties they occur on sandy, muddy, pebbly, or rocky coasts; in clear, muddy, or brackish water; in rocky crevices and tide pools; or afloat and adrift on the high seas. They are found from the Arctic to the tropics. Some mollusks not included in this chapter inhabit freshwater lakes and streams, live underground, or are found on trees.

Habits

Many, perhaps most, mollusks are slow-moving or sedentary. A few, however, such as the squids, are among the swiftest travelers. Various species can burrow in mud, wood, stone, or even lead cables; some are parasitic. A number can swim; others can crawl or jump; and the sea arrow, or flying squid, can even travel in the air. As in most great aggregations of living forms, some are carnivorous, some are vegetarians, and some are scavengers.

Reproduction

The phylum exhibits great diversity in reproduction. In some species the sexes are separate; some species are hermaphroditic; in others the same individual changes, sometimes even alternates, its sex. An interesting example is the Virginia Oyster, which indulges in an alternation of the sexes. The Quahog practices sex reversal, changing from male to female as it grows older.

Some species, such as the oyster, cast their eggs and sperm adrift on the surrounding sea. Higher forms are more careful. The male octopus has a tentacle designed to gather and deliver the sperm into the mantle cavity of the female. To make sure it arrives safely, the male argonaut takes the precaution of detaching and leaving the tentacle along with the sperm.

The shelled mollusks are essentially sedentary, although some scallops are active swimmers. Breeding seasons vary among different species; the custom of eating oysters only in months with an *r* coincides roughly with protection for the bivalve during the breeding season.

Evolution

Geologically speaking, mollusks go back to the Cambrian period. Because of their calcareous shells they are commonly preserved as fossils, and the rocks of former ages often abound with remains. Because of their abundance and the ease with which shallow-water forms have spread throughout the world in past ages of widespread equable temperatures, many have become index fossils and are used by the paleontologist to correlate geologic formations in different regions.

Under conditions of a relatively stable environment, such as the sea, the rate of evolutionary change slows down and is influenced more by competition with other animals than by the environment. So slowly does change proceed among certain well-adapted species living in such a near-stable environment that the same species of seashells living today may often be found in deposits 50,000,000 years old. This is a situation unknown among birds or mammals, which are subject to the wide variations of rapidly changing conditions on land. Molluscan genera living today may often be recognized in Mesozoic and occasionally even in Paleozoic strata.

Conservation

Mollusks which are eaten, such as clams and oysters, often suffer from pollution and overharvesting. However, much government research and legislation have gone into the problem of helping ten clams live and grow where only one did before. Commercial interest in the production of these shellfish is so great that noncommercial conservationists are more disposed to worry about other mollusks.

Most marine mollusks are not at the moment in danger of extinction or even significant depletion. However, this may change with excessive collecting. Even now, along some favored Florida beaches, the near-shore flats have been so well combed by amateurs that few individuals of the most sought-after species remain. The student, naturalist, and collector should remember to put back in its original position every rock or underwater shelter under which he looks. Any one of these may be a mollusk habitat, and without shelter the mollusks may go elsewhere or disappear.

Shell Collecting Techniques

The most productive place to collect shells is the seashore at low tide. Strolling along a sandy stretch of beach can prove fruitful in finding shells that have been washed up by a late tide. Much of such shelly material, however, will be wave-eroded, broken, or damaged. The valves of a bivalve shell are commonly found separated, and the lips of a univalve usually show broken edges and worn shoulders.

When no longer content with beach-worn specimens, a collector should search for living invertebrates in their natural habitats. This opens up unlimited possibilities for acquiring a worthwhile shell collection. It is thus important to pay attention to tides, seasonal moods of the ocean, and the effects of wind and current.

Some states may require possession of a regular collecting license before collection of living seashells is permitted. Additionally, for certain species, there are stringent protective regulations. The collector should understand these laws before collecting living mollusks.

The best time to collect and look for living shellfish is at low tide, when exploration of the exposed tidal zone is easiest. Because many marine creatures hide under objects to await the return of the tide, any stone, plank, or piece of driftwood should be turned over, and then replaced as originally found. Bivalves can often be detected in sand or on mud flats by small holes which reveal the siphon; and weight above may cause the animal to withdraw its siphon with a sharp squirt of water. Snails, which usually bury themselves just beneath the surface, reveal their locations by small telltale mounds or by their trails.

Shell collecting, of course, is only part of the fascination of malacology. The study of the living animal offers much more. Many of the life histories of almost all of the 50,000 marine mollusks still await investigation. In this field science has taken only its first steps.

Nomenclature

In this chapter every effort has been made to use the latest acceptable scientific nomenclature. The scientific name (genus and species) tells much that is known about a particular mollusk from its most distant ancestry to the present. Common or popular names are an attempt at standardization, and to many scientists translations seem inappropriate. It is suggested that the student concentrate on mastering the scientific name. However, the common names are included here, as an aid to the beginning malacologist.

Range and Scope
Of the more than 50,000 marine mollusks living in the world today, there are 6000 known in North America, most of them found in deep water. In this chapter 171 shallow-water species are discussed. They represent all species above ¼ inch (0.64 cm) in size that occur regularly to a depth of 10 fathoms (18.3 m) from the southern Canadian border south along the Atlantic Coast of the United States to southern Florida. Land and freshwater mollusks are not included; nor are those normally found in waters less than 10 fathoms deep, which are under ¼ inch in size.

Illustrations
Virtually every species discussed in this chapter is illustrated. The illustrations are captioned on the plates by common names of the individual species. The name is followed by a number indicating the size of the illustration relative to the actual life size of the species.

USEFUL REFERENCES

Abbott, R. T., 1968. *Seashells of North America.* New York: Golden Press.

———.1974. *American Seashells.* 2nd ed. New York: Van Nostrand Reinhold Co.

Emerson, W. K., and Jacobson, M. K. 1976. *The American Museum of Natural History Guide to Shells, Land, Freshwater and Marine, from Nova Scotia to Florida.* New York: Alfred A. Knopf.

Warmke, G. L., and Abbott, R. T. 1975. *Caribbean Seashells.* (Reprint of 1961 edition.) New York: Dover Publishing Co.

GLOSSARY

Aperture Opening or mouth of a snail.

Beak Noselike angle near hinge of bivalve shell, where growth started.

Byssus Strong threads that attach some clams to their supports.

Callus Shelly deposit sometimes covering umbilicus.

Canal A narrow extension of the lip of the aperture containing the siphons in many snails.

Columella Central pillar of snail shell.

Cord A band, found on shells with whorls.

Frondose Thin, flattened, often wavy, leaflike projection on a rib.

Ganglion (*pl.* ganglia) Knot or collection of nerve cells.

Hinge Area where the valves of a bivalve join.

Ligament Horny, elastic structure joining two shells of a bivalve.

Lunule Impressed or modified area in front of the beak on the outside of many bivalve shells.

Mantle A membranous flap or outer covering of the soft parts of a mollusk; it secretes the shelly material.

Operculum In some snails, a "trap door" that closes the aperture when animal's foot is withdrawn into shell.

Pallial line A mark on the inner surface of a bivalve shell, more or less parallel with the margin, caused by attachment of the mantle.

Parietal wall The inner lip area.

Periostracum Thin, noncalcareous outer layer of shells, often worn off those found on beach.

Radula Ribbonlike, saw-edged "tongue" of snails, used in feeding.

Sinus A deep cut.

Siphon Organ for admitting and expelling water from body cavity of bivalves.

Spire Upper whorls from the apex to the body whorl.

Suture Spiral line of spire in snails, separating the whorls.

Turbinate Top-shaped.

Umbilicus Hole in center of snail shell, at base of body whorl.

Umbo (*pl.* umbones) Swollen part of bivalve shell, near beak.

Varix (*pl.* varices) One of the prominent ridges across each whorl of certain gastropods showing a former position of the outer lip of the aperture; often called a "rib."

Whorl One of the turns of a snail shell.

CHITONS

Class Polyplacophora

Chitons, or coat-of-mail shells, are sluggish, primitive saltwater mollusks, believed to be close in structure to the common ancestor of all mollusks. Typically they are flat and oblong and have eight armorlike overlapping plates held together by a surrounding girdle, often bearing spiny hairs. The animal uses the foot beneath to move about on or to hold fast to a rock by suction. Most chitons prefer shallow water and darkness. They are often found on the underside of rocks, where they feed on small algae. Some 600 species are distributed worldwide except in polar seas.

True Chitons

Order Chitonida

CHAETOPLEURID CHITONS

Family Chaetopleuridae

EASTERN CHITON

Chaetopleura apiculata **94:4**

Description
Size, ⅓–¾ in. (0.9–1.9 cm). Shape oval to oblong; plates raised, covered with beads, which concentrate in rows toward the center; girdle narrow, mottled, slightly grained, and with scattered short hairs. Color gray or pale yellow, sometimes tinged with red.
Habitat
Low tide to 15 fathoms (27.4 m).
Range
Cape Cod to Fla.

COMMON CHITONS

Family Chitonidae

TUBERCULATE CHITON

Chiton tuberculatus **94:1**

Description
Size, 2–3 in. (5.1–7.6 cm). Oval; valves with strong radial riblets on lateral areas; girdle with small overlapping scales. Color of valves various shades of grayish green, with girdle showing alternate patches of pale gray-green and black.
Habitat
Common on rocky shores.
Range
S. Fla. and W. Indies.

FUZZY CHITON

Acanthopleura granulata **94:3**

Description
Size, 2–3 in. (5.1–7.6 cm). Rather broadly ovate, granular sculpture; valves usually lack color because of erosion; girdle rather broad, with coarse hairlike spines, pale grayish color, sometimes with black splotches.

Habitat
Common on rocks.
Range
S. Fla. and W. Indies.

SLENDER CHITONS
Family Ischnochitonidae

The shells of this family have comparatively large plates which are not covered by the girdle. The front and back plates are attached in the same manner and the back plate has no radiating ribs. These chitons are found on rocks between tide levels.

WHITE CHITON
Ischnochiton albus **94:5**

Description
Size, ½ in. (1.3 cm). Plates smooth with some growth ridges; girdle rough, scaly. Outside white, cream, or pale orange, occasionally with brown marks; inside white.
Similarities
Red Chiton is slightly larger and has a convex anterior-valve front slope.
Habitat
Common in cold, shallow water.
Range
Arctic seas to Mass.

MOTTLED CHITON
Tonicella marmorea **94:6**

Description
Size, 1 in. (2.5 cm). Oval to oblong, without hairs. Color creamy brown, mottled heavily with deep red.
Habitat
Common in 1–50 fathoms (1.8–91.5 m).
Range
Greenland to Mass.

RED CHITON
Tonicella ruber **94:2**

Description
Size, ½–1 in. (1.3–2.5 cm). Oblong, valves smooth; girdle with minute, elevated scales. Color pale brown with orange-red blotches or overall red.
Similarities
White Chiton is slightly smaller, with front slope of anterior valve straight to slightly concave.
Habitat
Common in 1–80 fathoms (1.8–146.3 m).
Range
Arctic seas to Conn.

SNAILS

Class Gastropoda

There are worldwide some 40,000 species of marine snails. All those that are treated in this book have a single, usually spiral, shell. In the limpets this spiral feature is more apparent in the early stages. The nudibranchs, or sea hares, which are not discussed, are naked sea slugs without external shells.

As with other mollusks, the soft mantle secretes the shell, which grows around the mantle and increases in size as the animal grows. Truly marine snails breathe with gills and move by means of an extensible foot. Most snails are quite mobile, traveling about at will over the ocean floor. Some, such as the floating violet snail, are lifelong wanderers on the open sea.

Snails may be carnivorous, herbivorous, or scavengers. The Flat Slipper is typical of those that feed on tiny organisms suspended in sea water; ceriths are typical of the detritus feeders that strain their nourishment out of the mud.

Snails have a low order of sight, hearing, and presumably taste; but the senses of smell and touch are well developed. Various collections of ganglia control the movements of the animal and correspond to a brain in higher forms. In attacking their prey most snails use a long ribbon-shaped "tongue," or radula, well toothed and remotely resembling a file. This is protruded through the opening and is worked back and forth over the shell of the snail's prey, such as a clam. Finally, a hole is bored into the interior of the prey and the snail sucks out its meal. Most snails are nocturnal.

Ancient Stomach-footed Gastropods

Order Archaeogastropoda

KEYHOLE LIMPETS

Family Fissurellidae

These are limpets with a hole or slit near the summit. The shell is cone-shaped and has an oval base and many radiating ribs.

LINNAEUS' KEYHOLE LIMPET

Puncturella noachina **94:8**

Description
Size, ½ in. (1.3 cm). Shell white, glossy within; 21–26 main ribs with smaller secondary ribs between them, part way down; small slit in front of apex.
Habitat
Low-tide mark to depth of 1 mi. (1.6 km).
Range
Arctic and s. to Cape Cod.

LITTLE KEYHOLE LIMPET
Diodora cayenensis **94:9**

Description
Size, 1–2 in. (2.5–5.1 cm). Shell whitish to dark gray and pinkish; inside white to bluish gray; every fourth radiating rib larger; hole keyhole-shaped.
Habitat
Intertidal to moderate depths on stones, shells, or seaweeds; common.
Range
Chesapeake Bay to Brazil.

LISTER'S KEYHOLE LIMPET
Diadora listeri **94:7**

Description
Size, 1–2 in. (2.5–5.1 cm). Shell whitish to gray with every second radiating rib larger and the concentric ribs stronger, forming a more distinctly cancellate (crisscrossed ridges) sculpture than in *D. cayenensis*; intersections of ribs often nodulose or scaly.
Habitat
Intertidal; common.
Range
S. Fla. to Brazil.

LIMPETS
Family Acmaeidae

ATLANTIC PLATE LIMPET
Collisella testudinalis **94:10**

Description
Size, to 1½ in. (3.8 cm). Shell oval, conical, depressed, smoothish. Outside blue-gray to cream with irregular axial bars checkered brown; inside rich brown with wide light gray border narrowly checkered with brown at the outer margin. Top of spire usually nearly central. Unlike Family Fissurellidae, has no opening at top of shell.
Habitat
Tide pools and shallow-water rocks.
Remarks
By developing the power of suction on rocks limpets have learned to survive and flourish in the zone between the tides.
Range
Arctic seas to Long Island.

TOP SHELLS
Family Trochidae

The top shells are snails with pearly insides. They are generally cone-shaped and have a horny operculum. These common snails feed on vegetation.

GREENLAND TOP SHELL
Margarites groenlandicus **94:11**

Description
Size, to ¾ in. (1.9 cm). Shell glossy cream to brown; 4–5 whorls flattened above, wavy at sutures; smooth, or with about 12 low, smooth spiral ridges; initial whorl smooth, shiny; umbilicus broad, funnel-shaped.
Habitat
Dredged to 150 fathoms (274 m).
Range
Arctic seas to Mass.

MOTTLED TOP SHELL
Calliostoma jujubinum **94:12**

Description
Size, ½–1¼ in. (1.3–3.2 cm). Shell pyramidal, brownish or reddish, usually with a pale zone below the suture and often with irregular darker or lighter splotches near lower part of whorls, which are weakly beaded; last whorl with swollen, rounded periphery.
Habitat
Under rocks on reefs.
Range
N.C. to Brazil.

TURBAN SHELLS
Family Turbinidae

These are solid shells with rounded, usually sculptured whorls, and with a shelly operculum.

KNOBBY TURBAN
Turbo castaneus **95:3**

Description
Size, 1–1½ in. (2.5–3.8 cm). Shell nodulose, usually brownish, grayish, or orange, and often with flamelike markings; operculum gray-greenish with a white border.
Habitat
Common in shallow water.
Range
N.C. to Fla. and Tex., and s. to Brazil.

LONG-SPINED STAR SHELL
Astraea phoebia **95:2**

Description
Size, 1½–2½ in. (3.8–6.4 cm). Shell depressed, almost flattened on its underside, whorls strongly sculptured, with flattened triangular spines on the periphery; aperture silvery. Color grayish to yellowish green; young shells with long spines and with iridescent green and yellow color.
Habitat
Common in shallow water.
Range
Fla. to W. Indies and Brazil.

AMERICAN STAR SHELL
Astraea americanum **95:1**

Description
Size, 1–1½ in. (2.5–3.8 cm). Shell with high conical spire, flat sides, and swollen, rounded periphery; sculptured with numerous fine axial riblets; color grayish to yellowish white.
Habitat
Common in shallow water.
Range
Fla. to Mexico and W. Indies.

NERITES
Family Neritidae

These snails have solid, usually broadly oval to flattened, smooth to spirally ribbed shells with a broad columella which may be smooth, tuberculate, or ribbed and has teeth at the inner edge. The operculum is shelly, with an internal peglike projection at one end.

BLEEDING TOOTH
Nerita peloronta **95:4**

Description
Size, ¾–1½ in. (1.9–3.8 cm). Shell broadly oval, weakly spirally sculptured, yellowish white with zigzag black and red markings. Rather broad white callus in columellar area marked with an orange-red splotch near the 2 teeth at the margin.
Habitat
Abundant on rocky ocean shores.
Range
S. Fla. through W. Indies.

VARIEGATED NERITE
Nerita versicolor **95:7**

Description
Size, ¾–1 in. (1.9–2.5 cm). Shell spirally ribbed, grayish with irregular spiral rows of black and red spots. Columellar callus with usually 4 strong teeth.
Habitat
Found with Bleeding Tooth, on rocky ocean shores.
Range
Fla. through W. Indies.

TESSELLATED NERITE
Nerita tessellata **95:6**

Description
Size, ½–¾ in. (1.3–1.9 cm). Shell sculptured with strong spiral cords of varying width separated by narrow grooves, the ribs irregularly marked by alternate spots of black and white; the broad, concave columellar area white, with 2 small teeth at apertural edge.
Habitat
Common under rocks in shallow water.
Range
Fla., W. Indies, to Brazil.

ZEBRA NERITE
Puperita pupa **95:5**

Description
Size, ⅓–½ in. (0.9–1.3 cm). Shell smooth, white with irregular rather narrow black axial lines; inside of outer lip and columellar callus yellow; operculum smooth.
Habitat
In pools above tide line.
Range
Fla. through W. Indies.

Middle Stomach-footed Gastropods
Order Mesogastropoda

PERIWINKLES
Family Littorinidae

These are shore dwellers, some living in brackish and some in salt water. Several species are amphibious and can survive long periods without water. The shell is spiral, turbinate, or globular, and it has few whorls and no umbilicus. Eggs are hatched inside or outside the body.

COMMON PERIWINKLE
Littorina littorea **95:13**

Description
Size, ½–1 in. (1.3–2.5 cm). Shell thick, smoothish; 6–7 whorls (seem fewer in beach-worn specimens) with very fine threads and wrinkles. Brownish to dark gray; base of columella white; inside chocolate-brown; white border of outer lip often checkered with brown.
Habitat
Common on algae and seaweed clinging to rocks.
Range
Labrador to N.J.

SMOOTH PERIWINKLE
Littorina obtusata **95:9**

Description
Size, ⅓–½ in. (0.9–1.3 cm). Shell stout, smooth, globular; 3–4 whorls; point low, dull. Color variable, usually bright yellowish brown; sometimes with bold brown spiral band; columella white.
Habitat
Common on rocks and rockweed.
Remarks
In the conquest of new habitats this periwinkle has ventured farthest from its ancestral home, the sea. It gives birth to living young and thus, independent of the water for reproduction, is able to fight drought and the hazards of land life at the very edge of the spring-tide saltwater splash line.
Range
Labrador to Cape May.

SALT MARSH PERIWINKLE

Littorina irrorata **95:10**

Description
Size, 1 in. (2.5 cm). Shell heavy, plump; top strongly pointed; 5 whorls with fine ridges. Grayish white with streaks of brownish dots on spiral ridges; opening oval, whitish; outer lip strong, sharp, slightly flaring.
Habitat
Marine vegetation between tides.
Range
N.Y. to cen. Fla. and Tex.

ROUGH PERIWINKLE

Littorina saxatilis **95:8**

Description
Size, ½ in. (1.3 cm) or less. Shell ovate, coarse; spire moderately elevated; 4–5 well-defined convex whorls. Color ashy or yellowish gray; surface markings show growth lines, revolving grooves. Opening oval, operculum horny.
Habitat
On rocks, often above high-tide mark.
Range
New England and n.

KNOBBY PERIWINKLE

Tectarius muricatus **95:11**

Description
Size, ½–1 in. (1.3–2.5 cm). Shell gray, rather broadly conical, whorls with spiral rows of regularly spaced beads.
Habitat
Common on rocks well above the water line.
Range
S. Fla. through W. Indies.

CHINK SHELLS

Family Lacunidae

The stoutly conical chink shells are smooth, rather fragile periwinkles with thin shells, a half-moon aperture, and a shelflike columella alongside which is the characteristic lengthened groove, or chinklike umbilicus.

LITTLE CHINK SHELL

Lacuna vincta **95:12**

Description
Size, to ⅜ in. (0.95 cm). Typical of its family, this shell is fragile, plump, and conical, with a half-moon–shaped aperture. Identify it by the lengthened groove or chink beside the columella. Shell smooth, with small spiral scratches; dingy white to tan, often banded with purple; groove of columella (central pillar of spiral whorls) deep, umbilicus very small.
Habitat
On marine growths; often found washed ashore after storms.
Range
Arctic Ocean to R.I.

SUNDIAL SHELLS
Family Architectonicidae

COMMON SUNDIAL
Architectonica nobilis **96:1**

Description
Size, 1–2 in. (2.5–5.1 cm) dia. Shell solid, flattened, almost discus-shaped, with strongly sculptured whorls above and below; deep, moderately narrow umbilicus with a strongly nodulose edge. Color whitish yellow with spiral rows of brown spots particularly strong below the suture.
Habitat
In sand to 20 fathoms (36.6 m).
Range
N.C. to W. Indies.

SCREW SHELLS
Family Turritellidae

These are long, slender shells of many whorls, with a generally rounded aperture. In some forms the later whorls become uncoiled and very irregular, as in the following species.

COMMON WORM SHELL
Vermicularia spirata **96:7**

Description
Size, 2–7+ in. (5.1–17.8+ cm). Early part of shell coiled for about ¼–1 in. (0.6–2.5 cm) in regular elongate spiral; then growth becomes highly irregular and usually marked by irregular axial ridges; color grayish to yellowish brown.
Habitat
In shallow water, often partially embedded in sponges or ascidians.
Range
S. Fla. through W. Indies.

HORN SHELLS
Family Cerithiidae

The horn shells are distinguished by their long shape, numerous whorls, and oblique opening. They are generally found on grass and seaweed in fairly shallow water.

FLORIDA HORN SHELL
Cerithium atratum **96:3**

Description
Size, 1–2 in. (2.5–5.1 cm). Shell with variable sculpture, whorls more or less angulate, sometimes with strong nodes, or else with spiral beaded sculpture dominant. Color dark brown to pale yellowish white with brown markings.
Habitat
In shallow water in bays or coral reefs.
Range
N.C. to Brazil; W. Africa.

DWARF HORN SHELL
Cerithium lutosum **96:6**

Description
Size, ⅓–¾ in. (0.9–1.9 cm). Shell small, whorls sculptured with spiral rows of even-sized beads and fine spiral threads. Color pale to dark brown, grayish white with brown markings or pure white.
Habitat
Common and often in great numbers in intertidal zones.
Range
N.C. through W. Indies and Tex. to Venezuela.

VARIABLE BITTIUM
Bittium varium **96:4**

Description
Size, ⅛ in. (0.3 cm). Shell with spiral rows of beads, occasionally with nodulose axial riblets; a thickened varix is usually present.
Habitat
In eelgrass below low tide.
Range
Md. to Brazil.

AWL-SHAPED CERITHS
Family Cerithiopsidae

EMERSON'S CERITH
Cerithiopsis emersoni **96:5**

Description
Size, ½–¾ in. (1.3–1.9 cm). Shell strong, slim; chocolate-brown but may become ash-gray from erosion; 14–15 whorls with flattish sides, each having 3 rows of rounded beads of lighter brown. Aperture small, inner lip twisted.
Habitat
1–33 fathoms (1.8–60.4 m).
Range
Mass. to W. Indies and Brazil.

ADAMS' MINIATURE CERITH
Seila adamsi **96:2**

Description
Size, ¼–½ in. (0.6–1.3 cm). Shell long, slim; somewhat smooth and concave at base; about 12 flat whorls each with 3 spiral ridges, giving impression of threads on a screw; last whorl has 4 spiral ridges. No beads as in Emerson's Cerith. Dark brown to orange brown; aperture small; lip sharp, fragile, undulating.
Habitat
Common to 40 fathoms (73.2 m).
Range
Mass. to W. Indies.

MODULUS
Family Modulidae

ATLANTIC MODULUS
Modulus modulus **96:10**

Description
Size, ½–¾ in. (1.3–1.9 cm). Shell with a low, conical, flat-sided spire; low axial ribs; and a rounded base, with spiral cords, last whorl with a sharp peripheral keel; small toothlike projection at base of columella. Color grayish white.
Habitat
In algae and eelgrass in shallow water.
Range
N.C. to Brazil.

WENTLETRAPS
Family Epitoniidae

Wentletraps are easily recognized by their high-spired, white shells consisting of many ribbed whorls increasing in size from top to bottom. The shells of these carnivorous snails are often called "staircase shells."

ANGLED WENTLETRAP
Epitonium angulatum **96:8**

Description
Size, ¾–1 in. (1.9–2.5 cm). Shell stout, glossy; 8 convex whorls with 9–10 strong, thin, somewhat angled ribs. China-white; outer lip thick, flaring backward; no umbilicus.
Habitat
Common in shallow water to 25 fathoms (45.7 m).
Range
N.Y. to Tex.

BROWN-BANDED WENTLETRAP
Epitonium rupicolum **96:9**

Description
Size, ½–1 in. (1.3–2.5 cm). The only wentletrap in the range with 1 or 2 brown spiral bands on the body whorl. Shell has 8–11 rounded whorls with 12–18 low ribs; white, pinkish, or yellow to purplish brown.
Habitat
Common from low water to 20 fathoms (36.6 m) and found deeper.
Range
Cape Cod to Tex.

PURPLE SEA SNAILS

Family Janthinidae

This is a small group of pelagic snails, usually purple, sometimes brown, that float on the surface of the sea by means of a raft of bubbles. Shells often cast up on beaches after an onshore blow.

COMMON PURPLE SEA SNAIL

Janthina janthina **96:11**

Description
Size, 1–1½ in. (2.5–3.8 cm). Shell thin, whitish above, purple below, last whorl bluntly angular; outer lip sinuate.
Habitat
Normally pelagic but often cast up on shore in great numbers.
Range
Worldwide tropical.

BOAT SHELLS

Family Calyptraeidae

These stationary snails are distinguished by their round or oval shape and horizontal or cuplike platform inside, which helps keep the animal in the shell.

STRIATE CUP-AND-SAUCER

Crucibulum striatum **97:1**

Description
Size, 1 in. (2.5 cm). Caplike without; has only ⅓ of the cuplike platform attached to the main shell. Conical; top smooth, somewhat twisted, may or may not be central; rest of outside has light raised radial lines. Outside, yellowish or pinkish white, streaked with brown; inside glossy, yellowish, translucent.
Habitat
Common on rocks below low-water line.
Range
N.S. to S.C.

FLAT SLIPPER SHELL

Crepidula plana **97:4**

Description
Size, ½–1½ in. (1.3–3.8 cm). Shell flat; platform less than ½ length of shell, top pointed and slightly off-center. Shell translucent, glossy within.
Habitat
Shallow water; on coastal mud flats many are found attached to large empty shells.
Range
Canada to Tex.

COMMON BOAT SHELL

Crepidula fornicata **97:2**

Description
Size, ¾–2 in. (1.9–5.1 cm). Shell varies from nearly flat to high and arched, often corrugated from attachment to a ribbed surface;

platform concave, wavy at edge; apex drawn to one side. Color dingy white to tan, often flecked with brown; inside, white to purplish brown, glossy; platform whitish.

Habitat

Common in shallow water, often adhering to each other in clusters or to other shells.

Range

Canada to Tex.

LITTLE BOAT SHELL

Crepidula convexa **97:3**

Description

Size, ¼–½ in. (0.6–1.3 cm). Shell varies from sturdy to fragile, sometimes lightly wrinkled; small muscle scar under right outer corner or deck. Similar to Common Boat Shell but much smaller and has a straight edge to the platform. Color reddish to purple brown, sometimes spotted; inside including platform, pale bluish to chestnut.

Habitat

Attached to stones, shells, and seaweeds.

Range

Mass. to W. Indies.

COFFEE BEAN SHELLS

Family Eratoidae

COFFEE BEAN TRIVIA

Trivia pediculus **97:9**

Description

Size, ½ in. (1.3 cm). Shell small, broadly oval, brownish pink to rosy pink with three pairs of irregular brown spots on back. Sculpture consists of strong transverse ribs on back and base, with a depression in the center of back.

Habitat

Intertidal to 25 fathoms (45.7 m).

Range

N.C. to Brazil.

FOUR-SPOTTED TRIVIA

Trivia quadripunctata **97:13**

Description

Size, ⅛–¼ in. (0.3–0.6 cm). Shell small, oval, pink; four small irregular brown spots on back; fine riblets crossing back, narrow shallow furrow in center of back.

Habitat

In shallow water.

Range

S. Fla. through W. Indies.

CONCHS
Family Strombidae

FLORIDA FIGHTING CONCH
Strombus alatus **97:10**

Description
Size, 3–4 in. (7.6–10.2 cm). Shell reddish brown or yellowish brown, spire usually lighter in color, inner lip darker. Last whorl has pointed knobs on shoulder; outer lip winglike with deep notch on lower portion.
Habitat
In sand in shallow water.
Range
N.C. to Fla. and Tex.

MOON SHELLS
Family Naticidae

These carnivorous snails have rounded shells, a sharp-edged aperture, and a large foot which in some species covers the whole shell when extended. The sand dollars so often seen on beaches are the egg cases of these handsome and ferocious snails.

SHARK EYE
Polinices duplicatus **97:5**

Description
Size, 1-2½ in. (2.5–6.4 cm). Shell generally broader than high, giving a flattened appearance; 4–5 whorls. Outside chestnut-brown tinged with blue, sometimes slate-gray or tan; inside brownish. Brown lobe wholly or partially covering umbilicus.
Habitat
Mud and sand flats at low tide.
Remarks
The thin, brown operculum of this species (and of the Common Moon Shell) may be known as the "dead sailor's toenail."
Range
Cape Cod to the Gulf of Mexico.

COMMON MOON SHELL
Lunatia heros **97:6**

Description
Size, 3–4 in. (7.6–10.2 cm). The largest moon shell in the range. Shell dirty white to ashy brown, 4–5 whorls, body whorl broadly expanded behind; aperture large, oval, glossy white to brownish; umbilicus large, round, extending almost to top of shell; periostracum yellowish brown, thin.
Habitat
Moist sand at low tide.
Range
Gulf of St. Lawrence to off N.C.

COLORFUL MOON SHELL

Natica canrena **97:8**

Description
Size, 1–2½ in. (2.5–6.4 cm). Shell smooth except for wrinkles near suture, with wavy axial brown lines and four rows of small squarish or arrow-shaped brown spots on spiral white bands usually on a light brown background. Umbilicus rather large. Calcareous operculum with spiral grooves.
Habitat
In sand in shallow water.
Range
N.C. through W. Indies to Brazil.

MINIATURE NATICA

Natica pusilla **97:7**

Description
Size, ¼–⅓ in. (0.6–0.9 cm). Shell glossy, yellow-white, occasionally with faint bands; oval in shape, and often with an open chink next to the umbilical callus.
Habitat
2–18 fathoms (3.7–32.9 m).
Range
Cape Cod to Fla., the Gulf states, and W. Indies.

COMMON BABY'S EAR

Sinum perspectivum **97:11**

Description
Size, 1–1½ in. (2.5–3.8 cm); height less than ½ in. (1.3 cm). The flattest moon shell in the range. Shell white; 3–4 whorls with many spiral lines and a few growth lines. Aperture large, wide, flaring; inside glossy; no umbilicus; periostracum yellowish, thin.
Habitat
Sandy beaches and shallow water.
Range
Va. to W. Indies.

COWRIES

Family Cypraeidae

This is a large family. The shells are usually shiny, brightly colored, and oval. Generally, the last whorl envelops most of the spire, and the opening is long and narrow with teeth on both lips. It is a favorite with collectors.

DEER COWRY

Cypraea cervus **97:14**

Description
Size, 3–7 in. (7.6–17.8 cm). A relatively large shell, brown with white spots on the back. It is large, inflated, with numerous small spots.
Habitat
In water at or near low tide.
Range
Off N.C. to Bermuda and Cuba.

GRAY COWRY
Cypraea cinerea **97:12**

Description
Size, ¾–1½ in. (1.9–3.8 cm). Shell broadly ovate; color pinkish brown when fresh, yellowish brown when faded, often with small black spots, or showing obscure pale transverse bands; base creamy white.
Habitat
In shallow water under rocks on reefs.
Range
N.C. through W. Indies to Brazil.

EGG SHELLS
Family Ovulidae

SINGLE-TOOTHED SIMNIA
Neosimnia uniplicata **98:2**

Description
Size, ½–¾ in. (1.3–1.9 cm). A long, slender, cylindrical shell with a narrow, slitlike aperture. Shell smooth, glossy; 1 twisted columellar fold, which folds under upon itself at the opening. Color pink or purplish, sometimes yellowish white.
Habitat
In shallow water, generally attached to sea fans and sea whips.
Habits
Color adapts to color of host.
Range
Va. through W. Indies to Brazil.

FLAMINGO TONGUE
Cyphoma gibbosum **98:1**

Description
Size, ¾–1¾ in. (1.9–4.5 cm). Shell elongate, glossy, somewhat flattened; strong transverse ridge in middle of back; aperture without teeth on lips. Color pinkish or yellowish orange, white area in center of back. The living animal pale pinkish with numerous black squarish rings.
Habitat
On sea fans and gorgonians in shallow to moderately deep water.
Range
N.C. to Brazil.

HELMET SHELLS
Family Cassidae

SCOTCH BONNET
Phalium granulatum **98:6**

Description
Size, 1½–3 in. (3.8–7.6 cm). Shell usually with numerous grooves which may be weakly beaded on body whorl, but sometimes smooth; outer lip often thickened with many teeth on inner side; inner lip with a white shieldlike callus which is usually more or less granulose. Color pinkish yellow to whitish with irregular pale brown spots spirally arranged.

Habitat
In shallow to moderately deep water.
Range
N.C. to Brazil.

TRITON SHELLS
Family Cymatiidae

HAIRY TRITON
Cymatium pileare **98:8**

Description
Size, 1½–4 in. (3.8–10.2 cm). Shell elongate-ovate; moderately high, broad spire; whorls sculptured with weakly beaded spiral cords of irregular size; usually has a periostracum with flattened hairlike projections; varices prominent. Aperture pale brown with inner lip dark brown between the white teeth.
Habitat
Common in shallow water.
Range
S.C. to Brazil.

WHITE-MOUTHED TRITON
Cymatium muricinum **98:7**

Description
Size, 2 in. (5.1 cm). Shell strong, with 5 whorls, sharp apex. Color spotty, may be banded with brown. Opening large, notched at upper end; canal long and open. Surface ribbed.
Habitat
Common in shallow water.
Range
S. Fla. to W. Indies.

FROG SHELLS
Family Bursidae

The frog shells are ovate, laterally somewhat compressed, and bear two rows of continuous varices, one row on each side.

GRANULAR FROG SHELL
Bursa granularis **98:9**

Description
Size, ¾–2 in. (1.9–5.1 cm). Shell flattened laterally, sides marked by varices on each whorl; whorls sculptured by spiral cords of varying size which are adorned with uniformly sized beads and nodules. Color brownish with the larger nodes appearing white.
Habitat
In shallow water.
Range
S. Fla. to Brazil.

New Stomach-footed Gastropods
Order Neogastropoda

ROCK SHELLS
Family Muricidae

These shells are strong and thick, usually spiny, and with three varices per whorl. The snails are carnivorous and active and prefer shallow rocky or pebbly bottoms.

APPLE MUREX
Phyllonotus pomum **98:4**

Description
Size, 2–4½ in. (5.1–11.4 cm). Shell stout, grayish or whitish to yellowish white, often irregularly spotted with brown; 3 nodose varices per whorl, 1–3 axial ribs between the varices, all crossed by spiral cords of varying strength; siphonal canal narrow, almost covered by a broad rim, giving strength; inner lip with broad, marginate callus; outer lip usually spotted with brown within.
Habitat
Common in shallow water.
Range
N.C. to Brazil.

LACE MUREX
Chicoreus florifer dilectus **98:3**

Description
Size, 1–3 in. (2.5–7.6 cm). Shell light brown or whitish with 3 stout and strongly and densely frondose varices per whorl; varix at outer lip reaches to end of rather long siphonal canal; strong knob between the varices; aperture small and oval.
Habitat
Common in shallow water.
Range
S.C. to s. Fla.

GIANT EASTERN MUREX
Muricanthus fulvescens **98:5**

Description
Length, 5–7 in. (12.7–17.8 cm). Shell stout, solid, white to grayish; body whorl broad with moderately long siphonal canal and 7–9 varices with separated stout triangular spines.
Habitat
Common seasonally in 5–40 fathoms (9.1–73.2 m).
Range
N.C. to Fla. and Tex.

THICK-LIPPED DRILL
Eupleura caudata **98:12**

Description
Size, ½–1 in. (1.3–2.5 cm). The rock shell with the most prominent ridges and knobs. Shell has 5–6 whorls, flattened at shoulders, with

9–11 raised ridges; the 2 diametrically opposite ridges have knobs. Color red-tinged to grayish white; outer lip thick, with 6 small teeth; edge of lip often bluish; canal straight, narrow.

Habitat
Rocks at low tide.

Remarks
An enemy of the oyster.

Range
Cape Cod to Fla.

FLORIDA DRILL

Calotrophon ostrearum **98:10**

Description
Size, ¾–1 in. (1.9–2.5 cm). Shell grayish to grayish brown in color with strongly angulate ribs crossed by spiral cords. Aperture pale yellowish orange to purplish brown, with outer lip with fine incised lines within.

Habitat
Common in shallow water to 30 fathoms (54.9 m).

Range
W. coast of Fla.

OYSTER DRILL

Urosalpinx cinerea **98:11**

Description
Size, ½–1 in. (1.3–2.5 cm). Shell dirty gray or yellowish, often with brownish spiral bands; 5–6 whorls with many raised lines and wavy folds. Opening oval, purple to tan; lip slightly thickened, sharp, sometimes with 2–6 small teeth; canal short.

Habitat
Oyster beds; very common.

Remarks
An enemy of the oyster.

Range
N.S. to Fla.

ROCK PURPLE

Nucella lapillus **98:13**

Description
Size, 1–2 in. (2.5–5.1 cm). The most variously colored rock shell in the range. Spire short, sharp; 5 whorls; body whorl large, furrowed, wrinkled; spines and varices not prominent. Color ranges from dull white to yellowish orange and purplish brown, with or without dark spiral bands; aperture oval, lip arched.

Habitat
Crevices in rocks between tide levels.

Range
Labrador to N.Y.

FLORIDA ROCK SHELL

Thais haemastoma floridana **98:14**

Description
Size, 2–3 in. (5.1–7.6 cm). Shell stout with angulate whorls, sometimes knobbed on the angle, usually rather finely spirally grooved or almost smooth. Color grayish to yellowish with

irregular markings of brown, the spiral threads often whitish, aperture orange-pink.
Habitat
Common in shallow water.
Food
Bivalves and barnacles.
Range
N.C. to Brazil.

DOVE SHELLS
Family Columbellidae

These small shells are somewhat long and narrow, colorful and glossy, with rather narrow apertures and thickened outer lips.

MOTTLED DOVE SHELL
Columbella mercatoria **99:6**

Description
Size, ½–¾ in. (1.3–1.9 cm). Shell solid, broad, with shallow spiral grooves; outer lip thick with strong teeth on inner edge. Color variable; whitish, yellow, or orange, variously spotted and blotched with brown or white.
Habitat
Common in shallow water under rocks.
Range
E. Fla. to Brazil.

GREEDY DOVE SHELL
Anachis avara **99:8**

Description
Size, ⅜–½ in. (0.95–1.3 cm). Shell with 6–7 whorls; upper half of body whorl has 12 smooth ribs interrupting the revolving lines found on entire shell. Color straw, sometimes dull gray, with irregularly placed large white spots; aperture small, oval.
Habitat
Low tide; common.
Range
N.J. to Tex.

WHELKS
Family Buccinidae

The shells of this northern carnivorous family are large and thick and have few whorls. They are usually pear-shaped and have a pointed spire. The aperture is large.

WAVED WHELK
Buccinum undatum **99:4**

Description
Size, 2–4 in. (5.1–10.2 cm). Shell solid; about 6 whorls form a sharp spire with 12 vertical ribs on each, becoming weaker on the body whorl. Many raised revolving lines, together with the ribs, give the shell a wavy appearance. Color gray to reddish or yellowish brown, aperture white to yellow. Aperture oval, extends to ½ length of shell; lip sharp, canal short.

Habitat
Offshore to several fathoms.
Food
A scavenger.
Remarks
The egg capsules of this snail once were used for soap by sailors, who called them "sea wash balls." The snail is edible.
Range
Arctic seas to N.J.

STIMPSON'S WHELK
Colus stimpsoni **99:14**

Description
Size, 3–5 in. (7.6–12.7 cm). The most spindle-shaped whelk in the range. Shell gray to chalk white; 6–8 whorls with many slightly raised spiral lines; aperture oval, about half length of shell; lip not flaring; periostracum thin, grayish green.
Habitat
1–640 fathoms (1.8–1170 m).
Range
Labrador to off N.C.

BROWN-CORDED NEPTUNE
Neptunea decemcostata **99:10**

Description
Size, 3–4½ in. (7.6–10.2 cm). Shell heavy, pointed, grayish white; 6–7 whorls; 7–10 red-brown spiral cords on body whorl, 2–3 on upper whorls; aperture white, canal short.
Habitat
Cold water offshore.
Range
N.S. to Mass.

LARGE WHELKS
Family Melongenidae

CHANNELED WHELK
Busycon canaliculatum **99:2**

Description
Size, 5–7½ in. (12.7–19.1 cm). Has 5–6 whorls; is distinguished by deep channel at the spiral seam of each whorl, forming a terrace. Shell pear-shaped; body whorl large, narrowing below, ending in a long straight canal. Outside creamy gray, inside yellowish; periostracum thick, gray, velvety.
Habitat
Common in sandy shallow water.
Food
Clams, oysters.
Reproduction
Egg ribbon contains capsules with sharp edges; often found on beach.
Remarks
Edible.
Range
Cape Cod to St. Augustine.

KNOBBED WHELK

Busycon carica **99:3**

Description
Size, 4–9 in. (10.2–22.9 cm). Shell pear-shaped, usually right-handed; spire low, knobbed; 6 whorls; large, broad body whorl has blunt knobs on shoulder. Color gray to yellowish, young with purplish streaks; aperture oval, yellow or orange-red within; outer lip sharp, inner lip arched and twisted; canal long, open.
Habitat
Common in shallow water with pebbly bottom.
Food
Clams, oysters.
Reproduction
Egg ribbon contains capsules with flat edges; often found on beach.
Remarks
Edible.
Range
Cape Cod to cen. Fla.

LIGHTNING WHELK

Busycon contrarium **99:1**

Description
Size, 4–16 in. (10.2–40.6 cm). Shell left-handed; short, triangular knobs at shoulder; long, tapering siphonal canal. Color grayish white with rather widely spaced axial brown lines and spirally arranged large splotches of pale brown.
Habitat
Rather abundant in shallow to moderately deep waters.
Range
N.J. to Fla. and Gulf of Mexico.

CROWN CONCH

Melongena corona **99:11**

Description
Size, 2–5¼ in. (5.1–13.3 cm). Shell with 1, sometimes 2, rows of hollow spines below the flattened shoulder, and a smaller row of spines near base whorl with more or less obscure spiral threads. Color whitish with dark brownish spiral bands of varying width, the surface covered when fresh with a yellowish brown periostracum.
Habitat
Common in shallow water, especially on shores of Gulf of Mexico.
Range
Fla. and Gulf Coast.

DOG WHELKS

Family Nassariidae

These scavenging snails have small, plump shells with pointed spires, an oval aperture, a short notchlike canal, and usually a distinct columellar callus. They inhabit all seas.

MOTTLED DOG WHELK

Nassarius vibex **99:9**

Description
Size, ½ in. (1.3 cm). Shell heavy; about 6 whorls with many folds, beaded ridges, and distinct spiral bands; body whorl large, with 12

folds, 10 lines. Color grayish brown to white, often mottled or with broken bands of dark brown; aperture notched at both ends; outer lip thick, toothed; heavy callus on inner lip. Shell shorter, chunkier than in New England Dog Whelk.

Habitat

Common on sandy shores, mud flats.

Range

Cape Cod to W. Indies.

NEW ENGLAND DOG WHELK

Nassarius trivittatus **99:7**

Description

Size, ½–¾ in. (1.3–1.9 cm). Shell with 8–9 whorls, slightly flattened at shoulder and lightly grooved; 4–5 rows of beading on whorls; spire sharp. Color gray to yellowish white, sometimes with a few brown bands; aperture oval; outer lip thin, scalloped. Shell larger, taller than in Mottled Dog Whelk.

Habitat

Common between tides to 45 fathoms (82.3 m).

Range

N.S. to S.C.

MUD SNAIL

Nassarius obsoletus **99:5**

Description

Size, ¾–1 in. (1.9–2.5 cm). The commonest and the darkest dog whelk in the range. Shell has 6 whorls with uneven lines crossed by light growth lines and oblique fold lines toward base; spire usually worn away. Color dark brown to purplish black; aperture oval; outer lip thin; inner lip arched, folded.

Habitat

Mud flats.

Remarks

Their keen sense of smell makes these snails move actively after their food. In so doing they leave groovelike trails in the mud, often seen by the bay-shore beachcomber.

Range

Gulf of St. Lawrence to ne. Fla.

TULIP SHELLS AND HORSE CONCHS

Family Fasciolariidae

TULIP SHELL

Fasciolaria tulipa **99:13**

Description

Size, 3–8 in. (7.6–20.3 cm). Shell almost smooth or with obscure spiral threads but always with a few strong, somewhat crinkled spiral threads below the suture. Color whitish, grayish, or pinkish white with irregular orange, brown, or greenish brown spots and with many interrupted spiral chestnut lines; sometimes shell is wholly suffused with orange or brown.

Habitat

Common intertidally in bays and estuaries.

Range

N.C. to Brazil.

FLORIDA BANDED TULIP
Fasciolaria lilium hunteria **99:12**

Description
Size, 2½–3 in. (6.4–7.6 cm). Shell yellowish white or grayish with 5–6 red-brown spiral lines on last whorl, and with irregular orange or bluish gray axial flammules.
Habitat
Common in quiet shallow waters.
Range
N.C. to Fla. and Ala.

FLORIDA HORSE CONCH
Pleuroploca gigantea **99:15**

Description
Size, 1–2 ft. (0.3–0.6 m). Shell with high spire and moderately long siphonal canal; strong nodose ribs on whorls, crossed by rather strong spiral cords. Color whitish to pale reddish orange, usually covered by a thick periostracum which flakes off when dry.
Habitat
On sandy bottom in 3–20 ft. (0.9–6.1 m).
Range
N.C. to Fla. and Mexico.

OLIVE SHELLS
Family Olividae

The shells of this family are more or less cylindrical in shape, since the much-inflated body whorl tends to conceal all the earlier whorls. Widely distributed in warm seas, the shells are often brightly colored and usually smoothly polished.

LETTERED OLIVE
Oliva sayana **100:1**

Description
Size, 2–2½ in. (5.1–6.4 cm). Shell glossy, elongate, sides moderately convex. Color gray to grayish yellow with irregular brownish zigzag or tentlike axial markings of light and dark brown, the latter usually occurring in transverse bands.
Habitat
In sand in shallow water.
Range
N.C. to Brazil.

MINUTE DWARF OLIVE
Olivella mutica **100:2**

Description
Size, ¼–⅝ in. (0.6–1.6 cm). Shell glossy, variously banded with gray and brown, sometimes with irregular spots below suture or completely brown. Parietal and columellar callus rather heavy, extending above top of aperture to next whorl; lower part of columella with obliquely entering folds on a raised callus.
Habitat
In sand or mud in shallow water.
Range
N.C. to s. Fla. and Bahamas.

RICE DWARF OLIVE
Olivella floralia **100:3**

Description
Size, ¼–½ in. (0.6–1.3 cm). Shell slender, glossy, white; spire often colored yellow, grayish, brown, or rosy; body whorl sometimes mottled with brown. Parietal callus thin, extending beyond the aperture.
Habitat
Common in shallow water.
Range
N.C. through Fla. and W. Indies to Brazil.

MITER SHELLS
Family Mitridae

BEADED MITER
Mitra nodulosa **100:8**

Description
Size, ¾–1½ in. (1.9–3.8 cm). Shell rather elongate, with narrow axial ribs crossed by spiral grooves resulting in a beaded sculpture; columella with 4 folds, the upper one strongest. Color yellowish to orange-brown, often with the top of the whorls whitish.
Habitat
Under rocks at or below low tide.
Range
N.C. to Brazil.

VOLUTE SHELLS
Family Volutidae

JUNONIA
Scaphella junonia **100:14**

Description
Size, 5–6 in. (12.7–15.2 cm). Shell solid, smooth, whitish to pinkish with spiral rows of squarish dark brown spots; 4 rather strong folds on columella.
Habitat
1–30 fathoms (1.8–54.9 m).
Remarks
A striking shell, much prized by Florida collectors and rarely washed ashore. There are several varieties differing in coloring and spotting.
Range
N.C. to Fla., Gulf states, and Mexico.

NUTMEG SHELLS
Family Cancellariidae

COMMON NUTMEG
Cancellaria reticulata **100:7**

Description
Size, 1–1¾ in. (2.5–4.5 cm). Solid, ovate shell, sculptured by axial and spiral threads giving a reticulate appearance; strong fold on

columella, with a weaker one below it. Color grayish to yellowish white irregularly mottled with brown.

Habitat
Common in sand in shallow water.

Range
N.C. to Brazil.

MARGINELLA SHELLS
Family Marginellidae

COMMON ATLANTIC MARGINELLA
Prunum apicinum **100:9**

Description
Size, ¼–⅓ in. (0.6–0.9 cm). Shell glossy, smooth, subtriangular; rather thick, smooth outer lip, 4 folds at base of inner lip. Color yellowish or grayish, with usually 3 obscure dark bands on body whorl and 2 or 3 small orange-brown spots on outer margin of outer lip.

Habitat
Common in shallow water.

Range
N.C. to W. Indies.

CONE SHELLS
Family Conidae

This is a large family of conical tropical shells with a generally narrow aperture and thin outer lip. Cone shells are carnivorous and are provided with retractile harpoonlike teeth that can be venomous in some species.

ALPHABET CONE
Conus spurius atlanticus **100:12**

Description
Size, 2–3 in. (5.1–7.6 cm). Shell smooth except for lines of growth. Color whitish or yellowish with spiral rows of irregular squarish brownish orange blotches.

Habitat
Common in shallow water.

Range
Fla. and Gulf of Mexico.

JASPER CONE
Conus jaspideus **100:13**

Description
Size, ½–¾ in. (1.3–1.9 cm). Spire of shell conically elevated, last whorl sometimes smooth but usually with incised lines, and some forms (i.e., form *verrucosus*) with spiral rows of nodules. Color variable, generally mottled with reddish brown or yellowish orange, occasionally pinkish.

Habitat
In shallow to moderately deep water, generally under stones and coral.

Range
S. Fla. to Brazil.

AUGER SHELLS

Family Terebridae

Members of this family typically are very long and slender and have many whorls.

ATLANTIC AUGER

Terebra dislocata **100:4**

Description
Size, 1½–2 in. (3.8–5.1 cm). Shell long and slender; white, pinkish, or brown. Small aperture; lip notched below. Has 12–15 whorls, each divided by a deep spiral groove, giving the impression of double this number of whorls; some 25 ribs per whorl; columella twisted.
Habitat
Sandy bottoms in shallow water.
Range
N.C. to W. Indies.

SALLÉ'S AUGER

Impages salleana **100:5**

Description
Size, 1–1⅓ in. (2.5–3.4 cm). A slender, pointed, straight-sided shell; strong axial folds on whorls below the suture. Color bluish or brownish gray, white band below suture, dark brown spots between the folds.
Habitat
Burrowing in intertidal sand.
Range
Fla. to Mexico.

Bubble Shells and Sea Hares

Order Cephalaspidea

In this order the shell may or may not be present. If present, it is in a reduced condition and becomes thinner as the size of the snail increases and envelops more of the shell.

SMALL BUBBLE SHELLS

Family Haminoeidae

These slugs have shells which are nearly cylindrical, and the large body whorl almost engulfs the spire. The spiral line is channeled and forms a continuous groove. The shell gives the impression of having been turned on a lathe, hence its other name of "lathe shell."

EASTERN PAPER BUBBLE

Haminoea solitaria **100:6**

Description
Size, ½ in. (1.3 cm). Shell white or yellowish, translucent, oval; faint, fine growth lines and revolving ridges.
Habitat
In shallow water with muddy or sandy bottoms.
Range
Cape Cod to N.C.

Notch-banded Gastropods
Order Pyramidellida

PYRAMS
Family Pyramidellidae

Members of this family of tiny augerlike shells are parasites on worms or bivalves. They hold on by suction, pierce the host's body with a tiny dagger, and suck its juices.

INTERRUPTED TURBONILLE
Turbonilla interrupta **100:10**

Description
Size, ¼ in. (0.6 cm). Tall, slender, with 8 broad whorls and many ribs; color pale yellow, luster waxy; aperture roundish.
Habitat
2–107 fathoms (3.7–195.7 m).
Range
Maine to W. Indies.

IMPRESSED ODOSTOME
Odostomia impressa **100:11**

Description
Size, ¼ in. (0.6 cm). Shell elongate, pear-shaped, with about 8 whorls carrying flattened revolving ridges; body whorl surrounded by ridges; color milky white; aperture oval.
Habitat
Shallow water.
Range
Mass. Bay to Gulf of Mexico.

TUSK SHELLS
Class Scaphopoda

This small class of sluggish univalves is distinguished by a single hollow tusklike shell with an opening at each end. The animals live in the sand and feed on minute marine organisms in the sea water, which is sucked in through the small rear end and expelled through the front. Tiny filaments surrounding the internal mouth extract this food from the water. Tusk shells have an extensible wedge-shaped foot. They breathe through the mantle, not through the gills, and they live in clean water at various depths.

TUSK SHELLS
Family Dentaliidae

The shells of this family are elongate, curved somewhat like a miniature elephant's tusk, open at both ends with the greatest diameter at the aperture.

STIMPSON'S TUSK

Dentalium entale stimpsoni **101:3**

Description
Size, 1–2 in. (2.5–5.1 cm). Shell long, slender, tubular, and moderately curved; round in cross section; top usually worn. Color ivory-white with buffy or tan tones.
Habitat
Common in 8–1200 fathoms (14.6–2,194.6 m).
Range
N.S. to Cape Cod.

IVORY TUSK

Dentalium eboreum **101:4**

Description
Size, 1–2½ in. (2.5–6.4 cm). Shell slender, glossy, gently curved, smooth except for concentric growth lines and obscure microscopic longitudinal ridges at apex, which has a narrow slit; color whitish.
Habitat
In sand in moderately deep water.
Range
N.C. to Tex. and W. Indies.

CLAMS

Class Bivalvia

There are worldwide some 12,000 species of marine clams. These are bivalves; that is, they have two shells, or valves, which are joined at the hinge by a ligament and held together by one or two strong muscles. Except in a short larval stage, most are sedentary throughout their lives. Some, however, have a fleshy and extensible foot and can move about, and the scallops can swim by clapping their valves together.

Clams breathe both through their gills and through their mantle, which, as in the snails, secretes the calcium carbonate which makes the shell. Growth lines on the shell show seasons of relative quiet, such as winter. Clams, in general, possess no sense of taste or smell and are apparently limited to identifying edible bits of food and to closing their valves if a starfish goes by. They have no head; three somewhat enlarged sets of ganglia take the place of a brain. Some species, however, such as scallops, have little eyespots and can see.

Most clams are vegetarians. They live by extracting minute particles of food from the sea water that passes over their gills or is pumped over them by their siphons. These clams are called suspension feeders, as they get their food from particles suspended in the water. Others are deposit feeders and suck up food from the mud with their siphons. Certain clams and oysters are the source of pearls, but few of commercial importance are produced by the species in the East.

Awning Clams

Order Solemyoida

VEILED CLAMS

Family Solemyacidae

These clams are primitive in form and are fairly uncommon in the eastern range. Their two valves are equal in size, somewhat long and fragile, and are covered by a shiny, horny brown periostracum which extends beyond the shell. There is a gap between the valves both in front and behind.

ATLANTIC AWNING CLAM

Solemya velum **101:2**

Description
Size, ½–1 in. (1.3–2.5 cm). Shell oblong, delicate, smooth, yellowish brown; 15 slightly indented wide radiating lines; inside grayish blue. Periostracum hangs in scallops beyond edges of valves.
Habitat
Mud and sand beyond low water.
Range
N.S. to Fla.

Nut Clams

Order Nuculoida

NUT SHELLS

Family Nuculidae

These are small, three-cornered or ovate shells with pearly interiors, finely denticulated ventral margins, and a row of teeth on each side of the beak cavity but no ligamental pit between them.

ATLANTIC NUT CLAM

Nucula proxima **101:7**

Description
Size, ⅓ in. (0.9 cm). Shell whitish, inflated, thin, translucent; inside grayish; periostracum light olive.
Habitat
Mud just offshore.
Range
N.S. to Tex.

YOLDIAS

Family Nuculanidae

This family has small, oval valves that are equal in size and have teeth at the hinge, with a pit for the ligament. The shell is usually prominently angled behind and rounded in front.

POINTED NUT CLAM
Nuculana acuta **101:6**

Description
Size, ¼–⅜ in. (0.6–0.95 cm). Shell white, rounded in front, sharply pointed behind, strongly and evenly grooved; many teeth at hinge, pit central; periostracum thin, yellowish to brown.
Habitat
Common in shallow waters with sandy bottoms.
Range
Cape Cod to W. Indies.

FILE YOLDIA
Yoldia limatula **101:5**

Description
Size, 1–2½ in. (2.5–6.4 cm). Shell shiny green to light brown, faintly lined; long, oval, narrowing toward back; 20–22 prominent teeth in filelike arrangement; umbo nearly central; inside bluish white.
Habitat
Common below low-tide mark.
Habits
Active; can swim and jump.
Range
Maine to Cape May.

Arks and Bittersweets
Order Arcoida

ARK SHELLS
Family Arcidae

Most arks are found in warm seas, but some are found in cooler waters. The shells are somewhat box-shaped and heavily ribbed, with a narrow hinge line bearing many teeth on both sides. The umbones are toward the back of the shell, which usually carries a heavy periostracum.

TURKEY WING
Arca zebra **101:12**

Description
Size, 2–3¼ in. (5.1–8.3 cm). Shell with a broad ligamental area between the beaks; ribs of varying widths; surface covered with a coarse brown matted periostracum, generally more or less worn off.
Habitat
Attached to rocks and corals by byssus in shallow water.
Range
N.C. to Brazil.

MOSSY ARK
Arca imbricata **101:9**

Description
Size, 1½–2½ in. (3.8–6.4 cm). Shell smaller and relatively broader than Turkey Wing, with a fine beaded sculpture and a larger byssus opening; suffused with dark brown.

Habitat
Attached by byssus to underside of rocks.
Range
N.C. to Brazil.

TRANSVERSE ARK
Anadara transversa **101:10**

Description
Size, ½–1½ in. (1.3–3.8 cm). The smallest of the ark shells in the range. Shell oblong, dull white, with heavy gray-brown periostracum; 30–35 ribs on each valve; left valve overlaps right valve, beaks curve inward.
Habitat
Sandy bottoms offshore.
Range
Cape Cod to Tex.

BLOOD ARK
Anadara ovalis **101:8**

Description
Size, 1½–2½ in. (3.8–6.4 cm). One of the very few mollusks that have red blood. Shell white, thick, solid, oval; umbones knobby. Each valve has 26–35 broad radiating ribs separated by narrow grooves; lower part of valves covered with thick brown periostracum.
Habitat
Sandy bottoms in shallow water.
Range
Cape Cod to W. Indies.

HEAVY ARK
Noetia ponderosa **101:11**

Description
Size, 2–2½ in. (5.1–6.4 cm). Shell creamy white, heavy, inflated; valves equal in size, symmetrically curved with about 32 radiating ribs on each; prominent ridge from beak to hind margin with valve falling off almost straight behind. Umbones placed toward front of valves and bend backward; periostracum blackish, heavy.
Habitat
Sandy bottoms.
Remarks
These are often found as fossils on Wildwood and Cape May beaches in New Jersey. They come from an interglacial deposit laid down when the sea there was as warm as it is in Virginia today.
Range
Va. to Gulf of Mexico.

BITTERSWEET CLAMS
Family Glycymerididae

This family consists of a small group of heavy, usually orbicular, equivalve, porcellaneous shells, generally with a soft velvety periostracum; the beaks are incurved, the hinge heavy and with many small teeth, and the ligament external with grooves diverging from the area. The largest muscle scar is at the anterior end.

COMB BITTERSWEET
Glycymeris pectinata **101:1**

Description
Size, ½–1¼ in. (1.3–3.2 cm). Shell orbicular in outline, rather flattened, with numerous smooth, flattened radial ribs crossed by fine crowded concentric threads. Color grayish or yellowish gray, irregularly spotted with brown.
Habitat
Common in shallow water.
Range
N.C. to Brazil.

Mussels
Order Mytiloida

MUSSELS
Family Mytilidae

The shells of this family are long, narrow, and oval, with valves of equal size and shape and often with a pearly lining. The umbo is close to the front of the shell. The periostracum is heavy, dark, and sometimes hairy. Some species burrow; others attach themselves to rocks or pilings with a byssus, or set of threads, which they spin.

BLUE MUSSEL
Mytilus edulis **102:5**

Description
Size, 1–3 in. (2.5–7.6 cm). Shell long, a flattened oval, blue-black, often with pale violet patches; no ribs, but often with fine growth lines. Umbo at extreme top, 4 teeth; inside white and blue-violet; periostracum thin, shiny.
Habitat
Very common clustered on rocks and pilings at low tide.
Range
Arctic Ocean to S.C.

HOOKED MUSSEL
Ischadium recurvus **102:4**

Description
Size, 1–2½ in. (2.5–6.4 cm). Shell green-brown to gray-black; rather flat, wide, strongly hooked at front end; many wavy ribs, edge of shell at umbo has 3–4 small, longish teeth; inside pink to purplish brown with narrow grayish blue border.
Habitat
Clustered on rocks and shells in shallow water.
Range
Cape Cod to W. Indies.

PAPER MUSSEL
Amygdalum papyrium **102:3**

Description
Size, 1–1½ in. (2.5–3.8 cm). Shell smooth, long, oval, very fragile, shiny; green-blue to yellowish brown; inside white, shiny.

Habitat
Common in bays in mud at roots of grass.
Reproduction
This species builds a nest with its byssal threads.
Range
Md. to Tex.

HORSE MUSSEL
Modiolus modiolus **102:1**

Description
Size, 2–6 in. (5.1–15.2 cm). Shell coarse, oval, dark blue to mauve-white, with concentric lines; beaks slightly to one side of short, narrow front end. Hind end broad and round, inside white; periostracum thick, dark.
Habitat
Burrows in sand or rock crevices in cool, moderately deep water; common.
Range
Arctic seas to ne. Fla.

RIBBED MUSSEL
Geukensia demissus **102:2**

Description
Size, 2–4 in. (5.1–10.2 cm). Shell shiny, pinkish or black-brown to greenish yellow; coarse radiating ribs at back, becoming finer toward front; bottom margin straight in young, curved in old; old often have top outside of shell eroded away toward front. Umbo to one side of top; inside bluish white, often purplish white toward back; periostracum dark brown.
Habitat
Common in mud and sand at low tide.
Range
Gulf of St. Lawrence to S.C.

PEN SHELLS
Family Pinnidae

The members of this family are generally elongately triangular with rather thin shells, the pointed end embedded in sandy mud where it anchors itself to a hard object by means of a strong byssus composed of many fine threads.

FLESH PEN SHELL
Pinna carnea **102:10**

Description
Size, 4–11 in. (10.2–27.9 cm). Shell rather thin, more or less translucent, with 8–12 ribs which may or may not bear erect, hollow spines. Color light brown or pinkish to orange-brown.
Habitat
In shallow water in sandy mud.
Range
Fla. to Brazil.

STIFF PEN SHELL
Atrina rigida **102:8**

Description
Size, 5–11 in. (12.7–27.9 cm). Shell moderately stout with usually 10–25 radial ribs generally having rather long tubular spines. Color light brown to dark grayish brown.
Habitat
In shallow water.
Range
N.C. to W. Indies.

Scallops and Oysters
Order Pterioida

TREE OYSTERS
Family Isognomonidae

Some species in this family of flattened bivalves are commonly found attached by their byssus to the roots of mangrove trees, others to rocks and coral reefs.

FLAT TREE OYSTER
Isognomon alatus **102:6**

Description
Size, 2–3 in. (5.1–7.6 cm). Compressed shells, purplish brown to grayish, irregularly oval in outline.
Habitat
Attached in clumps by byssus to mangrove roots.
Range
Fla. to Brazil.

SCALLOPS
Family Pectinidae

The valves of scallops may or may not be equal. The shells have radiating ribs and scalloped edges; they lack teeth. There is an ear-shaped projection on either side of the umbones. Many of the animals are edible.

Scallops are one of the relatively few mollusks that swim. They have eyespots on the edge of their body, which give them some power of sight. On favored beaches or after violent storms, thousands of their shells will strew the strand. Almost no two are alike.

DEEP SEA SCALLOP
Placopecten magellanicus **103:9**

Description
Size, 5–8 in. (12.7–20.3 cm). Shell almost round; upper valve yellow to red-brown, slightly convex; lower valve flat, pinkish white. Ears symmetrical; valves gap at front and have many fine, raised threads and grooves, crossed by growth lines; inside shiny white.
Habitat
10–100 fathoms (18.3–182.9 m).
Range
Labrador to N.C.

LION'S PAW
Nodipecten nodosus **103:10**

Description
Size, 3–6 in. (7.6–15.2 cm). Shell rather stout, with about 9 broad, nodose ribs, which are finely sculptured with whitish radial threads or riblets and separated by deep interspaces; nodes often protuberant and hollow. Color orange or dark red, often irregularly spotted with white.
Habitat
Rather abundant in moderate depths.
Range
N.C. to Brazil.

BAY SCALLOP
Argopecten irradians **103:8**

Description
Size, 2–3 in. (5.1–7.6 cm). The well-known edible scallop. Shell varies greatly in color; white, gray, or bluish to yellowish or reddish brown, shading to deep violet at hinge; weathered shells slaty. Valves equal in size, each with 17–21 strongly rounded ribs with narrower grooves; ears slightly unequal and ridged; edge evenly scalloped; inside purplish white or mottled.
Habitat
Common among eelgrass in shallow water of bays.
Range
Cape Cod to N.J.

CALICO SCALLOP
Argopecten gibbus **103:7**

Description
Size, 1–2½ in. (2.5–6.4 cm). Shell smaller than Bay Scallop, more inflated, with ribs closer together and rather squarish in cross section. Lower valve usually white; upper valve variously and brightly colored in rose or red, or white with red or purplish mottlings.
Habitat
Common in shallow to moderately deep water.
Range
Md. to Brazil.

SPINY OYSTERS
Family Spondylidae

This is a family of usually rather large, thick-shelled, radially ribbed, and generally more or less spiny shells with a ball-and-socket hinge and a central internal ligament. The margin of the interior is often distinctly colored. The shells are always attached to the substrate by the lower valve.

SPINY OYSTER
Spondylus americanus **102:7**

Description
Size, 3–4 in. (7.6–10.2 cm). Shell often quite thick and heavy; many irregular and scabrous radial riblets; several larger ribs often with spines which may be long, slender, and flattened. Color white, reddish orange, or yellow.

Habitat
Attached to hard substrate in shallow to rather deep water.
Range
N.C. to Brazil.

CAT'S PAW OYSTERS
Family Plicatulidae

KITTEN'S PAW
Plicatula gibbosa **102:9**

Description
Size, ¾–1 in. (1.9–2.5 cm). Shell rather compressed with 6–9 strong, occasionally divergent ribs, resulting in a wavy margin; hinge with 2 strong interlocking teeth in each valve. Color whitish to gray with short, fine reddish lines on the ribs.
Habitat
Attached to rocks or coral in shallow to moderately deep water.
Range
N.C. to Brazil.

JINGLE SHELLS
Family Anomiidae

This family is characterized by the hole in the lower valve, which gives the anchoring threads an outlet to attach themselves to a rock, shell, or other support. The shells are fragile, roundish, waxy in luster, and have unequal-sized valves. The upper valve is left-handed and dome-shaped; the lower one with the hole is right-handed, smaller, and concave. The upper one has three muscle scars inside and is the "jingle shell" most commonly found washed up on shore.

COMMON JINGLE SHELL
Anomia simplex **103:4**

Description
Size, 1–2 in. (2.5–5.1 cm). Shell varies from silvery to yellow, red, or blue-black, often translucent; shape varies with habitat, usually unevenly rounded, thin, smooth; edges irregular, sometimes jagged; upper valve plump, lower valve flat, hole near top.
Habitat
Common; attached to shells, rocks, boats, in moderately deep water.
Range
Maine to W. Indies.

PRICKLY JINGLE SHELL
Anomia squamula

Description
Size, ½–¾ in. (1.3–1.9 cm). Shell opaque, yellowish white, unevenly oval. Upper valve rough, convex, with rows of small prickly scales. Lower valve flat, small hole near hinge; inside whitish.
Habitat
Attached to rocks, shells, and seaweed in cold water.
Range
N.S. to N.C.

OYSTERS
Family Ostreidae

Oyster shells are all asymmetrical and are greatly varied, depending largely upon the bottom to which they are affixed.

VIRGINIA OYSTER
Crassostrea virginica **103:5**

Description
Size, 2–6 in. (5.1–15.2 cm). Shell dingy gray, coarse-textured, thick, somewhat elongated and curved. Upper valve smaller and less rounded than lower; beaks long, curved, and grooved by growth of the ligament. Inside white; muscle scar almost central and deep purple.
Habitat
Attached to hard surfaces.
Reproduction
Up to 50,000,000 eggs are released by 1 animal at 1 spawning.
Remarks
Perhaps the most valuable commercial mollusk in the range; widely cultivated. "Blue Points" are round, "Lynnhavens" broad and long.
Range
Gulf of St. Lawrence to W. Indies.

COON OYSTER
Lopha frons **103:6**

Description
Size, 1–2½ in. (2.5–6.4 cm). Shell oval to elongate with numerous irregular ridged ribs. Has shelly projections or claspers for attaching to support.
Habitat
Attached to corals in shallow to moderately deep water.
Range
Fla. to Brazil.

Lucines, Clams, and Razors
Order Veneroida

CARDITAS
Family Carditidae

BROAD-RIBBED CARDITA
Carditamera floridana **104:2**

Description
Size, 1–1½ in. (2.5–3.8 cm). Shell elongate, solid, with about 20 strong, flattened, radial ribs; elongate beads on earlier part of ribs, becoming fine, crowded threads near margin. Color whitish gray with oblong reddish brown spots on ribs arranged in concentric rows.
Habitat
Common on sandy bottom, moderate depths.
Range
Fla. to Mexico.

LUCINES

Family Lucinidae

The equivalve shells of this family are orbicular, strong, and laterally compressed and have small but definite beaks. Most species are white.

PENNSYLVANIA LUCINE

Lucina pensylvanica **103:2**

Description
Size, 1–2 in. (2.5–5.1 cm). Shell ovate, rather inflated, stout; conspicuous, fine, concentric ridges; strong furrow running from beak to posterior ventral edge. Color white, occasionally orange.
Habitat
May be locally common in shallow water.
Range
N.C. to W. Indies.

GREAT WHITE LUCINE

Codakia orbicularis **103:3**

Description
Size, 1½–3¾ in. (3.8–9.5 cm). Shell slightly broader than high; stout, compressed, with numerous coarse radial riblets crossed by fine, concentric threads that make the riblets beaded. Color white inside and outside; occasionally tinged with yellow, often suffused with rose near hinge or along margins.
Habitat
Common in sandy, shallow water.
Range
Fla. to Brazil.

CROSSHATCHED LUCINE

Divaricella quadrisulcata **103:1**

Description
Size, ½–1 in. (1.3–2.5 cm). Shell white, glossy, solid, and circular; valves with 5 or more concentric grooves and many fine radial lines, which form a unique pattern running obliquely toward the end of each valve; right valve has 1 small tooth, left has 2; inside margin with fine scalloping.
Habitat
Common in 10–50 fathoms (18.3–91.4 m).
Range
Mass. to W. Indies.

JEWEL BOXES

Family Chamidae

These thick-shelled bivalves are attached by one valve to the substrate, possess a strong hinge with few large teeth, and are sculptured with frondose, concentric ridges or spiny radial ribs.

LEAFY JEWEL BOX

Chama macerophylla **104:3**

Description
Size, 1½–3 in. (3.8–7.6 cm). Shell sculptured with concentric rows of leafy ridges; ridges in specimens from quiet deep water with delicate scaly spines. Color may be white, yellow, orange, to purplish or combinations; internally often more or less suffused with purple.
Habitat
In shallow to deeper water.
Range
N.C. to Brazil.

ASTARTES

Family Astartidae

The shells of this family have prominent umbones which are almost central. They are small, usually brownish, triangular, and with well-developed cardinal teeth at the hinge. They all have conspicuous concentric grooves and growth lines.

BOREAL ASTARTE

Astarte borealis **104:1**

Description
Size, 1–2 in. (2.5–5.1 cm). Shell white, oval, only slightly raised, smooth, with low concentric growth lines; valves equal; inside smooth, margin not scalloped, periostracum brownish.
Habitat
Common offshore to 30 fathoms (54.9 m).
Range
Arctic seas to Mass. Bay.

COCKLES

Family Cardiidae

The shells of this family have valves of equal size, which often gape behind; the prominent beaks are almost central. The margins are toothed or scalloped; the hinge teeth are arched; the pallial line is wavy behind; and the general shape of the shell, viewed from the side, is that of a heart. These cockles are mobile and have no byssus.

GREAT HEART COCKLE

Dinocardium robustum **104:5**

Description
Size, 3–6 in. (7.6–15.2 cm). The largest and most colorful cockle in the range. Shell yellowish brown, streaked and spotted with chestnut and purplish brown; plump, abruptly flattening toward

the back; 30–35 ribs radiating from the high, rounded beak. Valve margins toothed, inside salmon-pink.
Habitat
Sandy bottoms at moderate depth.
Range
Va. to Mexico.

COMMON COCKLE
Trachycardium muricatum **104:4**

Description
Size, 1¼–2 in. (3.2–5.1 cm). Shell nearly circular in outline with many radial ribs, more or less spiny except at the center. Color whitish or yellowish; interior white or tinged with yellow.
Habitat
Common in shallow water.
Range
N.C. to Brazil.

VENUS CLAMS
Family Veneridae

These clams are distinguished by an unusual symmetry. The valves are equal, the teeth interlocking, the colors often arresting. The lunule is clear and deep. The shell is thick and strong, the pallial line is wavy, and the muscle scars are oval. The inside edge is often ridged or scalloped.

NORTHERN QUAHOG
Mercenaria mercenaria **106:8**

Description
Size, 3–5 in. (7.6–12.7 cm). Shell dingy white, thick, rounded; prominent concentric lines fairly widely spaced at the beaks, becoming very close toward margins. Valves have smooth central area; beaks are toward the short front; back long with an oval sweep; lunule ¾ as wide as long. Inside chalky white, violet muscle scars, inside margin lightly ridged.
Habitat
Mud or sand at low tide.
Remarks
This is the most common hard-shelled clam.
Range
Gulf of St. Lawrence to Gulf of Mexico.

DISK DOSINIA
Dosinia discus **106:4**

Description
Size, 2–3 in. (5.1–7.6 cm). The flattest Venus clam in the range; has finer and closer concentric lines than the other species. Shell snowy white, valves round, beaks shallow and small, hinge strong and thick; periostracum thin, yellowish.
Habitat
In sand, moderate depths.
Range
Va. to the Bahamas.

CROSS-BARRED CHIONE

Chione cancellata **106:3**

Description
Size, 1–1¾ in. (2.5–4.5 cm). Shell solid, rather stout, roughly triangular in outline, with widely spaced lamellar concentric ribs and rather regular axial riblets. Color whitish or grayish, generally maculated with reddish brown to brown, often in axial rays; interior often tinged with bluish purple.
Habitat
Common in shallow water.
Range
N.C. to Brazil.

SUNRAY SHELL

Macrocallista nimbosa **106:9**

Description
Size, 4–6 in. (10.2–15.2 cm). Shell elongately oval, rather stout, smooth, glossy; pale grayish yellow with grayish purple radial rays, generally more or less interrupted; surface covered by a thin brownish periostracum.
Habitat
Rather common in sandy areas in shallow water.
Range
N.C. to Fla. and Tex.

CHECKERBOARD CLAM

Macrocallista maculata **106:10**

Description
Size, 1½–3 in. (3.8–7.6 cm). Shell smaller, more broadly oval than in Sunray Shell and covered with irregular spots arranged in concentric and often radial patterns; otherwise similar.
Habitat
Moderately common in shallow water.
Range
N.C. to Brazil.

ROCK DWELLERS

Family Petricolidae

These bivalves are long, are rounded in front and narrowing behind, and have a weak hinge, which is almost toothless.

FALSE ANGEL WING

Petricola pholadiformis **106:1**

Description
Size, 2–2¼ in. (5.1–5.7 cm). Shell chalky white, fragile, rectangular. Umbones raised, placed close to short, sharply rounded front end. Valves have 2 teeth, many radiating ribs crowded and weak behind, strong and widely spaced in front; growth lines emphasized at intervals.
Habitat
Burrows in clay between tides.
Remarks
Usually found with the valves attached together in pairs.
Range
Gulf of St. Lawrence to Gulf of Mexico.

RAZOR CLAMS

Family Solenidae

The shells of this family are very long and narrow, and gape at each end. The valves are equal in size and shape. Distributed worldwide in the sandy bottoms of shallow coastal waters, all species are considered edible.

RIBBED POD

Siliqua costata **104:8**

Description
Size, 2–2½ in. (5.1–6.4 cm). Shell oblong, greenish purple, smooth, fragile, somewhat flat; beaks very small. Inside purplish white, glossy, with white raised rib from hinge to halfway across valve; periostracum olive-brown, iridescent, smooth.
Habitat
Common in sandy bottoms of shallow water.
Range
Gulf of St. Lawrence to N.J.

COMMON RAZOR

Ensis directus **104:10**

Description
Size, 6–10 in. (15.2–25.4 cm). Shell white, thin, long, narrow, slightly curved; hinge, with sharp plate of 1 tooth on 1 valve between other valve which has 2 teeth and a double plate. Ends sharp, squarish, and gaping, inside white, stained purple; periostracum dark greenish brown, thin, and suggesting varnish.
Habitat
Common in sand at low water.
Remarks
These active animals burrow rapidly in the sand if disturbed. If captured they will attempt to jump away with amazing speed.
Range
Labrador to S.C.

TELLINS

Family Tellinidae

These bivalves are usually polished and colorful. The valves are equal, rounded in front, but sharp and slightly folded behind. They are fairly flat, and the edges close evenly. Tellins are among the most beautiful shells.

LINED TELLIN

Tellina alternata **105:1**

Description
Size, 2–3 in. (5.1–7.6 cm). Shell compressed, rather solid, surface sculptured with evenly spaced grooves. Color whitish or yellowish or tinged with pink.
Habitat
In sand in shallow to moderately deep water.
Range
N.C. to Fla. and Tex.

NORTHERN TELLIN
Tellina agilis **105:3**

Description
Size, ⅓–¾ in. (0.9–1.9 cm). Shell iridescent white or rose, inside and out, fragile, flat, translucent, with many very fine growth lines; front rounded, back slopes to a rounded point. Valves have 2 cardinal teeth; glossy but not enameled; no radial grooves; pallial sinus large, rounded; ventral margin slightly curved.
Habitat
Sandy bottoms below tides.
Range
Gulf of St. Lawrence to N.C.

DeKAY'S TELLIN
Tellina versicolor **105:5**

Description
Size, ½ in. (1.3 cm). Shell white, opalescent, with pink rays widening toward edge. Pallial sinus close to muscle scar; ventral margin straight, not curved. Similar to Northern Tellin but slightly longer.
Habitat
In sand in shallow water to 50 ft. (15.2 m).
Range
N.Y. to W. Indies.

CHALKY TELLIN
Macoma calcarea **105:6**

Description
Size, 1½–2 in. (3.8–5.1 cm). Shell chalky white, thin, and with fine concentric lines; front evenly rounded, back narrow and slightly twisted; valves have 2 hinge teeth; periostracum olive-gray.
Habitat
Common in cold water, 5–40 fathoms (9.1–73.2 m).
Range
Greenland to Long Island.

BALTIC MACOMA
Macoma balthica **105:4**

Description
Size, ½–1½ in. (1.3–3.8 cm). Shell thin and white or rosy in a sandy location; thick, bluish red in mud; slightly triangular, rounded in front; back angular and with a rounded tip. Beaks nearly central, fairly prominent; inside white or rose. Remnants of thin grayish periostracum often seen on margins.
Habitat
Common between tides and up creeks and rivers.
Range
Arctic seas to off Ga.

TENTA MACOMA
Macoma tenta **105:2**

Description
Size, ½–¾ in. (1.3–1.9 cm). Shell shiny, thin, oval, elongate, pinkish white; many sharp, fine growth lines; front long, rounded; back narrow, twisted, and gaping. Inside white, tinged with yellow, with fine radiating lines; 2 teeth on right valve, 1 on left.
Habitat
Muddy or sandy bottoms in shallow water; common.
Range
Cape Cod to W. Indies.

WEDGE SHELLS
Family Donacidae

These small wedge-shaped clams are long and rounded at the front, short and straight in back. The beaks are toward the back. The edges of the valves are usually ridged.

SOUTHERN COQUINA
Donax variabilis **105:10**

Description
Size, ½–1 in. (1.3–2.5 cm). Color varies ranging from shades of white, yellow, pink, and blue to mauve, with darker rays sometimes crossed with bands of other colors as in a plaid. Front end of shell smooth, occasionally minutely lined; small radial lines toward the center, becoming larger toward blunt back end. Inside glossy, varying in color, with finely toothed edges.
Habitat
Sandy beaches between tides.
Range
Va. to Tex.

SANGUIN CLAMS
Family Psammobiidae

FALSE COQUINA
Heterodonax bimaculatus **105:8**

Description
Size, ½–1 in. (1.3–2.5 cm). Shell broadly oval, smooth except for fine growth lines; 2 small teeth in each valve. Color variable; pure white, purplish, or whitish, suffused with rose or orange, or with internal rose rays, or maculated with purplish gray.
Habitat
Common intertidally on sandy beaches.
Range
S. Fla. to W. Indies.

SEMELES

Family Semelidae

Chiefly warm-seas animals, the Semeles have rounded oval shells that are little inflated, with the posterior end characterized by relatively obscure folds.

WHITE SEMELE

Semele proficua **105:7**

Description
Size, ¾–1½ in. (1.9–3.8 cm). Shell subcircular, rather stout, with fine concentric ridges and minute radial striations. Color whitish to yellowish white, internally white or yellowish often spotted with purple or pink.
Habitat
Common offshore.
Range
N.C. to Brazil.

LONG SIPHON CLAMS

Family Solecurtidae

These clams are somewhat similar to the tellins but larger. The sides of the shell are more nearly parallel and are often marked with fine concentric lines.

PURPLISH RAZOR

Tagelus divisus **105:9**

Description
Size, 1½ in. (3.8 cm). Shell fragile, smooth, translucent, purplish; oblong, with round end; back narrower than front. Beaks nearly central; purplish rib stripe runs from beaks toward margin; 2 cardinal teeth in each valve; inside purplish, glossy; periostracum thin glossy brown.
Habitat
Sandy bottoms in shallow water.
Range
Cape Cod to the Caribbean.

BLACK CLAMS

Family Arcticidae

BLACK CLAM

Arctica islandica **106:5**

Description
Size, 3–5 in. (7.6–12.7 cm). Shell almost round, dull, chalky, smoothish, with low, concentric growth lines; inside white; periostracum thick, blackish.
Habitat
Sandy mud, 5–80 fathoms (9.1–146.3 m).
Range
Newfoundland to N.C.

SURF CLAMS

Family Mactridae

These clams have equal valves, usually tightly closed, but sometimes with a very small gap. The hinge has two cardinal teeth.

ATLANTIC SURF CLAM

Spisula solidissima **104:6**

Description
Size, 4–7 in. (10.2–17.8 cm). Shell yellowish white, thick, oval; smoothish with many growth lines. Beaks large, nearly central, hinge strong, ligament internal with large, fairly straight spoon-shaped pit, cardinal tooth small; lateral teeth long, with notched edges.
Habitat
Under sand in surf and farther out; abundant just offshore.
Remarks
This is probably the best-known single shell along the North Atlantic Coast. The shells are conspicuous and common, so the first-time visitor to the ocean often takes some home as souvenirs. The meat makes a good chowder. Since the animals are abundant just offshore, a real northeaster will pile them up on the beach.
Range
N.S. to S.C.

DWARF SURF CLAM

Mulinia lateralis **104:7**

Description
Size, ⅓–¾ in. (0.9–1.9 cm). Shell whitish, inflated, smoothish, triangular and convex, with concentric wrinkles and a radial ridge toward the back. Beaks prominent, nearly central, pointing forward; hinge strong; marginal teeth V-shaped; periostracum thin, yellowish brown.
Similarities
Resembles young Atlantic Surf Clam, which has a larger spoon at hinge and tiny saw teeth along the larger side hinge teeth.
Habitat
Abundant in sand in warm, shallow water.
Range
Maine to Tex.

WEDGE RANGIA

Rangia cuneata **104:9**

Description
Size, 1–2½ in. (2.5–6.4 cm). Shell heavy, stout, smooth; high, strong beaks; whitish but usually covered by a grayish brown periostracum; interior white, glossy.
Habitat
Common in mud in fresh to brackish water of coastal rivers.
Range
Md. to Tex.

Clams, Rock Borers, Piddocks, and Shipworms

Order Myoida

SOFT-SHELLED CLAMS

Family Myidae

A distinctive feature of these shells is the spoon-shaped tooth in the left valve which fits into a corresponding pit in the right valve. The valves are usually unequal and gape at both ends.

COMMON SOFT-SHELLED CLAM

Mya arenaria **106:7**

Description
Size, 1–6 in. (2.5–15.2 cm). Shell gray or chalky white, thick, wrinkled by growth lines; elliptical. Spoon-shaped tooth long, shallow; pallial sinus somewhat V-shaped; periostracum thin, gray to straw-colored.
Habitat
Common in shallow bottoms between tides.
Habits
Reveals presence by abrupt, vertical squirt of water from suddenly indrawn siphon.
Remarks
Highly edible, especially when steamed.
Range
Labrador to off N.C.

BASKET CLAMS

Family Corbulidae

Formerly the Family Aloididae, the shells are small, solid, inequivalve with one valve usually overlapping the other. Commonly ribbed centrally, the valves may gape slightly at the anterior end. Each valve has one upright conical tooth. They occur worldwide in temperate waters.

COMMON BASKET CLAM

Corbula contracta **106:6**

Description
Size, ½ in. (1.3 cm). Shell white, plump, irregularly oval with smooth, regular, concentric ridges. Valves unequal, rounded in front, pointed behind, lower back of right valve overlaps left. Angular fold from beak to margin; left valve has V-shaped notch in front of beak; hinge tooth upright, slender; large conical tooth fits into corresponding pit in the other valve; periostracum thin, brownish.
Habitat
Common in sandy or muddy shallows.
Range
Cape Cod to W. Indies.

ROCK BORERS

Family Hiatellidae

The nestling and burrowing habits of these clams cause irregularities and considerable variation in their shells. They are chalky, are toothless, and have an interrupted pallial line.

ARCTIC ROCK BORER

Hiatella arctica **106:2**

Description
Size, 1–3 in. (2.5–7.6 cm). Shell chalky white, oblong, oval, or distorted, with heavy, uneven growth lines. Valve margins almost parallel in front and behind, occasionally gaping behind; radial rib toward rear; beaks close somewhat behind top of valves; inside whitish; periostracum grayish, thin, flaky.
Habitat
Common, burrowing in mud or rock in cold water.
Range
Arctic seas to deep water in W. Indies.

PIDDOCKS

Family Pholadidae

These clams are able to bore through hard rocks. The long, narrowed shells have sharp toothlike ridges in front. They gape at both ends and lack both ligament and hinge teeth.

COMMON ANGEL WING

Cyrtopleura costata **107:9**

Description
Size, 4–8 in. (10.2–20.3 cm). Shell snowy white, rather fragile, widely gaping; valves wing-shaped, each with about 30 strong radial ribs crossed by growth lines, forming cuplike scales; beaks close to front with internal spoon-shaped brace; on the inside, sculpture is reversed; periostracum thin, gray.
Habitat
Burrows to depth of 12 in. (30.5 cm) in sandy mud.
Range
Mass. to W. Indies.

FALLEN ANGEL WING

Barnea truncata **107:2**

Description
Size, 2–2½ in. (5.1–6.4 cm). Shell white, delicate, oblong; triangular, scaly, and pointed in front; smoothish, bluntly cut off behind; both ends gaping. Growth lines prominent, thin; broken into scales at radial ribs, giving beaded appearance; long, shelly plate reinforces hinge. Beaks prominent, leaning toward back, siphons can extend 3 times length of shell.
Habitat
Burrows in clay, rock, or wood between tides.
Range
Mass. Bay to Fla.

GREAT PIDDOCK

Zirfaea crispata **107:5**

Description
Size, 1–2 in. (2.5–5.1 cm). Shell grayish, pointed in front, rounded behind; valves widely gaping at ends, touching at center. Surface equally divided by deep furrow from beak through middle of valves; front area has tooth ridges, entire valve has coarse concentric growth lines; inside whitish.
Habitat
Bores into soft rock in cold water.
Range
Newfoundland to N.J.

STRIATE WOOD PIDDOCK

Martesia striata **107:7**

Description
Size, ¾–1½ in. (1.9–3.8 cm). Shell elongately ovate, front end broadly rounded, gaping at hind end; area of beaks covered by a roundish shieldlike plate. Front end with fine, curved, denticulate, concentric riblets; shallow radial groove separating it from posterior end, which has smoothish concentric ridges that at very end become irregular wrinkles.
Habitat
Bores into wood.
Range
N.C. to Brazil.

WEDGE-SHAPED PIDDOCK

Martesia cuneiformis **107:6**

Description
Size, ½–¾ in. (1.3–1.9 cm). Shell smaller, less elongate, more oval than in Striate Wood Piddock; the plate on top triangular with a groove in the center.
Habitat
Bores into wood.
Range
N.C. to Brazil.

SHIPWORMS

Family Teredinidae

Shipworms are an aberrant group of clams in which the shell is used for boring into saltwater-soaked wood, and the body, which maintains the connection with the outside world of sea water, becomes greatly lengthened, sometimes to forty times the length of the shell. Numerous species are recognized throughout the world. They are distinguished by differences in the featherlike pallets, composed of a series of cones, at the far end of the body, which control the admission of sea water into the siphons.

One female lays millions of eggs. The free-swimming young soon attach themselves to anything wooden and start boring in. They never leave this hole but keep on boring and lengthening their bodies the rest of their lives.

NAVAL SHIPWORM

Teredo navalis **107:8**

Description
Size of shell, ½ in. (1.3 cm); length of pallets, to 2 in. (5.1 cm). Valves thin, sharp, globular; pallets pointed, slightly compressed, and symmetrical, with shallow cones, widening from a slender stalk and tapering to a somewhat hollow tip. Extreme ⅓ of pallet covered by yellowish brown periostracum.
Habitat
Common in wood soaked by sea water.
Food
Plankton from sea water; cellulose from the wood it excavates.
Range
Arctic to tropic seas.

Note: The **GOULD'S SHIPWORM,** *Bankia gouldi*, has ½-inch (1.2-cm) pallets of deep-cupped cones.

Paper Shells, Slender Clams, and Spoon Shells

Order Pholadomyoida

PAPER SHELLS

Family Lyonsiidae

These shells have small fragile valves that are unequal in size and somewhat rectangular.

GLASSY LYONSIA

Lyonsia hyalina **107:3**

Description
Size, ½–¾ in. (1.3–1.9 cm). Shell very fragile, white or beige, pearly translucent; rounded in front; long, narrow somewhat cut off behind and gaping. Hinge weak, toothless; beaks small, pointed toward front. Valves covered with thin periostracum having radial wrinkles bearing minute fringes, which catch sand with which to coat shell.
Habitat
Low water to 34 fathoms (62.2 m); common.
Range
N.S. to S.C.

SLENDER CLAMS

Family Pandoridae

These small shells are very thin and flat. The valves are white, unequal, and pearly inside. The beaks are not noticeable, and there are two teeth at the hinge which fit into matching grooves.

GOULD'S PANDORA

Pandora gouldiana **107:4**

Description
Size, 1 in. (2.5 cm). Shell chalky white, roughly oval, rounded in front; back cut off almost straight, curving strongly at lower edge. Right valve flat, left valve convex; hinge line concave, curving up toward back; inside pearly; brownish periostracum on edges of valves.
Habitat
Rocky bottoms to 20 fathoms (36.6 m).
Range
Gulf of St. Lawrence to Cape May.

SPOON SHELLS

Family Periplomatidae

These shells are oval, gape slightly, and have a faint pearly glow. A spoon-shaped tooth at the hinge is supported by a small triangular process beneath, which is often lost.

UNEQUAL SPOON SHELL

Periploma margaritaceum **107:1**

Description
Size, ¾–1¾ in. (1.9–4.5 cm). Shell with left valve more inflated than and slightly overlapping the right; surface smooth, except front end minutely granulose and separated from rest of shell by a strong ridge and groove from beak to lower margin.
Habitat
Common on certain beaches.
Range
S.C. to Tex.

CEPHALOPODS

Class Cephalopoda

Cephalopods are carnivorous mollusks with the shell, when present, lengthened and usually inside, not outside, the animal. The region of the head is surrounded by tentacles or arms, which have prehensile suckers. The mouth is round with a parrotlike beak and there are two eyes, often large. The body is either cylindrical or rounded and frequently has terminal fins. The outside has colored spots which, when contracted or expanded, change the color of the surface of the animal. With the exception of the Paper Nautilus, all cephalopods can discharge an inky fluid as a means of defense.

This class includes some of the most highly organized, swift-moving, and intelligent of the invertebrates. Some octopi in tests have shown an intelligence greater than that of many vertebrates.

Cuttlefish

Order Sepioidea

RAM'S HORNS

Family Spirulidae

RAM'S HORN

Spirula spirula **108:2**

Description
Size, ¾–1 in. (1.9–2.5 cm). Shell fragile, in the form of a flat, few-whorled spiral, the coils not touching; internally divided into chambers by concave partitions; the pearly concave surface of last partition, visible at aperture, has a tiny hole near inner edge. Color yellowish white with the internal partitions showing through as opaque white lines.
Habitat
Pelagic, living in depths of 60–500 fathoms (109.8–914.5 m).
Range
Cast up on beaches, Mass. to Brazil.

BOB-TAILED SQUIDS

Family Sepiolidae

ATLANTIC BOB-TAILED SQUID

Rossia tenera **108:1**

Description
Size, including arms, 3–4 in. (7.6–10.2 cm). The smallest squid in the range; characterized by the comparatively large suckers at the middle of the side arms. Body short, soft; pinkish with small brownish spots (papillae). Short arms about ½ length of long arms; fins about ½ body length. Internal shell small, soft, slender.
Habitat
7–250 fathoms (12.8–457.2 m).
Range
N.S. to se. Fla.

Squids
Order Teuthoidea

COMMON SQUIDS
Family Loliginidae

These animals have a long, tapering, cylindrical body with ten arms and large triangular terminal fins. The arms bear two rows of suckers; each sucker is surrounded by a horny, dentated ring. The tentacular arms bear four rows of suckers on small clubs. The horny internal pen is slightly broader at its base and tapers to a point, with a keel on the underside.

COMMON SQUID
Loligo pealei **108:3**

Description
Size, 1–1½ ft. (30.5–45.7 cm). Distinguished by long triangular fins over ½ length of body, with their tips meeting at the end. Body long, narrow; gray with reddish to purple spots; large; 8 short arms with 2 rows of suckers; 2 long arms with 4 rows on clublike ends.
Habitat
Common from low water to 50 fathoms (91.4 m).
Remarks
This species includes mackerel in its diet. If some of them attack a school of mackerel of the same size, a rugged battle may ensue. The squid try to fasten on to the neck of the mackerel; the mackerel try to sink their jaws in the squid. Sometimes the mackerel win. These squid are used commercially for food and bait.
Range
S. Labrador to Fla.

DWARF SQUID
Lolliguncula brevis **108:5**

Description
Size, 5–10 in. (12.7–25.4 cm). The short, rounded fins which meet to form an oval at the base of the body are characteristic of this animal. Body short, with large purplish spots. Upper arms very short; long arms equal to length of body, with 4 rows of suckers at ends; fins white underneath.
Habitat
Warm waters.
Range
Del. Bay to Brazil.

SEA ARROWS
Family Ommastrephidae

SEA ARROW
Illex illecebrosus **108:4**

Description
Size, 12–18 in. (30.5–45.7 cm). Body long, cylindrical. Notch in front of eyes; head ridged transversely behind the eyes, narrows abruptly to neck, and has smooth, deep depression underneath to

hold siphon. Superficially similar to the Common Squid, but easily distinguished by shorter fins and small eyes.

Habitat
Common in open sea; during summer in schools close to shore in New England.

Habits
Can skim through air.

Remarks
Used commercially for food and bait.

Range
Greenland to Gulf of Mexico.

Octopods

Order Octopoda

The octopods are characterized by having only eight arms, lacking the tentacular arms of the squids. The arms bear suckers which, unlike those of the squids, are not on stalks and lack horny rings. There is no internal pen or shell.

OCTOPI

Family Octopodidae

ATLANTIC OCTOPUS

Octopus vulgaris **108:7**

Description
Size, with longest arm, 1–2½ ft. (30.5–76.2 cm). Body short, thick, depressed, rounded behind. Skin smoothish, translucent, blue-white with brown dots; 8 arms connected partway by a web, each has 2 rows of suckers; third right arm in males serves as sexual intromittent organ. Head broad, enlarged, and warty around eyes; depressed between.

Habitat
Rocks and niches close to shore.

Range
Gulf of Maine to W. Indies.

ARGONAUTS

Family Argonautidae

PAPER NAUTILUS

Argonauta argo **108:6**

Description
Size, 4–8 in. (10.2–20.3 cm). Dorsal arms of the female secrete a thin, spiral, few-whorled papery shell, which is used to contain the eggs and into which the female can retreat when disturbed. Shell symmetrical, compressed, with branching radial ridges on surface terminating in sharp nodes on twin median keels. Keels tinged with brownish black in the early parts of shell; remainder of shell white.

Habitat
Pelagic, in tropical seas.

Range
Found on beaches in tropical Fla.

Other Marine Invertebrates

Consulting Editor
Robert D. Barnes
Department of Invertebrate Zoology, Gettysburg College

Illustrations
Plates 109–110 Klarie Phipps
Text Illustrations by John Cameron Yrizarry and
Nancy Lou Gahan

Other Marine Invertebrates

Marine invertebrates are sea animals which possess no backbone. In this book they have been divided into two groups. The phylum Mollusca, or mollusks, includes the study of the chapter preceding this one. Included in this chapter are the most conspicuous species of eleven other phyla of marine invertebrates. Because of the great number of species this includes, it is not possible to treat them with the completeness given to the vertebrates in other sections of this book. However, the bare essentials of habits and life histories given below reveal at least a little of the fascinating world that lies between the tides and beneath the waves.

Habitat

The diversity of animal life in the oceans is enormous. Many species live in deep water and are therefore not easily seen. However, there is a rich fauna at the edge of the sea, between the tide marks, and in water no deeper than knee level, which can be studied and enjoyed by all who are interested. Unfortunately, surf beaches, which are frequented by the largest number of beachcombers and visitors to the shore, are one of the most inhospitable and impoverished marine habitats. Few animals can live in sand pounded by waves.

The naturalist who wants to see marine animals must explore other kinds of habitats. At low tides, one should comb the protected beaches of inlets, bays, and sounds, looking beneath driftwood and algae cast up by spring tides. Tide pools on rocky shores are menageries; ways to seek out the inhabitants include poking beneath and between clumps of shells on oyster reefs (wearing gloves), examining wharf pilings and rock jetties at low tide, rambling over exposed sand and mud flats at low tide, and digging beneath surface markings.

Conservation

Many of these habitats and species are protected because they are in state parks, and collecting or damaging specimens is prohibited by law. In some states, all marine life is protected and may not be collected without a permit. As with most wildlife, the formerly abundant supply of marine invertebrates has been reduced by vandals, collectors, students, overfishing, insecticides, and loss of habitat.

Range and Scope

This chapter describes the most common, typical, or spectacular marine invertebrates (aside from the mollusks) that occur along the Atlantic coastline, principally from southern Florida to Canada, seaward to ten fathoms. The selection includes most of the species that the amateur field naturalist is likely to encounter. All of the species included are marine, or saltwater, invertebrates. The size given is usually the maximum growth of the species.

Illustrations

Each species illustrated is shown as it would be seen by the naked eye, just as it would be found on the seashore, rather than with the use of a high-powered microscope. Common names, for those species for which they are accepted, have been included. When there is no widely used common name, the common name of the group to which the species belongs has been captioned in parentheses on the illustrated plates, numbers 94 and 95, and in the text illustrations to facilitate species identification.

USEFUL REFERENCES

Barnes, R. D. 1980. *Invertebrate Zoology*, 4th ed. Philadelphia: Saunders.

Gosner, K. L. 1971. *Guide to Identification of Marine and Estuarine Invertebrates.* New York: Interscience Publishers.

Kaestner, A. 1967–1970. *Invertebrate Zoology.* New York: Interscience Publishers.

MacGinitie, G. E., and MacGinitie, N. 1968. *Natural History of Marine Animals.* New York: McGraw-Hill.

Pennak, R. W. 1978. 2nd ed. *Freshwater Invertebrates of the United States.* New York: John Wiley and Sons.

Smith, F. G. W. 1971. *Atlantic Reef Corals.* Coral Gables, Fla.: Univ. of Miami Press.

Smith, R. I., ed. 1964. *Keys to Marine Invertebrates of the Woods Hole Region.* Contribution No. 11. Systematics–Ecology Program. Woods Hole: Marine Biology Lab.

GLOSSARY

Aboral The side away from the mouth.

Ampulla (pl. ampullae) A membranous sac.

Anterior Front; forward part.

Branchiae Gills.

Capitulum Barnacle plate.

Carapace Shieldlike plate covering above and on sides.

Chelipeds Claw-bearing appendages.

Cilia Short, movable hairlike processes of cell surface.

Cirri Tendrils or fleshy appendages.

Cloaca Common duct into which digestive, excretory, and reproductive systems empty.

Detritus Fine particles of organic or inorganic origin.

Dorsal Pertaining to the upper side of the body.

Flagellum Long, movable hairlike process of cell surface.

Ganglion (pl. ganglia) A mass of nerve cell bodies.

Gnathopods Crustacean appendages which manipulate food.

Gonads Organs producing the sex cells (ovary and testis).

Hermaphroditic Having both male and female reproductive organs in one individual.

Hydranth Individual polyps in a cnidarian colony.

Interradius Space between radii.

Interray Space between rays.

Lobule Part of a lobe.

Madreporic plate Water intake disk.

Mantle An enveloping external covering of the body.

Medusa Jellyfish.

Mesenteries Folds of tissue in the body cavity.

Nephridia Excretory and internal water regulatory systems.

Palp A small, often fleshy, feeler.

Papilla (pl. papillae) A small, nipplelike projection.

Parapodia Side feet.

Peduncle Stalk.

Peristome Region around the mouth.

Polyp Attached tentaculate animal like hydra and sea anemone. Some species colonial.

Prehensile Grasping; capable of suspending body.

Proboscis Anterior body extension or tubular organ for feeding.

Radius (pl. radii) An imaginary radial plane dividing the body of a radially symmetrical animal into similar parts.

Ray Any of the radiating divisions of the body of an echinoderm with all its included parts.

Rostrum Beak or beaklike part of an animal.

Scutum Barnacle plate.

Sessile Attached directly by its base.

Setae Bristles.

Spicule Rod or needlelike skeletal unit.

Spongin A horny substance.

Stolon Extension of body wall.

Telson Posterior projection of last body segment.

Test Hard internal skeleton.

Thoracic Pertaining to the thorax, or middle division of the body.

Umbo Prominence above hinge in bivalves.

Ventral Pertaining to the underside of the body.

Zooid Individual animal in a colony.

SPONGES

Phylum Porifera

Sponges are very different from most other animals. They are attached, and there is little discernible movement of any part of the body. There is no front end, no head, no mouth or gut. Essentially a sponge is a mass of tissue organized around a system of water canals. The in-flowing canals are provided with microscopic openings, from which the name *Porifera*, or "pore bearer," is derived. The out-flowing canals open to the surface through one or more large openings, or oscula. Seawater is driven through the canals by internal flagellated cells, the water stream providing the animal with food and oxygen and removing wastes.

Sponges range in size from species that can just be seen with the naked eye to others that would fill a bushel basket. The growth form differs depending upon the species—encrusting sheets, rounded masses, erect branches, tubes, and vases—and there may be some variation even within a species. The body is supported by a skeleton of minute mineral rods (spicules) or organic fibers (spongin) or both.

Sponges reproduce sexually to form a free-swimming ciliated larva, which after a few days attaches permanently to the bottom or to some object underwater and develops into a mature sponge.

There are a few freshwater sponges, but the vast majority of some 5,000 species live in the sea, from the intertidal zone to deep ocean basins.

CALCAREOUS SPONGES

Class Calcarea

SCYPHA CILIATA

Fig. 123

Description
Length, to 1 in. (2.54 cm). Cigar-shaped, covered with thin surface projections; excurrent canal surrounded by palisade of long, slender, needlelike limy spicules. Color yellowish white. Budded individuals separate completely, or almost completely, from the parent, giving the colony the appearance of a bunch of tiny bottles.
Habitat
Seaweed, shells, stones, and dock piles, low water to 300 ft. (91.4 m).
Range
Greenland to Long Island.

SILICEOUS AND HORNY SPONGES

Class Demospongiae

BORING SPONGE

Cliona celata

Fig. 123

Description
Sponge fills channels excavated in old mollusk shells and appears at the surface as yellow plugs ⅛ in. (0.32 cm) in dia., or less. In shells which have been extensively riddled, the sponge may overgrow the surface as a thick irregular crust as much as 8 in.

(20.3 cm) across. Excavation begins when the larva settles to the bottom. Signs of this sponge are the finely perforated clam and oyster shells it has attacked.

Habitat
Gravel or shell bottoms, low tide to 100 ft. (30.5 m).

Remarks
Boring sponge is an important agent in the breakdown of shell and coral.

Range
Arctic to the Carolinas.

Fig. 123

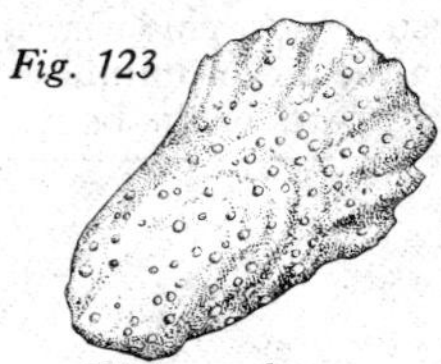

Cliona celata
Boring Sponge

Scypha ciliata
(Sponge)

Haliclona oculata
Deadman's Fingers

RED SPONGE

Microciona prolifera **109:6**

Description
Size, irregular masses to 1 ft. (30.48 cm) high and 6 in. (15.2 cm) square at base. Commonly limited to small, thin encrusting sheets. Surface has scattered oscula, numerous pores. Color bright red.

Habitat
Grows on rocks, oyster shells, piles, old bottles, from low tide to 60 ft. (18.28 m).

Range
Maine to the West Indies.

DEADMAN'S FINGERS

Haliclona oculata *Fig. 123*

Description
Size, fingers 3–6 in. (7.6–15.2 cm) long, oscula $\frac{1}{10}$ in. (0.25 cm) dia. Colonies compact, large; fingers close-set on slender stem; oscula numerous. Color pale orange-red.

Habitat
From low tide to 500 ft. (152.4 m); common.

Remarks
The fingerlike branchings from the central stem give this sponge its name. Bleached skeletons are often found on eastern beaches.

Range
Arctic to Cape May.

HYMENIACIDON HELIOPHILA

Description
Size, erect peaked tufts up to several inches across. Color bright orange.

Habitat
From low tide to 150 ft. (45.7 m) on shells and stones; common in some areas.

Range
Bay of Fundy through the Carolinas.

CNIDARIANS

Phylum Cnidaria

This numerous and largely marine phylum is distinguished by one or more rings of tentacles encircling the mouth, which is the only opening into the interior of the body. The symmetry is thus radial. Some cnidarians (polyps) have a tubular, attached body, with the oral tentaculate end directed upward; others (medusae, or jellyfish) are bowl- or bell-shaped and free-swimming, with the mouth directed downward. Many polypoid cnidarians are colonial, with the individuals connected together by living tissue. Most cnidarians feed on other animals, which they capture with the aid of unique stinging cells located on their tentacles and other parts of the body.

HYDROIDS

Class Hydrozoa

Members of the class Hydrozoa exhibit polypoid, medusoid, or both body forms in the course of their life cycle. Most marine polypoid species are colonial and are known as hydroids. They are common on pilings, jetties, and rocks, where their attached plantlike growth form may cause them to be dismissed as "seaweed." There are also species which grow on the surfaces of the larger seaweeds. The medusae, or jellyfish, of this class are quite small, usually less than one inch in diameter, and are therefore less conspicuous than the larger, more familiar jellyfish of the next class. Hydrozoan jellyfish are usually a reproductive stage arising from the hydroid colony. Hydroids may produce a painful sting. Meat tenderizer quickly rubbed on jellyfish stings is very effective in reducing painful reaction.

Only a few of the many species of hydroids are described below.

Uncupped Hydroids

Suborder Gymnoblastea

In this group the polyps do not form a protective cup (hydrotheca) around themselves, as those of the suborder Calyptoblastea. However, the stems of the colony may have a firm protective covering.

TUBULARIA CROCEA

109:5, *Fig. 124*

Description
Size, to 4 in. (10.2 cm) high. Straight, stalked clumps with tangled and spreading bases, growing in colonies. Stalks gray, tentaculate heads (hydranths) pink.
Habitat
Submerged rocks, pilings, cables.
Range
N.B. to Fla.

HYDRACTINIA ECHINATA

Fig. 124

Description
Lives on surface of old snail shells that are inhabited by hermit crabs. Size depends upon size of shell. Hydroid colony gives shell a furry appearance; color brown to cream.

Habitat
Shallow water, on shells inhabited by hermit crabs.
Range
Labrador to the Carolinas.

Fig. 124 **Hydroids**

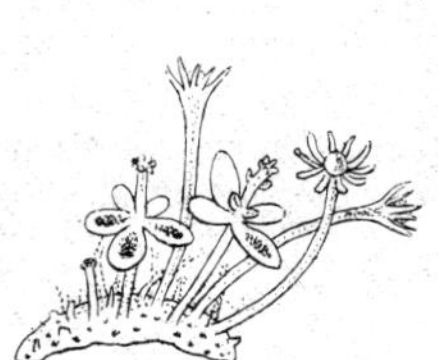

Hydractinia echinata

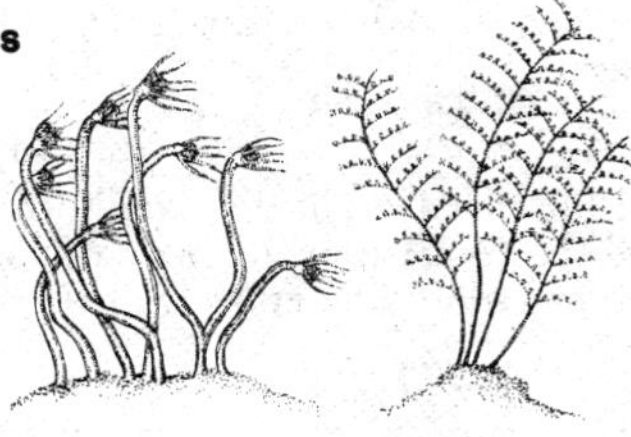

Tubularia crocea *Pennaria tiarella*

PENNARIA TIARELLA

Fig. 124

Description
Size, up to 6 in. (15.2 cm) long. Featherlike or fernlike growth form. Pale orange.
Habitat
Common on pilings and rocks in shallow water.
Range
Cape Cod to Fla.

EUDENDRIUM RAMOSUM

Description
Size, colony to 4 in. (10.2 cm) long. Colonies form irregular branching tufts. Color bright orange.
Habitat
On pilings, jetties, rocks, from shallow to deep water.
Range
Bay of Fundy to Fla.

Cupped Hydroids

Suborder Calyptoblastea

In this group of hydroids the transparent protective covering not only occurs on the stems but also extends up around the polyps to form a protective cup, or hydrotheca, into which the polyp can withdraw.

OBELIA (Species)

Description
Size, colony to 1 in. (2.5 cm) long. Small, delicate, irregular branching colonies. Oral end of polyps within a protective cup but difficult to see without magnification. Color pink or translucent gray.
Habitat
On pilings, rocks, and seaweeds to 200 ft. (60.9 m).
Remarks
There are many other hydroids with growth form similar to *Obelia*. They can be distinguished only with the aid of a microscope and knowledge of structural details.
Range
Bay of Fundy to Fla.

SERTULARIA PUMILA

Description
Size, up to 1½ in. (3.8 cm) long. Branching stalks resembling jigsaw blades placed back to back, the "teeth" being the protective houses of the polyps. Color silvery.
Habitat
Common on seaweed.
Remarks
All of the members of this family (Sertulariidae) have a similar growth pattern, although size differs. Many different species are distributed along the entire East Coast.
Range
Bay of Fundy to N.J.

PORTUGUESE MAN-OF-WAR
Physalia physalis **109:3**

Description
Floating colony up to 1 ft. (30.5 cm) in length, surmounted by an elongate purplish-blue gas-filled sac. Long tentacles, 2 to many feet in length.
Habitat
Warm tropical waters.
Habits
Tentacles sting small animals and convey captured food to mouth by slow contraction. Some small fish, immune to sting, live and travel among tentacles.
Remarks
Has a painful sting, which can be extremely dangerous and should be carefully avoided by swimmers.
Range
Gulf of Mexico, West Indies, and Gulf Stream. Blown ashore in Fla. and the tropics and occasionally in N.J.

JELLYFISHES
Class Scyphozoa

These are the large jellyfishes that are seen swimming near the shore or washed up on beaches. They are found in all seas and vary in diameter from about one inch (2.5 cm) to seven and one-half feet (2.3 m). Scyphozoan jellyfish lack the flaplike shelf, or velum, within the bell margin that is found in the little hydrozoan jellyfish, and four long flaps, or oral arms, hang down from around the mouth. The eggs of the sexually reproducing adult jellyfish develop into an inconspicuous polypoid stage, the scyphistoma, which attaches to seaweeds or to hard bottoms. It is not colonial and is usually less than one-half in. (1.3 cm) high. When budding, it may look like a stack of tiny plates. The top plate, showing tentacles, detaches at regular intervals and swims away as a tiny juvenile jellyfish.

SEA NETTLE
Dactylometra quinquecirrha **109:1**

Description
Size, bell to 8 in. (20.3 cm), tentacles to 3 ft. (91.4 cm). Oral arms long, ruffled. Tentacles of bell margin, 40, located between lobes on the bell margin; every sixth position between lobes lacks a

tentacle, the area being occupied by an inconspicuous sense organ. Stinging cells on both bell and tentacles. Bell milky white or rosy pink to tan, sometimes with rosy stripes; oral arms milky white or pink.

Habitat

Common in less saline waters of bays and estuaries.

Range

Cape Cod to Fla.; especially common in the Chesapeake Bay and certain parts of the Carolina sounds.

CYANEA CAPILLATA 109:2

Description

Size, bell dia. to 3 ft. (91.4 cm), rarely to 8 ft. (24.4 m). Bell margin notched, forming 32 lobes. Long tentacles arranged in 8 clusters around bell margin; number of tentacles increases with bell diameter, reaching as many as 800 in large specimens. Bell brownish to purplish-red; margin translucent.

Habitat

Cool coastal waters.

Range

Bay of Fundy to Cape Hatteras.

MOON JELLY

Aurelia aurita 109:4

Description

Size, bell dia. 6–10 in. (15.2–25.4 cm). Marginal lobes of bell, 8, broad, flattened. Tentacles numerous, short, having the appearance of a fringe around the bell margin. Color bluish-white; male gonads yellow, female gonads pink, having a four-leaf-clover pattern when viewed through bell.

Habitat

Coastal waters.

Range

Bay of Fundy to Cape Hatteras.

STOMOLOPHUS MELEAGRIS *Fig. 125*

Description

Size, bell dia. 6 in. (15.2 cm). Bell rigid, shaped like mushroom cap. No tentacles on bell margin; oral arms form short, wide, frilly mass.

Habitat

Coastal waters.

Habits

Strong swimmers, moving horizontally through water when swimming rapidly.

Range

Va. and s.

Fig. 125

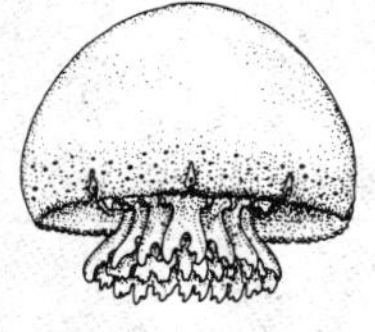

Stomolophus meleagris (Jellyfish)

Leptogorgia virgulata Whip Coral, p. 632

SEA ANEMONES AND CORALS

Class Anthozoa

Members of the class Anthozoa are entirely polypoid; there is no free-swimming medusa stage in the life cycle. The mouth leads through a short gullet to the internal gut cavity, which is partially compartmented by vertical partitions (mesenteries).

In contrast to most other anthozoans, sea anemones are solitary; that is, individuals are not attached together in colonies. They also attain the largest size of any anthozoan polyps. (The colonies of other groups may be quite large, but the individual polyps are generally small.) The heavy cylindrical body of sea anemones, called the column, is surmounted by the oral disc bearing the tentacles and mouth. In many sea anemones a collarlike fold around the top of the column closes over the oral disc when the sea anemone is contracted.

The name *coral* most commonly refers to those anthozoans known as stony, madreporarian, or scleractinian corals. They are closely related to sea anemones, but differ in being colonial and in secreting a skeleton of calcium carbonate. The living colony covers or rests on top of the skeleton. The position of the polyps is indicated in the skeleton of most corals by the presence of cuplike depressions.

The *Astrangia danae* is the only common shallow-water coral along the East Coast, but in the Florida Keys and West Indies there are many species of reef corals.

WHIP CORAL

109:13

Leptogorgia virgulata

Fig. 125

Description

Size, colony 1–3 ft. (30.5–91.4 cm) long. Colonies have form of whiplike rods the diameter of drinking straws. Small polyps embedded in the common tissue of the rod. Colony supported by an internal organic core and microscopic calcareous spicules. Color yellow, orange, or purple.

Habitat

Attached to rocks and shells in shallow water.

Range

Va. to Brazil.

SEA PANSY

Renilla reniformis

Description

Size, colony about 2 in. (5.1 cm) dia. Colony rounded, leaflike, attached by a short stem, fleshy but containing microscopic spicules; small polyps embedded on upper side of "leaf." Color purple.

Habitat

Attached on sandy bottoms in shallow water to 40 ft. (12.2 m).

Range

Va. and s.

BURROWING OR TUBE-DWELLING SEA ANEMONE

Ceriantheopsis americanus *Fig. 126*

Description
Size, extended body 6 in. (15.2 cm) long, ½ in. (1.3 cm) dia. Mouth surrounded by two circlets of 125 or more tentacles. Color brown.
Habitat
Soft bottoms of sand and mud, from intertidal zone to 60 ft. (18.3 m).
Habits
Lives within a soft tube buried in mud or sand. Only oral end emerges to surface.
Range
Cape Cod and s.

Fig. 126 **Sea Anemones**

Diadumene leucolena

Ceriantheopsis americanus
Burrowing Sea Anemone

Metridium senile
Brown Sea Anemone, p. 634

AIPTASIA PALLIDA AND ERUPTAURANTIA

Description
Size, *A. pallida* to 1½ in. (3.8 cm) dia., *A. eruptaurantia* smaller. Short sea anemones with one circlet of about 50 tentacles. Color highly variable: *A. pallida* white to dark brown with white stripes on tentacles; *A. eruptaurantia* may be orange, also with white stripes on tentacles.
Habitat
On rocks and shells in shallow water.
Range
Cape Hatteras and s.

DIADUMENE LEUCOLENA *Fig. 126*

Description
Size, to 1 in. (2.5 cm) long. Body long, cylindrical, translucent; mesenteries visible. Tentacles, 40–60 in 4 circlets, delicate, tapering; inner ones are "catch tentacles," longer, thicker, and more opaque (a distinguishing feature); central series longest. Color white, pinkish, or greenish.
Habitat
Oysters, dark rock overhangs, ships' bottoms, under stones on rocky beaches; common.
Range
Cape Cod to Cape Hatteras.

BROWN SEA ANEMONE

Metridium senile — *Fig. 126*

Description
Size, column to 4 in. (10.2 cm) high, to 3 in. (7.6 cm) dia. Adult, column dark brown blotched with white, orange, or straw, and streaked with brown. Disc, when expanded, is convex, very broad, with up to 1000 short, white, tapered tentacles, which distinguish it from *Diadumene.*
Habitat
Docks or rocks from low water to 300 ft. (91.4 m); common.
Range
Arctic seas to N.J.

ASTRANGIA DANAE — 109:14

Description
Size, colony up to 4 in. (10.2 cm) wide. Skeleton with distinct cups about $\frac{3}{16}$ in. (0.48 cm) dia. Colony more or less encrusting; few tentacles. Living colony pinkish or white, translucent.
Habitat
Attached to stones and shells in shallow water. Broken and wave-worn skeletons often found on beaches.
Range
Cape Cod to Fla.

COMB JELLIES

Phylum Ctenophora

Members of the phylum Ctenophora are somewhat similar to jellyfish in having transparent jellylike bodies. The distinguishing feature is the eight bands of comb plates composed of fused cilia, which are used in locomotion. Comb jellies are usually globular in shape with a bilateral symmetry, rather than the radial symmetry of the Cnidaria. Some have retractile tentacles but without stinging cells. The digestive system starts with a slitlike mouth that leads to a tubular stomach. This branches and leads to tubes which lie directly beneath the comb plates.

Almost all comb jellies are free-swimming or -floating and occur in vast numbers when conditions are favorable. They are carnivorous, feeding on plankton. Some species feed on other comb jellies.

SEA WALNUT

Pleurobrachia pileus — *Fig. 127*

Description
Size, body ¾ in. (1.8 cm), tentacles to 15 in. (38.1 cm). Oval, plump; 2 branched tentacles. Color pink, yellow, or clear.

Fig. 127

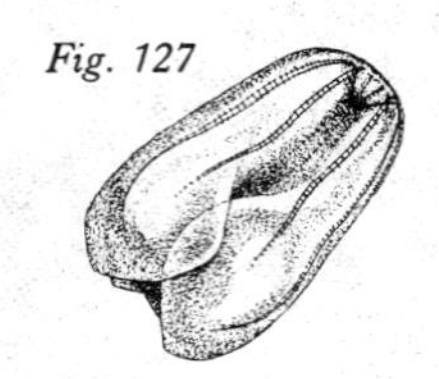

Mnemiopsis leidyi
(Comb Jelly)

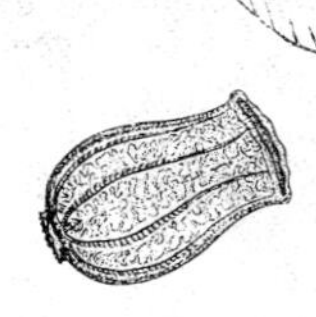

Beroe cucumis
(Comb Jelly)

Pleurobrachia pileus
Sea Walnut

Habitat
Coastal water to deepest ocean; frequent in winter.
Range
Arctic seas to Long Island.

MNEMIOPSIS LEIDYI

Fig. 127

Description
Size, to 6 in. (15.2 cm) long. Oval, with oral end divided into two large lobes. Combs produce blue-green luminescence when animal is disturbed at night.
Habitat
Shallow water to deep sea; often extremely abundant on bathing beaches.
Range
Cape Cod and s.

BEROE CUCUMIS AND OVATA

Fig. 127

Description
Size, to 4 in. (10.2 cm) long, to 3 in. (7.6 cm) wide. Body bell-shaped with neither tentacles nor lobes. Color pinkish or translucent; comb plates iridescent.
Habitat
Shallow water to open ocean.
Range
B. cucumis, Arctic seas to S.C.; *B. ovata,* Cape Cod to Cape Hatteras.

RIBBON WORMS

Phylum Nemertea

Most ribbon worms are small, inconspicuous animals. They are greatly flattened, but the body is elongate and more wormlike than flatworms. They possess a long eversible proboscis that can be shot out of the anterior to capture prey. In addition to many small species which occupy the same sorts of habitats as the flatworms, there are a few very large ribbon worms, such as the one described below, that live beneath stones or burrow in sand.

RIBBON WORM

Fig. 128

Cerebratulus lacteus

Description
Size, rarely to 6 ft. (1.8 m) long, to 1 in. (2.5 cm) wide. Head lance-shaped; front end of worm thicker, narrower than rear end, which has a small, slender projection, often lost. Color pink or creamy, with internal organs showing as a brownish band.

Fig. 128

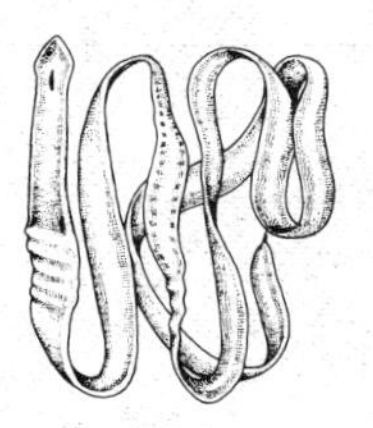

Ribbon Worm

Habitat
Sandy or muddy bottoms in shallow water. May be found under flat rocks on beaches at low tide.
Remarks
Popular bait for hand-line fishing.
Range
Maine to Fla.

MOSS ANIMALS

Phylum Bryozoa

The phylum Bryozoa contains many species of common colonial marine animals, but they are unknown to most beachcombers because of their minute size. The attached erect or encrusting colonies may reach several inches high or across, but the individuals composing the colony are less than 1/16 in. (0.16 cm) in length. Each is enclosed within a horny or calcareous exoskeleton. Bryozoans feed on plankton and detritus, which is collected by means of a crown of tentacles protruded through an opening in the skeletal encasement. However, the tentacles and other body structures can be seen only with magnification.

A few of the more common conspicuous species are described below. The description is based on aspects of the colony which may be seen with the naked eye.

CRISIA EBURNEA — *Fig. 129*

Description
Size, colony ½–¾ in. (1.3–1.8 cm) high. Erect branching colony; each branch composed of minute, white, outward-curving cylinders arranged in 2 alternating rows.
Habitat
Attached to solid objects in shallow water to 100 ft. (30.5 m).
Range
Arctic to Long Island.

Note: The related species *C. denticulata* ranges from Bay of Fundy to Florida and reaches one in. (2.5 cm) in height. It possesses a fluffier colony with black jointed basal sections.

Fig. 129
Moss Animals

Bugula turrita

Crisia eburnea

BUGULA TURRITA and NERITINA *Fig. 129*

Description
Size, colony 1–6 in. (2.5–15.2 cm) high, sometimes higher. Colony erect, composed of branching tufts; superficially looks like seaweed, but finely sectioned structure of each branch and brown or purplish-brown color are distinctive.
Habitat
Species of *Bugula* are very common on pilings, jetties, buoys, and other objects in shallow water.
Range
B. turrita mostly n. of Cape Hatteras and *B. neritina* s.

Fig. 130
Moss Animals

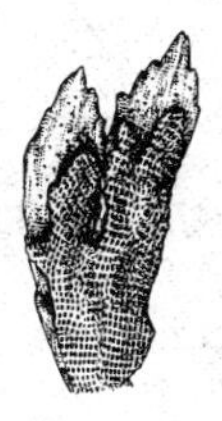

Electra species *Membranipora* species

ELECTRA AND MEMBRANIPORA (Species) *Fig. 130*

Description
Size, colony usually less than 1 in. (2.5 cm) long. Species of these two genera form white, lacy encrusting colonies. Separation of species requires observation of microscopic characteristics.
Habitat
On algae, stones, and shells from the intertidal zone to over 1000 ft. (304.8 m).
Range
Entire East Coast. *Membranipora tuberculata* abundant on floating *sargassum* (gulf weed), which is sometimes washed up on southern beaches, more rarely in the north.

SCHIZOPORELLA UNICORNIS *Fig. 131*

Description
Size, varies from small to very large since species is colonial. Heavy encrusting colonies, wide spreading, sometimes vertically expanded. Color whitish-orange or brick-red.
Habitat
On stones, shells, pilings, from low-tide mark to 100 ft. (30.5 m).
Range
Bay of Fundy to Fla.

Fig. 131

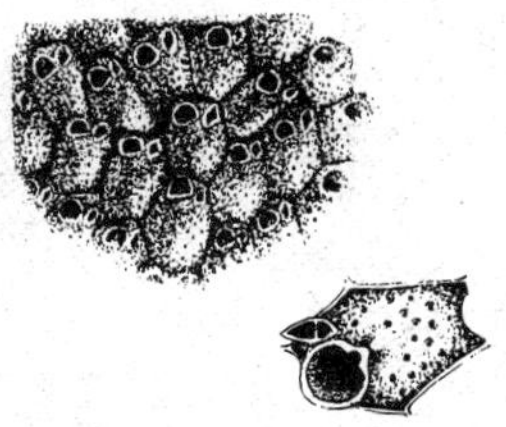

top view

side view

Schizoporella unicornis (Moss Animal)

Terebratulina septentrionalis (Lamp Shell), p. 638

LAMP SHELLS

Phylum Brachiopoda

The bivalved animals of this phylum bear a superficial resemblance to clams. However, the valves cover the upper and lower surfaces of the animal instead of the sides as in clams; thus the valves are bilaterally symmetrical, not symmetrical to each other. In articulate brachiopods (the type represented below), the upper, dorsal valve is smaller and usually flatter and the ventral valve extends behind the interlocking hinge joint. Most brachiopods are attached to the bottom by a stalk, which distinguishes them from any mollusk. A passage for the stalk is provided either by a notch or circular opening in the ventral valve. They usually are attached upside down.

TEREBRATULINA SEPTENTRIONALIS — *Fig. 131*

Description
Size, ½ in. (1.3 cm) long. Shell oval, thin, semitransparent. Ventral valve points toward rear; dorsal valve with calcareous loop inside. Hinge line curves up in middle. Color yellowish white.
Habitat
Rocky bottoms, shallow water to 120 ft. (36.6 m).
Range
Arctic seas to Cape Cod.

ACORN WORMS

Phylum Hemichordata

Acorn worms, which are the most conspicuous members of this small phylum of exclusively marine animals, burrow in sand and mud or live beneath stones or algae. The soft body is divided into three primary regions: the proboscis, the collar, and the trunk. The proboscis is a muscular structure attached to the collar by a narrow neck. The mouth opens widely at the anterior ventral margin of the collar, which is partly overlapping the proboscis. The trunk is the longest region of the body. It has paired gill slits in rows on either side in its anterior part and tapers toward the rear, ending with the terminal anus.

SACCOGLOSSUS KOWALEVSKII — 110:13

Description
Size, 6 in. (15.2 cm) long. Body slender, proboscis carrot-shaped, collar narrow. Odor faintly as of iodine. Color orange-yellow to brown, proboscis white to pink, collar red-orange.
Habitat
Fine, muddy sand in shallow water.
Remarks
Burrow may be marked in the intertidal zone by small coils of castings on the sand surface.
Range
Cape Cod and s.

BALANOGLOSSUS AURANTIACUS

Fig. 132

Description
Similar to *Saccoglossus kowalevskii*, but proboscis short and rounded; color often paler and size larger. Smells very strongly of iodine.
Habitat
In sand in the intertidal zone and shallow water.
Remarks
Large piles of castings may be very conspicuous on a sand flat at low tide.
Range
Va. and s.

Fig. 132

Castings of *Balanoglossus aurantiacus*

ECHINODERMS

Phylum Echinodermata

The members of this exclusively marine phylum are all relatively large invertebrates and include such familiar animals as starfish, sea urchins, and sand dollars. A five-part radial symmetry is a distinguishing feature of echinoderms, although there are some species of starfish in which there are more than five arms and some other echinoderms in which the original radial symmetry has become less apparent. All echinoderms contain an internal skeleton of calcareous pieces, which usually includes surface spines of various shapes and sizes. The name *echinoderm* means "spiny skin." Also peculiar to echinoderms is the water vascular system, a system of internal canals connected to small, hollow surface appendages called tube feet. The tube feet are used for crawling, feeding, or other functions and are located in five long areas (ambulacral areas), one for each of the radial divisions of the body.

STARFISH

Class Asteroidea

Starfish, or sea stars, have bodies composed of five tapered arms, or sometimes more, arranged around a central disk. The lower side of each arm bears a longitudinal ambulacral groove that contains the tube feet. The five ambulacral grooves converge on the mouth, located in the center of the underside of the disk. A buttonlike structure, the madreporite, is usually conspicuous in an interradial position to one side of the upper surface of the disk.

Starfish crawl by means of their tube feet. Some species live in rocky habitats; others are adapted for life on sandy bottoms. Most are carnivorous, eating clams, snails, and other echinoderms. They feed either by swallowing their prey whole or by everting their stomachs out of their mouths over the prey. Some species, such as the common *Asterias* of the Atlantic coast, are able to slither their everted stomachs between the valves of oysters and clams.

ASTROPECTEN ARTICULATUS

Fig. 133

Description
Size, 4–12 in. (10.2–30.5 cm) dia. Distinguished by large, pale marginal plates that border the arms and bear fringelike spines. Surface smooth. Color variable, often orange to mauve or blue above, pale below.
Habitat
Sandy bottoms from 20–400 ft. (6.1–121.9 m).
Range
N.J. to Mexico.

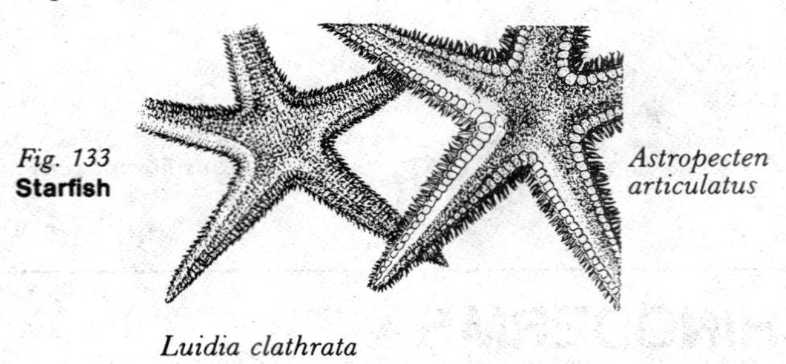

Fig. 133
Starfish

LUIDIA CLATHRATA

Fig. 133

Description
Size, 4–10 in. (10.2–25.4 cm) dia. Like *Astropecten,* has a smooth surface, but lacks the conspicuous marginal plates. Color rosy orange to gray.
Habitat
Sandy bottoms, shallow water to 500 ft. (152.4 m).
Range
N.J. to Brazil.

HENRICIA SANGUINOLENTA

109:9

Description
Size, 2½–4 in. (6.5–10.2 cm) dia. Radius of disk ⅕ length of arms. Arms slender, smoothly rounded. Upper parts usually blood-red or purple, often mottled, sometimes orange or creamy; underparts yellow.
Habitat
Rock crevices from shore to 600 ft. (182.9 m).
Range
Greenland to Cape Hatteras.

PURPLE SUN STAR

Solaster endeca

Fig. 134

Description
Size, 8½–16 in. (21.6–40.6 cm) dia. Arms, 7–13, usually 10, slender, as long as disk is wide. Upper surface red-purple; madreporic plate and margins yellow; underside white.
Habitat
Shallow water to 600 ft. (182.9 m).
Range
Greenland to Cape Cod.

COMMON SUN STAR

Solaster papposus *Fig. 134*

Description
Size, 8–15 in. (20.3–38.1 cm) dia. Arms, 8–14, shorter than disk dia. Disk large; dorsal spines large, tufted. Single series of marginal spines very prominent. Disk with concentric color bands often of scarlet, crimson, or orange, sometimes with purple tones; arms banded pink or white, tips crimson; underside whitish.
Habitat
Shoreline to depth of 1 mi. (1.609 km).
Range
Arctic seas to N.J.

Fig. 134

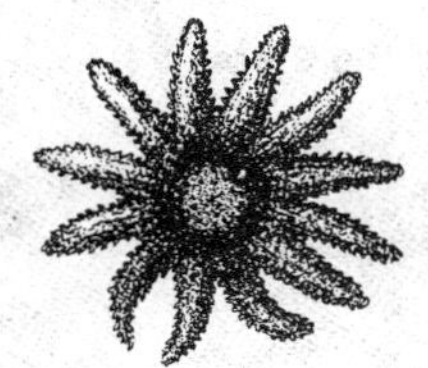

Solaster papposus
Common Sun Star

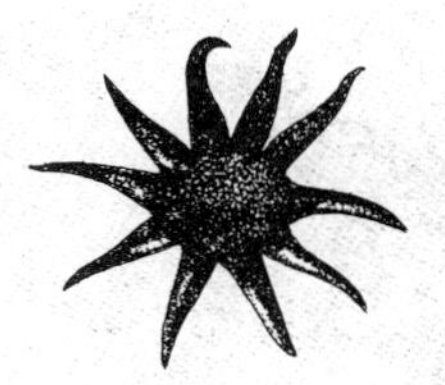

Solaster endeca
Purple Sun Star

ASTERIAS FORBESII AND VULGARIS 109:10

Description
Size, 6–12 in. (15.2–30.5 cm) dia. Arms, 4–7, about twice as long as disk dia.; tuberculate spines give rough surface. Color brown to purple, green, or orange; madreporite of *A. forbesi* red-orange, of *A. vulgaris* light yellow.
Habitat
Rocky and shelly bottoms, shallow water to over 150 ft. (45.7 m).
Range
A. forbesi n. of Cape Cod, *A. vulgaris* s.

BRITTLE STARS

Class Ophiuroidea

Brittle stars, or serpent stars, are similar to starfish in possessing arms, but the arms are long, slender, and jointed and sharply distinct from the small, round central disk. In the basket stars, which are not included here, the arms are branched. There are no ambulacral grooves on the underside of the arms, and the small, inconspicuous tube feet are used in feeding and respiration rather than in locomotion. The madreporite is on the lower side of the disk and is not easily seen.

Brittle stars move by pushing with their flexible arms and are the most agile of all echinoderms. They will readily sever one or more arms if seized. Although common shallow-water animals, they hide beneath stones, shells, and other objects and are thus not as conspicuous as many other echinoderms. Brittle stars are scavengers or feed on fine bottom detritus.

Species of brittle stars are differentiated by fine structural details, most of which cannot be seen with the naked eye. The following four species are commonly encountered in shallow water.

DAISY BRITTLE STAR
Ophiopholis aculeata *Fig. 135*

Description
Size, disk ½–1 in. (1.3–2.5 cm) wide, arms 3–3½ in. (7.6–8.9 cm) long. Disk bulges between the arms, giving it an almost circular outline. Spines on arms conspicuous, stand out from the arm. Colors vary from rusty-red to blue, green, or brown. Disk usually mottled; arms banded.
Habitat
Low water to depth of 1 mi. (1.6 km).
Range
Arctic seas to Long Island.

Fig. 135 **Brittle Stars**

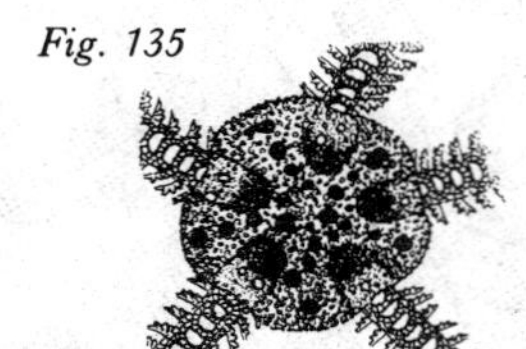

Ophiopholis aculeata
Daisy Brittle Star

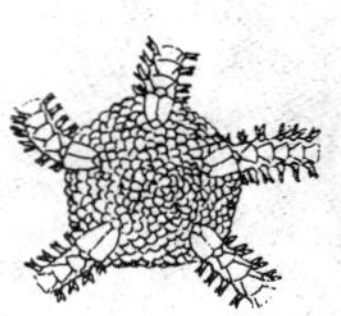

Amphipholis squamata

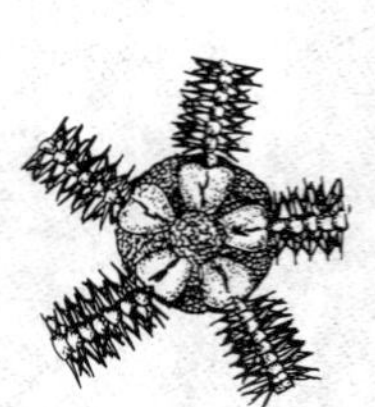

Ophiothrix angulata

AMPHIPHOLIS SQUAMATA *Fig. 135*

Description
Size, disk ⅕ in. (0.5 cm) wide, arms 1¼ in. (3.1 cm) long. Arms with prominent lateral spines. Color gray; white spot on base of each arm.
Habitat
Shoreline to 350 ft. (106.7 m).
Range
Arctic seas to Long Island.

GREEN BRITTLE STAR
Ophioderma brevispinum **109:12**

Description
Size, disk ½ in. (1.3 cm) dia., arms 3 in. (7.6 cm) long. Arms, 4–6, usually 5; disk 5-sided; spines very short, usually lying flat against arm. Color dark greenish-brown to clear green, often banded.
Habitat
Tide pools, sand flats; to depth of 600 ft. (182.9 m).
Range
Cape Cod to Brazil.

OPHIOTHRIX ANGULATA **109:11,** *Fig. 135*

Description
Size, disk ½ in. (1.3 cm) dia. Arms with long spines. Color highly variable; brown, red, green, mottled and banded; white stripe on upper surface of each arm, especially toward tip.
Habitat
Beneath stones and shell clumps from shallow to deep water.
Range
Va. and s.

SEA URCHINS AND SAND DOLLARS

Class Echinoidea

The body of echinoids is not drawn out into arms but rather is globular and disk-shaped. The skeletal pieces are fused together to form a rigid internal shell, or test, parts of which are commonly picked up on beaches or sand flats. The surface of the test bears tubercles upon which are mounted movable spines, long and heavy in sea urchins and minute in sand dollars. The thousands of tiny spines give sand dollars a furry look. The five ambulacral areas containing the tube feet run over the body of echinoids as five meridians, converging on the mouth in the center below and on the anus above. Each tube foot connects to the interior parts of the water vascular system by a pair of canals through the test, which can be seen in a dried clean sea urchin test as paired perforations.

Most sea urchins are adapted for life on hard bottoms and, like starfish, move by means of their tube feet. They may also push with their spines and erect them in defense. Sea urchins are scavengers, scraping the bottom with their complex jaw apparatus, called Aristotle's lantern, which can be seen projecting slightly from the mouth.

Sand dollars are adapted for life on soft bottoms and use their minute spines to burrow partially or completely below the sand surface. The tube feet are used in respiration and in moving food. The five conspicuous petallike areas on the upper surface of the test mark the position of the respiratory tube feet. Sand dollars feed on fine detritus sifted out between the spines and transported by cilia and the tube feet to the mouth. Grooves for food transport can be seen on the underside of the test. The calcareous piece found in the center of the sand dollar when it is broken open is the vestigial jaw apparatus.

PURPLE SEA URCHIN

Arbacia punctulata **109:8**

Description
Size, test, 2 in. (5.1 cm) wide, spines, 1 in. (2.5 cm) high. Color purple to dark purplish brown. Test domed; center of dome and surrounding area bare; rest of test almost hidden by sturdy spines.
Habitat
Seaweed in tide pools; rocky and shelly bottoms in shallow water.
Range
Cape Cod to Gulf of Mexico.

KEYHOLE URCHIN

Mellita quinquiesperforata **109:7**

Description
Size, test to 4½ in. (11.4 cm) wide. Round, flattened on rear margin. Pattern of tube feet clear, somewhat irregular. Test has 5 slots (from above); 4 stretch from petal tips to margin; 1 is between petals.
Habitat
Sandy bottoms in shallow coastal waters.
Range
Md. to the West Indies.

GREEN SEA URCHIN

Strongylocentrotus droebachiensis *Fig. 136*

Description

Size, test to 3½ in. (8.9 cm) wide, spines ½ in. (1.3 cm) long. Test somewhat flattened; spines equal in length, crowded, completely covering test. Test greenish brown, spines bright green.

Habitat

Tide pools and rocky bottoms to 4000 ft. (1219.2 m); common.

Range

N.J. (deep water); Cape Cod (shallow water) and n.

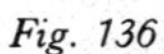

Fig. 136

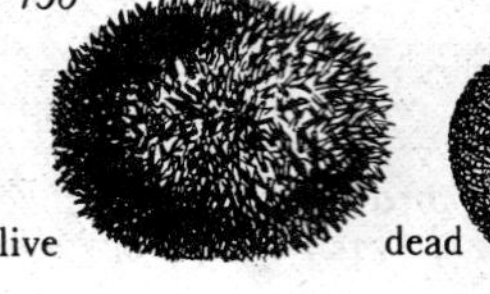

alive

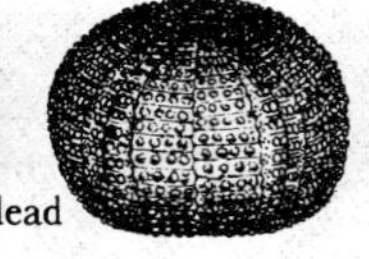

dead

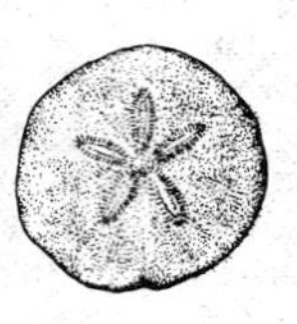

Strongylocentrotus droebachiensis
Green Sea Urchin

Echinarachnius parma
Sand Dollar

WHITE SEA URCHIN

Lytechinus variegatus

Description

Size, test to 2 in. (5.1 cm) wide, spines about ½ in. (1.3 cm) or less. Spines shorter and more slender than those of *Arbacia.* Color white, sometimes greenish (in the West Indies, very dark purple).

Habitat

Shelly and rocky bottoms, shallow water to 150 ft. (45.8 m).

Range

Cape Hatteras to the West Indies.

SAND DOLLAR

Echinarachnius parma *Fig. 136*

Description

Similar to the sand dollar above but lacks slots.

Habitat

Sandy bottoms in shallow water.

Range

Bay of Fundy to Cape Hatteras.

HEART URCHIN

Moira atropos *Fig. 137*

Description

Size, 2¼ in. (5.7 cm) long. Spines short and numerous as in sand dollars but body not flattened and slightly longer than wide. Color brown.

Habitat

Burrows beneath sand surface from low-tide mark to 450 ft. (137.2 m).

Range

Va. (s. of Chesapeake Bay) to Fla.

SEA CUCUMBERS

Class Holothurioidea

Sea cucumbers are elongated along the radial axis, which runs between the mouth and anus, giving the body a cucumber- or worm-like shape. These animals lie on their side rather than on the oral end as do sea urchins. The body has a leathery texture, for the skeleton is reduced to microscopic pieces.

Sea cucumbers creep over the bottom using the rows of tube feet, which run down the length of the body, or they burrow in sand and mud. They feed on bottom detritus or plankton collected by means of a crown of tentacles representing modified tube feet, which can be extended from around the mouth. The tentacles are usually contracted and not easily seen when the animal is collected.

PENTAMERA PULCHERRIMA — 109:16

Description
Size, 1½–2 in. (3.8-5.1 cm) long. Gherkin-shaped, narrowing at ends. Tentacles, 10, branched; lower 2 shorter. Tube feet in 2 rows on each of 5 tracts. Color yellowish or white.

Habitat
Shallow, muddy bottom with eelgrass.

Range
Cape Cod to the Carolinas.

SCLERODACTYLA BRIAREUS — *Fig. 137*

Description
Size, 6 in. (15.2 cm) long. Tentacles, 10, treelike. Body saclike, with tube feet scattered all over the surface, giving it a furry appearance. Color greenish, brown, or black.

Habitat
Buried in muddy sand in shallow water.

Range
Cape Cod and s.

Fig. 137

Leptosynapta inhaerens (Sea Cucumber)

Moira atropos Heart Urchin

Sclerodactyla briareus (Sea Cucumber)

Caudina arenata, p.646 (Sea Cucumber)

LEPTOSYNAPTA INHAERENS — *Fig. 137*

Description
Size, 5–9 in. (12.7–22.8 cm) long, and to ⅜ in. (0.96 cm) dia. Delicate wormlike body. Tube feet absent; ambulacral areas marked by five longitudinal lines extending length of body. Tentacles, 12, each having 5–7 side branches. Color transparent to whitish.

Habitat
Within muddy sand, shallow water to 600 ft. (182.9 m).

Range
Bay of Fundy to the Carolinas.

CAUDINA ARENATA *Fig. 137*

Description
Size, 4–7 in. (10.2–17.8 cm) long, ½–1 in. (1.3–2.5 cm) dia. Forward portion of body chunky; terminal ⅓ slender, taillike. Skin smooth. Mouth surrounded by 15 short, stubby tentacles, split crosswise at tips, each forming 4 fingerlike projections. Tube feet absent; 5 clusters of papillae around anus. Color translucent pinkish to purple.
Habitat
Burrowing, with tail protruding, in mud or sand from shoreline to 100 ft. (30.5 m).
Range
Bay of Fundy to Cape Cod.

SEGMENTED WORMS

Phylum Annelida

Members of the phylum Annelida have the body divided into a linear series of similar parts, or segments. At the anterior end of the body is the nonsegmental prostomium, which makes up most of the head and lies just in front of the ventral mouth. Behind the prostomium lies the peristomium (surrounding the mouth), which is in turn followed by the trunk segments. The trunk segments usually carry similar external and internal structures. Annelid worms are found in the sea, in fresh water, and in soil. Earthworms are familiar examples of terrestrial species.

MARINE SEGMENTED WORMS

Class Polychaeta

Of approximately 9000 described species of segmented worms, two-thirds live in the sea. Most of these marine annelids belong to the class Polychaeta, which not only is the largest class but contains the greatest diversity of forms. Polychaetes are distinguished from other annelids by the pair of fleshy lateral projections, or parapodia, carried by each trunk segment. The parapodia contain bundles of fine horny rods, or bristles, called setae, from which the name *polychaeta,* or "many setae," is derived. A head is often well developed in polychaetes and may bear eyes, antennae, and other sensory projections. In some polychaetes the head carries elaborate feeding appendages, such as tentacles or featherlike projections.

Many different life styles are found among polychaetes. Some species live beneath stones, in rock crevices, on algae, or among other sessile animals; some burrow in sand or mud; and others live within tubes either secreted by the worm or composed of sand grains or other materials cemented together. The tubes are planted in mud or sand or attached to various objects, and they are usually much more conspicuous than their inhabitants. Polychaetes use their parapodia for crawling or gripping the sides of their burrow

The feeding habits of polychaetes vary greatly. Some are carnivores or algae eaters and possess jaws for seizing and tearing. Many consume detritus or even mud and sand from which the detritus is digested. Some are filter feeders, removing plankton by passing a current of water over some part of the body (the filter).

The following species are a few of the larger, more conspicuous polychaetes that might be encountered. Identification has been limited to family where that category is more easily recognized than species.

SEA MOUSE
Aphrodita hastata *Fig. 138*

Description
Size, to 7 in. (17.8 cm) long, 2 in. (5.1 cm) wide. Body elliptical. Head end bears 1 tentacle and 2 long palps. Back slightly arched, with dense coat of long setae. Setae gray on top of back, golden green and iridescent on sides.
Habitat
In mud.
Range
Gulf of St. Lawrence to Chesapeake Bay.

Fig. 138

Aphrodita hastata
Sea Mouse

Lepidonotus squamatus
(Scale Worm)

SCALE WORMS
Families Polynoidae and Sigalionidae *Fig. 138*

Description
Size, ¼–3 in. (0.64–7.6 cm) long. Upper surface of body covered by flat overlapping scales. Color varies; black, grays, browns common.
Habitat
Beneath stones, in rock crevices, among oyster or mussel shells, in algae.
Range
Entire E. Coast.

CLAM WORMS
Family Nereidae **110:12**
Fig. 139

Description
Size, 2 in. (5.1 cm) to many inches long. Distinguished by large, flat parapodia and a distinct head structure. (The latter, however, may be difficult to discern without a hand lens.)
Habitat
Burrows in muddy sand in shallow water.
Range
Bay of Fundy to Va.

Note: The **GREEN CLAM WORM,** *Nereis virens,* is a large greenish species reaching a length of eighteen inches, and is commonly used as bait.

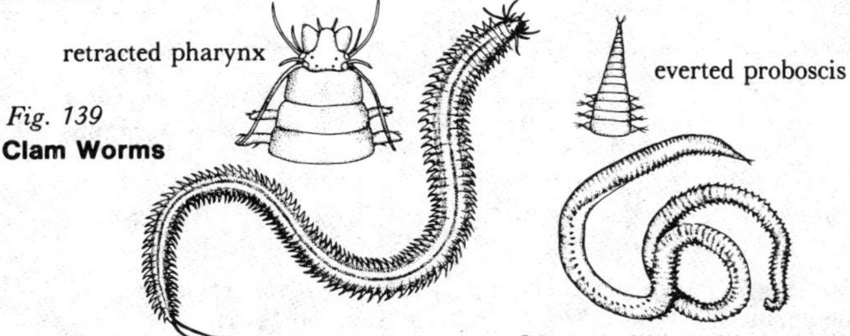

Fig. 139
Clam Worms

Nereis virens

Glycera dibranchiata, p. 648

DIOPATRA CUPREA 110:14

Description
Size, 12 in. (30.5 cm.) long, ½ in. (1.3 cm) wide. Body with 5 tentacles in front, tapering rows of feathery gills on front ⅓ of body. Color bluish green with metallic iridescence, tentacles pale, gills bright red.
Habitat
Shallow water or intertidal zone.
Remarks
The tube of this worm is commonly abundant and conspicuous in the intertidal zone or protected beaches or sand flats. About the size of a pencil, it is shaped like a ship's funnel and is adorned with bits of algae and shell. About an inch or so of the tube sticks up above the sand surface.
Range
Cape Cod and s.

BLOOD WORMS
Glycera dibranchiata and *americana* *Fig. 139*

Description
Size, 6–8 in. (15.3–20.3 cm) long. Cylindrical body, pointed at anterior end. Easily recognized by long proboscis that is periodically shot out of the mouth when the worm is handled. Proboscis armed at tip with 4 black jaws.
Habitat
Muddy sand from intertidal zone to deep water; live in a relatively complex burrow system.
Remarks
Widely used for bait.
Range
Cape Cod to the Carolinas.

LUGWORMS
Arenicola (species) *Fig. 140*

Description
Size, 6–12 in. (15.2–30.5 cm) long, dia. up to that of a finger. Body large, cylindrical; parapodia reduced to ridges; 11–13 pairs of lateral gills. Color light tan, green, greenish black.
Habitat
Intertidal zone. Live in L-shaped burrows in sand; top of burrow often marked by castings. Large transparent gelatinous egg masses, oval or elongate, commonly attached to top of burrow.
Range
Entire E. Coast.

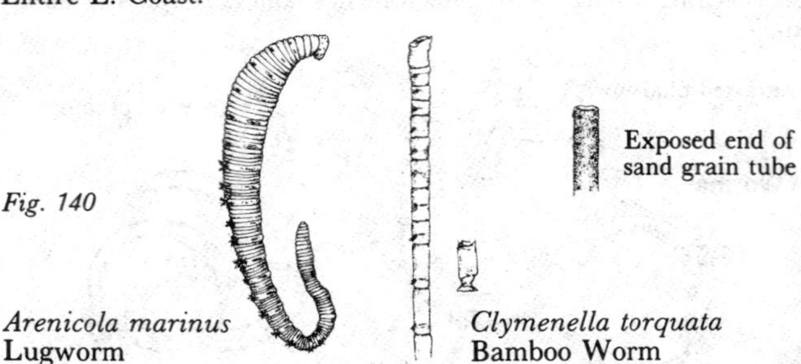

Fig. 140

CHAETOPTERUS VARIOPEDATUS — *Fig. 141*

Description
Size, 6 in. (15.2 cm) long. Strange-looking, flaccid worm; body with 3 sections, flattened anterior region, middle region with semicircular parapodia, posterior section with rather elongate parapodia. Most easily recognized by secreted parchmentlike tube, U-shaped, 1–1½ ft. (30.5–45.7 cm) long, about 1½ in. (3.8 cm) dia. at middle; ends, about size of large drinking straws, stick ½–1 in. (1.3–2.5 cm) above sand surface, about 1 ft. (0.3 m) apart.
Habitat
Sandy protected bottoms from the low-tide mark to 20 ft. (6.1 m).
Range
Cape Cod to Fla.

Note: Living in the tube with *Chaetopterus* are often found both sexes of the little commensal crabs *Pinnixa chaetopterana* or *Polyonyx gibbesi.* The latter species can be recognized by the very small fifth pair of legs.

Marine Segmented Worms

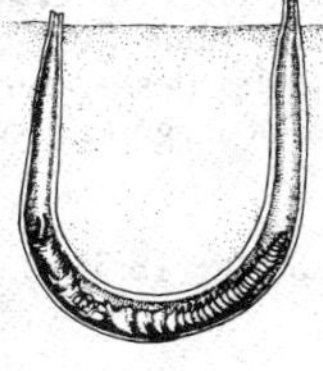

Chaetopterus variopedatus

Fig. 141

Pectinaria gouldii
Trumpet Worm, p. 650

Sabellaria vulgaris, p. 650

BAMBOO WORMS

Clymenella torquata and *Axiothella mucosa* — *Fig. 140*

Description
Size, to 4 in. (10.2 cm) long. Elongated segments give worm the appearance of a bamboo cane. Straight sand-grain tubes up to size of large matchsticks planted upright in sandy bottoms; end of tube projects about ½ in. (1.3 cm) above sand surface.
Habitat
Sandy bottoms in intertidal zone and shallow water. *C. torquata* found to 300 ft. (91.4 m).
Range
C. torquata, Bay of Fundy to the Carolinas; *A. mucosa,* Cape Hatteras and s.

AMPHITRITE ORNATA — **110:15**

Description
Size, 12–15 in. (30.5–38.1 cm) long. Body bears great mass of coiling tentacles at anterior end and 3 pairs of branched gills on dorsal side just behind tentacles. Color pinkish, gills red.
Habitat
Lives in mucus-lined burrow in muddy bottoms from shallow water to intertidal zone.

Remarks
Amphitrite belongs to the family Terebellidae, whose members can easily be recognized by the mass of coiling tentacles at the anterior end.
Range
Cape Cod and s.

Note: The similar *johnstoni* has a brownish body with about 24 pairs of setae clusters in anterior body region. It occurs from the Arctic to Cape Cod.

TRUMPET WORM

Pectinaria gouldii *Fig. 141*

Description
Size, tube to 1½ in. (3.8 cm) long, ¼ in. (0.64 cm) wide; worm slightly longer than tube in which it lives. Worm with 2 feathery gills on each side of head. Head with oblique double coronet of stiff bristles for digging; when withdrawn form a defensive phalanx. Tentacles numerous and in a cluster; each flattened, folded lengthwise to form a groove underneath. Body flesh-colored, mottled; gills bright red; bristles golden; tentacles flesh-colored. Tube trumpet-shaped, formed of single layer of large and small sand grains.
Habitat
Buried in sand and mud from low water to 60 ft. (18.3 m).
Range
Maine to Fla.

SABELLARIA VULGARIS

Fig. 141

Description
Size, less than 1 in. (2.5 cm) long. Worm with concentric rows of yellow setae at anterior end. Firm sand-grain tubes plastered on shells and stones, frequently massed in colonies.
Habitat
Intertidal zone and shallow water.
Range
Cape Cod to Va.

FEATHER DUSTER WORMS, FAN WORMS, PEACOCK WORMS

Families Sabellidae and Serpulidae

Description
Two families of tube-dwelling polychaetes in which the anterior end bears a crown of long featherlike projections, or radioles. (The radioles are arranged in a funnel in many species and as two spirals in others.) Most easily distinguished by their tubes.
Habitat
Most members of both families attach their tubes to firm objects in shallow water.
Habits
The sabellids secrete membranous tubes or ones composed of foreign particles; serpulids secrete white tubes of calcium carbonate, and one radiole is modified as a stopper.
Range
Both worldwide.

Note: Two common serpulids are *Hydroides dianthus* (*Fig. 142*), which builds twisted chalky tubes on shells or stones, and *Spirorbis borealis,* which constructs little spiral tubes resembling snail shells attached to seaweeds. *S. borealis* (*Fig. 142*) occurs from Bay of Fundy to Cape Cod, but other species of *Spirorbis* are found beyond that range.

Fig. 142

Hydroides dianthus
(Segmented Worm)

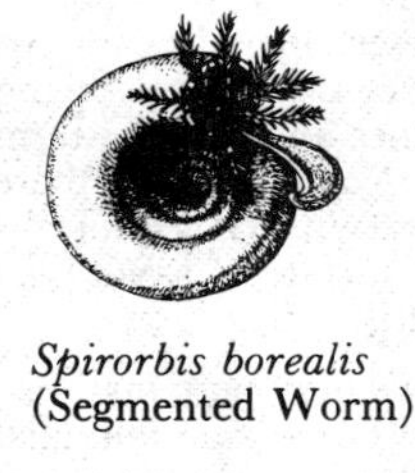

Spirorbis borealis
(Segmented Worm)

Echiurus pallasii
(Echiuran Worm)

ECHIURANS

Phylum Echiura

The Echiura is a small phylum of worms that differ from their relatives the annelids in lacking segmentation and in possessing a fleshy, shovel-shaped anterior end, called the proboscis. The trunk is somewhat sausage-shaped. Echiurans burrow in mud and sand or inhabit rock crevices. One species along the east coast of the United States lives within old sand-dollar tests.

ECHIURUS PALLASII

Fig. 142

Description
Length, 1–12 in. (2.5–30.5 cm). Proboscis fleshy, spoon-shaped. Body gray, yellow, or orange; proboscis light yellow with orange grooves.
Habitat
Mud, sand, and rock bottoms in shallow water.
Range
Arctic to the Carolinas.

ARTHROPODS

Phylum Arthropoda

The phylum Arthropoda contains the spiders, mites, crabs, shrimps, insects, centipedes, and millipedes and is the largest phylum of animals. There are more species of arthropods than of all other animals combined. The body of arthropods is covered by an external skeleton containing a hard substance called chitin. This exoskeleton, or carapace, is the so-called shell of crabs and shrimps. Periodic molting, or shedding, of the exoskeleton permits growth to occur. The phylum is characterized by jointed legs, from which the name *arthropod*, or "jointed feet," is derived. Arthropods are segmented, but the segmentation is often modified and not as distinct as in annelids.

HORSESHOE CRABS

Class Merostomata

The members of this little group of arthropods are not crabs at all but are related to the terrestrial scorpions and spiders. Like their relatives, they lack antennae, and the head and anterior part of the trunk are fused together as a cephalothorax. The cephalothorax has the outline of a horseshoe and bears a pair of large lateral eyes. Behind the cephalothorax is the large triangular abdomen, which carries short lateral spines and a large, terminal, spikelike telson. On the underside of the cephalothorax there are six pairs of leglike appendages, the first five with claws and the sixth modified for clearing the body undersurface of sand. On the underside of the abdomen are leaflike gills.

Horseshoe crabs live on sandy bottoms in shallow water, where they feed on soft-bodied burrowing animals. They crawl with their legs but may use the telson for pushing or righting themselves if turned over. Horseshoe crabs are harmless animals, and the telson makes a convenient handle with which to pick them up or carry them.

HORSESHOE CRAB

Limulus polyphemus *Fig. 143*

Description
Size, to 28 in. (71.1 cm) long or more. Adults dark brown, young yellowish.
Habitat
Sandy bottoms in shallow water.
Range
N.S. to Fla.

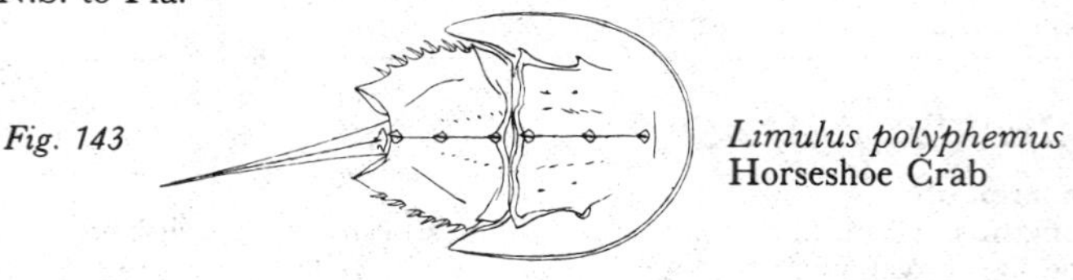

Fig. 143 *Limulus polyphemus* Horseshoe Crab

CRUSTACEANS

Class Crustacea

Crustaceans are distinguished from other arthropods by their two pairs of antennae. They are largely aquatic animals and are the principal representatives of the phylum in the sea. The great diversity in form and size ranges from microscopic wormlike species which live between sand grains to certain crabs with 10-ft. (3.05-m) leg spans.

BARNACLES

Subclass Cirripedia

Barnacles are the only crustaceans that are attached, living on stones, pilings, and other objects. The larva, which is free-swimming, attaches at the end of its planktonic life by means of a cement gland at the base of the antennae. The fixed animal then becomes encased within protective calcareous plates, another distinctive feature of the group. Two pairs of these plates are movable and guard the opening into the mantle cavity, which houses the legs and other parts of the body. The legs, or cirri, bear long bristles (setae) and can be extended out of the opening like scoops, collecting plankton as food.

Not surprisingly, there are species of barnacles which attach to other animals, such as crabs, turtles, and whales, and species which are parasitic. There are also some barnacles which bore into coralline rock or old shells.

Stalked Barnacles

Suborder Lepadomorpha

GOOSE BARNACLES

Lepas (species) **110:9**

Description
Size, about 2 in. (5.1 cm) long. Only 5 calcareous plates; encased part of body laterally compressed. Attach by a long, heavy, naked stalk.
Habitat
Attached to floating objects, such as timbers and bottles, often in dense clusters.
Range
Entire E. Coast; most species cosmopolitan.

Acorn Barnacles

Suborder Balanomorpha

The barnacles below lack a stalk and are called sessile or acorn barnacles. The two pairs of movable plates are surrounded by a ring of vertical plates forming a rigid circular wall. The attachment surface is called the basis.

CHTHAMALUS FRAGILIS

Description
Size, to ⅜ in. (0.96 cm) dia. Basis membranous. Plates brown or gray.
Habitat
Marsh grass, rocks, pilings in intertidal zone.
Range
Cape Cod to the Carolinas.

BALANUS BALANOIDES

Fig. 144

Description
Size, to ⅜ in. (0.96 cm) dia. Basis membranous. Plates white, smooth.
Habitat
Rocks, pilings in intertidal zone.
Range
Bay of Fundy to Cape Hatteras.

BALANUS EBURNEUS

Fig. 144

Description
Size, to 1 in. (2.5 cm) dia. Basis calcareous. Plates white, smooth.
Habitat
Rocks, pilings in intertidal zone; will tolerate low salinities.
Range
Maine to West Indies.

Fig. 144
Acorn Barnacles

Balanus eburneus

Balanus balanoides, p. 653

Balanus improvisus

Balanus crenatus

Balanus amphitrite

BALANUS CRENATUS

Fig. 144

Description
Size, to 1 in. (2.5 cm) dia. Basis calcareous. Wall plates white, rough; edges of wall sharp, saw-toothed.
Habitat
Shells, rocks from tide mark to 300 ft. (91.4 m).
Range
Bay of Fundy to Long Island Sound.

BALANUS IMPROVISUS

Fig. 144

Description
Size, to ½ in. (1.3 cm) dia. Basis calcareous. Wall plates white, surface and edges smooth.
Habitat
Rocks, pilings in intertidal zone; tolerates water of low salinity.
Range
N.S. to Patagonia.

BALANUS AMPHITRITE

Fig. 144

Description
Size, to ⅜ in. (0.96 cm) dia. (greater in south). Basis calcareous. Wall plates smooth, tinged or striped with pink, red, or purple.
Habitat
Rocks, pilings in intertidal zone to 150 ft. (45.7 m).
Range
Cape Cod to the West Indies.

Isopods

Order Isopoda

Members of this order have somewhat dorsoventrally flattened bodies. Viewed from above, the anterior half, or thoracic region, bearing the seven pairs of legs, is not sharply demarcated from the posterior half, or abdomen, which carries flattened respiratory appendages. The young are brooded in a marsupium beneath the thorax.

Sow bugs, pill bugs, and wood lice are familiar land forms, but the far greater number of isopods are marine, either free-living or parasitic. The latter are often twisted or distorted with one side shorter than the other.

The following species are a few common forms that may be encountered in the intertidal zone or in shallow water.

FISH LOUSE

Lironeca ovalis — *Fig. 145*

Description
Size, 1 in. (2.5 cm) long, ½–¾ in. (1.3–1.8 cm) wide. Body oval, flattened, asymmetrical; foremost pair of antennae widely separated. Thoracic segments, 7; head compactly set in the first; legs, doubly curved, hooked, prehensile.

Habitat
Coastal waters where bluefish and other hosts abound.

Habits
Parasitic on the gills and in the mouths of fish.

Range
Cape Cod to Tex.

Fig. 145
Isopods

Gammarus locusta
Seaweed Hopper, p. 657

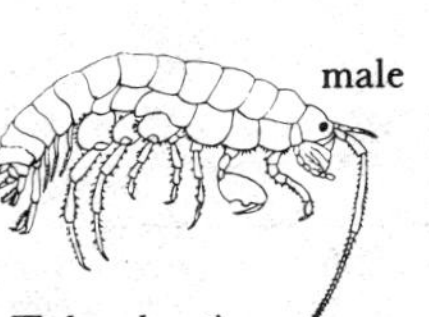

Talorchestia longicornis, p. 657

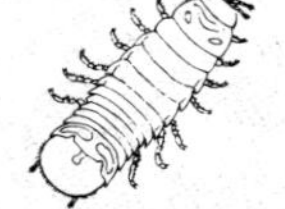

Lironeca ovalis
Fish Louse

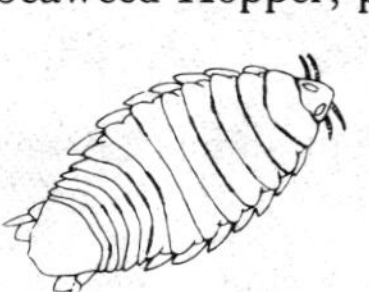

Limnoria lignorum
Gribble

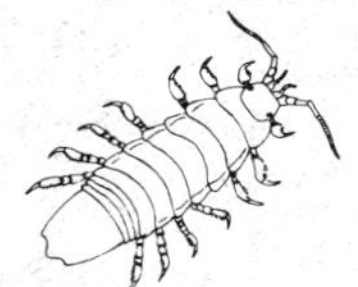

Idotea baltica, p. 656

GRIBBLES

Limnoria lignorum and *tripunctata* — *Fig. 145*

Description
Size, $\frac{3}{16}$ in. (0.48 cm) long. Body 3 times longer than wide. Color gray.

Habitat
Wood borers, attacking the surface layers of pilings and other submerged timbers in shallow water.

Range
L. lignorum, Newfoundland to N.C.; *L. tripunctata,* se. U.S.

IDOTEA (Species) *Fig. 145*

Description
Size, to ¾ in. (1.8 cm) long. Abdomen appears from above to be composed of three segments counting terminal telson. Eyes round, large. Color gray, brown, greenish.
Habitat
In algae, eelgrass, gravel bottoms, swimming in shallow to deep water.
Range
Entire E. Coast.

SPHAEROMA QUADRIDENTATUM *Fig. 146*

Description
Size, ⅜ in. (0.96 cm) long. Body oval in outline; abdomen from above with only two visible segments counting terminal telson. Color gray.
Habitat
On pilings, in old barnacle shells, on algae, beneath stones, from intertidal zone to 3 ft. (91.4 cm).
Habits
Capable of rolling up into a ball.
Range
Cape Cod to Key West, Fla.

Fig. 146
Isopods

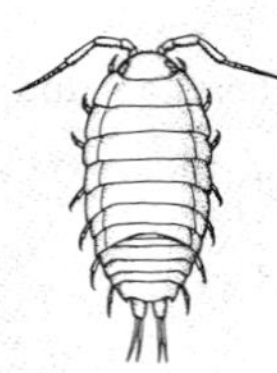

Lygia oceanica

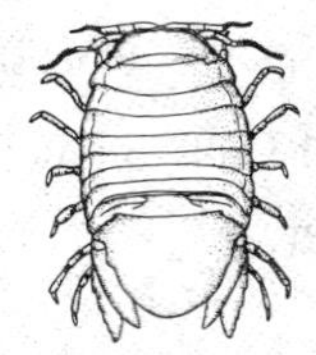

Sphaeroma quadridentatum

LYGIA OCEANICA AND EXOTICA *Fig. 146*

Description
Size, to 1 in. (2.5 cm) long. Resemble large pill bugs. Color gray.
Habitat
Intertidal and supratidal zone, but animals always stay above water level.
Habits
Run rapidly over docking, pilings, and stones above the water line.
Range
L. oceanica, n. of Cape Cod; *L. exotica*, Del. and s.

Amphipods

Order Amphipoda

In members of this order the body is usually laterally compressed and the axis strongly curved. They may be distinguished from small shrimps and prawns by the absence of a carapace, by the body that is jointed and flexible along its entire length, and by their smaller size (most are less than one-half inch long). Some of the legs (usually the first two, called gnathopods) are subchelate; that is, the last segment, the finger, closes against the end of the preceding segment. As in isopods, the young are carried in a marsupium beneath the thorax.

There are over 5500 species in this order, almost exclusively marine. Many are found in the beach wrack or climbing on hydroids and ascidians attached to rocks and pilings. Identification of species is difficult without magnification and knowledge of technical characters. The following groups can be recognized most easily by their habitat.

SEAWEED HOPPERS, SCUDS

Fig. 145

Family Gammaridae

Description
Size, ¼–1 in. (0.64–2.5 cm) long. Color brown, olive, mottled.
Habitat
Common within algae and under stones in intertidal zone and very shallow water, i.e. *Gammarus locusta*.
Habits
Scurry rapidly, listing to one side.
Range
Entire E. Coast.

BEACH FLEAS, SAND FLEAS, BEACH HOPPERS

109:15
Fig 145

Family Talitridae

Description
Size, ¼–1 in. (0.64–2.5 cm) long. Of the two genera, *Orchestia* and *Talorchestia*, the latter reaches the largest size; species of *Orchestia* are no longer than ½ in. (1.3 cm). Color waxy-white to brown or gray.
Habitat
High-tide mark; semiterrestrial, living beneath seaweed and driftwood washed up on beaches by extremely high tides.
Habits
Ability to jump is a distinctive feature (very noticeable when drift is turned over).
Range
Entire E. Coast.

WOOD-BORING AMPHIPOD

Chelura terebrans

Fig. 147

Description
Size, ¼–⅜ in. (0.48–0.64 cm) long.
Habitat
Bores into pilings and other submerged marine timbers much like the isopod *Limnoria*.
Range
Cape Cod to Cape Hatteras.

Fig. 147

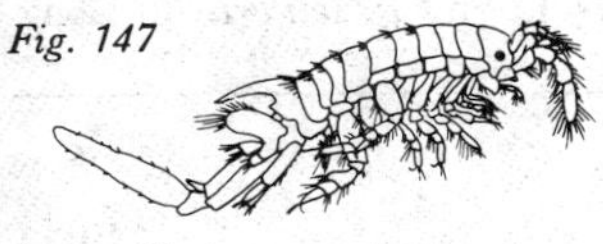

Chelura terebrans
Wood-boring Amphipod

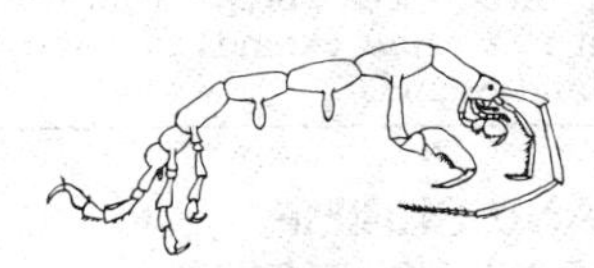

Aeginella longicornia, p. 658
(Skeleton Shrimp)

SKELETON SHRIMP
Family Caprellidae *Fig. 147*

Description
Size, to 1 in. (2.5 cm.) long, but usually less. Strange-looking aberrant amphipods with long, narrow, rather cylindrical bodies with a greatly reduced abdomen.
Habitat
On attached organisms in shallow and deep water; may be especially abundant on hydroids and such erect bryozoans as *Bugula*.
Habits
Climb about various attached organisms, using the grasping claws at the ends of their legs.
Range
Entire E. Coast.

Shrimps, Lobsters, and Crabs
Order Decapoda

These are the most highly organized crustaceans. The head and thorax are rigidly fused to each other and form the cephalothorax, which is covered by the carapace. The anterior tip of the head often projects forward as a rostrum (snout), like the prow of a ship. The ten pairs of legs give this order its name; although some terminate in pincers, most of them are adapted for walking or swimming. The majority of this group are marine, but some have invaded fresh water, and a few live on land. Decapods include familiar shrimps, prawns, lobsters, crayfishes and crabs. This diverse order is divided into a number of sections and tribes, encompassed by the suborders Natantia and Reptantia.

Swimming Decapods
Suborder Natantia

EDIBLE SHRIMP
Penaeus aztecus, setiferus, and *duorarum*

Description
Size, body to 6 in. (15.2 cm) long, antennae to 12 in. (30.5 cm). The largest shrimps on the East Coast. Rostrum ⅔ length of carapace, forms a low keel on the carapace with lateral grooves. Color brownish green.
Habitat
Shallow water to about 300 ft. (91.4 m); tolerate low salinities.
Range
All three species occur from the Chesapeake to the Gulf of Mexico, but *P. aztecus* extends to Cape May, N.J., and *P. setiferus* to Long Island Sound.

SAND SHRIMP
Crangon septemspinosa **110:11**

Description
Size, to 2 in. (5.1 cm) long. Body somewhat tubular, tapers to tail; gray-green, translucent on sand, darker on mud. Rostrum short, front end blunt; this and the comparatively large size help identify

this animal when swimming, although its translucency makes it hard to see.

Habitat

Mud, sand, surf, drainage ditches.

Remarks

This species is the food of many fishes. Other members of the genus are eaten by man, more so on the West Coast and abroad than in eastern North America.

Range

Labrador to N.C.

GRASS SHRIMP

Palaemonetes (species) *Fig. 148*

Description

Size, 1 in. (2.5 cm) long. The three eastern species are difficult to separate and have similar ranges. Tip of rostrum forked or unforked, with teeth on underside; two sets of small claws, the second slightly larger; first segment of abdomen partly hidden by plate of second segment. Color more or less transparent, spotted with brown.

Habitat

Eelgrass, in ditches, salt marshes.

Range

Maine to Tex.

Fig. 148

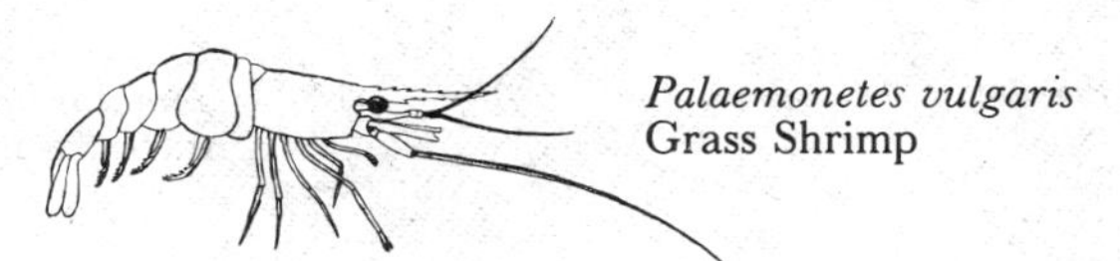

Palaemonetes vulgaris
Grass Shrimp

SNAPPING SHRIMP

Family Alphaeidae

Description

Size, to 2 in. (5.1 cm) long, but usually smaller. Distinguished by the single greatly enlarged claw.

Habitat

Among oyster shells, within rock crevices.

Remarks

The structure and closing mechanism of the claw is such that it produces a popping sound. The claw functions in threat displays toward other snapping shrimp of the same species, apparently ensuring that pairs of shrimp are spaced out within available habitats.

Range

The family is largely tropical and semitropical, but some species extend as far north as Va. on the E. Coast.

Creeping or Walking Decapods

Suborder Reptantia

AMERICAN LOBSTER

Homarus americanus *Fig. 149*

Description

Size, to 12 in. (30.5 cm) long, sometimes more; occasionally over 2 ft. (0.6 m) and up to 44 lb. (24 kg). The largest crustacean in the range. Claws, 2, very large, asymmetrical; one has crushing edge, other is a sharp grasper.

Habitat

Rocky or sandy bottoms from shallow to very deep water.

Remarks

The greatest commercial catch is made between Nova Scotia and Long Island. The excellent flavor of the lobster is a perpetual threat to its numbers, which in many areas have been sharply reduced. A wise conservation program rigidly enforced is essential in order to guard against the further depletion of this natural resource.

Range

Labrador to the Carolinas.

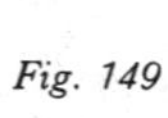

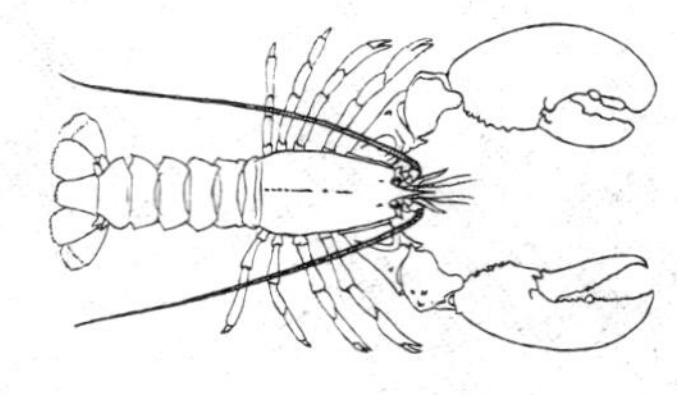

Fig. 149 *Homarus americanus* American Lobster

MOLE CRAB, SAND BUG

Emerita talpoida **110:2**

Description

Size, 1½ in. (3.8 cm) long. Carapace olive-shaped, shiny white with faint purple markings. First antenna small, second long and feathery. First pair of legs directed forward; second, third, and fourth strong, leaflike at tips; fifth pair threadlike, hidden.

Habitat

Clean sand between tides on surf beaches.

Habits

Animals reburrow rapidly after being washed out by incoming waves.

Remarks

One of the few conspicuous inhabitants of ocean-front beaches.

Range

Cape Cod to Fla.

MUD SHRIMP, BURROWING SHRIMP

Upogebia affinis *Fig. 150*

Description

Size, to 4 in. (10.2 cm) long. Shrimplike, but abdomen dorsoventrally flattened as in lobsters and crayfish. First pair of legs with equal-sized claws. Exoskeleton rather soft. Color pale gray.

Habitat
Lives in burrows in muddy sand or shelly mud.
Range
Cape Cod and s.

Note: The similar *Callianassa atlantica* has claws of unequal size and burrows in cleaner sand than does *Upogebia*. It occurs from N.C. to La.

Fig. 150

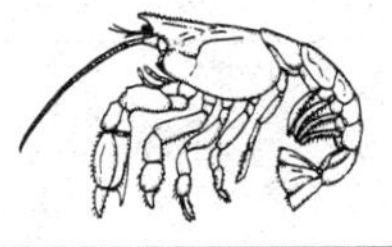

Upogebia affinis
Mud Shrimp

HERMIT CRABS AND MOLE CRABS

Section Anomura

Hermit crabs house themselves in the empty shells of snails, whelks, conchs, or other gastropod mollusks. They cannot live without a shell and must locate another empty shell when they outgrow the old one. The shell is carried with the fourth or fifth pair of legs, which is jammed against the inner shell wall, and with a hook-shaped appendage at the end of the abdomen. The abdomen is asymmetrical and soft. Mole crabs do not house themselves in empty mollusk shells.

PAGURUS POLLICARIS 110:1

Description
Size, carapace to 1½ in. (3.8 cm) long. Claws covered with fine tubercles; right much larger than left, wide and flat, acting as a "door" to the shell when animal is withdrawn. Color reddish brown or tan, claws white or gray.
Habitat
Shallow water over sand or mud bottoms.
Range
Mass. to Tex.

PAGURUS LONGICARPUS

Description
Size, carapace ⅜ in. (0.96 cm) long. Larger right claw subcylindrical. Color variable, upper surface of claws and legs may be iridescent.
Habitat
Shallow water to 150 ft. (91.4 m).
Range
Mass. to Tex.

STRIPED HERMIT CRAB

Clibanarius vittatus

Description
Size, carapace to 1¼ in. (5.2 cm) long. Claws equal. Color greenish to dark brown; legs with white, gray, or light orange stripes.
Habitat
Low-tide mark to 6 ft. (1.8 m).
Range
Va. to Brazil.

TRUE CRABS

Section Brachyura

A short body form has evolved in decapods as a consequence of reduction and folding of the abdomen beneath the cephalothorax. These decapods are called crabs. The cephalothorax is as wide as or wider than it is long, and a sideways gait is usually well developed.

SPIDER CRAB

Libinia emarginata *Fig. 151*

Description
Size, carapace to 4 in. (10.2 cm) long. Body somewhat heart-shaped, rostrum long and forked; carapace with tubercles and with 9 spines on middle line. Legs long, slender, giving crab a spiderlike appearance. Color brownish or whitish; algae growing on shell may vary this appearance.
Habitat
Mud, rocks, and oyster beds; common.
Range
N.S. to Fla.

Note: The similar species, *L. dubia,* has a longer, more deeply divided rostrum and only 6 spines on its middle line. It occurs from Cape Cod south.

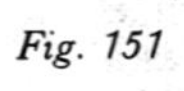
Fig. 151

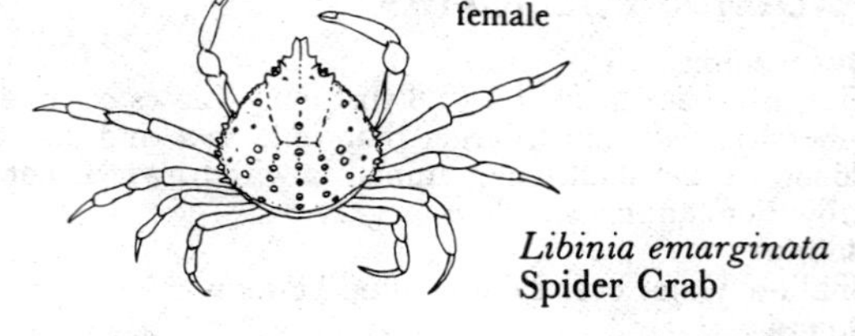

Libinia emarginata
Spider Crab

GREEN CRAB

Carcinus maenas *Fig. 152*

Description
Size, carapace 2¼ in. (5.7 cm) long. Claws short, sturdy. Last legs slightly flattened. Color orange, red, or dark green, mottled with black and yellow.
Habitat
Rocky bottoms in shallow water.
Range
Maine to N.J. Since 1900, has extended its range n. from Cape Cod to N.S.

Fig. 152

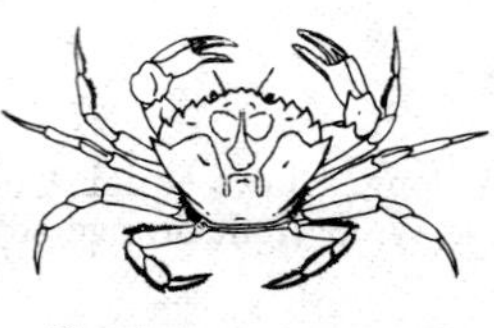

Carcinus maenas
Green Crab

Pinnotheres ostreum
Oyster Crab, p. 664

LADY CRAB

Ovalipes ocellatus **110:3**

Description
Size, 2 in. (5.1 cm) long, 2¼ in. (5.7 cm) wide. Last legs equipped with swimming paddles. Color bluish white with red or purple spots.
Habitat
Sandy bottoms; common.
Range
Cape Cod to Gulf of Mexico.

BLUE CRAB

Callinectes sapidus **110:4**

Description
Size, 3 in. (7.6 cm) long, 7 in. (17.8 cm) wide. Carapace wide, having front margin with spines and ending with a large spike on either side; 4 teeth between eye sockets; last pair of legs paddle-shaped. Carapace gray- or blue-green, legs blue.
Habitat
Grassy, sandy, and muddy bottoms from low-tide mark to 100 ft. (30.5 m). Can tolerate very low salinities. Common in bays, inlets, and estuaries.
Remarks
Members of the family Portunidae, to which the Lady Crab and Blue Crab belong, have the last pair of legs paddle-shaped and adapted for swimming. There are other species having the general form of the Blue Crab but differing in structural details and in coloration.
Range
Cape Cod to Tex.

ROCK CRAB

Cancer irroratus **110:7**

Description
Size, carapace 3 × 5 in. (7.6 × 12.7 cm). Carapace transversely oval, granulated but smooth, yellowish with bronzy-purple spots; sometimes with light spots in circular design, and light borders to front spines.
Habitat
Among rocks between tide marks to a depth of several hundred feet.
Range
Labrador to Fla.; in deeper water south of Cape Cod.

Note: The similar **JONAH CRAB,** *C. borealis,* has the large marginal teeth of the carapace subdivided into smaller teeth. It occurs from Nova Scotia to Florida.

SESARMA RETICULATUM

Description
Size, ⅞ in. (2.2 cm) long, 1⅛ in. (2.9 cm) wide. Carapace a short transverse rectangle. Color dark olive to black or purple.
Habitat
Amphibious; lives in burrows in mud in salt marshes.
Range
Woods Hole to Tex.

SESARMA CINEREUM

Description
Size, ¾ in. (1.8 cm) long. Carapace almost square. Color brown.
Habitat
Semiterrestrial; lives beneath drift at the high-tide mark on protected beaches; sometimes found at the high-tide mark on pilings and jetties.
Range
Chesapeake Bay to British Honduras.

MUD CRABS
Family Xanthidae

Description
Size, usually less than 1 in. (2.5 cm) wide. Carapace transversely oval with teeth on the laterofrontal margin. Legs and claws stout. Color drab, mostly dark gray and brown; claws with black, white, and brown.
Habitat
Beneath rocks and in oyster clumps and reefs on sandy and muddy bottoms; very common. Some species tolerate very low salinities. The six species which inhabit eastern shores are difficult to distinguish.
Remarks
Panopeus herbstii is the easiest of the six species to distinguish. It reaches a length of 1 in. (2.5 cm) and possesses a large, blunt tooth near the base of the movable finger of the large claw. It may be found as far north as Massachusetts.
Range
Entire E. Coast.

OYSTER CRAB OR PEA CRAB
Pinnotheres ostreum *Fig. 152*

Description
Size, to ¾ in. (1.8 cm). Carapace round, pinky white, smooth.
Habitat
The eastern crab that is found inside the shells of living oysters. Females spend their lives commensally within oysters. Males are more free-living.
Range
Cape Cod to Tex.

Note: The similar but hairy **MUSSEL CRAB,** *P. maculatus,* of similar range, lives in the shells of mussels, clams, and scallops.

GHOST CRAB
Ocypode quadrata **110:5**

Description
Size, carapace 1½ in. (3.8 cm) long, 2 in. (5.1 cm) wide. Carapace rectangular. Pale and ghostly looking, the color of dry sand. Eye stalks very prominent.
Habitat
Terrestrial; lives in burrows in dunes above tide mark.
Habits
Comes out at night to feed.
Range
N.J. to Fla.

MUD FIDDLER CRAB
Uca pugnax — *Fig. 153*

Description
Size, carapace 1 in. (2.5 cm) wide. Carapace trapezoidal. Inner surface of great claw ridged diagonally from base of fixed finger toward hind margin of hand or palm. Color dark olive; great claw paler than carapace. Distinguished from *U. minax* by lack of red joints.
Habitat
Salt marshes.
Range
Cape Cod to Mexico.

Note: Members of the genus *Uca* are amphibious. They make their home in burrows excavated in the intertidal zone, from which they emerge at low tide to feed on fine organic matter mixed with surface sand or mud. Male fiddlers, in contrast to the female, in which both claws are small and of equal size, have one very large claw.

Fig. 153 *Uca minax*
Red-jointed Fiddler Crab

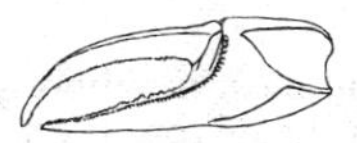

Uca pugilator
Sand Fiddler Crab

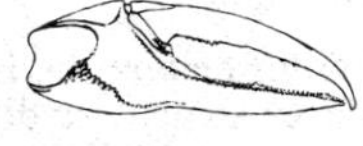

Uca pugnax
Mud Fiddler Crab

SAND FIDDLER CRAB
Uca pugilator — **110:6** *Fig. 153*

Description
Size, to 1¼ in. (3.2 cm). Similar to *U. pugnax* but slightly larger, and rear of great claw more or less evenly pebbled on all sides, with no strong oblique ridge as in other fiddlers; no red markings at joints of great claw as in *U. minax*.
Habitat
On protected sand or sand-mud beaches where the salinity of the water is relatively high.
Habits
Often occur in large aggregations.
Range
Boston to Tex.

RED-JOINTED FIDDLER CRAB
Uca minax — *Fig. 153*

Description
Size, carapace 1½ in. (3.8 cm) across. The only fiddler in the range with red joints on the great claw. Forward ⅓ light, rear ⅔ dark; H in middle with red crossbar. Great claw diagonally ridged within.
Habitat
From beach inland almost to fresh water.
Range
Cape Cod to Tex.

Mantis Shrimps

Order Stomatopoda

The members of this small order of crustaceans are not decapods, although they are about the same size as true shrimps. The distinctive feature of the group is the large, powerful second pair of legs, resembling those of the praying mantis. The terminal piece of this pair of legs folds into the penultimate segment like the blade of a pocket knife. These predatory crustaceans are most widely distributed in tropical and subtropical waters.

MANTIS SHRIMP

Squilla empusa **110:10**

Description
Size, to 7 in. (17.8 cm) long. Thorax short, abdomen long. Second legs large, with terminal piece folding into penultimate segment. Color yellowish green, segments edged with yellow, tail rosy.
Habitat
Burrows in sandy bottoms in shallow water.
Range
Cape Cod to Fla.

CHORDATES

Phylum Chordata

To this phylum belong the largest and most highly organized animals. Three structures distinguish them from the members of other phyla: (1) a notochord, a cyclindrical stiff supporting rod of tissue dorsal to the gut; (2) a dorsal hollow nerve cord (the spinal cord), which lies above the notochord; and (3) lateral pharyngeal clefts, or gill slits.

The vertebrate chordates—fish, reptiles, birds, and mammals—are very familiar. The notochord in most of these chordates is found only in the embryo, and the vertebrae develop around the embryonic notochord. In higher terrestrial vertebrates the gill slits are formed in embryonic life but become modified or lost in the course of later development. The outer ear canal of humans is a modified part of a gill slit.

There are also some chordates without backbones. These invertebrate chordates are all marine and include some very common animals, but ones which are unfamiliar to most laymen.

TUNICATES OR SEA SQUIRTS

Class Ascidiacea

Sea squirts are a very specialized group of marine chordates. In general they are baglike sedentary animals, with an incurrent and an excurrent siphon. The body is covered by a transparent or opaque envelope, or tunic, which contains tunicin, a substance very similar to the cellulose found in plants. Sea squirts may be either solitary or colonial. In colonial species, the zooids (individuals) are usually very small and are embedded in a common tunic, which may form a tough, flexible, attached mass.

The incurrent siphon brings water and food particles to the perforated pharynx, through which the water is filtered. The food particles are trapped in films of mucus and so carried to the mouth. When sea squirts contract at low tide, they often eject a stream of water from the siphons. The incurrent siphons of colonial forms are separate, but the excurrent siphons may open into a common cloaca.

The tiny free-swimming larva of these animals resembles a tadpole and possesses all of the chordate characteristics. Following attachment, the notochord and dorsal hollow nerve cord degenerate.

STYELA PARTITA

110:8

Description
Size, to 1 in. (2.5 cm) long. Irregularly oval; attaches at posterior end or along entire ventral surface. Outer covering yellowish brown, tough, with an irregular surface. Openings of siphons rectangular, terminating prominent tubes marked with triangular spots of purple and white.
Habitat
On stones, shells, pilings.
Range
Maine to West Indies.

SEA GRAPES

Molgula manhattensis

Fig. 154

Description
Size, to 1 in. (2.5 cm) dia. The only eastern sea squirt with a round ball-shaped body, usually occurring in groups or colonies (hence its common name). Tunic ashy gray, rough, slightly translucent; often covered with sand, bits of seaweed, eelgrass, etc. Siphons long, incurrent with 6 lobes at margin, excurrent with 4.
Habitat
Piles, stones, and eelgrass.
Range
Maine to La.

Fig. 154
Sea Squirts

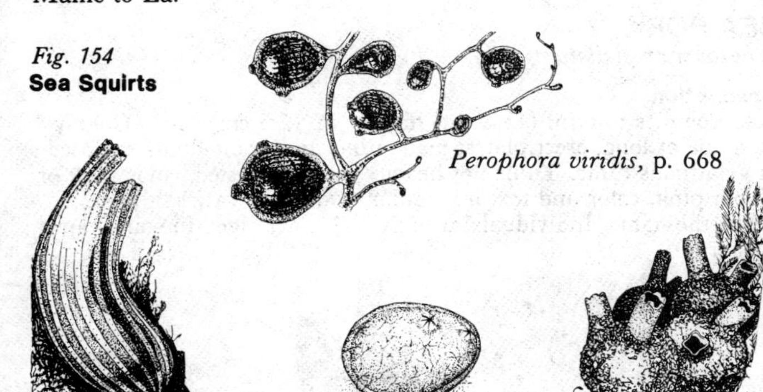

Perophora viridis, p. 668

Ciona intestinalis, p. 668
Sea Vase

Ascidia callosa, p. 668

Molgula manhattensis
Sea Grapes

ASCIDIA CALLOSA AND INTERRUPTA

Fig. 154

Description
Size, 2 × 1 in. (5.1 × 2.5 cm). Body oval, length twice width; flattened, usually attaches on one side. Outer covering translucent; greenish, yellowish, or gray; thick, fleshy, firm. Siphons short.
Habitat
Shallow to deep water on pilings, stones, and shells.
Range
A. callosa, Arctic to Cape Cod; *A. interrupta*, s. of Cape Hatteras.

SEA VASE

Ciona intestinalis

Fig. 154

Description
Size, to 5 in. (12.7 cm) high. Body slender, attaches at rear. Outer tunic translucent, yellow, spotted with orange or red; spots vanish on death. Strong longitudinal muscle bands may be seen through the tunic; 2 siphons at upper end give animal a vaselike shape.
Habitat
Buoys, timbers, rocks.
Range
N.S. to Long Island.

PEROPHORA VIRIDIS

Fig. 154

Description
Size, zooids ⅛ in. (0.32 cm) across. Semicolonial; individual zooids transparent, soft green, or chartreuse; connected together by a vinelike stolon.
Habitat
Rocks, pilings, kelp, or other firm base.
Range
Cape Cod to Fla.

SEA PORK

Amaroucium stellatum

Fig. 155

Description
Size, colonies to 6 in. (15.2 cm) high, 1 in. (2.5 cm) thick. Colony in shape of long, erect plates; made up of tiny individuals enclosed in gelatinous tunic. Tunic not heavily sand-encrusted, color pink or bluish pink; color and texture similar to those of raw salt pork, hence the name. Individuals, usually 6–18, arranged in star-shaped clusters.

Fig. 155

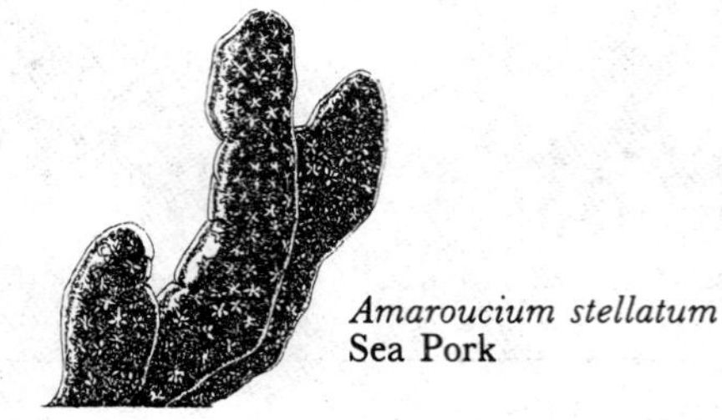

Amaroucium stellatum
Sea Pork

Habitat
Coastal waters, 10–20 ft. (3.1–6.1 m) deep.
Range
Maine to Cape Hatteras.

Note: The similar *A. constellatum* forms pear-shaped masses 1½ to 2 inches (1.3–5.1 cm) high and has a cream-colored tunic through which the reddish-orange individuals may be clearly seen. It occurs from Cape Cod to Florida.

Life Lists

BIRDS

_______ Acadian Flycatcher
_______ Alder Flycatcher
_______ American Avocet
_______ American Bittern
_______ American Coot
_______ American Golden Plover
_______ American Goldfinch
_______ American Kestrel
_______ American Redstart
_______ American Robin
_______ American White Pelican
_______ American Wigeon
_______ American Woodcock
_______ Anhinga
_______ Arctic Loon
_______ Arctic Tern
_______ Audubon's Shearwater

_______ Bachman's Sparrow
_______ Bachman's Warbler
_______ Baird's Sandpiper
_______ Baird's Sparrow
_______ Bald Eagle
_______ Bank Swallow
_______ Barn Owl
_______ Barn Swallow
_______ Barred Owl
_______ Barrow's Goldeneye
_______ Bay-breasted Warbler
_______ Bell's Vireo
_______ Belted Kingfisher
_______ Bewick's Wren
_______ Black-and-White Warbler
_______ Black-backed Woodpecker
_______ Black-bellied Plover
_______ Black-billed Cuckoo
_______ Black-billed Magpie
_______ Blackburnian Warbler
_______ Black-capped Chickadee
_______ Black-crowned Night Heron
_______ Black Duck
_______ Black Guillemot
_______ Black-headed Grosbeak
_______ Black-headed Gull
_______ Black-legged Kittiwake
_______ Black-necked Stilt
_______ Blackpoll Warbler
_______ Black Rail
_______ Black Scoter
_______ Black Skimmer
_______ Black Tern
_______ Black-throated Blue Warbler
_______ Black-throated Gray Warbler
_______ Black-throated Green Warbler
_______ Black-throated Sparrow
_______ Black Vulture
_______ Black-whiskered Vireo
_______ Blue-gray Gnatcatcher
_______ Blue Grosbeak
_______ Blue Jay
_______ Blue-winged Teal
_______ Blue-winged Warbler
_______ Boat-tailed Grackle
_______ Bobolink
_______ Bobwhite
_______ Bohemian Waxwing
_______ Bonaparte's Gull
_______ Boreal Chickadee
_______ Boreal Owl
_______ Brant
_______ Brewer's Blackbird
_______ Brewer's Sparrow
_______ Brewster's Warbler
_______ Bridled Tern
_______ Broad-tailed Hummingbird
_______ Broad-winged Hawk
_______ Brown Creeper
_______ Brown-headed Cowbird
_______ Brown-headed Nuthatch
_______ Brown Pelican
_______ Brown Thrasher
_______ Buff-breasted Sandpiper
_______ Bufflehead
_______ Burrowing Owl

_______ Canada Goose
_______ Canada Warbler
_______ Canvasback
_______ Canyon Wren
_______ Cape May Warbler
_______ Cardinal
_______ Carolina Chickadee
_______ Carolina Wren
_______ Caspian Tern
_______ Cassin's Kingbird
_______ Cassin's Sparrow
_______ Cattle Egret
_______ Cedar Waxwing
_______ Cerulean Warbler
_______ Chestnut-collared Longspur
_______ Chestnut-sided Warbler
_______ Chimney Swift
_______ Chipping Sparrow

_______ Chuck-will's Widow
_______ Clapper Rail
_______ Clark's Nutcracker
_______ Clay-colored Sparrow
_______ Cliff Swallow
_______ Common Crow
_______ Common Eider
_______ Common Flicker
_______ Common Gallinule
_______ Common Goldeneye
_______ Common Grackle
_______ Common Ground Dove
_______ Common Loon
_______ Common Merganser
_______ Common Murre
_______ Common Nighthawk
_______ Common Oystercatcher
_______ Common Puffin
_______ Common Raven
_______ Common Redpoll
_______ Common Snipe
_______ Common Teal
_______ Common Tern
_______ Common Wheatear
_______ Common Yellowthroat
_______ Connecticut Warbler
_______ Cooper's Hawk
_______ Cory's Shearwater
_______ Crested Caracara
_______ Curlew Sandpiper

_______ Dark-eyed Junco
_______ Dickcissel
_______ Dipper
_______ Double-crested Cormorant
_______ Dovekie
_______ Downy Woodpecker
_______ Dunlin
_______ Dusky Flycatcher

_______ Eared Grebe
_______ Eastern Bluebird
_______ Eastern Kingbird
_______ Eastern Meadowlark
_______ Eastern Phoebe
_______ Eastern Wood Pewee
_______ Eurasian Wigeon
_______ European Tree Sparrow
_______ Evening Grosbeak

_______ Ferruginous Hawk
_______ Field Sparrow
_______ Fish Crow
_______ Forster's Tern
_______ Fox Sparrow
_______ Franklin's Gull
_______ Fulvous Whistling-Duck

_______ Gadwall
_______ Gannet
_______ Glaucous Gull
_______ Glossy Ibis
_______ Golden-crowned Kinglet
_______ Golden Eagle
_______ Golden-winged Warbler
_______ Grasshopper Sparrow
_______ Gray Catbird
_______ Gray-cheeked Thrush
_______ Gray-crowned Rosy Finch
_______ Gray Jay
_______ Gray Kingbird
_______ Gray Partridge
_______ Great Black-backed Gull
_______ Great Blue Heron
_______ Great Cormorant
_______ Great Crested Flycatcher
_______ Great Egret
_______ Greater Prairie Chicken
_______ Greater Scaup
_______ Greater Shearwater
_______ Greater Yellowlegs
_______ Great Gray Owl
_______ Great Horned Owl
_______ Great-tailed Grackle
_______ Green Heron
_______ Green-tailed Towhee
_______ Gull-billed Tern
_______ Gyrfalcon

_______ Hairy Woodpecker
_______ Harlequin Duck
_______ Harris' Sparrow
_______ Hawk Owl
_______ Henslow's Sparrow
_______ Hermit Thrush
_______ Herring Gull
_______ Hoary Redpoll
_______ Hooded Merganser
_______ Hooded Warbler
_______ Horned Grebe
_______ Horned Lark
_______ House Finch
_______ House Sparrow
_______ House Wren
_______ Hudsonian Godwit

_______ Iceland Gull
_______ Indigo Bunting
_______ Ivory-billed Woodpecker

_______ Ivory Gull

_______ Kentucky Warbler
_______ Killdeer
_______ King Eider
_______ King Rail
_______ Kirtland's Warbler

_______ Ladder-backed Woodpecker
_______ Lapland Longspur
_______ Lark Bunting
_______ Lark Sparrow
_______ Laughing Gull
_______ Lawrence's Warbler
_______ Lazuli Bunting
_______ Leach's Storm-Petrel
_______ Least Bittern
_______ Least Flycatcher
_______ Least Sandpiper
_______ Least Tern
_______ Le Conte's Sparrow
_______ Lesser Black-backed Gull
_______ Lesser Goldfinch
_______ Lesser Prairie Chicken
_______ Lesser Scaup
_______ Lesser Yellowlegs
_______ Lewis' Woodpecker
_______ Limpkin
_______ Lincoln's Sparrow
_______ Little Blue Heron
_______ Little Gull
_______ Loggerhead Shrike
_______ Long-billed Curlew
_______ Long-billed Dowitcher
_______ Long-billed Marsh Wren
_______ Long-eared Owl
_______ Long-tailed Jaeger
_______ Louisiana Heron
_______ Louisiana Waterthrush

_______ McCown's Longspur
_______ MacGillivray's Warbler
_______ Magnificent Frigatebird
_______ Magnolia Warbler
_______ Mallard
_______ Mangrove Cuckoo
_______ Manx Shearwater
_______ Marbled Godwit
_______ Merlin
_______ Mew Gull
_______ Mississippi Kite
_______ Mountain Bluebird
_______ Mourning Dove
_______ Mourning Warbler
_______ Mute Swan

_______ Nashville Warbler
_______ Neotropical (Olivaceous) Cormorant
_______ Northern Fulmar
_______ Northern Goshawk
_______ Northern Harrier
_______ Northern Mockingbird
_______ Northern Oriole
_______ Northern Parula
_______ Northern Phalarope
_______ Northern Pintail
_______ Northern Shoveler
_______ Northern Shrike
_______ Northern Skua
_______ Northern Waterthrush

_______ Oldsquaw
_______ Olive-sided Flycatcher
_______ Orange-crowned Warbler
_______ Orchard Oriole
_______ Osprey
_______ Ovenbird

_______ Painted Bunting
_______ Palm Warbler
_______ Parasitic Jaeger
_______ Pectoral Sandpiper
_______ Peregrine Falcon
_______ Philadelphia Vireo
_______ Pied-billed Grebe
_______ Pileated Woodpecker
_______ Pine Grosbeak
_______ Pine Siskin
_______ Pine Warbler
_______ Pinyon Jay
_______ Piping Plover
_______ Pomarine Jaeger
_______ Poor-will
_______ Prairie Warbler
_______ Prothonotary Warbler
_______ Purple Finch
_______ Purple Gallinule
_______ Purple Martin
_______ Purple Sandpiper
_______ Pygmy Nuthatch

_______ Razorbill
_______ Red-bellied Woodpecker
_______ Red-breasted Merganser
_______ Red-breasted Nuthatch
_______ Red-cockaded Woodpecker
_______ Red Crossbill
_______ Reddish Egret
_______ Red-eyed Vireo
_______ Redhead
_______ Red-headed Woodpecker

______ Red Knot
______ Red-necked Grebe
______ Red Phalarope
______ Red-shouldered Hawk
______ Red-tailed Hawk
______ Red-throated Loon
______ Red-winged Blackbird
______ Reeve
______ Ring-billed Gull
______ Ring-necked Duck
______ Ring-necked Pheasant
______ Roadrunner
______ Rock Pigeon
______ Rock Ptarmigan
______ Rock Wren
______ Roseate Spoonbill
______ Roseate Tern
______ Rose-breasted Grosbeak
______ Ross' Gull
______ Rough-legged Hawk
______ Rough-winged Swallow
______ Royal Tern
______ Ruby-crowned Kinglet
______ Ruby-throated Hummingbird
______ Ruddy Duck
______ Ruddy Turnstone
______ Ruff
______ Ruffed Grouse
______ Rufous-crowned Sparrow
______ Rufous-sided Towhee
______ Rusty Blackbird

______ Sabine's Gull
______ Sage Thrasher
______ Sanderling
______ Sandhill Crane
______ Sandwich Tern
______ Savannah Sparrow
______ Saw-whet Owl
______ Say's Phoebe
______ Scarlet Tanager
______ Scissor-tailed Flycatcher
______ Screech Owl
______ Scrub Jay
______ Seaside Sparrow
______ Semipalmated Plover
______ Semipalmated Sandpiper
______ Sharp-shinned Hawk
______ Sharp-tailed Grouse
______ Sharp-tailed Sparrow
______ Short-billed Dowitcher
______ Short-billed Marsh Wren
______ Short-eared Owl
______ Short-tailed Hawk
______ Smith's Longspur
______ Smooth-billed Ani
______ Snail Kite
______ Snow Bunting
______ Snow Goose
______ Snowy Egret
______ Snowy Owl
______ Snowy Plover
______ Solitary Sandpiper
______ Solitary Vireo
______ Song Sparrow
______ Sooty Shearwater
______ Sooty Tern
______ Sora
______ Spot-breasted Oriole
______ Spotted Sandpiper
______ Sprague's Pipit
______ Spruce Grouse
______ Starling
______ Steller's Jay
______ Stilt Sandpiper
______ Summer Tanager
______ Surf Scoter
______ Swainson's Hawk
______ Swainson's Thrush
______ Swainson's Warbler
______ Swallow-tailed Kite
______ Swamp Sparrow

______ Tennessee Warbler
______ Thayer's Gull
______ Thick-billed Murre
______ Three-toed Woodpecker
______ Townsend's Solitaire
______ Tree Sparrow
______ Tree Swallow
______ Tufted Titmouse
______ Turkey
______ Turkey Vulture

______ Upland Sandpiper

______ Varied Thrush
______ Veery
______ Vesper Sparrow
______ Violet-green Swallow
______ Virginia Rail

______ Warbling Vireo
______ Water Pipit
______ Western Flycatcher
______ Western Grebe
______ Western Kingbird
______ Western Meadowlark
______ Western Sandpiper
______ Western Tanager
______ Western Wood Pewee
______ Whimbrel
______ Whip-poor-will
______ Whistling Swan
______ White-breasted Nuthatch

_____ White-crowned Pigeon
_____ White-crowned Sparrow
_____ White-eyed Vireo
_____ White-faced Ibis
_____ White Ibis
_____ White-fronted Goose
_____ White-rumped Sandpiper
_____ White-tailed (Gray Sea) Eagle
_____ White-tailed Kite
_____ White-throated Sparrow
_____ White-throated Swift
_____ White-winged Crossbill
_____ White-winged Scoter
_____ Whooping Crane
_____ Willet
_____ Willow Flycatcher
_____ Willow Ptarmigan
_____ Wilson's Phalarope
_____ Wilson's Plover
_____ Wilson's Storm-Petrel
_____ Wilson's Warbler
_____ Winter Wren
_____ Wood Duck
_____ Wood Stork (Wood Ibis)
_____ Wood Thrush
_____ Worm-eating Warbler

_____ Yellow-bellied Flycatcher
_____ Yellow-bellied Sapsucker
_____ Yellow-billed Cuckoo
_____ Yellow-breasted Chat
_____ Yellow-crowned Night Heron
_____ Yellow-headed Blackbird
_____ Yellow Rail
_____ Yellow-rumped Warbler
_____ Yellow-throated Vireo
_____ Yellow-throated Warbler
_____ Yellow Warbler

MAMMALS

_____ Arctic Fox
_____ Arctic Hare
_____ Arctic Shrew
_____ Atlantic Spotted Dolphin
_____ Atlantic White-sided Dolphin

_____ Badger
_____ Beach Vole
_____ Bearded Seal
_____ Beaver
_____ Big Brown Bat
_____ Big Free-tailed Bat
_____ Bison
_____ Black Bear
_____ Black-footed Ferret
_____ Black Rat
_____ Black Right Whale
_____ Black-tailed Jackrabbit
_____ Black-tailed Prairie Dog
_____ Blue Whale
_____ Bobcat
_____ Bottle-nosed Dolphin
_____ Bowhead Whale
_____ Brazilian Free-tailed Bat
_____ Brown Lemming

_____ Caribou
_____ Cave Myotis
_____ Coati
_____ Collared Lemming
_____ Common Dolphin
_____ Common Pilot Whale
_____ Cotton Mouse
_____ Coyote

_____ Deer Mouse
_____ Desert Shrew
_____ Dwarf Sperm Whale

_____ Eastern Chipmunk
_____ Eastern Cottontail
_____ Eastern Harvest Mouse
_____ Eastern Mole
_____ Eastern Pipistrelle
_____ Eastern Spotted Skunk
_____ Eastern Woodrat
_____ Elk
_____ Ermine
_____ European Hare
_____ Evening Bat

_____ False Killer Whale
_____ Fin Whale
_____ Fisher
_____ Florida Mouse
_____ Fox Squirrel
_____ Franklin's Ground Squirrel
_____ Fulvous Harvest Mouse

_____ Gaspé Shrew
_____ Gervais' Beaked Whale
_____ Golden Mouse
_____ Goose-beaked Whale
_____ Grampus
_____ Gray Fox

_______ Gray Myotis
_______ Gray Seal
_______ Gray Squirrel
_______ Gray Wolf

_______ Hairy-tailed Mole
_______ Harbor Porpoise
_______ Harbor Seal
_______ Harp Seal
_______ Heather Vole
_______ Hispid Cotton Rat
_______ Hispid Pocket Mouse
_______ Hoary Bat
_______ Hog-nosed Skunk
_______ Hooded Seal
_______ House Mouse
_______ Hump-backed Whale

_______ Indiana Myotis

_______ Keen's Myotis
_______ Killer Whale

_______ Labrador Collared Lemming
_______ Least Chipmunk
_______ Least Shrew
_______ Least Weasel
_______ Little Brown Myotis
_______ Long-tailed Shrew
_______ Long-tailed Weasel
_______ Lynx

_______ Manatee
_______ Marsh Rabbit
_______ Marsh Rice Rat
_______ Marten
_______ Masked Shrew
_______ Meadow Jumping Mouse
_______ Meadow Vole
_______ Mink
_______ Minke Whale
_______ Moose
_______ Mountain Lion
_______ Mule Deer
_______ Muskox
_______ Muskrat

_______ Narwhal
_______ New England Cottontail
_______ Nine-banded Armadillo
_______ North Atlantic Beaked Whale
_______ North Atlantic Bottle-nosed Whale
_______ Northern Bog Lemming
_______ Northern Flying Squirrel
_______ Northern Grasshopper Mouse
_______ Northern Pocket Gopher
_______ Northern Pygmy Mouse
_______ Northern Yellow Bat
_______ Norway Rat
_______ Nutria

_______ Oldfield Mouse
_______ Olive-backed Pocket Mouse
_______ Ord's Kangaroo Rat

_______ Plains Harvest Mouse
_______ Plains Pocket Gopher
_______ Plains Pocket Mouse
_______ Polar Bear
_______ Porcupine
_______ Prairie Vole
_______ Pronghorn
_______ Pygmy Shrew
_______ Pygmy Sperm Whale

_______ Raccoon
_______ Rafinesque's Big-eared Bat
_______ Red Bat
_______ Red Fox
_______ Red Squirrel
_______ Red Wolf
_______ Richardson's Ground Squirrel
_______ Ringed Seal
_______ Ringtail Cat
_______ River Otter
_______ Rock Squirrel
_______ Rock Vole
_______ Rough-toothed Porpoise
_______ Round-tailed Muskrat

_______ Sei Whale
_______ Seminole Bat
_______ Short-finned Pilot Whale
_______ Short-tailed Shrew
_______ Silver-haired Bat
_______ Small-footed Myotis
_______ Smoky Shrew
_______ Snowshoe Hare
_______ Southeastern Myotis
_______ Southeastern Pocket Gopher
_______ Southeastern Shrew
_______ Southern Bog Lemming
_______ Southern Flying Squirrel

_______ Southern Plains Woodrat
_______ Southern Red-backed Vole
_______ Southern Short-tailed Shrew
_______ Sperm Whale
_______ Spotted Ground Squirrel
_______ Star-nosed Mole
_______ Striped Porpoise
_______ Striped Skunk
_______ Swamp Rabbit
_______ Swift Fox

_______ Texas Mouse
_______ Thirteen-lined Ground Squirrel
_______ Townsend's Big-eared Bat
_______ Tropical Beaked Whale
_______ True's Beaked Whale

_______ Virginia Opossum

_______ Walrus
_______ Water Shrew
_______ Western Harvest Mouse
_______ White-beaked Dolphin
_______ White-footed Mouse
_______ White-tailed Deer
_______ White-tailed Jackrabbit
_______ White Whale
_______ Wild Boar
_______ Wolverine
_______ Woodchuck
_______ Woodland Jumping Mouse
_______ Woodland Vole

REPTILES

_______ Alabama Map Turtle
_______ Alabama Red-bellied Turtle
_______ Alligator Snapping Turtle
_______ American Alligator
_______ Atlantic Ridley

_______ Banded Water Snake
_______ Barbour's Map Turtle
_______ Black Kingsnake
_______ Black-knobbed Map Turtle
_______ Black Racer
_______ Black Rat Snake
_______ Black Swamp Snake
_______ Blanding's Turtle
_______ Blotched Water Snake
_______ Blue Racer
_______ Bog Turtle
_______ Broad-banded Copperhead
_______ Broad-banded Water Snake
_______ Broad-headed Skink
_______ Brown Snake
_______ Brown Water Snake
_______ Bullsnake
_______ Butler's Garter Snake
_______ Buttermilk Racer

_______ Cagle's Map Turtle
_______ Carolina Salt Marsh Snake
_______ Central Plains Milk Snake
_______ Checkered Garter Snake
_______ Chicken Turtle
_______ Coachwhip
_______ Coal Skink
_______ Collared Lizard
_______ Common Garter Snake
_______ Common Kingsnake
_______ Cooter
_______ Copper-bellied Water Snake
_______ Copperhead
_______ Corn Snake
_______ Cottonmouth
_______ Cumberland slider

_______ Desert Kingsnake
_______ Diamondback Terrapin
_______ Diamondback Water Snake

_______ Eastern Box Turtle
_______ Eastern Coral Snake
_______ Eastern Diamondback Rattlesnake
_______ Eastern Fence Lizard
_______ Eastern Garter Snake
_______ Eastern Glass Lizard
_______ Eastern Hognose Snake
_______ Eastern Kingsnake
_______ Eastern Milk Snake
_______ Eastern Mud Turtle
_______ Eastern Painted Turtle
_______ Eastern Ribbon Snake
_______ Eastern Yellow-bellied Racer

_______ False Map Turtle
_______ Five-lined Skink
_______ Flat-headed Snake
_______ Florida Box Turtle

______ Florida Kingsnake
______ Florida Red-bellied Turtle
______ Florida Scrub Lizard
______ Florida Snapping Turtle
______ Florida Softshell
______ Florida Water Snake
______ Fox Snake

______ Glossy Snake
______ Glossy Crayfish Snake
______ Gopher Tortoise
______ Graham's Crayfish Snake
______ Gray Rat Snake
______ Great Plains Rat Snake
______ Great Plains Skink
______ Green Anole
______ Green Turtle
______ Green Water Snake
______ Ground Skink
______ Ground Snake
______ Gulf Coast Box Turtle
______ Gulf Coast Salt Marsh Snake

______ Hawksbill

______ Indigo Snake
______ Island Glass Lizard

______ Key Ringneck Snake
______ Kirtland's Water Snake

______ Lake Erie Water Snake
______ Leatherback
______ Lesser Earless Lizard
______ Lined Snake
______ Loggerhead
______ Loggerhead Musk Turtle
______ Long-nosed Snake
______ Louisiana Milk Snake

______ Many-lined Skink
______ Map Turtle
______ Massasauga
______ Mexican Milk Snake
______ Midland Painted Turtle
______ Midland Water Snake
______ Milk Snake
______ Mississippi Ringneck Snake
______ Mole Kingsnake
______ Mud Snake

______ Night Snake
______ Northern Copperhead
______ Northern Pine Snake
______ Northern Ringneck Snake
______ Northern Water Snake

______ Osage Copperhead
______ Ouachita Map Turtle

______ Painted Turtle
______ Pale Milk Snake
______ Pigmy Rattlesnake
______ Pine Snake
______ Pine Woods Snake
______ *Pituophis melanoleucus sayi*
______ Plain-bellied Water Snake
______ Plains Black-headed Snake
______ Plains Garter Snake
______ Prairie Kingsnake
______ Prairie Ringneck Snake
______ Prairie Skink

______ Queen Snake

______ Racer
______ Rainbow Snake
______ Rat Snake
______ Razor-backed Musk Turtle
______ Red Milk Snake
______ Red-bellied Snake
______ Red-bellied Turtle
______ Red-bellied Water Snake
______ Red-eared Slider
______ Red-sided Garter Snake
______ Ringed Map Turtle
______ Ringneck Snake
______ River Cooter
______ Rough Earth Snake
______ Rough Green Snake

______ Scarlet Kingsnake
______ Scarlet Snake
______ Short-headed Garter Snake
______ Short-horned Lizard
______ Short-tailed Snake
______ Six-lined Racerunner
______ Slender Glass Lizard
______ Slider
______ Smooth Earth Snake
______ Smooth Green Snake
______ Smooth Softshell
______ Snapping Turtle

_____ Southeastern Crowned Snake
_____ Southeastern Five-lined Skink
_____ Southern Copperhead
_____ Southern Hognose Snake
_____ Southern Painted Turtle
_____ Southern Ringneck Snake
_____ Southern Water Snake
_____ Speckled Kingsnake
_____ Spiny Softshell
_____ Spotted Turtle
_____ *Sternotherus minor minor*
_____ Stinkpot
_____ Striped Mud Turtle
_____ Striped Crayfish Snake
_____ Stripe-necked Musk Turtle

_____ Texas Blind Snake
_____ Texas Horned Lizard
_____ Texas Map Turtle
_____ Texas Rat Snake
_____ Three-toed Box Turtle
_____ Timber Rattlesnake

_____ Western Box Turtle
_____ Western Hognose Snake
_____ Western Painted Turtle
_____ Western Rattlesnake
_____ Western Ribbon Snake
_____ Western Terrestrial Garter Snake
_____ Wood Turtle
_____ Worm Snake

_____ Yellow-bellied Slider
_____ Yellow-bellied Water Snake
_____ Yellow-blotched Map Turtle
_____ Yellow Mud Turtle
_____ Yellow Rat Snake

AMPHIBIANS

_____ Alabama Waterdog
_____ American Toad

_____ Barking Treefrog
_____ Bird-voiced Treefrog
_____ Black-bellied Salamander
_____ Black Mountain Salamander
_____ Blue-spotted Salamander
_____ Brimley's Chorus Frog
_____ Bullfrog

_____ Canadian Toad
_____ Carpenter Frog
_____ Cave Salamander
_____ Cheat Mountain Salamander
_____ Couch's Spadefoot
_____ Crayfish Frog

_____ Dark-sided Salamander
_____ Dusky Salamander
_____ Dwarf Salamander
_____ Dwarf Siren
_____ Dwarf Waterdog

_____ Eastern Narrow-mouthed Toad
_____ Eastern Spadefoot

_____ Flatwoods Salamander
_____ Four-toed Salamander

_____ Gray Treefrog
_____ Greater Siren
_____ Great Plains Narrow-mouthed Toad
_____ Great Plains Toad
_____ Green Frog
_____ Green Salamander
_____ Green Toad
_____ Green Treefrog
_____ Grotto Salamander
_____ Gulf Coast Toad

_____ Hellbender

_____ Jefferson Salamander
_____ Jordan's Woodland Salamander

_____ Lesser Siren
_____ Little Grass Frog
_____ Long-tailed Salamander

_____ Mabee's Salamander
_____ Many-lined Salamander
_____ Many-ribbed Salamander
_____ Marbled Salamander
_____ Mink Frog
_____ Mole Salamander
_____ Mountain Chorus Frog

_______ Mountain Dusky Salamander
_______ Mudpuppy
_______ Mud Salamander

_______ Newt
_______ Northern Cricket Frog
_______ Northern Leopard Frog

_______ Oak Toad
_______ Oklahoma Salamander
_______ Ornate Chorus Frog

_______ Pickerel Frog
_______ Pig Frog
_______ Pigmy Salamander
_______ Pine Barrens Treefrog
_______ Pine Woods Treefrog
_______ Plains Leopard Frog
_______ Plains Spadefoot

_______ Ravine Salamander
_______ Red-backed Salamander
_______ Red Salamander
_______ Red-spotted Toad
_______ Ringed Salamander
_______ River Frog

_______ Seal Salamander
_______ Seepage Salamander
_______ Shovel-nosed Salamander
_______ Silvery Salamander
_______ Slimy Salamander
_______ Small-mouthed Salamander
_______ Southern Chorus Frog
_______ Southern Cricket Frog
_______ Southern Dusky Salamander
_______ Southern Leopard Frog
_______ Southern Red-backed Salamander
_______ Southern Toad
_______ Spotted Chorus Frog
_______ Spotted Salamander
_______ Spring Peeper
_______ Spring Salamander
_______ Squirrel Treefrog
_______ Strecker's Chorus Frog
_______ Striped Chorus Frog
_______ Striped Newt

_______ Tennessee Cave Salamander
_______ Texas Toad
_______ Three-toed Amphiuma
_______ Tiger Salamander
_______ Tremblay's Salamander
_______ Two-lined Salamander
_______ Two-toed Amphiuma

_______ Valley and Ridge Salamander

_______ Wehrle's Salamander
_______ Weller's Salamander
_______ Western Spadefoot
_______ Western Toad
_______ West Virginia Cave Salamander
_______ White-spotted Salamander
_______ Wood Frog
_______ Woodhouse's Toad

_______ Yonahlossee Salamander

_______ Zigzag Salamander

FISHES

_______ African Pompano
_______ Alabama Shad
_______ Albacore
_______ Alewife
_______ Alligator Gar
_______ American Eel
_______ American Plaice
_______ American Sand Lance
_______ American Shad
_______ Angel Shark
_______ Arctic Grayling
_______ Atlantic Bonito
_______ Atlantic Cod
_______ Atlantic Croaker
_______ Atlantic Cutlassfish
_______ Atlantic Halibut
_______ Atlantic Herring
_______ Atlantic Mackerel
_______ Atlantic Manta
_______ Atlantic Moonfish
_______ Atlantic Needlefish
_______ Atlantic Salmon
_______ Atlantic Silverside
_______ Atlantic Stingray
_______ Atlantic Sturgeon
_______ Atlantic Tomcod
_______ Atlantic Torpedo
_______ Atlantic Wolffish

_______ Banded Darter
_______ Banded Drum
_______ Banded Killifish
_______ Banded Rudderfish
_______ Banded Sunfish
_______ Bantam Sunfish
_______ Barn-door Skate
_______ Basking Shark

_______ Bay Anchovy
_______ Bermuda Chub
_______ Big-eyed Scad
_______ Bigmouth Buffalo
_______ Bitterling
_______ Blackbanded Sunfish
_______ Black Buffalo
_______ Black Bullhead
_______ Black Crappie
_______ Black Drum
_______ Blackfin Cisco
_______ Blackfin Tuna
_______ Blacknose Dace
_______ Black Sea Bass
_______ Blackstripe Topminnow
_______ Blackside Darter
_______ Bloater
_______ Blueback Herring
_______ Blue Catfish
_______ Bluefin Tuna
_______ Bluefish
_______ Bluegill
_______ Blue Marlin
_______ Blue Runner
_______ Bluespot Sunfish
_______ Bluespotted Cornetfish
_______ Blue Sucker
_______ Bluntnose Minnow
_______ Bonefish
_______ Bowfin
_______ Brook Silverside
_______ Brook Stickleback
_______ Brook Trout
_______ Brown Bullhead
_______ Brown Trout
_______ Bullnose Ray
_______ Bull Shark
_______ Burbot
_______ Butterfish

_______ Carp
_______ Central Mudminnow
_______ Cero
_______ Chain Pickerel
_______ Channel Catfish
_______ Chub Mackerel
_______ Cisco
_______ Clearnose Skate
_______ Cobia
_______ Common Shiner
_______ Conger Eel
_______ Cownose Ray
_______ Creek Chub
_______ Creek Chubsucker
_______ Crevelle Jack
_______ Cunner
_______ Cusk
_______ Cutlips Minnow
_______ Cutthroat Trout

_______ Deepwater Cisco
_______ Dolphin
_______ Dusky Shark

_______ Eastern Mudminnow
_______ Eastern Sand Darter
_______ Emerald Shiner
_______ Eyed Flounder

_______ Fallfish
_______ Fantail Darter
_______ Fathead Minnow
_______ Flathead Catfish
_______ Florida Pompano
_______ Flying Gurnard
_______ Fourspot Flounder
_______ Freshwater Drum

_______ Gafftopsail Catfish
_______ Gizzard Shad
_______ Golden Redhorse
_______ Golden Shiner
_______ Goldeye
_______ Goldfish
_______ Goosefish
_______ Great Amberjack
_______ Great Barracuda
_______ Great Blue Shark
_______ Greenland Halibut
_______ Greenland Shark
_______ Greenside Darter
_______ Green Sunfish
_______ Grubby

_______ Haddock
_______ Hagfish
_______ Halfbeak
_______ Hickory Shad
_______ Highfin Carpsucker
_______ Hogchoker
_______ Horse-eyed Jack

_______ Johnny Darter

_______ King Mackerel
_______ Kiyi

_______ Ladyfish
_______ Lake Chubsucker
_______ Lake Sturgeon
_______ Lake Trout
_______ Lake Whitefish
_______ Largemouth Bass
_______ Leatherjacket
_______ Lined Sea Horse
_______ Little Skate
_______ Little Tuna
_______ Logperch
_______ Longear Sunfish
_______ Longhorn Sculpin
_______ Longjaw Chub

______ Longnose Dace
______ Longnose Gar
______ Longnose Sucker
______ Lookdown

______ Mackerel Scad
______ Margined Madtom
______ Menhaden
______ Mooneye
______ Mosquitofish
______ Mottled Sculpin
______ Mountain Sucker
______ Mountain Whitefish
______ Mummichog
______ Muskellunge

______ Naked Goby
______ Nipigon Cisco
______ Northern Cavefish
______ Northern Hog Sucker
______ Northern Kingfish
______ Northern Pike
______ Northern Pipefish
______ Northern Puffer
______ Northern Sea Robin
______ Northern Sennet
______ Northern Stargazer
______ Northern Studfish
______ Northern Swamp Darter

______ Ocean Sunfish
______ Ohio Redhorse
______ Orange Filefish
______ Orange Spotted Sunfish
______ Oyster Toadfish

______ Paddlefish
______ Pallid Sturgeon
______ Palometa
______ Pearl Dace
______ Permit
______ Pigfish
______ Pinfish
______ Pirate Perch
______ Plains Carpsucker
______ Planehead Filefish
______ Pollack
______ Porbeagle Shark
______ Pumpkinseed

______ Quillback

______ Rainbow Darter
______ Rainbow Smelt
______ Rainbow Trout
______ Redbreast Sunfish
______ Red Drum
______ Redear Sunfish
______ Redfin Pickerel
______ Redfish
______ Red Hake
______ Red Goatfish
______ Red Grouper
______ Redside Dace
______ River Carpsucker
______ River Chub
______ River Redhorse
______ Roanoke Hog Sucker
______ Rockbass
______ Rock Gunnel
______ Rosa's Blindfish
______ Rosy Dace
______ Rough Scad
______ Roughtail Stingray
______ Round Herring
______ Round Scad
______ Round Whitefish
______ Rustyside Sucker

______ Sandbar Shark
______ Sand Tiger
______ Satinfin Shiner
______ Sauger
______ Scrawled Cowfish
______ Scup
______ Sea Catfish
______ Sea Lamprey
______ Sea Raven
______ Sharksucker
______ Sheepshead
______ Sheepshead Minnow
______ Shortfin Mako Shark
______ Shorthead Redhorse
______ Shorthorn Sculpin
______ Shortjaw Cisco
______ Shortnose Chub
______ Shortnose Gar
______ Shortnose Sturgeon
______ Shovelnose Sturgeon
______ Silver Chub
______ Silver Hake
______ Silverjaw Minnow
______ Silver Lamprey
______ Silver Perch
______ Silver Redhorse
______ Silvery Minnow
______ Skilletfish
______ Skipjack Herring
______ Skipjack Tuna
______ Smallmouth Bass
______ Smallmouth Buffalo
______ Smallmouth Flounder
______ Smooth Butterfly Ray
______ Smooth Dogfish
______ Smooth Flounder
______ Smooth Hammerhead
______ Smooth Puffer
______ Sockeye Salmon
______ Southern Cavefish
______ Southern Kingfish

_______ Southern Redbelly Dace
_______ Spanish Mackerel
_______ Spanish Sardine
_______ Spiny Butterfly
_______ Spiny Dogfish
_______ Spot
_______ Spotfin Shiner
_______ Spottail Shiner
_______ Spotted Bass
_______ Spotted Eagle Ray
_______ Spotted Gar
_______ Spotted Hake
_______ Spotted Sea Trout
_______ Spotted Sucker
_______ Spotted Sunfish
_______ Springfish
_______ Starhead Topminnow
_______ Stonecat
_______ Stoneroller
_______ Striped Anchovy
_______ Striped Bass
_______ Striped Burrfish
_______ Striped Killifish
_______ Striped Mullet
_______ Striped Sea Robin
_______ Suckermouth Minnow
_______ Summer Flounder
_______ Swampfish
_______ Swordfish

_______ Tadpole Madtom
_______ Tarpon
_______ Tautog
_______ Thorny Skate
_______ Thresher Shark
_______ Tiger Shark
_______ Tilefish
_______ Tonguetied Chub
_______ Torrent Sucker
_______ Tripletail
_______ Troutperch
_______ Trunkfish

_______ Virginia Lancelet

_______ Walleye
_______ Warmouth
_______ Weakfish
_______ White Bass
_______ White Catfish
_______ White Crappie
_______ White Hake
_______ White Marlin
_______ White Mullet
_______ White Perch
_______ White Shark
_______ White Sucker
_______ Windowpane
_______ Winter Flounder
_______ Winter Skate
_______ Witch Flounder
_______ Wrymouth

_______ Yellow Bass
_______ Yellow Bullhead
_______ Yellowfin Tuna
_______ Yellow Jack
_______ Yellow Perch
_______ Yellowtail Flounder

MOLLUSKS

_______ Adams' Miniature Cerith
_______ Alphabet Cone
_______ American Star Shell
_______ Angled Wentletrap
_______ Apple Murex
_______ Arctic Rock Borer
_______ Atlantic Auger
_______ Atlantic Awning Clam
_______ Atlantic Bob-tailed Squid
_______ Atlantic Modulus
_______ Atlantic Nut Clam
_______ Atlantic Octopus
_______ Atlantic Plate Limpet
_______ Atlantic Surf Clam

_______ Baltic Macoma
_______ Bay Scallop
_______ Beaded Miter
_______ Black Clam
_______ Bleeding Tooth
_______ Blood Ark
_______ Blue Mussel
_______ Boreal Astarte
_______ Broad-ribbed Cardita
_______ Brown-banded Wentletrap
_______ Brown-corded Neptune

_______ Calico Scallop
_______ Chalky Tellin
_______ Channeled Whelk
_______ Checkerboard Clam
_______ Coffee Bean Trivia
_______ Colorful Moon Shell
_______ Comb Bittersweet
_______ Common Angel Wing
_______ Common Atlantic Marginella
_______ Common Baby's Ear
_______ Common Basket Clam
_______ Common Boat Shell
_______ Common Cockle
_______ Common Jingle Shell
_______ Common Moon Shell
_______ Common Nutmeg
_______ Common Periwinkle
_______ Common Purple Sea Snail

______ Common Razor
______ Common Soft-shelled Clam
______ Common Squid
______ Common Sundial
______ Common Worm Shell
______ Coon Oyster
______ Cross-barred Chione
______ Crosshatched Lucine
______ Crown Conch

______ Deep Sea Scallop
______ Deer Cowry
______ DeKay's Tellin
______ Disk Dosinia
______ Dwarf Horn Shell
______ Dwarf Squid
______ Dwarf Surf Clam

______ Emerson's Cerith
______ Eastern Chiton
______ Eastern Paper Bubble

______ Fallen Angel Wing
______ False Angel Wing
______ False Coquina
______ File Yoldia
______ Flamingo Tongue
______ Flat Slipper Shell
______ Flat Tree Oyster
______ Flesh Pen Shell
______ Florida Banded Tulip
______ Florida Drill
______ Florida Fighting Conch
______ Florida Horn Shell
______ Florida Horse Conch
______ Florida Rock Shell
______ Four-spotted Trivia
______ Fuzzy Chiton

______ Giant Eastern Murex
______ Glassy Lyonsia
______ Gould's Pandora
______ Gould's Shipworm
______ Granular Frog Shell
______ Gray Cowry
______ Great Heart Cockle
______ Great Piddock
______ Great White Lucine
______ Greedy Dove Shell
______ Greenland Top Shell

______ Hairy Triton
______ Heavy Ark
______ Hooked Mussel
______ Horse Mussel

______ Impressed Odostome
______ Interrupted Turbonille
______ Ivory Tusk

______ Jasper Cone
______ Junonia

______ Kitten's Paw
______ Knobbed Whelk
______ Knobby Periwinkle
______ Knobby Turban

______ Lace Murex
______ Leafy Jewel Box
______ Lettered Olive
______ Lightning Whelk
______ Lined Tellin
______ Linnaeus' Keyhole Limpet
______ Lion's Paw
______ Lister's Keyhole Limpet
______ Little Boat Shell
______ Little Chink Shell
______ Little Keyhole Limpet
______ Long-spined Star Shell

______ Miniature Natica
______ Minute Dwarf Olive
______ Mossy Ark
______ Mottled Chiton
______ Mottled Dog Whelk
______ Mottled Dove Shell
______ Mottled Top Shell
______ Mud Snail

______ Naval Shipworm
______ New England Dog Whelk
______ Northern Quahog
______ Northern Tellin

______ Oyster Drill

______ Paper Mussel
______ Paper Nautilus
______ Pennsylvania Lucine
______ Pointed Nut Clam
______ Prickly Jingle Shell
______ Purplish Razor

______ Ram's Horn
______ Red Chiton
______ Ribbed Mussel
______ Ribbed Pod
______ Rice Dwarf Olive
______ Rock Purple
______ Rough Periwinkle

______ Sallé's Auger
______ Salt Marsh Periwinkle
______ Scotch Bonnet
______ Sea Arrow
______ Shark Eye
______ Single-toothed Simnia

_______ Smooth Periwinkle
_______ Southern Coquina
_______ Spiny Oyster
_______ Stiff Pen Shell
_______ Stimpson's Tusk
_______ Stimpson's Whelk
_______ Striate Cup-and-saucer
_______ Striate Wood Piddock
_______ Sunray Shell

_______ Tenta Macoma
_______ Tesselated Nerite
_______ Thick-lipped Drill
_______ Transverse Ark
_______ Tuberculate Chiton
_______ Tulip Shell
_______ Turkey Wing

_______ Unequal Spoon Shell

_______ Variable Bittium
_______ Variegated Nerite
_______ Virginia Oyster

_______ Waved Whelk
_______ Wedge Rangia
_______ Wedge-shaped Piddock
_______ White Chiton
_______ White-mouthed Triton
_______ White Semele

_______ Zebra Nerite

OTHER MARINE INVERTEBRATES

_______ *Aiptasia eruptaurantia*
_______ *Aiptasia pallida*
_______ American Lobster
_______ *Amphipholis squamata*
_______ *Amphitrite johnstoni*
_______ *Amphitrite ornata*
_______ *Ascidia callosa*
_______ *Ascidia interrupta*
_______ *Asterias forbesii*
_______ *Asterias vulgaris*
_______ *Astrangia danae*
_______ *Astropecten articulatus*

_______ *Balanoglossus aurantiacus*
_______ *Balanus amphitrite*
_______ *Balanus balanoides*
_______ *Balanus crenatus*
_______ *Balanus eburneus*
_______ *Balanus improvisus*
_______ Bamboo Worms
_______ Beach Fleas
_______ Beach Hoppers
_______ *Beroe cucumis*
_______ *Berod ovata*

_______ Blood Worms
_______ Blue Crab
_______ Boring Sponge
_______ Brown Sea Anemone
_______ *Bugula neritina*
_______ *Bugula turrita*
_______ Burrowing Sea Anemone
_______ Burrowing Shrimp

_______ *Callianassa atlantica*
_______ *Caudina arenata*
_______ *Chaetopterus variopedatus*
_______ *Chthamalus fragilis*
_______ Clam Worms
_______ Common Sun Star
_______ *Crisia denticulata*
_______ *Crisia eburnea*
_______ *Cyanea capillata*

_______ Daisy Brittle Star
_______ Deadman's Fingers
_______ *Diadumene leucolena*
_______ *Diopatra cuprea*

_______ *Echiurus pallasii*
_______ Edible Shrimp
_______ *Electra* (species)
_______ *Eudendrium ramosum*

_______ Fish Louse

_______ Ghost Crab
_______ Goose Barnacles
_______ Grass Shrimp
_______ Green Brittle Star
_______ Green Clam Worm
_______ Green Crab
_______ Green Sea Urchin
_______ Gribbles

_______ Heart Urchin
_______ *Henricia sanguinolenta*
_______ Horseshoe Crab
_______ *Hydractinia echinata*
_______ *Hydroides dianthus*
_______ *Hymeniacidon heliophila*

_______ *Idotea* (species)

_______ Jonah Crab

_______ Keyhole Urchin

_______ Lady Crab
_______ *Leptosynapta inhaerens*
_______ Lugworms
_______ *Luidia clathrata*
_______ *Lygia exotica*

OTHER MARINE INVERTEBRATES

______ *Lygia oceanica*

______ Mantis Shrimp
______ *Membranipora* (species)
______ *Mnemiopsis leidyi*
______ Mole Crab
______ Moon Jelly
______ Mud Crabs
______ Mud Fiddler Crab
______ Mud Shrimp
______ Mussel Crab

______ *Obelia* (species)
______ *Ophiothrix angulata*
______ Oyster Crab

______ *Pagurus longicarpus*
______ *Pagurus pollicaris*
______ *Panopeus herbstii*
______ Peacock Worms
______ Pea Crab
______ *Pennaria tiarella*
______ *Pentamera pulcherrima*
______ *Perophora viridis*
______ *Pinnixa chaetopterana*
______ *Polyonyx gibbesi*
______ Portuguese Man-of-War
______ Purple Sea Urchin
______ Purple Sun Star

______ Red-jointed Fiddler Crab
______ Red Sponge
______ Ribbon Worm
______ Rock Crab

______ *Sabellaria vulgaris*
______ *Saccoglossus kowalevskii*
______ Sand Bug
______ Sand Dollar
______ Sand Fiddler Crab
______ Sand Fleas
______ Sand Shrimp
______ Scale Worms
______ *Schizoporella unicornis*
______ *Sclerodactyla briareus*
______ Scuds
______ *Scypha ciliata*
______ Sea Grapes
______ Sea Mouse
______ Sea Nettle
______ Sea Pansy
______ Sea Pork
______ Sea Vase
______ Sea Walnut
______ Seaweed Hoppers
______ *Sertularia pumila*
______ *Sesarma cinereum*
______ *Sesarma reticulatum*
______ Skeleton Shrimp
______ Snapping Shrimp
______ *Sphaeroma quadridentatum*
______ Spider Crab
______ *Spirorbis borealis*
______ *Stomolophus meleagris*
______ Striped Hermit Crab
______ *Styela partita*

______ *Terebratulina septentrionalis*
______ Trumpet Worm
______ Tube-dwelling Sea Anemone
______ *Tubularia crocea*

______ Whip Coral
______ White Sea Urchin
______ Wood-boring Amphipod

Index

Animal species are indexed in general by common name. A number of species, however, especially the marine invertebrates, do not have common names, and these species are indexed by their Latin (genus-species) nomenclature. Higher taxa, such as classes, orders and families, are indexed by both common and scientific names. A number in *italics* indicates a text page on which the species is illustrated; a reference in **bold** type indicates the number of the color plate on which the species is illustrated and the position of the species on the color plate.

G

Q

R

S

NOTES

NOTES

NOTES

NOTES

NOTES